THEORETICAL AND APPLIED MECHANICS 1996

INTERNATIONAL UNION OF THEORETICAL
AND APPLIED MECHANICS

THEORETICAL AND APPLIED MECHANICS 1996

1996

Proceedings of the XIXth International Congress of Theoretical and Applied Mechanics,
Kyoto, Japan, 25-31 August 1996

Edited by

Tomomasa TATSUMI

Vice Director
International Institute for Advanced Studies
Kyoto, Japan

Eiichi WATANABE

Department of Civil Engineering
Kyoto University, Japan

and

Tsutomu KAMBE

Department of Physics
University of Tokyo, Japan

1997
ELSEVIER
AMSTERDAM • LAUSANNE • NEW YORK • OXFORD • SHANNON • TOKYO

ELSEVIER SCIENCE B.V.
Sara Burgerhartstraat 25
P.O. Box 211, 1000 AE Amsterdam, The Netherlands

ISBN: 0 444 82446 4

This book is printed on acid-free paper.

Printed in The Netherlands

PREFACE

This book contains the Proceedings of the XIXth International Congress of Theoretical and Applied Mechanics, held at the Kyoto International Conference Hall, Kyoto, August 25–31, 1996. The Congress (ICTAM Kyoto 1996) was organized under the auspices of the International Union of Theoretical and Applied Mechanics (IUTAM) by invitation of the Japan National Committee of Theoretical and Applied Mechanics, with the sponsorship of the Science Council of Japan, the Japan Society of Civil Engineers, the Japan Society of Mechanical Engineers, the Architectural Institute of Japan, and the support of other academic, public institutions and corporations.

This volume includes the full texts of the two General Lectures, of introductory lectures of the six Minisymposia and of the Sectional Lectures, according to the list on pages xi and xii. The contributed papers presented at the congress are listed by author and title; most of them will be published in appropriate scientific journals.

The publication of these Proceedings has been handled promptly and capably by Elsevier Science and their editors, to whom we are very grateful.

Tomomasa Tatsumi
(Vice Director,
International Institute
for Advanced Studies)

Eiichi Watanabe
(Kyoto University)

Tsutomu Kambe
(University of Tokyo)

Kyoto
November 1996

Opening Ceremony

Professor Tomomasa Tatsumi, President of the Congress

Professor Leen van Wijngaarden, President of IUTAM

Professor Takuji Kobori

Professor Sir James Lighthill

Audience at the Opening Ceremony

Professor L. van Wijngaarden presenting the Bureau Prize to Mr. J. Vollmann

CONTENTS

OPENING AND CLOSING LECTURES

INTRODUCTORY LECTURES OF MINISYMPOSIA

"Vorticity Dynamics and Turbulence"

"Non-Newtonian Fluid Flow"

"Aero- and Hydroacoustics"

"Mechanics of Heterogeneous and Composite Solids"

CONGRESS COMMITTEE OF IUTAM

Chairman: L. van Wijngaarden* (The Netherlands)
Secretary: N. Olhoff* (Denmark)

A. Acrivos* (USA)
H. Aref (USA)
S. Bodner* (Israel)
B.A. Boley (USA)
C.R. Calladine (UK)
C. Cercignani (Italy)
G.G. Chernyi (Russia)
J. Engelbrecht (Estonia)
P. Germain (France)
Z. Hashin (Israel)
M.A. Hayes (Ireland)
N.J. Hoff (USA)
J.W. Hutchinson (USA)
T. Inoue (Japan)
J. Jimenez (Spain)
S. Kaliszky* (Hungary)
B.L. Karihaloo (Australia)
A.N. Kounadis (Greece)

Y.H. Ku (USA)
S. Leibovich (USA)
B. Lundberg (Sweden)
G.E.A. Meier (Germany)
H.K. Moffatt* (UK)
Z. Mroz (Poland)
B.C. Nakra (India)
J.T. Oden (USA)
T.J. Pedley (UK)
M. Sayir (Switzerland)
W. Schiehlen (Germany)
B. Tabarrok (Canada)
T. Tatsumi (Japan)
Ren Wang (China)
Zemin Zheng (China)
F. Ziegler (Austria)

* Members of Executive Committee

LOCAL ORGANIZING COMMITTEE

Chairman: Y. Yamamoto* (Univ. of Tokyo, Prof. Emer.)
Vice-Chairmen: M. Ito* (Univ. of Tokyo, Prof. Emer.)   K. Kawata* (Univ. of Tokyo, Prof. Emer.)
    H. Ohashi*(President of Kogakuin Univ.)   T. Tatsumi* (Int'l. Inst. for Adv. Studies)
Secretary General: E. Watanabe* (Kyoto Univ.)

H. Abe (President, The Japan Society
    of Mechanical Engineers)
Y. Aihara (Tokai Univ.)
T. Funaki (Osaka Univ.)
K. Gotoh (Osaka Pref. Coll. Tech.)
Y. Hangai (Univ. of Tokyo)
T. Inoue* (Kyoto Univ.)
T. Kakutani (Kinki Univ.)
T. Kambe* (Univ. of Tokyo)
S. Kida (Nat. Inst. for Fusion Sci.)
S. Kobayashi (Kyoto Univ.)
T. Kosaka (former President, The Japan
    Society of Civil Engineers)
T. Masuda (Kyoto Univ.)
M. Matsumoto (Kyoto Univ.)

M. Matsuo (Nagoya Univ., President, The Japan
    Society of Civil Engineers)
S. Murakami* (Nagoya Univ.)
H. Nakagawa* (Ritsumeikan Univ.)
T. Nakamura (Kyoto Univ.,President,The Archi-
    tectural Institute of Japan)
R. Nakano (Gifu Univ.)
F. Ogino* (Kyoto Univ.)
N. Satofuka* (Kyoto Inst. of Tech.)
T. Tamura* (Kyoto Univ.)
Y. Tomita (Kyoto Univ.)
K. Uetani* (Kyoto Univ.)
G. Yagawa* (Univ. of Tokyo)

* Members of Executive Committee

LOCAL EXECUTIVE COMMITTEE

Chairman and President of XIXth ICTAM:

T. Tatsumi (Vice Director of International

Institute for Advanced Studies)

Vice-Chairman: H. Nakagawa (Ritsumeikan Univ.)

Secretary General: E. Watanabe (Kyoto Univ.)

T. Funaki (Osaka Univ.)

K. Gotoh (Osaka Pref. Coll. Tech.)

T. Inoue (Kyoto Univ.)

T. Kakutani (Kinki Univ.)

T. Kawahara (Kyoto Univ.)

S. Kida (Nat. Inst. for Fusion Sci.)

H. Kitagawa (Osaka Univ.)

S. Kobayashi (Kyoto Univ.)

M. Matsumoto (Kyoto Univ.)

S. Morioka (Kyoto Univ., Prof. Emer.)

S. Murakami (Nagoya Univ.)

T. Nakamura (Kyoto Univ.)

F. Ogino (Kyoto Univ.)

N. Satofuka (Kyoto Inst. of Tech.)

K. Suzuki (Kyoto Univ.)

T. Tamura (Kyoto Univ.)

Y. Tomita (Kobe Univ.)

K. Uetani (Kyoto Univ.)

FINANCE COMMITTEE

Chairman: M. Ito (Univ. of Tokyo, Prof. Emer.)

Vice-Chairmen: H. Ohashi (President of Kogakuin Univ.)

H. Nakagawa (Ritsumeikan Univ.)

T. Nakamura (Kyoto Univ.)

Members: Y. Aihara (Tokai Univ.)

T. Inoue (Kyoto Univ.)

A. Kobayashi (Science Univ. of Tokyo)

G. Yagawa (Univ. of Tokyo)

The XIXth International Congress of Theoretical and Applied Mechanics
(XIXth ICTAM 1996) is held under the auspices of

The International Union of Theoretical and Applied Mechanics (IUTAM)

by invitation of

The Japan National Committee of Theoretical and Applied Mechanics

under the sponsorship and support of

- The Science Council of Japan
- The Japan Society of Civil Engineers
- The Japan Society of Mechanical Engineers
- The Architectural Institute of Japan

and with the support of the following academic and public organizations:

- The Physical Society of Japan
- Mathematical Society of Japan
- The Japanese Society of Irrigation, Drainage and Reclamation Engineering
- The Japan Society of Applied Physics
- The Japan Society of Aeronautical and Space Sciences
- The Mining and Materials Processing Institute of Japan
- The Society of Naval Architects of Japan
- The Japan Society of Fluid Mechanics
- Atomic Energy Society of Japan
- The Society of Rheology, Japan
- The Society of Chemical Engineers, Japan
- Meteorological Society of Japan
- The Seismological Society of Japan
- The Japanese Geotechnical Society
- Japan Concrete Institute
- Japanese Society of Steel Construction
- The Japan Society for Precision Engineering
- Society of Automotive Engineers of Japan, Inc.
- The Japan Society for Technology of Plasticity
- Gas Turbine Society of Japan
- Turbomachinery Society of Japan
- The Geothermal Research Society of Japan
- The Society of Instrument and Control Engineers
- The Visualization Society of Japan
- Japan Society of Computational Fluid Dynamics
- The Japan Society of Multiphase Flow
- Japan Society for Industrial and Applied Mathematics
- Japan Society for Simulation Technology
- The Society of Polymer Science, Japan
- The Ceramic Society of Japan
- The Society of Materials Science, Japan
- The Japanese Society for Strength and Fracture of Materials
- The Japan Society for Composite Materials
- Japan Society of Medical Electronics and Biological Engineering

The **financial support** has been made from the following public organizations and corporations:

- The Commemorative Association for the Japan World Exposition(1970)
- Maeda Memorial Engineering Foundation
- The Mitsubishi Foundation
- The Kajima Foundation

- Japan Federation of Construction Contractor
- Japan Automobile Manufacturers Association, Inc.
- The Japan Iron and Steel Federation
- The Federation of Electric Power Companies
- The Tokyo Bankers Association, Inc.
- The Japan Electrical Manufacturers' Association
- Japan Association of Steel Bridge Construction
- The Kozai Club
- The Federation of Pharmaceutical Manufacturers' Association of Japan
- Offshore Steel Structures Contractors Association
- The Japan Gas Association
- Kyoto Industrial Association
- Association of Tokyo Stock Exchange Regular Members
- The Life Insurance Association of Japan
- The Marine and Fire Insurance Association of Japan, Inc.
- Regional Banks Association of Japan
- Trust Companies Association of Japan
- The Japanese Shipowners' Association

- Nippon Steel Corporation
- Mitsubishi Heavy Industries, Ltd.
- Nippon Telegraph and Telephone Corporation
- Ishikawajima-Harima Heavy Industries Co., Ltd.
- Kawasaki Heavy Industries, Ltd.
- Asahi Glass Co., Ltd.
- Fuji Xerox Co., Ltd.
- Hitachi Zosen Corporation
- Kao Corporation
- Komatsu Ltd.
- Mitsui Engineering and Shipbuilding Co, Ltd.
- NEC Corporation
- Sumitomo Heavy Industries, Ltd.
- The Furukawa Electric Co., Ltd.
- Fujitsu Ltd.
- Kandenko Co., Ltd.
- Kinden Corporation
- Sumitomo Chemical Co., Ltd.
- Ricoh Company, Ltd.
- Oki Electric Industry Co., Ltd.
- Sanyu Consultants Inc.

LIST OF PARTICIPANTS

AUSTRALIA (11 participants)

Boger, David V.
Griffiths, Ross W.
Grimshaw, Roger H.J.
Hosking, Roger J.
Jenkins, David R.
Kerr, Ross C.
Muehlhaus, Hans B.
Phan-Thien, Nhan
Soria, Julio
Tanner, Roger I.
Zhang, Liangchi

AUSTRIA (10 participants)

Buryachenko, Valeri A.
Heuer, Rudolf
Kluwick, Alfred
Mang, Herbert A.
Rammerstorfer, Franz G.
Steiner, Wolfgang
Steinrueck, Herbert
Troger, Hans
Wohlhart, Karl B.
Ziegler, Franz F.

BELARUS (2 participants)

Migoun, Nikolai P.
Zhuravkov, Michail A.

BELGIUM (8 participants)

Boulanger, Philippe S.
Crochet, Marcel J.
Dehombreux, Pierre B.
Dimova, Vesselina I.
Duysinx, Pierre J.
Fisette, Paul
Geradin, Michel
Samin, Jean-Claude

BRAZIL (2 participants)

Mamiya, Edgar N.
Mazzilli, Carlos E.N.

BULGARIA (2 participants)

Bontcheva, Nikolina
Ivanov, Tzolo P.

CANADA (16 participants)

Ballyk, Peter D.
Chuang, Jim
Epstein, Marcelo
Graham, G.A.C.
Hoa, Suong Van
Jain, Mukesh K.
Kalamkarov, Alexander L.
Modi, Vinod J.
Paidoussis, Michael P.
Qin, Zhong
Rimrott, Friedrich P.J.
Schiavone, Peter
Sigurdson, Lorenz W.
Szyszkowski, Walerian
Tabarrok, Bez
Trochu, Francois

CHINA (BEIJING) (19 participants)

Chang, Huai-Xin
Chen, Yi-Heng
Cheng, Gengdong
He, You-Sheng
Hwang, Keh-Chih
Liu, Gao-Lian
Wan, Decheng
Wang, Daozeng
Wang, Ren
Xu, Jian-Xue
Xu, Kai-Yu
Yang, Wei
Ye, Qu-Yuan

Yin, Xiao Chun
Yu, Da-Bang
Zhang, Ruo-Jing
Zheng, Quan-Shui
Zhou, You-He
Zhu, Wenhui

CHINA (TAIPEI) (19 participants)

Chao, Ching-Kong
Chen, Chao-Hsun
Chen, Falin
Chen, Kuo-Ching
Chen, Tungyang
Chiang, Chun-Ron
Chiang, Dar-Yun
Chu, Chin-Chou
Chu, Shi-Sheng
Hsu, Tze-Chi
Lei, U.
Ma, Chien-Ching
Miau, Jiun-Jih
Tarn, Jiann-Quo
Wang, An-Bang
Wang, Li-Sheng
Wen, Chih-Yung
Yang, Cheng-Ying
Yang, Yeong-Bin

CZECH REPUBLIC (4 participants)

Dvorak, Rudorf
Klimes, Frantisek
Marsik, Frantisek
Okrouhlik, Miloslav

DENMARK (25 participants)

Andreasen, Jens H.
Bendsoe, Martin P.
Brons, Morten
Condra, Thomas J.
Hansen, John M.
Hansen, Michael R.
Hassager, Ole
Hauggaard, Anders B.

Jacobsen, Christian B.
Jensen, Henrik M.
Jensen, Jens C.
Karihaloo, Bhushan L.
Larsen, Poul S.
Mikkelsen, Lars P.
Niordson, Frithiof
Olhoff, Niels
Pedersen, Niels L.
Pedersen, Pauli
Ravn, Peter
Sigmund, Ole
Sorensen, Jens N.
Terndrup Pedersen, Preben
Thomsen, Ole T.
True, Hans C.G.
Tvergaard, Viggo

EGYPT (2 participants)

Abo-Elezz, Aly E.S.
Ismael, Mohamed Kholoussi

ESTONIA (2 participants)

Engelbrecht, Juri
Lellep, Jaan

FINLAND (4 participants)

Koski, Juhani
Miettinen, Antero
Mikkola, Martti J.
Virtanen, Simo S.

FRANCE (50 participants)

Abdul-Latif, Akrum A.
Anselmet, Fabien
Bacri, Jean-Claude
Baptiste, Didier
Barthelet, Pierre
Barthes-Biesel, Dominique
Bataille, Jean
Ben Hadid, Hamda
Bernadou, Michel J.

Besnard, Didier C.
Boehler, Jean-Paul
Calloch, Sylvain
Chaboche, Jean L.
Chedmail, Patrick
Chen, Gang
Chomaz, Jean-Marc
Choquin, Jean Philippe
Darve, Felix A.
Dias, Frederic
Fabre, Jean
Francois, Marc
Fressengeas, Claude
Germain, Paul
Grediac, Michel
Hopfinger, Emil J.
Huerre, Patrick
Ibrahimbegovic, Adnan
Iooss, Gerard M.
Izrar, Boujema I.
Laadhari, Faduzi
Lagarde, Alexis
Lazarus, Veronique
Leorat, Jacques
Lexcellent, Christian
Licht, Christian
Metais, Olivier
Moes, Nicolas
Moreau, Rene
Perkins, Richard J.
Potier-Ferry, Michel
Pouliquen, Olivier
Pouquet, Annick G.
Risso, Frederic
Rossi, Maurice
Salencon, Jean C.
Sellier, Antoine
Tran-Cong, Sabine
Weiss, Daniel A.
Zaleski, Stephane
Zarka, Joseph

GERMANY (50 participants)

Auerbach, David E.
Bechert, Dietrich W.

Bohn, Dieter E.
Buellesbach, Juergen
Dallmann, Uwe
Deml, Markus
Dickopp, Christian
Ehret, Thorsten O.W.
Erk, Patrick P.
Fauser, Jurgen
Fiebig, Martin
Fuerst, Daniel
Gabbert, Ulrich
Gaul, Lothar
Gersten, Klaus
Greve, Ralf
Grill, Heiner
Hagedorn, Peter
Hofmann, Bernd
Hutter, Kolumban
Kalkuehler, Kathrin
Koschel, Wolfgang W.
Krause, Egon
Kreuzer, Edwin J.
Kuhn, Guenther R.
Lippmann, Horst
Maisser, Peter
Meier, Gerd E.A.
Mitra, Nimai
Mohring, Willi
Mueller, Ulrich
Niebergall, Mathias
Obrecht, Hans
Pfeiffer, Friedrich G.
Popp, Karl
Rozvany, George
Schiehlen, Werner O.
Schnack, Eckart
Schnerr, Guenter H.
Schweizerhof, Karl
Stein, Erwin
Vasanta Ram, Venkatesa I.
Warnatz, Juergen
Wauer, Joerg
Wechsler, Klaus F.
Wedig, Walter V.
Weyh, Bernhardt
Wittenburg, Jens

Wunderlich, Walter
Zhou, Ming

GREECE (2 participants)

Kotsovinoy, Despina E.
Kounadis, Anthony N.

HONG KONG (7 participants)

Chau, Kam T.
Tang, Shiu K.
Tong, Pin
Vladimirov, Vladimir A.
Wong, Chun Hung
Wu, Charles C.K.
Yu, Tong-Xi

HUNGARY (4 participants)

Beda, Peter B.
Kaliszky, Sandor
Stepan, Gabor
Tarnai, Tibor

INDIA (4 participants)

Agrawal, Sharad Chandra
Bhargava, Raj Rani
Narasimha, Roddam
Singh, Shamsher B.

IRELAND (1 participant)

Hayes, Michael A.

ISRAEL (19 participants)

Benveniste, Yakov
Bodner, Sol R.
Chiskis, Alexander
Elad, David
El-Boher, Arik
Elperin, Tov
Elyada, Dov I.
Hashin, Zvi

Igor, Ragachevskii
Muravskii, Grigori
Nir, Avinoam
Parnes, Raymond
Pismen, Leonid M.
Rubin, Miles B.
Shliomis, Mark I.
Solan, Alexander
Tumin, Anatoli
Volokh, Konstantin Y.
Weinstein, Mordechai

ITALY (18 participants)

Benedettini, Francesco C.
Bigoni, Davide
Brancaleoni, Sabio
Carpinteri, Alberto
Cercignani, Carlo
Chiaia, Bernardino M.
Chiandussi, Giorgio
Contro, Roberto
Corigliano, Alberto
D'Acquisto, Leonardo
Luciano, Raimondo
Maier, Giulio
Nardinocchi, Paola
Nova, Roberto
Podio-Guidugli, Paolo
Sacchi-Landriani, Giannantonio
Schrefler, Bernhard
Vatta, Furio

JAPAN (333 participants)

Abe, Hiroyuki
Abe, Masato
Abe, Takeji
Adachi, Junji
Adachi, Taiji
Aihara, Yasuhiko
Aizawa, Tatsuhiko
Akamatsu, Teruaki
Akishita, Sadao
Andoh, Hiroteru
Aoki, Kazuo

Araki, Keisuke
Araki, Yoshikazu
Arimitsu, Yutaka
Asai, Masahito
Ashida, Fumihiro
Biwa, Shiro
Chen, Xinzhong
Ding, Li
Egami, Yasuhiro
Fujii, Fumio
Fujii, Ikuya
Fujikawa, Shigeo
Fujimura, Kaoru
Fujino, Yozo
Fujita, Hajime
Fujita, Takafumi
Fukushima, Takayuki
Fukuyu, Akio
Funaki, Toshihiko
Funakoshi, Mitsuaki
Furukawa, Tomonari
Furuta, Hitoshi
Furuya, Hiroshi
Furuya, Osamu
Goto, Seiji
Goto, Tomonobu
Gotoh, Kanefusa
Goya, Moriaki
Hagiwara, Yoshimichi
Hanazaki, Hideshi
Hangai, Yasuhiko
Hara, Toshiaki
Hasebe, Norio
Hasegawa, Tomiichi
Hashiba, Kunio
Hashimoto, Yoshio
Hasimoto, Hidenori
Hata, Toshiaki
Hayase, Toshiyuki
Hayashi, Kunio
Henderson, Le Roy F.
Higashino, Fumio
Hikihara, Takashi
Hildebrand, Brian G.
Hino, Mikio
Hirahara, Hiroyuki

Hirasawa, Toku
Hirayama, Hirohumi
Honda, Akihiro
Hori, Muneo
Horii, Hideyuki
Ichihara, Mie
Ichijo, Makoto
Iemura, Hirokazu
Igarashi, Akira
Iida, Akiyoshi
Iida, Oaki
Ikawa, Nozomu
Ikeda, Toru
Imai, Isao
Imanishi, Etsujiro
Inamuro, Takaji
Inokuti, Hiroo
Inoue, Hirotsugu
Inoue, Tatsuo
Ishihara, Masayuki
Ishihara, Takashi
Ishii, Katsuya
Ishii, Ryuji
Ishii, Yoshio
Ito, Hidesato
Ito, Manabu
Itoh, Nobutake
Itoh, Sumiko
Iwai, Takashi
Ju, Dong-Ying
Kaji, Shojiro
Kakutani, Tsunehiko
Kambe, Tsutomu
Kameda, Masaharu
Kamiya, Kunio
Kamiyama, Shinichi
Kasagi, Nobuhide
Kato, Chisachi
Kato, Yukari
Katsuyama, Tomoo
Kawaguchi, Akihisa
Kawahara, Genta
Kawahara, Takuji
Kawai, Masamichi
Kawamura, Ryusuke
Kawashima, Koichiro

Kawata, Kozo
Kawatate, Kazuo
Kida, Sigeo
Kida, Teruhiko
Kimura, Kichiro
Kimura, Yoshifumi
Kinoshita, Takeshi
Kishimoto, Kikuo
Kitagawa, Tetsuya
Kiya, Masaru
Kobayashi, Ryoji
Kobayashi, Shoichi
Kobori, Takuji
Koga, Tatsuzo
Kogiso, Nozomu
Kohama, Yasuaki
Komori, Satoru
Kondo, Jiro
Kong, Lingzhe
Kosaka, Ikuo
Krishna Rao, Juvva V.S.
Kubota, Yuzuru
Kumagai, Teruo
Kumazaki, Ikutaro
Kundu, Sourav
Kunoh, Takahiko
Kurino, Haruhiko
Kuwahara, Kunio
Li, Hsi-Shang
Li, Xin-Zen
Liu, Yufu
Maruta, Eizo
Matsui, Goichi
Matsumoto, Akira
Matsumoto, Masaru
Matsumoto, Tatsuji
Matsumoto, Yoichiro
Matsunaga, Yasuhiro
Matsuo, Minoru
Matsuo, Yoshimasa
Meguro, Toshikatsu
Meisner, Mark J.
Miki, Mitsunori
Miwa, Masataka
Miyajima, Mitsuharu
Miyazaki, Takeshi

Mizuno, Mamoru
Mizuta, Yo
Mochizuki, Osamu
Monji, Hideaki
Motogi, Shinya
Murakami, Masahide
Murakami, Sumio
Murakami, Youichi
Murakami, Yukitaka
Murotsu, Yoshisada
Naert, Antoine
Nagai, Ken-ichi
Nagai, Kenichiro
Nagaki, Shigeru
Nagano, Katsutoshi
Nagata, Kazutoshi
Nagata, Kouji
Nakabayashi, Koichi
Nakagawa, Hiroji
Nakajima, Fumio
Nakamachi, Eiji
Nakamura, Ikuo
Nakamura, Shigehisa
Nakamura, Tsuneyoshi
Nakamura, Yoshiya
Nakanishi, Masato
Nakano, Ryoki
Namba, Masanobu
Narumi, Takatsune
Nemat-Alla, Mahmound
Nishijima, Yasunori
Nishimura, Fumihito
Nishimura, Kouichi
Nishimura, Naoshi
Nishinari, Katsuhiro
Nishioka, Kazumi
Nishioka, Michio
Noda, Naotake
Noguchi, Hirohisa
Nohguchi, Yasuaki
Nonaka, Taijiro
Noto, Katsuhisa
Ogino, Fumimaru
Ohaba, Motoyoshi
Ohashi, Hideo
Ohba, Kenkichi

Ohji, Kiyotsugu
Ohkitani, Koji
Ohkubo, Sadaji
Ohno, Nobutada
Ohta, Yoshiki
Oka, Fusao
Oka, Kenji
Oka, Yasushi
Okada, Takehide
Onishi, Yoshimoto
Ootao, Yoshihiro
Osyczka, Andrzej
Ota, Yukio
Poovarodom, Nakhorn
Qian, Jun
Saijo, Kenji
Saitou, Masatoshi
Saka, Masumi
Sakai, Chifuyu
Sakai, Yasuhiko
Sakamoto, Makoto
Sano, Osamu
Sanyal, Dipayan
Sasaki, Katsuhiko
Satake, Masao
Satake, Shinichi
Satofuka, Nobuyuki
Sawada, Toshio
Sha, Weiming
Shao, Shaowen
Shibata, Heki
Shibutani, Yoji
Shibuya, Toshikazu
Shigeishi, Mitsuhiro
Shindo, Yasuhide
Shintani, Kazuhito
Shirato, Hiromichi
Shunei, Mekaru
Sone, Yoshio
Song, Wusheng
Sudo, Seiichi
Sugihara-Seki, Masako
Sugimori, Tadayuki
Sugimoto, Hiroshi
Sugimoto, Nobumasa
Sugimoto, Takeshi

Sugiura, Kunitomo
Sumi, Yoichi
Sunada, Shigeru
Suzuki, Katsuhiro
Suzuki, Kenjiro
Suzuki, Shinichi
Suzuki, Yuji
Tadano, Shigeru
Takagi, Shohei
Takagi, Shu
Takahashi, Kiyoshi
Takahira, Hiroyuki
Takai, Shinro
Takano, Naoki
Takano, Yasunari
Takaoka, Masanori
Takasugi, Nobuhide
Takayama, Kazuyoshi
Takewaki, Izuru
Takezono, Shigeo
Tamagawa, Masaaki
Tamaki, Toshihiro
Tamura, Takeshi
Tamura, Yukio
Tanabe, Yuji
Tanaka, Eiichi
Tanaka, Kikuaki
Tanaka, Mitsuru
Tani, Junji
Tanifuji, Katsuya
Tanigawa, Yoshinobu
Tanishita, Kazuo
Tao, Katsumi
Tatsumi, Tomomasa
Terada, Kazuko
Terao, Ken
Tobushi, Hisaaki
Tokuda, Masataka
Tokugawa, Naoko
Tokura, Ikuo
Tomita, Yoshihiro
Touhei, Terumi
Toyoda, Yukihiro
Tsuchida, Eiichiro
Tsumori, Fujio
Tsuta, Toshio

Tsuyuki, Koji
Ueda, Yoshisuke
Uehara, Takuya
Ueno, Kazuyuki
Uetani, Koji
Ujihashi, Sadayuki
Umeda, Akira
Umeda, Yoshikuni
Umeki, Makoto
Utsunomiya, Tomoaki
Venkataramana, Katta
Wada, Akira
Wakui, Takashi
Warkentin, David J.
Washio, Toshikatsu
Watanabe, Eiichi
Watanabe, Kazuhiro
Watanabe, Osamu
Wells, John C.
Willmott, Alexander P.
Woods, William P.
Wu, Lin Z.
Wu, Xu
Yagawa, Genki
Yamada, Hiroshi
Yamada, Kazuhiko
Yamada, Michio
Yamada, Seishi
Yamada, Takahiro
Yamaguchi, Akimasa
Yamaguchi, Hiroki
Yamakawa, Hiroshi
Yamamoto, Kyoji
Yamamoto, Yoshiyuki
Yamamura, Masato
Yamashita, Minoru
Yamazaki, Koetsu
Yanase, Shinichiro
Yokota, Rikizo
Yokoyama, Takashi
Yoshida, Fusahito
Yoshikawa, Nobuhiro
Yoshimura, Hiroaki
Yoshimura, Shinobu
Yoshinaga, Takao
Yoshizawa, Masatsugu

KOREA (11 participants)

Choi, Deok-Kee
Choi, Hang S.
Im, Jong S.
Im, Seyoung
Kim, Sang-Hwan
Kim, Seung J.
Lee, Duck-Joo
Park, Jinho
Park, Young-Sun
Suh, Yong-Kweon
Yoo, Seung-Hyun

LATVIA (2 participants)

Blums, Elmars
Tamuzs, Vitauts

LITHUANIA (1 participant)

Vasauskas, Vytautas

MEXICO (1 participant)

Davalos-Orozco, Luis A.

NETHERLANDS (33 participants)

Aarts, Annemarie C.T.
Baaijens, Frank P.
Biesheuvel, Arie
Boomkamp, Paul A.M.
Bulthuis, Hindrik F.
Coene, Rene
de Bruin, Gerrit J.
Dieterman, Harry A.
Groen, Arend E.
Hoeijmakers, Hendrik W.M.
Kalker, Joost J.
Koppens, Willy D.
Koren, Barry
Krasnopolskaia, Tatiana S.
Li, Zi-Li
Meijaard, Jacob P.
Meijers, Pieter

Meleshko, Viatcheslav V.
Mellema, Jorrit
Periard, Frederic J.
Poorte, Edwin
Sluys, Lambertus J.
Spelt, Peter D.M.
Steenbrink, Alexander C.
van Campen, Dick H.
van de Ven, Alphons A.F.
van Heijst, G.J.F.
van Mier, Jan G.M.
van Wijngaarden, Leen
Verschueren, Maykel
Vosse, Frans N.
Woering, Arend A.
Wolfert, A.R.M.

NEW ZEALAND (3 participants)

Collins, Ian F.
Remennikov, Alexander M.
Sneyd, Alfred D.

NORWAY (2 participants)

Dysthe, Kristian B.
Irgens, Fridtjov

POLAND (21 participants)

Basista, Michal
Bauer, Jacek
Blajer, Wojciech
Bodnar, Adam
Chrzanowski, Marcin
Chudzikiewicz, Andrzej
Dems, Krzysztof
Gutkowski, Witold
Jarzebowski, Andrzej
Korbel, Andrzej
Kowalewski, Tomasz A.
Maciejewski, Jan
Mroz, Zenon
Muc, Aleksander
Nowacki, Wojciech K.
Pecherski, Ryszard B.

Perzyna, Piotr
Petryk, Henryk
Szefer, Gwidon
Tylikowski, Andrzej
Waniewski, Maciej

PORTUGAL (7 participants)

Ambrosio, Jorge A.C.
Camotim, Dinar R.Z.
Martins, Joao A.C.
Mascarenhas, Maria-Luisa
Negrao, Joao H.J.O.
Oliveira, Eduardo A.
Trabucho, L.

RUSSIA (27 participants)

Asmolov, Evgeny S.
Chashechkin, Yuli D.
Chernyi, Georimir
Chernyshenko, Sergei I.
Chugunov, Vladimir A.
Fedorchenko, Alexander T.
Fomin, Sergei A.
Frolov, Konstantin V.
Golub, Victor
Kazakov, Alexander V.
Kedrinskii, Valery K.
Kopiev, Victor F.
Korobeinikov, Victor P.
Kozlov, Viktor V.
Kuryachii, Alexander P.
Liberzon, Mark R.
Manuilovich, Sergei V.
Mikhailov, Gleb K.
Neyland, Vera M.
Nezlin, Mikhail V.
Nikitin, Nikolai V.
Obrezanova, Olga A.
Prostokishin, Valerii M.
Salamatin, Andrey N.
Selyugin, Sergei V.
Sorokin, Sergey V.
Tazioukov, Faruk Kh.

SERBIA (1 participant)

    Milosevic, Vesna O.

SINGAPORE (1 participant)

    Wang, Yao-Ping

SLOVAKIA (1 participant)

    Brilla, Jozef

SLOVENIA (1 participant)

    Mejak, George

SOUTH AFRICA (4 participants)

    Du Toit, Charl G.
    Ibragimov, Nail H.
    Mason, David P.
    Momoniat, Ebrahim

SWEDEN (16 participants)

    Adolfsson, Erik M.
    Andersson, Lars-Erik V.
    Bostroem, Anders E.
    Drugge, Lars J.
    Enelund, Mikael
    Gupta, Ram B.
    Hallstroem, Stefan
    Jansson, Per A.
    Krenk, Steen
    Lundberg, Bengt
    Mahler, Lennart
    Nygren, Tomas
    Sjostrom, Soren
    Stensson, Annika E.
    Storakers, Bertil M.
    Wirdelius, Hakan

SWITZERLAND (13 participants)

    Beguelin, Philippe
    Chen, Thomas

    Dual, Jurg
    Huet, Christian
    Isay, Michael P.
    Mazza, Edoardo
    Monkewitz, Peter A.
    Nakkasyan, Alice
    Roussopoulos, Kimon
    Sayir, Mahir B.
    Schultz, Marcus
    Schwab, Christoph
    Vollmann, Johannes

THAILAND (2 participants)

    Senjuntichai, Teerapong
    Wijeyewickrema, Anil C.

UK (66 participants)

    Babitsky, Vladimir I.
    Bajer, Konrad
    Bernasconi, Daniel J.
    Blachut, Jan
    Blake, John R.
    Calladine, Christopher R.
    Carpenter, Peter W.
    Choi, Kwing-So
    Chui, Atta Y.K.
    Crighton, David G.
    Croll, James G.A.
    Davidson, Peter A.
    Davies, Christopher
    Dellar, Paul J.
    Dowling, Ann P.
    Duck, Peter W.
    Economou, Chrisanthi
    Ffowcs Williams, John E.
    Fradkin, Larissa J.
    Fu, Yibin
    Gordon, Timothy J.
    Harlen, Oliver G.
    Haughton, David M.
    Healey, Jonathan J.
    Hogan, Stephen J.
    Hoyle, Rebecca B.
    Hughes, David W.

Hunt, Julian C.R.
Jones, Mark C.W.
Kazakoff, Alexander B.
Ker, Robert F.
Leslie, Frank M.
Lighthill, James
Linden, Paul F.
Lingwood, Rebecca J.
Lucey, Anthony D.
Luo, Xiao-Yu
Mak, Vincent W.S.
Moffatt, H.Keith
Movchan, Alexander B.
Movchan, Natalia V.
Nagata, Masato
Pan, Jingzhe
Parker, David F.
Peake, Nigel
Pedley, Timothy J.
Pellegrino, Sergio
Price, William G.
Proctor, Michael R.E.
Raghunathan, Srinivasan
Reid, Stephen R.
Ricca, Renzo L.
Rucklidge, Alastair M.
Smith, Frank T.
Smyshlyaev, Valery P.
Soldatos, Kostas P.
Soward, Andrew M.
Spencer, Anthony J.M.
Stronge, William J.
Thomas, Peter J.
Toropov, Vassili V.
Tutty, Owen R.
Walters, Ken
Watterson, John K.
Webster, Mike F.
Willis, John R.

UKRAINE (4 participants)

Berbyuk, Victor E.
Guz', Alexander N.
Kostenko, Jewgenij J.
Lvov, Guennadi

USA (107 participants)

Achenbach, Jan D.
Acrivos, Andreas
Agrawal, Sunil K.
Akylas, Triantaphyllos R.
Aref, Hassan
Atassi, H.M.
Bajaj, Anil K.
Baker, A.J.
Balachandran, Bala
Bammann, Douglas J.
Banerjee, Sanjoy
Batra, Romesh C.
Bau, Haim H.
Bauchau, Olivier A.
Bogy, David B.
Boley, Bruno A.
Brady, John F.
Brennen, Christopher E.
Brown, Robert A.
Buckingham, Michael J.
Buckmaster, John D.
Chaudhry, Hans R.
Chen, Chuan F.
Childress, W.Stephen
Christensen, Richard M.
Colonius, Timothy E.
Crandall, Stephen H.
Crocker, Malcolm J.
Dempsey, John P.
Denda, Mitsunori
Findley, Thomas W.
Freeman, Jeffrey S.
Freund, L.B.
Gidaspow, Dimitri
Grenestedt, Joachim L.
Gu, Pei
Haftka, Raphael T.
Hanagud, Satnya V.
Hara, Tetsu
Hetnarski, Richard B.
Hilburger, Mark W.
Hodge, Philip G.
Hornung, Hans G.
Hsieh, Tsuying

Huang, T.C.
Hueckel, Tomasz
Jensen, Soren S.
Jimenez, Javier
Joseph, Daniel D.
Kerr, R.M.
Kim, Kwang Y.
Knio, Omar M.
Koplik, Joel
Krajcinovic, Dusan
Laursen, Tod A.
Leal, L.Gary
Leibovich, Sidney
Leissa, Arthur W.
Lele, Sanjiva K.
Li, Keyu
Li, Ming
Lin, Xiao
Mal, Ajit K.
Markenscoff, Xanthippi
Masri, Sami F.
Matalon, Moshe
McCulloch, Andrew D.
Melville, W.Kendall
Milton, Graeme W.
Moon, Francis C.
Moser, Robert D.
Mura, Toshio
Nemat-Nasser, Sia
Nosenchuck, Daniel M.
Oden, J. Tinsley
Orlandea, Nicolae V.
Ostoja-Starzewski, Martin
Pan, Yu
Phillips, Owen M.

Phillips, William R.C.
Ponte-Castaneda, Pedro
Prosperetti, Andrea
Rajapakse, Yapa D.S.
Ravichandran, G.
Renardy, Michael
Renardy, Yuriko
Rodin, Gregory J.
Schreyer, Howard L.
Shaqfeh, Eric S.G.
Shtern, Vladimir N.
Stadler, Wolfram
Steele, Charles R.
Stewart, Donald S.
Talreja, Ramesh R.
Tamura, Shigeyuki
Tan, Benkui
Tortorelli, Daniel A.
Tryggvason, Gretar
Tsitverblit, Naftali A.
Voyiadjis, George Z.
Vu-Quoc, Loc
Walker, J.David A.
Wright, Paul K.
Wu, Chien H.
Yong, Zhou
Zhang, Ruichong
Zikry, Mohammed A.

YUGOSLAVIA (1 participant)

Ruzic, Dobroslav D.

TOTAL 939 participants

REPORT ON THE CONGRESS

The decision to accept the invitation from Japan to hold the XIXth International Congress in Kyoto was taken by the Congress Committee of IUTAM during its meeting in Haifa in August 1992. It was decided that the main elements of the program would be the same as those established for the Congress at Lyngby in 1984, at Grenoble in 1988 and at Haifa in 1992, namely Opening and Closing General Lectures, 15 Sectional Lectures and a large number of Contributed Papers contained in parallel Lecture Sessions and in parallel Seminar Presentation Sessions. Although the pattern of the program for the XIXth Congress is the same as that of the previous three Congresses, the Congress Committee has implemented some new ideas and concepts in its planning for the Congress in order to promote broader participants, in particular by scientists from outside the host country. Thus, for the XIXth Congress the Committee agreed to increase the number of Mini-Symposia from three to six, and to select about 40 Pre-nominated Sessions on typical topics of Mechanics which are not covered by the Mini-Symposia.

The Congress Committee selected the following six topics for the Mini-Symposia.

MF1. Vorticity Dynamics and Turbulence
MF2. Non-Newtonian Fluid Flow
MF3. Aero- and Hydroacoustics
MS1. Mechanics of Heterogeneous and Composite Solids
MS2. Structural Optimization
MS3. Solid Mechanics in Manufacturing

Like those of the Mini-Symposia, the topics of the Pre-nominated Sessions were listed in the Announcement of the Congress along with the names of the Chairpersons for the Sessions.

The Congress Committee also selected two general lecturers, Professor Takuji Kobori (Japan) to present the Opening Lecture and Professor Sir James Lighthill (UK) to present the Closing Lecture, as well as 15 Sectional Lecturers. The Chairpersons of the Mini-Symposia further selected 21 Introductory Lectures for their symposia.

The contributed papers were presented in parallel sessions, either as 20 minutes Lectures or in Seminar-Presentation Sessions which were scheduled separately from the lectures. There were lively discussions both after the lectures and in the Seminar-Presentation Sessions, where the second half of the Sessions were devoted to general discussions guided by the Chairpersons.

The scientific program of the Congress was presented from Monday afternoon till Friday evening in 7-8 parallel sessions. The Seminar-Presentation Sessions were held in the Event Hall of KICH in 9 partitioned spaces on Tuesday afternoon and Thursday afternoon, at times when no other Lecture Sessions were in progress. This resulted in good attendance and very lively discussions.

The detailed statistics of the Congress are presented on the following page. It is noted that the proportion of acceptances to submissions was almost 1/2. The meaning of the symbols used in the following listing by countries is as follows.

S   Submitted abstracts
A-L   Accepted as Lecture
A-SP   Accepted as Seminar-Presentation
L   Presented as Lecture
SP   Presented as Seminar-Presentation
IL   Invited Lecture

COUNTRY	PAPERS						PARTICIPANTS
	S	A-L	A-SP	L	SP	IL	
ALGERIA	1	1					
ARMENIA	2						
AUSTRALIA	21	7	4	7	2	1	11
AUSTRIA	10	7	3	7	3		10
AZERBAIJAN	4		1				
BAHRAIN	2						
BELARUS	11	3	4	1	1		2
BELGIUM	7	3	3	3	3	1	8
BRAZIL	8	1	1	1	1		2
BULGARIA	4	2	2	1	1		2
CANADA	31	11	7	8	5	2	16
CHILE	1						
CHINA - BEIJING	116	15	21	9	9	1	19
CHINA - TAIPEI	32	8	9	7	9		19
CZECH REPUBLIC	10	1	2	1	2		4
DENMARK	24	14	6	13	5	1	25
EGYPT	2						2
ESTONIA	4	1	2	1			2
FINLAND	5		1		1		4
FRANCE	77	31	18	27	13	3	50
GERMANY	86	42	20	33	12	2	50
GREECE	3		1		1		2
HONG KONG	15	3	5	2	4		7
HUNGARY	5	1	1	1	1		4
INDIA	40	3	5		3	1	4
IRAN	3						
IRELAND							1
ISRAEL	30	18	5	13	4	1	19
ITALY	30	12	8	8	3	1	18
JAPAN	308	101	102	98	90	4	333
KAZAKHSTAN	3						
KOREA	19	1	9	1	6		11
KUWAIT	1						
LATVIA	11	2	2	1	0	1	2
LITHUANIA	4		1		1		1
MACAU	5						
MACEDONIA	2						
MALAYSIA	1						
MEXICO	3	1		1			1
NETHERLANDS	39	21	7	21	6	1	33
NEW ZEALAND	2	1	1	1	1		3
NORWAY	4	2	1	2			2
PAKISTAN	1						
POLAND	32	10	9	8	8	1	21
PORTUGAL	10	3	4	3	4		7
ROMANIA	10		2				
RUSSIA	186	19	38	14	9		27
SERBIA	1		1		1		1
SINGAPORE	2		1		1		1
SLOVAKIA							1
SLOVENIA	1	1		1			1
SOUTH AFRICA	6	2	1	1	1		4
SPAIN	4	1	1				
SWEDEN	14	3	4	2	4	1	16
SWITZERLAND	20	13	4	12	3		13
THAILAND	3		2		2		2
TURKY	2						
UKRAINE	56	1	13		3		4
UNITED KINGDOM	110	54	15	44	10	4	66
USA	182	88	26	68	11	12	107
UZBEKISTAN	4		1				
VIETNAM	3						
YEMEN	1						
YUGOSLAVIA	8						1
TOTAL	1642	508	374	421	244	38	939

└ 882 ┘ └ 665 ┘

└── 703 ──┘

OPENING CEREMONY

The Opening Ceremony of the Congress was held in the Main Hall of the Kyoto International Conference Hall (KICH), at 10 am. on Monday, 26th August, 1996. Professor Eiichi Watanabe, Secretary General of the Congress, served the chair of the Opening Ceremony.

"Koto" (Japanese Harp) performance

Before the opening addresses, "Koto" (Japanese Harp) performance has been held for about fifteen minutes by five lady players from Osaka.

Opening by Professor Yoshiyuki Yamamoto, Chairman of Local Organizing Committee

Professor Yoshiyuki Yamamoto, the Chairman of the Local Organizing Committee, opened the Opening Ceremony as follows.

"Chairman, Distinguished Guests, Ladies and Gentlemen. It is my great honor to open the 19th International Congress of Theoretical and Applied Mechanics and to greet this outstanding gathering of more than one thousands scientists and engineers, which will ensure the success of the Congress. On behalf of the Local Organizing Committee, I welcome you all to Japan and Kyoto. It is a great pleasure to inform that this is the first International Congress of Theoretical and Applied Mechanics ever held in the East Asia. Kyoto had been the capital of Japan for eleven centuries since its foundation in 794, until 1869.

Taking this opportunity, I would like to introduce the Dais to you. Professor Leen van Wijngaarden, President of IUTAM. Professor Paul Germain, Vice-President of IUTAM. Professor Bruno A. Boley, Treasurer of IUTAM. Professor Franz Ziegler, Secretary General of IUTAM. Professor Niels Olhoff, Secretary of the Congress Committee of IUTAM. Professor Tomomasa Tatsumi, President of the Congress. Professor Hiroo Inokuchi, Representative of the Science Council of Japan. Mr. Yorikane Masumoto, Mayor of Kyoto City. Professor Manabu Ito, Chairman of the Finance Committee of the Congress.

Ladies and Gentlemen. The hottest season in Kyoto is over. So, I believe, you would like this lovely city very much, because there are so many interesting places to visit, thousands of shrines and temples, beautiful gardens, for example. We are honored to have you here and happy to host the 19th ICTAM in this traditional city of Kyoto. I sincerely hope that you will enjoy yourselves at the Congress. Thank you very much."

Opening Address by Professor Tomomasa Tatsumi, President of the XIXth ICTAM

Professor Tomomasa Tatsumi, the Chairman of the Local Executive Committee and the President of the 19th International Congress of Theoretical and Applied Mechanics opened the XIXth ICTAM by the following words.

"Dr. President, Distinguished Guests, Honorable Delegates, Ladies and Gentlemen. It is my great honor and privilege to be able to announce the opening of the 19th International Congress of Theoretical and Applied Mechanics. I wish to extend my hearty welcome to you, who have come from all parts of the world in order to attend this Congress.

As you know, ICTAM has a long and wonderful history of more than 70 years. In 1922, an informal conference on Applied Mechanics was organized by a young Hungarian scientist, Theodore von Karman and an Italian mathematician, Tullio Levi-Civita in Innsbruck, Austria. Shortly after this conference, in 1924, the first International Congress of Applied Mechanics was held in Delft, the Netherlands, with a partnership of Dutch scientist, Jan M. Burgers. We can easily observe in these first conferences a strong wish of the organizers to identify Applied Mechanics as a new sector of scientific activity, bordering on Physics and Mathematics on one hand and Engineering on the other. The succeeding long history of ICTAM seems to prove how this wish and scope of the ancestors of Applied Mechanics were far reaching and fruitful. Up to the present Congress, we can count nineteen ICTAMs, ever increasing in their scales and contents. Also a number of IUTAM Symposia and Summer Schools are held every year.

The huge volume of outcome of these IUTAM conferences provide us with precious and important sources of information for all fields of science and technology. Now, Theoretical and Applied Mechanics does not only constitute a central core of modern science and technology but it provides us with a powerful driving force for developments of modern production industries.

In the latter half of this Century, Theoretical and Applied Mechanics has become to have less commitment with military engineering and to be more involved in problems of human welfare, such as protection of global environments or reduction of natural disasters. You may easily notice such a trend in the subject of the Opening and Closing Lectures of this Congress. The former is "Structural Control for Large Earthquakes" by Professor Takuji Kobori, and the latter is "Typhoons, Hurricanes and Fluid Mechanics" by Professor Sir James Lighthill.

Actually, Mechanics itself is an old branch of science, which can dates back to the days of Newton, Galileo, or even Aristotle. But its recent developments are quite new. Discovery of "Soliton" of water waves has clearly shown that the double image of "particle and wave" is not necessarily limited to quantum-mechanical microscale phenomena. Study of "Chaos" in dynamical systems has led to a drastic change in philosophical concepts of "determinism" and "randomness." Observation of "Fractal" structure of turbulent motions has resulted in finding the beauty of "virtual reality." Now, Theoretical and Applied Mechanics seems to have infinite potentiality of evolution.

We are looking forward to listening and learning in this Congress the whole outcome of such developments in Theoretical and Applied Mechanics. In conclusion, I wish to express my hearty thanks to all people and institutions, who kindly helped us in all possible ways for materializing this Congress in Kyoto. Thank you."

Opening Address by Professor Leen van Wijngaarden, President of IUTAM

Professor Leen van Wijngaarden, the President of IUTAM addressed the gathering.

"Dear Mayor of Kyoto, Dear colleagues representing Japanese scientific authorities, Dear colleagues from the IUTAM Bureau, Dear colleagues from all parts of the world, Ladies and Gentlemen. On behalf of IUTAM, it is my pleasure to welcome you here in Kyoto, where we are looking forward to a nice and fruitful Congress in Mechanics. As Professor Tatsumi already pointed out, Mechanics is a very old science going back to Newton, Galileo and even further, and Japan is very old country with large and very impressive traditions. So I think that it is not only Mechanics, that is attracting us to Japan, but also beautiful temples, shrines, and the country as a whole, and thus IUTAM has come to Japan. Earlier there have been many IUTAM symposia on specific subjects in Japan but never such a big gathering as we have now.

Now, you will discover that if you are looking into the Japanese characters and traditions, the Japanese style is shy. The symbol of this is a story about Mt. Fuji. If you perhaps traveled from Tokyo to Kyoto by Shinkansen, you passed along Mt. Fuji. But sometimes, Mt. Fuji is in the clouds and you cannot see the top. The Japanese say that it is because Mt. Fuji is very shy and does not want to be seen. And so shy I feel today as representing IUTAM because we are only 50 years old. You will find in your registration case a small booklet giving you some of the details of how IUTAM was born 50 years ago. This is nothing in comparison with the age of the monuments which you will find here. So IUTAM feels very shy to be here in such a big hall in such an old and impressive country.

But we can make ourselves visible in Japan and in the world by having a nice and good Congress. Over the last four years, the Congress Committee and the Executive Committee have been working very hard and we have new things like Prenominated Sessions, like six in stead of three Mini-Symposia, and a many other new aspects to this Congress and the people that have organized and prepared these have been working very hard. I remember that at our last meeting in Vienna last year we said to each other that we are looking forward to good wish for August 1996. Now it is August 1996 and we are here, but this will not be a success because of all the preparations that has been done, but this can only be a success by the collective work of all of you participating in the congress and bringing good papers on new developments in Mechanics. Apart from that, enjoy yourself in the city of Kyoto and in the country of Japan. It is again my pleasure to welcome you, expressing a hope and full expectation that we will have a very nice Congress here in Japan. Thank you."

Greetings by Professor Hiroo Inokuchi, Representative of Science Council of Japan

Professor Hiroo Inokuchi, the Representative of the Science Council of Japan, brought the greetings to the Congress participants.

"Good morning, Mr. Chairman, Ladies and Gentlemen. It is a great honor for me to have an opportunity to speak on behalf of the Science Council of Japan to people gathering from over 50 countries for the 19th International Congress of Theoretical and Applied Mechanics.

The Science Council of Japan was established in 1949 as a governmental organization representing qualified Japanese scientists, covering all fields of humanity, social and natural sciences. The International Union of Theoretical and Applied Mechanics is one of the international academic organizations where the Council is affiliated to, and the Liaison Committee for Mechanics of the Council is acting as the local committee of IUTAM.

We have organized the present Congress jointly with the Japan Society of Civil Engineers, the Japan Society of Mechanical Engineers, and the Architectural Institute of Japan. It is my great pleasure to meet distinguished mechanical scientists from all over the world and to have an opportunity to attend fascinating lectures and seminar presentations. We regard it a very delightful and meaningful event to have the present Congress for the first time in Japan since its first Congress held in Delft in 1924.

The IUTAM is the sole international academic organization dedicated to promotion of research in Mechanics, which have been continuously expanding and diversifying year after year. In the present Congress, it is expected that a number of research outcomes will be reported across various research fields, and intensive discussions will take place on future directions of research, problems of primary importance, international collaboration and so on.

The site of the Congress, Kyoto is located in the center of Japan and flourished as a political, economic and cultural center of Japan for over 1,000 years since 794. Kyoto is provided not only with beautiful natural scenery and cultural heritage but also with the tradition of science and advanced technology. I hope you will bring home some elements of Japanese traditional culture as well as the latest achievements in Mechanics through your participation in the Congress.

In conclusion, I sincerely wish a great success of the 19th ICTAM and hope that the present Conference will become a really memorable one for all the participants. Thank you very much."

Message from Mr. Ryutaro Hashimoto, Prime Minister of Japan

Professor Watanabe presented to the audience the following message from the Prime Minister, Mr. Ryutaro Hashimoto.

"I am pleased to extend a hearty welcome to all delegates of the world at the opening of the 19th International Congress of Theoretical and Applied Mechanics in Kyoto. I wish a great success in this International Congress for the creative advancement of Mechanics."

Welcome Address by Mr. Yorikane Masumoto, Mayor of Kyoto City

Mr. Yorikane Masumoto, the Mayor of Kyoto City, made his welcome address as follows.

"Ladies and Gentlemen. I wish to express my sincere congratulation to the opening of the 19th International Congress of Theoretical and Applied Mechanics attended by a large number of participants from all over the world. It is my great pleasure and honor to be able to extend my hearty welcome to all of you on behalf of citizens of Kyoto city. I also wish to thank you for that our city of Kyoto has been chosen as the venue for the present 1996 Congress.

Kyoto City has a long history of twelve hundred years as the capital of Japan, and has been developed as the cultural core of this country. Now Kyoto City enjoys beautiful natural environments and inherits from its long history a large number of cultural heritage, including fourteen World Cultural Heritages nominated by UNESCO. Indeed, the City is not only a center of traditional Japanese culture but also the flourishing place of modern advanced engineering. Kyoto includes in its civic area, thirty-nine Universities, including several centers of excellence, and representative companies and enterprises of high technology. Thus, it seems quite adequate that the present Congress, which is the first IUTAM Congress in Asia and Pacific regions, has been decided to take place in Kyoto, and I take it a great honorable decision for us.

Theoretical and Applied Mechanics is the most important research field, which composes not only the central core of natural science but also the fundamental basis for developments of modern industries. We eagerly hope that the Kyoto Congress will be carried out as a most fruitful and wonderful Congress, which will certainly contribute to a world-wide progress in Mechanics and further developments in science and technology for human welfare.

It is still hot in Kyoto towards the end of August, but I sincerely hope that you will really enjoy the elegance of passing summer in Kyoto, that is the spiritual home of Japanese people. In conclusion, I wish to express my sincere thanks to the President, Professor Leen van Wijngaarden and other members of International Union of Theoretical and Applied Mechanics, and the President, Professor Masao Ito and other members of Science Council of Japan for their devoted efforts for realizing the 19th International Congress of Theoretical and Applied Mechanics at our City of Kyoto. I heartily wish the great success of this Congress and health and happiness of all you participants."

Closing Remarks by Professor Manabu Ito, Chairman of Finance Committee

Finally, Professor Manabu Ito, the Chairman of the Finance Committee, made the closing remarks for the Opening Ceremony.

"Ladies and Gentlemen, as a Vice-Chairman of the Local Organizing Committee and also a member of the Science Council of Japan sponsoring this Congress, I am very much pleased to welcome such large number of participants to Kyoto. I appreciate in particular those participants who have come far away from overseas. As closing remarks to this session, I would like to speak a few words as the Chairman of the Finance Committee. Because the registration fees cannot cover all the budget of the Congress, we had to ask for a fairly large amount of donation to a number of Japanese industries, corporations and organizations which are listed in the program of the Congress. In spite of difficult situations of economy in these years, they have kindly responded to our request. Therefore, I wish to express our cordial thanks to them for their financial support. I am also grateful to Professor Watanabe who is chairing this Ceremony and his colleagues Dr. K. Sugiura and Dr. T. Utsunomiya, for their very hard work as the secretariat of this Kyoto Congress.

Usually it is very hot and humid in Kyoto and sometimes affected by Typhoons in this season. But fortunately, after passing of a cold front, we have now a relatively comfortable weather. I hope that all participants from overseas may enjoy pleasant days in Kyoto and in Japan. I wish that this Congress would be fruitful for all the participants. Thank you for your attention."

SOCIAL PROGRAMS

The Social Programs included Get-together Party on Sunday, Reception on Monday evening, Banquet on Thursday evening, and other programs such as Afternoon Excursions on Wednesday, Accompanying Persons' Programs and a Post Congress Tour.

The Reception of a buffet dinner style was held on Monday, 26 August, 7-9 pm. in the Garden of KICH and a welcome address was given by Professor Yosunori Nishijima, Vice-President of the Science Council of Japan. The optional Banquet Dinner was held on Thursday, 29 August, 8-10 pm. in the Event Hall of KICH and a greeting address and toast were given by Professor Jiro Kondo, Former President of the Science Council of Japan.

CLOSING CEREMONY

The Closing Ceremony was held in Room A of KICH from 6.00 pm. on Friday, 30 August 1996, immediately after the Closing Lecture by Professor Sir James Lighthill. Professor Watanabe chaired the Closing Ceremony.

Closing Address by Professor Tomomasa Tatsumi, President of the XIXth ICTAM

Professor Tomomasa Tatsumi made the Closing Address as follows.

"Dr. Chairman, Ladies and Gentlemen. The present 19th International Congress of Theoretical and Applied Mechanics, which started on the last Sunday has now finished its scientific programs with the marvelous Closing Lecture by Sir James Lighthill. Now, it is my happy privilege to announce the closing of the 19th International Congress of Theoretical and Applied Mechanics.

Numerical statistics of the present Congress is being given by Professor Niels Olhoff in his Congress Report which will follow soon. I can only mention that the total registered number of participants of this Congress is about 940. This number is considerably larger than those of previous Congresses. As a reason for that I should mention the devoting efforts of the Chairmen of Mini-Symposia and Pre-nominated Sessions. They have done their best in stimulating interests of scientists and engineers in respective subjects and encouraged them to participate in this Congress.

Now, I would like to express my hearty thanks and respects to all participants, including Session Chairmen, Lecturers, in particular those of Opening and Closing Lectures, Seminar presenters, discussants and audiences for their invaluable efforts to make the present Congress one of the most successful, scientifically fruitful and enjoyable Congresses in the long history of IUTAM and IUAM. I wish to extend my sincere gratitude to all members of the Bureau, Congress Committee and General Assembly of IUTAM, who decided to have the 19th ICTAM in Kyoto, Japan, and for their overwhelming support for the present Congress. My hearty gratitudes are also due to the Science Council of Japan, which co-sponsored the present Congress and a number of Academic Societies, Foundations and Companies, which supported financially and in other forms our Congress.

Lastly, but by no means leastly, I would like to express my hearty thanks to all people related with the Local Organizing Committee headed by Professor Yoshiyuki Yamamoto, the Local Financial Committee headed by Professor Manabu Ito, and the Local Executive Committee headed by myself, in particular, its General Secretary Professor Eiichi Watanabe and his staffs including Dr. Tomoaki Utsunomiya for their long and still continuing excellent works in support of our Congress.

Now, it was announced by Professor Leen van Wijngaarden in the Banquet last night that the next 20th IUTAM Congress of the Year 2000 is to be held at Chicago, USA. Until then, let us work more on Theoretical and Applied Mechanics with the strong belief that it is actually the most advanced and central core of our modern science and technology. I am looking forward to meeting you again at Chicago in the Year of the Turn of the Century. Thank you."

Congress Report by Professor Niels Olhoff, Secretary of the Congress Committee of IUTAM

Professor Niels Olhoff made the Congress Report as follows.

"Ladies and Gentlemen. I am convinced that you agree with me that this Congress in Kyoto will be long remembered as the one that was very successful and marvelously organized one. It has been absolutely perfect and our Japanese hosts have almost spoiled us by making such an excellent meeting and using these excellent Congress facilities we have here.

We owe our sincere thanks to the Local Organizing Committee for their very hard and dedicated work toward making this Congress a considerable success. On behalf of all the participants, I am sure, I would like to convey our deep gratitude to in particular, the President of this Congress, Professor Tomomasa Tatsumi, and the Secretary General, Professor Eiichi Watanabe, and Dr. Tomoaki Utsunomiya with whom I have had the most smooth, efficient and frictionless cooperation. I have been very grateful for that.

I also wish to acknowledge substantial financial support that the Congress has received from the Japanese funds, from the Japanese Science Council and from various other Japanese sources, and I think it should be mentioned here that actually the amount of the received supports acquired by the Local Organizing Committee actually superseded total amounts of received registration fees, which means that more than half of the total budgets of the Congress has been covered by sheer Japanese funds.

Professor Tatsumi informed about the total number of participants in the congress was 939 and I can add that they come from no less than 46 different countries. I can also mention the number of accompanying persons 94, which means that the total participation is more than 1,000. A little more statistics which might interest you, the number of contributed papers presented as Lectures is 421, and the number of papers in the Seminar Presentation Sessions is 244, which makes the total number of 665 contributed papers presented here.

And then, of course, we have had two enlightening Opening and Closing Lectures, and excellent Introductory Lectures of Mini-Symposia and Sectional Lectures, and all these nice things with papers in the Lecture and Seminar Sessions make the total up to 703 papers.

Unfortunately, there have been some no-shows. Conditions in the world are not as used to be or as should be. It means that, of course, a little more than these papers, the number of 882 contributed papers were actually accepted. I can mention for your information that the total number of contributed papers submitted to the Congress was almost 1650 and this is clearly more than the number of papers submitted to the previous Congresses. I can say in this case that about a half of the number of papers was accepted. I would like to stress here that this rejection rate reflects the IUTAM's policy to maintain the highest scientific standard at its conferences and I think this 19th Congress has, in my opinion, reflected this scientific quality requirement very well.

I will not take your time since Professor Tatsumi thanked general Plenary Lecturers, Sectional Lecturers, invited Introductory Lecturers to Mini-Symposia and all the other Lecturers and Seminar presenters and the Chairmen of the Sessions and of course, all the Participants. But I would like to mention a few things which may be noteworthy. More or less, the pattern of the program of this Congress has followed the pattern of the earlier Congresses but with some differences that we increased the number of Mini-Symposia from three to six, and also introduced a concept of Pre-nominated Sessions. Actually we introduced 40 of those on typical topics of Mechanics which are not covered by the Mini-Symposia. The idea here is that as a recent trend in scientific meetings people prefer specialized meetings and certainly there is a strong demand for this in every IUTAM Congresses. So this is a kind of adaptation of these ideas that we have introduced Pre-nominated Sessions. Of course, it helps very much the whole procedures of selecting papers and structuring the program at an earlier stage. It is very important that we select very good chairmen for these Sessions and that we involve them in our work in soliciting papers, in organizing and setting up the Sessions, and also in communicating with the International Papers Committee (IPC) in endorsing papers and so forth.

So, I would like to say that generally speaking this new feature has been going very well, and hope we will continue that at the next Congress. At least I am very sure that this structure of having Mini-Symposia and also Pre-nominated Sessions is the reason why we have been able at this Congress to draw about 200 out of country participants to Japan than in the previous three Congresses. So it is this Congress which was held in Japan that attracted certainly all people in Mechanics and really has been a very international conference. I think this is very nice.

A few things the IUTAM desires are to have the role as an international umbrella organization for Theoretical and Applied Mechanics, and they wish to do this among other things, by strengthening the IUTAM relationship and collaboration with the Affiliated Organizations. They do this in two ways or maybe in three ways. The first way is that they have asked the Congress Committee to adopt members of the representatives from the Affiliated Organizations in the Congress Committee. For the current Congress, we have also followed the wishes of the Bureau and the General Assembly to arrange Pre-nominated Sessions and also Mini-Symposia jointly with some of our Affiliated Organizations. I am very pleased here and want to acknowledge the cooperation. I should mention the International Association for Computational Mechanics (IACM), with which we have had three Pre-nominated Sessions together, and we have had one with the International Association for Vehicle System Dynamics (IAVSD), and one with the International Association for Hydromagnetic Phenomena and Applications (HYDROMAG). We have had a Mini-Symposium with a new Affiliated Organization, the International Society for Structural and Multidisciplinary Optimization (ISSMO). So we wish to thank very much for very valuable cooperation with these Affiliated Organizations. We hope in the Year 2000, we can certainly expand this collaboration with our Affiliated Organizations.

Well, I am afraid, time is running. I would finally have two reports on the retirement of 10 members of the Congress Committee of IUTAM and the election of 10 new members. The retiring members are Professor Boley, Professor Chernyi, Professor Germain, Professor Hashin, Professor Hutchinson, Professor Karihaloo, Professor Leibovich, Professor Mroz, Professor Tabarrok, and Professor Zemin Zheng. I wish to thank all, these very nice persons, for their cooperation and work in the Congress Committee. In particular, I would like to thank Professor Germain who was the past President of IUTAM, has served also as the President and the Chairman of the Congress Committee for four years. I would also like to thank

Professor Boley who for eight years had the job to act as the Secretary of the Congress Committee and also as the Executive Committee. I should mention for correctness that one of the members I mentioned who retires, does not really retire. It is Professor Karihaloo who was former from Australia but now in Denmark. Professor Karihaloo suggested to relinquish his duties in the Congress Committee to give the seat to an Australian representative, and I am pleased to announce that the Congress Committee and then the General Assembly has appointed Professor Tanner from Australia to replace Professor Karihaloo.

The other 9 members of the Congress Committee which were appointed by the General Assembly are the following: Professor Bogy from USA, Professor Freund also from USA, who is now also the Treasurer of IUTAM, Professor Gladwell from Canada, Professor Lugner from Austria, Professor Moreau from France, Professor Nigmatulin from Russia, Professor Pearson from UK, Professor Suquet from France, and Professor Fen-gan Zhuang from the People's Republic of China. I wish to welcome these new members, whose appointment will take effect from the 1st of November, this year. I would like to mention that two of the new members represent Affiliated Organizations: Professor Lugner, the International Association for Vehicle System Dynamics, and Professor Pearson, International Committee on Rheology, and actually also Professor Moreau representing HYDROMAG, if we decide that so. This would imply that the number of people representing the Affiliated Organizations is increased from three to six then over the last four years.

Finally, I would like to mention that the Congress Committee yesterday appointed its new Executive Committee also from the 1st of November, and the members are: Professor Schielen, for reasons which I lay for a moment, Professor Acrivos, Professor Bodner, Professor Moreau, Professor Patrick, and myself as a new secretary. Well, I would like to conclude this report, by once again, thanking our Japanese hosts, and assure that we will return to our home countries with warm memories of this very exceptional hospitality which you have shown us with a warm memories of well organized and successful Congress which we have had here in Kyoto. Thank you very much."

Concluding Address by Professor Leen van Wijngaarden, President of IUTAM

Professor Leen van Wijngaarden made the following concluding address.

"It is getting late, so, I will be short. First of all, I have the pleasant duty to announce the IUTAM Prizes. You may have read in the announcement that since 1988, IUTAM awards two prizes for outstanding presentations by younger scientists either in Poster or Lecture Session. To be young is a flexible concept but here it is understood to be younger than around 35 years old. The prize winning paper in Solid Mechanics has as the title: "Wave Propagation in Cylindrical Structures Containing Viscoelastic Components," and the author is Mr. J. Vollmann. Then the prize winning paper in Fluid Mechanics has the title "An Experimental Study of Gravity Currents in a Rotating Frame of Reference," and the author is Dr. P. J. Thomas from the UK. The both authors are given a check of 500 US dollars each.

Then, the second piece of information has already been disclosed by Tomomasa Tatsumi and I told it to those who were present at Banquet of last night, the place of the 20th ICTAM in the Year 2000 will be Chicago, and it will be supported by the Consortium of Universities located in Chicago area. The local president will be Professor Hassan Alef. I can say that IUTAM has in mind to increase for Chicago the number of prizes from two to four, two in Solid Mechanics and two in Fluid Mechanics.

Then, it is a pleasure to announce that the General Assembly of IUTAM has elected at its meeting of Wednesday afternoon new Bureau members starting from 1st of November, this year. The President will be Professor Werner Schielen from Stuttgart, Germany. The Secretary General Elect is Professor Michael Hayes from Ireland. The Treasurer, another important officer in IUTAM, which for the last four years has been fulfilled by Professor Boley, will be Professor Ben Freund from USA. And I myself will continue for another four years as the Vice President. Four other members, normal officers of the Bureau that have been elected are Professor Keith Moffatt from Cambridge, UK, Professor Tomomasa Tatsumi from Japan, Professor Juli Engelbrecht from Estonia, and Professor Ren Wang from Beijing, China.

Another piece of information which might interest you is that apart from those who are delegates of national Adhering Organizations, our General Assembly has the so-called Members at Large. We have elected a new Member at Large, that is Professor Paul Germain from France. With that I have come to an end of information that I would like to convey to you.

So what rests now is a sort of thanks to one man that has not been mentioned in all the sort of thanks, that is, Professor Niels Olhoff. Because both he and Professor Watanabe have done a tremendous lot of work and I was expressing on Monday when we started the hope that they would not fall apart in this week because they have had still more work to do. But they have brought this Conference to a good success and thanks of us are very earned by them. We have already thanked to our hosts, but I think it is good thing to thank you participants, because whatever the surface of every body is, a Congress success is made by the participants, and as far as I am able to see, we have had high quality Congress with many very good lectures and very good questions and answers. So that every body whoever I have spoken are well satisfied and you may remember that at Olympic games, the President Samaranch gave some qualifications at Atlanta, he did not say that it was the best Olympic games, but he said it was most extraordinary. Accordingly I think we cannot say that Kyoto was the best because we have no exact criteria to measure that, but I am on the safe side that if I conclude that it is one of the best. I think you know that IUTAM will organize the next ICTAM in Chicago. I hope you all come and convince others who are not here to come also in four years to Chicago. Then, I join you in envisioning good trip to home and still staying a couple of days here to enjoy Kyoto and Japan, and I say a Japanese word, Sayonara."

OPENING AND CLOSING LECTURES

Theoretical and Applied Mechanics 1996
T. Tatsumi, E. Watanabe and T. Kambe (Editors)
© 1997 Elsevier Science B.V. All rights reserved.

Structural control for large earthquakes

T. Kobori

Professor Emeritus of Kyoto University
KI Bldg. 6-5-30 Akasaka, Minato-ku, Tokyo 107, Japan

This paper describes changes in seismic structural design in Japan based on data from past earthquakes and damage they have caused. Both the extremely complicated seismic motions and the inherent uncertainty of earthquakes has led to the introduction of active structural control in civil engineering. Technological innovations of the recent years have significantly advanced this concept, and in the beginning of 1985 full-scale research was launched for its practical application. Currently proposed control strategies are categorized considering the principle of structural control, and theoretical research and applications are reported. The future direction of this research and development is described and a practical approach to control systems for large earthquakes is indicated.

1. EARTHQUAKE RESISTANT STRUCTURE TO STRUCTURAL CONTROL

Changes in seismic structural design in Japan based on data from past earthquakes and damage they have caused are described. Structural control has been introduced because both the extremely complicated seismic motions and the inherent uncertainty make it impossible to guarantee structural safety by conventional design methods. The principle of structural control is considered and currently proposed control strategies are categorized.

1.1. Earthquakes and damage in Japan

Most of Japan have suffered from large earthquakes since ancient times. Each of nearly twenty earthquakes and tsunamis has claimed over one hundred lives since the Meiji era when Japan became a modern country.

In 1891 the Nobi Earthquake extensively damaged Aichi and Gifu prefectures. The Neo-dani Fault ran approximately 80 km from north to south, and shifted the earth's surface six meters vertically and two meters horizontally in Mizutori village. This was recognized as one of the largest earthquakes to occur in an intraplate. Five years later, an earthquake occurred off the coast of Sanriku in the Tohoku district, and the resulting tsunami took more than twenty-two thousands lives. This is recognized as the largest earthquake to occur along an interplate.

On September 1, 1923, the Great Kanto Earthquake, the earthquake of the century, hit the southern part of the Kanto district and caused extensive damage in Odawara, Yokohama and Tokyo. Now the day is memorialized in Japan as Disaster Prevention Day. The first epicenter was located in Sagami bay, and several successive faults slipped to the east. The conflagration spread rapidly and increased the number of victims in many cities, since the earthquake occurred at the lunch time. This caused civil engineers to reconsider the existing technology. It was even found that damage to structures depended on soil characteristics. In 1944, the Toh-Nankai Earthquake caused tremendous damage in the Chubu and Kinki districts extensively. The rate of destroyed building structures was especially high in reclaimed coastal area and on alluvial soil along rivers.

Date	Name	M_{JMA}	Lost lives	Total and half collapse of buildings and houses
Mar. 14, 1872	Hamada Earthquake	7.1	552	over 5,000
Oct. 28, 1891	Nobi Earthquake	8.0	7,273	over 220,000
Oct. 22, 1894	Syonai Earthquake	7.0	726	6,255
June 15, 1896	Sanriku Earthquake Tsunami	8.5	21,959	over 10,000
Aug. 31, 1896	Riku-u Earthquake	7.2	209	5,792
Sep. 1, 1923	The Great Kanto Earthquake	7.9	142,000	over 254,000
May 23, 1925	Kita-Tajima Earthquake	6.8	428	1,295
Mar. 7, 1927	Kita-Tango Earthquake	7.3	2,925	12,584
Nov. 26, 1930	Kita-Izu Earthquake	7.3	272	2,165
Mar. 3, 1933	Sanriku-oki Earthquake	8.1	3,064	5,851
Sep. 10, 1943	Tottori Earthquake	7.2	1,083	13,643
Dec. 7, 1944	Toh-Nankai Earthquake	7.9	1,223	57,248
Jan. 13, 1945	Mikawa Earthquake	6.8	2,306	32,963
Dec. 21, 1946	Nankai Earthquake	8.0	1,330	36,529
June 28, 1948	Fukui Earthquake	7.1	3,769	48,000
May 23, 1960	Chili Earthquake Tsunami		142	over 3,500
May 26, 1983	Nihon-kai Chubu Earthquake	7.7	104	3,101
July 12, 1993	Hokkaido Nansei-oki Earthquake	7.8	231	over 600
Jan. 17, 1995	Hyogo-ken Nanbu Earthquake	7.2	over 6,300	over 207,000

The 1946 Nankai Earthquake was the first shock to the western part in the reconstruction of damaged cities after World War II. The 1948 Fukui Earthquake caused great damage locally because the focal depth was 20 km in a shallow intraplate. The extremely strong ground motion destroyed all houses in several villages built on the alluvial soil of the Fukui plain. It is well known that the shock effect of this earthquake led the Japan Meteorological Agency (JMA) to add a level of "7" to the Seismic Intensity Scale. Later, we are reminded of the liquefaction of saturated sand that occurred in the 1964 Niigata Earthquake, the collapse of reinforced concrete structures designed according to the Japanese earthquake resistant design code in the 1968 Tokachi-oki Earthquake, and the tsunami disasters in both the 1983 Nihon-kai Chubu and the 1993 Hokkaido Nansei-oki Earthquakes. These earthquakes resulted in fewer deaths than the Fukui one because they occurred below the ocean floor.

Early in the morning of January 17, 1995, the Hyogo-ken Nanbu Earthquake, centered near Awaji Island, suddenly hit Kobe, causing a rare tragedy in this modern city. This great disaster had a social impact as well as scientific one. Ways of preventing urban disaster and encouraging the spread of new technologies and the revival of old-fashioned cities must be studied as well as earthquake engineering and seismology.

1.2. Changes in seismic structural design

In Japan, it is still a matter of concern whether or not we live in a seismically safe house. Research on earthquake resistant structures began with the establishment of the Imperial Earthquake Investigation Committee just after the 1891 Nobi Earthquake that destroyed most brick buildings copied easily from both Europe and the United States.

Prof. Sano and Prof. Naito each pioneered a theory for earthquake resistant design before the Great Kanto Earthquake occurred. Prof. Sano opened a course for simple design of structures, introducing a design seismic coefficient that expressed the ratio of seismic force to total weight of a structure. Prof. Naito discovered the efficiency of shear wall in suppressing structural deformations and proposed their optimal arrangement. It is well known that the old Nihon-Kogyo Bank building to which he had applied his theory suffered no damage in the Kanto Earthquake. However, most buildings built in the western style were severely damaged. As a result, both brick and stone were replaced as structural materials by reinforced concrete and steel, and the design seismic coefficient was legally determined to be more than 0.1 in cities. This dire calamity produced many competent researchers in structural engineering and seismology. The resulting wealth of research determined the design method for low-storied and medium-rise buildings designed as rigid structures, based on the force method that is useful even now. There was discussion on whether or not a rigid structure was better than a flexible structure, though the nature of earthquakes was still unclear.

The 1948 Fukui Earthquake stimulated the research and development which had been stagnant since before World War II. Two years later, the design seismic coefficient was revised to more than 0.2. The strong motions recorded in the United States had installed accelerometers in Japan since 1953. In addition, advances of modern computers made it easy to analyze both earthquake records and structural responses. Results of the new research fruit showed that seismic motions in hard soil include many short-period waves and few long-period waves. In other words, on hard soils seismic forces are large in a rigid structure and small in a flexible structure. Thus, it is concluded that high-rise buildings can be built in Japan if they have enough deformation capacity. This conclusion is based on the theoretical research on nonlinear vibration of civil engineering structures conducted since the 1950's. In 1963 the maximum legal height of a structure, 31.0 meters, was increased, and in 1968 a 36-story high-rise building, the Kasumigaseki Building, was completed for the first time in our country. Now there are about 250 high-rise buildings more than 100 meters high in Japan.

The flexible structure design method became widely accepted, especially after the 1968 Tokachi-oki Earthquake in Japan and the 1971 San Fernando Earthquake in the United States, because these disasters clearly demonstrated problems with the rigid structure design method. A five-year research project revised the old structural calculation standard in 1981. The revised building code assumes two earthquake scales and requires that designers attend not only to strength but also to ductility for many kinds of structures. This proved to be mostly appropriate in the 1995 Hyogo-ken Nanbu Earthquake.

Table 1.2. The history of seismic structural design in Japan.

1891	The Nobi Earthquake occurred.
1892	The Imperial Earthquake Investigation Committee was established.
1916	Prof. Sano proposed the design seismic coefficient.
1922	Prof. Naito proposed shear walls and their optimal arrangement.
1923	The Great Kanto Earthquake occurred. The Nihon-Kogyo Bank building was undamaged, although most buildings in the western style were damaged.
1924	The design seismic coefficient was legally determined to be more than 0.1 in cities. The maximum allowable building height was confirmed to be 30.3 meters.
1927	"Rigid structure versus flexible structure" was discussed until 1936. The rigid structure design method was mostly completed by the 1940's.
1948	The Fukui Earthquakes occurred.
1950	The building code modified the design seismic coefficient to more than 0.2. Theoretical research on nonlinear vibration was conducted after the 1950's, using newly developed digital computers.
1953	Accelerometers for strong ground motion were first installed.
1963	The maximum legal building height of 31.0 meters was legally lifted.
1968	The Kasumigaseki Building was completed. The Tokachi-oki Earthquake demonstrated some problems with the rigid structure design method.
1981	The building code was revised, with great importance attached to strength and ductility.
1985	Base isolation was first applied to a two-story private house.
1989	An AMD was first installed in a ten-story office building.

1.3. Nature of earthquakes

Geophysics describes that motion of crustal plates causes accumulation of strain energy in rock beds, and a fault motion occurs when the energy exceeds a certain limit. The shock propagates through the earth's crust, rock beds and surface soil with reflection and refraction to civil engineering structures. Body waves propagate three-dimensionally in the form of the longitudinal wave and the transversal wave, and the surface waves propagate two-dimensionally in the surface soil in the form of Rayleigh wave and Love wave. The fault motion, the soil nature and the topography influence wave propagation, and affect frequency dependency. These waves excite structures, while the structures affect the ground motion. Furthermore, the earth is an inhomogeneous inelastic material. As a result, predicting the characteristics of seismic motion becomes extremely complicated because of the highly uncertain information about the earth. We cannot precisely predict when, where and how a fault will occur or how a seismic motion will arrive at structures.

Figure 1.1 shows three observed accelerograms with running response spectra (h=1%). Figure (a) indicates the accelerogram observed at El Centro in Southern California during the Imperial Valley Earthquake in 1940. The observation is well known as the first recorded strong seismic motion. This record has been customarily used in structural design in Japan, because we had few seismic records in the past for structural response analysis. The large amplitudes include the dominant frequencies from 1.0 Hz to 3.0 Hz during the initial five

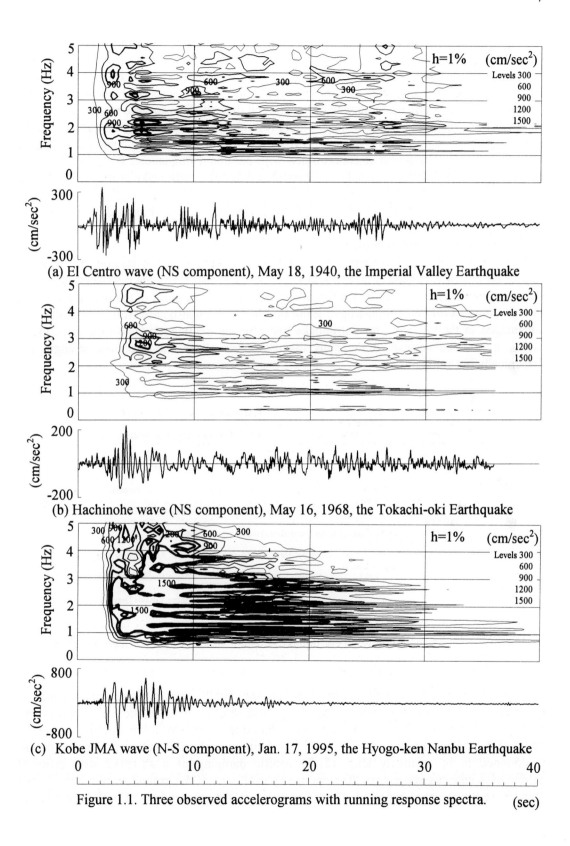

(a) El Centro wave (NS component), May 18, 1940, the Imperial Valley Earthquake

(b) Hachinohe wave (NS component), May 16, 1968, the Tokachi-oki Earthquake

(c) Kobe JMA wave (N-S component), Jan. 17, 1995, the Hyogo-ken Nanbu Earthquake

Figure 1.1. Three observed accelerograms with running response spectra. (sec)

8

seconds. Just after twelve seconds, the seismic motion suddenly increased again. Figure (b) indicates the accelerogram observed at Hachinohe in the northeastern (Tohoku) district in Honshu, when the 1968 Tokachi-oki Earthquake occurred, with its epicenter in its Pacific Ocean. The dominant frequency varies from high to low as time goes on. However, 0.4 Hz is recognized as the dominant frequency of the alluvial deposit. This seismic motion excites tall buildings more than the El Centro wave. Figure (c) indicates the accelerogram observed at the Kobe Ocean Meteorological Agency (OMA), when the Hyogo-ken Nanbu Earthquake occurred in 1995. The power is concentrated with dominant frequencies from 1.0 Hz to 3.0 Hz during the initial fifteen seconds. It is considered that the earthquake destroyed most civil engineering structures in the initial ten seconds at the longest. These figures show that seismic motion has the characteristic of high nonstationarity in amplitude and frequency.

1.4. Principle of structural control

Figure 1.2 indicates the progressive transition of principle in the relationship between earthquake resistant structures and structural control. The first earthquake resistant structures were designed as rigid structures with small deformations. "The proof of the pudding is in the eating." was that the old Nihon-Kogyo Bank building, a rigid structure with shear walls, was undamaged in the Great Kanto Earthquake just after its completion. The rigid structure is useful in practice because it demands that designers check the structural strength not dynamically but statically.

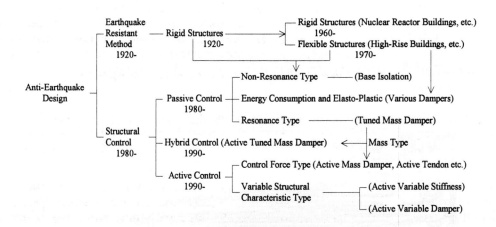

Figure 1.2. Relationship between earthquake resistant structures and structural control.

However, it became increasing apparent by the 1950's that a skyscraper cannot be designed as a rigid structure, and flexible design was introduced as an alternative philosophy for structural design. This new philosophy was based upon the two following theories: (1) A structure is excited less by an earthquake if it is built on hard soil and the first natural period is designed to be relatively long. (2) Hysteretic damping of a structure with efficient deformation capacity suppresses the response by itself even if the ground motion has a larger first-resonant wave than expected. These are based on the assumptions of both hard bearing ground and a predicted ground motion. These conclusions have also been applied to base-isolated structures.

In spite of seismologist' tireless efforts, seismic ground motions at the base of a structure will continue to be uncertain and unclear even in the future. For example, a structure with a safety factor of approximately two cannot efficiently utilize the fuzzy information from seismology for safety. As a result, there is only one means for achieving high safety in which a building controls its response itself. This is the origin of my expectation for structural control against earthquakes. The best way to conquer uncertain seismic motions must be the Real-time Integrated Earthquake Mitigation System. Active structural control is performed at the second stage of the system as shown in Figure 1.3. The system could read the information near the epicenter and transmit it to cities just before the earthquake motion arrived. This first stage would make the structural control system stand-by before the event. For example, it preferably offers the power plant in a building to a control system, and the control system selects the most appropriate control law. The principal civil engineering structures could accept this new approach to hazard mitigation in the modern society.

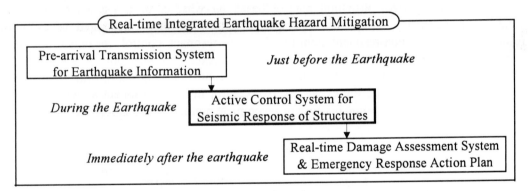

Figure 1.3. General concept of the Real-time Integrated Earthquake Hazard Mitigation System.

Structural control means providing a structure with characteristics that suppress its response under dynamic loadings. The five principles of structural control are as follows.
(1) Cut off the transmission of the earthquake ground motion to the structure.
(2) Isolate the natural period of the structural system from the predominant frequency domain of the earthquake ground motion.
(3) Achieve a non-resonant state by providing nonlinear characteristics.
(4) Apply control forces to the structure.
(5) Utilize the energy absorption mechanism.
If the first principle were perfectly achieved under earthquakes, the others would became unnecessary. However, in reality more than two principles must be satisfactorily combined to protect structures. There is no better structural control than utilizing ordinary structural members rather than special control devices. However, this is possible only if the earthquake ground motion can be predicted such as the time history.

Structural control can be classified in various ways. From the viewpoint of control energy supply, control strategies are categorized as passive, active and hybrid. Base isolation and many dampers are defined as passive control systems that do not demand external energy supply. Base isolation is based on items (2) and (5), and utilizes laminated rubbers and dampers, respectively. (5) is directly applied to passive dampers such as tuned mass damper (TMD) and elasto-plastic damper. Furthermore, this principle is indirectly applied to most active control systems such as active mass damper/driver (AMD) and active tendon, although they are based on (4). The fourth is indirectly applied to TMD (passive type) although it is based on (5). Hybrid control is defined as either the conversion of passive control into active to save the energy supply or the combination of active and passive control. The active tuned mass damper (ATMD), a so-called hybrid mass damper (HMD), tunes the period of a mass to one of the natural periods of the structure.

From the viewpoint of the control device, active control is categorized whether or not an auxiliary mass is necessary. The mass-damper type operates a mass to produce a control force. The non-mass damper type includes active tendon, active variable stiffness (AVS), active variable damper and so on. Active tendon applies a control force to a structure as the braces. AVS and active variable damper change the characteristics of a structure, and they are categorized as "semi-active" system that demands little energy. AVS is based on (3).

Figure 1.4 categorizes structural control based on whether or not a control principle depends on frequency. Frequency-dependent control is further categorized in as either resonant control or non-resonant control. Whereas resonant control consumes the input energy in an installed system, non-resonant control makes it smaller outside a structure. In other words, the difference seems to be whether a control system deals with input energy "after the event" or "before the event". Active control in a non-resonant system inevitably analyzes the disturbance on-line because of incomplete prediction of the seismic motion.

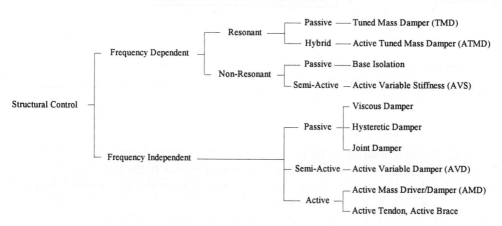

Figure 1.4. Frequency dependence of structural control systems.

2. STATE-OF-THE ART IN ACTIVE CONTROL

In the 1950's the author proposed a basic concept for active seismic-response-controlled structures for attaining non-resonance under nonstationary excitations like earthquake ground motions by providing nonlinearity to itself [1-3]. This concept was based upon the discovery of a principle that both energy dissipation with nonlinearity and the variety of the natural periods in a structure suppress its response under large earthquakes. Meanwhile, J.P.Yao showed the way to the present active control research in civil engineering in the 1970's when modern control theory was recognized as a useful procedure [4]. He described the concept that civil engineering structures counter earthquakes and winds not through their structural members but through a control force. Based on his concept, modern control theory has been applied to civil engineering structures mainly in the United States. It became necessary to restudy the applicability of modern control theory, because civil engineering structures are large and moreover the external forces acting on them are both severe and highly uncertain. Research produced many kinds of control devices and research and development was promoted, both theoretical and experimental. Some ideas seem to be impractical at present and require further development. However, a lot of research indicates that active control is efficient for civil engineering structures, because it provides instantaneous responses.

In this new research field, LQ/LQG (Linear Quadratic Gaussian) control theory was applied to control-force-type systems such as AMD and active tendon. In particular, a lot of research and development was conducted on mass dampers, and they are now being applied. In structural engineering, vibration problems are also examined again from the standpoint of control engineering. As a result, active vibration control has become a large field in structural engineering. However, the research has not yet reached its ultimate target, which is to ensure structural safety during large earthquakes.

The following classifies research on active structural control into theoretical research and application development, and reports the state of the art, because it is important to understand the present situation.

2.1. Theoretical research

Most active/hybrid control systems have been developed based upon LQ/LQG control theory. This theory has been widely accepted in many engineering fields and is extremely practical because it places importance on a balance between control effect and effort. The balance must be first required in design. Strain energy is considered in the control law when the displacement relative to the base is weighted in the performance function. Kinematic energy is considered when the relative velocity is weighted. Relative velocity is generally weighted for a civil engineering structure with low damping. This weighting efficiently adds high damping to a structure according to the resulting velocity feedback control law. A high-damped system makes the response less sensitive to the nature of the earthquake than a low-damped one. Stationary feedback gain is generally implemented in most installed control systems, because the input excitation cannot be predicted and is assumed to be Gaussian white noise. This means that the acceleration of an excitation is neglected in the relative coordinate, and both its velocity and its displacement are neglected in the absolute coordinate.

LQ theory

State equation: $\dot{x}(t) = Ax(t) + Bu(t) + Dw(t)$ (2.1)

where the excitation $w(t)$ is generally assumed to be white noise.

Performance index: $\quad J = \int_0^\infty \{x^T(t)Qx(t) + u^T(t)Ru(t)\}dt$ $\hspace{2cm}$ (2.2)

Control law: $\quad u(t) = Kx(t) = -R^{-1}B^T \Lambda x(t)$ $\hspace{3cm}$ (2.3)

$\hspace{2cm}$ where Λ is the solution of the stationary Riccati equation as follows.

Riccati equation: $\quad Q - \Lambda BR^{-1}B^T\Lambda + \Lambda A + A^T\Lambda = 0$ $\hspace{2cm}$ (2.4)

J.N.Yang et al. derived Instantaneous optimal control by removing the integral symbol from the performance index (2.2) [5]. The feedback control law is independent of the structure's characteristics, and the feedback gain is proportional to the weighting matrices. Eq.(2.6) indicates that the feedback response is closely related to the location of the controllers. Feedback control can be converted into either feedforward control or feedback-feedforward combined control with the same control efficiency. This conversion directs attention to the nonstationarity of earthquakes. One example of this attention was that the performance index was extended to include the input energy [6]. The control law could be easily applied to nonlinear systems because the Riccati equation no longer needs to be solved.

Instantaneous optimal control [5]

$\hspace{1cm}$ Performance index: $\quad J = x^T(t)Qx(t) + u^T(t)Ru(t)$ $\hspace{2.5cm}$ (2.5)

$\hspace{1cm}$ Feedback control law: $\quad u(t) = -\dfrac{\Delta t}{2}R^{-1}B^T Qx(t)$ $\hspace{2.5cm}$ (2.6)

$\hspace{1cm}$ Feedforward control law: $\quad u(t) = -\dfrac{\Delta t}{2}[\dfrac{(\Delta t)^2}{4}B^T QB + R]^{-1}B^T Q[T(t - \Delta t) + \dfrac{\Delta t}{2}Dw(t)]$

$\hspace{12cm}$ (2.7)

$\hspace{2cm}$ where $\quad T(t - \Delta t) = \exp(A\Delta t)[x(t - \Delta t) + \dfrac{\Delta t}{2}\{Bu(t - \Delta t) + Dw(t - \Delta t)\}]$

Recently, J.Suhardjo et al. generally extended the higher-ordered weighting for response to the performance index and introduced the nonlinear feedback control law in a polynomial form [7]. The gains are also determined by solving the Riccati equation off-line. It is applied to active tendon in a linear SDOF structure after application to the Duffing equation, a nonlinear system [8]. Comparison with linear control strategy indicates that while nonlinear control exerts a larger peak control force than linear control, it also significantly reduces the total control energy. Z.Wu et al. similarly proposed a higher-ordered performance index to reduce peak response [9].

Generalized optimal control [7,8]

$\hspace{1cm}$ Performance index: $\quad J = \int_{t_0}^{t_1}\{Q_{ij}x^{ij}(t) + R_{ij}u^{ij}(t) + Q_{ijk}x^{ijk}(t) + Q_{ijkl}x^{ijkl}(t) + \cdots\}dt$ $\hspace{0.5cm}$ (2.8)

$\hspace{1cm}$ Control law: $\quad u^i(t) = K_j^i x^j(t) + K_{jk}^i x^{jk}(t) + K_{jkl}^i x^{jkl}(t) + \cdots$ $\hspace{1.5cm}$ (2.9)

$\hspace{2cm}$ where u^i is the ith control force, $K_{jkl\cdots}^i$ is the ith gain of the nth-order,

$\hspace{2cm}$ and $x^{ijk\cdots}$ is the response of nth-order.

The assumption of state feedback in these theories requires that both the observer and suboptimal control subjected to control structure constraints, because it is very difficult to apply it to a high-ordered MDOF civil engineering structure. The minimum error excitation method and the minimum norm method are useful for establishing output feedback based on stability [10].

The H^∞ control theory has also been studied in civil engineering with new development in control engineering. The theory reflects robust control stability in implicit forms, which is important for practical use, since errors inevitably occur in modeling an objective structure. Two output equations (2.11) and (2.12) can separate the controlled variable z from the observed variable y, independently of the state variable. These two equations also imply the direction from velocity and/or displacement feedback to acceleration feedback. The weighting for both the response and the controllers can be implemented in the frequency domain. This control combines LQ/LQG control in the time domain with classical control in the frequency domain, and simultaneously takes both state feedback control and state estimation into consideration. However, the H^∞ theory disadvantageously produces high-ordered controllers when it is applied to an MDOF structure with constricted observation and high-ordered weighting. It would become more practical if the dimension were reduced in modeling the structure. In other words, H^∞ control works best when the first natural mode becomes more dominant than any other modes and it is mainly controlled. For example, it is better when an HMD is applied to a flexible structure. The μ-synthesis is now searching for an application to civil engineering. If linearity is assumed, new control strategies will be proposed considering stability based on LQ/LQG theory. From this viewpoint, LQ/LQG theory has become the most effective judgment standard for newly developed control laws.

H^∞ control

$$\dot{x}(t) = Ax(t) + B_1 w(t) + B_2 u(t) \qquad (2.10)$$

System equation: $\quad z(t) = C_1 x(t) + D_{11} w(t) + D_{12} u(t) \qquad (2.11)$

$$y(t) = C_2 x(t) + D_{21} w(t) + D_{22} u(t) \qquad (2.12)$$

Performance index: $\left\| G_{zw}(s) \right\|_\infty = \left\| G_{11} + G_{12} K (I - G_{22} K)^{-1} G_{21} \right\|_\infty < \gamma \qquad (2.13)$

where $G_{ij}(s)$ and $K(s)$ are the transfer functions of a structure and a controller, respectively, and $G_{ij}(s) = C_i (sI - A)^{-1} B_j + D_{ij}$ $(i,j=1,2)$.

Control law: $u(s) = K(s)y(s) \qquad (2.14)$

where the closed-loop system (G, K) is designed to be internally stable.

Modeling error introduces research in both system identification and adaptive control. Precise identification evidently makes control effect higher than modeling error compensation by control algorithm. Therefore, identification techniques continue to be studied from the viewpoint of active control. This technique also serves as a background to adaptive control. It will be impossible in the near future to understand the characteristics of earthquake ground motions with a precision equal to the structural ones. This recognition opens the way to adaptive control that changes structural parameters moment by moment. The gain schedule method is useful for practice, although it is theoretically not classified as adaptive control. Some realized control systems select a feedback gain from some prepared gains by judging the response level on-line [11,12]. Continuous change of gain has been recently proposed because discontinuous change suddenly causes a spike-shaped noise, especially in response acceleration [13]. Adaptive control also introduces both feedforward and nonlinear controls.

Feedforward control seems to be neglected in theoretical research, because it provides only one-way control with no correlation to a structure's response and its effect depends only on the modeling of the phenomena. However, it has significant potential because it reduces the disturbance level directly. J.Suhardjo et al. proposed a feedback-feedforward control with the same quadratic performance index as LQ/LQG control and Instantaneous optimal control [14]. The combined control augments the equation of motion for the structural system with an appropriate model of the earthquake excitation based on filtering the Gaussian white noise process. The information from the structure and earthquake excitation model is utilized in the feedforward control law with an observer designed to estimate the states of the earthquake model based upon the base acceleration measurements. The feedback and feedforward gains can be determined off-line with some additional computational effort. The result indicates that the proposed control offers advantages in performance over feedback control. In general, the extended equation of motion is introduced when the seismic motion and/or the controller are considered in modeling a total system. Figure 2.1 shows an example of a structural system with dynamic feedback and feedforward controllers.

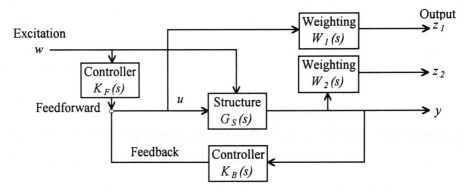

Figure 2.1. Block diagram showing one example of an extended system.

The control laws for nonlinear and hysteretic systems were reported based on Instantaneous control theory. After the theory was extended to incorporate the specific hysteretic model, its applicability to base-isolated structures with a control force was studied [15,16]. This control law requires measurement of the velocity, the displacement and the hysteretic component. Its application to AVS indicates that it is important to add damping to a structure even if the stiffness is changed to achieve non-resonance [17]. Another application is the control with constraints of maximum control input [18]. The nonlinear control law was introduced based on the minimum principle of optimality as the linear control law. The limitation of control forces was also considered based on bang-bang control that produces a control force only with both maximum and minimum values [19]. The applicability of the continuous sliding mode control, a nonlinear control theory, was recently studied [20]. The proposed strategy continuously changes the control force to evade the undesirable chattering effect at the switching surface. The response noise causes a severe problem in building structures that attach importance to living comfort.

Control in the absolute coordinate has a high potential for improving living comfort and safety of vital equipment. When control aims at suppressing the absolute acceleration, feedforward control is proposed in the form of measuring both the base velocity and base displacement. This feedforward control can ideally cut off transmission of input excitations if the control force is large enough. The control might become most effective in application to the soft-storied structure such as the base-isolation, since the control force is relatively smaller [21]. The absolute velocity feedback control law has been studied in civil engineering since it was successfully applied to automobile suspension systems. The control completely suppresses the acceleration in the frequency domain, though it necessitates a larger control force than the relative velocity feedback control in the low frequency domain (described in Section 3.1). It is a so-called "sky-hooked damper" because it performs as if the damper were hooked to the sky in the absolute coordinate. Active control produces this advantageous characteristic when the state vector $x(t)$ is defined in the absolute coordinate and the velocity is weighted in the performance function. However, the relative velocity feedback control suppresses the acceleration less than the absolute feedback control in the high frequency domain. It is introduced not only by active control but also by passive control, although the sky-hooked damper is introduced only by active control. J.Inaudi et al. proposed predictive control for base-isolated structures, in which the performance index includes absolute values predicted to compensate for sensing delay [22].

Consideration is also given to compensation of control lag. Phase lag should generally be considered to be a more severe problem than time lag, because the natural periods in civil engineering are much longer than in electrical and mechanical engineering. The phase lag problem is closely related to the H^∞ theory that leads the dynamics of controllers.

The location of controllers greatly influences the control effect. For example, a mass damper works best at the location that vibrates with the largest amplitude of the objective mode. An active tendon works best in the story that indicates the largest relative amplitude of the objective mode. The location problem seems to be easy when only one controller aims at suppressing only one vibration mode. However, it suddenly becomes complicated with the interaction of controllers when plural controllers aim at plural modes. Most of the control theories cannot directly offer the optimal location for controllers, since the matrix for the controllers B is determined a priori. This problem has been discussed as the optimization problem even in the passive control field under the assumption of input excitation. In active control, the optimal location in plane was recently discussed when only one controller simultaneously reduces vibrations in both transverse and torsional directions [23].

Fuzzy prediction and neural network are categorized as non-mechanical (non-physical) models. Fuzzy prediction is a procedure for solving difficult problems such as ambiguity in situation comprehension, incomplete information and diversity in assessments that cannot be evaded in practice. The applicability was studied in selection of the control law under the complex control target, prediction of disturbance level, and identification of structural parameters [24-27]. The main future problem is to clarify both the optimal membership function and the relation to the proposed control theories from physical viewpoints. The neural network is a new optimization procedure which models information processing in the human brain. The back-propagation method has been applied to structural identification and structural control with identification [28-30]. The methodology will be expected to indicate the stability, how the various formulations relate to each other, and how the scale of judgment processing affects control. They could be applied not only to reduction of vibration but also to sensing of abnormal situations such as control system malfunction. The non-physical models shows us that it is important to attempt physical consideration for the proposed control strategy independently of the applied theory.

Recent aspects of theoretical research are well-reflected in the proceedings of the First World Conference on Structural Control (1WCSC), that was held in Los Angeles 1994 [31].

2.2. Applications & developments in Japan

Interest in active structural control has grown remarkably in the civil engineering field all over the world in the past ten years. Japan is one step ahead of other countries, especially in practical application to building structures. Most applications in Japan are described in the proceedings of both the International Workshop on Structural Control (1993) and 1WCSC (1994) [31,32]. In 1989 an AMD system was installed in an actual ten-story building for the first time to suppress vibrations caused by frequent earthquakes and winds. Its efficiency has been verified by observation records and simulation [33,34]. This success has greatly influenced subsequent research and development. Building structures with active/hybrid control system now number about twenty, as shown in Table 2.1. R & D in Japan has been promoted from the standpoint of practical application unlike that in other countries. Most installed systems were designed using the output feedback control law based upon LQ/LQG control theory.

The AMD system, the most popular mass-damper type, operates an auxiliary mass, and produces a control force to counter disturbances to a structure. This is a classical strategy. The HMD system tunes the dominant period of the mass to one of the natural periods of the structure. It is also called an ATMD, since the idea is to convert the passive-type TMD into an active one [35]. Two systems, Nos.7 and 18, are also proposed as the derivations of HMD, that is, an AMD mounted on a TMD. The AMD effectively moves the TMD to cope with inharmonic excitations, because a TMD cannot react instantaneously. Both systems are designed to save the external energy supply by utilizing resonance. In general, whereas an AMD is efficient for all vibration modes, an HMD is efficient for only one mode. However, it is efficient for most civil engineering structures in controlling the lowest (first) mode, and HMD systems utilize this property.

All developed mass damper systems add damping to structures and aim at improving living comfort during small/medium earthquakes and frequent winds such as typhoons. They cannot yet protect a structure from large earthquakes because of the huge mass and energy supply required to drive them. One idea for the huge auxiliary mass has been suggested which is to utilize facilities such as heliports and heat storage tanks placed on building roofs (Systems Nos.4 & 9). Table 2.1 draws attention to the fact that most systems aim to reduce complicated vibrations in both lateral and torsional directions. AMD/HMD have already been applied to several bridges to suppress vibrations under wind and traffic loadings. Bridges often require control systems even under construction.

The AVS (No.2) changes the stiffness of a structure based on the feedforward control law

[36]. It measures the base acceleration and analyzes it in the frequency domain with band-pass filters. It then changes the first natural period of the structure from 0.4 to 1.0 second. The non-resonance-type system is expected to reduce seismic loads of large earthquakes with only a small external energy supply.

Table 2.1 Active/hybrid control systems applied to building structures in Japan.

No.	System	Applied Building (story, total weight, 1st period)	Developer Date	Control law, Controlled direction & Scale of devices
1	AMD	Kyobashi Seiwa Bldg. in Tokyo (10 stories, 400tf, 0.9s)	Kajima 1989	Output velocity feedback (LQ/LQG theory), AMD1 (4.2tf) for transverse +AMD2 (1.2 tf) for torsional
2	AVS	KaTRI No.21 Bldg. in Tokyo (3 stories, 400tf, 0.4-1.0s)	Kajima 1990	Feedforward (Observe and analyze the base acceleration), 3 braces for transverse
3	AMD	Sendagaya INTES in Tokyo (11 stories, 3,300tf, 1.7s)	Takenaka 1991	Output feedback (LQ/LQG theory) 2 AMDs (36tf*2) for transverse & torsional
4	HMD	Applause Tower in Osaka (32 stories, 14,000tf, 3.8s)	Takenaka 1992	Output feedback (LQ/LQG theory) HMD (480tf) for 2 horizontal
5	HMD	Kansai Airport Control Tower in Osaka (86m, 2,570tf, 1.3s)	MHI 1992	State feedback (LQ/LQG theory) 2 HMDs (5tf*2) for 2 horizontal
6	HMD	ORC200 Symbol Tower in Osaka (50 stories, 56,680tf, 3.9s)	Shimizu 1993	Output feedback (LQ/LQG theory) 2 HMDs (110tf*2) for transverse & torsional
7	HMD	Ando Nishikicho Bldg. in Tokyo (14 stories, 2,600tf, 1.3s)	Kajima 1993	Acceleration & velocity feedback (Fixed-points theory) 2 AMDs (2tf*2) on TMD (20tf) for 2 horizontal
8	HMD	Landmark Tower in Yokohama (70 stories, 260,610tf, 5.2s)	MHI 1993	Output feedback (LQ/LQG theory) 2 HMDs (170tf*2) for 2 horizontal
9	HMD	Long Term Credit Bank Bldg. in Tokyo (21 stories, 39,800tf, 3.0s)	Nikken Sekkei & Prof.Fujita, 1993	Output feedback (LQ/LQG theory) HMD (120tf) for transverse
10	HMD	Porte Kanazawa (29 stories, 2.9s)	Takenaka 1994	Output feedback (LQ/LQG theory) 2 HMDs (50tf*2) for transverse & torsional
11	HMD	Shinjuku Park Tower in Tokyo (52 stories, 130,000tf, 5.2s)	Kajima & IHI 1994	Output feedback (LQ/LQG theory) 3 HMDs (110tf*3) for transverse & torsional
12	HMD	ACT Tower in Hamamatsu (46 stories, 4.7s)	MHI 1994	Output feedback (LQ/LQG theory) 2 HMDs (70tf*2) for transverse
13	AMD	Riverside Sumida in Tokyo (33 stories, 52,000tf, 2.9s)	Obayashi 1994	State feedback for simplified model (LQ/LQG theory) 2 AMDs (36tf*2) for transverse & torsional
14	HMD	Osaka World Trade Center (52 stories, 60,000tf, 5.8s)	MHI 1994	Output feedback (LQ/LQG theory) 2 HMDs (50tf*2) for 2 horizontal
15	HMD	Hotel Ocean 45 in Miyazaki (43 stories, 83,650tf, 3.0s)	Shimizu 1994	Output feedback (LQ/LQG theory) 2 HMDs (171tf*2) for weak-axis dir. & torsional
16	HMD	Hirobe Miyake Bldg. in Tokyo (9 stories, 270tf, 0.8s)	Fujita & Prof.Fujita, 1994	Output feedback (LQ/LQG theory) HMD(2tf) for transverse
17	HMD	MKD8 Tokyo Hikarigaoka Bldg. (24 stories, 2.0s)	Maeda 1994	Output feedback (LQ/LQG theory) 2 HMDs (22tf*2) for 2 horizontal
18	HMD	Dowa Kasai Phoenix Tower in Osaka (29 Stories, 26,800tf, 3.8s)	Kajima & KRC 1995	Acceleration & velocity feedback (Fixed-point theory) 2 sets of AMD (6tf) on TMD (36tf) for 2 horizontal
19	HMD	Rinku Gate Tower Bldg. in Osaka (56 stories, 4.4s)	MHI 1995	Output feedback (LQ/LQG theory) 2 HMDs (110tf*2) for transverse & torsional

18

The developed systems are expected to be verified under nonstationary excitations such as earthquakes and winds. Both a harmonic excitation test and a free vibration test are first conducted to directly compare the controlled and uncontrolled responses. These tests provide data for structural identification in the second phase. Next, the structure and control system combined model is identified again by utilizing the measured controlled response under nonstationary excitations. Last, the observed controlled response is compared with the uncontrolled one that the identified model estimates, because both controlled and uncontrolled responses are not measured simultaneously. Structural identification is recognized as an important technique in both application and development. Indirect verification can be conducted even if either the analytical model were incompletely identified or the excitation were unmeasured. The control effect can be verified by comparing the two transfer functions with the control system on and off after the responses and excitations are measured [37]. The effect can be easily estimated while control is repeatedly interrupted under winds, since this kind of disturbance continues for a long time.

3. AN APPROACH TO CONTORL OF LARGE EARTHQUAKE FORCES

Structural control ultimately aims at an undamaged structure even under large earthquakes. However, most active/hybrid control systems are presently designed to simply improve residential amenity during medium earthquakes and strong winds. In general, they require a huge amount of energy and installation space for the control in large earthquakes. A semi-active control system is useful, because it changes the characteristics of a structure with minimal energy. This chapter presents one approach to control of large earthquake forces for a high-rise building. An active variable damping (AVD) system has been developed, which actively controls the damping force of a variable hydraulic damper installed in a structure. It operates with low power and without any additional energy supply or equipment such as pumps. Many control devices can be installed in each story in combination with braces or walls without requiring special installation spaces. A large-scale shaking table experiment using a three-story test model and a verification test for the performance of a prototype model have been conducted [38, 39].

This chapter describes the control performance of the AVD system during large earthquakes, based on results of seismic response analyses for a typical tall building. First, LQ control theory introduces feedback control to an SDOF system, and its advantages are considered. Second, the actively controlled response is compared with the passively controlled and uncontrolled responses, when the AVD system is applied to a tall building model. Furthermore, the control performance against large earthquakes is verified from the viewpoint of practical applicability, considering the maximum damping force of the control device. The analytical results imply that the AVD system would improve the structural safety and functional security of computers and other important equipment during large earthquakes.

3.1 Basic study on SDOF system
The AVD system actively controls with a computer the damping force of the device to minimize the structure's response. Commands are sent from the computer to the device to

generate the optimal control force. Various control theories can be applied to calculate the optimal force. In particular, significant progress has been made recently in robust control theories such as H^∞ control. However, if control performance is to be considered more important than robustness, LQ control is an appropriate choice.

Figure 3.1 shows an SDOF undamped system with an active control device. m is the mass of the structure, k is the stiffness, and $u(t)$ is the optimal control force. $y(t)$ is the displacement relatively to the base, and $y_o(t)$ is the ground displacement. When LQ control theory is applied, the state vector $x(t)$ can be defined in either the relative coordinate or the absolute coordinate as described in Section 2.1. Figure 3.2 compares the controlled responses with uncontrolled ones (NC) in the frequency domain, weighting Q and R as follows.

$$Q = \begin{bmatrix} 1 & 0 \\ 0 & 0 \end{bmatrix}, \quad R = r \tag{3.1}$$

This weighting evidently introduces velocity feedback control. The circular frequency of a harmonic excitation ω is normalized by the undamped natural circular frequency of the system $\omega_0 = \sqrt{m/k}$.

When the state vector $x(t)$ is defined in the absolute coordinate, the absolute acceleration and the relative displacement are completely suppressed under the uncontrolled response in the higher frequency-ratio range. However, when the vector is defined in the relative coordinate, both responses are completely suppressed in the lower frequency-ratio range. This absolute velocity feedback control makes the control force smaller in the higher frequency-ratio range. This difference becomes more significant with increasing control force, suggesting that the control law must be changed in accordance with the ratio of the excitation frequency to the natural frequency of the structure.

3.2. Application to a tall building

Figures 3.3(a) and (b) show a 100-meter-high, 26-story, steel-framed building. Because the AVD system was to be installed, the building frame was statically designed using the allowable stress design with a base shear coefficient of 0.08. This coefficient is the ratio of shear force in the first story to total building weight. The total weight of the structure is about 31,000tonf. Four control devices were planned to be installed in each of the first to the 25th stories. Figure 3.3(b) depicts two devices in two frames for each horizontal direction, and Figure 3.4(a) depicts the installation of a device in each frame.

3.2.1. Analytical model

The structure was modeled as an MDOF system with two bending-shear deforming elements in the transverse direction: the outside frame A with a device and the inner frame B without one. The device and brace were modeled as shown in Figure 3.4(b). The device was given as a Maxwell model, in which a variable damping element with damping coefficient $c(t)$ and a spring element with stiffness k_v were joined in series. The device model and a spring element with brace stiffness were joined in series and incorporated in the building model. The first undamped natural period of the building is 3.45sec., the second is 1.19sec., and the third is 0.71sec.. The internal viscous damping is assumed to be 2% of the critical

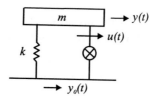

Figure 3.1. SDOF system with active control.

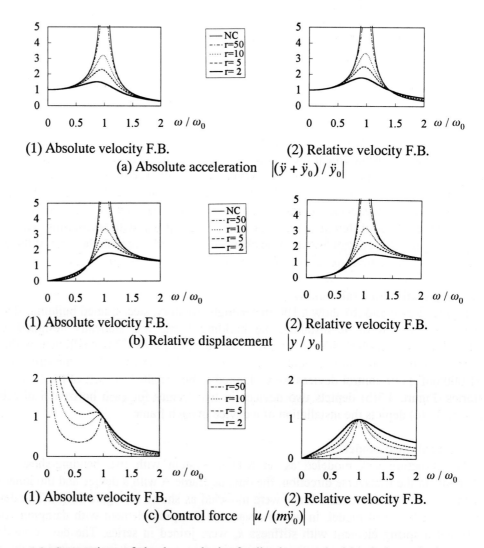

(1) Absolute velocity F.B. (2) Relative velocity F.B.

(a) Absolute acceleration $\left|(\ddot{y} + \ddot{y}_0)/\ddot{y}_0\right|$

(1) Absolute velocity F.B. (2) Relative velocity F.B.

(b) Relative displacement $\left|y/y_0\right|$

(1) Absolute velocity F.B. (2) Relative velocity F.B.

(c) Control force $\left|u/(m\ddot{y}_0)\right|$

Figure 3.2. Comparison of absolute velocity feedback control with relative velocity feedback control in the frequency domain.

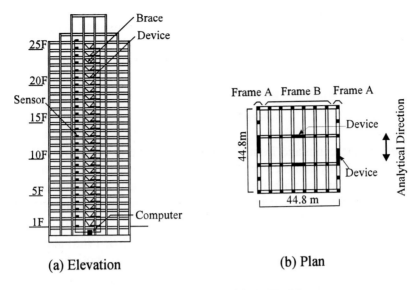

(a) Elevation

(b) Plan

Figure 3.3. Objective building with AVD system.

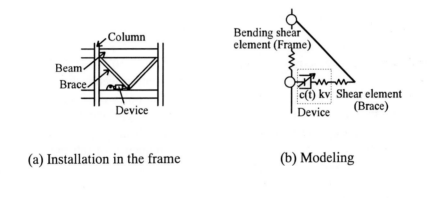

(a) Installation in the frame

(b) Modeling

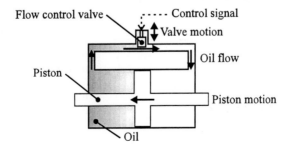

(c) Outline of the device

Figure 3.4. Installation, modeling and outline of control device.

22

value at the first natural period.

The control device receives commands from the computer, and generates a damping force by adjusting the opening of the internal flow control valve as shown in Figure 3.4(c). The damping force $f_v(t)$ is assumed to comply with the following equation:

$$f_v(t) = \begin{cases} c(t)v(t) = u(t) & \textit{where } u(t)v(t) > 0 \\ 0 & \textit{where } u(t)v(t) \le 0 \end{cases} \tag{3.2}$$

In which $u(t)$ is the command from the computer to the device (optimal control force), $v(t)$ and $c(t)$ are the velocity and the variable damping coefficient of the device, respectively. The device becomes a damper with fixed damping coefficient c if passive control is assumed, where damping force $f_p(t) = cv(t)$. The passive system has advantages of simplicity and cost efficiency.

The El Centro wave (NS, 1940), the Hachinohe wave (NS, 1968) and the Kobe wave (NS, 1995) described in Section 1.3 were used as input earthquake motions.

3.2.2. Control system design

The weighting matrices in the performance index Eq.(2.2) when the LQ theory is applied to the system are set as follows:

$$Q = \begin{bmatrix} [Q_v] & [0] \\ [0] & [0] \end{bmatrix}, \quad [Q_v] = diag(1), \quad R = diag(r) \tag{3.3}$$

As a result, only the parameter r is changed from 10^{-1}, 10^{-2}, 10^{-3} to 10^{-4} in the design.

The state feedback control is simplified to velocity output feedback control, since the gains for displacement only slightly affect the response. The poles confirm that this control adds only high damping to a structure, as shown in Figure 3.5. The absolute velocity feedback law is adopted as an appropriate method for tall buildings. It has high control performance and reduces the control force, because the first natural frequency of tall buildings is lower than the dominant earthquake frequencies, and its power is generally concentrated in the higher frequency range. A digital computer was used in the practical application as presented in Reference [38]. Feedback gain based on LQ control theory for a discrete-time system was used for the seismic response analysis. A computer sampling time of 10^{-3} sec. was assumed.

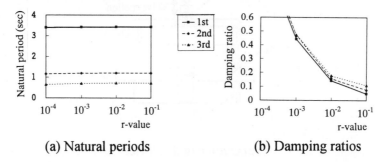

(a) Natural periods   (b) Damping ratios

Figure 3.5. Natural periods and damping ratios of the controlled system.

3.2.3. AVD system versus passive system

Figures 3.6.1(a) to (c) show the maximum controlled story drift angles with the uncontrolled response, when the linear structure is excited by the El Centro wave normalized to a maximum velocity of 25cm/sec (255.4cm/sec^2). This input level demands that a structure vibrate within the elastic range according to the Japanese building code. Figure 3.6.1(a) show that the responses with the AVD system decrease with decreased weighting r. The control effect is not proportional to r and is limited to around $r=10^{-3}$. Figures 3.6.1(b) show the results for the passive damper using four damping coefficients c=122.5, 245, 490 and 980 kN sec/cm. The responses decrease with increasing damping forces, and are similarly limited to around c=490 kN sec/cm. Figure 3.6.1(c) compares the responses for both the AVD system with $r=10^{-3}$ and the passive damper with c=490 kN sec/cm with the uncontrolled response (NC). The story drift angle at intermediate stories for the passive system is reduced by 1/4, and that for the AVD system is reduced by 1/2. Figure 3.6.2(a) to (c) show the maximum shear forces of Frame B corresponding to Figures 3.6.1. The shear force at intermediate stories for the passive system is reduced by 1/4, and that for the AVD system is reduced by 1/2.

3.2.4. Control performance against large earthquakes

Elasto-plasticity for the building frame and the limit of the control device's damping force are considered in practical design for large earthquakes. A normal trilinear-type shear force-deformation curve was assumed on the basis of static incremental non-linear analysis, and modeled according to the following equation in the damping force [40].

$$f_v(t) = \begin{cases} f_{max}sign(v(t)) & where \quad u(t)v(t) > 0 \quad and \quad |u(t)| > f_{max} \\ c(t)v(t) = u(t) & where \quad u(t)v(t) > 0 \quad and \quad |u(t)| \le f_{max} \\ 0 & where \quad u(t)v(t) \le 0 \end{cases} \tag{3.4}$$

where f_{max} is assumed to be 1,960kN, as the limit of the device's damping force on the basis of mockup test results [39].

Figure 3.7 shows the resulting maximum responses for the El Centro wave where maximum velocity is normalized to 50cm/sec. Elastic limit shear force (E. Lim) is also shown. The story drift angles are maintained at under 1/200, and the shear forces are maintained within the elastic range of the structure, although the damping forces are limited. The uncontrolled response is decreased due to additional hysteretic damping of the frame in the nonlinear range. Figure 3.8 shows the control forces in both the first and 20th stories in the time histories, corresponding to Figure 3.7. The damping forces reach the limitation of the devices (1,960kN$\times$2). It is confirmed in practical application that this level requires only 30 to 50W electric power per device.

Figures 3.9 compares the maximum uncontrolled responses when three waves are input. The maximum velocities of three waves are normalized to 50 cm/sec. The corresponding maximum accelerations are 510.8cm/sec^2 for the El Centro wave, 338.0cm/sec^2 for the Hachinohe wave and 453.0cm/sec^2 for the Kobe wave.

Figure 3.10 compares the controlled responses. The uncontrolled story drift angle clearly exceeds 1/200 and the ductility factor exceeds 1 under all waves. The ductility factor is defined as the ratio of actual deformation to maximum elastic deformation.

24

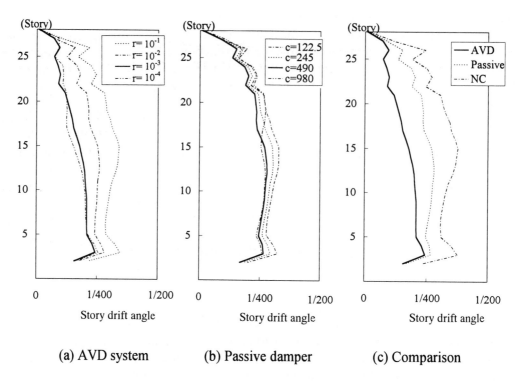

(a) AVD system   (b) Passive damper   (c) Comparison

Figure 3.6.1. Maximum story drift angles under the El Centro wave (25cm/sec.).

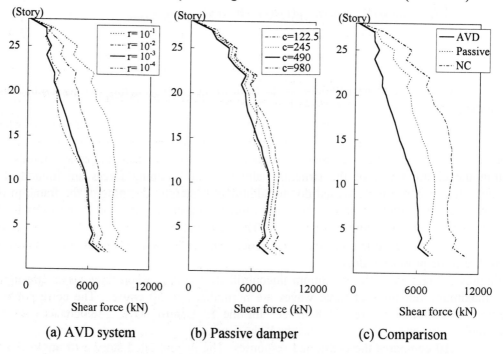

(a) AVD system   (b) Passive damper   (c) Comparison

Figure 3.6.2. Maximum shear forces under the El Centro wave (25cm/sec.).

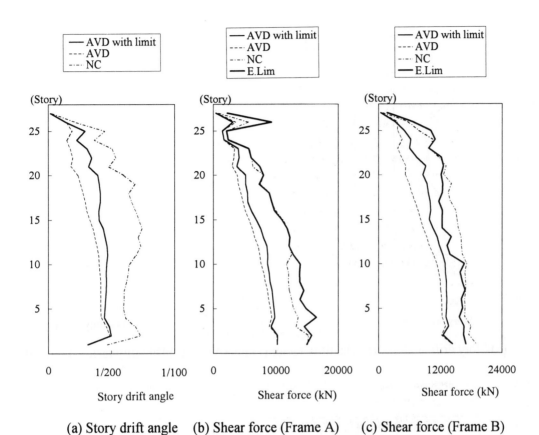

(a) Story drift angle   (b) Shear force (Frame A)   (c) Shear force (Frame B)

Figure 3.7. Maximum responses under the El Centro wave (50cm/sec.).

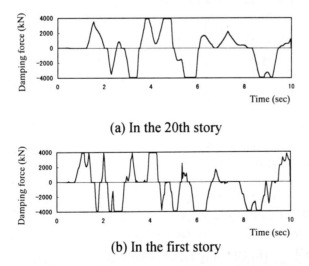

(a) In the 20th story

(b) In the first story

Figure 3.8. Damping forces with the limitation.

26

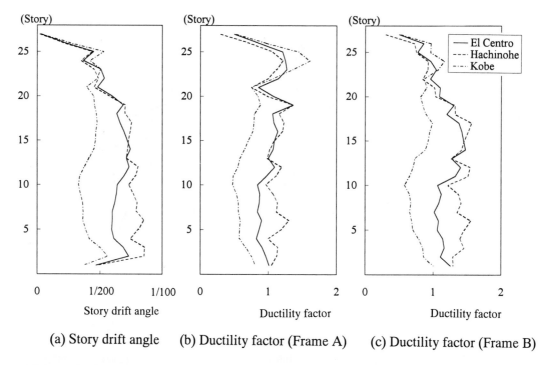

(a) Story drift angle   (b) Ductility factor (Frame A)   (c) Ductility factor (Frame B)

Figure 3.9. Comparison of the uncontrolled responses under three earthquakes (50cm/sec).

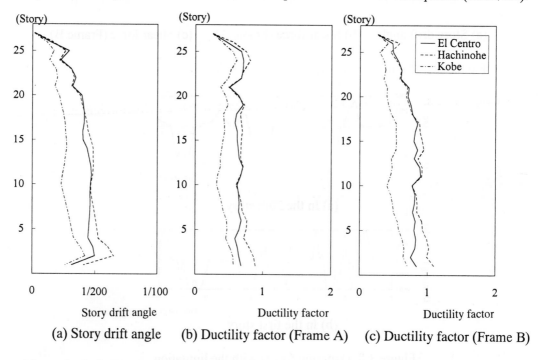

(a) Story drift angle   (b) Ductility factor (Frame A)   (c) Ductility factor (Frame B)

Figure 3.10. Comparison of the controlled responses under three earthquakes (50cm/sec).

For the AVD system, however, small responses were achieved with a story drift angle of around 1/200 and a ductility factor of about 1. These controlled results indicate the structural responses maintain within the elastic range and the building structure remains intact even during large earthquakes.

4. CONCLUSIONS

Active/hybrid vibration control of structures has been attracting a growing interest worldwide as an innovative technology in earthquake engineering for the last decade. This technology has been advanced thanks to new developments in both mechanical and control engineering. The state of the art is reported from the viewpoints of theoretical research and application development, after classifying currently proposed control strategies. The research situation conquered the first phase in which structural responses are effectively suppressed under moderate earthquakes and strong winds. Now it faces the second phase in which responses are maintained within an elastic range under large earthquakes. One approach to the control of large earthquake forces is described by applying the AVD system to a typical tall building structure. The analytical results imply the direction of the near future; the semi-active control system has the advantages of low external energy supply and compact devices.

REFERENCES

1. T.Kobori and R.Minai, J. Struct. Contr. Eng. AIJ No.52 (1956) 41 (in Japanese).
2. T.Kobori and R.Minai, *ibid.* No.66 (1960) 253 (in Japanese).
3. T.Kobori and R.Minai, *ibid.* No.66 (1960) 257 (in Japanese).
4. J.T.P.Yao, J. Struct. Div. Proc. ASCE Vol.98 No.ST7 (1972) 1567.
5. J.N.Yang et al., J. Eng. Mech. ASCE Vol.113 No.9 (1987) 1369.
6. T.Sato et al., Trans. Japan Nat. Sym. Active Struct. Res. Contr. (1992) 109 (in Japanese).
7. J.Suhardjo et al., Int. J. Nonlinear Mech. Vol.27 No.2 (1992) 157.
8. D.P.Tomasula et al., J. Eng. Mech. ASCE Vol.122 No.3 (1996) 218.
9. Z.Wu et al., Proc. 1WCSC Vol.1 (1994) TP2-50.
10. R.L.Kosut, Trans. Automatic Contr. IEEE Vol.AC-5 No.5 (1970) 557.
11. T.Suzuki et al., Proc. Annual Meeting AIJ Vol.Struct-1 (1993) 753 (in Japanese).
12. K.Tamura et al., Proc. 1WCSC Vol.3 (1994) FA2-13.
13. I.Nagashima et al., J. Struct. Constr. Eng. AIJ No.483 (1996) 39 (in Japanese).
14. J.Suhardjo et al., Struct. Safety No.8 (1990) 69.
15. J.N.Yang et al., J. Eng. Mech. ASCE Vol.118 No.7 (1991) 1423.
16. J.N.Yang et al., *ibid.* Vol.118 No.7 (1991) 1441.
17. Y.Ikeda and T.Kobori, J. Struct. Contr. Eng. AIJ No.435 (1992) 51 (in Japanese).
18. S.Noda, Trans. Japan Nat. Sym. Active Struct. Res. Contr. (1992) 65 (in Japanese).
19. B.Bhartia et al., Proc. 1WCSC Vol.2 (1994) TP2-17.
20. J.N.Yang et al., J. Eng. Mech. ASCE Vol.121 No.12 (1995) 1330.
21. M.Kageyama et al., Trans. Japan Nat. Sym. Active Struct. Res. Contr. (1992) 181 (in Japanese).

22. J.Inaudi et al., Earthquake Eng. Struct. Dyn. Vol.21 No.6 (1992) 471.
23. Y.Ikeda and T.Kobori, Proc.1WCSC Vol.3 (1994) FP1-3.
24. H.Furuta et al., *ibid.* Vol.1 (1994) WP1-3.
25. M.Yamada et al., *ibid.* Vol.1 (1994) WP1-13.
26. K.Goto et al., *ibid.* Vol.1 (1994) WP1-41.
27. A.Joghataie and J.Ghaboussi, *ibid.* Vol.1 (1994) WP1-21.
28. T.Sato et al., Trans. Japan Nat. Sym. Active Struct. Res. Contr.(1992) 109 (in Japansese).
29. T.Sasaki and K.Yoshida, Proc. Annual Meeting AIJ Vol.Struct.-1 (1993) 845 (in Japanese).
30. L.Faravelli and T.Yao, Proc. 1WCSC Vol.1 (1994) WP1-49.
31. Proc. 1WCSC (1994, Los Angeles).
32. Proc. Int. Workshop on Struct. Contr. (1993, Hawaii).
33. T.Kobori et al., Earthquake Eng. Struct. Dyn. Vol.20 No.2 (1991) 133.
34. T.Kobori et al., *ibid.* Vol.20 No.2 (1991) 151.
35. J.C.H.Chang and T.T.Soong, J. Eng. Mech. Div. ASCE Vol.106 No.EM6 (1980) 1091.
36. T.Kobori et al. Earthquake Eng. Struct. Dyn. Vol.22 No.11 (1993) 925.
37. M.Yamamoto and S.Aizawa, Proc. 1WCSC Vol.3 (1994) FP1-13.
38. N. Kurata et al., *ibid.* Vol. 2 (1994) TP2-108.
39. T. Mizuno et al., PVP-Vol. 229 ASME (1992) 163.
40. N. Kurata et al., Proc. 1ECSC (1996) (in press).

Theoretical and Applied Mechanics 1996
T. Tatsumi, E. Watanabe and T. Kambe (Editors)

29

Typhoons, Hurricanes and Fluid Mechanics

James Lighthill

Mathematics Department, University College London,
Gower St., London WC1E 6BT, England

The worldwide knowledge and experience of IUTAM in fluid mechanics − including the mechanics of a pair of strongly interacting fluids such as the atmosphere and the ocean − contributes, in a close collaboration with ICSU partners and with the World Meteorological Organisation, to fruitful efforts directed at reducing the human impact of possible Tropical Cyclone Disasters.

1. INTRODUCTION: CYCLONE AS CYCLOPS

On receiving the Congress Committee's greatly appreciated invitation to give this Closing Lecture 'on a topic within Fluid Mechanics', and after careful thought, I decided to devote the lecture to the fluid mechanics of Tropical Cyclones. The scientific term Tropical Cyclone describes [1,2] a single well defined natural phenomenon, which in Japan and other areas of the Northwest Pacific is called 'typhoon' (literally, 'great wind') while in the Northwest Atlantic it is known by the name 'hurricane' (of Caribbean origin); this latter name being used also in the Northeast Pacific, whereas the scientific expression Tropical Cyclone (or, more simply, just cyclone) is preferred in tropical areas of the Southwest Pacific and of the Indian Ocean.

In meteorology, of course, the word cyclone on its own is used [3] to denote any centre of low pressure towards which, by the well known Coriolis effect, surface winds spiral inwards 'cyclonically'; that is, anticlockwise in the Northern, but clockwise in the Southern, Hemisphere. In addition to satisfying this general requirement for a cyclone, however, the Tropical Cyclone proper possesses an altogether special feature − clearly exhibited (Figure 1) in satellite imagery − which is absent from other cyclones. This is the famous 'eye of the storm': a calm region of circular shape which is often nearly free of clouds, although it is surrounded by a wall of extremely dense cloud (the 'eyewall') where wind speeds are as high as 50 m/s or more.

The fierce one-eyed monster of Greek myth was called Cyclops. Now a Tropical Cyclone, with its formidable threats to human life and property, is undoubtedly fierce; it has moreover a single circular eye (actually, the word Cyclops in Greek signifies 'circular eye'); and, above all, its huge diameter of several hundreds of kilometres makes it truly a monster among meteorological phenomena. So three special features of a Tropical Cyclone − its immense size, its one circular eye, its terrifying ferocity − are summed up in the alliterative phrase 'Cyclone as Cyclops'.

30

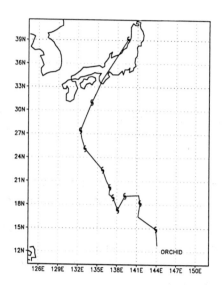

Figure 1. Satellite picture of a Tropical Cyclone (the only one, out of 9 such pictures shown in the Lecture itself, for which there is space in a written text limited to 26 pages).

Figure 2. On this observed track of Supertyphoon Orchid, the symbols show the eye's position at 0000Z on 20 September 1994 and at 24-hourly intervals thereafter.

It is always over a tropical Ocean (Atlantic, Pacific or Indian) that a Tropical Cyclone is formed, but human beings experience its ferocity when it approaches land. Then great destruction may result, either from the direct force of extreme winds on manmade structures, or else from coastal inundation by a storm surge; as can occur when nearshore water, acted on by such winds, piles up against a coastline. The worst recorded storm surge, produced [4] by a 1970 Tropical Cyclone in the Bay of Bengal, caused 300,000 deaths in Bangladesh.

Even a city as far from the equator as 35°N, like this beautiful city of Kyoto, is by no means immune from the threat of extreme Tropical Cyclone winds. For example, on 29 September 1994, "Supertyphoon Orchid" passed close to Kyoto [5] with some seriously damaging consequences; especially, to structures which, having been built before the introduction of modern building codes for wind-hazard resistance, had not been adapted later to take those into account.

A map (Figure 2) showing the track followed by the eye of supertyphoon Orchid recalls the severe problems which forecasters face in predicting such a track. The track of a Tropical Cyclone (TC) commonly involves just as many sudden changes of direction as were observed for Orchid. Accordingly, any approach to TC track prediction by a simple process of extrapolation of the path followed in the last 24 or 48 hours (meteorologists call this a 'persistence' approach to forecasting) can achieve only relatively poor reliability. Here is one of the reasons why fluid mechanics needs to play an important role in TC forecasting.

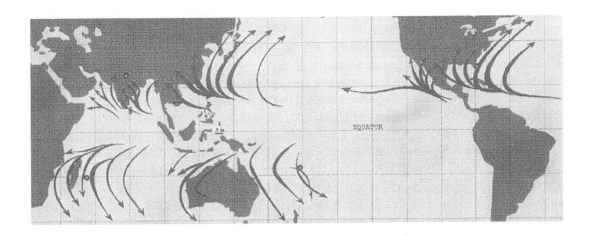

Figure 3. Schematic diagram for the world's oceans of (drastically smoothed) 'frequent tracks' followed by Tropical Cyclones.

In another map, designed to give a summary view of the TC as a worldwide phenomenon, the highly irregular character of individual TC tracks is altogether suppressed as a result of a sort of statistical 'smoothing'. This map (Figure 3) shows 'frequent tracks' (drastically smoothed) of TCs commonly appearing in the Caribbean, in the Pacific, around Australasia, and in the Indian Ocean. Very large numbers of islands, and extremely extensive coastal regions of continents, are seen from this map to be seriously threatened by TC hazards. Formidable hazards may arise moreover from TCs with clockwise rotation (in the Southern Hemisphere, of course); one of those, for example, having destroyed three-quarters of Darwin, Australia in 1974.

The map reconfirms also that all TCs are formed over oceanic regions (even though the hazards they generate arise only later, when they approach land). Their study, then, involves not simply fluid mechanics but specifically the mechanics of (at least) two interacting fluids: ocean and atmosphere.

2. GLOBAL HAZARDS AND INTERNATIONAL UNIONS

So TCs are global hazards, crying out for study by responsible International Unions such as IUTAM. This Union, indeed, first entered the field when it organised, along with the International Union of Geodesy and Geophysics (IUGG), the Joint IUTAM/IUGG Symposium (Reading, UK, 1980) entitled 'Intense Atmospheric Vortices'[6].

Actually, it had been our revered ex-President Maurice Roy who had forcefully and persuasively argued that IUTAM needed to give attention to the mechanics of such geophysical phenomena as develop exceptionally great kinetic energies; actually, those in a TC are commonly of order 10^{18}J or more. In response to these arguments I willingly agreed to organise a Symposium on this topic in a close collaboration with IUGG (just as, in 1977, we had jointly organised [7] the 'Monsoon Dynamics' Symposium in New Delhi). Moreover, a further topic − not only TCs but also tornadoes − was included in the subject matter of the 'Intense Atmospheric Vortices' Symposium, and the resulting book has been found a valuable introduction to both topics.

Then resolutions of the UN General Assembly designated the 1990's an International Decade for Natural Disaster Reduction (IDNDR), with a well defined aim: to reduce the human impact of natural disasters — and, above all, their impact on relatively poor countries. With this objective, an efficient IDNDR Secretariat has been set up in Geneva as part of the UN's Department of Humanitarian Affairs.

It was at the 1989 meeting of IUTAM's Bureau that I agreed to investigate (as fellow Bureau members suggested) whether a useful contribution to IDNDR might be made by IUTAM; again, jointly with IUGG. For this investigation I acted with an IUGG colleague, Professor R.P. Pearce, to call together a suitable Workshop, comprising ourselves and seven other persons: Drs. K.A. Emanuel, B. Johns, Y. Kurihara, M.E. McIntyre, M.P. Singh, G.B. Tucker and Zheng Zhemin. This small group made an intensive study of the possibility of such a contribution to IDNDR when it met in Vienna (during the week in August 1990 which immediately preceded IUTAM's General Assembly).

The influential Report issued by the Vienna Workshop stressed [8] that very many relatively poor tropical countries are vulnerable to disasters associated with a TC's extreme wind speeds and huge horizontal length scale. In the context of IDNDR, therefore, the Report proposed an extensive programme of new interdisciplinary studies aimed above all at improving the quality of TC forecasts and warnings (mainly with regard to timeliness and reliability). On the other hand those other Intense Atmospheric Vortices, the tornadoes, which are on a far smaller scale (of several kilometres only instead of several hundred kilometres) and mainly appear in countries that are 'not so poor', might appropriately be omitted from the IDNDR programme.

The Report of the Vienna Workshop emphasized also

(i) the key roles of ocean science — and of air/sea interaction mechanics — in TC study;

(ii) the importance of IUTAM for fluid mechanics, of IUGG for geophysics, and of SCOR for oceanography, as ICSU bodies that could work well with the UN's World Meteorological Organisation (WMO) on TC disasters;

(iii) the vital need, if forecasting was to be improved, for better knowledge of initial conditions; and

(iv) the great value to be attached to ever closer links between TC experts from Asia and from other continents.

Finally the Report recommended that all these objectives should be actively pursued at a Joint ICSU/WMO Symposium to be held in Beijing, China. I should gratefully add that the Report's proposals were backed (August 1990) by IUTAM's General Assembly and Bureau, and later supported by IUGG, SCOR, ICSU and WMO; while arrangements for the proposed Beijing Symposium were finalised, again in Vienna, during IUGG's own General Assembly (August 1991).

All of the important matters (i) to (iv) were highlighted, along with several others, during the Joint ICSU/WMO Symposium on Tropical Cyclone Disasters (held during

October 1992 in Beijing, with the book of the Symposium being published in the following year [9] by Peking University Press); and, in this Lecture, I propose to lay stress on every one of them while referring also to some more recent developments. These latter include the published report [10] of yet another Symposium co-sponsored by ICSU and WMO, which I chaired during 22 November - 1 December 1993 in Huatulco, Mexico, on the subject 'Global Climate Change and Tropical Cyclones'. But first of all I must return to the theme of 'Cyclone as Cyclops', and ask the key question: 'Why is it that intense Tropical Cyclones possess a circular eye?'

3. WET-AIR THERMODYNAMICS AND A DISAPPEARING SPIRAL

In every cyclone, of course, the surface winds spiral inward cyclonically; yet in Tropical Cyclones, on the other hand, this spiral motion's inward component performs an extraordinary 'disappearing trick' at the eyewall: that circular wall of extremely dense cloud (of the type known to meteorologists as 'convective') which surrounds an eye often nearly free of cloud. What causes the inward component of the spiral motion to disappear there?

The answer is somewhat astonishing: in the eyewall, fast upward motions are able to lift the air right up to the base of the stratosphere (situated at about 15 km altitude), where it then spirals outward in a broadly anticyclonic (although not usually very symmetrical) motion. Thus, the spiral 'disappears' because fast upward motions lift surface air to stratospheric altitudes.

Yet, under ordinary circumstances, air simply cannot rise like this! This is because rising air, as its pressure drops, expands and therefore cools; actually (see below) by 1°C per 100m through energy lost in the work of expansion. Now, in any stable atmosphere, the surrounding air is not so cold; in other words, the atmosphere's temperature drop with height is less than 1°C per 100m. The rising air, then, being colder than its surroundings, necessarily falls back.

The simple thermodynamics of air rising 'under ordinary circumstances' yields such a conclusion from the easily derived equation (1). Because the pressure p drops with height z at a rate ρg, equal to the weight of air per unit volume, it follows that, during any height increase dz,

$$-gdz = \frac{dp}{\rho} = c_p dT \tag{1}$$

for a perfect gas in an adiabatic change. Here, c_p is the specific heat of air at constant pressure (about 1000J/kg per 1°C), so that heat input at constant pressure is $c_p dT$; on the other hand, heat input at constant temperature equals the work of expansion pdV if specific volume V increases by dV, where by Boyle's law $pdV = -Vdp = -\rho^{-1}dp$; and these two heat inputs must cancel under adiabatic conditions (of no exchange of heat with the surroundings). Equation (1) gives the rate of temperature drop with height as

$$-\frac{dT}{dz} = \frac{g}{c_p} = \frac{10\text{ms}^{-2}}{1000\text{J/kg per 1°C}} = 1°\text{C per 100m}, \tag{2}$$

just as stated in the previous paragraph.

3.1. Wet-air thermodynamics

By contrast, any corresponding conclusions from wet-air thermodynamics are altogether different. By wet air I mean air that is 100% humid; in other words, saturated with water vapour. Now, whereas rising air that is wet (in this sense) does cool still, nevertheless this cooling causes some condensation of water vapour into rain drops; in which process latent heat is released, so that the degree of cooling is less.

In wet-air thermodynamics the equation governing rising air is changed from (1) to

$$-gdz = \frac{dp}{\rho} = c_p dT + Ldq, \qquad (3)$$

where the latent heat L means the excess energy per unit mass of water vapour over water in condensed form, while q is the concentration (by mass) of water vapour. The crucial fact when wet air rises is that q continues to take its saturated value q_s which is a steeply increasing function of temperature; this, of course, is why cooling air experiences that reduction in vapour concentration q which necessarily implies some condensation.

Very roughly indeed (see below for precise details), the effect of the added term Ldq appearing in equation (3), with q rising so steeply as function of T, is to double the coefficient of dT on the right-hand side from its simple value c_p occurring in (1). For rising wet air, then, the resulting rate of temperature drop with height,

$$\Gamma = -\frac{dT}{dz}, \qquad (4)$$

assumes very roughly half the former value (2), becoming about $\frac{1}{2}°$C per 100m. Accordingly, if the surrounding fluid's rate of temperature drop with height exceeds this (lying, then, somewhere between $\frac{1}{2}°$C per 100m and its maximum possible value of 1°C per 100m for a stable atmosphere), rising wet air will always remain warmer than its surroundings and therefore can continue to rise. Such a capability, nonetheless, exists only for wet air in the strict sense of 100% relative humidity; by contrast, equation (1) is satisfied by air 'under ordinary circumstances'; that is, with relative humidity less than 100%.

It is necessary, then, to interpret a TC's eyewall as that location where the very long spiral path pursued by winds directly over the surface of the ocean has finally caused them to become saturated with water vapour (100% humidity). Then the resulting wet air is indeed able to rise if, in the sense sketched above, the surrounding atmosphere is 'not too stable'.

Far more precision can be injected into the above arguments if equation (3) is used with the density ρ given by the perfect-gas law as p/RT while q takes its saturated value

$$q_s = 0 \cdot 62 \frac{p_v(T)}{p}. \qquad (5)$$

Here, 0·62 is just the water/air molecular-weight ratio, while the vapour pressure $p_v(T)$ may be plotted (Figure 4) as a function of temperature in two alternative curves, representing partial pressure of water vapour (i) over liquid water and (ii) over ice. In the lower part of the eyewall, of course, we are concerned with vapour condensing to liquid water. Higher up, however, vapour condenses to ice crystals; which, indeed, are found to be abundant in the upper air over a Tropical Cyclone. Transition from one curve to the

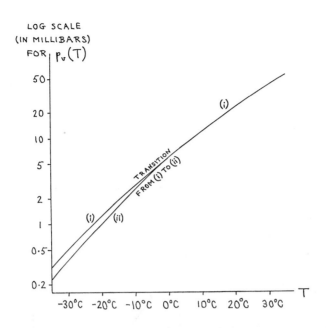

Figure 4. The partial pressure $p_v(T)$ of water vapour (i) over liquid water and (ii) over ice.

other takes place somewhere between 0°C and −10°C, depending on the amount of supercooling of condensed water.

Values of Γ, the rate (4) of temperature drop with height for wet air, were obtained from equations (3) and (5) by List [11]; and are here plotted (Figure 5) against temperature for a range of different pressures, with both $p_v(T)$ and L given their values for vapour over liquid water when $T > 0°C$ while values for vapour over ice are used when $T < -10°C$. Dotted lines indicate possible transitions between the two sets of curves.

Equations (3) and (5) also specify the ratio dp/dT as a function of p and T in the form of a differential equation which can be solved given a boundary condition at (say) the base of the eyewall. When typical base values of $p = 950$mb and $T = 30°C$ are used to derive this solution, the corresponding values of Γ are as shown by the broken line. (They are a little less than the very rough approximation $\frac{1}{2}°C$ per 100m in the lower part of the eyewall, and a little more in the upper part.) Air in the eyewall can rise when rates of temperature drop in the surrounding atmosphere exceed these Γ values.

The corresponding heights z at which the Γ values would be found if wet air rose (as assumed) in a strictly adiabatic process are readily deduced from the broken-line values of Γ by numerically integrating equation (4) from a base value of $T = 30°C$ at $z = 0$. This derives the height z as the function of T shown in the dash-dotted line.

Real processes within the rising fluid of the eyewall include, of course, some departures from strictly adiabatic wet-air behaviour; the most important one arising, perhaps, from mixing caused by entrainment of ambient fluid. Under these circumstances the value of T reached at an altitude of 15km typical of the base of the stratosphere will not be nearly so far above a typical ambient temperature of $-70°C$ as the value derived $(-35°C)$ on

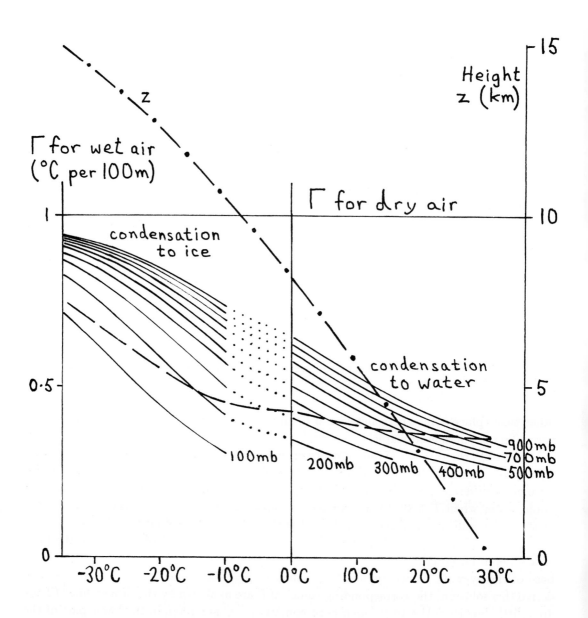

Figure 5. Solid curves show Γ, the rate of temperature drop (4) with height for adiabatically rising wet air, computed [11] as a function of temperature T in °C and pressure p in millibars (mb). Dotted curves suggest possible transitions between values for vapour condensing (i) into liquid water and (ii) into ice. The broken-line curve plots values of Γ in an eyewall where wet air is assumed to rise adiabatically from a base at 30°C and 950mb, while the dash-dotted line gives corresponding values of the height z (again assuming a strictly adiabatic process).

the assumption of a strictly adiabatic process. Nonetheless the adiabatic curves for wet air given an impressive indication of those lifting forces that power a TC.

3.2. The TC viewed as a heat engine

Emanuel [12,13] has pointed out how wet-air thermodynamics allows us to view the TC as a heat engine where the working fluid is just that mix of dry air with water in all its forms (vapour, droplets, ice crystals) which appears in the atmosphere. From this standpoint all the heat intake occurs over the ocean, and essentially consists of latent heat of evaporation transferred during the long spiral path pursued by winds before they reach saturation. Somewhat remarkably, all this heat intake occurs at a practically constant temperature (that of the ocean surface). This is because most of the cooling of air which could be expected to result from provision of latent heat is cancelled by the vigorous processes of radiative and turbulent heat transfer which take place at the interface between ocean and atmosphere. After that the heat engine's nearly adiabatic work-output phase is concentrated in the eyewall; while, finally, the heat-loss phase takes place at a nearly stratospheric temperature.

Essentially, the heat-engine cycle so described is close to a Carnot cycle: one with heat-intake and heat-loss phases occurring at different constant temperatures, and separated by an adiabatic work-output phase. Here, the large difference between the heat-intake temperature T_1 and the heat-loss temperature T_0 suggests a substantial value for the 'thermal efficiency' η: that proportion of the heat intake which appears in the output of mechanical work. Thus, the classical efficiency value

$$\eta = \frac{T_1 - T_0}{T_1} \tag{6}$$

(with temperature in kelvins) for an ideal Carnot cycle could take values of order $\frac{1}{3}$ if T_1 were about 300K (a typical sea-surface temperature) and T_0 about 200K (a typical stratospheric temperature).

Whatever a realistic value for η may be, that proportion η of the heat intake at the ocean surface which generates mechanical energy — above all in the form of extreme winds — is required in a TC to balance all the frictional dissipation of energy in these winds occurring near the ocean surface itself. Study of this important balance explains [12,13] why the TC is a tropical phenomenon: the heat intake per unit mass of air depends critically on the concentration q_s (by mass) of water vapour under saturated conditions, which increases steeply with temperature (Figure 4), while there is no such dependence on temperature in the dissipation rate per unit mass. (But see also section 4.3 below.)

4. AIR AND OCEAN IN STRONG INTERACTION

A TC's energetics, then, depend critically upon extremely strong interactions between atmosphere and ocean; 'extremely' strong in the sense that the interactions take place at extreme wind speeds. They include

   (i) that transfer of water vapour from ocean to atmosphere which is necessary to allow 100% humidity to be reached — so that air in the eyewall can rise to great heights; along with

(ii) such heat transfer from ocean to nearby air as is required to keep their temperatures equal; yet opposed by

(iii) a transfer of momentum from air to ocean associated with its frictional resistance to surface winds.

Thus, while in any TC the ocean gives energy to the atmosphere, nonetheless the atmosphere in turn 'gives back' momentum to the ocean — with some potentially disastrous results of just two main kinds, each whipped up by extreme surface winds. On the deep ocean these tend to produce intense surface waves (see section 4.1) while in shallow water (see section 5.3) they may generate those powerful bulk motions which are known as storm surges, and which can gravely threaten coastal populations.

4.1. Exceptional ocean waves

Ocean waves of exceptional height may of course be of immediate local importance, as severe hazards to shipping — while being capable also, 'at long range', of augmenting the flooding threat to a coast. Now a well known requirement for a high wind to generate intense waves is for it to act over a sufficiently long distance: the so-called 'fetch' needed for wave amplitudes to reach large values. Similarly, any exceptional wave heights may be attainable only when extreme winds are applied over a long distance — so how is it that the relatively localised extreme winds around a TC's eyewall can act over a long enough fetch?

An answer to this question is suggested by the fact that the energy in a group of ocean waves travels at the group velocity (half the velocity of their crests). Accordingly, extreme TC winds can continue to energize one and the same wave group over a long distance or fetch provided that the travel velocity of the TC as a whole coincides with the group velocity of the waves [14]. For a typical TC travel velocity around 10 m/s this condition identifies waves of length around $\lambda = 250$m as those which might continue to be excited; and, certainly, an extended fetch would be required for such long waves to build up to near maximum heights.

A schematic diagram of a travelling TC shows (Figure 6) which part generates waves moving in the direction of the travel velocity. Energy in a wave group there generated, at wavelengths satisfying the above condition, must travel at this velocity. Therefore, these waves can continue to be amplified by extreme winds, so that they are able to reach very great heights; as with the wave thus produced with height 32m (from crest to trough) which was observed in the early hours of 11 September 1995 by the crew of the celebrated liner Queen Elizabeth II.

It goes without saying, of course, that TC winds generate waves with a most diverse range of lengths — not just those very big wavelengths which I've stressed up to now (because they allow such exceptional wave heights to be attained). Moreover, many far shorter waves are quite important; for example, as helping to determine those roughness lengths which affect values of transfer coefficients for momentum, heat and water vapour. In addition, the continual breaking of these waves can be described as creating between air and ocean enormous volumes of 'a third fluid' — spray — which I shall need to re-emphasize later (section 4.3) as a principal feature of intense interactions between air and ocean.

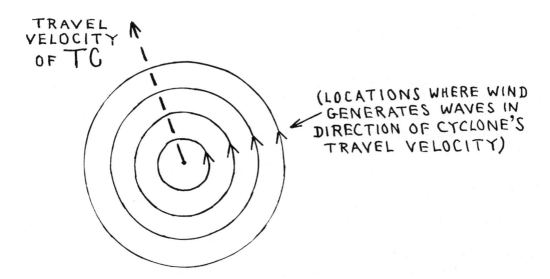

Figure 6. Schematic diagram of a travelling TC, showing which part generates ocean waves moving in the direction of the travel velocity.

4.2. Conditions for TC formation and intensification

Before that, however, I must enumerate those conditions under which such strong interactions have been observed to produce TC formation and intensification. In an important series of papers (see [6], p.3 and [9], p.116) W.M. Gray, after summarising and analysing very fully the observational record, concluded that TC initiation — often called 'cyclogenesis' — requires all six of the following conditions to be satisfied (here, Gray's well accepted list of 6 necessary conditions is accompanied with brief bracketed comments that relate each condition to physical discussions in section 3 above):

(i) latitudes, whether N. or S. of the equator, must be at least 5° (for Coriolis effect to yield a cyclonic spiralling of surface winds);

(ii) rates of temperature drop with height in surrounding air must exceed the value Γ appropriate to wet air rising adiabatically (for ascent of air in the eyewall to be adequately powered);

(iii) temperatures T at the ocean surface must be at least 26°C (for the saturated water-vapour concentration (5) to provide sufficient latent-heat input to cyclonically spiralling winds);

(iv) vertical shears, i.e. gradients of wind with height in surrounding air, must be relatively small (for disruption of the TC flow structure's axisymmetry and vertical coherence to be avoided);

(v) relative humidities in the middle troposphere must be sufficiently high (for prevention of any possible 'drying out' of the eyewall upflow by entrained air); and, finally,

(vi) some rather substantial amounts of cyclonic vorticity must previously be present at low altitude (for that process which is described in section 3 as the Carnot-cycle heat engine to be able to start).

I emphasize again that all six conditions are shown by Gray from the observational record to be necessary for TC formation and intensification.

In particular, condition (vi) refutes any idea that a TC arises from some sort of 'instability to small-amplitude disturbances.' On the contrary, only a pre-existing atmospheric disturbance of substantial amplitude can, by means of strong interaction between air and ocean, be amplified into a Tropical Cyclone. Among examples of how this can happen, I briefly mention two.

In the western Pacific, a commonly observed phenomenon is the 'monsoon trough': a long line of low pressure formed in the aftermath of a monsoon. At the eastern end of a monsoon trough, a TC is often formed through strong interaction of local cyclonic vorticity with the tropical ocean.

Over the Atlantic, on the other hand, the meteorology of the Sahara acts to generate waves with continually growing amplitude which steadily travel westward in the upper atmosphere. Then, after sufficient amplification of these westward-travelling waves, a 'crest' may break away to form a 'cut-off low' with strong cyclonic rotation which may then penetrate down to the sea surface and 'ignite the Carnot engine.' For further discussion of the fluid mechanics of TC initiation, see section 5.1.

4.3. Observing air-sea interaction at extreme wind speeds

In the meantime, this section 4 has amply shown how air-sea interaction at extreme wind speeds can exert many different influences on a TC's formation and development, as well as on its possibly hazardous consequences; so that great importance must be attached to direct observations of the detailed nature of that interaction. It has above all been the gifted and courageous crews of Russian research ships who have contributed such direct observations by deliberate navigation through typhoons − and those from ICSU who planned the 'Tropical Cyclone Disasters' Symposium (Beijing, 1992) were pleased to be able to draw the attention of the world's meteorologists to these Russian achievements. In particular, the excellent Symposium paper given by V.D. Pudov (see [9], p.367 − along with other papers referred to therein) summarised that valuable research programme.

Here, I mainly emphasize Dr. Pudov's 1988 observations, made with great care near typhoons 'Tess' (8830) and 'Skip' (8831), of a steep increase with surface wind speed in the concentration of water droplets ('spray'), found alongside a closely parallel fall in air temperature below that of the ocean surface (Figure 7). These observations suggest a slight correction to the thermodynamic picture painted in section 3. Thus, if

(a) much of the vapour transfer to surface winds came from spray droplets, then

(b) cooling from the corresponding latent heat transfer might not be fully made up by heat transfer from ocean surface, so that

(c) air temperature (as observed) would reach an equilibrium value below that of the ocean surface; and, in consequence,

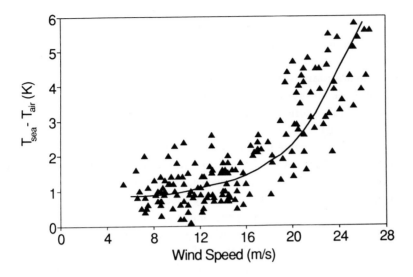

Figure 7. The 1988 measurements by Pudov (see [9], p.374 and [15], p.123), demonstrating for increasing wind speed an increasing difference (in °C) between temperatures at the sea surface and in neighbouring air — as ascribed [15] to air cooling by spray evaporation.

(d) the average temperature of saturated air around the base of the eyewall would be less than the sea-surface temperature.

From the thermodynamic viewpoint, the importance of such a correction (d) to the heat-intake temperature T_1 lies, of course, not in the rather modest resulting drop in the Carnot efficiency (6), but in the much more significant reduction in the overall latent heat intake per unit mass of air associated with the very steep dependence (Figure 4) of saturated water-vapour concentration q_s on the heat-intake temperature T_1.

Fairall, Kepert and Holland, in a careful study [15] of Pudov's data, have given a theoretical analysis in support of the above interpretation. Breaking waves generate spray in two ways: surface bursting of bubbles of entrapped air makes smaller droplets (radii up to $60\mu m$), while spume formation at whitecaps makes larger droplets (radii over $40\mu m$). An increase in the fraction of ocean surface covered by whitecaps is viewed as the main influence of increasing wind speed on spray formation; nonetheless these authors recognize that, at wind speeds above the value (around 40m/s) for which that whitecap fraction approaches unity, spray formation may yet continue to increase 'in an unknown manner' and so they discourage application of their theory at greater wind speeds. Very briefly, their predictions at 40 m/s are that the mass density of spray should reach only 0.008 kgm^{-3} (less than 1% of the air density), and yet that vapour transfer from spray to air should exceed direct transfer from the ocean surface by an order of magnitude. Thus, hypothesis (a) wins from this analysis quite strong support — which, moreover, would not be qualitatively altered even if rate of generation of spray droplets had been overestimated by up to a factor of 2 as suggested in some other studies [16].

At relatively high, although not 'extreme', wind speeds, the major international programme HEXOS (humidity exchange over the sea) gained much valuable information [17] on water vapour transfer rates. In particular, as wind speeds increased to 18m/s, North Sea measurements on the fixed platform Meetpost Noordwijk found no statistically significant increase in the standard vapour transfer coefficient. In this range of wind speeds, on the other hand, changes are absent also in Pudov's above-noted data. For the TC environment, then, there remains considerable confidence in hypothesis (a) and its suggested implications (b), (c) and (d).

Future observations of air-ocean interaction at extreme wind speeds will depend increasingly on satellite imagery. Already, a large amount of information on sea-surface roughness, including directional spectra of waves, is obtained from the earth resources satellite ERS-1 of the European Space Agency by means of a C-band radar (at 5.3GHz) which measures Bragg back-scatter from the rough ocean surface. Such 'scatterometer' data [18] are concerned, not with 'exceptional' waves developed over a long fetch (section 4.1), but with the general statistical distribution of wave roughness (as a function of length and orientation) associated with local winds. Accordingly, they are often used to suggest an approximate distribution of surface wind vectors. These data may increase in value if, as proposed by Institut Français de recherche pour l'exploitation de la mer (IFREMER), current scatterometer measurements at 25km resolution can be made routinely available.

4.4. Global Climate Change and Tropical Cyclones

Yet another problem closely related to strong interactions between air and ocean is the question, posed at an ICSU/WMO Symposium which I chaired from 22 November to 1 December 1993, of whether expected directions of global climate change are liable to produce significant effects on the frequency or on the intensity of Tropical Cyclones. Very briefly, the report on the Symposium concluded [10] that, even after an expected doubling of CO_2 in the atmosphere, those effects were likely to be small relative to the usual large year-by-year variations.

Indeed, models of climate change following such a doubling suggest a mean surface temperature rise in the tropical oceans of only 1°C (because ocean temperatures respond less than land temperatures, and tropical oceans less than oceans in general). Admittedly, in relation to the six conditions of section 4.2, a rise of 1°C helps to satisfy condition (iii), which might be thought to make TC formation and intensification significantly more likely. Nonetheless all consideration of other influences, including those of conditions (ii), (iv), (v) and (vi), tends to work the other way and diminish expectations of a rise in TC frequency.

As to any effect on TC intensities, the thermodynamic argument of section 3.2 does as it stands indicate a positive influence of mean sea-surface temperature. This, however, is where considerations about spray droplets outlined in section 4.3 may yet again tend to diminish any such influence. Indeed, if any increase in TC intensity produces an enhancement of effects (a) to (d), then this increase may readily become self-limiting because saturated air rising from the base of the eyewall may fall short, to an increasing extent, of gaining that full latent-heat intake (for powering the TC) which would be associated with the sea-surface temperature.

Next, the report gives detailed consideration to TC statistics, stressing not only the large year-to-year variation which has already been mentioned but also the absence of any discernible trends in recent decades. Finally, it studies scatter diagrams in which each TC is represented by a single point related to when and where it reached maximum intensity, that intensity being plotted against the actual sea-surface temperature found there during the month in question. The absence of significant correlation when the data are plotted in this way is seen as confirming the report's broadly negative conclusions about climate-change influences on TC statistics.

5. FLUID MECHANICS AND TROPICAL CYCLONE DISASTERS

Tropical Cyclone Disasters can of course arise quite soon after TC formation, when a small oceanic island may suffer the impact of TC winds. Far more frequently, however, major disasters occur when a TC reaches a substantial land mass, where extreme winds may cause severe damage either directly by their action on manmade structures or indirectly through storm-surge flooding; and where, in addition, river valleys can be massively flooded if the TC's huge water content is released over their catchments areas.

Scientific studies of all types of TC disaster —and of their possible mitigation — are founded primarily on fluid mechanics. Moreover, many different aspects of the mechanics of what the late A.E. Gill [19] called 'the earth's fluid envelope' (atmosphere, ocean, rivers, lakes, groundwater) contribute to such studies, in ways which are briefly outlined in section 5.

5.1. Fluid Mechanics and TC formation

Even though knowledge of Gray's six necessary conditions for TC formation (see section 4.2) is of very great value, nonetheless some severe difficulties still stand in the way of any attempt to forecast just when and where a TC will appear. Admittedly, it can be argued that obstacles to forecasting TC initiation present no huge embarrassment; after all, satellite pictures soon pinpoint a TC, and for very many purposes (see section 5.2) a good forecast of future movements, along with any intensity changes, for a TC already observed by satellite may be quite sufficient.

Most TC forecasters, on the other hand, who also take into account the above-noted dangers to oceanic islands, feel strongly committed to watch out for likely TC originators. These include, of course, those monsoon troughs and westward-moving waves which were mentioned in section 4.2. Recently, moreover, a simple idea from fluid mechanics — potential vorticity — has begun [20] to prove useful in this context. Potential vorticity for the atmosphere was defined by H. Ertel [21] in 1942, and it turns out that anomalously large cyclonic values of potential vorticity can be viewed as 'having the potential' to produce strong cyclonic rotation; which, after gaining energy from air-sea interaction as described in section 4.2, may initiate a TC.

From the fluid-mechanics standpoint, potential vorticity may be interpreted as follows. The vertical component of the atmosphere's absolute vorticity may be written $f + \zeta$, where the Coriolis expression

$$f = 2\Omega \sin \theta \tag{7}$$

(with θ as latitude) gives the contribution from earth's rotation at angular velocity Ω, while ζ represents the vertical vorticity component of the winds themselves (air motions relative to the rotating earth). If now h stands for the vertical spacing between two nearby surfaces of constant entropy S — these are surfaces tending to move with the fluid — then the absolute vorticity $f + \zeta$ tends to respond to any vertical stretching of fluid elements by a variation in direct proportion to h. This implies a tendency for the quantity

$$(f + \zeta)\frac{\partial S}{\partial z}, \tag{8}$$

defined as potential vorticity, to remain constant from each particle of fluid.

The stress laid by modern meteorologists on distributions of potential vorticity recalls that earlier revolution in low-speed aerodynamics which arose from Prandtl's committed recognition of how a vorticity field both uniquely determines the flow around an aircraft and can display its key features. Indeed the likeness runs deep: an analogous 'inversion theorem' exists (see [9], p.143) for the meteorological potential-vorticity field; which, furthermore, may help to indicate key cyclonic developments. Forecasters are able to utilise such indications because modern programs (see section 6.2) which apply computational fluid dynamics to weather prediction are able to give printouts of potential-vorticity distributions.

5.2. Fluid mechanics of TC tracks

In the general context of TC disaster prediction on the other hand the primary need is to forecast, for a TC already observed by satellite, its future 'track'. This means the path pursued by the eye of the storm.

An associated need, second only to the tracking requirement, may be to forecast any changes in TC intensity. Admittedly, while a TC travels over a tropical ocean, there exists an approximate balance between energy gain from the Carnot cycle and energy loss by friction at the surface. Nonetheless, this balance can only be approximate; accordingly, questions about intensity changes need to be put even over a tropical ocean. Furthermore, rates of energy gain (but not of loss) diminish over colder ocean areas, where the TC therefore weakens — while over land there remains no energy supply at all (and no physical basis for the presence of an eyewall) so that weakening is even more rapid. Yet these considerations tend perhaps to strengthen the key demand for track prediction, just because the gravest threats of disaster may be centred around where a TC first reaches land.

The track which a TC follows is often extremely complicated (Figure 2); primarily, because it is influenced by the TC's interactions with the entire surrounding field of atmospheric motions. Such far-ranging influences can be fully taken into account only in a global numerical model (section 6.2) of those motions. Nevertheless a preliminary analysis by the methods of theoretical mechanics may help both in interpreting the outputs of numerical models and in suggesting (see section 7.2) how a model might be further improved.

The fluid mechanics of TC tracks views them as influenced to a similar extent both by the general ambient-flow pattern (including both first and second horizontal derivatives of ambient flow velocities) and by the well known 'beta effect'. Here, β stands for the northward gradient in the Coriolis parameter (7); evidently, horizontal gradients in ambient values of the absolute vorticity $f + \zeta$ are fully specified by this mix of data.

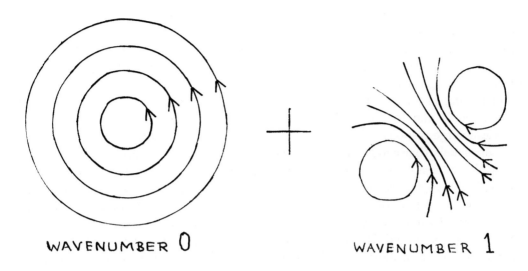

WAVENUMBER 0 WAVENUMBER 1

Figure 8. For a TC wind field, a much better approximation than a purely axisymmetrical pattern of azimuthal winds is a linear combination of such a component (with wavenumber zero) and another component (with wavenumber 1) which tends to convect the whole TC pattern.

Those influences, taken together, can perturb the TC's inherently axisymmetric structure, in a manner highlighted by the 1987 paper [22] of J.C.L. Chan and R.T. Williams. Also, studies by R.B. Smith (see [9], p.264) have still further elucidated the nature of this perturbation; yet I emphasize here the contribution of Dr. Chan as a rather striking example of good theoretical mechanics which, later on, would lead him directly to propose valuable practical improvements to TC track forecasting (see section 7.2).

The departures from axisymmetry may be described in terms of Fourier components having different wavenumbers with respect to an azimuthal angle; then a purely axisymmetric TC pattern is represented by a component of wave number zero. A key discovery by the authors cited is that a component of wavenumber 1 represents (Figure 8) by far the biggest perturbation to that axisymmetric motion. Furthermore, the form of this component near the TC's centre is such that it tends to convect the TC pattern. Here, then, is at least one of the mechanisms underlying the nature of TC tracks — a subject to which I return in sections 6 and 7.

5.3. Wind forcing of shelf waters

Strong winds in general (including TC winds) may force shallow waters over a continental shelf into large 'bulk motions', nearly uniform with respect to depth except in a bottom boundary layer. Such a response is called barotropic.

The especially extensive shelf waters off Bangladesh — where depths of order 10m extend over almost 100km — can massively respond to TC winds. During the years since the 1970 storm surge (mentioned in section 1 as the worst ever), another mega-disaster hit Bangladesh on 29 April 1991, when the combination [23] of storm-surge flooding and extreme winds led to 138,000 deaths. This was a case when, although TC warnings had been widely disseminated from 26 April, some of the human action called for in response was tragically delayed until intense winds began to make such action difficult. Here I should add that many TC forecasters acknowledge a special problem in winning public confidence in storm-surge warnings; which at present suffer from enhanced false-alarm rates because of extreme sensitivity to TC track-forecast timing (briefly, storm surges are much more damaging if a TC's arrival coincides with high tide).

Wind forcing of shelf waters is important too in a non-TC context. For example, the Netherlands and Britain need to be always on the alert against a possibly disastrous storm-surge response of the North Sea. This is an area where increasingly reliable numerical models (barotropic, with bottom friction and an appropriate horizontal eddy viscosity) have been developed to assist decision-making: a model developed [24] at the Proudman Oceanographic Laboratory near Liverpool determines when the Thames Barrier must be raised to protect London, while one developed [25] at the Delft Hydraulics Laboratory (with Rijkswaterstaat) has a key role for Dutch coastal defences. Both laboratories have collaborated [26,27] in developing a new, highly promising Wave-Tide-Surge coupled model (which allows e.g. for the influence of wind waves on augmenting both surface friction coefficient and bottom dissipation).

For TC forcing around US shores, NOAA operates an excellent model pioneered [28] by C.P. Jelesnianski. Fine models have been developed also in China and India (see [9], p.423 and p.442) − but this afternoon I especially wish to praise the work of Kyoto University's Disaster Prevention Research Institute with its highly effective model [23] for TC-generated surges: one of those (see also [29]) that stood up well to the test of being applied retrospectively to the April 1991 disaster in Bangladesh. Yet again, however, I remark that future operational use even of outstanding models in the Bay of Bengal may need to await improved accuracy in TC track forecasts.

Some workers in this field regard CFD alone as a useful guide, but I myself believe that theoretical fluid mechanics can make a contribution; above all, in deciding whether or not shelf waters exhibit a resonant type of response (as opposed to one that is mainly determined by a balance between wind forcing and dissipative effects). In a joint paper at the Beijing Symposium (see [9], p.410), Bryan Johns and I did identify a resonant response on any rather long continental shelf − taking the east coast of India as an example. Very briefly, the 'transient forcing' [30] produced by a TC crossing the shelf at an acute angle may generate a short 'group' (in the wave-theory sense) of shelf waves, having a phase velocity equal to the along-shore component of TC travel velocity, provided that the wavenumber so determined matches the TC's dimensions.

Actually, coastal responses to wind forcing have long been observed, both in the North Sea [31] and around Australia [32], to include shelf waves. These, however, are Kelvin waves; that is, they propagate along a shelf 'anticyclonically relative to the coast' with a radian frequency ω which is less than the Coriolis parameter f.

The general theory of shelf waves, on the other hand, shows [33] that cyclonic (as well as anticyclonic) propagation is possible when $\omega > f$. For the Bay of Bengal, we found that the relevant frequencies have $\omega >> f$ (making Coriolis effect negligible) and established that a TC is able to generate a group of shelf waves which propagates relative to the east coast of India 'cyclonically' (that is, in the N.E. direction).

There is one highly simplified case, that of a shelf with uniform bottom slope, when such shelf waves become the famous 'edge waves' discovered by Stokes, and it is a pleasure to acknowledge that the possibility of these being excited by a hurricane had been suggested 40 years ago [34] by H. Greenspan. Moreover I myself outlined a nonlinear theory of edge waves, and applied it to the propagation of a group of such waves, in my lecture contributed to 18th ICTAM (Haifa, 1992); however, while preparing that lecture for publication, I completed a literature search which revealed that, in extending Stokes's edge-wave theory to a nonlinear analysis, I had been preceded [35] by none other than my old friend and erstwhile pupil G.B. Whitham!

5.4. Kinematics of river-valley floods

And that reminds me to mention how, in relation to those other flood disasters which may arise in river valleys — often from extreme precipitation — Whitham and I collaborated long ago [36] to emphasize how the speed, with which a flood wave travels down the river, is determined above all by fluid kinematics. Essentially, at each place along the river, any local level of the water surface determines simultaneously both the cross-sectional area of water A and the downstream volume-flow rate Q (the latter arising from a local balance between gravity forcing and frictional dissipation); thus permitting for each place the construction of a plot of Q against A, with a slope

$$C = \frac{\partial Q}{\partial A} \text{ for } x \text{ constant.} \tag{9}$$

Here, x stands for distance down the river.

Simple kinematics now shows that C represents the local speed with which the flood wave propagates downstream from where new runoff into the river is occurring. This is because equation (9) gives

$$\frac{\partial Q}{\partial t} = C\frac{\partial A}{\partial t} = -C\frac{\partial Q}{\partial x}, \tag{10}$$

where the last two expressions are identical by the equation of continuity (law of conservation of volume). Equation (10), confirming that values of the flow rate Q travel down river at speed C, is yet another useful insight from theoretical fluid mechanics.

6. FORECASTS AND WARNINGS

The UN World Conference on Natural Disaster Reduction (Yokohama, 23 to 27 May 1994) issued a strong Yokohama Message [37], stressing above all the contribution that preparedness can make to Natural Disaster Reduction of all kinds. Such preparedness includes proper use of (and provision of)

(i) Hazard Resistant Structures, and

(ii) Forecasts and Warnings;

which were the titles of two of the Conference's Technical Sessions. Both (i) and (ii) are specially important for reducing the human impact of TC Disasters — and indeed both involve fluid mechanics! — but I offer here just a brief summary of the TC-related recommendations [38] from the Technical Session (which I chaired) on Hazard Resistant Structures.

Half of these recommendations are concerned with structures able to resist extreme winds. Happily, local building codes in countries subject to TC threats do offer good advice, so our key recommendation was necessarily that BUILDERS COMPLY WITH CODES. (Usually, such compliance demands only rather small additional costs, yet it is vital that these should not be avoided.) For example, a new roof needs to be of 'hipped' construction (alternatively, an existing gable roof must have a parapet added) and to be anchored e.g. by 'hurricane straps'; masonry walls need bracing; metal sheeting must have proper thicknesses and fastenings; etc. etc.

Again, for construction to resist strong storm surges, just one main recommendation suffices: build at least 10m above mean sea level. For example, on threatened Bangladesh islands, low artificial hills are being created in the impressive 'Multipurpose Cyclone Shelter Project'. It is called multipurpose because the structure built on such a hill acts primarily as either a school or a health centre − or as whatever else the islanders themselves feel is most needed − and secondarily as a shelter to house the local population safely (with their livestock crowded nearby) after receipt of a storm-surge warning.

6.1. From forecasts to effective warnings

Of course any such reference to how an effective warning may be of vital importance for disaster mitigation reminds us that two problems of closely comparable difficulty are faced by forecasting experts:

(a) to apply good science (including good fluid mechanics) to making timely and reliable forecasts of patterns of wind, wave, surge, flood; and

(b) on the basis of such forecasts, to disseminate useful warnings that will convincingly advise people under threat about actions needed for their protection.

The essential processes which link (a) to (b) represent a key part of TC Disaster Reduction and one that is taken particularly seriously by the World Meteorological Organisation. Also, in the context of the International Decade (IDNDR), I can report with some pride that the UK's own so-called IDNDR Flagship Project, entitled 'Forecasts and Warnings', is directed entirely towards achieving a further strengthening of those links within tropical countries. In this lecture about TC Fluid Mechanics on the other hand, while having felt the need to lay proper stress on (b), I must now devote to (a) alone all of the time which remains.

6.2. Advances in global weather forecasting

For weather forecasting in general, the big improvements achieved during the past two decades came above all [39] from combining better fluid mechanics with better atmospheric physics in a well constructed numerical model of global atmospheric processes. Such a model for numerical weather prediction (NWP), then, uses refined CFD methods into which the physics of air/water mixtures, and also of radiative heat transfer, have been comprehensively injected.

Here, before I describe just one of the many good global NWP models that are in operational use, I must pause to answer a question which I know will be forming in the minds of many in this audience: a question on whether those big improvements actually resulted from meteorological satellites. But the answer to this question is 'no', because satellites do 'now-casting': they tell us about the weather at this moment, not about the future. Improvements in forecasting, which is concerned of course with weather in the future, have come almost entirely from NWP developments.

Admittedly, initial conditions are highly important for any CFD model, and for NWP in particular. Every NWP model, indeed, starts from initial data (comprehensively 'smoothed' in a certain sense to ensure compatibility with the model) which are internationally derived for the global atmosphere each day at 0000Z and 1200Z (midnight and noon at longitude zero). Such data are obtained

    (i) by weather stations over land from regularly released radiosonde balloons (telemetering data from all altitudes); and also

    (ii) from local weather radars; and, especially,

    (iii) from geostationary and polar-orbiting satellites; as well as

    (iv) from ships and (above all) aircraft;

these keen users (iv) of global forecasts being happy to feed valuable data — alongside data from (i) and (iii) — into the WMO's massive World Weather Watch system.

One of today's many good NWP models is that used [40,41] by the UK Meteorological Office (UKMO), both to give global advice to aircraft and shipping and other industries, and to issue public forecasts and warnings. It is this model that I choose (partly for a reason which emerges in section 7.2) to describe below in brief summary.

The UKMO model uses a CFD grid with 19 different levels in the vertical and with horizontal spacings of five-sixths degrees in latitude and five-fourths degrees in longitude. The numerical scheme is 'conservative' (in the technical sense that quantities which physically should be conserved remain conserved even in the finite-difference representation) and includes parameterisations for

    (a) type of boundary layer (dependent on Richardson number),

    (b) radiative characteristics of land surfaces (as affected by soil type, vegetation, snow depth, etc.),

    (c) radiative characteristics of atmosphere (as affected by greenhouse gases),

    (d) both large-scale cloud cover and 'convective' clouds,

    (e) air-topography interaction (with correct localisation of level for loss of air momentum due to gravity-wave drag), and of course

    (f) air-ocean interaction; as well as

    (g) horizontal and vertical eddy diffusion.

The model is run twice daily, with initial data derived (see above) at 0000Z and 1200Z, in each case leading to global forecasts for regularly spaced instants up to 6 days ahead.

This, like the world's many other excellent models, has achieved big measurable improvements in forecasting accuracy. For all of them, on the other hand, there may be a need to ask what limits, if any, exist on the possible extent of further improvements.

6.3. Weather predictability in general

The past decade has brought increasingly precise knowledge about 'predictability horizons' for systems like the global weather that, in a now well defined sense, exhibit 'chaotic' properties. Furthermore, the existence of several good NWP models has made possible the quantitative determination of such horizons; in other words, of effective limits on how far ahead prediction is possible.

Every NWP model, indeed, allows 'predictability experiments' − in which the model is run with a range of different sets of initial conditions varying only slightly among themselves. This generates what is called [42] an 'ensemble' of different forecasts; which, as the time t increases, diverge more and more from one another. After a certain time, forecast differences resulting from quite small shifts in initial conditions are seen to have reached levels where no confidence can be placed in any of the forecasts. Moreover, the limits on predictability identified in this way are increasingly identified as 'real' limits for the actual global atmosphere, primarily because NWP models have become so good.

In the general theory of 'deterministic chaos', such sensitivity to initial conditions is described [43] by Lyapunov exponents, of which the largest characterizes an exponential growth with time t in solution differences resulting from shifts in initial conditions. Similarly, refined studies of NWP models have determined such exponents; furthermore, they did so in a way that is wholly consistent with what is known about the real atmosphere's sensitivity to small disturbances (e.g. the 'baroclinic instability' phenomenon arising over mid-latitude oceans).

Both for models and for the real atmosphere, the existence of an exponential growth in solution differences explains first of all why limits on predictability, though by no means precisely defined, may nonetheless possess useful approximate values. And a second, perhaps even more vital, inference is that − given the good models now available − any attempt to move still closer towards a predictability horizon may above all demand YET MORE ACCURATE INITIAL CONDITIONS.

All these remarks apply with undiminished force when NWP is used for TC forecasting. Admittedly, unusual difficulties in numerical representation may then arise, e.g. from steepness of gradients near an eyewall, and forecasting improvements are being vigorously sought in two ways (notably, by members of the Geophysical Fluid Dynamics Laboratory at Princeton − see [44]; and [9], p.190):

(i) with specialised local refinements to the model, and

(ii) with more accurate initial conditions.

Yet in a single lecture I cannot include both; so, for the reasons just indicated, I concentrate in my final section 7 on the big improvements which (ii) on its own may be able to make.

7. IMPROVED INITIAL DATA FOR TC FORECASTING

Improved initial data for TC forecasting may be of two kinds: either specially measured (the best kind, as described in section 7.1) or specially estimated (as argued in section 7.2 to constitute 'a good second-best').

7.1. Possible futures for measured initial data

Measured initial data for TC forecasting need imperatively to be three-dimensional (3D) in character; see, for example, section 4.2 on the subject of how important for TC behaviour are questions of how winds, temperatures, humidities, etc. vary initially with height. This essential need for 3D data is recalled here to emphasize the point that the required data cannot come just from 2D satellite imagery. Yet neither can other regular sources of initial data (section 6.1) help: the oceanic areas where TCs appear tend to lack radiosonde stations (and even weather radars); while aircraft, in general, avoid TCs!

An honourable exception to this last generalisation is offered in the eastern USA by NOAA's Aircraft Operations Center, from which manned aircraft fly regularly into Caribbean hurricanes. Since 1982, moreover, such aircraft have deployed (see [45], p.6) a most valuable instrument for deriving 3D initial data; namely, the OMEGA Dropsonde, able to telemeter to the aircraft variations in physical quantities from any level below it. Use of such initial data has been shown to produce a statistically significant reduction in forecast errors.

In all other tropical regions, however, very high costs have continued to offer an overwhelming obstacle to TC reconnaissance with manned aircraft. It was against this background that the Joint ICSU/WMO Symposium (Beijing, 1992) adopted as a major goal the identification of a new and satisfactorily cost-effective means of acquiring 3D initial data for TC forecasting. Then extensive debate during the Symposium produced full agreement on what has been described (see [9], p.585) as the meeting's 'chief conclusion'; namely, that participants should give strong support to the vigorously proposed development of the small computer-piloted TC reconnaissance aircraft Aerosonde.

An impressive feature of the piston-engined Aerosonde, with its 3m wing span and 12-15kg all-up weight, is its combination of low cost with high technology — that is, with an advanced computer as its pilot, with an advanced GPS navigation system (yielding wind vectors by comparison of aircraft motions relative to ground and to air), and with meteorological instrumentation as advanced as in modern radiosondes. Low cost, of course, makes tolerable the occasional loss of an aircraft, although it should be emphasized that the 24 Aerosonde flights during 1995 involved only one such loss (from engine failure). In another of these flights, lasting over 8 hours, the aircraft monitored conditions at altitudes 0 to 4 km, flying into a heavy thunderstorm (which had an intense gust front) yet returning safely to base. Future developments of Aerosonde by Sencon Environmental Systems, in association with Australia's Bureau of Meteorology, are aimed at a doubling of endurance by 1997, followed by introduction of a supercharged powerplant which will progressively extend the aircraft's ceiling to 16 km.

7.2. Fruitful uses of estimated initial data

In the meantime, ingeniously estimated initial data have been shown, where measured data were not available, to constitute 'a good second-best.' Yet another happy result of the Beijing Symposium was increased co-operation of TC experts from Asia with those from other continents — including a much valued visit by the able Hong Kong meteorologist Dr. J.C.L. Chan to the U.K. Meteorological Office.

There he worked with UKMO's gifted NWP specialists on developing a major improvement to existing schemes for applying 'bogus' (or, more correctly, estimated) initial data. This improvement uses the fluid-mechanical models outlined in section 5.2, including those [22] of Dr. Chan himself, to yield appropriate 3D interpretations of satellite imagery; winds at four heights, and at four radii from the storm centre, being estimated [46] as the vector sum of

(i) an expected axisymmetrical (that is, azimuthal) wind field and

(ii) the TC's observed travel velocity.

Here, the wind-field component with wavenumber 1 can appropriately be approximated by its central value (ii) in the region where initial data are most needed, and the blend of this addition with good choices for the four radii has generated a major improvement over previous 'bogusing' methods.

Indeed, the new initial-data scheme was assessed experimentally [46] in detailed trials, lasting from 25 August to 12 September 1994, on that UKMO model which was briefly described in section 6.2; and the striking successes observed (including, for example, a reduction from 201km to 123km in mean TC tracking error − as averaged over forty-five 24-hour forecasts) led to a decision to introduce the scheme operationally from 25 October 1994. Furthermore, its good success has since been maintained; for example, a recent independent assessment of 10 different NWP models for all the Atlantic hurricanes in 1995 gave the UKMO model lowest mean tracking errors for 1-day forecasts (143 km) and for 2-day forecasts (225 km) and for 3-day forecasts (335 km).

7.3. Concluding remarks

I mentioned in section 7.2 just one example among many fruitful 'post-Beijing' collaborations, after outlining in section 7.1 that promising Aerosonde development which had also been stimulated by discussion in Beijing, and in section 4.3 some advances in understanding air-sea interaction at extreme wind speeds that were influenced by an unusual interdisciplinary blend of contributors to the 'Tropical Cyclone Disasters' Symposium. I believe, then, that IUTAM (after this Closing Lecture of its Kyoto Congress) may conclude that its collaboration on TC Disasters, both with ICSU partners and with WMO, has fruitfully stressed the key contributions of

(a) oceanography in TC science; of

(b) improved initial conditions in TC forecasting; and of

(c) growing links between TC experts from Asia and from other continents.

Moreover, at the WMO/ICSU International Workshop on Tropical Cyclones (Hainan, China, January 1998), all of these points will be emphasized still further − in the ever exciting context of 'Typhoons, Hurricanes and Fluid Mechanics'.

ACKNOWLEDGEMENTS

I offer heartfelt thanks for assistance with TC studies to the International Council of Scientific Unions, to the World Meteorological Organisation, to the UK's Meteorological Office and Proudman Oceanographic Laboratory, to l'Institut Français de recherche pour l'exploitation de la mer, to the Bureau of Meteorology Research Centre (Melbourne) and to the Geophysical Fluid Dynamics Laboratory (Princeton).

REFERENCES

1. R.A. Anthes, *Tropical Cyclones: their Evolution, Structure and Effects*, Amer. Meteorol. Soc., Boston, 1982.
2. World Meteorological Organisation, *Fifteen Years of the WMO Tropical Cyclone Programme*, WMO, Geneva, 1995.
3. J.R. Holton, *An Introduction to Dynamical Meteorology*, 2nd Ed., Acad. Press, New York, 1979.
4. N.L. Frank and S.A. Husain, *Bull. Amer. Meteorol. Soc.* **52** (1971), 438.
5. ESCAP/WMO, *Typhoon Committee Annual Review for 1994*, World Meteorological Organisation, Geneva, 1995.
6. L. Bengtsson and J. Lighthill (Eds.), *Intense Atmospheric Vortices*, Springer, Berlin, 1982.
7. J. Lighthill and R.P. Pearce (Eds.), *Monsoon Dynamics*, Cambridge Univ. Press, 1982.
8. J. Lighthill, Tropical Cyclone Disasters, Appendix 2 to *First Report of the ICSU Special Committee for IDNDR*, Internat. Council of Scientific Unions, Paris, 1990.
9. J. Lighthill, Zheng Zhemin, G. Holland and K. Emanuel (Eds.), *Tropical Cyclone Disasters*, Peking Univ. Press, Beijing, 1993.
10. J. Lighthill, G. Holland, W. Gray, C. Landsea, G. Craig, J. Evans, Y. Kurihara and C. Guard, *Bull. Amer. Meteorol. Soc.* **75** (1994), 2147.
11. R.J. List, *Smithsonian Meteorological Tables*, Smithsonian Institution Press, Washington, 1951.
12. K.A. Emanuel, *J. Atmos. Sci.* **43** (1986), 585.
13. K.A. Emanuel, *Ann. Rev. Fluid Mech.* **23** (1991), 179.
14. The WAMDI Group, *J. Phys. Oceanogr.* **18** (1988), 1775.
15. C.W. Fairall, J.D. Kepert and G.J. Holland, *Global Atmos. Ocean System* **2** (1994), 121.
16. K.B. Katsaros and G. de Leeuw, *J. Geophys. Res.* **97** (1994), 14339.
17. J. De Cosmo, K.B. Katsaros, S.D. Smith, R.J. Anderson, W.A. Oost, K. Bumke and H. Chadwick, Air-sea exchange of water vapor and sensible heat: the HEXOS results, *J. Geophys. Res.* (to appear, 1996).
18. Y. Quilfen, K. Katsaros and B. Chapron, Structures des champs de vent de surface dans les cyclones tropicaux observés par le diffusiomètre d'ERS-1 et relations avec les vagues et les précipitations, *Mém d l'Inst. océanogr. Monaco* no. 18 (1994).
19. A.E. Gill, *Atmosphere-Ocean Dynamics*, Acad. Press, New York, 1982.
20. B.J. Hoskins, M.E. McIntyre and A.W. Robertson, *Quart. J. Roy. Meteorol. Soc.* **111** (1985), 877.
21. H. Ertel, *Meteorol. Zeits.* **59** (1942), 271.
22. J.C.L. Chan and R.T. Williams, *J. Atmos. Sci.* **44** (1987), 1257.
23. J. Katsura, T. Hayashi, H. Nishimura, M. Isobe, T. Yamashita, Y. Kawata, T. Yasuta and H. Nakagawa, *Storm Surge and Severe Wind Disasters caused by the 1991 Cyclone in Bangladesh*, Japanese Group for the Study of Natural Disaster Science, Kyoto, 1992.
24. R.A. Flather, R. Proctor and J. Wolf, Oceanographic forecast models, *Computer modelling in the environmental sciences* (D.G. Farmer and M.J. Rycroft, Eds.) 15,

Oxford Univ. Press, 1991.

25. G.K. Verboom, J.G. de Ronde and R.P. van Dijk, *Contin. Shelf Res.* **12** (1992), 213.

26. C. Mastenbroek, G. Burgers and P.A.E.M. Janssen, *J. Phys. Oceanogr.* **23** (1993), 1856.

27. X. Wu, R.A. Flather and J. Wolf, *A third generation wave model of European continental shelf seas with depth and current refraction due to tides and surges and its validation using GEOSAT and buoy measurements*, Proudman Oceanogr. Lab., Merseyside, UK, 1994.

28. C.P. Jelesnianski, *Mon. Weath. Rev.* **93** (1965), 343.

29. R.A. Flather, *J. Phys. Oceanogr.* **24** (1994), 172.

30. J. Lighthill, *Dynam. Atmos. Oceans* **23** (1996), 3.

31. R.L. Gordon and J.M. Huthnance, *Contin. Shelf Res.* **7** (1987), 1015.

32. C.B. Fandry, L.M. Leslie and R.K. Steedman, *J. Phys. Oceanogr.* **14** (1984), 582.

33. J.M. Huthnance, *J. Fluid Mech.* **69** (1975), 689.

34. H. Greenspan, *J. Fluid Mech.* **1** (1956), 574.

35. G.B. Whitham, *J. Fluid Mech.* **74** (1976), 353.

36. M.J. Lighthill and G.B. Whitham, *Proc. Roy. Soc. A* **229** (1955), 281.

37. O. Elo, Outcome of the Conference: Yokohama Strategy, including a Plan of Action for Natural Disaster Reduction, *STOP Disasters* no. 19-20 (1994), 8, United Nations IDNDR Secretariat, Geneva.

38. J. Lighthill, report of Technical Committee B: Hazard Resistant Structures, *STOP Disasters,* no. 19-20 (1994), 16, United Nations IDNDR Secretariat, Geneva.

39. S. Manabe, *Issues in Atmospheric and Ocean Modeling: Part B, Weather Dynamics,* Acad. Press, New York, 1985.

40. T. Davies and J.C.R. Hunt, New developments in Numerical Weather Prediction, *Proc. of ICFD Conf. on Numerical Methods for Fluid Dynamics* (K.W. Morton and M.J. Baines, Eds.), 1, Oxford Univ. Press, 1995.

41. M.J.P. Cullen, *Meteorol. Mag.* **122** (1993), 81.

42. T.N. Palmer, F. Molteni, R. Murea, R. Buizza, P. Chapelet and J. Trebbia, *Ensemble Prediction,* ECMWF Tech. Memo 188, European Centre for Medium-range Weather Prediction, Reading, UK, 1992.

43. J. Argyris, G. Faust and M. Hause, *An Exploration of Chaos,* North-Holland, Amsterdam, 1994.

44. Y. Kurihara, *Mon. Weath. Rev.* **118** (1990), 2185.

45. American Meteorological Society, *21st Conference on Hurricanes and Tropical Meteorology,* AMS, Boston, 1995.

46. J.T. Heming, J.C.L. Chan and A.M. Radford, *Meteorol. Appl.* **2** (1995), 171.

INTRODUCTORY LECTURES OF
MINISYMPOSIA

Theoretical and Applied Mechanics 1996
T. Tatsumi, E. Watanabe and T. Kambe (Editors)
© 1997 Elsevier Science B.V. All rights reserved.

Euler Singularities and Turbulence

Robert M. Kerr[a]

[a] Geophysical Turbulence Program, National Center for Atmospheric Research
P.O. Box 3000, Boulder, CO 80307, USA

For a calculation of anti-parallel vortices which has already shown strong singular dynamics consistent with well-known bounds on the peak, local vorticity and strain, new evidence of singular behavior and structure is shown. For the same projected singular time of T, this includes reemphasizing the numerical evidence consistent with a blowup in the global enstrophy production as $1/(T-t)$ in addition to the expected blowups of peak vorticity and strain as $1/(T-t)$. Blowup of enstrophy production of this form implies logarithmic blowup of the enstrophy and analysis in progress of the structure of vorticity involved in singular dynamics suggests that this is consistent with the existence of two length scales, one collapsing as $\rho_i \sim (T-t)$ and the other as $r \sim (T-t)^{1/2}$. Using two-dimensional slices, it is shown that these two length scales can be incorporated into a model of the three-dimensional structure of vorticity by concentrating vorticity on the surface of a collapsing, blunt cone. Following from this, weaker numerical evidence for a blowup of peak velocity going as $1/(T-t)^{1/2}$ is presented. New tools for looking for singular dynamics in more general flows are discussed and finally for more general flows a curious coincidence is noted where the ratio of the time when pausible singular behavior is indicated and the time when the enstrophy peaks, leading to turbulence, is always about 2.

1. INTRODUCTION

It has recently been shown [1] that with sufficient resolution and care to numerical details that evidence for a singularity of the incompressible, three-dimensional Euler equations can be obtained for at least one idealized initial condition, anti-parallel vortices with a localized perturbation. The strength of the conclusion comes from detailed comparisons with analytic bounds on the growth of the peak vorticity and strain first introduced by Beale, Kato, and Majda (BKM) [2]. This contribution will discuss in more detail the physical space structure identified earlier [1], including its relationship to other calculations claiming singular behavior and some new hints on how this structure might play a role in generating turbulence. In addition, while consistency with the BKM result is essential, here with $\omega_p = ||\omega||_\infty \sim 1/(T-t)$, it will be stressed that several other tests are necessary in order to make the case for singular dynamics convincing.

In order to encapsulate the three-dimensional structure, a black and white version of the 1996 color cover figure of Nonlinearity [3] is given in Figure 1. The flow is at $t = 17$ in the Euler calculation [1], where the singular time is currently estimated to be $T = 18.7$, which is consistent with the limits found before [1]. One half of one of the anti-parallel vortices, cut through the symmetry plane of maximum vorticity, is shown. The image vortex is below the lower, or dividing, plane in the figure. The coordinates used to describe the structure will be horizontal x, vertical z, and in the direction of the initial vorticity y. In Figure 1 the z direction is expanded by a factor of 4. The dominant feature in this visualization is an isosurface set at $0.6 \times \omega_p$, inside of which bright curves, with brightness indicating vortex strength, follow the visualization mesh, which is much coarser than the original computational mesh. Outside the surface, vortex lines that originate from within the surface are shown. The starting points of the vortex lines are chosen by a random algorithm weighted by vortex strength and to avoid confusion between the bright inner mesh lines and the vortex lines, none of the starting points for the vortex lines are allowed to originate near the symmetry plane.

Three visualization procedures are used in Figure 1, that is mesh lines, isosurface, and vortex lines, in order to illustrate how, based on the analysis to be described, the physical space structure can be divided into three regions, inner, intermediate, and outer. The brightest mesh lines define the inner region, which analysis here and earlier [1] show is collapsing in the x and z directions with a characteristic length that goes as $\rho_i \sim (T - t)$ and contains the peak vorticity, which increases as $\omega_p \approx 20/(T - t)$. The lower contour mesh lines indicate that they are not collapsing uniformly in x and z, but are being strongly deformed. Where that deformation starts, then out to the isosurface, is what will be called the intermediate region. Rather than being described by a single length scale, the intermediate region is best described by two, these going as the width and length of the lobes of the isosurface in the symmetry plane. From a fully three-dimensional perspective the vorticity in the inner and intermediate regions, within the isosurface, can be imagined as sitting on the outer surface of an elongated cone with a blunt tip. The analysis to be described shows that the thickness of vorticity on the surface of the cone goes as $\rho_c \sim (T - t)$, the extent of the cone goes as $r_c \sim (T - t)^{1/2}$, the vorticity in this layer is $\omega \sim \omega_p$, and that all of the vorticity in this layer is actively involved in the singular dynamics. Therefore the inner and intermediate regions can also be called the active regions.

The flow in the third region, outside the isosurface, is shown by the vortex lines to be helical. It will be argued that this region, which will contain the majority of enstrophy, circulation, and energy as the estimated singular time is approached, does not actively participate in the singular dynamics. However, it could play a crucial role in the generation of turbulence following viscous reconnection and annihilation of the inner regions. Weak evidence consistent with velocity growing as $(T - t)^{-1/2}$ on the outside edge of the intermediate region, that is a distance $r \sim (T - t)^{1/2}$ from the peak vorticity, will be described.

The outline of the paper is as follows. First, there will be two-dimensional slices illustrating in more detail the structure, especially with respect to the two length scales. It will be argued that the observation of three distinct regions and two length scales is consistent with observations of the blowup of enstrophy production $d\Omega/dt = \int \omega_i e_{ij} \omega_j dV \sim (T-t)^{-1}$ and peak velocity $v_p^2 \sim (T-t)^{-1}$, although current mathematical tools are unable to produce meaningful bounds on the behavior of either quantity. Qualitative arguments for the role of the outer region in allowing universal singular dynamics by serving as a disposal

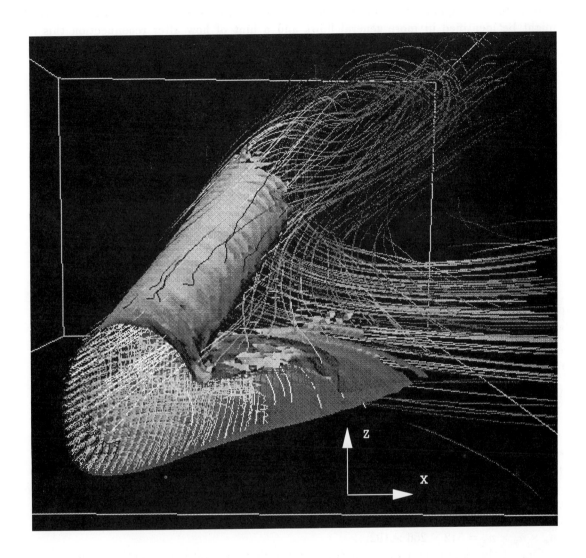

Figure 1. Three dimensional visualization of the singular collapse of anti-parallel vortex tubes in the incompressible Euler equations at $t = 17$. One half of one of the anti-parallel vortices, cut through the symmetry plane of maximum vorticity with z expanded by 4 is shown. The dominant feature is an isosurface set at $0.6 \times \omega_p$. Bright curves within the isosurface indicate vortex strength in an inner region containing ω_p. Out to the surface is an intermediate region and outside the surface is an outer region indicated by vortex lines that originate from within the surface.

for non-universal properties will be mentioned.

Following this qualitative analysis, there will be a discussion of how singular behavior might be identified in more general flows and a hint of how after reconnection this is related to the turbulent cascade. In the following discussion, peak values will refer to the L^∞ norm of a quantity, for example $\omega_p = ||\omega||_\infty$. Quadratic global quantities are the integral over space, for example enstrophy is $\Omega = \int \omega^2 dV$, which in the mathematical literature is the square of the L^2 norm, $||\omega||_2^2$.

2. STRUCTURAL ANALYSIS

To complement the overview of the three dimensional structure in Figure 1, a number of slices through the structure will be used. These visualizations will be taken at $t = 17.5$, whereas Figure 1 is at $t = 17$. Times later than $t = 17$ were not used before [1] because numerical noise tended to obscure the results. In subsequent analysis it has become apparent that the major source of this noise was a build-up of energy in the highest modes, which can be seen in the spectra [1]. Much smoother plots and more consistent analysis can be obtained to later times by simply filtering out these high wavenumber modes smoothly using a small quartic hyperviscous term of the form $\exp(-\nu_4 k^4)$. Removal of these higher modes in analysis changed the kinetic energy, peak vorticity, and enstrophy by less than a per cent. To produce the contour plots in Figures 2 and 3, filtering was done only in x, the direction with the worst resolution problems, with $\nu_4 = 8 \times 10^{-7}$, where the minimum wavenumber for the 4π domain in x was $\min(k_x) = 0.5$.

The ability to extend the analysis to $t = 17.5$ has played a significant role in producing some of the analysis to be discussed. This is because some properties do not lock onto consistent singular behavior until $t = 17$, and at least one more reasonably later time is needed to confirm this behavior. Failure to lock strongly onto singular behavior until $t = 17$ was noted before for strain [1] and will be true of the velocity here. To complete the analysis one earlier time displaying singular behavior in most of its properties is included, $t = 15$. Details of each of these calculations is given in the original paper [1]. For a computational domain of $4\pi \times 2\pi \times 2\pi$, using symmetries in y and z, the mesh for $t = 15$ and 17 was $n_x \times n_y \times n_z = 512 \times 256 \times 128$ and for $t = 17.5$ the mesh was $n_x \times n_y \times n_z = 512 \times 256 \times 192$.

One of the dominant features noticed in the first calculations of anti-parallel vortex reconnection on a mesh [4–7] was the strong flattening and deformation of the vortex cores into vortex sheets. Only at the leading edge of the vortex structure was there thickening, a feature termed the head, where the following vortex sheet was called the tail. Graphics in the original analysis of these calculations [1] was designed to demonstrate that the head should more accurately be described as another vortex sheet perpendicular to the tail. The lower left frame in Figure 2 represents the xz slice in the symmetry plane studied earlier [1], but at the later time of $t = 17.5$. The other frames in Figure 2 represent different xz slices, all at $t = 17.5$, in order to show more of the 3D structure. The two perpendicular sheets in the lower left frame are the thinnest dark regions. The original analysis [1] concentrated on the scaling in the vicinity of the juncture between the two sheets, which is the location of ω_p. It was shown that the maxima of all the components of the stress tensor in the symmetry plane are near this corner, with their positions converging to a single point as $\rho_i \sim (T-t) \to 0$ and their magnitudes growing as $1/(T-t)$. This paper will refer to the region where all positions scale as $\rho_i \sim (T-t)$ as the scaling of the inner region. The small x in the lower right frame of Figure 2 locates one of these maxima, that of strain along the vorticity. That all components of the stress, and therefore all components of the vorticity, must grow as $1/(T-t)$ is consistent with mathematical requirements [8,9] which must be satisfied by any calculation claiming singular behavior.

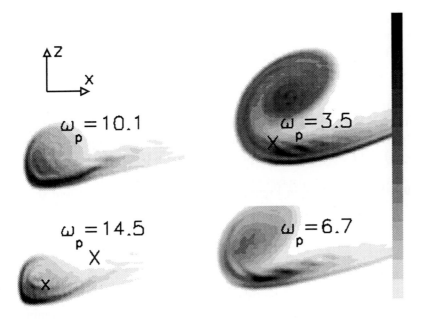

Figure 2. Contour plots of vorticity for 4 xz slices in y. The xz domain is $0.7\pi \times 0.6\pi$ with y spacing of 0.06π. The vertical coordinate uses the Chebyshev mesh. That is, the non-uniform Chebyshev mesh is plotted with uniform spacing. This reduces noise in interpolating to a uniform mesh and resolves the sharp vertical structures better. The maximum vorticity in each slice is shown. Small x indicates position of $e_{yy,p}$ and large X represents the xz position of the extent of the intermediate scaling region and the peak vortical velocity v_p. The X in upper right represents the approximate y position of these.

By taking profiles in x and z of the vorticity across the perpendicular sheets of the vortex structure in Figure 2, it can be shown that the thickness of the leading edge of these perpendicular sheets scales as $\rho \sim (T-t)$, the inner length scale. This is done by scaling x by $\Delta x = x - x|_{\omega_p}$, z by $\Delta z = z - z|_{\omega_p}$ and vorticity by $(T-t)^{-1}$, and taking only $(\Delta x, \Delta z) < 0$. A second feature to note about the lower right frame in Figure 2 is that the vorticity throughout the perpendicular vortex sheets is the order of the peak vorticity, that is $\omega \sim \omega_p \sim (T-t)^{-1}$. To see this quantitatively and how vorticity scales along the vortex sheets, profiles have been taken through the perpendicular sheets that show that the extent of the perpendicular sheets from their juncture goes as a second length scale $r \sim (T-t)^{1/2}$ and that the vorticity everywhere in the perpendicular sheets, appropriately rescaled by ρ_i and r, is $\omega \sim \omega_p \sim (T-t)^{-1}$.

In an incompressible flow, there cannot be simultaneous collapse in all three directions, there must be at least one direction of stretching. However, there is no reason the scaling regime could not collapse in all directions if in the directions of stretching there is strong curvature of field lines. So far, all the analysis has been in the xz symmetry plane. Along with the outer xz slices in Figure 2, Figures 3 and 4 show slices in yz and xy respectively in order to show the structure out of the symmetry plane. Figure 3 shows vorticity intensity in yz planes starting in the lower left with a plane running through the peak vorticity in the lower left frame of Figure 2 and moving to greater positions in x, that is to the right in Figure 2. In general, the region contain large vorticity in these figures, that is $\omega \sim \omega_p$, increases in y and z as the slices move back away from the peak vorticity. The large X's in the upper right frame of Figure 2 and in Figures 3 and 4 represent approximately the same (x, y, z) position. Quantitative analysis of vorticity scaling in these figures shows that $\omega \sim \omega_p$ between the positions of ω_p, $(x_p, 0, z_p)$, and $X(t)$, which goes as $(x - x_p, y, z - z_p) \sim (T-t)^{1/2}$. More specifically, the y position of $X(t)$ goes as $y/(T-t)^{1/2} \approx 0.5$.

For $y/(T-t)^{1/2} > 0.5$, beyond the X in Figure 3, swirl becomes significant and scaling with respect to $(T-t)^{1/2}$ begins to break down. The upper right frame of Figure 2 also shows swirl starting at $y/(T-t)^{1/2} \approx 0.5$. The structural significance to $y/(T-t)^{1/2} \approx 0.5$ seems to be that from $y = 0$ out to this point, xz vorticity slices in Figure 3 can be characterized as two vortex sheets meeting at a corner. In Figure 1 this region is enclosed by the isosurface and within that isosurface vortex lines are relatively straight. Therefore, in y, the beginning of the outer region will be taken as the upper right frame of Figure 2 and beyond the large X in Figure 3. This is where helical vortex lines begin to dominate in Figure 1.

In order to illustrate the vortex structure in xy, Figure 4 contours vorticity in an xy slice through $z/(T-t)^{1/2} = 0.15$, with the arrows representing the horizontal velocity in this plane. An xy slice through the peak vorticity at $z/(T-t)^{1/2} = 0.05$ would show similar structures for both velocity and vorticity, except the vorticity would be more concentrated in the region immediately around the peak vorticity, extending less in both x and y. This would be consistent with the trend in Figures 3 , where contours spread out as x moves away from the first slice through the peak vorticity. Note that in Figure 4, only at the leading edge in x is there a structure whose thickness is the order of $\rho_i \sim (T-t)$. Moving back in x and away from the symmetry plane (lower boundary) in y, the only relevant length scale appears to be $r \sim (T-t)^{1/2}$. Crude analysis on the radius of curvature r_c of the strongest vortex lines, which go through the leading edge in Figure 4, suggests that $r_c \sim (T-t)^{1/2}$.

A structural model can help in the visualization of vorticity. Assume that vorticity of order ω_p is concentrated on the surface of an elongated, blunt cone. The extent of the cone from its tip will be the order of $r \sim (T-t)^{1/2}$ and the thickness of the vorticity layer will be the order of $\rho \sim (T-t)$. Then, the volume δV occupied by $\omega \sim \omega_p$ will be the order of $r^2 \rho \sim (T-t)^2$ and the area δA of δV through the symmetry plane, that is the exposed plane in Figure 1, is the order of $r\rho \sim (T-t)^{3/2}$. Since this volume is

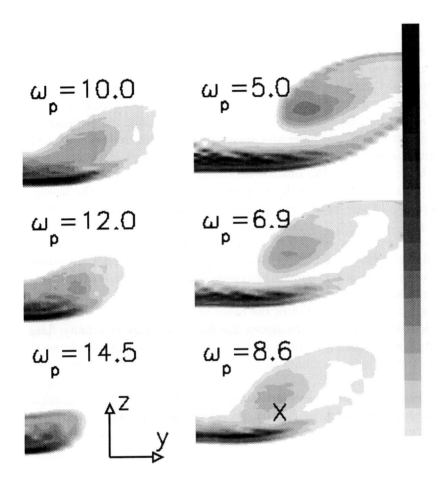

Figure 3. Contour plots of vorticity for 6 yz slices in x. yz domain is $0.32\pi \times 0.59\pi$ with x spacing of 0.07π. The vertical coordinate uses the Chebyshev mesh. The lower right frame passes through ω_p and the maximum vorticity in each slice is shown. X represents the position of the extent of intermediate scaling region and the peak vortical velocity v_p.

shrinking with time, there is no reason it should contain all of any quantity. For example, the circulation in this volume $\delta\sigma \sim \omega_p \delta A \sim (T-t)^{1/2}$. The rest is left behind in what is being called the outer region as the volume active in the singular dynamics collapses, that is the region represented by the swirling vortex lines outside the isosurface in Figure 1.

In this model, the enstrophy contained in δV will go as $\omega_p^2 \delta V \sim 1$. The numerical results [1] suggest logarithmic growth of enstrophy. Direct analysis of the time-dependence of Ω is rather weak in this regard because logarithmic growth is so small, but the evidence for $1/(T-t)$ behavior of $d\Omega/dt$ is much stronger and energy spectra seem to be approaching k^{-3}, which gives logarithmic growth of Ω if the upper wavenumber cutoff grows as $k_{max} \sim (T-t)^{-1}$. This might be equivalent to the smallest length scale in the problem decreasing as $\rho \sim (T-t)$, which is the observed time-dependence of ρ_i. Therefore, for the present purposes it will be taken that $\Omega \sim -\log(T-t)$ and this can be consistent with enstrophy in the active regions being $O(1)$ if most of the enstrophy produced by the singular dynamics is left behind in the outer region as the active regions collapse. Thus, the outer region might be characterized as the rubbish bin of the singular dynamics, containing everything that is no longer necessary to maintain singular behavior. This could include circulation, enstrophy, kinetic energy, and helicity.

Another quantity that could be deposited in the outer region is vortex line length. If vorticity blows up, it is trivial to see that the vortex line length must go to infinity. Line length going to infinity could be accommodated in only two ways, the line could go to infinity or it could curl up. The vortex lines in the outer region in Figure 1 would be consistent with the vortex lines curling up as a way of accommodating infinite line length.

One quantity that this model says is contained almost entirely in the active regions is the global enstrophy production $d\Omega/dt = \int \omega_i e_{ij} \omega_j dV$. Assuming that the strain along the vorticity is also the order of ω_p, then the enstrophy production $\int_{\delta V} \omega_i e_{ij} \omega_j dV$ will go as $\omega_p^3 \delta V \sim (T-t)^{-1}$, as observed [1].

3. VELOCITY SCALING

The large X's in Figures 2, 3, and 4, besides indicating the boundary to what are being called the intermediate and outer regions, also represent the position of the peak velocity at $t = 17.5$. If only the plane in z indicating the boundary between the intermediate and outer regions is used, new analysis suggests a blowup in velocity going as $v_p \sim (T-t)^{1/2}$. Note that along the symmetry plane, at the bottom, the velocities are relatively small and for $t = 17.5$, small velocities when compared to the peak value in Figure 4 are found for all values of z in the symmetry plane. Let us consider how small velocities on the symmetry plane, with a peak velocity at the boundary between the active, intermediate region and the outer region, would be consistent with the structural model just proposed.

Small velocities on the symmetry plane, that is $O(1)$, would be consistent with the structural model in two ways. First, the velocity in the symmetry plane u should go as the shear in one of the lobes in Figure 2 times its thickness, or $u \sim \omega_p \times \rho \sim (T-t)^{-1} \times (T-t) \sim O(1)$. Second, u should go as the circulation in the active region divided by a radius, the outer length r in this case, so one would expect that $u \sim \delta\sigma/r \sim (T-t)^{1/2}/(T-t)^{1/2} \sim O(1)$ again.

Given these arguments for velocity of $O(1)$, how might the velocity blow up away from

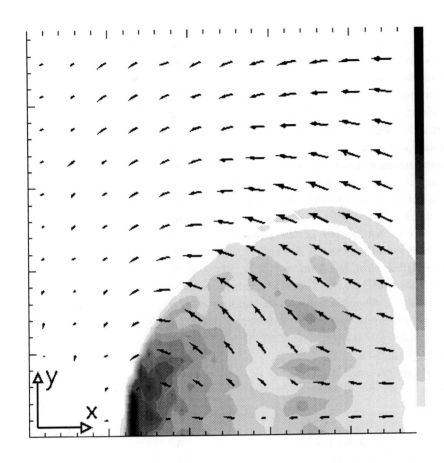

Figure 4. Contours of vorticity and arrows representing horizontal velocity through $z = 0.06\pi$, the plane containing the peak vortical velocity v_p and the boundary in z between the intermediate and outer scaling regions at $t = 17.5$.

the symmetry plane? It might if there is some concentration of vorticity of the order of ω_p such that the thickness of the shear layer is the order of r instead of ρ, or if $\omega \sim \omega_p$ is concentrated in a radius $r \sim r$ such that the circulation across this distance r was the order of the original circulation σ_o. While the detailed analysis has yet to be done, the region of swirl in the upper right frame in Figure 2 seems to be the type of concentration of vorticity that would satisfy these properties. Note that the large X's in the upper right frame of Figure 2 and in Figure 3, which represent the position of maximum velocity, are located on the edges of this swirl.

However, there is an important caveat on this evidence for velocity blowup. Detailed analysis shows that for the early times discussed here, from $t = 15$ to 17, the peak velocity u_p over the entire domain is located where the two symmetry planes meet at the bottom of Figures 1 and 2 and obeys $u_p \sim (T^* - t)^{1/2}$, but T^* does not equal the singular time T estimated from the time dependence of the peak vorticity. However, since the peak velocity off the symmetry plane v_p that is consistent with the singular time T overtakes u_p by $t = 17.5$, it is possible that the behavior of u_p up to $t = 17$ is a transient of the initial conditions that does not have any direct bearing on the singular dynamics. The time dependence of the peak velocities chosen in the two ways, u_p from over the entire domain and v_p from only over the specified z plane, are shown in Figure 5.

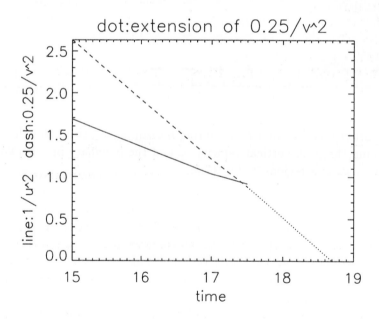

Figure 5. Time dependence of inverse squares of maximum streamwise and vortical velocities, $1/u_p^2$ (solid) and $0.25/v_p^2$ (dotted) and its extension to the estimated singular time (dot).

4. SCALING TESTS FOR SINGULARITY

The calculation of anti-parallel vortices discussed here [1] is not the only numerical calculation for which claims of singular behavior have been made. For this reason, it is important to compare the tests done here and earlier [1] with what has been done, or could be done, for other flows with pausible singular behavior. Five other geometries for which arguments for singular behavior could be made, roughly in order of how strong the arguments are in the opinion of this author, are the interaction of orthogonal vortices [10], axisymmetric flow with swirl [11], the Taylor-Green vortex [12,13], Kida flow [14], and decay from random initial conditions [15].

None of what will be discussed should be taken as proof that there is a singularity in any of these flows, which seems beyond current analytic or numerical methods. Instead, where agreement with singular dynamics is discussed it will mean consistency between the numerical solution, analytic constraints [2,16,9], and other tests to be proposed here.

The first essential test for singular behavior is whether the peak vorticity obeys the analytic restriction of BKM. Given singular dynamics, $\omega_p \sim (T-t)^{-1}$ was expected based on assumptions of scaling behavior of the form $\omega_p \sim (T-t)^{-\gamma_\omega}$, where BKM tells us that $\gamma_\omega \geq 1$, and dimensional analysis [1]. Mathematics can also tell us that similar behavior for all components of the stress, strain and vorticity is necessary in order for there to be singular dynamics [8,9], and as discussed was used as one of the tests of singular dynamics in the original analysis [1]. In particular, consistency in the time scaling of ω_p and of the maximum of the component of strain along vorticity in the symmetry plane $e_{yy,p} = ||e_{yy}||_\infty \sim (T-t)^{-1}$ was shown [1].

The second test of strain is needed because by itself very strong growth in the peak vorticity such as superexponential could be misinterpreted as singular [7]. For anything other than singular growth, $e_{yy,p}$ will quickly flatten out. In addition to the tools suggested by mathematical analysis, the analysis of the calculation [1] has shown singular growth for the enstrophy production $\int \omega_i e_{ij} \omega_j dV \sim (T-t)^{-1}$. This was the third strong test for singular dynamics used in the original analysis [1].

The importance of using several tests suggested by mathematical analysis and theory to check scaling behavior in numerical solutions is by no means limited to finding singularities of 3D Euler. In heat convection, the Rayleigh-Bénard problem, no current numerical method can simulate a long enough range of Rayleigh number Ra scaling to determine convincingly whether the normalized heat flux or Nusselt number Nu obeys the classical result $Nu \sim Ra^\gamma$ with $\gamma = 1/3$ or the hard convective turbulence result of $\gamma \approx 2/7$. However, using several different measures of boundary layer scaling, more convincing numerical evidence can be obtained [17].

In addition to the quantitative tests on ω_p, $e_{yy,p}$, and $\int \omega_i e_{ij} \omega_j dV$, the final strength of the earlier analysis showing singular dynamics [1] was the structural imformation provided. That is, it was shown how thin vortex structures and sharp corners developed as the singular time was approached. As has been discussed here, this type of information is essential if one wants to construct a convincing physical model of how a singularity could develop. An important property that comes out of this structural analysis is that there must be some region into which vortex lines escape as the vorticity in the active regions grows. It has been shown here that this escape hatch is along the vortex lines into the swirling outer region in Figure 1. Another property hinted at by the analysis here is that the minimum radius of curvature of vortex lines in all directions in the active regions will decrease as the outer length scale $r \sim (T - t)^{1/2}$. While strong curvature in at least two directions is now an analytic result [16], this analysis suggests more, that curvature in three directions could be required.

Since the calculation discussed here is inviscid, and most of the other calculations that could be singular in the inviscid limit are actually viscous calculations, it would be important to know how viscosity affects the proposed tests. What is usually done to ascertain singular behavior with viscous calculations is to take a series of calculations at increasing Reynolds numbers. This is very dangerous and has led to misleading results for several different problems. To date, a viscous calculation of anti-parallel vortex reconnection, with resolution equivalent to the Euler calculation [1], and showing strong trends consistent with a singularity in the inviscid limit, has not been done. Therefore, definitive conclusions cannot be made. The best current example of such a viscous calculation [5] used crude initial conditions and only moderate resolution by current standards. The only test from that calculation that might have more general use is that the circulation in the symmetry plane decreased abruptly at the most likely singular time. Nonetheless, it is important to find out which of the above tests used on the anti-parallel Euler calculation [1] has been applied to other flows and what the results are.

How these tests have been applied to the other calculations mentioned will only be sketched. For some, the evidence consistent with singular growth is unpublished analysis done by this author after generating equivalent numerical data. For example, published calculations of inviscid orthogonal vortices [10] and viscous Taylor-Green [12] on the equivalent of 256^3 meshes (without symmetries) have been recalculated. What is discussed here can be seen in those earlier publications, but is not necessarily obvious. For these recalculations and data for decaying random initial conditions [15], the peak vorticity and a measure of peak strain, $\sup(\omega_i e_{ij} \omega_j)/\omega_p^2$, as functions of time have been found. For Taylor-Green and decaying random initial conditions, this analysis was reported in an earlier conference proceeding [18] along with the time dependence of enstrophy production and structural information for how the prevailing view that vortex sheets dominate the singular properties of these flows might break down [19].

For inviscid orthogonal vortices it is found that ω_p grows by a factor of 8 over the range where $1/\omega_p$ is almost linear and with consistent strain behavior. For Taylor-Green, ω_p grows by only a factor of 4 over the regime of linear behavior in $1/\omega_p$, consistent strain behavior, and growth in enstrophy production. The estimated singularity time for Taylor-Green is what was once predicted based on time series expansions [20]. A factor of 4 for growth of ω_p is hardly convincing, and further analysis comparing ratios of stress components to the anti-parallel and orthogonal vortex cases will be necessary to support the evidence for a singularity of Taylor-Green.

The evidence for singularities for axisymmetric flow with swirl and Kida flow quoted here come from the published results. For the axisymmetric with swirl calculation [11] both the ω and strain tests have been done, but this author distrusts the results because of the analytic requirement that all components of the vorticity should have this scaling is broken at an early time. However, the published structures do show thinning and angles in vortex sheets which would be consistent with the structures for the anti-parallel case.

Based on these observations and the analysis to date, this calculation should be redone with greater care to the numerics. Unlike the published analysis [11], my prediction is that if axisymmetric with swirl does have a singularity, it should occur on the axis and the vortex structure should approach the axis as $r \sim (T-t)^{1/2}$. This is based on the belief that curvature in all three directions is necessary, that the axis is the only place where this would be allowed in an axisymmetric flow, and the crude result from the anti-parallel calculation that the curvature goes as $r_c \sim (T-t)^{1/2}$.

For Kida flow[14], none of the proposed tests has been done and while there is an interesting sharp growth in ω_p, its location lacks some of the structural properties that seem to be essential for singular growth in the anti-parallel case. This does not rule out singular behavior since in Taylor-Green, the initial location of ω_p is also not where it might eventually go singular. In particular, the initial location of ω_p in Taylor-Green [18] is on one of the obvious symmetry planes, but the collision of vortex pairs suppresses this initial trend and any signs of singular development eventually move off this plane. Kida flow is also characterized by colliding pairs of vortices, so it would not be surprizing if once all the recommended tests have been done, the location of growing vorticity showing the greatest consistency between all the tests will also be off the obvious symmetry planes.

For analysis of decaying random initial conditions [18], none of the proposed tests shows anything except a slight peak in ω_p and $\int \omega_i e_{ij} \omega_j dV$ at roughly the same time. However, as a final note, if this is a sign of a singular time, then in published work for decaying random initial conditions flow [15], Taylor-Green [12], and Kida flow [14], it appears that independent of Reynolds number, the peak in the enstrophy always occurs at roughly twice the time that has been estimated for a possible singularity in each case. And immediately after this peak in enstrophy is when all the usual properties of homogeneous turbulence appear. This coincidence suggests that reconnection associated with a finite-time singularity might be what triggers so new type of dynamics, again in some sense occurring in a finite time and therefore at least nearly singular, that leads to a more traditional cascade mechanism. How this might occur and how it is related to fully turbulent dynamics should be some of the next major questions to be addressed.

ACKNOWLEDGEMENTS

NCAR is sponsored by the National Science Foundation. Discussions with P. Constantin, J.D. Gibbon and A. Majda and interactions while the author was a visitor at the Isaac Newton Institute for Mathematical Sciences, Cambrige, United Kingdom are acknowledged. The new scaling analysis was begun while the author was a visitor at the Weizmann Institute of Science, Israel. Graphics advice by Å. Nordlund is appreciated.

REFERENCES

1. R.M. Kerr, "Evidence for a singularity of the three-dimensional incompressible Euler equations." *Phys. Fluids A* **5**, 1725 (1993).
2. J.T. Beale, T. Kato and A. Majda, "Remarks on the breakdown of smooth solutions of the 3-D Euler equations." *Commun. Math. Phys.* **94**, 61 (1984).
3. R.M. Kerr, "Cover illustration: vortex structure of Euler collapse." *Nonlinearity* **9**, 271 (1996).

4. A. Pumir and R. M. Kerr, "Numerical simulation of interacting vortex tubes." *Phys. Rev. Lett.* **58**, 1636–1639 (1987).

5. R.M. Kerr and F. Hussain, "Simulation of vortex reconnection." *Physica D* **37**, 474–484 (1989).

6. M. V. Melander and F. Hussain, "Cross-linking of two antiparallel vortex tubes." *Phys. Fluids A* **1**, 633–636 (1989).

7. A. Pumir and E. D. Siggia, "Collapsing solutions to the 3-D Euler equations." *Phys. Fluids A* **2**, 220–241 (1990).

8. G. Ponce, "Remark on a paper by J.T. Beale, T. Kato and A. Majda", *Commun. Math. Phys.* **98**, 349 (1985).

9. P. Constantin, C. Fefferman, and A.J. Majda, "Sufficient conditions for regularity for the 3D incompressible Euler equations." Preprint (1995).

10. O.N. Boratav, R.B. Pelz, and N.J. Zabusky, "Reconnection in orthogonally interacting vortex tubes: Direct numerical simulations and quantifications." *Phys. Fluids A* **5**, 581 (1993)

11. R. Grauer and T. C. Sideris, "Finite time singularties in ideal fluids with swirl." *Physica D* **88**, 116–132 (1995).

12. M. E. Brachet, D. I. Meiron, S. A. Orszag, B. G. Nickel, R. H. Morf and U. Frisch, "Small-scale structure of the Taylor-Green vortex." *J. Fluid Mech.* **130**, 411–452 (1983).

13. M. E. Brachet, "Direct simulation of three-dimensional turbulence in the Taylor-Green vortex", *Fluid Dyn. Res.* **8**, 1–8 (1991).

14. O.N. Boratav and R.B. Pelz, "Direct numerical simulation of transition to turbulence from a high-symmetry" inital conditions. *Phys. Fluids* **6**, 2757 (1994).

15. J. R. Herring & R.M. Kerr, "Development of enstrophy and spectra in numerical turbulence." *Phys. Fluids A* **5**, 2792 (1993).

16. P. Constantin, "Geometric statistics in turbulence", *SIAM Review* **36**, 73–98 (1994).

17. R. M. Kerr, "Rayleigh number scaling in numerical convection", *J. Fluid Mech.* **310**, 139 (1996).

18. R.M. Kerr, "The role of singularities in turbulence." In *Unstable and Turbulent Motion of Fluid.* (ed. S. Kida). World Scientific Publishing, Singapore (1993).

19. M.E. Brachet, M. Meneguzzi, A. Vincent, H. Politano and P.L. Sulem, "Numerical evidence of smooth self-similar dynamics and possibility of subsequent collapse for three-dimensional ideal flows." *Phys. Fluids A* **4** 2845 (1992).

20. R.H. Morf, S.A. Orszag, and U. Frisch, "Spontaneous singularity in three-dimensional inviscid, incompressible flow." *Phys. Rev. Lett.* **44**, 572 (1980).

Theoretical and Applied Mechanics 1996
T. Tatsumi, E. Watanabe and T. Kambe (Editors)

Vortices in turbulent mixing layers

Robert D. Moser[a]

[a]Department of Theoretical and Applied Mechanics,
University of Illinois at Urbana-Champaign,
104 S. Wright St., Urbana, IL 61801, USA

The importance of coherent vortices in the evolution of the mixing layer that forms between two fluid streams of different velocities has long been recognized. In transitional (i.e. preturbulent) mixing layers, there are well-known vortex structures (spanwise "rollers" and streamwise "ribs"), the dynamics and stability of which are largely understood and are discussed. Similar vortex structures have been observed in a variety of experiments in fully turbulent mixing layers, though they are not observed in all circumstances. The conditions necessary for these transitional structures to exist in the turbulent flow appears to be the continued occurrence of organized "pairings" of the spanwise rollers. This and the impact the structures have when they do appear are explored.

1. INTRODUCTION

The mixing layer that develops between two fluid streams moving at different velocities is an important model problem for the study of turbulence in free shear layers, which occur in a variety of technological applications. Transitional and turbulent mixing layers have consequently been the subject of many studies over the years. In particular, one of the first observations of coherent structures in turbulence by Brown & Roshko (1974) was in a mixing layer. They observed large, predominantly two-dimensional, spanwise vortex structures (rollers), whose diameters are approximately the thickness of the layer. It was also observed (Brown & Roshko 1974; Bernal & Roshko 1986) that predominantly streamwise counter-rotating vortices (rib vortices after Hussain 1983) populate the so-called braid region between the corotating rollers. These rib vortices line up with the extensional strain in this region.

The presence of these large-scale structures necessarily has a profound effect on the dynamics of the mixing layer. Therefore, the dynamics of these large-scale structures is of great interest. Most of our understanding of the large-scale mixing layer vortices has come from the observation that these structures are similar to, and often behave like structures in transitional mixing layers, particularly the spanwise rollers that arise from the initial Kelvin-Helmholtz instability of the mixing layer, and the structures resulting from the two- and three-dimensional instabilities of these Kelvin-Helmholtz rollers. There have been many studies of these instabilities and the resulting transitional structures (usually at low-Reynolds number) with the ultimate aim of understanding the turbulent flow (e.g. Winant & Browand 1974; Corcos & Lin 1984). These efforts have been aided by flow-visualization experiments in turbulent mixing layers in which some of the processes occurring in the transitional flows (*e.g.* pairing) have been observed.

While it has been widely assumed that the mechanisms governing the Kelvin-Helmholtz rollers and the three-dimensional rib vortices between them dominate the behavior of the turbulent mixing layer, this is by no means certain. There have been various experimental (Chandrsuda, Mehta, Weir & Bradshaw 1978) and computational (Comte, Lesieur & Lamballais 1992; Rogers & Moser 1994) studies that have suggested that other effects are important, though in several cases these too are related to stability properties of the transitional layer (*e.g.* helical pairing, Pierrehumbert & Widnall 1982).

In this paper, the properties of what might be called the standard transitional mixing layer structures (ribs and rollers) are reviewed (§2). Then the extent to which the evolution of these structures are relevant to the analogous structures in the turbulent layer is explored (§3). Finally, the impact of the vortical structures on the bulk properties of the turbulent layer will be discussed (§4).

2. VORTICES IN TRANSITIONAL MIXING LAYERS

In studying the evolution of the mixing layer, the usual experimental setup uses a splitter plate mounted in a wind tunnel to separate the two streams moving at different velocities. The mixing layer then forms starting from the trailing edge of the plate and evolves down stream; this is the spatially evolving mixing layer. In contrast, theoretical and computational studies often use a slightly different configuration; at the initial time, the free-stream velocities are prescribed, with a given profile (possibly a sharp jump) specified between them. The layer then evolves in time, forming a time-evolving mixing layer. Asymptoticly, for $\Delta U/U_c \ll 1$, where ΔU is the velocity difference and U_c is the average of the velocities of the two streams, the spatially evolving mixing layer viewed in a reference frame moving at U_c is equivalent to a temporally evolving mixing layer. By comparing computed spatially and temporally evolving mixing layers, Rogers, Moser & Buell (1990) found that the evolution of the vortices is virtually identical to quite large values of $\Delta U/U_c$ (up to 0.3). Thus, in the discussion to follow, we will generally use data from spatial and temporal mixing layers interchangeably.

2.1. Two-Dimensional Structure

When the splitter plate boundary layers or the initial velocity fields are laminar, the evolution of the mixing layer involves the development of instabilities and often a transition to turbulence. The initial instability is a Kelvin-Helmholtz instability due to the presence of an inflection point in the velocity profile. The linear stability of simple analytic mixing layer velocity profiles, such as an error function (the self-similar solution for the time-developing laminar mixing layer) or a hyperbolic tangent, are well established (Michalke 1964; Monkewitz & Huerre 1982; Huerre & Monkewitz 1985). As a Kelvin-Helmholtz instability mode develops and becomes nonlinear, the spanwise vorticity in the mixing layer rolls up forming an array of concentrated spanwise vortices all of the same sign separated by braid regions, which are nearly devoid of spanwise vorticity (see Figure 1). This rollup process has been studied extensively using numerical simulation (Acton 1976; Patnaik, Sherman & Corcos 1976; Riley & Metcalfe 1980; Corcos & Sherman 1984; Jacobs & Pullin 1989; Rogers & Moser 1992), and has been observed in flow

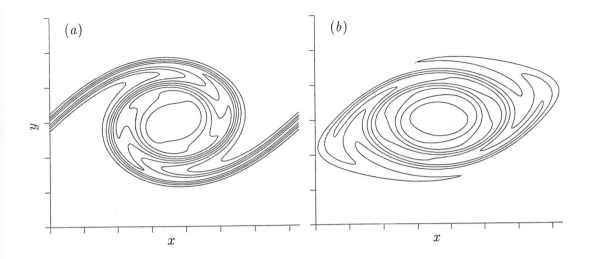

Figure 1. Contours of ω_z in a two-dimensional rolling up mixing layer (from Rogers & Moser 1992, same as 2D0P in Moser & Rogers 1993). Shown is one period of a periodic array of rollers. The contour increment is -0.2 and tic marks are at δ_ω^0 intervals. (a) $t = 9.8$ (completion of rollup). (b) $t = 14.6$ (beginning of oversaturation).

visualization experiments (Browand & Weidman 1976). Depending on the wavelength of the disturbance and the Reynolds number, the resulting large-scale vortex can have a complicated internal structure. Further, as the flow evolves past the initial rollup, the vortices become elongated, and reintroduce vorticity into the braid region (Figure 1).

The Kelvin-Helmholtz rollers are themselves unstable to a variety of disturbances (Pierrehumbert & Widnall 1982; Corcos & Lin 1984; Metcalfe, Hussain, Menon & Hayakawa 1987). One of these is a two-dimensional disturbance at twice the Kelvin-Helmholtz wavelength. This instability usually (depending on relative phase) leads to the amalgamation of pairs of roller vortices in a process called pairing, as observed experimentally by Winant & Browand (1974). The result is a new array of larger vortices at twice the original wavelength. Winant & Browand (1974) observed that in their low Reynolds number, transitional mixing layer, pairing was responsible for most of the growth of the layer. As the roller vortices pair, they rotate one above the other (Figure 2), as would be expected of an isolated pair of like-sign vortices. This has the effect of enlarging the remaining braid region and drawing more vorticity out of it (Moser & Rogers 1993). As the paired roller evolves further, the individual vortices coalesce and two arms of weaker vorticity are ejected from the core (Figure 2), which is required for energy conservation (Martel, Mora & Jimenez 1989).

2.2. Three-Dimensional Structure

The two-dimensional rollers are also unstable to three-dimensional disturbances (Pierrehumbert & Widnall 1982; Metcalfe et al. 1987; Corcos & Lin 1984). The most studied three-dimensional instability in this flow is the so-called translative instability (Pierrehumbert & Widnall 1982), which results in the bending of the rollers and the creation

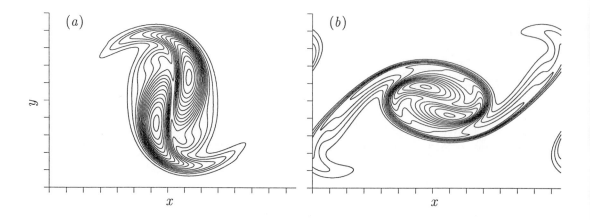

Figure 2. Contours of ω_z in a two-dimensional mixing layer undergoing its first pairing (flow 2D1P in Moser & Rogers 1993). Shown is one period of a periodic array of paired rollers at (a) the pairing time ($t = 21.5$)and (b) the time at which the arm of vorticity enters the braid region ($t = 30.4$).

of rib vortices. There has been much discussion in the literature as to what instability mechanism underlies the translative instability observed by Pierrehumbert & Widnall (1982). One candidate is the instability of strained or elliptic vortices(J. 1988; Waleffe 1990; Pierrehumbert 1986; Landman & Saffman 1987), of which the mixing layer rollers are an example. This instability arises because there are inertial waves on the vortex that do not rotate with the vortex. Thus the vorticity associated with the wave can stay aligned with the extensional direction of the strain on the vortex so that it can be stretched. This instability leads to a bending of the vortex. The other possibility often discussed with regard to the translative instability is one associated with the braid region (Ashurst & Meiburg 1988; Lasheras & Choi 1988; Bell & Mehta 1990). It is not obvious what this instability is, since an analysis similar to that for the elliptic flow has not yet been done. However, it is not difficult to imagine a mechanism whereby predominantly spanwise vorticity in the strain-dominated braid region is distorted and then stretched by the mean strain. Since the mean flow is not rotational, the disturbance vorticity can stay aligned with the extensional strain. This would lead to the observed rib vortices.

With this mechanism, it is clear that if the spanwise vorticity in the braid region is zero, then the strength (circulation) of the rib vortices cannot increase. For particularly simple mixing layers with spanwise symmetries as described by Rogers & Moser (1992), the evolution of the rib circulation (Γ) can be expressed:

$$\frac{\partial \Gamma}{\partial t} = \int_y u\omega_z \, dy \Big|_{z=z_0}, \tag{1}$$

where z_0 is the spanwise location at which the streamwise vorticity is zero due to the symmetry. Indeed, Rogers & Moser (1992) and Moser & Rogers (1993) observed that the growth in circulation of the rib vortices nearly ceased when the rollup or pairing of the mixing layer caused the spanwise vorticity in the braid regions to become small. This is

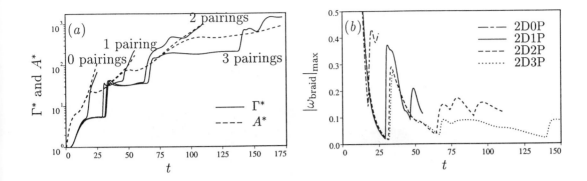

Figure 3. Time development of (a) the rib circulation $\Gamma^* = \Gamma(t)/\Gamma(0)$ and the disturbance amplitude $A^* = A(t)/A(0)$ of three-dimensional linear perturbations and (b) the maximum magnitude of of the spanwise vorticity associated with the two-dimensional base flow in the middle of the surviving braid region. The two-dimensional base flow undergoes 0, 1, 2, or 3 pairings (flows 2D0P, 2D1P, 2D2P, and 2D3P respectively, in Moser & Rogers 1993).

well illustrated in Figure 3, where the maximum magnitude of spanwise vorticity in the braid region of a mixing layer undergoing various numbers of pairing, and the circulation of the rib vortices in a three-dimensional linear perturbation are plotted. Note that whenever the maximum spanwise vorticity takes a jump, so does the rib vortex circulation. These jumps in braid vorticity occur when either the arms of spanwise vorticity associated with a pairing enter the braid region (see above and Figure 2), or pairing stops and the roller vortex becomes increasingly elliptical, bringing vorticity back into the braid region (see above and Figure 1).

Rogers & Moser (1992) concluded that both the elliptic vortex instability (core instability) and the braid instability are active in the three-dimensional translative instability of the rolled-up mixing layer, and that they are coupled. In fact it is not correct to discuss them independently. The fact that they are coupled is easily seen in Figure 3, where the integrated amplitude of the three-dimensional disturbance (square root of the energy), which has its largest contribution in the roller cores, is observed to plateau when the rib circulation stops growing.

The three-dimensional structure that arises from the nonlinear evolution of the translative instability in a non-pairing mixing layer as computed by Rogers & Moser (1992) is shown in Figure 4. Note the predominantly streamwise rib vortices in the braid regions. These vortices are nearly axisymmetric, and near the center of the braid region are very similar to a burgers vortex in vorticity profile. This is rather remarkable since the braid region is a region dominated by plane strain (not axisymmetric strain). There is a threshold rib circulation, depending on viscosity, rib spacing and strain, below which such "collapsed" vortices cannot form (Lin & Corcos 1984). It is such collapsed nearly circular rib vortices that have been observed in flow visualization experiments (Bernal & Roshko 1986). The cup-like concentrations of spanwise vorticity apparent in the figure occur at the "bends" of the now bent spanwise roller. Vorticity becomes con-

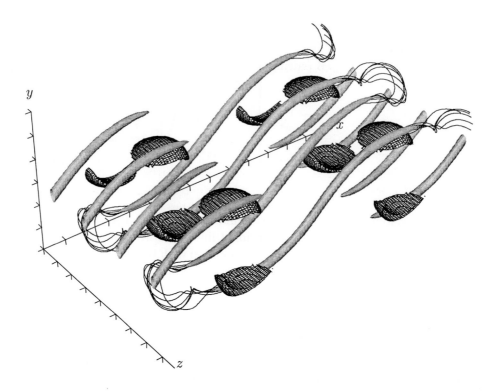

Figure 4. Surfaces of constant vorticity magnitude and vortex lines in a three-dimensional rolled-up mixing layer (flow ROLLUP from Rogers & Moser 1992 at $t = 12.8$). Cross-hatched surfaces represent $\omega_z = -4.0$ and shaded surfaces show $\sqrt{\omega_x^2 + \omega_y^2} = 4.0$. The "rib" structures contain ω_x and ω_y of the same sign and this sign alternates in z (negative for the closest rib). The same vortex lines go through both of the counterrotating rib vortex pairs (concealed by the rib surface contour). Tic marks are at δ_ω^0 intervals.

centrated there because of the combined spanwise-extensive strain induced by the rib vortices bending over the rollers and the "legs" of the bent spanwise rollers.

When a three-dimensional mixing layer similar to that shown in Figure 4 undergoes a pairing, there is a complex interaction between the cup-like structures of spanwise vorticity in the two pairing rollers and the rib vortices that are engulfed during the pairing (Moser & Rogers 1993). The result is that the roller core becomes increasingly complex until after a second paring, the flow is apparently turbulent. However, the rib vortices in the surviving braid region remain.

There are several other scenarios by which the mixing layer becomes three dimensional (Pierrehumbert & Widnall 1982; Nygaard & Glezer 1994; Collis, Lele, Moser & Rogers 1994). The one that has evoked the most interest arises from the so-called helical pairing instability identified by Pierrehumbert & Widnall (1982). In this instability the three-dimensional perturbation is subharmonic to the array of spanwise rollers. As in the translative instability, one of the effects is to bend the roller vortices, except in this case, the bending in adjacent rollers is 180 degrees out of phase. The result is that a section of a roller will appear to begin pairing with a section of its down-stream neighbor, while the next section will appear to begin pairing with its upstream neighbor. If the respective sections were to actually rotate about each other and amalgamate, the resulting vortex lines would form an alternating twisted pattern (note that "helical" is really a misnomer here since the topology is not that of a helix). This instability is thought to be responsible for the twisted flow visualization patterns observed in mixing layers by Chandrsuda et al. (1978). Similar evidence of twisted rollers has also been found in computations(Comte et al. 1992; Rogers & Moser 1994). Using numerical simulations with controlled initial condition, Collis et al. (1994) showed that, provided the spanwise wavelength of the disturbance was sufficiently large (larger than the streamwise wavelength of the roller vortices), such local pairing resulting in twisted rollers would indeed occur. However, local amalgamation into a single vortex did not occur due to local self induction.

3. VORTICES IN TURBULENT MIXING LAYERS

If vortical features of the transitional mixing layers are to occur in fully turbulent layers, they clearly must be features that are robust to the presence of turbulence. For example, it is clear that we should not expect the cup-shaped spanwise vorticity concentrations that occur in the three dimensional roller (see Figure 4) to be present in the turbulent case. The reason is that these do not survive even a single pairing. Thus from the above discussion, the Kelvin-Helmholtz rollers, their paring and the rib vortices are features that could potentially appear in turbulent mixing layers. And indeed, these features have been observed experimentally in turbulent layers. The issues concerning us here are the conditions required for these features to occur and how their evolution might differ from that of the transitional structures.

The experiments of Brown & Roshko (1974) indicate that the familiar mixing layer rolers can exist and pair in a fully turbulent layer. But, other experiments in different facilities (Chandrsuda et al. 1978) suggest that these features are not present. It is not clear which of the subtle often uncontrolled differences in these and other experiments might lead to such a qualitative difference in the flow evolution. One possibility that might account for these variations is the uncontrolled disturbance environment in which the mixing layer forms. In particular, the tip of the splitter plat in an experiment can introduce vortical disturbances into the layer in response to acoustic fluctuations. The mechanism is viscous and occurs even at high Reynolds numbers due to the extremely small size of the tip (there may even be a singularity if the tip is sharp). If the acoustic environment is dominated by essentially one-dimensional longitudinal resonance modes as might occur in a wind tunnel, then the splitter plat tip will be a source of two-dimensional disturbances. This will obviously depend on the details of the experimental facility.

To investigate the impact of introducing such preferentially two-dimensional disturbances, Rogers & Moser (1994) performed a series of three numerical simulations with varying strengths of two-dimensional disturbances added to a turbulent initial condition representing the boundary layer turbulence that would be present in an experiment when the splitter plate boundary layers are turbulent. The simulations included an unforced, a weakly forced and a strongly forced case, where the added two-dimensional energy is 0., 0.47 and 10.1 times the fluctuation energy of the turbulent boundary layers. Since it has

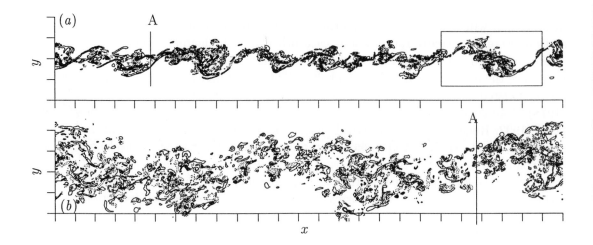

Figure 5. Contours of spanwise vorticity in x–y planes at (a) $z = 18.8\delta_m^0$ and $t = 78.5$, (b) $z = 0$ and $t = 187.5$ in the unforced flow of Rogers & Moser (1994). Contour increments are ± 0.25, positive contours are dotted, and tic marks are at $5\delta_m^0$ increments. The vertical lines mark the locations of the z–y planes depicted in Figure 6. The rollers in the region marked by the box in (a) are undergoing a pairing

generally not been possible to measure in experiments the magnitude of the disturbances introduced by the the splitter plate tip, it is difficult to say whether these two-dimensional forcings are reasonable, though it does seem that in the strongly forced case the forcing is quite large. In the following, only the unforced and strongly forced flows will be discussed. The behavior of the weakly forced flow is intermediate.

Visualizations of the spanwise vorticity of the unforced mixing layer at two times are shown in Figure 5. At the earlier time one can distinguish rollers and braid regions, though the internal structure is quite complicated. Some of the braid regions have particularly simple structure, consisting of thin filaments of spanwise vorticity. Furthermore, there appears to be a pairing occurring in the region marked with a box in Figure 5a. In the braid regions we can look for evidence of rib vortices by visualizing a passive scalar in a y-z plane cutting through a braid. For example, scalar in the plane marked "A" in Figure 5a is shown in Figure 6a. The commonly observed rollups of the scalar interface and "mushroom" shaped structures, similar to those observed by Bernal & Roshko (1986) and others, indicate that there are indeed predominantly streamwise organized vortices, and more detailed visualizations (not shown here) indicate that they line up with the extensive strain and extend from the bottom of one "roller" to the top of the next.

To visualize the spanwise coherence of the rollers, Rogers & Moser (1994) devised an indicator function

$$\mathcal{R}(x, z) = \int_{x'} \int_y \int_{z'} G(x - x', z - z') \frac{\partial v}{\partial x'}(x', y, z') \, dx' \, dy \, dz' \tag{2}$$

where v is the vertical velocity component and G is a bivariat Gaussian filter kernel, with filter width equal to the momentum thickness of the layer. It is shown by Rogers &

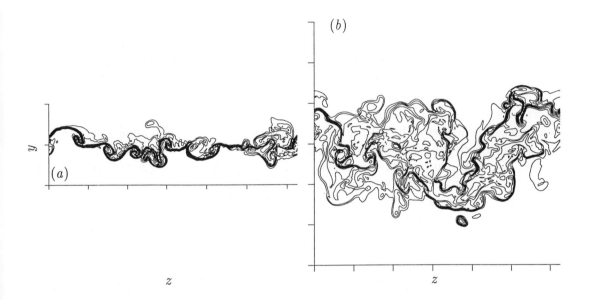

Figure 6. Contours of passive scalar in the z–y plane at the lines marked A (braid region) in Figure 5, at (a) $t = 78.5$ and (b) $t = 187.5$ in the unforced flow of Rogers & Moser (1994). The contour increment is 0.1. Tic marks are at $5\delta_m^0$ intervals.

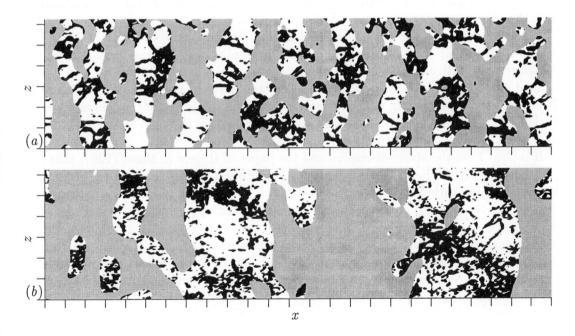

Figure 7. Locations in x and z of the rollers (gray), as determined by $\mathcal{R} < 0$ at (a) $t = 78.5$, and (b) $t = 187.5$ in the unforced flow of Rogers & Moser (1994). Also shown (in black) are the locations where the maximum enstrophy exceeds $2.4\Delta U^2/\delta_m^{0^2}$. Tic marks are at $5\delta_m^0$ intervals.

Moser (1994) that this quantity measures the predominance of large-scale strain over large scale rotation and should be negative in a roller. Regions where this quantity is negative are shown as gray in Figure 7. Note that at the early time, the "rollers" have minimal spanwise coherence, and indeed the pairing identified in Figure 5a appears to only be occurring locally. Also shown in Figure 7 are regions of large enstrophy (black). The existence of long thin regions of large enstrophy spanning several of the "braid regions" is further evidence of rib vortices.

At later time however, the picture is quite different. There are still apparently "rollers" and braid regions as evidenced by the indicator visualized in Figure 7b, but now the "rollers" span the entire spanwise domain of the computation. This is only because the size and separation of the rollers has become of the same order as the spanwise domain size. Furthermore, as shown in Figure 5b, the rollers and braids really just correspond to regions where the vortical turbulent layer is relatively thicker or thinner. The braid regions are full of turbulence. As a consequence, there is no discernible evidence of rib vortices in either of Figures 5b or 7b. The reason for this change of behavior is that after the local pairing shown in Figure 5a, there are no further pairings, local or otherwise. Instead, the rollers grow and the streamwise spacing increases by "nibbling" away at neighboring rollers, in a process that is in some ways similar to the "tearing" amalgamation described by Moore & Saffman (1975). Without the pairing process, which extracts vorticity from the braid regions as discussed in §2.1, there is no mechanism to keep the braid regions essentially devoid of turbulence as was observed at earlier time. Apparently, the turbulence in the braid regions causes any rib vortices that may be present to loose coherence. The cessation of pairing also has other consequences as discussed in §4.

The main effects of the two-dimensional forcing on the vortex structure are that the rollers are essentially two-dimensional from the beginning and that the pairings continue for longer. In particular, in the strongly forced case, pairings occur until the end of the simulations. Late time visualizations of the strongly forced case are shown in Figure 8. It is apparent that a paring is in progress and that the braid regions are largely free of turbulence. Also, the long predominately streamwise vortices shown in Figure 8b, indicate that there are quite coherent rib vortices.

It is clear from the above discussion and the paper of Rogers & Moser (1994) that the paring of the Kelvin-Helmholtz rollers is the key process that determines the structural character of the turbulent mixing layer. However, the role of pairing in preserving the rib vortices is somewhat curious. On the one hand, pairing removes turbulence from the braid region so rib vortices can remain coherent, but on the other hand the transitional studies suggest that pairing also removes the spanwise vorticity from the braid region, which inhibits the *growth* of the rib vortices. Further, in the turbulent case, no arms of vorticity have been observed to leave paired rollers and reenter the braid region as they do in two-dimensional (transitional) pairings. The evolution of the rib vortices and there circulation have not been tracked in the strongly forced turbulent layer of Rogers & Moser (1994), so how or whether the rib vortices in the turbulent case do grow is not known.

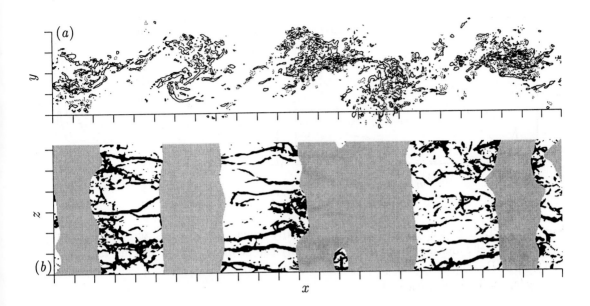

Figure 8. (a) Contours of spanwise vorticity in an x–y plane and (b) locations in x and z of the rollers (gray) as determined by $\mathcal{R} < 0$ at $t = 101.5$ in the strongly forced flow of Rogers & Moser (1994). In (a), contour increments are $\pm 0.25\Delta U/\delta_m^0$ and positive contours are dotted. Also shown in (b)(black) are the locations where the maximum enstrophy exceeds $2.4\Delta U^2/\delta_m^{0\,2}$. Tic marks are at $5\delta_m^0$ increments.

4. IMPACT OF VORTEX STRUCTURES

With such large differences in the qualitative structure of the forced and unforced mixing layers, one expects that there will be differences in the bulk properties of the layer as well. Rogers & Moser (1994) cite several, of which only two will be discussed here.

The first is the most basic property, the growth rate r of the layer. In the time developing mixing layer this is measured as

$$r = \frac{1}{\Delta U}\frac{d\delta_m}{dt} \tag{3}$$

where $\Delta U = U_2 - U_1$ is the difference between the free-stream velocities on the two sides of the layer with $U_2 > U_1$, and δ_m is the momentum thickness defined by

$$\delta_m = \frac{1}{\Delta U^2}\int_{-\infty}^{\infty}(U_2 - U)(U - U_1)\,dy. \tag{4}$$

Here, U is the mean velocity, which in the time developing layer depends only on time and y. The equivalent growth rate in the spatially evolving mixing layer is

$$r = \frac{U_c}{\Delta U}\frac{d\delta_m}{dx}, \tag{5}$$

where $U_c = (U_1 + U_2)/2$ is the convection velocity of large structures in the layer. Among experiments, this growth rate varies by almost a factor of two (0.014 to 0.022, Dimotakis

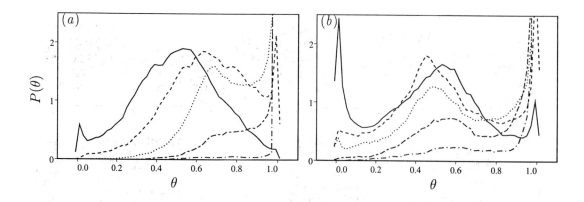

Figure 9. PDFs of the passive scalar at equally spaced y-locations across the layer for (a) the unforced mixing layer at $t = 150$, (b) the strongly forced mixing layer at $t = 119.6$ from Rogers & Moser (1994). —— $y/\delta_m \approx 0$, ---- $y/\delta_m \approx 1$, ········ $y/\delta_m \approx 2$, —·— $y/\delta_m \approx 3$, —··— $y/\delta_m \approx 4$.

1991). In the numerical simulations of Rogers & Moser (1994), r varies from 0.014 in the unforced case to 0.017 in the strongly forced case. Thus, unless even stronger two-dimensional forcing is present in some experiments than in the strongly forced simulations, it seems that unintended two-dimensional forcing (due to the splitter plate tip or other causes) and the resulting continued paring of the mixing layer rollers are not the cause of most of the variation in mixing layer growth rate observed in experiments. This is an unexpected result since in transitional mixing layer experiments, pairing has been found to be responsible for much of the growth of the layer (Winant & Browand 1974).

The second bulk property of interest here is associated with scalar mixing. In many technological applications, a mixing layer forms between two streams containing different scalar constituents (e.g. chemical reactants) and the mixing of these constituents is of great importance (e.g. to facilitate chemical reactions). Each time mixing layer rollers pair, they engulf a large slug of free-stream fluid into the vortical part of the layer, bringing whatever scalar constituents are carried. Further, as has been observed in several experiments (Brown & Roshko 1974; Konrad 1976), these incursions of irrotational free stream fluid penetrate almost the entire width of the mixing layer, as is evident in Figure 8a. As a consequence, free-stream fluid and the scalar it carries get mixed into the layer in a two-step process. First it is engulfed into the layer during pairing, then it is mixed into the rollers. This characteristic of a pairing mixing layer was used by Broadwell & Breidenthal (1982) to construct a model for scalar mixing for this flow.

With this sort of process, one expects that the concentration of scalar would be uniformly distributed through the roller core, with the result that the probability density function (PDF) of scalar concentration should be the same throughout the roller. The scalar PDF including both rollers and braid regions should thus be tri-modal, with a broad peak centered at some intermediate scalar concentration representing the the roller contribution and a spike (though possibly small) at both of the free stream concentrations representing the incursion of free-stream fluid. In particular, the location of the central peak of the PDF (i.e. the most probable concentration in the roller) should be independent of y location. This has been called a "nonmarching" PDF, and it does in fact occur in the strongly forced flow (Figure 9b). But at late time, the scalar PDF in the unforced flow (Figure 9a) has a different behavior. Here the location of the central peak changes with y location so that as one approaches one free stream or the other, the most probable scalar concentration approaches the value in that free-stream (a marching PDF). This is indicative of an entirely different mixing process in which free-stream fluid is brought gradually into the layer by small-scale turbulence at the edge of the vortical region. It is consistent with an eddy diffusivity model of the scalar mixing, while the nonmarching PDF is not. The association of nonmarching scalar PDFs with continued pairing is further supported by the observation that at early times, before the last pairing shown in Figure 5, the scalar PDFs in the unforced case were also nonmarching.

This difference in mixing character could be of great importance in devices involving chemical reaction (e.g. combustion). It would certainly change the local rates at which the reaction occurs and the distribution of reagents and products in the layer. However, there is currently no credible evidence of a marching scalar PDF in an experimental fully turbulent mixing layer (Karasso & Mungal 1996). Indeed, based on their experiments, Karasso & Mungal (1996) suggest that in its asymptotic state, a fully turbulent mixing layer should always exhibit a nearly nonmarching scalar PDF (they call it "tilted", specifically the location of the central peak of the PDF is only weakly dependent on y). They have also conclusively determined that the main experimental study that reported a marching PDF (Batt 1977) obtained that result due to inadequate spatial resolution of the scalar concentration. Whether the unforced simulation of Rogers & Moser (1994) yields a marching PDF because it has not reached its asymptotic state[1] or because its disturbance environment is fundamentally different from the experiments (e.g. no splitter plate tip) is not clear at this point. The issue is of some importance since it determines the robustness of nonmarching PDFs and all that they imply to changes in flow environment.

Regardless of the reasons for the discrepancy in mixing character, the simulations of Rogers & Moser (1994) suggest that the character of the scalar mixing is controlled by the dynamics of the large-scale vortical structures. If they pair and engulf free-stream fluid, nonmarching PDFs should result.

REFERENCES

ACTON, E. 1976 The modelling of large eddies in a two-dimensional shear layer. *J. Fluid Mech.* **76**, 561–592.

ASHURST, W. T. & MEIBURG, E. 1988 Three-dimensional shear layers via vortex dynamics. *J. Fluid Mech.* **189**, 87–116.

BATT, R. G. 1977 Turbulent mixing of passive and chemically reacting species in a low-speed shear layer. *J. Fluid Mech.* **82**, 53–95.

BELL, J. H. & MEHTA, R. D. 1990 Development of a two-stream mixing layer from tripped and untripped boundary layers. *AIAA J.* **28**, 2034–2042.

---

[1]Karasso & Mungal (1996) and Rogers & Moser (1994) note that the simulations have only evolved long enough for the thickness to grow by a factor of 10, rather than several hundred as in the experiments.

84

BERNAL, L. P. & ROSHKO, A. 1986 Streamwise vortex structure in plane mixing layers. *J. Fluid Mech.* **170**, 499–525.

BROADWELL, E. J. & BREIDENTHAL, R. E. 1982 A simple model of mixing and chemical reaction in a turbulent shear layer. *J. Fluid Mech.* **125**, 397–410.

BROWAND, F. K. & WEIDMAN, P. D. 1976 Large scales in the developing mixing layer. *J. Fluid Mech.* **76**, 127–144.

BROWN, G. L. & ROSHKO, A. 1974 On density effects and large structure in turbulent mixing layers. *J. Fluid Mech.* **64**, 775–816.

CHANDRSUDA, C., MEHTA, R. D., WEIR, A. D. & BRADSHAW, P. 1978 Effect of free-stream turbulence on large structure in turbulent mixing layers. *J. Fluid Mech.* **85**, 693–704.

COLLIS, S. S., LELE, S. K., MOSER, R. D. & ROGERS, M. M. 1994 The evolution of a plane mixing layer with spanwise nonuniform forcing. *Phy. Fluids* **6**, 381–396.

COMTE, P., LESIEUR, M. & LAMBALLAIS, E. 1992 Large- and small-scale stirring of vorticity and a passive scalar in a 3-d temporal mixing layer. *Phys. Fluids A* **4**, 2761–2778.

CORCOS, G. M. & LIN, S. J. 1984 The mixing layer: deterministic models of a turbulent flow. part 2. the origin of the three-dimensional motion. *J. Fluid Mech.* **139**, 67–95.

CORCOS, G. M. & SHERMAN, F. S. 1984 The mixing layer: Deterministic models of a turbulent flow. part 1. introduction and the two-dimensional flow. *J. Fluid Mech.* **139**, 29–65.

DIMOTAKIS, P. E. 1991 Turbulent free shear layer mixing and combustion. In *High-speed flight propulsion systems* (ed. S. N. B. Murthy & E. T. Curran), Number 137 in Progress in Astronautics and Aeronautics, pp. 265–340. AIAA.

HUERRE, P. & MONKEWITZ, P. A. 1985 Absolute and convective instabilities in free shear layers. *J. Fluid Mech.* **159**, 151–168.

HUSSAIN, A. K. M. F. 1983 Coherent structures–reality and myth. *Phys. Fluids* **26**, 2816–2850.

J., B. B. 1988 Three-dimensional instability of elliptical flow. *Phys. Rev. Lett.* **57**, 2160–2163.

JACOBS, P. A. & PULLIN, D. I. 1989 Multiple-contour-dynamic simulation of eddy scales in the plane shear layer. *J. Fluid Mech.* **199**, 89–124.

KARASSO, P. S. & MUNGAL, M. G. 1996 Scalar mixing and reaction in plane liquid shear layers. *J. Fluid Mech.* **323**, 23–63.

KONRAD, J. H. 1976 An experimental investigation of mixing in two-dimensional turbulent shear flows with applications to diffusion-limited chemical reactions. Tech. Rep. CIT-8-PU, Calif. Inst. Technol., Pasadena, CA.

LANDMAN, M. J. & SAFFMAN, P. G. 1987 The three-dimensional instability of strained vortices in a viscous fluid. *Phys. Fluids* **30**, 2339–2342.

LASHERAS, J. C. & CHOI, H. 1988 Three-dimensional instability of a plane free shear layer: an experimental study of the formation and evolution of streamwise vortices. *J. Fluid Mech.* **189**, 53–86.

LIN, S. J. & CORCOS, G. M. 1984 The mixing layer: Deterministic models of a turbulent flow. part 3. the effect of plane strain on the dynamics of streamwise vortices. *J. Fluid*

Mech. **141**, 139–178.

MARTEL, C., MORA, E. & JIMENEZ, J. 1989 Small scales generation in 2-d mixing layers. *Bull. Am. Phys. Soc.* **34**, 2268.

METCALFE, R. W., HUSSAIN, A. K. M. F., MENON, S. & HAYAKAWA, M. 1987 Coherent structures in a turbulent mixing layer: A comparison between direct numerical simulation and experiments. In *Turbulent Shear Flows 5* (ed. F. D. et al), Berlin, pp. 110–123. Springer-Verlag.

MICHALKE, A. 1964 On the inviscid instability of the hyperbolic-tangent velocity profile. *J. Fluid Mech.* **19**, 543–546.

MONKEWITZ, P. A. & HUERRE, P. 1982 Influence of the velocity ratio on the spatial instability of mixing layers. *Phys. Fluids* **25**, 1137–1143.

MOORE, D. W. & SAFFMAN, P. G. 1975 The density of organized vortices in a turbulent mixing layer. *J. Fluid Mech.* **69**, 465–473.

MOSER, R. D. & ROGERS, M. M. 1993 The three-dimensional evolution of a plane mixing layer: pairing and transition to turbulence. *J. Fluid. Mech.* **247**, 275–320.

NYGAARD, K. J. & GLEZER, A. 1994 The effect of phase variation and cross-shear on vortical structures in a plane mixing layer. *J. Fluid Mech.* **276**, 21–59.

PATNAIK, P. C., SHERMAN, F. S. & CORCOS, G. M. 1976 A numerical simulationof kelvin-helmholtz waves of finite amplitude. *J. Fluid Mech.* **73**, 215–240.

PIERREHUMBERT, R. T. 1986 Universal short-wave instability of two-dimensional eddies in an inviscid fluid. *Phys. Rev. Lett.* **57**, 2157–2159.

PIERREHUMBERT, R. T. & WIDNALL, S. E. 1982 The two- and three-dimensional instabilities of a spatially periodic shear layer. *J. Fluid Mech.* **114**, 59–82.

RILEY, J. J. & METCALFE, R. W. 1980 Direct numerical simulation of a perturbed turbulent mixing layer. paper 80-0274, AIAA.

ROGERS, M. M. & MOSER, R. D. 1992 The three-dimensional evolution of a plane mixing layer: the kelvin-helmholtz rollup. *J. Fluid. Mech.* **243**, 183–226.

ROGERS, M. M. & MOSER, R. D. 1994 Direct simulation of a self-similar turbulent mixing layer. *Phys. Fluids* **6**, 903–923.

ROGERS, M. M., MOSER, R. D. & BUELL, J. C. 1990 A direct comparison of spatially- and temporally-evolving mixing layers. *Bull. Am. Phys. Soc.* **35**, 2294.

WALEFFE, F. 1990 On the three-dimensional instability of strained vortices. *Phys. Fluids A* **2**, 76–80.

WINANT, C. D. & BROWAND, F. K. 1974 Vortex pairing: the mechanism of turbulent mixing layer growth at moderate reynolds number. *J. Fluid Mech.* **63**, 237–255.

Theoretical and Applied Mechanics 1996
T. Tatsumi, E. Watanabe and T. Kambe (Editors)
© 1997 Elsevier Science B.V. All rights reserved.

Vortices in rotating and stratified turbulence

O. Métais [a] *

[a]Laboratoire des Écoulements Géophysiques et Industriels / IMG
B.P. 53, 38041 Grenoble Cédex 9, FRANCE

We first review numerical and experimental studies of stably-stratified flows aimed at the investigation of coherent vortices. It is shown that the late stage of strongly-stratified (homogeneous) turbulence correspond to scattered pancake-shaped vortex patches lying in almost horizontal planes. These are associated with intense vertical shear at their interface leading to an important energy dissipation. We next consider the dynamics of coherent vortices present in free and wall-bounded shear flows submitted to solid-body rotation. It is shown that a cyclonic rotation always makes the coherent vortices more two-dimensional. Anticyclonic vortices are however highly three-dimensionalized for moderate rotation rates and two-dimensionalization is recovered for fast rotation. For destabilizing anticyclonic rotation, a very efficient mechanism to create intense longitudinal vortices is provided, thanks to a linear longitudinal instability followed by a vigorous stretching of absolute vorticity.

1. Introduction

Rotation and stable density stratification modify the turbulence dynamics in many geophysical situations and on a large range of scales. Many industrial turbulent flows are also submitted to rotation and/or thermal stratification effects (turbomachinary, thermohydraulics). The important dynamical processes are strongly related to the coherent vortices imbedded within these flows. We here summarize the results of numerical simulations aimed at investigating the effects of stable-stratification and/or solid-body rotation on turbulence and coherent vortices. In a first part, we review numerical and experimental studies showing the particular vortex topology in the late stage of strongly-stratified (homogeneous) turbulence. The second part will be devoted to the dynamics of coherent vortices present in free and wall-bounded shear flows submitted to solid-body rotation.

2. General structure of strongly-stratified flows

The presence of a strong density stratification yields the inhibition of vertical motion. Riley et al. (1981) and Lilly (1983) have therefore suggested that, in the limit of small Froude numbers stably-stratified turbulence could obey a two-dimensional turbulence dynamics. The numerical studies by Herring and Métais (1989) and Métais and Herring (1989) have indeed shown that the horizontal motion tends to dominate in a strongly

---

*LEGI is sponsored by CNRS, INPG and UJF.

88

stably-stratified environment, since the growth of the vertical velocity fluctuations is inhibited. However, the flow reorganizes itself into decoupled horizontal layers leading to a strong vertical variability. Very intense vertical shear of the horizontal velocity is thus generated at the interface between the layers, which leads to energy dissipation, and prevents the turbulence from exhibiting the characteristics of two-dimensional turbulence.

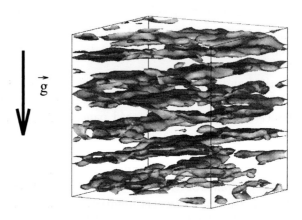

Figure 1. Two-dimensionally forced strongly-stratified turbulence; $N = 4\pi$ isosurface of one component of the horizontal vorticity ω_y (24% of the instantaneous maximum, from Métais *et al.*, 1994).

Figure 1 shows an isosurface of one component of the horizontal vorticity ω_y (24% of the instantaneous maximum) obtained in a numerical simulation of the three-dimensional Navier-Stokes equations within the Boussinesq approximation. Energy injection is applied in the small scales (see Métais *et al.*, 1994; Métais *et al.*, 1995a, for details). The Brünt-Vaissälä frequency N is here chosen in such a way that the Froude number Fr is small (strong stratification). It is clear from Figure 1 that the vertical shear of the horizontal velocity gives rise to intense horizontal vorticity generation at the interface of the layers. Similar flow layering is also observed in freely-decaying turbulence: Métais and Herring (1989) showed that the dissipation of the total energy is primarily attributable to the vertical variability of the horizontal vortical velocity component. The recent simulations by Kimura and Herring (1996) have demonstrated that the vorticity vector is more and more horizontal as the stratification is increased. Their three-dimensional simulations have also displayed scattered pancake-shaped vortex patches lying almost in horizontal planes. The long term behaviour (when the system has reached a state in which vertical fluctuations have been suppressed) for the High Reynolds number regime has been experimentally studied by Fincham and Maxworthy (1995). The full field measurements was made possible through the use of digital partical image velocimetry (DPIV). These experiment consisting in towing a rake of vertical flat-plates through a linearly stratified medium, have shown that the vertical shear between the horizontal "pancakes" is responsible for over 90% of the flow dissipation.

3. Coherent vortices in rotating shear flows

Turbulent or transitional shear flows in a rotating frame have been extensively studied due to their importance in many geophysical and engineering applications. Within these flows, the local Rossby number, which characterizes the relative importance of inertial and Coriolis forces, can vary significantly. Typical values of the Rossby number are on the order of 0.05 in mesoscale oceanic eddies and in Jupiter's Great Red Spot, 0.3 for large synoptic-scale atmospheric perturbations, and 2.5 for the atmospheric wake of a small island. Turbulence in rotating fluids finds numerous industrial applications in turbo-machinery; e.g., the turbulent characteristics of the flow in blade passages of radial pumps and compressor impellers determine the efficiency of these devices. Turbulence is also of great importance for the cooling by the fluid inside the blades. Depending upon the magnitude of the radial velocity, the Rossby number within rotating machines can range from values close to unity to very small ones (of the order 0.05).

Laboratory experiments (see Bidokhti and Tritton, 1992, for the mixing layer; Witt and Joubert, 1985 and Chabert d'Hières et al., 1988, for the wake), theoretical works, numerical simulations and atmospheric and oceanic observations (see Etling 1990) show that there are three basic effects associated with rotating free-shear flows. (i) If the shear vorticity is parallel and of same sign as the rotation vector (cyclonic rotation), the flow is made more two-dimensional. (ii) If the two vectors are anti-parallel (anticyclonic rotation), destabilization is observed at moderate rotation rates (high Rossby numbers), while (iii) two-dimensionalization is recovered for fast rotation. For the channel fow, former experimental and numerical studies at moderate rotation rates (Johnston et al., 1972 ; Kristofferson et Andersson, 1993; Piomelli and Liu, 1995) have shown that the flow becomes strongly asymmetric with respect to the channel center. They indicate a reduction (resp. increase) of the velocity fluctuations in the cyclonic (resp. anticyclonic) region. Note that various other terms are used to designate the cyclonic and anticyclonic sides of the rotating channel. The names suction and pressure sides originate from the pressure gradient due to the Coriolis force, and the terms trailing and leading sides are borrowed from the turbo-machinery literature. It is easy to show that the asymmetry between the cyclonic and anticyclonic cases can only be explained by considering the influence of rotation on the growth of three-dimensional perturbations. In order to describe the early stage of the development, we first present results of a three-dimensional linear-stability analysis of planar free-shear flows. The nonlinear regime will be investigated through three-dimensional simulations of the Navier-Stokes equations for free-shear flows and the channel flow. We will show that the key dynamical mechanism in the case of destabilizing anticyclonic rotation is the concentration and longitudinal stretching of absolute vorticity.

3.1. Linear-stability analysis

In order to describe the early stage of the development, Yanase et al. (1993) have performed a three-dimensional, viscous, linear-stability analysis of two planar free-shear flows subject to rigid-body rotation oriented along the span: the mixing layer and the plane wake. The mean ambient velocity is oriented in the longitudinal direction x and varies with y, $[\bar{u}(y), 0, 0]$. The rotation vector is the spanwise direction z, $\vec{\Omega} = (0, 0, \Omega)$. The Rossby number is here based upon the maximum ambient vorticity of the basic

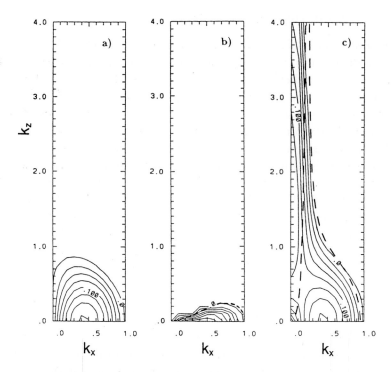

Figure 2. Contours of constant amplification rates in the three-dimensional linear-instability analysis of a rotating mixing layer at Reynolds number 150; a) non-rotating case, b) cyclonic case at $R_o = 1$, c) anticyclonic case at $R_o = -20$ (from Lesieur and Métais, 1996).

profile, that is, the vorticity at the inflexion point(s), $-(d\bar{u}/dy)_i$, i.e.,

$$R_o = -(d\bar{u}/dy)_i/2\Omega \quad . \tag{1}$$

In the mixing layer, R_o is positive for cyclonic rotation and negative for anticyclonic rotation. For a wake, one has to consider the modulus $|R_o|$ of the Rossby number. In order to examine the (temporal) linear stability of these flows, we follow the usual procedure consisting in developing the velocity and pressure fields as sums of the basic fields plus a small perturbation, $\underline{u}'(x, y, z, t)$ and $\pi'(x, y, z, t)$. We are only seeking for normal modes solutions (see, e.g. Drazin and Reid, 1981) of the form:

$$\underline{u}'(x, y, z, t) = \underline{U}'(y) \, exp \, i(k_x x + k_z z - \omega t)$$

$$\pi'(x, y, z, t) = \Pi'(y) \, exp \, i(k_x x + k_z z - \omega t) \quad ,$$

where $\underline{U}'(y) = (U'(y), V'(y), W'(y))$. After substituting into the equations of motion, one obtains after linearization and nondimensionalization the generalized three-dimensional Orr-Sommerfeld equations, which have been solved numerically by Yanase *et al.* (1993) making use of Galerkin methods based upon Chebychev polynomials.

In order to illustrate the very distinct effects of cyclonic and anticyclonic rotation, we examine the maximum amplification rate, $\sigma_m = Im(\omega_m)$, as a function of wave vector (k_x, k_z). Results are presented for the mixing-layer (hyperbolic tangent profile) at a Reynolds number $Re = 150$. Figure 2 shows σ_m contour plots in the (k_x, k_z)-plane, for various Rossby numbers: $R_o = \infty$ (figure 2a), $R_o = 1$ (figure 2b, cyclonic) and $R_o = -20$ (figure 2c, anticyclonic). In the non-rotating case, the overall maximum amplification occurs for $k_z = 0$ (which corresponds to a two-dimensional disturbance, in good agreement with Squire's theorem). The maximum value is $\sigma_m = 0.180$ at a wave number $k_x = 0.433$, close to the inviscid theory predictions. In the cyclonic case, the instability region is clearly reduced by the rotation and concentrated towards the k_x axis: the rotation has a stabilizing influence on the three-dimensional motions, and plays a two-dimensionalizing role. The most-amplified wavenumber is unchanged. For even moderate anticyclonic rotation, the flow stability is dramatically modified. Along the k_x-axis, the peak corresponding to Kelvin-Helmholtz instability is still present, with the same location and amplification rate as in the non-rotating case. However, a second peak appears along the k_z-axis (for $k_z = 1.51$ and $\sigma_m = 0.147$) corresponding to a purely streamwise instability. It will be designated as the shear/Coriolis instability. Yanase *et al.* (1993) have shown that the streamwise mode is maximally amplified for a Rossby number ≈ 2.5: its wavelength is 4.6 times smaller than Kelvin-Helmholtz's mode, and its amplification rate 1.8 times larger. At $R_o = -1$ a dramatic change occurs in the flow stability: the shear/Coriolis instability disappears and the flow is stabilized for $R_o \geq -1$, that is for high rotation rates.

Notice that, for purely longitudinal modes ($k_x = 0$) and in the limit $Re \to \infty$, the stability problem reduces to:

$$\left(\frac{d^2}{dy^2} - k_z^2\right) V' - \frac{k_z^2}{s^2} 4\Omega^2 \left(1 + Ro(y)\right) V' = 0 \tag{2}$$

where $s = -i\omega$ and $R_o(y) = -(d\bar{u}/dy)/2\Omega$ is a local Rossby number. This is a classical eigenvalue problem of the Stürm-Liouville type for which well-known theorems have been derived (see e.g. Drazin and Reid, 1981). In particular, it can be shown that a necessary and sufficient condition for instability is that the coefficient

$$[1 + Ro(y)] < 0$$

should be negative somewhere in the flow. This is essentially the result previously found by Pedley (1969) and Hart (1971), consistent with the above linear-stability analysis. Notice that, for a purely longitudinal perturbation, the eigenvalue equation is similar to the one governing the radial velocity when studying the inviscid linear centrifugal instability in the limit of axisymmetric disturbances.

3.2. The non-linear regime

The previous analysis allows to describe the early linear stage of the perturbations growth. Further insight in the nonlinear regime, can be obtained by performing three-dimensional simulations of the full Navier-Stokes equation.

In a former study (Lesieur *et al.*, 1991) have emphasized the importance of considering the absolute vorticity, and not just the relative vorticity, since Kelvin's circulation theorem directly applies to the absolute vorticity. For example, if the relative vorticity is written as the sum of the ambient $-(d\bar{u}/dy)\underline{z}$ and fluctuating $\underline{\omega}'$ components. The absolute vorticity then writes $(2\Omega - d\bar{u}/dy)\,\underline{z} + \underline{\omega}'$. If the flow is locally cyclonic (i.e., Ω and $-(d\bar{u}/dy)$ have the same sign), then the absolute vortex lines are closer to the spanwise direction than the corresponding relative ones. Therefore, as compared to the non-rotating case, the effectiveness of vortex turning and stretching is reduced. Conversely, if the flow is locally anticyclonic, especially for the regions where 2Ω has a value close to $d\bar{u}/dy$ (weak absolute spanwise vorticity), absolute vortex lines are very convoluted, and will be very rapidly stretched out all over the flow, as a dye would do. It was thus predicted that in rotating shear flows, the vortex filaments of Rossby number -1 (hence anticyclonic) would be stretched into longitudinal alternate vortices. This phenomenological argument will be referred to as the weak-absolute vorticity stretching principle.

3.3. Numerical simulations of rotating free-shear flows
3.3.1. Temporally-growing mixing layer

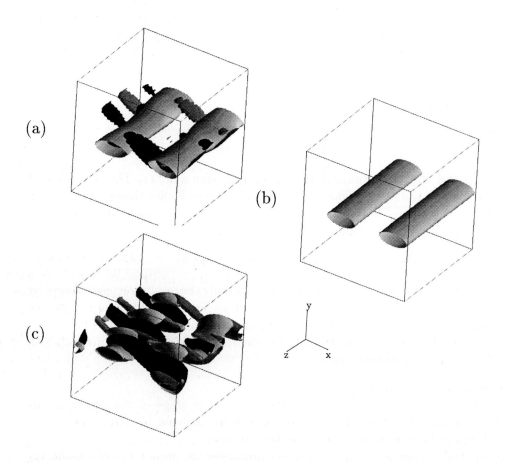

Figure 3. Mixing layer. Relative vorticity isosurfaces at $t = 17.8\delta_i/|U_0|$. a) nonrotating case: $\omega_z = 45\%$ of $|\omega_{2D}^{(i)}|$ (vorticity maximum associated with the initial mean profile), light gray; $\omega_l = \sqrt{\omega_x^2 + \omega_y^2} = 4.5\%$ of $|\omega_{2D}^{(i)}|$ coloured by the sign of ω_x, black $\omega_x < 0$, dark gray $\omega_x > 0$. b) $R_o^{(i)} = -1$. $\omega_z = 45\%$ of $|\omega_{2D}^{(i)}|$, light gray. c) $R_o^{(i)} = -5$. $\omega_z = 45\%$ of $|\omega_{2D}^{(i)}|$, light gray; $\omega_l = 22.5\%$ of $|\omega_{2D}^{(i)}|$ coloured by the sign of ω_x, black $\omega_x < 0$, dark gray $\omega_x > 0$ (from Métais et al., 1995b).

We here concentrate on the rotating mixing layer case and we will show, through numerical simulations, how the rotation modifies the three-dimensional flow topology. The reader can refer to Lesieur *et al.* (1991), Métais *et al.* (1992), and Métais *et al.* (1995b) for more details. Temporal shear flows are here considered with periodicity in the streamwise direction. We consider a mixing layer associated with a hyperbolic-tangent mean-velocity profile: $\bar{u}(y) = U_0 \tanh y/\delta$ where $2U_0$ is the velocity difference across the layer, and $\delta = \delta_i/2$, with δ_i initial vorticity thickness. We define the Rossby number as, $R_o^{(i)} = -U_0/2\Omega\delta$. $R_o^{(i)}$ is positive for cyclonic rotation (U_0 and Ω of opposite sign) and negative for anticyclonic rotation. The Reynolds number $Re^{(i)} = |U_0|\delta/\nu$ is here taken equal to 50. Initially, a low-amplitude random noise is superposed upon the ambient velocity profiles. Two different types of perturbations are considered. Firstly, a quasi-two-dimensional one consisting of the superposition of a purely two-dimensional perturbation (z independent) of kinetic energy $\epsilon_{2D}U_0^2$ and a three-dimensional perturbation of energy $\epsilon_{3D}U_0^2$, with $\epsilon_{2D} = 10\epsilon_{3D} = 10^{-4}$. The perturbation peaks at the fundamental Kelvin-Helmholtz mode. This case will be referred to as the "forced transition" case. The second type of perturbations which has been considered is purely three-dimensional, with $\epsilon_{3D} = 10^{-4}$ and $\epsilon_{2D} = 0$. The noise is now a white-noise which does not favor any mode, and the most amplified one can freely emerge: this case will be called the "natural transition" case.

We will show that the simulations have confirmed the global trends observed in the experiments and predicted by the linear-stability analysis: the Kelvin-Helmholtz vortices are two-dimensionalized by the rotation when these are cyclonic; this is also true for rapid anti-cyclonic rotation. Conversely, a moderate anti-cyclonic rotation disrupts the primary vortices. The latter are stretched into intense longitudinal alternate vortices of Rossby number -1, as predicted by the weak absolute vorticity stretching principle.

In the early stage of the development, the results have corroborated the linear-stability predictions for quantities such as the critical Rossby for maximum destabilization, found to be ≈ -2. A special care is here given to the coherent-vortex dynamics. Now we look at the three-dimensional flow structure, in the forced transition case. We focus on the relative vorticity iso-surfaces at $t = 17.8\delta_i/|U_0|$ obtained in the non-rotating case ($R_o^{(i)} = \infty$), and for anticyclonic rotation at $R_o^{(i)} = -5$ and $R_o^{(i)} = -1$.

(1) $R_o^{(i)} = \infty$ (Figure 3a). Here, one observes quasi-two-dimensional Kelvin-Helmholtz billows, slightly distorted in the spanwise direction. Weak longitudinal vortices are stretched between the primary rolls: those are identified through isosurfaces of weak longitudinal vorticity.

(2) $R_o^{(i)} = -1$ (Figure 3b) displays the spanwise vorticity field with the same iso-contour value as in the non-rotating case. Anticyclonic and cyclonic flows are similar at this Rossby number, and a strong two-dimensionalization is observed in both cases. The longitudinal vortices have disappeared. Furthermore, the two-dimensionalization tendency can be observed in the cyclonic case even for large positive $R_o^{(i)}$. This agrees well with both the predictions of the linear stability analysis by Yanase *et al.* (1993) and the phenomelogical theory proposed by Lesieur *et al.* (1991).

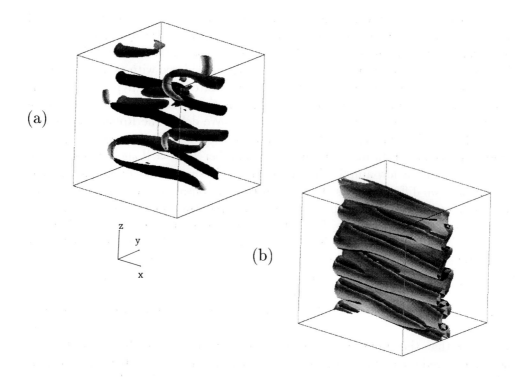

Figure 4. Mixing layer. Relative vorticity isosurfaces at $t = 26.8\delta_i/|U_0|$. (a) forced-transition; (b) natural transition. ω_z, light gray; ω_l (longitudinal vorticity) coloured by the sign of ω_x, black $\omega_x < 0$, dark gray $\omega_x > 0$ (from Lesieur and Métais, 1996).

(3) $R_o^{(i)} = -5$ (Figure 3c). The Kelvin-Helmholtz vortices are now highly distorted and exhibit strong oscillations along the spanwise direction. The longitudinal vorticity is much higher than in the non-rotating case: we observe the simultaneous formation of Kelvin-Helmholtz vortices and longitudinal hairpin vortices which are stretched inbetween. As time goes on, this produces an important increase of the longitudinal vorticity component: by the end of the run the longitudinal vorticity is approximately twice the one associated with the initial mean velocity profile. By the end of the run (Figure 4a), the Kelvin-Helmholtz vortices have been totally dislocated and the flow is entirely composed of hairpin-shaped longitudinal vortices. A similar sequence had been proposed in this case by Lesieur et al. (1991), using the weak absolute vorticity stretching mechanism: weak absolute vorticity in the stagnation region between the Kelvin-Helmholtz rollers would be stretched longitudinally between the latter, yielding longitudinal alternate vortices which should destroy the primary vortices. It is worth noting that these vorticity structures originate from the growth of the longitudinal mode predicted by the linear stability analysis (see Yanase et al. 1993).

In the case of natural transition, we observe the rapid formation of purely longitudinal structures corresponding to regions of high *relative* vorticity (Figure 4b). These exhibit some analogies with the Görtler vortices observed in the boundary layer over a con-

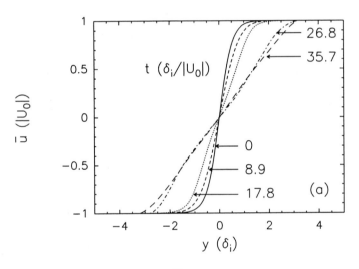

cave wall. These structures have a spanwise wavelength λ_s corresponding to the fastest shear/Coriolis mode predicted by the linear-stability analysis of Yanase *et al.* (1993). Close examination of the time evolution of the *absolute* vortex lines shows that the flow undergoes very distinct stages. In the first stage, the vorticity dynamics are dominated by quasi-linear mechanisms yielding absolute vortex lines inclined at 45° with respect to the horizontal plane. These are in phase in the longitudinal direction. In a second stage, nonlinear stretching mechanisms yield a concentration of quasi-horizontal longitudinal hairpins of absolute vorticity. The dynamics are then dominated by a strong quasi-horizontal stretching of longitudinal *absolute* spanwise vorticity. During that stage, we observe that the ambient velocity profile exhibits a long range of nearly constant shear slope 2Ω (see Figure 5). It is associated with a local Rossby number $R_o(y) = -1$, where

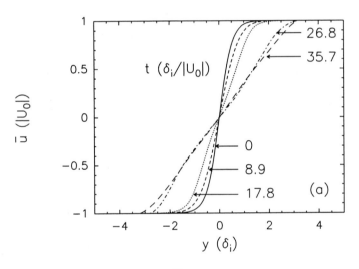

Figure 5. Anticyclonic mixing layer at $R_o^{(i)} = -5$, natural transition. Time evolution of the mean velocity profile $\bar{u}(y,t)$, normalized by $|U_0|$. y is non-dimensionalized by δ_i and the time unit is $\delta_i/|U_0|$ (from Métais *et al.* 1995b).

$R_o(y) = -(d\bar{u}(y)/dy)/2\Omega$. In this region, the mean absolute vorticity is close to zero. As suggested by the phenomenological theory of Lesieur *et al.* (1991), in such a range of near-zero absolute spanwise vorticity, the production of longitudinal vorticity is favored. Thus a very efficient mechanism to create intense longitudinal vortices in rotating anticyclonic shear layers is provided, thanks to a linear longitudinal instability followed by a vigorous stretching of absolute vorticity.

3.3.2. Spatially-growing wake

As opposed to the previous study, the periodicity assumption in the longitudinal direction is dropped and we consider here spatially growing flows. We have used a three-dimensional code developed by M.-A. Gonze (1993) combining pseudo-spectral methods

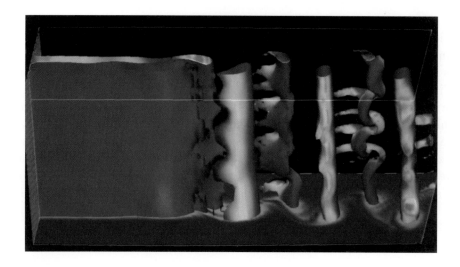

Figure 6. Spatially growing wake at $R_o = 5$, vorticity modulus isosurfaces $\omega_n \sim 0.31\omega_{max}^{(up)}$, colored by the pressure field, $t = 76r_m/U_m$.

in the periodicity directions (spanwise and shear directions; noted y and z) and compact finite difference schemes (Lele, 1992) in the longitudinal one (noted x). This numerical algorithm allows one to reach a precision comparable to spectral methods. An open-boundary condition is applied at the downstream boundary. The mean velocity imposed at the inflow boundary corresponds to a gaussian velocity profile:

$$U(y) = U_m \left[1 - \exp\left(\ln 2 \, \frac{y^2}{r_m^2} \right) \right] \quad, \tag{3}$$

where r_m is the half deficit velocity width. A three-dimensional perturbation of amplitude $10^{-2}U_m^2$ is superposed to the upstream mean flow (white noise in time). The maximum vorticity $\omega_{max}^{(up)} = (-dU/dy)_{max} \approx 0.7U_m/r_m$. Thus, the upstream Rossby number is defined as

$$R_o^{(up)} = 0.7 \, \frac{U_m}{|2\Omega|r_m} \quad. \tag{4}$$

Calculations are performed with $100 \times 40 \times 40$ resolution points for a domain size of $L_x = 40r_m, L_y = 16r_m, L_z = 16r_m$. The Reynolds number is fixed $R_e = U_m r_m/\nu = 200$ and we vary $R_o^{(up)}$. For each case, we focus, at time $t = 76r_m/U_m$, on vorticity modulus iso-surfaces, colored by the pressure.

At $R_o = \infty$, the alternate sign quasi-two-dimensional primary vortices of the Karman street appear (not shown here). The development of the three-dimensionality induce slight oscillations of the rolls along the span.

The picture is completely different when $R_o^{(up)} = 5$ (rotation axis along the spanwise direction). Figure 6 shows that the Karman street symmetry breaks down: the cyclonic vortices three-dimensionality is decreased as compared to the non-rotating case, whereas

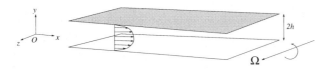

Figure 7. Rotating channel flow configuration.

the anticyclones are destabilized and stretched into longitudinal hairpin vortices. The ratio between their spanwise wavelength λ_z and the longitudinal wavelength of the Karman rolls λ_x agrees well with linear stability predictions of Yanase *et al.* (1993): $\lambda_x/\lambda_z \approx 2$. Furthermore, in accordance with the mechanism proposed by Lesieur *et al.* (1991), these vortices result from the stretching of vortex lines of weak absolute vorticity leading to a longitudinal vorticity maximum $\approx \omega_{max}^{(up)}$.

We find a critical value $R_o^{(up)} \approx 2.5$ for which a maximum anticyclonic destabilization is achieved. The Rossby number based upon the local mean-velocity profile $(R_o^{(l)})$ decreases with the downstream position: a saturation of the spanwise velocity fluctuations is observed for $R_o^{(l)} \approx 1$ and the flow is subsequently restabilized. However, the longitudinal length of the computational domain does not allow addressing the issue of an eventual reappearance of well defined spanwise anticyclones.

When a fast rotation is applied (low Rossby number régime), two-dimensional cyclonic and anticyclonic vortices are observed in agreement with Taylor-Proudman theorem. In that case, since geostrophic equilibrium holds, anticyclones and cyclones respectively correspond to high- and low- pressure zones.

3.4. Numerical simulations of rotating channel flow

We here investigate, via three-dimensional direct numerical simulations, the influence of spanwise rotation on the vortex topology in turbulent channel flow (see Lamballais *et al.*, 1995, 1996; for further details). The flow configuration is shown on Figure 7.

The non-dimensional parameters are the Reynolds number $Re = U_c h/\nu$ (h half-channel height and U_c centerline velocity of the initial laminar velocity profile), and $|R_o^{(i)}| = U_c/\Omega h$ the Rossby number associated with the vorticity maxima of the initial basic velocity profile. The Reynolds number is taken equal to 3750 when the Rossby number is varied: $R_o^{(i)} = \infty$ (no rotation) , 18, 6 and 2. The computational domain size is $(L_x, L_y, L_z) = (4\pi h, 2h, 2\pi h)$ corresponding to the following resolution $128 \times 129 \times 180$ for $R_o^{(i)} = 18$ and 6 ; $128 \times 129 \times 128$ for $R_o^{(i)} = \infty$ and $96 \times 129 \times 128$ for $R_o^{(i)} = 2$. The statistics are obtained through temporal and spatial (in the x, z directions) averaging. The upper (resp. lower) channel wall, corresponding to $y > 0$ (resp. $y < 0$), will be called cyclonic (resp. anticyclonic) since the relative mean vorticity is parallel (resp. anti-parallel) to the rotation vector $\vec{\Omega}$.

The initial conditions consist in a field issued from a non-rotating calculation on which we impose a spanwise rotation, We then integrate till the flow has reached a statistically steady state. Previous experimental and numerical studies (Johnston *et al.*, 1972 ; Kristofferson et Andersson, 1993) have shown that the turbulence intensity is amplified

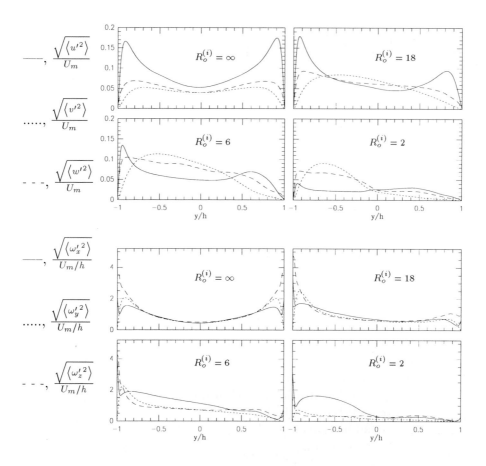

Figure 8. R.m.s. velocity and vorticity fluctuations for various rotation regimes (from Lamballais *et al.*, 1996).

on the anticyclonic side while relaminarization occurs on the cyclonic side. They have also exhibited large-scale longitudinal rolls in the anticyclonic region, which have been assimilated with Taylor-Görtler vortices. The statistics shown on Figure 8 for $R_o^{(i)} = 18$ confirm these findings. For higher rotation rates, however, the longitudinal velocity fluctuations intensity is reduced everywhere (including in the anticyclonic region) as compared with the non-rotating case. This indicates a decrease of the intensity of the low- and high-speed streaks close to the wall. The latter have totally disappeared at $R_o^{(i)} = 2$, since the longitudinal velocity component has become negligable in front of the spanwise and vertical components.

For $R_o^{(i)} \leq 6$, the longitudinal vorticity component dominates the other two components within most of the anticyclonic region (cf Figure 8). At $R_o^{(i)} = 6$, the longitudinal vorticity is far more important than in the non-rotating case, indicating an enhancement of longitudinal stretching.

We now show that the vortex topology is strongly affected by the rotation. The figure 9 displays a three-dimensional visualization of a relative-vorticity modulus isosurface. The eddies activity is dramatically reduced by the rotation on the cyclonic side. Conversely,

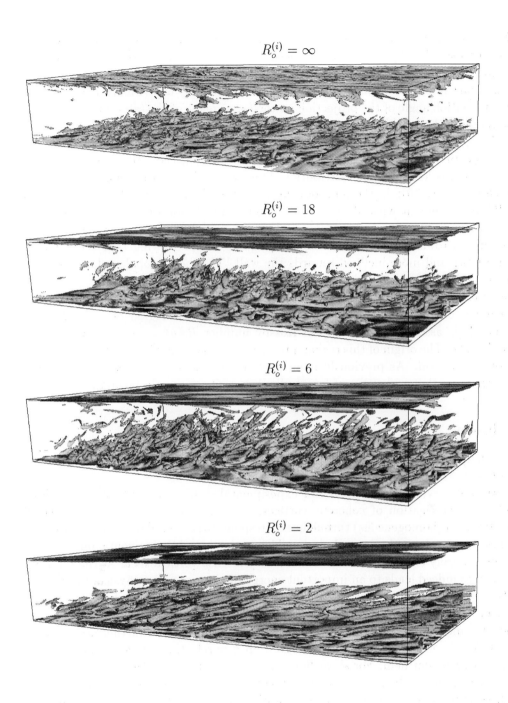

Figure 9. Vorticity modulus isosurface : $\omega = 4.5$ for $R_o^{(i)} = \infty, 18, 6$; $\omega = 3.4$ for $R_o^{(i)} = 2$ (from Lamballais *et al.*, 1996).

well-defined hairpin vortices are clearly visible in the anticyclonic region. They reach a maximum intensity for $R_o^{(i)} = 6$ and the best defined organization for $R_o^{(i)} = 2$ with a weak inclination with respect to the wall.

We next examine the distribution of the angle $\theta = tan^{-1}(\omega_y/\omega_x)$ (Figure 10). It represents the inclination angle of the projection on the (x, y)-plane of the (absolute or relative) vorticity vector with respect to the wall. We here follow the procedure developed by Moin & Kim (1985). Each contribution to the distribution is weighted by the magnitude of the vorticity projection. We here consider a flow region far from the wall corresponding to $y = -0.6h$. At this location, the distribution differs appreciably in the non-rotating and rotating cases (figure 10): in the non-rotating case, the histogram for θ presents a peak around 45° in agreement with Moin & Kim (1985). As the rotation rate increases, the peak for θ becomes narrower and narrower indicating a much a greater flow organization. It is also shifted towards smaller and smaller angular values showing that the vorticity vectors are more and more inclined towards the wall. The most probale values for θ are close to 45°, 35°, 25° and 10° for $R_o^{(i)} = \infty, 18, 6, 2$ respectively.

Similarly to the mixing-layer case mentioned above, a striking feature of the flow is that the mean velocity profile exhibits a characteristic linear region of slope 2Ω (see Figure 11). It is associated with a local Rossby number $R_o(y) = -1$, where $R_o(y) = -(d\bar{u}(y)/dy)/2\Omega$. The origin of this region, previously observed in laboratory experiments, is still not understood. As previously mentioned, in such a range of near-zero absolute spanwise vorticity, the production of longitudinal vorticity is greatly favored.

4. Conclusion

We have here demonstrated the very distinct effects of stable-stratification and rotation on the coherent vortex dynamics.

First, we have reviewed numerical and experimental studies of stably-stratified flows aimed at the investigation of coherent vortices. It was shown that the late stage of strongly-stratified (homogeneous) turbulence correspond to scattered pancake-shaped vortex patches lying in almost horizontal planes. The motion is mainly horizontal but strongly differs from a purely two-dimensional motion: very intense vertical shear at the interface of the pancakes indeed leads to an important energy dissipation. The physical mechanism setting the thickness of these horizontal pancakes is still a question of debate.

Conversely, a solid-body rotation tends to preserve coherence along the rotation axis direction in some regions of the flow. Bartello et al. (1994) have furthermore showed that this type of coherence can freely emerge from a purely three-dimensional isotropic homogeneous turbulent field. For shear flows, it is shown that a cyclonic rotation always makes the coherent vortices more two-dimensional. Anticyclonic vortices are however highly three-dimensionalized for moderate rotation rates and two-dimensionalization is recovered for fast rotation. For destabilizing anticyclonic rotation, the linear-stability studies shows a new type of instability, the shear Coriolis instability consisting in the amplification of purely longitudinal modes. For this regime, one striking feature of the rotating free-shear flows is that the mean velocity profile exhibits a characteristic linear region of slope 2Ω. In such a range of near-zero absolute spanwise vorticity, the concentration and production by stretching of longitudinal vorticity is highly favored and

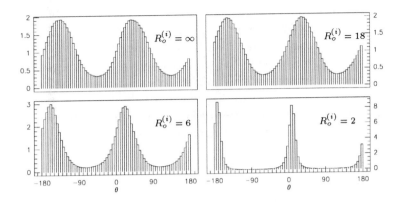

Figure 10. Probability density function of the angle θ at $y = -0.6\,h$ for various rotation regimes (from Lamballais *et al.*, 1996).

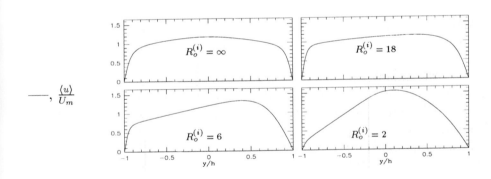

Figure 11. Mean velocity profiles for various rotation regimes (from Lamballais *et al.*, 1996).

leads to the formation of longitudinal hairpin vortices. This seems to corroborate the prediction of the phenomenological theory proposed by Lesieur *et al.* (1991) based on the weak-absolute vorticity stretching principle. The progresses in experimental measurements (DPIV) should render possible the identification of these longitudinal vortices in laboratory experiments studying rotating mixing layers, wakes or channel flows.

Acknowledgments Computations were carried out at the IDRIS (Institut du Développement et des Ressources en Informatique Scientifique, Paris).

REFERENCES

1. Bartello, P., Métais, O. and Lesieur, M., 1994. Coherent structures in rotating three-dimensional turbulence. *J. Fluid Mech.*, **273**, pp 1–29.
2. Bidokhti, A.A. and Tritton, D.J. 1992. The structure of a turbulent free shear layer in a rotating fluid. *J. Fluid Mech.*, **241**, pp. 469–502.
3. Chabert d'Hières, G., Davies, P.A. and Didelle, H., 1988. Laboratory studies of pseudo-periodic forcing due to vortex shedding from an isolated solid obstacle in an homogeneous rotating fluid. In *20th Int. Liège Colloquium on Ocean Dynamics*, Elsevier Publisher.
4. Drazin, P.G. and Reid, W.H., 1981. *Hydrodynamic stability*, Cambridge University Press.
5. Etling, D., 1990. Mesoscale vortex shedding from large islands: a comparison to laboratory experiments in rotating stratified flows. *Meteorol. Atmos. Phys.*, **43**, 145.
6. Fincham, A.M., Maxworthy, T. and Spedding G.R., 1995. Energy dissipation and vortex structure in freely decaying stratified grid turbulence. *Dynamics of Atmospheres and Oceans*, **23**, pp. 155–169.
7. Gonze, M.-A., 1993. Simulation numérique des sillages en transition à la turbulence. PhD thesis, Institut National Polytechnique de Grenoble, France.
8. Hart, J.E. 1971. Instability and secondary motion in a rotating channel flow. *J. Fluid Mech.* **45**, pp. 341–351.
9. Herring, J.R. and Métais, O., 1989. Numerical experiments in forced stably stratified turbulence. *J. Fluid Mech.*, **202**, pp. 97-115.
10. Johnston, J.P., Halleen, R.M. and Lezius, D.K., 1972. Effects of spanwise rotation on the structure of two-dimnesional fully developped turbulent channel flow. *J.Fluid Mech.*, **56**, pp. 533–557.
11. Kimura, Y. and Herring, J.R., 1996. Diffusion in stably stratified turbulence. *submitted to the J. Fluid Mech.*.
12. Kristoffersen, R. and Andersson, H.I., 1993. Direct simulations of low-Reynolds-number turbulent flow in a rotating channel. *J. Fluid Mech.*, **256**, pp. 163-197.
13. Lamballais, E., Lesieur, M. and Métais, O., 1995. Effects of spnawise rotation on the vorticity stretching in transitional and turbulent channel flow. *Selected Proceedings of the Tenth Symposium on Turbulent Shear Flows, International Journal of Heat and Fluid Flow*, **17**.
14. Lamballais, E., Lesieur, M. and Métais, O., 1996. Influence d'une rotation d'entraînement sur les tourbillons cohérents dans un canal. *C.R. Acad. Sci. Paris*,

bf 323, Série IIb, pp. 95–101.

15. Lele, S. K., 1992. Compact finite difference schemes with spectral-like resolution. *J. Comput. Phys.*, **103**, pp. 16–42.

16. Lesieur, M., and Métais, O., 1996. Turbulence and coherent vortices. In *Computational Fluid Dynamics*, Les Houches, Session LIX, 1993. M. Lesieur, P. Comte and J. Zinn-Justin, eds. Elsevier Science, pp. 111–163.

17. Lesieur, M., Yanase, S. and Métais, O., 1991. Stabilizing and destabilizing effects of a solid-body rotation upon quasi-two-dimensional shear layers. *Phys. Fluids A*, **3**, pp. 403-407.

18. Lilly, D.K., 1983. Stratified turbulence and the mesoscale variability of the atmosphere. *J. Atmos. Sci.*, **40**, pp. 749-761.

19. Métais, O., Bartello, P., Garnier, E., Riley, J.J. and Lesieur, M., 1995a. Inverse cascade in stably-stratified rotating turbulence. *Dynamics of Atmospheres and Oceans*, **23**, pp. 193–203.

20. Métais, O., Flores, C., Yanase, S., Riley, J.J. and Lesieur, M., 1995b. Rotating free-shear flows. Part 2: numerical simulations. *J. Fluid Mech.*, **293**, pp. 47–80.

21. Métais, O. and Herring, J.R., 1989. Numerical simulations of freely evolving turbulence in stably stratified fluids. *J. Fluid Mech.*, **202**, pp. 117-148.

22. Métais, O., Riley, J.J., and Lesieur, M., 1994. Numerical simulations of stably-stratified rotating turbulence. In *Stably-stratified flows- flow and dispersion over topography*, I.P. Castro and N.J. Rockliff eds., Oxford University Press, pp. 139-151.

23. Métais, O., Yanase, S., Flores, C., Bartello, P. & Lesieur, M., 1992. Reorganization of coherent vortices in shear layers under the action of solid-body rotation. Selected proceedings of the *Turbulent Shear Flows VIII*, Springer-Verlag, pp.415-430

24. Pedley, T.J., 1969. On the stability of viscous flow in a rapidly rotating pipe. *J. Fluid Mech.*, **35**, pp. 97–115.

25. Piomelli U. and Liu J., 1995. Large-eddy simulation of rotating channel flows using a localized dynamic model. *Phys. Fluids* **7** (4), pp. 839–848

26. Riley, J.J., Metcalfe, R.W. and Weissman, M.A. (1981). Direct numerical simulations of homogeneous turbulence in density stratified fluids. In: B.J. West (Editor), AIP Conf. Proc. N° 76, New York, Nonlinear Properties of Internal Waves, pp. 79-112.

27. Witt, H.T. and Joubert, P.N. 1985. Effect of Rotation on a Turbulent Wake. In *Proc. 5th Symp. on Turbulent Shear Flows, Cornell*, pp. 21.25-21.30.

28. Yanase, S., Flores, C., Métais, O., and Riley, J.J. 1993 Rotating free shear flows Part 1: Linear stability analysis, *Phys. of Fluids. A.* **5** (11), pp. 2725-2737.

Theoretical and Applied Mechanics 1996
T. Tatsumi, E. Watanabe and T. Kambe (Editors)

Laboratory modelling of geophysical vortices

G.J.F. van Heijst and R.R. Trieling

Fluid Dynamics Laboratory, Department of Physics, Eindhoven University of
Technology, P.O. Box 513, 5600 MB Eindhoven, The Netherlands.

Vortices emerging in geophysical or two-dimensional turbulence may undergo serious
deformations due to the non-uniform ambient flow induced by neighbouring vortices.
In order to investigate the behaviour of vortices in a strain or shear flow, laboratory
experiments were carried out in a rotating and a stratified fluid. Certain kinematic aspects
of the observed vortex evolution are explained by simple point-vortex models.

1. INTRODUCTION

Large-scale geophysical flows are essentially affected by density stratification and background rotation. Together with geometrical constraints (flatness of the domain), these
effects cause the large-scale motions in the atmosphere and oceans to be approximately
two-dimensional. It is a well-known fact that two-dimensional turbulence is characterized
by the inverse energy cascade, which implies a spectral flux of energy to the larger scales
of motion. This property of self-organization was nicely illustrated by numerical simulations of decaying two-dimensional turbulence on a periodic domain by *e.g.* McWilliams [1]:
vortex structures were observed to emerge gradually from a randomly generated initial
turbulent field. The emerging vortices were all of the monopole-type, *i.e.* consisting of a
single nested set of closed streamlines. In a later high-resolution simulation of forced two-
dimensional turbulence, Legras *et al.* [2] found that the flow at some stage also contained
dipolar and tripolar vortices, in addition to monopolar structures. Similar vortices were
observed in laboratory experiments in stratified and rotating fluids [3,4]. Later studies
have revealed important characteristics of dipolar, tripolar and even higher-mode vortex
structures [5–7].

In order to better understand the evolution of two-dimensional turbulence towards
an 'organized' state of a number of isolated vortices, it is instructive to consider one
particular single vortex embedded in the 2D flow field and follow it in space and time.
In fact, this was done by McWilliams [8] in a numerical simulation of two-dimensional
turbulence. In the initial stage of the flow evolution, smaller and larger coherent vortex
structures appear to emerge from the initial random field. Because of their large number
and their close proximity at that stage, the vortices undergo serious deformations, in
some cases even to such an extent that they are completely torn apart. Some time later,
the flow field consists of a cloud of intense vortices drifting around in the plane, once
in a while interacting with neighbouring vortices. When like-signed vortices come close
together, they may start to rotate around each other, possibly resulting in a merger.

Although the merging process results in a single vortex containing a large portion of the vorticity originally in the two individual vortices, many long vorticity filaments are formed during this fusion [9,10]. Another type of interaction that may be encountered is when two oppositely-signed vortices are brought close together: depending on the vorticity distribution of the individual monopolar vortices such an interaction may result in the formation of a dipolar structure. As a result of these various types of interactions, the number of vortices gradually decreases, and they become more and more 'isolated' in the sense that the vortices are only weakly affected by each other.

In an earlier stage, a vortex structure may undergo serious deformations due to a non-uniform ambient flow induced by neighbouring vortices that are too distant to start a direct interaction. As a first approximation, such a non-uniform flow may be considered as a linear shear. A vortex patch subjected to shear may exhibit different types of behaviour, as described analytically by Chaplygin (see [11]) and Kida [12]. Recent contour-dynamics simulations [13] have provided important insight in the (partial) destruction of such vortices, through a process called 'vortex stripping'. The present paper describes laboratory experiments on vortices that are subjected to such a non-uniform background flow. Two cases of deforming background flow will be considered: a pure strain flow and an irrotational annular shear flow. For both cases, the observed vortex behaviour is explained by using relatively simple point-vortex models.

2. MONOPOLAR VORTEX IN A STRAIN FLOW

The dynamics of monopolar vortices in a strain flow have been investigated by performing laboratory experiments in a stratified fluid. Some kinematic aspects of the observed vortex evolution, such as the characteristic formation of two filaments, appear to be well captured by a simple point vortex model.

2.1. Laboratory experiments

The laboratory experiments were carried out in a square tank, with inner dimensions 1×1 m and a depth of 0.3 m, which was filled with a linearly salt-stratified fluid. The background strain flow was established by four rotating horizontal disks, positioned at the corners of a square in the mid-plane of the tank (see Figure 1). The disks (diameter 10 cm, thickness 0.5 cm, at diagonal distance D) were rotated at a constant angular speed Ω_d, in the directions indicated in Figure 1. Quantitative flow measurements were carried out by tracking small polystyrene particles floating at the mid-plane. Their motions were recorded by a video camera that was mounted above the tank. Afterwards, the digitized video images were analysed by using the package *DigImage* developed by Dalziel [14]. The measurements revealed that after a while a perfect planar strain flow was established in the central part of the domain: the velocity components u and v in the Cartesian x, y-directions, taken along the symmetry axes, were found to be very close to $(u, v) = (ex, -ey)$, with e the strain rate. Once a steady strain flow was obtained, a monopolar vortex was created by immersing a small solid sphere (diameter 2.5 cm) and letting it rotate at constant speed Ω_s for a while in the centre of the strain field (this vortex generation technique was also used in [15]). After carefully removing the spinning sphere, a monopolar vortex flow results, the motion being confined to a thin pancake-shaped region. It was found that the vorticity distribution of the flow at the symmetry

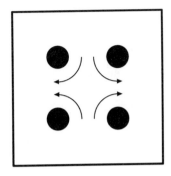 

Figure 1. Schematic drawing of the experimental arrangement used to generate a planar strain flow in a linearly stratified fluid.

plane is closely approximated by that of a so-called Gaussian vortex:

$$\omega_z^* = [1 - \tfrac{1}{2}(r^*)^2]\exp\{-\tfrac{1}{2}(r^*)^2\} , \tag{1}$$

with $\omega_z^* = \omega_z/\hat{\omega}_z$ the scaled vorticity and $r^* = r/\hat{r}$ the scaled radial coordinate. Here $\hat{\omega}_z$ represents the maximum vorticity, while $\hat{r}$ is the radial position of the maximum azimuthal velocity. Owing to appreciable diffusion in radial and vertical directions, the vortex expands while decaying. However, careful measurements (see [16]) have revealed that the vortex structure is well represented by (1) even during later stages of the decay. A characteristic feature of Gaussian vortices is the presence of a ring of weaker, oppositely-signed vorticity around the core (the total circulation is zero).

The typical evolution of such a vortex subjected to a background strain is illustrated by the sequence of vorticity contour plots shown in Figure 2. At $t = 75$ s (a) the core of the vortex is circular, but the negative vorticity initially present in a concentric ring is already concentrated into two separate patches, located on opposite sides of the vortex core. Note that these patches are oriented at an angle of approximately 45° with respect to the strain axes, as can be clearly seen in (b). Subsequently, they are rapidly advected by the strain flow (c), and a single-signed elliptical vortex remains (d). In the next stage (e)-(g), one observes a fast elongation and an overturning motion of the vortex towards the horizontal strain axis. Finally, after the vortex has completely been torn apart along the horizontal strain axis, only the basic strain flow remains, see (h). It should be stressed here that the decay of the vortex structure is a complicated process, in which both diffusion (in horizontal and vertical directions) and 'erosion' by the ambient flow play an important role. The effects of diffusion have been studied in absence of a background flow [15,16], and are now understood quite well. The presence of ambient strain or shear flow implies the removal of low-amplitude vorticity at the edge of the vortex, a process usually referred to as 'vortex stripping' (see e.g. [17]). Contour-dynamics simulations have demonstrated that the 'peeling off' of low-amplitude vorticity results in gradient intensification at the edge of the vortex, while the low-level vorticity is advected by the ambient strain flow in the form of two long filaments. In such inviscid simulations the gradually steepening vorticity gradients at the vortex edge increasingly inhibit the mechanism of stripping,

108

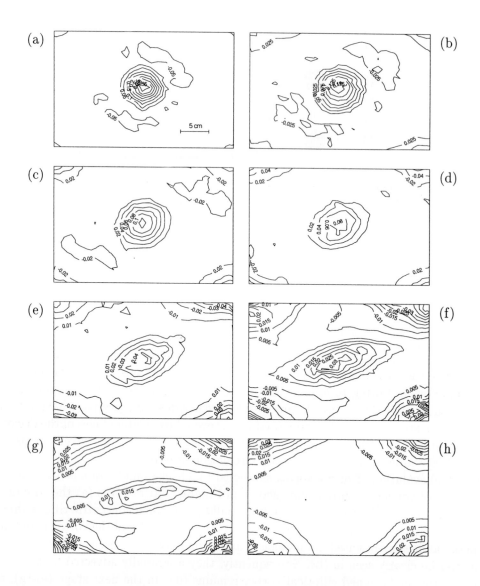

Figure 2. Measured vorticity distributions $\omega_z(x,y)$ during the evolution of a monopolar vortex in a strain flow with strength $e = 0.47 \times 10^{-2}$ s^{-1} : $t = 75$ s (a), 135 s (b), 195 s (c), 255 s (d), 375 s (e), 495 s (f), 615 s (g) and 855 s (h). Experimental parameters: $\Omega_d = 3.0$ rpm, $D = 40$ cm, $N = 2.8$ rad s^{-1}. The vortex was generated by a sphere spinning at $\Omega_s = 162$ rpm during $\delta t = 15$ s.

and an elliptical sharp-edged vortex survives [13]. In the case of viscous flows, however, diffusion effects are promoted by sharp gradients, so that the low-level vorticity that is removed by stripping (advection by ambient flow) is simultaneously replaced by horizontal diffusion. Obviously, the combination of diffusion and advective processes results in a very efficient decay of the vortex. Similar features were observed by Mariotti *et al.* [18] in their numerical study of monopolar vortices in a background shear flow.

In order to gain some insight in the advection characteristics of this particular flow configuration, experiments were performed in which the initial vortex was coloured with a passive dye (fluorescein). The typical evolution of the dye-visualized vortex is illustrated by the sequence of photographs shown in Figure 3. The photographs clearly show that the initially circular vortex (a) gradually becomes elliptic (b), while in the next stage (c,d) two filaments of dye are shed from the vortex. Note that at this stage the elliptical vortex remains oriented at an angle of approximately 45° with respect to the strain axes. The vortex becomes more and more elongated while the stripping continues (c-e), until eventually it is completely torn apart (f). Of course, one should be cautious when comparing the dye visualizations of the vortex evolution (Figure 3) with the measured vorticity distributions (Figure 2), since the diffusivities of these properties differ by one or two orders of magnitude. However, the dye visualizations provide important information about the (coloured) fluid that was initially inside the vortex. The two tails visible in Figure 3(c,d) are indicative of the vortex stripping process: they show that material is removed from the vortex edge and subsequently advected by the ambient flow. As mentioned before, such filaments have been found in contour-dynamics simulations, but in most practical situations the rapid viscous diffusion will quickly smear out the vorticity gradients, so that vorticity filaments will be hard to observe. In fact, only some weak traces of vorticity filaments are visible in Figure 2(f,g).

2.2. Point-vortex model

Figure 4 shows a detailed picture of the digitized flow due to the vortex in the strain field. One can clearly observe the presence of two stagnation points, the line through which is oriented at an angle of approximately 45° to the horizontal strain axis. In a recent study [19], the present authors investigated the kinematic aspects of this vortex configuration by the simplest possible model, *viz.* a point vortex of strength γ placed in the centre of the strain flow $u = ex, v = -ey$. The stream function Ψ associated with this flow field, expressed in polar coordinates r and θ, takes the following form:

$$\Psi = \tfrac{1}{2}er^2 \sin 2\theta - \frac{\gamma}{2\pi} \ln r + C , \tag{2}$$

with C an arbitrary constant. After a straightforward analysis, the positions (r_s, θ_s) of the stagnation points are found to be given by

$$r_s = \left(\frac{\gamma}{2\pi e}\right)^{1/2} , \quad \theta_s = \frac{\pi}{4} + k\pi , \quad k = \text{integer.} \tag{3}$$

As in the experiment, the stagnation points lie on a line oriented at 45° with respect to the strain axes.

The advection properties of this flow were investigated by following an arbitrary patch of passive tracers, initially placed somewhere in the flow. Because of conservation of mass

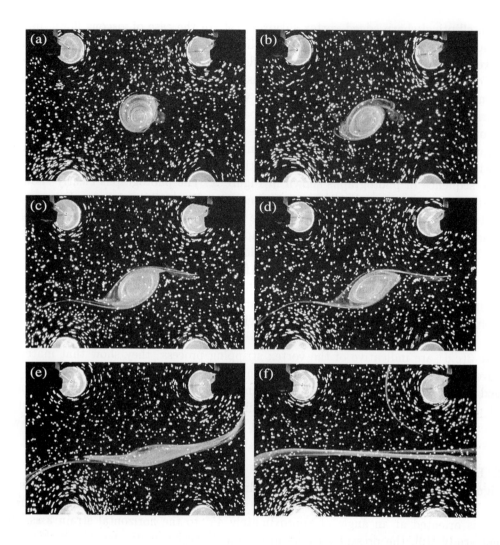

Figure 3. Evolution of a dye-visualized vortex in a strain flow with $e = 2.0 \times 10^{-2}$ s^{-1}. The photographs were taken at $t = 66$ s (a), 115 s (b), 160 s (c), 180 s (d), 280 s (e) and 430 s (f). Experimental parameters: $\Omega_d = 11.4$ rpm, $D = 40$ cm, $N = 3.2$ rad s^{-1}, $\Omega_s = 320$ rpm, $\delta t = 10$ s.

Figure 4. Close-up of the flow due to a vortex in a strain flow: (a) digitized particle paths (monitored during 40 s, starting at $t = 420$ s), (b) measured velocity vecotrs, (c) interpolated velocity field and (d) corresponding streamline pattern. Experimental parameters: $\Omega_d = 1.9$ rpm, $D = 50$ cm, $\Omega_s = 150$ rpm, $\delta t = 15$ s, $N = 2.8$ rad s^{-1}.

(*i.e.* area), it suffices to only trace the boundary of the patch. In most simulations, the initial contour was defined by 60 nodes. Subsequent deformation (stretching and folding) may cause considerable increase of the contour length, which requires additional nodes (up to a maximum of 5000) to be placed on the contour in regions of large stretching or folding. This method of contour kinematics is discussed in [20]. In one of the simulations the viscous decay of the vortex was modelled by an exponentially decreasing point-vortex strength

$$\gamma(t) = \gamma_0 \exp(-t/\tau) , \tag{4}$$

with γ_0 the initial strength and τ a characteristic e-folding time. A typical result of a contour kinematics simulation is presented in Figure 5. A circular contour of passive tracers is initially placed within the separatrix of the flow (the dotted lines indicate the streamlines intersecting in the stagnation points; the enclosed area is usually referred to as the separatrix area or the 'atmosphere'), see (a). If there would be no time dependence (constant vortex strength), the separatrix would act as a solid boundary, separating the fluid within it from the exterior fluid. In other words, fluid would never escape from the separatrix area. When time-dependence is introduced, however, the flow changes drastically. In the case of a decreasing strength $\gamma(t)$, the topological structure of the flow field does not change, but one observes a gradual shrinking of the separatrix area: the stagnation points gradually move towards the vortex, while the shape of the streamline pattern remains self-similar. Although in the first stages, when the passive contour is still inside the separatrix, only filaments are observed that are wrapped around the contained core area (b), at some stage the separatrix area has decreased to such an extent that the passive tracers start to leak away from the atmosphere (c,d). At this stage, the beginning of filamentation can be clearly seen. Eventually all the tracers of the contour lie outside the separatrix (e,f), and the contour takes on a shape similar to that observed in the laboratory experiment (Figure 3).

Close inspection of the contour reveals the occurrence of multiple filaments (see Figure 5d,e). Although these are hard to see in the dye visualizations of the flow (Figure 3), some evidence of multiple filamentation was found in a number of experiments.

It can be concluded that, despite its simplicity, the point-vortex model is a useful tool in gaining understanding of the basic advection properties of a (decaying) vortex in a strain flow.

3. MONOPOLAR VORTEX IN AN IRROTATIONAL ANNULAR SHEAR FLOW

In contrast to the experiment described in the previous section, the behaviour of a monopolar vortex in an irrotational annular shear flow has been investigated experimentally in a rotating, homogeneous fluid. As in the previous case, a point-vortex model will be applied to obtain a basic understanding of certain kinematic aspects of the vortex evolution.

3.1. Laboratory experiments

The laboratory experiments were carried out in a large circular tank (diameter 98 cm, working depth 30 cm) that was placed on a turntable that can rotate steadily at an angular speed Ω about its vertical axis (see Figure 6). The tank was filled with water (kinematic viscosity ν), with a fluid depth $H = 16$ cm, and the rotation speed of the table was kept at 0.70 rad s^{-1}. A paraboloidal bottom was used in order to avoid any topographic vorticity production (by stretching or squeezing of fluid columns) associated with the curved free surface.

The basic annular shear flow was generated by the following source-sink configuration: fluid is sucked away at flow rate Q through a tube positioned in the centre of the tank, while it is returned through a ring-shaped line source in the bottom corner. When the flux from the source to the sink is small enough, it is entirely carried by the Ekman layer at the tank bottom, which has a thickness of typically 1 mm. In the geostrophic flow region above the Ekman layer, the motion is purely azimuthal, its velocity distribution being given by

$$v_\theta(r) = \frac{\gamma_s}{2\pi r} \tag{5}$$

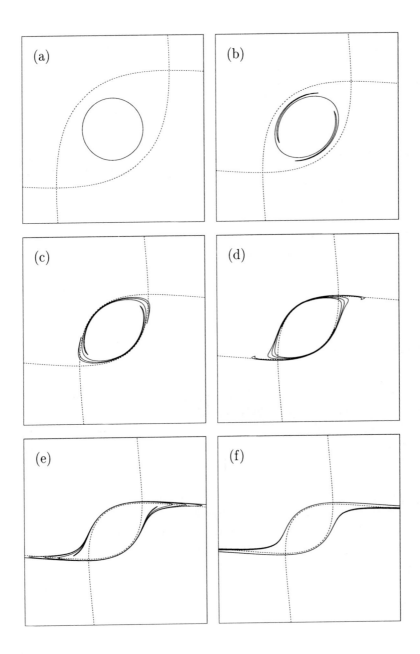

Figure 5. Calculated evolution of an initially circular contour of passive tracers, initially placed inside the separatrix (dashed line) associated with an exponentially decaying point vortex in a strain flow: $t^* = 0$ (a), 0.5 (b), 1.0 (c), 1.5 (d), 2.0 (e) and 3.0 (f), with $t^* = et$. Parameter values: initial contour radius $r_0 = 0.38(\gamma_0/2\pi e)^{1/2}$, decay timescale $\tau = 4/e$.

114

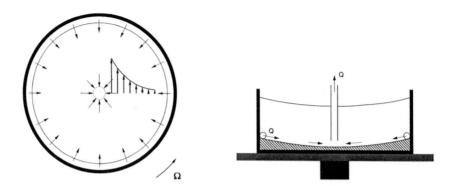

Figure 6. Schematic drawing of the experimental arrangement for the generation of an annular shear flow in a rotating fluid.

with

$$\gamma_s = 2Q \left(\frac{\Omega}{\nu} \right)^{1/2} \tag{6}$$

Obviously, the geostrophic flow driven by the source-sink arrangement corresponds with that of a potential vortex with strength γ_s. Hence, this basic annular flow is irrotational.

A cyclonic vortex could be created by locally withdrawing fluid (at rate q) through a perforated tube that was placed vertically in the fluid. As can be understood from conservation of angular momentum, the radial motion that is forced by this sink is deflected, and effectively results in a cyclonic swirling flow. Once the forcing was stopped and the tube was lifted out of the fluid, a well-defined vortex was established within a few rotation periods (see also [21]). The flow field of this vortex (either in presence or in absence of the background flow) could be determined accurately by tracking small tracer particles floating at the free surface. It was found that the vorticity distribution of these sink-induced vortices are closely represented by the so-called Lamb-vortex:

$$\omega(r) = \frac{\gamma_v}{\pi R^2} \exp(-r^2/R^2) , \tag{7}$$

with γ_v the total circulation and R a typical length scale. This scale is a measure of the vortex radius, and it is proportional to the radius $\hat{r}$ of the maximum azimuthal velocity. As a result of the Ekman layer action, the vortex gradually spins down. Previous experiments have demonstrated that this decay of the vortex flow is close to exponential.

Figure 7 shows the evolution of the sink-induced cyclonic vortex in the annular basic shear flow. Soon after the vortex forcing was stopped, the vortex flow was visualized by adding fluorescent dye. As is apparent from Figure 7, the vortex is advected in cyclonic (anti-clockwise) direction, while it is gradually deformed: the vortex acquires an oval shape, and two dye filaments are observed. Initially, see (a), one filament is expelled into the ambient fluid at the rear side of the vortex. Later, see (b,c), a second filament is formed at the front side. In the course of the experiment, the filaments become longer and longer, while the vortex core gradually shrinks (d-g). Note that the oval-shaped vortex

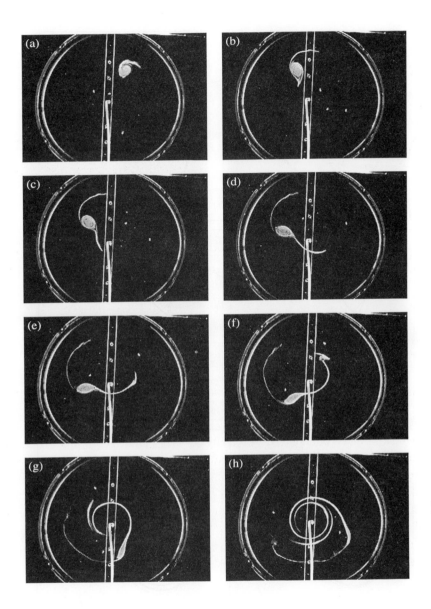

Figure 7. Sequence of photographs showing the dye-visualized evolution of a monopolar cyclonic vortex in a cyclonic annular shear flow at successive times after the vortex forcing was stopped: $t/T_E = 0.13$ (a), 0.24 (b), 0.29 (c), 0.34 (d), 0.39 (e), 0.45 (f), 0.55 (g) and 0.71 (h), with $T_E \equiv H/(\nu\Omega)^{1/2} = 190$ s the Ekman timescale. Experimental parameters: $Q = 10.5$ ml s^{-1}, $q = 40$ ml s^{-1} and forcing duration $\delta t = 10$ s.

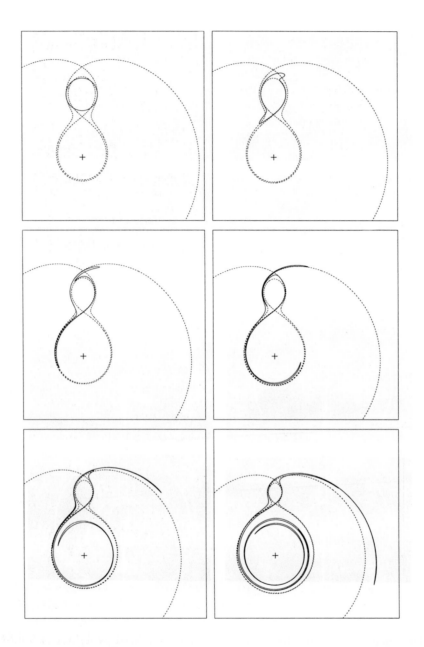

Figure 8. Calculated evolution of an initially circular contour of passive tracers, initially placed outside the outer separatrix of an exponentially decaying 'free' vortex in an annular shear flow induced by a point vortex fixed in the origin (indicated by a cross). The separatrices of the co-rotating flow pattern are shown by dashed lines.

core is oriented radially, *i.e.* perpendicular to the direction of the background shear flow. In the final stage (h), the vortex is completely torn apart in a long filament that is being wrapped around the sink in the tank centre.

3.2. Point-vortex model

In order to gain insight in the observed stripping process, the flow is modelled by a combination of two potential vortices: the background flow is modelled by a potential vortex of constant strength γ_s fixed in the origin, whereas the monopolar vortex is represented by a point vortex of decaying strength $\gamma_v(t)$ located at (x_0, y_0). The latter point vortex is passively advected by the flow due to the vortex in $(0,0)$.

In a co-rotating frame, the stream function of this flow configuration is given by:

$$\Psi(x,y) = -\frac{\gamma_s}{2}\ln\left[\frac{x^2+y^2}{x_0^2+y_0^2}\right] - \frac{\gamma_v}{2}\ln\left[\frac{(x-x_0)^2+(y-y_0)^2}{x_0^2+y_0^2}\right] + \frac{\gamma_s}{2}\frac{x^2+y^2}{x_0^2+y_0^2}. \tag{8}$$

Close inspection of the streamline pattern in the co-rotating frame reveals the existence of two hyperbolic points (stagnation points). Each of these points is associated with the self-intersection of a streamline: as can be seen in Figure 8, the inner separatrix has a figure-eight shape and encloses each point vortex separately, while the outer separatrix encloses both vortices. The stagnation points lie on a radial line through the origin.

When the 'free' point vortex in (x_0, y_0) is prescribed to decay exponentially, in accordance with the experimental observation, the structure of the streamline pattern is not so much affected, although the area of both separatrices decreases gradually. This has a major effect on the advection of individual tracers, however. The decrease of the separatrix area implies that fluid 'leaks' away, into the ambient area. This process is illustrated by the sequence of plots presented in Figure 8, which show the evolution of a contour of passive scalars. Initially the contour had a circular shape, and was positioned just outside the outer separatrix, enclosing the 'free' vortex. As time proceeds, two long tails are seen to develop, remarkably similar to the dye filaments visible in Figure 7. Also, the shape and the orientation of the inner region enclosed by the contour shows a remarkable resemblance with that observed in the dye-visualized vortex.

4. CONCLUDING REMARKS

The laboratory experiments discussed in the previous sections clearly illustrate the role of vortex stripping in the case that a two-dimensional vortex is subjected to strain. In particular the dye visualizations demonstrate the existence of long filaments, as was also predicted by (both viscous and inviscid) numerical simulations. It was shown that simple point-vortex models (with prescribed strengths variations in order to mimick viscous decay) capture the main kinematic characteristics of the observed vortex evolutions remarkably well. A more detailed description of the laboratory experiments and the modelling of vortices in strain and shear can be found in [22] and in a few future publications of the present authors.

Acknowledgements

The authors are grateful to Marcel Beckers and André Linssen for their contributions to the experimental work.

One of us (R.R.T.) gratefully acknowledges financial support from the Netherlands Foundation for Fundamental Research on Matter (FOM).

REFERENCES

1. J.C. McWilliams – The emergence of isolated coherent vortices in turbulent flows. *J. Fluid Mech.* **146**, 21–43 (1984).
2. B. Legras, P. Santangelo & R. Benzi – High-resolution numerical experiments for forced two-dimensional turbulence. *Europhys. Lett.* **3**, 811–818 (1988).
3. G.J.F. van Heijst & J.B. Flór – Dipole formation and collisions in a stratified fluid. *Nature* **340**, 212–215 (1989).
4. G.J.F. van Heijst & R.C. Kloosterziel – Tripolar vortices in a rotating fluid. *Nature* **338**, 569–571 (1989).
5. G.J.F. van Heijst, R.C. Kloosterziel & C.W.M. Williams – Laboratory experiments on the tripolar vortex in a rotating fluid. *J. Fluid Mech.* **225**, 301–331 (1991).
6. J.B. Flór & G.J.F. van Heijst – An experimental study of dipolar structures in a stratified fluid. *J. Fluid Mech.* **279**, 101–133 (1994).
7. G.F. Carnevale & R.C. Kloosterziel – Emergence and evolution of triangular vortices. *J. Fluid Mech.* **259**, 305–331 (1994).
8. J.C. McWilliams – The vortices of two-dimensional turbulence. *J. Fluid Mech.* **219**, 361–385 (1991).
9. M.V. Melander, N.J. Zabusky & J.C. McWilliams – Symmetric vortex merger in two dimensions: causes and conditions. *J. Fluid Mech.* **195**, 303–340 (1988).
10. M.V. Melander, N.J. Zabusky & J.C. McWilliams – Asymmetric vortex merger in two dimensions: Which vortex is "victorious"? *Phys. Fluids* **30**, 2610–2612 (1987).
11. V.V. Meleshko & G.J.F. van Heijst – On Chaplygin's investigations of two-dimensional vortex structures in an inviscid fluid. *J. Fluid Mech.* **272**, 157–182 (1994).
12. S. Kida – Motion of an elliptic vortex in a uniform shear flow. *J. Phys. Soc. Japan* **50**, 3517–3520 (1981).
13. D.G. Dritschel – Contour dynamics and contour surgery: numerical algorithms for extended, high-resolution modelling of vortex dynamics in two-dimensional, inviscid, incompressible flows. *Comput. Phys. Rep.* **10**, 77–146 (1989).
14. S. Dalziel – *DigImage. Image Processing for Fluid Dynamics.* Cambridge Environmental Research Consultants Ltd. (1992).
15. J.B. Flór & G.J.F. van Heijst – Stable and unstable monopolar vortices in a stratified fluid. *J. Fluid Mech.* **311**, 257–287 (1996).
16. R.R. Trieling & G.J.F. van Heijst – Decay of monopolar vortices in a stratified fluid. Submitted to *Phys. Fluids* (1996).
17. B. Legras & D.G. Dritschel – Vortex stripping and the generation of high vorticity gradients in two-dimensional flows. *Appl. Sci. Res.* **51**, 445–455 (1993).
18. A. Mariotti, B. Legras & D.G. Dritschel – Vortex stripping and the erosion of coherent structures in two-dimensional flows. *Phys. Fluids* **6**, 3954–3962 (1994).

19. R.R. Trieling & G.J.F. van Heijst – Kinematic properties of monopolar vortices in a strain flow. Submitted to *Fluid Dyn. Res.* (1996).

20. V.V. Meleshko & G.J.F. van Heijst – Interacting two-dimensional vortex structures: point vortices, contour kinematics and stirring properties. *Chaos, Solitons & Fractals* **4**, 977–1010 (1994).

21. E.J. Hopfinger & G.J.F. van Heijst – Vortices in rotating fluids. *Ann. Rev. Fluid Mech.* **25**, 241–289 (1993).

22. R.R. Trieling – Two-Dimensional Vortices in Strain and Shear Flows. PhD thesis, Eindhoven University of Technology, The Netherlands (1996).

Theoretical and Applied Mechanics 1996
T. Tatsumi, E. Watanabe and T. Kambe (Editors)
© 1997 Elsevier Science B.V. All rights reserved.

Development of wall turbulence structure in transitional flows

Masahito Asai[a] and Michio Nishioka[b]

[a]Department of Aerospace Engineering, Tokyo Metropolitan Institute of Technology, Hino, Tokyo 191, Japan

[b]Department of Aerospace Engineering, Osaka Prefecture University, Sakai, Osaka 593, Japan

This paper describes the development of wall turbulence structure in boundary-layer transition triggered by artificially-introduced high-intensity hairpin vortices at subcritical Reynolds numbers.   In this transition occurring beyond a momentum-thickness Reynolds number of 130, a key event for the wall turbulence generation is found to be successive evolution of wall shear layers with streamwise vortices into hairpin vortices, which is almost the same as the near-wall phenomena observed in the final stage of the ribbon-induced transition in plane Poiseuille flow.   In fact, recent detailed observations of the development of turbulent patch initially caused by orifice-generated hairpin vortices reconfirm the near-wall activities.

1. INTRODUCTION

Detailed experimental observations of the boundary-layer transition have much contributed to our understanding of a sequence of flow instabilities leading to the wall turbulence. We can now draw a transition scenario up to the first appearance of wall turbulence structure, in particular, for the case of the so-called ribbon-induced transition where a single Tollmien-Schlichting wave is introduced as the initial disturbance; see the reviews by Morkovin and Reshotko [1], Saric [2] and Kachanov [3]. The present experimental study is concerned with the development of transitional wall turbulence structure which can be observed at the stage just prior to the developed turbulent stage.   Needless to say, to clarify the key mechanism responsible for generating and sustaining wall turbulence is one of the most important subjects in transition research as well as in turbulence research.

In the ribbon-induced transition, the important phenomenon responsible for the wall turbulence generation occurs at the stage of the high-frequency secondary instability, i.e., the breakdown of internal high-shear layer into high-frequency hairpin vortices. Through a series of the experiments in plane Poiseuille flow which has almost the same stability and transition characteristics as Blasius boundary layer [4], Nishioka and Asai first identified the disturbance directly triggering the near-wall bursts (observed in the final stage of the transition) to be hairpin vortices coming from upstream [5, 6]. Namely, after the passage of hairpin vortices resulting from the high-frequency secondary instability[7], the near-wall fluid is lifted up away from the wall to form a wall shear layer, which soon evolves (or breaks down) into hairpin vortices (or wall hairpins), emphasizing the importance of hairpin vortices and their regeneration. The important role of hairpin vortices in generating and sustaining wall turbulence is also understood through many recent studies on the near-wall turbulence; see the review by Robinson [8].

The successive occurrence of hairpin vortices has been observed in the development of turbulent spots too; see the detailed flow visualization by Matsui [9]. Acarlar and Smith [10, 11] examined the response of boundary layer to hairpin vortices excited behind a hemisphere on the wall, and suggested that the hairpin vortices were responsible for the wall turbulence generation. The regeneration process of hairpin vortices was further studied by Smith et al. [12], and Haidari and Smith [13]. Singer and Joslin [14] and Singer [15] also showed through direct Navier-Stokes simulations that the similar regeneration process of hairpin vortices leads to the formation of turbulent spot.

To clarify the critical condition for wall turbulence generation, Asai and Nishioka examined the boundary layer transition caused by high-intensity hairpin vortices at subcritical Reynolds numbers experimentally. They found that when the near wall flow is disturbed by the energetic hairpin vortices (initially introduced at the leading edge), uplifted wall shear layers are generated and soon evolve into hairpin vortices [16–18]. The successive growth of hairpin vortices is found to occur only beyond the critical Reynolds number (based on the momentum thickness of laminar boundary layer) $R_\theta = 130$; about 65 % of the critical R_θ (= 200) for the linear instability (based on the parallel flow theory). The development of wall turbulence structure was further examined for the case of the so-called lateral turbulent contamination initially caused by hairpin vortices periodically excited through a small wall orifice. Importantly, the lateral growth of the turbulent patch at the subcritical Reynolds numbers is rather weak, but quite periodic in the growth of wall turbulence structure, compared with those observed at high Reynolds numbers [19, 20].

In the present paper, we review our important findings obtained in a series of experiments on the ribbon-induced transition and the subcritical boundary-layer transition triggered by high-intensity hairpin vortices. Furthermore, new results are presented for the disturbance development in the periodic lateral contamination process at subcritical Reynolds numbers.

2. FINAL STAGE OF RIBBON-INDUCED TRANSITION

First we would like to describe the flow structure observed at the final stage of the ribbon-induced transition in plane Poiseuille flow. The experiment was conducted by using a rectangular channel with large aspect ratio (about 27.4). The Reynolds number R_h based on the channel half-depth h (= 14.6 mm) and the center-plane velocity U_c is from 3500 to 6200: Note that the critical R_h for the linear instability is 5772. The detailed observation was made mainly for the flow at $R_h = 5000$, with $U_c = 9.8$ m/s.

The transition is governed by a sequence of flow instabilities; that is, the primary viscous (T-S) instability [4], the secondary instability (of fundamental resonance) leading to the three-dimensional wave growth [21, 22], and the subsequent high-frequency instability generating high-frequency hairpin vortices [5–7]. The product of the secondary instability is the so-called peak-valley structure characterized by the three-dimensional, Λ-shaped vorticity field, which soon develops high-shear layers away from the wall at spanwise peak positions. The development of high-shear layer is governed by the vorticity stretching and viscous diffusion effects, and its thickness is determined by the local balance between these effects. Such viscosity-conditioned thickness is much smaller than the channel half depth h for the transition Reynolds numbers, being one-tenth of h at $R_h = 5000$. The subsequent high-frequency instability is essentially of an inviscid inflectional nature which is free from viscous wall effects, and dominated by the shear-layer thickness and the velocity at the inflection point. The high-shear layer is highly three-dimensional so that the high-frequency disturbances amplified by the inflectional instability rapidly evolve into hairpin-shaped vortices. Figure 1 illustrates a typical waveform of u-fluctuation at the stage of the wave breakdown into hairpin vortices. We see four low-speed spikes (of 400 – 500 Hz) in each cycle of the primary wave ($f_{TS} = 72$ Hz); we call this stage the 4-spike stage. These spikes correspond to the passage of the hairpin vortices. Here, we notice that fluctuations of much higher frequency appear on each spike signal. The higher-frequency component superposed to each spike is filtered out by a high-pass filter (with a cut-off frequency of 1500 Hz), and its squared signal u_H^2 is also shown in Figure 1. Note that the higher-frequency component above 1500 Hz is in the viscous dissipation range at the present Reynolds number $R_h = 5000$. Interestingly, the smallest-scale fluctuations appear riding on the stretched legs and the heads of hairpin vortices. This fact enables us to identify the shape of the hairpin vortices by means of the periodic sampling of u_H^2 signal; the periodic sampling provides the envelop of the u_H^2 waveform, which is denoted by $\langle u_H^2 \rangle$. For instance, Figures 2(a) – 2(c) illustrate instantaneous equi-intensity contours of $\langle u_H^2 \rangle^{1/2}$ at the spanwise peak position ($z/h = 0$) and in the z-t cross-sections at $y/h = 0.70$ and 0.51 respectively. These figures demonstrate the development of three hairpin-shaped vortices clearly. The heads of hairpin vortices are located at heights of $1.1h$, $0.9h$ and $0.7h$ respectively and the spanwise distance between their legs is about $0.3h$ ($= 2.0 - 2.5$ mm).

At the beginning of the breakdown of the high-shear layer, the hairpin vortices develop away from the wall. However, the intensity of the hairpin vortices is so high, being more than 30 % of U_c in terms of u-fluctuation, that they can induce high-intensity fluctuations near the wall. In fact, as a result of being disturbed by the passage of hairpin vortices resulting from the high-frequency instability, the near-wall flow develops streaky structures, a pair of high-speed lumps of fluid with a low-speed lump in between, as shown in Figure 3 which illustrates the instantaneous equi-velocity contours at $y/h = 0.06$ at the 4-spike stage (the time t/T corresponds to that in Figure 2). The spanwise distance between the streaks is about $0.6h$ ($= 4.5$ mm), twice the distance between the legs of the above-passing hairpin vortices. As the streaks develop, a low-speed lump lifts up from the low-speed streak to form a wall shear layer, which soon evolves into a hairpin vortex. In terms of instantaneous equi-shear contours

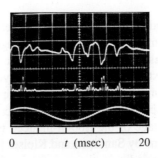

$$0 \qquad t \text{ (msec)} \qquad 20$$

Figure 1. Waveform of u-fluctuation at the 4-spike stage ($R_h = 5000$, $2\pi f_{TS} h/U_c = 0.334$). From upper trace; total u, high-pass filtered u_H^2 and ribbon current (72 Hz).

124

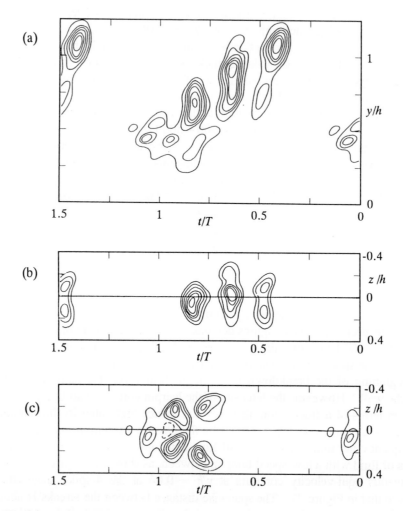

Figure 2. Equi-intensity contours of high-frequency component (> 1500 Hz) at the 4-spike stage (R_h = 5000); $\langle u_H^2 \rangle^{1/2}/U_\infty$ = const. lines. (a) z/h = 0, (b) y/h = 0.70, (c) y/h = 0.51. Contour levels range from 0.4% to 1.0 % in (a), and from 0.4 % to 0.9 % in (b) and (c).

at a spanwise peak (z/h = 0) and equi-velocity contours near the wall (y/h = 0.06), Figure 4 illustrates such near-wall phenomena at the multi-spike stage where the high-shear layer has broken down into hairpin vortices almost completely. The head of hairpin vortices moves at higher speed than the near-wall structures (legs of hairpin vortices and near-wall shear layer), so the breakdown process is strongly affected by the overlapping hairpin vortices generated in the next cycle. The near-wall activities observed may be called the wall burst. The similar near-wall activities in the final stage of the ribbon-induced transition are also observed by the recent direct numerical simulations by Sandham and Kleiser [23] and Rist and Fasel [24].

The flow remains almost periodic in time as far as the primary and secondary high-frequency fluctuations are concerned. Nevertheless, the time-mean velocity over the streaky region exhibits the wall-law characterized by the log-law profile for the developed wall turbulence. The spanwise scale of the wall streaks is determined to be about 0.6h or about 120

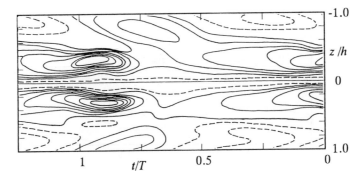

Figure 3. $(U+u)/U_c$ = const. lines at y/h = 0.06 at the 4-spike stage (R_h = 5000). Contour levels range from 0.1 to 0.5.

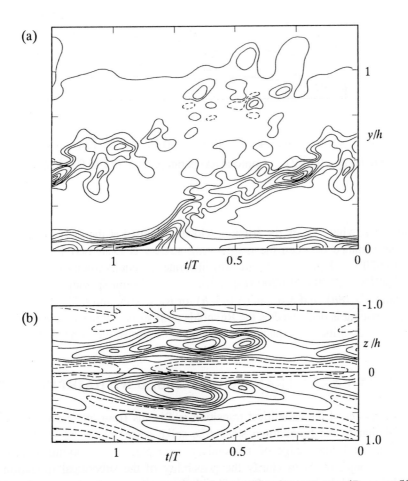

Figure 4. Instantaneous flow structure at the multi-spike stage (R_h = 5000). (a) $(h/U_c)\partial(U+u)/\partial y$ = const. lines at spanwise peak (z/h = 0), (b) $(U+u)/U_c$ = const. lines at y/h = 0.06. Contour levels range from 0.0 to 5.0 in (a), and from 0.1 to 0.65 in (b).

126

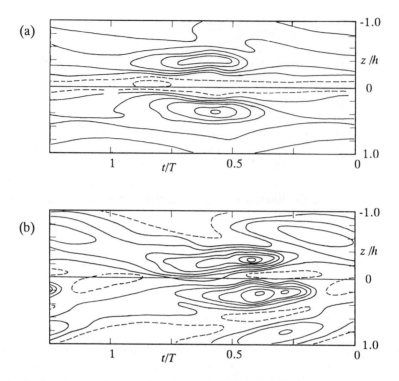

Figure 5. $(U+u)/U_c$ = const. lines at $y/h = 0.06$. (a) $R_h = 3500$ ($2\pi f_{TS}h/U_c = 0.331$) and (b) $R_h = 6200$ ($2\pi f_{TS}h/U_c = 0.322$). Contour levels range from 0.1 to 0.4 in (a), and from 0.1 to 0.5 in (b).

in wall units v/u_τ, where u_τ is the friction velocity obtained at the developed turbulent stage. The spanwise distance between the neighboring streaks is examined for Reynolds numbers from 3500 to 6200. Figures 5(a) and 5(b) illustrate the equi-velocity contours near the wall ($y/h = 0.06$) at $R_h = 3500$ and 6200 respectively. The lateral spacing between the streaks is about 5.7 mm (= 0.78h) and 3.8 mm (= 0.52h) for $R_h = 3500$ and 6200 respectively. Thus, the spanwise spacing is decreased as the Reynolds number is increased, and vice versa. But, in terms of the wall units, both spacings are found to be almost the same as that for $R_h = 5000$.

3. BOUNDARY-LAYER TRANSITION CAUSED BY HAIRPIN VORTICES

In the experiment of the subcritical transition in Blasius flow, we have tried to realize the flow stage similar to the breakdown stage (the spike stage) observed in the ribbon-induced transition, near the leading edge of a boundary layer plate, by applying a periodic acoustic forcing. The objective is to clarify the possibility of the subcritical transition, the related critical condition, and the key event for wall turbulence generation, through observing the first appearance of wall turbulence structure.

The experiment is conducted in a wind tunnel of open jet type, 200 mm × 200 mm in cross section. A flat plate set in the test section is 600 mm long, 3 mm thick and 195 mm in span, and has a sharp leading edge. The freestream velocity U_∞ is fixed at 4 m/s. To excite vortices at the leading edge acoustically, a loudspeaker of 30 cm woofer is used: The forcing frequency is 50 Hz. The acoustic forcing is applied almost perpendicularly to the boundary-layer plate, and therefore can cause local unsteady flow separation at the sharp leading edge in each forcing cycle. The separation bubble is accompanied by an intense thin shear layer away from the wall. The thin bubble shear layer is quite unstable to high-frequency disturbances and soon breaks down into small-scale discrete vortices. Thin plastic tapes of 0.05 mm thick in 3 mm wide are glued around the leading edge every 6 mm in the spanwise direction, so that the shear layer rolls up into hairpin-shaped vortices. Even the seemingly negligible spanwise periodicity (due to the thin tape) well controls the exact lateral positions for the hairpin generation. This indicates that under the combined effects of the background normal-to-wall shear and the Biot-Savart law the near-wall concentration of the spanwise vorticity may easily be disturbed to develop into hairpin-shaped vortices. By adjusting the intensity of the acoustic forcing, we can observe flow stages, which, as far as the waveforms of u-fluctuation are concerned, are similar to the spike (breakdown) stage observed in the ribbon-induced transition.

The instantaneous u-fluctuation induced by the leading-edge-generated hairpin vortices (which we call the leading-edge hairpins in short) is more than 30 % of U_∞ near the leading edge as shown in Figure 6 which illustrates the waveforms of u-fluctuation at $x = 10$ mm. However, the leading-edge-hairpins evolving from the thin bubble shear layer are so small in scale (their frequency is about 400 Hz) that they soon decay downstream due to viscous dissipation and diffusion. Figure 7 shows a time sequence of the disturbance development visualized by means of smoke-wire with a high-speed camera. Each photo is taken at an interval of 1/200 sec, one fourth of the period of the acoustic forcing (at 50 Hz). Here, the smoke is released from the two y-positions, i.e., at the boundary layer edge (y = 3.0 mm) and near the wall (y = 0.5 mm) to give information on the decay of the leading-edge hairpins as well as the associated near-wall activity. Up to $x = 100$ mm, the leading-edge hairpins are very active and can lift up the near-wall smoke released from $y = 0.5$ mm; note that the height of the momentum thickness is about 0.5 mm at $x = 100$ mm. Beyond $x = 100$ mm, the leading-edge hairpins get weak. The rolled-up smoke associated with the leading-edge hairpins only convects downstream without disturbing the near-wall flow. It should be noted that when the smoke is released from $x = 90$ mm, no appreciable near-wall activity is observed. With increasing the forcing, however, the leading-edge hairpins become more energetic and survive

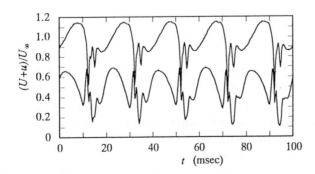

Figure 6. Waveforms of u-fluctuation at $x = 10$ mm. Upper; $y = 1.6$ mm, lower; $y = 0.35$ mm.

128

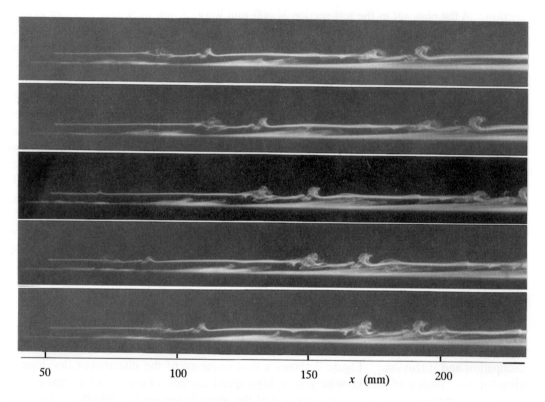

Figure 7. Smoke-wire visualization of development of the leading-edge-generated hairpin vortices. Time interval between frames is 1/200 sec.

further downstream.   In that case, they can trigger the subcritical transition.   Figures 8(a) and 8(b) visualize the subcritical transition caused by such energetic leading-edge hairpin vortices.   Smoke is released from $x = 90$ mm (lower smoke) and 100 mm (upper smoke). Thus the leading-edge hairpins are still active as seen from the rapid roll-up of the upper smoke (streak) line, and importantly, the lifted-up smoke released from the lower y-position shows that being associated with the passage of the leading-edge hairpins, the shear layer near the wall starts to develop into discrete vortices around $x = 150$ mm.   Indeed, the cross-section view of the near-wall flow given in Figure 8(b), taken at an instance a half the period after the passage of the heads of the leading-edge hairpins, indicates the appearance of many mushroom-shape smoke patterns, which no doubt shows the development of three-dimensional wall shear layers with streamwise vortices. They evolve into hairpin vortices downstream, which we call "wall hairpins".   Here, the heads of the leading-edge hairpins convect at a speed of about $0.9U_\infty$, while the near wall vortices at $0.5U_\infty - 0.6U_\infty$.   Therefore, the newly generated vortices (wall hairpins) are soon caught up with by the leading-edge hairpins (of the next cycle) coming from upstream.   Figure 9 shows the disturbance growth by means of top view pictures of smoke-wire visualization.   Due to the successive generation of hairpin vortices, we see no laminar region over the whole spanwise extent beyond $x = 200$ mm.

In order to obtain the critical Reynolds number for the subcritical disturbance growth, we measured the coefficient of local skin friction (and the streamwise variation of the momentum thickness), and found that the friction coefficient starts to deviate from the laminar value at and around $x = 140$ mm where the momentum-thickness Reynolds number R_θ is about 130 for Blasius flow. The increase in the friction coefficient is caused by the momentum transfer due to wall hairpins. The wall shear layers with streamwise vortices (which appear after the passage of the leading-edge-generated hairpin vortices) become active and themselves evolve into wall hairpin in succession beyond $R_\theta = 130$. Here, it is worth noting that the Reynolds number R_θ at the critical station is almost the same as the minimum transition Reynolds number for plane Poiseuille flow [25].

The successive growth of wall hairpins is mainly due to the lift-up of three-dimensional wall shear layer by the streamwise vortices (probably induced by hairpin vortices coming from upstream), though the development of wall shear layer with streamwise vortices is under strong influence of viscosity at the subcritical Reynolds numbers. The successive growth of wall shear layer into wall hairpins starts to occur before being caught up with by the leading-edge hairpins (of the next cycle) coming from upstream, indicating the break-up of the disturbed wall shear layer by itself. Indeed, if the streamwise vortices are active, the wall shear layer is lifted up away from the wall and stretched in the lateral (spanwise) direction. The resulting intense wall shear layer is free from the wall effect and may become unstable due to the inflectional instability. For the regeneration of wall hairpins, however, the wall shear layer (with streamwise vortices) has to break down before decaying due to the effects of viscous diffusion and dissipation. In the present case, the time scale required for the generation of wall hairpins is found to be of the same order as the viscous diffusion time θ^2/ν around the critical station when we take the momentum thickness θ as the measure of wall-shear-layer thickness. And, the intensity of the near wall activity at the critical station is almost the same as that of fully turbulent cases in terms of u- and v-fluctuations.

Another important finding is as to the spanwise scale of near-wall vortices. Beyond the critical station, the spanwise scale of the active wall shear layers (i.e., spanwise distance between the neighboring mushrooms in the cross-section view of visualization) in wall units is increased downstream, eventually approaches the well-known value, $\lambda u_\tau/\nu = 100$ in terms of the mean spanwise distance (λ). In the following, the regeneration process of hairpin and/or

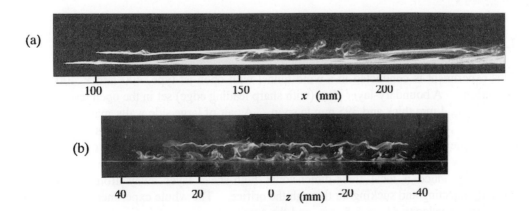

Figure 8. Smoke-wire visualization of the subcritical disturbance growth. (a) side view, (b) cross-section view (at $x = 150$ mm).

130

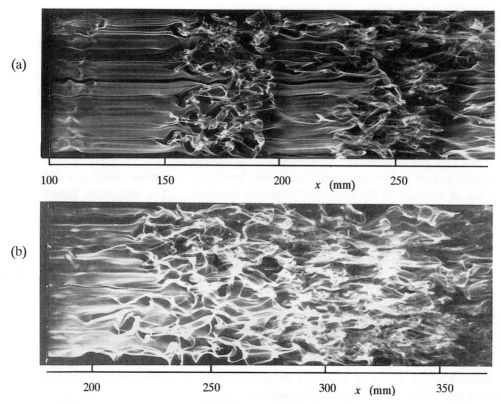

Figure 9. Top view photographs of smoke-wire visualization. (a) smoke from (x, y) = (100 mm, 3 mm), (b) smoke from (x, y) = (180 mm, 3.5 mm).

streamwise vortices is further examined for the case of the so-called lateral turbulent contamination initially caused by hairpin vortices periodically excited by using a wall orifice-loudspeaker system.

4. LATERAL CONTAMINATION CAUSED BY HAIRPIN VORTICES

This experiment is conducted in a wind tunnel with open jet type, 400 mm × 400 mm in cross section. A boundary-layer plate (with sharp leading edge) set in the open test section is 1000 mm long, 4 mm thick and 395 mm in span. A pair of large side walls maintain the two-dimensionality of the main stream. The orifice of 2.4 mm in diameter is drilled at the spanwise center position, 80 mm downstream of the leading edge, and is connected to a loudspeaker by a vinyl-hose. When the loudspeaker is driven at a single frequency, the forcing system can produce a local three-dimensional high-shear layer away from the wall by periodically injecting and sucking air through the orifice. The whole experiment is carried out at a free-stream velocity U_∞ = 5.5 m/s, and the forcing frequency f is fixed at 55 Hz. The free-stream turbulence is less than 0.1 % of U_∞ at the tunnel exit. Owing to the thin flat plate with sharp leading edge, the Blasius flow profile can develop close to the leading edge. The x-Reynolds number at the orifice station $(x_d = 80$ mm$)$ R_{xd} $(= x_d U_\infty / v$, where v is the kinematic

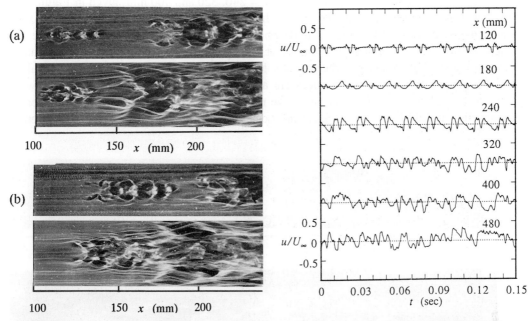

Figure 10. Visualization of orifice-generated hairpin vortices. (a) $t/T = 0$, (b) $t/T = 0.5$. smoke from $y = 2$ mm (upper) and 1 mm (lower).

Figure 11. Development of u-fluctuation; waveforms at $y = 1$ mm.

viscosity) is 3×10^4, where the momentum- and displacement-thickness Reynolds numbers, R_θ and R^* (for the Blasius flow without the forcing) are 113 and 295 respectively. At $x = 240$ mm and 560 mm, R^* is about 520 and 790 respectively: The critical R^* given by the Orr-Sommerfeld stability equation is 520. Without the forcing, the boundary layer remains laminar over the whole observation region ($x \leq 560$ mm).

Figure 10 shows the smoke-wire visualization of the orifice-generated hairpin vortices. A pair of top-view pictures (where smoke is released from $y = 1$ mm and 2 mm) are taken with the stroboscopic light synchronized with the forcing signal so as to obtain the flow structure at the same phase of the forcing. These pictures show that three or four hairpin vortices are excited immediately downstream of the orifice. As the primary orifice-generated hairpin vortices convect downstream, turbulent patch develops along a narrow strip extending from the orifice. Figure 11 shows the disturbance development at the spanwise center ($z = 0$) in terms of the waveforms of u-fluctuation at $y = 1$ mm. Beyond $x = 320$ mm, the waveforms are not unlike that of turbulent fluctuation. In fact, the y-distribution of mean velocity U (at $z = 0$) is found to approach the log-law profile at $x = 400$ mm ($R_x = 1.45 \times 10^5$) though the log-law region is rather limited. Here, the Reynolds number based on the momentum thickness in this transitional flow is about 275 and 410 at $x = 320$ and 480 mm ($R_x = 1.2 \times 10^5$ and 1.8×10^5) respectively.

The lateral growth of the turbulent (or active disturbance) region is also visualized by the smoke-wire technique. Figure 12 shows the side- and top-view pictures. The lateral disturbance growth occurs gradually through generating secondary vortices on both sides of the turbulent patch. The important feature of lateral contamination can be understood more

132

clearly from the cross-section view of smoke at x = 240, 320 and 400 mm, given in Figure 13. In each cross-section view, smoke is released from 60 mm upstream of the observation x to visualize the active vortical structures. In the photographs at x = 240 mm and 320 mm, we see the appearance of mushroom-shaped smoke at the spanwise edges of the turbulent patch, indicating a pair of streamwise vortices and/or hairpin-shaped vortices. Inside the turbulent patch, the generated (streamwise) vortices are aligned quite regularly with a mean spanwise spacing of about 5.9 mm at x = 400 mm; the mean spacing λ is obtained from a number of photographs. In this case too, in terms of the local wall unit v/u_τ (where u_τ denotes the friction

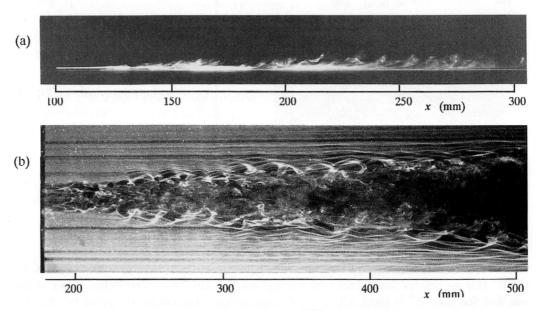

Figure 12. Smoke-wire visualization of the development of turbulent patch. (a) smoke from (x, y) = (100 mm, 1.5 mm), (b) smoke from (x, y) = (180 mm, 2 mm).

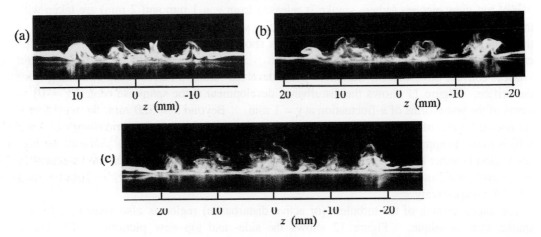

Figure 13. Cross-section view of smoke-wire visualization. (a) x = 240 mm, (b) x = 320 mm, (c) x = 400 mm.

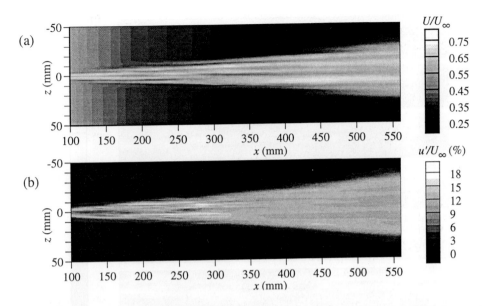

Figure 14. Lateral spreading of the turbulent patch. (a) mean velocity (U/U_∞) at $y = 1$ mm, (b) turbulent intensity (u'/U_∞) at $y = 1$ mm .

velocity at the spanwise center location), the spanwise spacing of the streamwise vortices ($\lambda u_\tau/\nu$) is about 110 at $x = 400$ mm.

In order to visualize the lateral spreading quantitatively, the spanwise distributions of the mean velocity and the r.m.s. value of u-fluctuation near the wall were measured at various x-stations from $x = 100$ to 560 mm. Figure 14(a) shows the equi-velocity contours at $y = 1$ mm. Inside the turbulent patch, we can clearly see the spanwisely alternating high- and low-speed regions extending in the streamwise direction, which are associated with the generation of the streamwise vortices.   The lateral spreading is also illustrated in Figure 14(b) in terms of the turbulent intensity, i.e., the r.m.s. value of u-fluctuation (at $y = 1$ mm). From the turbulent intensity map, the half angle of lateral spreading (defined by the contours with r.m.s. value larger than 5 % of U_∞ for instance) is estimated to be only about 2° for the subcritical Reynolds numbers $R_x < 0.9 \times 10^5$ ($R_\theta < 200$), which is much smaller than that of the turbulent spot or wedge observed at much higher Reynolds numbers, about 10°; see Schubauer and Klebanoff [26], Wygnanski et al. [27], Gad-el-Hak et al. [28]. The spreading angle increases as R_x increases downstream, for instance, about 5° around $R_x = 1.8 \times 10^5$ ($x = 480$ mm).

In the early stage of the disturbance development (at least up to $x = 240$ mm), the flow is quite periodic in time so that the ensemble-averaged flow structure can be obtained with good accuracy by the periodic sampling technique.   Figure 15 illustrates a time sequence of the disturbance development in terms of the ensemble-averaged streamwise velocity field ($U + u$)/U_∞ in the x-z plane at $y = 1$ mm. We see that immediately downstream of the orifice a distinct low speed streak develops along a streamwise strip at $z = 4$ mm, and grows longer as it convects downstream. The low-speed streak is no doubt due to streamwise vortices observed in the visualization picture (Figure 13). Subsequently, the breakdown occurs in the low-speed

134

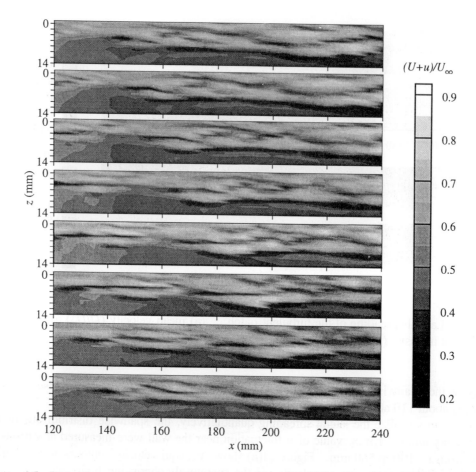

Figure 15. Instantaneous velocity field at $y = 1$ mm.   From top; $t/T = 0$, 1/8, 2/8, 3/8, 4/8, 5/8, 6/8, 7/8.

streak.   Figure 16 follows up the breakdown process in terms of the waveforms of the ensemble-averaged streamwise velocity at $y = 1$ mm and 2.3 mm along the $z = 4$ mm strip. On the streak at $z = 4$ mm, high-frequency spikes (not unlike the spikes observed in the ribbon-induced transition) appear in succession during the low-speed phase of the forcing cycle, indicating the breakdown of the lifted-up shear layer into small-scale (hairpin) vortices. During the process, a new low-speed streak together with neighboring high-speed lump of fluid is generated in the laminar region outside the primary streak (at $z = 4$ mm), along the $z = 8$ mm strip (downstream from $x = 130$ mm). In addition, after the primary streak sufficiently grows, another low-speed streak starts to develop along the $z = 5 - 6$ mm strip behind (upstream of) the primary streak. These newly-generated near-wall streaks observed along the $z = 5 - 6$ mm and 8 mm strips also grow longer upto 30 - 50 mm in streamwise length. Here, without the forcing, the boundary-layer thickness δ (= $3\delta^*$) of Blasius flow is about 3 mm and 4.3 mm at $x = 120$ mm and 240 mm respectively.   Therefore, the streamwise scale of the streaks observed is the order of 10δ or more.   On these low-speed streaks too, the lifted-up wall shear layer is formed and eventually breaks down into wall hairpins.   Figures 17 (a) and

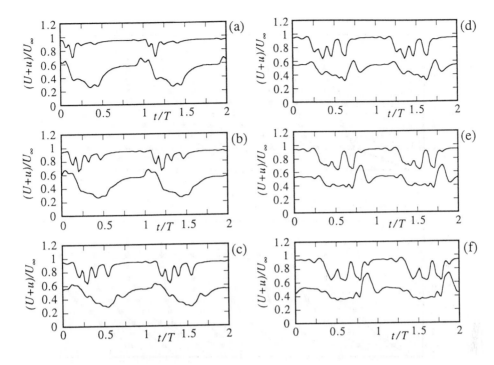

Figure 16. Development of u-fluctuation along $z = 4$ mm streak.   Waveforms at $y = 2.3$ mm (upper) and $y = 1$ mm (lower). From (a) to (f); $x = 120, 125, 130, 135, 140, 145$ mm.

17(b) show the formation and breakdown of the near-wall shear layer on the low speed streaks in terms of the instantaneous velocity distributions at $(x, z) = (180$ mm, 6 mm$)$ and $(180$ mm, 8 mm$)$ respectively.   On the both streaks, we can see inflectional velocity profiles over nearly a half of the forcing period.   In particular, the inflection point appears in the velocity profile around $t/T = 0.4$ close to the wall at $z = 8$ mm, indicating the formation of the near-wall shear layer.   On the $z = 5 - 6$ mm streak, the lifted-up wall shear layer has already evolved into hairpin vortices at this x-station.

Successive generation of hairpin vortices observed within the turbulent patch is partly attributed to the inflectional instability of the lifted-up wall shear layer which has been developed by the primary hairpin vortices. Another possible mechanism for generating hairpin (and/or streamwise) vortices may be a vortex-surface interaction which can excite secondary vortices immediately behind the legs of the primary hairpin vortex through the vortex-induced unsteady separation; see Peridier et al. [29, 30].   Smith et al. [12] and Haidari and Smith [13] examined the regeneration process of hairpin vortices by exciting a single hairpin vortex, and revealed that secondary hairpin vortices are generated between the legs of primary hairpin vortices as well as directly behind each leg through the process of vortex-surface interaction. They proposed that the adverse pressure gradient imposed by the primary hairpin vortices causes the eruption (lift-up) of low speed fluid to lead to the formation of secondary hairpin vortices, as first observed in our ribbon-induced transition [5, 6]. As for the vortex interaction with the wall, also see the review by Doligalski et al. [31].   In the present case, a new streak starts to develop from the rear portion of the primary low-speed streak, along the $z = 5 - 6$ mm

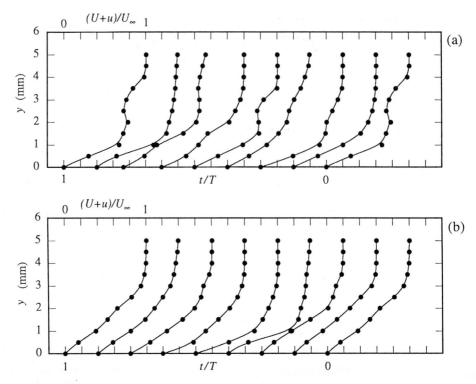

Figure 17. Instantaneous velocity distributions at $(x, z) = (180$ mm, 6 mm) in (a) and (180 mm, 8 mm) in (b) .

strip at $t/T = 5/8 - 7/8$ (see Figure 15), which is probably due to the secondary hairpin vortices generated immediately behind the leg of the primary hairpins. The generation mechanism for a pair of streamwise vortices observed at the spanwise edge of the turbulent patch, on the other hand, seems to be different.   Once a (main) spanwise vortex line deforms to hairpin-like shape, the bending propagates laterally due to combined effects of the Biot-Savart law and the velocity gradient (normal to the wall), as noted earlier and as illustrated schematically by Perry et al. [32].   The meandering (in the x-z plane) of streaks is of another importance for the evolution of the wall shear layer from the low-speed streak.   In fact, we see the bending (in the x-z plane) of the streaks, as well as their amalgamation in Figure 15.   These behaviors result from the interactions between neighboring streamwise and/or hairpin vortices.   The high-speed regions behind the low-speed streak observed along the $z = 4$ mm and the $z = 8$ mm strips are caused by the high-speed fluid associated with the neighboring hairpin (and/or streamwise) vortices developing along the $z = 5 - 6$ mm strip.   In fact, on the $z = 4$ mm streak at $t/T = 0.75 - 1.0$ in Figure 16, we can see a bursting-like waveform characterized by the occurrence of strong acceleration from negative to large positive fluctuation, the signature for the formation of intense near-wall shear layer $(\partial u/\partial y)$.   The wall shear layer formed on the $z = 8$ mm streak (around $t/T = 0.4 - 0.5$ in Figure 17b) is also intensified by the induced velocity due to the neighboring hairpin vortices (above the $z = 5 - 6$ mm streak).   It is added that recent numerical simulations of developed wall turbulence also emphasize the importance of bending and meandering of near-wall streaks in generating intense near-wall shear layers [33].

5. CONCLUDING REMARKS

We have examined the transition to wall turbulence in plane Poiseuille flow and Blasius flow. For the case of ribbon-induced transition in Poiseuille flow, the final stage is found to be governed by the wall activities, which are caused by the hairpin vortices (the product of high-frequency instability) and involve the appearance of high- and low-speed streaks (streamwise vortices), the growth of wall shear-layer and its evolution into wall hairpins (regeneration of hairpin vortices). The role played by the primary (T-S) and secondary instabilities is to set the stage for the high-frequency instability generating the primary hairpin vortices, which excite the near-wall flow to generate such wall turbulence structures as noted above. That the final, and thus real transition starts with the onset of the wall activities triggered by the primary hairpins is reconfirmed in Blasius flow through observing the transition process caused by leading-edge-generated and orifice-generated hairpin vortices. In these cases, the critical value of momentum-thickness Reynolds number for the wall activities to occur is found to be about 130. Below the critical, even if highly disturbed and once the wall streaks and/or streamwise vortices are created, they eventually decay downstream, so that the near-wall flow can not regenerate wall hairpins. These facts suggest that some flow instabilities are surely involved in the regeneration process of the wall turbulence structures. In this connection, it should be added that even beyond the critical $R_\theta = 130$ for the wall hairpins to be regenerated, the normal-to wall fluctuation v (associated with the near-wall activities) must satisfy the condition $v\theta/\nu > 8$, though this is our tentative conclusion [18].

ACKNOWLEDGMENT

The authors wish to express their sincere gratitude to Professors. S. Iida and H. Sato for their continual encouragement. This work was in part supported by the Grant-in-Aid for Scientific Research (No. 08651090 and No. 08455465), from the Ministry of Education, Science and Culture, Japan, as well as the Special Research Fund from the Tokyo Metropolitan Government.

REFERENCES

1. M.V. Morkovin and E. Reshotko, in Laminar-Turbulent Transition (D. Arnal and R. Michel), Springer (1990) 3.
2. W. Saric, Special course on Progress in Transition Modeling, AGARD Report 793 (1993).
3. Yu.S. Kachanov, Annu. Rev. Fluid Mech., 26 (1994) 411.
4. M. Nishioka, S. Iida and Y. Ichikawa, J. Fluid Mech., 72 (1975) 731.
5. M. Nishioka, M. Asai, and S. Iida : in Transition and Turbulence (ed. R.E. Meyer), Academic (1981) 113.
6. M. Nishioka and M. Asai, in Turbulence and Chaotic Phenomena in Fluids (ed. T. Tatsumi), North-Holland (1984) 87.
7. M. Nishioka, M. Asai, and S. Iida, in Laminar-Turbulent Transition (eds. R. Eppler and H. Fasel), Springer (1980) 37.
8. S.K. Robinson, Annu. Rev. Fluid Mech., 23 (1991) 601.

138

9. T. Matsui, in Laminar-Turbulent Transition (eds. R. Eppler and H. Fasel), Springer, (1980) 288.

10. M.S. Acarlar and C.R. Smith, J. Fluid Mech., 175 (1987) 1.

11. M.S. Acarlar and C.R. Smith, J. Fluid Mech., 175 (1987) 43.

12. C.R. Smith, J.D.A. Walker, A.H. Haidari and U. Sobrun, Phil. Trans. R. Soc. Lond. A, 336 (1991) 131.

13. A.H. Haidari and C.R. Smith, J. Fluid Mech., 277 (1994) 135.

14. B.A. Singer and R.D. Joslin, Phys. Fluids, 6 (1994) 3724.

15. B.A. Singer, Phys. Fluid, 8 (1996) 509.

16. M. Asai and M. Nishioka, in Laminar-Turbulent Transition (eds. D.Arnal and R.Michel), Springer (1990) 215.

17. M. Nishioka, M. Asai and S. Furumoto, Acta Mechanica (suppl.), 4 (1994) 87.

18. M. Asai and M. Nishioka, J. Fluid Mech., 297 (1995) 101.

19. M. Asai and M. Nishioka, in Laminar-Turbulent Transition (ed. R. Kobayashi), Springer (1995) 111.

20. M. Asai, K. Sawada and M. Nishioka, Fluid Dyn. Res., 18 (1996) 151.

21. M. Nishioka and M. Asai, in Laminar-Turbulent Transition (ed. V.V. Kozlov), Springer (1985) 173.

22. M. Asai and M. Nishioka, J. Fluid Mech., 208 (1989) 1.

23. N.D. Sandham and L. Kleiser, J. Fluid Mech., 245 (1992) 319.

24. U. Rist and H. Fasel, J. Fluid Mech., 298 (1995) 211.

25. M. Nishioka and M. Asai, J. Fluid Mech., 150 (1985) 441.

26. G.B. Schubauer and P.S. Klebanoff, NACA Rep. 1289 (1956).

27. M. Gad-el-Hak, R.F. Blackwelder and J.J. Riley, J. Fluid Mech. 110, (1981) 73.

28. I. Wygnanski, M. Sokolov, and D. Friedman, J. Fluid Mech., 78 (1976) 785.

29. V.J. Peridier, F.T. Smith and J.D.A. Walker, J. Fluid Mech., 232 (1991) 99.

30. V.J. Peridier, F.T. Smith and J.D.A. Walker, J. Fluid Mech., 232 (1991) 133.

31. T.L. Doligalski, C.R. Smith and J.D.A. Walker, Annu. Rev. Fluid Mech., 26 (1994) 573.

32. A.E. Perry, T.T. Lim and E.W. Teh, J. Fluid Mech., 104 (1981) 387.

33. A.V. Johansson, P.H. Alfredsson and J. Kim, J. Fluid Mech., 224 (1991) 579.

Theoretical and Applied Mechanics 1996
T. Tatsumi, E. Watanabe and T. Kambe (Editors)
© 1997 Elsevier Science B.V. All rights reserved.

Newtonian Elastic Liquids - A Paradox!

D.V. Boger and M.J. Solomon

Department of Chemical Engineering, The University of Melbourne, Parkville, Victoria 3052, Australia

ABSTRACT

Most scientists and engineers would regard the viscosity as being a unique physical property of the fluid. When the viscosity is constant, the fluid is regarded as Newtonian; when the viscosity is variable the fluid is regarded as non-Newtonian. This paper deals with a class of materials where the viscosity is constant and where the fluids exhibited elasticity to a varying degree. With such optically clear viscous polymeric liquids, observations of elastic effects in flows in the absence of varying viscosity effects (shear thinning) and fluid inertia can be made. Construction of such ideal elastic liquids is reviewed and their steady shear properties are used to establish the parameters in a hierarchy of constitutive equations (Newtonian, Maxwell, three-mode Maxwell, and three-mode KBKZ) for the international test fluid, M1.

Observations on the shape of an emerging jet from a reservoir for a Newtonian fluid and for Fluid M1 of the same constant viscosity are used to demonstrate the significant effect of elasticity on the flow and how effective predictions can be made of the shape of the jet for an elastic liquid.

The paper concludes with examination of Stokes flow (creeping flow around a sphere) where prediction of the influence of elasticity on the drag has not been successful. Here the drag is observed to increase and decrease relative to the known Newtonian drag for different ideal elastic fluids. The differences are argued to be related to the extensional characteristics of the molecule in the particular solution used for the observation. Clearly, in viscoelastic fluids one cannot divorce themselves from the basic chemistry of the particular system of interest.

1. INTRODUCTION

When theoretical and applied mechanics practitioners think of fluid mechanics they are usually thinking of Newtonian fluid mechanics with the non-Newtonian fluid being a small subset within the general framework of fluid mechanics. Such a view is a vast over-simplification because the challenges facing the research worker in Newtonian fluid mechanics and non-Newtonian fluid mechanics are very different. In Newtonian fluid mechanics no-one questions the validity of the Navier-Stokes equations, which are arrived at by use of the Newtonian constitutive equation with the general equations of motion. The challenge in Newtonian fluid mechanics is simply stated: to find solutions of the Navier-Stokes equations for all conceivable boundary value problems [1]. Figure 1 illustrates the process required for

140

the solution of a Newtonian fluid mechanics problem. There is only one area of Newtonian fluid mechanics where the conceptual problem of an appropriate constitutive equation arises and that is in turbulence. Here, although the Navier-Stokes equations are valid, one is forced to seek solutions in terms of the average velocities and pressures instead of their instantaneous values. In Newtonian fluid mechanics only one fluid property needs to be measured, viscosity.

NEWTONIAN FLUID MECHANICS

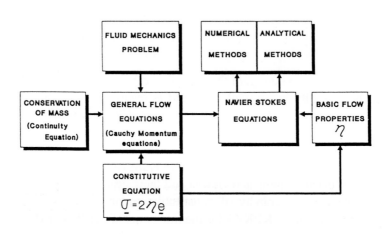

Figure 1. The process for solution of a Newtonian fluid mechanics problem.

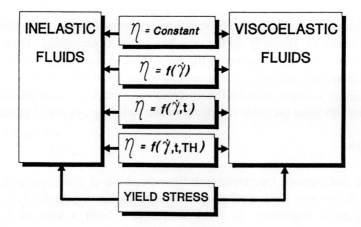

Figure 2. Classification of non-Newtonian fluid behaviour.

In contrast, as is illustrated in Figure 2, the viscosity for a non-Newtonian fluid can increase or decrease with deformation rate or time of deformation and flow may not occur until a critical stress (the yield stress) is exceeded. In addition, materials are classified as inelastic or viscoelastic dependent on the ratio of a characteristic time of the material (τ) to a characteristic time of the process of interest (t), the Deborah number

$$D_e = \frac{\tau}{t} \tag{1}$$

The Deborah number is in fact a measure of the relative importance of elasticity in a particular process. For example, if the characteristic time for the material molecular relaxation processes is of the same order as the characteristic time of the process, one can expect viscoelastic effects to be of importance. Even in so-called Newtonian fluids such as gases, where the characteristic time associated with the relaxation process may be of the order of 10^{-12}s, elastic-like flow phenomena would be observed if one were able to make observations on the same time scale. Figure 2 illustrates the range of behaviour which can be observed for rheologically complex fluids. Newtonian behaviour is only observed in the limit of low Deborah numbers and for fluids which exhibit a constant viscosity.

Rheology has provided a description of the mechanical behaviour in the idealised flows used for property measurement (viscometric flow) and has defined some, but perhaps not all, the flow properties that need to be measured in addition to the viscosity, $\eta(\dot\gamma)$, for viscoelastic fluids, i.e. the first normal stress difference, $N_{1(\dot\gamma)}$, and the second normal stress difference, $N_{2(\dot\gamma)}$, the loss and storage modulus, G'' (ω) and G' (ω), obtained from low strain sinusoidal deformation, and the extensional viscosity, h_e, obtained from steady elongational flows. Rheology has also provided some understanding of the relationship between molecular structure and mechanical properties, but has not provided a simple enough constitutive equation which describes all non-Newtonian fluids in all conceivable flow fields, like the Navier-Stokes equations. Thus, while the tools for solution of Newtonian fluid mechanics problems are available (Figure 1), non-Newtonian fluid mechanics is now in the era of developing and proving such tools for viscoelastic fluids.

Figure 3 illustrated the unknowns in the process loop for the solution of a viscoelastic fluid mechanics problem. In contrast to Newtonian fluid mechanics, non-Newtonian fluid mechanics has been concerned with the development of general constitutive equations for viscoelastic fluids. These constitutive equations should in principle lead to a definition of the flow properties that need to be measured to define the viscoelastic fluid and to the development of the equivalent of the Navier-Stokes equations for the solution of all possible boundary value problems. Unfortunately this idealistic program in rheology and non-Newtonian fluid mechanics has not been successful in defining a universally applicable constitutive equation. For example, from considering the process shown in Figure 3 for solution of a non-Newtonian fluid mechanics problem, it becomes immediately obvious that a dilemma exists. Without knowledge of a constitutive equation, a Navier-Stokes equivalent equation cannot be defined, nor can the basic flow properties that need to be measured be defined. Thus a problem in non-

Newtonian fluid mechanics becomes ill-posed unless a constitutive equation is first assumed. The process in the past has consisted of the assumption of a relatively simple viscoelastic constitutive equation, the generation of a solution (generally by numerical methods), and almost invariably, no experimental verification of the solution because no ideal enough material has been available to match the over-simplified constitutive equation assumed. In other words, a non-problem has been solved. On the other hand is the industrial practitioner who has made observations with materials whose properties are so complicated that no constitutive equation is applicable for the material, or if one seemingly is applicable to the material, the problem becomes so complex that even today, with the most sophisticated computers and numerical methods, a solution is not possible.

NON-NEWTONIAN FLUID MECHANICS

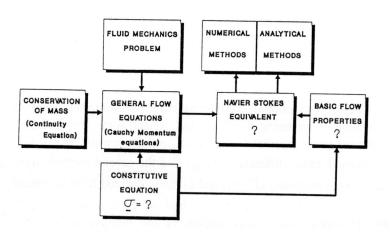

Figure 3. The process for solution of a non-Newtonian fluid mechanics problem.

Although many constitutive equations have been proposed, such an equation must fulfil two contradictory requirements: it must be complex enough to correctly describe the behaviour of some real materials or perhaps a class of materials in as wide a class of flows as possible, yet it must be simple enough for solutions to be possible for boundary value problems within the class of flows for which it is valid. Non-Newtonian fluid mechanics does not have the luxury of solving real problems that can be solved but has to be concerned with attempting to solve problems that may not be solvable with current tools. Such was the state of affairs in 1976 when, in first volume of the Journal of Non-Newtonian Fluid Mechanics, Professor G. Astarita addressed the question, "Is non-Newtonian fluid mechanics a culturally autonomous subject?". Since 1976 significant advances have been made in non-Newtonian fluid mechanics, largely as a result of three developments: the revolution in computation, the use of finite element techniques in non-Newtonian fluid mechanics [2], and the discovery of constant viscosity elastic liquids [3,4] In this paper the importance of ideal elastic liquids is illustrated using experimental observations with such materials in two flows - a gravity jet, and creeping flow around a sphere.

2. IDEAL ELASTIC LIQUIDS

Ideal elastic fluids are characterised by a constant viscosity in shear, while at the same time exhibiting most of the flow phenomena attributed to fluid elasticity: the rod climbing effect (Weissenberg effect), die swell in extrusion, the open channel syphon effect (Fano effect), stress relaxation and in entry flows, vortex enhancement and flow instabilities which result in melt fracture in polymer processing [5]. A discussion on viscoelastic flow phenomena and the constitutive equations developed to describe their behaviour can be found in the book by Bird [6] while an excellent introduction to the subject is available in Barnes et al. [7]. Viscoelastic fluids also exhibit an extensional viscosity which is strain and strain rate dependent, which is in contrast to a Newtonian fluid, where the extensional viscosity is constant, and three times the steady shear value (Trouton ratio).

Ideal elastic fluids are highly viscous constant viscosity fluids which exhibit the flow phenomena associated with fluid elasticity while their viscosity remains constant like that of a Newtonian fluid. Furthermore, optically clear fluids that exhibit these characteristics at room temperature can be constructed. Such fluids are aptly suited for use in the laboratory for observation of elastic effects in flow fields in the absence of any shear thinning and fluid inertia effects. A large number of elastic flow phenomena, in the absence of shear thinning and inertial effects, are illustrated in the book entitled "Rheological Phenomena in Focus" by Boger and Walters [5].

The development of constant viscosity elastic liquids (ideal elastic liquids) was reviewed in 1985 [4] and again in 1990 by Chai in his PhD thesis [8]. The first ideal elastic fluids consisted of polyacrylamide dissolved in maltose or corn syrup, using water as the second solvent [3,4 and 9]. Later the organic system of polyisobutylene in polybutene (using kerosene as the second solvent) has also shown such ideal properties [4, 10, 11 and 12]. An interesting and perhaps even more ideal fluid system is the polystyrene fluid constructed by Magda and Larson [13] and Solomon and Muller [14,15].

Thus a large number of organic and inorganic ideal elastic fluids have been constructed and used in laboratories around the world in the study of viscoelastic fluids and viscoelastic fluid mechanics. A co-ordinated study on a polyisobutylene/polybutene fluid (Fluid M1) was reported in Volume 35 of the Journal of Non-Newtonian Fluids. It is the observations with this fluid which are used in the first instance in this paper to demonstrate how ideal elastic fluids and their properties have helped to bridge the gap between prediction and observation as is illustrated in Figure 3. Fluid M1 was constructed in the laboratories of Professor T Sridhar in the Department of Chemical Engineering at Monash University (Australia) and distributed to a large number of laboratories worldwide for comparative study. The dynamic and steady shear properties of Fluid M1 measured in the laboratory at Melbourne University are shown in Figure 4. Five other laboratories measured the dynamic properties (the storage modulus, G', and the dynamic viscosity, η') while four other laboratories measured the first normal stress difference, N_1. The results were compared and discussed in some detail in the paper by Te Nijenhuis [16]. The comparative study concluded that the laboratories agreed on the measured storage modulus as a function of frequency on the observed first normal stress shear rate

144

relationship, and most importantly, that the simple fluid relationship from continuum mechanics

$$\underset{\dot{\gamma}\to 0}{Lim}\ \frac{N_1}{2\dot{\gamma}^2} = \underset{\omega\to 0}{Lim}\ \frac{G'}{\omega^2} \tag{2}$$

was valid.

Figure 4 illustrates the dynamic and steady shear properties of Fluid M1 at 21°C. Fluid M1 is a constant viscosity elastic liquid characterised by steady shear viscosity $\eta = 2.65$ Pa s and a Maxwell relaxation time of $\lambda_m = 0.28$ sec.

$$\lambda_m = \frac{N_1}{2\tau_{12}\dot{\gamma}_{12}} \tag{3}$$

where τ_{12} is the shear stress and $\dot{\gamma}_{12}$ is the shear rate. Parameters for describing M1 have been established by Chai [8, 17] for five constitutive equations: Newtonian, Maxwell, Oldroyd-B, Three-Mode Maxwell, and Three-Mode KBKZ. Details of how the various parameters were determined are available [8].

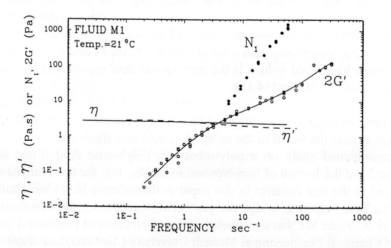

Figure 4. The steady shear (η, η_1) and dynamic (G', η',) flow properties of Fluid M1.

The predicted flow properties from the various constitutive equations in comparison to the measured steady and dynamic shear properties are shown in Figures 5 and 6. With reference to Figure 3 one has now defined five constitutive equations, which in turn have specified the flow parameters which must be measured. These parameters were all determined for Fluid M1 from the measured steady and dynamic shear properties, except for β in the KBKZ constitutive equation, which was extracted from elongational viscosity measurements of Walters on Fluid M1 [18]. Figure 5 shows that all of the constitutive equations, with the exception of the Maxwell model, predict the steady and dynamic viscosity fairly well, although the three-mode Maxwell Model (3M) and the KBKZ (K) appeared to be superior. In Figure 6 where N_1 and 2G' are examined, it is clear that the three-mode Maxwell Model and the KBKZ are superior, that the Newtonian and Maxwell constitutive equations are hopeless, with the Oldroyd-B being adequate up to moderate shear rate (a frequency of 10 sec^{-1}). We now have a simple elastic fluid with the flow properties specified for five constitutive equations. Each of these constitutive equations can now be used to develop the Navier-Stokes equivalent equation (see Figure 3).

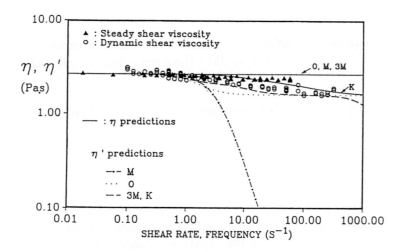

Figure 5. Constitutive equation predictions for the steady shear and dynamic viscosity: O - Oldroyd-B; M - Maxwell; 3M - Three-mode Maxwell; K - KBKZ.

It now remains to choose a fluid mechanics problem which is simple enough in order that it can be solved for the hierarchy of constitutive equations ranging from Newtonian to KBKZ, yet realistic enough to demonstrate a significant elastic effect on the flow field of practical interest where observations are easily made in the laboratory. In this way a comparison of laboratory observations made with a well-characterised elastic liquid can be compared with the theoretical prediction for a hierarchy of simple constitutive equations in non-Newtonian fluid mechanics - perhaps for the first time. This has been done by Chai and Yeow for a gravity jet flow [17].

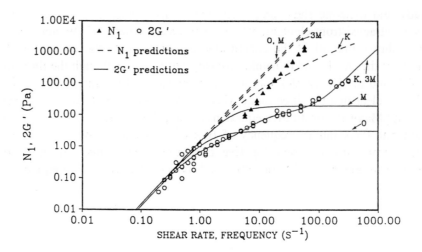

Figure 6. Constitutive equation predictions for the first normal stress difference and the storage modulus: O - Oldroyd-B; M - Maxwell; 3M - Three-mode Maxwell; K - KBKZ.

3. COMPARISON BETWEEN OBSERVATION AND PREDICTION FOR IDEAL ELASTIC LIQUIDS

3.1 Gravity Jet Flow

Figure 7 is the schematic diagram of the capillary rheometer used by Chai and Yeow [17] to generate a gravity jet. Basically, the test fluid (M1) was loaded into the reservoir of the capillary rheometer which had a reservoir diameter of 0.203 m, where the reservoir cylinder was pressurised with nitrogen gas. The test fluid flowed through 45° entry into the capillary tube with a $L_1/D_0 = 130$, which was sufficient to eliminate end effects in normal capillary rheometer operation. In a typical experiment the fluid left the capillary tube in creeping flow ($N_{Re} \leq 0.1$), stretching as it travelled downward under its own gravity; thus a moderate elongational flow is generated. The jet ended its travel in an overflow recovery beaker.

An image digitising system was used to record the shape of the jet from which the jet diameter was determined as a function of actual position. The actual technique was checked against a more conventional photographic technique. Furthermore, excellent agreement was obtained between the predicted and observed jet for an inelastic Newtonian jet [19]. The Newtonian prediction for the jet diameter at a wall shear rate of 83.3 sec^{-1} is shown in Figure 8 in comparison to the experimental results obtained for Fluid M1. Clearly a comparison of the confirmed Newtonian prediction and the experimental observation demonstrates that elasticity has a very significant effect on the jet diameter, increasing it by as much as six times at a dimensional axial distance from the exit of the rheometer of twenty.

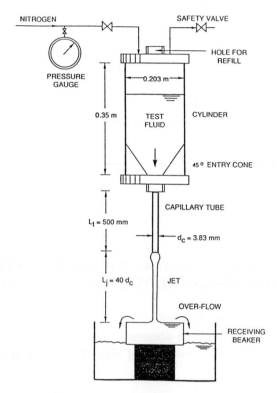

Figure 7. Schematic diagram of apparatus used to establish the gravity jet

The shape of the jet was also predicted for a Maxwell (M), Oldroyd-B (O), three-mode Maxwell (3M), and a three-mode KBKZ (K) fluid. The one-dimensional approximation normally used to describe fully developed jets was employed [20] by Chai and Yeow [17]. Solution details are available in Chai [8] where a similar method to that of Papanastasiou et al. [20] (used for spinning of polymer melts) was employed. Clearly, as is graphically demonstrated in Figure 8, elasticity dramatically affects the shape of the gravity jet. The Newtonian fluid prediction is completely inadequate to describe the behaviour of a jet of the same viscosity, as is the simplest of elastic fluid constitutive equations, the Maxwell model. The simplest constitutive equation which leads to an adequate prediction of the jet shape is the Oldroyd-B constitutive equation. The three-mode Maxwell prediction is adequate but worse than the Oldroyd-B model, while the three-mode KBKZ is virtually exact and does a great job predicting the observed expansion of the receiver, compared to any other of the models. The results reproduced in Figure 8 from the original work of Chai and Yeow [17] represent a serious attempt to compare the performance of a hierarchy of constitutive equations to careful experimental observation for a material whose properties are simple enough to be represented by the constitutive equations.

148

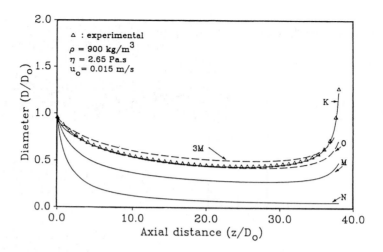

Figure 8. Comparison of the measured jet diameter with prediction at $\dot{\gamma}_w = 83.3$ s^{-1} for Fluid M1: Newtonian (N); Maxwell (M); Oldroyd-B (O); Three-mode Maxwell (3M); KBKZ (K).

In addition to the gravity jet, there have been other flows in which good agreement between experiment and numerical simulation has been obtained. The examples all deal with elastic flow instabilities: in a cone and plate [21] and parallel plate flow [22] and in flow between rotating concentric cylinders [23].

Such results lead one to be very optimistic about the ability to predict elastic effects in other flow fields. If this were only the case! As soon as a more significant elongational component enters the flow field, as is the case for tubular entry flows and creeping flow around a sphere, our ability to predict elastic effects with the simple constitutive equations just discussed disappear. The enigma is demonstrated for creeping flow around a sphere where the influence of fluid elasticity can result in both drag enhancement and drag reduction relative to the expected Newtonian Stokes drag.

3.2 Creeping Flow Around a Sphere

Creeping flow past a sphere in an unbounded medium, because of its classical origins and practical relevance to sedimentation in non-Newtonian fluids, has been the subject of extensive study. A related problem is the international numerical test problem of flow past a sphere in a cylinder where the sphere to cylinder ratio is one half [24]. Although the latter has also been often studied, we focus our attention on the unbounded case because the available experimental results highlight certain limitations in the constant-viscosity elastic liquid experimental strategy. Nevertheless, these experimental results also indicate that constant-viscosity ideal elastic liquids have played a role in uncovering how material properties other than those measured in steady shear affect flow behaviour.

Although both the velocity and stress fields of creeping flow past a sphere are of interest, the drag on the sphere is an easily measured quantity, since it can be computed from a measurement of a falling sphere's terminal velocity. While the drag for a Newtonian fluid is only a function of Reynolds number (C_d = 24/Re, where C_d is the drag coefficient of a sphere falling in an infinite medium and Re is the Reynolds number), the drag for a constant-viscosity elastic liquid also depends upon the Weissenberg number (We) of the flow. The Weissenberg number measures the relative strength of elasticity in the flow, and plays a role similar to the aforementioned Deborah number. In flow past a sphere, We = $(\lambda V_t)/R$ where λ is the material's characteristic relaxation time, V_t is the terminal velocity and R is the sphere radius. The effect of We on flow past a sphere in a constant viscosity elastic liquid can be examined by plotting the non-dimensional drag coefficient, X_e vs. We. X_e is the ratio of the measured drag to that of a Newtonian fluid of the same viscosity (X_e = (C_D Re)/24 for flow in an unbounded medium). Note that such an experimental approach is only possible for test fluids possessing a constant viscosity for all shear rates in the flow, since X_e is defined relative to that of a Newtonian fluid. Figure 9 plots X_e versus We for five different constant-viscosity elastic liquids. The data are as taken from the literature. Fluids 1,2 and 4 are from Solomon and Muller[14], fluid 3 from Tirtaatmadja et al. [25] and fluid 5 is from Chhabra et al.[26]. Measurements on constant-viscosity elastic liquids similar to fluids 3 and 5 have been reported by Chiemlewski et al. [27].

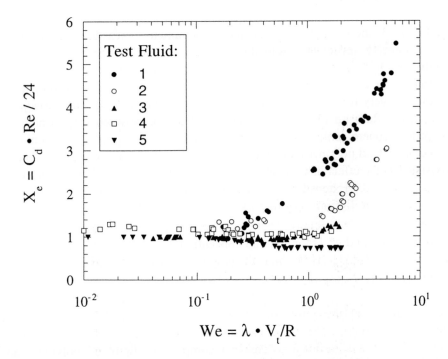

Figure 9. The effect of We on the non-dimensional drag coefficient is plotted for five constant viscosity elastic liquids. The data are taken from literature referenced in the text. The compositions and some model parameters for the test fluids are compiled in Table 1.

The significant effect of elasticity on flow past a sphere can be observed in Figure 9. If elasticity has no effect on the measured drag for a fluid, then $X_e = 1$, independent of We. In fact, in Figure 9, for We < 1 very little deviation from Newtonian behaviour is observed for any of the test fluids. This is in qualitative agreement with early perturbation solutions by Leslie and Tanner [28], Caswell and Schwarz [29], and Giesekus [30]. However, at high We, the constant-viscosity elastic liquids show drag reduction or drag enhancement relative to Newtonian fluids. What mechanism would cause these fluids, all of a constant-viscosity, to exhibit divergent drag behaviour? The magnitude of the effect is striking, with the fluid behaviour spanning from 500 percent drag enhancement to 30 percent drag reduction. Note that 500 percent drag enhancement corresponds to a factor of five reduction in the velocity of a sphere falling in the test fluid (relative to a Newtonian fluid of the same viscosity). This level of deviation from Newtonian behaviour is significantly greater than the predictions of current numerical simulations using different constitutive equations [31,32].

In addition, comparing the data for the five test fluids shows qualitative differences in drag behaviour among the fluids at high We. Since all the researchers report measurement errors to be less than 5% and all have taken care to account for wall effects on the measured drag, these differences are significant. They suggest that other quantities besides We are determinants of behaviour in constant-viscosity elastic liquids. Establishing what those differences are, and how they lead to the divergent drag behaviour of Figure 9, is the subject of current research in non-Newtonian fluid mechanics.

In terms of Figure 3, these results suggest that the parameters measured for constitutive modelling are not entirely sufficient. Something besides the steady-shear rheology of the constant-viscosity test fluids is required to understand trends in Figure 9. Since flow past a sphere is a complex flow containing an extensional component, the extensional properties of the test fluids could be very relevant here. Indeed, Chmielewski et al. [27], Tirtaatmadja et al. [25], and Chilcott and Rallison [31] have all suggested a link between drag behaviour in creeping flow past a sphere, extensional properties, and the molecular configuration and volume fraction of dissolved polymer molecules. The effect of molecular configuration can be more explicitly stated as the effect of extensibility, the degree to which a polymer molecule can be extended in a flow field. Extensibility is quantified by L, the ratio of the fully extended chain length (R_e) to the equilibrium root mean squared end-to-end vector (R_o). The volume fraction of dissolved polymer coil relative to the overlap concentration, φ ($\varphi = c/c^*$, where c is the polymer concentration and c* is the overlap concentration) is implicitly equivalent to the retardation parameter, η_p/η_s [31]. Here, η_p is the polymeric contribution and η_s is the solvent contribution to the viscosity.

It has been conjectured that these two quantities influence creeping flow past a sphere primarily at the stagnation points fore and aft of the sphere [31]. Since flow in the vicinity of a stagnation point contains a significant extensional component, dissolved polymer can be significantly deformed there. How this deformation affects the flow should depend on extensional properties. These properties are themselves controlled by the amount of dissolved polymer (governed by φ) and the extent to which it can be deformed (governed by L). Polymer physics indicates that L and φ depend on the polymer backbone structure, molecular weight (M_W) and polymer-solvent quality in the manner indicated in Solomon and Muller [14]. Table 1 compiles what is known about these quantities for the Figure 9 test fluids.

Table 1

Fluid	Polymer	Solvent[a]	concentration (wt%)	M_W or M_v (g/mole)	η (Pa·s)	η_p/η_s	λ[b] (s)	L[c]
1	PS	TCP/LMPS	0.16	2.0×10^7	5.0	1.2	6	97
2	PS	DOP/LMPS	0.16	2.0×10^7	4.0	1.0	2.5	160
3	PIB	PB/kerosene	0.24	3.8×10^6	3.16	0.87	0.14 - 0.20	–
4	PS	DOP/LMPS	0.23	2.0×10^6	12.0	0.25	0.3	50
5	PAA	glucose/H$_2$O	0.002 - 0.038	–	6.3 - 10	–	0.037 - 0.51	–

Notation: PS, polystyrene; PIB, polyisobutylene; PAA, polyacrylamide; TCP, tricresyl phosphate; LMPS, oligomeric PS; DOP, dioctyl phthalate, PB, polybutene.

(a) TCP and DOP systems assessed as good and poor solvents, respectively by Solomon and Muller [14].

Solvent quality of PIB systems assessed as poor relative to PAA system by Chmielewski et al. [27].

(b) relaxation times from steady-shear properties except fluid 4, from dynamic shear properties.

(c) from Solomon and Muller [15].

Estimating L apriori from molecular information requires a fluid consisting of monodisperse polymers of known M_W dissolved in a solvent of known quality. To estimate φ one must know the retardation parameter, η_p/η_s, or the polymer coil size (from light scattering or from the polymer M_W and solution solvent quality). From Table 1, it is clear that even a qualitative comparison in terms of volume fraction and extensibility effects among the five fluids is not possible, due to gaps in reported material properties. However, Solomon and Muller [14] have discussed the relative effect of L and φ on the behaviour of fluids 1, 2 and 4 and Chmielewski et al. [27] explain the relative drag enhancement/reduction of fluids 3 and 5 in terms of their solvent quality and extensibility. Nevertheless, a complete comparison is hampered by the fact that some measurements on the drag-reducing fluid (fluid 5) were performed in a shear-rate region wherein Ψ_1[1] decreases as a function of shear rate, whereas the four drag enhancing fluids show constant Ψ_1 over the experimental range of falling sphere shear rates.

---

[1] $\Psi_1 = \dfrac{N_1}{\gamma_{12}^2}$ is the first normal stress difference coefficient

The point here is not to argue that the ordering of the degree of drag enhancement/reduction in these five fluids is fully understood in terms of their molecular properties. Unresolved issues include the possible existence of flow instabilities, the substantial uncertainty in the assignment of λ in fluids containing a spectrum of relaxation times, the significance of a shear thinning $\Psi 1$, the competing effects of extensibility and volume fraction, and the difficulty of estimating L and φ in constant-viscosity elastic liquids. As further progress on these issues is made, a clearer picture of how molecular configuration directly affects creeping flow past a sphere will emerge. If this is the case, then constant-viscosity elastic liquids will have been used as a tool not only to understand the effect of elasticity, but also the effect of molecular configuration on this viscoelastic flow.

4. CONCLUSION

Returning to the central problem of constitutive modelling in non-Newtonian fluid mechanics, we see that in the case of flow past a sphere, the loop of Figure 3 has yet to be completed. This is certainly the case for other complex flows, particularly those containing a significant extensional component. Moreover, it is clear that work with traditional constant-viscosity elastic liquids is alone not enough to complete the experimental portion of the program. It appears (as previously noted by, for example, Rallison and Hinch [33]) that we require more molecular information about experimental fluids to advance understanding of these complex flows.

In addition to the characterisation of steady-shear properties of constant viscosity elastic liquids, the characterisation of fluid extensibility and volume fraction thus seems worthwhile. While the former can perhaps be assessed from elongational property measurements, the example of the international test fluid M1 calls attention to the difficulty of that approach. Another avenue involves computing these model parameters from microscopic solution properties such as polymer molecular weight, coil size, fully extended chain length and polymer solvent quality. This too can be difficult if not impossible to achieve in constant-viscosity elastic liquids, which, in all but a few cases, consist of polydisperse polymers dissolved in mixed solvents of poorly defined quality. We are thus led to conclude that if extensibility and extensional properties are indeed believed to be key model parameters in a flow, then perhaps the need to characterise them is more important than the need to prepare a constant-viscosity test fluid. For example, an extremely dilute test fluid, consisting of monodisperse polymer dissolved in a solvent of known quality (perhaps characterised by means of intrinsic viscometry) might in some instances be a better experimental tool than a constant-viscosity highly elastic liquid. That is, developing ideal materials with measurable model parameters relevant to the flow of interest still remains a challenge in non-Newtonian fluid mechanics.

REFERENCES

1. G. Astarita, J. Non-Newtonian Fluid Mech., 1 (1976) 203.
2. M. Crochet, A.R. Davies and K. Walters, Numerical Simulation of Non-Newtonian Flow, Elsevier, Amsterdam, 1984.
3. D.V. Boger, J. Non-Newtonian Fluid Mech., 3 (1977/78) 77.
4. R.J. Binnington and D.V. Boger, J. Rheol., 26(6) (1985) 887.
5. D.V. Boger and K. Walters, Rheological Phenomena in Focus, Elsevier, Amsterdam, 1993.
6. R.B. Bird. R.C. Armstrong and O. Hassager, Dynamics of Polymeric Liquids, 2nd Edition, Vol. 1, Wiley, New York, 1987.
7. H.A. Barnes, J.F. Hutton and K. Walters, An Introduction to Rheology, Elsevier, Amsterdam, 1989.
8. M.S. Chai, Elongational Flow of Boger Fluids, Ph.D. Thesis, The University of Melbourne, Parkville, 1990.
9. J.C. Chang and M.M. Denn, J. Non-Newtonian Fluid Mech., 5 (1979) 369.
10. G. Prilutski, R.K. Gupta, T. Sridhar and M.E. Ryan, J. Non-Newtonian Fluid Mech., 12 (1983) 233.
11. R.J. Binnington and D.V. Boger, Polymer Eng. Sci., 26 (1986) 133.
12. L.M. Quinzani, G.H. McKinley, R.A. Brown and R.C. Armstrong, J. Rheol., 34(5) (1990) 705.
13. J.J. Magda and R.G. Larson, J. Non-Newtonian Fluid Mech., 30 (1988) 1.
14. M.J. Solomon and S.J. Muller, J. Non-Newtonian Fluid Mech., 62 (1996) 81-94.
15. M.J. Solomon and S.J. Muller, J. Rheol., 40(5) (1996).
16. K. Te Nijenhuis, J. Non-Newtonian Fluid Mech., 35 (1990) 169.
17. M.S. Chai and Y.L. Yeow, J. Non-Newtonian Fluid Mech., 35 (1990) 459.
18. D.M. Binding, D.M. Jones and K. Walters, J. Non-Newtonian Fluid Mech., 35 (1990) 129.
19. C.T. Trang and Y.L. Yeow, J. Non-Newtonian Fluid Mech., 13 (1983) 203.
20. T.C. Papanastasiou, C.W. Mocosko, L.E. Scriven and Z. Chen, A.I.Ch.E.J., 33 (1987).
21. G.H. McKinley, A. Oztekin, J.A. Byars and R.A. Brown, J. Fluid Mech., 285 (1995) 123-164.
22. J.A. Byars, A. Oztekin, R.A. Brown and G.H. McKinley, J. Fluid Mech., 271 (1994) 173-218.
23. R.G. Larson, Es.G. Shaqfeh and S.J. Muller, J. Fluid Mech., 218 (1990) 573-600.
24. O. Hassager, J. Non-Newtonian Fluid Mech., 29 (1988) 2-5
25. V. Tirtaatmadja, P.H.T. Uhlerr and T. Sridhar, J. Non-Newtonian Fluid Mech., 35 (1990) 327-337.
26. R.P. Chhabra, P.H.T. Uhlherr and D.V. Boger, J. Non-Newtonian Fluid Mech., 6 (1980) 187-199.
27. C. Chmielewski, K.L. Nichols and K. Jayaraman, J. Non-Newtonian Fluid Mech., 35 (1990) 37-49.
27. C. Chmielewski, K.L. Nichols and K. Jayaraman, J. Non-Newtonian Fluid Mech., 35 (1990) 37-49.
28. F.M. Leslie and R.I. Tanner, J. Mech. Appl. Math., 14 (1961) 36-48.
29. B. Caswell and W.H. Schwarz, J. Fluid Mech., 13 (1962) 417-426.
30. H. Giesekus, Rheol. Acta, 3 (1963) 59-71.

154

32. B. Gervang, A.R. Davies and T.N. Phillips, J. Non-Newtonian Fluid Mech., 44 (1992) 281-306.

33. J.M. Rallison and E.J. Hinch, J. Non-Newtonian Fluid Mech., 29 (1988) 37-55.

Theoretical and Applied Mechanics 1996
T. Tatsumi, E. Watanabe and T. Kambe (Editors)
© 1997 Elsevier Science B.V. All rights reserved.

Computational and Experimental Studies of Polymeric Liquids in Mixed and Extensional Flow

L. G. Leal

Department of Chemical Engineering, University of California, Santa Barbara, Santa Barbara, CA 93106

1. INTRODUCTION

This paper describes a program of experimental and theoretical research which has, as its ultimate goal, the development of a predictive basis to describe the behavior of viscoelastic, *polymeric* liquids in nonhomogeneous, time-dependent, two- or three-dimensional flows.

This problem is much more difficult, both experimentally and theoretically, than the evaluation of material behavior in conventional shear-flow based *rheological* studies. Unlike the flow in *shear* rheometers, which is (approximately) unchanged by polymer contributions to the stress, virtually all other flows are *inhomogeneous* and thus strongly modified by the polymer, so that the velocity field is *a priori* unknown.[†] In experimental studies, it is crucial to measure *both* the polymer configuration (and/or stress) and the velocity gradient of the flow, and since the flow is inhomogeneous, these must be measured *locally* rather than globally as one does in standard shear rheometers. To evaluate the behavior of a constitutive model, one must generally compare experimental data with solutions of the full viscoelastic flow problem, including both the velocity field and the stress distributions. The only alternative is to use *measured* data for the flow, and compare predictions for the polymer configuration with experimental data, (e.g., birefringence). But this is only a partial test because it does not guarantee that the model will yield accurate predictions for the flow, e.g., stress-optic laws will not apply in a strongly nonlinear regime, nor if there are significant viscous (non-entropic) contributions to the stress.

The particular focus of this proposal is extension-dominated motions.[††] The importance of studying this class of flows has been widely recognized. Not only are regions of strong flow found in many flow systems of practical significance, but it is expected that the polymer behavior will differ fundamentally from its behavior in shear flows (there is a fundamental difference in the intrinsic ability to induce *stretching* of polymer chains). However, progress has been slow. Most attempts to measure extensional flow properties have failed to incorporate measurements of the velocity gradient, and thus produce widely scattered results. [4] The one recent attempt to incorporate measured flow data (apart from our own studies described below) utilized a converging channel flow in which the polymer contribution to the extensional stress was weak. [5] Computational studies, with the exception of a few very recent ones, [5-8] have not attempted quantitative comparisons with experimental data, and there is not a generally agreed upon optimal basis to determine constitutive model parameters. The few exceptional studies [5-8] have focused mostly on the velocity rather than velocity gradient field, and/or have not included stress or birefringence data. Further, most recent experimental work on non-

---

[†] Recent work[1] shows that this is true even of the stretching filament flows that have been recently devised by Sridhar,[2] and others in hopes of obtaining extensional viscosities without this complication.
[††] For planar geometries, this implies that the magnitude of the strain-rate exceeds the vorticity.[3]

viscometric flows has utilized relatively dilute (e.g., Boger) solutions, and thus does not provide insight into the behavior of strongly entangled systems.

Our *immediate* objectives are primarily fundamental. Thus, we focus on narrow molecular weight distribution solutions (ranging from highly entangled to dilute) of "linear" flexible polymers (polystyrene and polyisoprene), and on theoretical models derived from molecular theory. However, as stated earlier, our ultimate goal is a predictive capability for the two- or three-dimensional time-dependent flows that occur in many applications. One example of the importance of this objective is the sharp reduction in the enormous annual cost associated with the largely empirical development of new injection-molds that could be achieved if a predictive capability existed for time-dependent, three-dimensional *viscoelastic flows*—it is the influence of viscoelasticity on the final temperature and stress distributions in a molten part that is missing or over-simplified in current molding simulation software.

Experimental and theoretical tools (some unique to our lab) have evolved over the past several years to a point where many of the intrinsic deficiencies in the prior studies can be remedied. In this paper we briefly summarize recent progress in this direction.

2. BACKGROUND

We begin by describing our current experimental capabilities, as well as the general theoretical underpinning for our research. Following this, in Section III, we summarize recent research results.

2.1 Experimental Methodology

2.1.1 Flow Cells

This project is predicated on the assumption that it is important to study a broader class of flows than the viscometric flows of classical rheometry. The recent attempts to produce a pure uniaxial extension via filament stretching represent one effort to move in this direction.[1-2] In the present work, we seek to study a broader class of extension-dominated ("strong"[3]) flows, which are also easier to access for measurements of chain configuration, and which can be studied for a wider range of Deborah numbers. To access the parameter regime in which there are strong nonlinear effects, the polymer must experience both large strain rates (i.e., a Weissenberg number $\geq 0(1)$), and a large total strain in the strong flow region. We accomplish this latter objective by using a pair of flow devices known as the four-roll mill and the co-rotating two-roll mill, each of which contains a *stagnation* point (actually a line of stagnation points) at the center of the strong flow region where the polymer can be subjected to both large strain and strain-rate. The (new) four-roll mill in our laboratory can produce either a steady or time-dependent pure-strain, hyperbolic flow in the region between the four cylinders (all four cylinders are driven by a single DC stepping motor to insure "stability" of the stagnation point in transient flows). By using rollers of different size, the two-roll mill can produce a spectrum of planar flows ranging from near-shear to *mixed* flows that are between simple shear and the purely extensional, hyperbolic flow. Both devices are small, to avoid inertial instabilities, and to enable experiments with relatively small samples (~100 ml) of the expensive narrow molecular weight distribution solutions that we study.

2.1.2 Birefringence

We do two main kinds of measurement in these flow cells. The first is two-color birefringence, which is a technique to determine both the *magnitude* of birefringence, and the *orientation* of the principle axes of the refractive index tensor, in a *transient* flow where both Δn and the angle χ may be time-dependent. Temporal resolution is in the ms range (or better), which is far more than we need in the viscous systems that we study (see below), and spatial resolution is currently 50μm (though recent work at Bristol has shown that this can be reduced to 3-4μm with relatively simple modification using a CCD camera). [9] The gap width between cylinders is typically several millimeters. Although birefringence measurements can be made anywhere we wish in the flow, we generally focus on the stagnation point and its immediate neighborhood. The spatial localization of our birefringent measurement (50 μm) is small enough that the behavior of the polymer can be interpreted in terms solely of the velocity gradient history measured *at* the stagnation point. The most transparent evidence to support this claim actually comes from computational simulations of start-up flow for a dilute solution (the most difficult case because gradients of polymer configuration are largest in the dilute limit). In this case, the results obtained by *averaging* the predicted birefringence over a 50μm zone centered at the stagnation point, are virtually identical with the values predicted exactly at the stagnation point. This implies that the velocity gradient history experienced by chains that move through the birefringent zone is very close to that experienced exactly at the stagnation point, at least over the time span of a typical transient experiment.

We are fully aware of the potential ambiguity in interpreting intrinsic birefringence data in terms of polymer configuration, and of claims that large chain extensions do not occur in dilute solutions, as the usual interpretation would suggest. [10] However, recent direct visual observations of DNA molecules in extensional flow [11] demonstrate conclusively that chains *do* extend (and this is also consistent with recently measured stress levels in Boger fluids [1-2]). We believe that two-color birefringence is a useful, local probe of polymer configuration for the class of "slowly varying" two-dimensional flows that we consider. As discussed below, polarized dynamic light scattering can give very similar information to birefringence, [12] without some of the limitations, but at a cost of much more time-consuming experiments.

2.1.3 Dynamic Light Scattering

The second type of measurement developed recently in our laboratory is a *unique* adaptation of the classical method of polarized dynamic light scattering that is designed to provide a pointwise determination of components of the velocity *gradient* tensor in either steady [13] or transient [14] flows. *It is this capability (plus the choice of flow cells) that most strongly distinguishes our experimental capabilities from other laboratories.* Since we can measure both the stress components (via stress-optic laws in the regime where these are relevant) and the corresponding velocity gradient, we can determine the elastic contributions to the rheological properties for steady and transient flows ranging from hyperbolic, purely extensional flow in the four-roll mill to a near-shear flow in the two-roll mill. Indeed, as we will discuss below, we have used this approach in a recent study [14] to obtain steady "extensional" viscosities for an entangled polystyrene solution which initially show a strong decrease with increase of strain-rate as predicted via the reptation model of Doi and Edwards, [15] and then an upturn when chain stretching begins.

A major advantage of this method for determining $\nabla \mathbf{u}$, is that *no differentiation* of velocity data is required. Hence accuracy, and spatial resolution is greatly improved relative to methods that determine $\mathbf{u}$, and this is critical because of the large gradients that occur in many viscoelastic flows, and the small size of the flow cells that we use. Beyond this, however, another major bonus when the scattering is from the polymer chains (as opposed, for example, to "seed" particles), is that the initial amplitude of the correlation function depends upon the degree of optical anisotropy of the polymer. Since this changes as the polymer configuration is changed, the scattering experiment gives a *pointwise* measure of the chain configuration that is equivalent to birefringence, [12] but without the problem of multiple orders, and without integrating across the whole flow cell.

158

There are, in fact, very few groups that are presently equipped to make *simultaneous* measurements of polymer configuration *and* flow, and none with the advantage of *direct* measurements of the velocity gradient (or polymer configuration) at a *point* (via dynamic light scattering). Among the others, Quinzani *et al.*[5] have focused mainly on contraction flows, Schowalter and co-workers [16] have studied a wavy wall channel, while McKinley *et al.*,[6] Baajens *et al.* [7] and Muller [17] have all studied the wakes behind cylinders and spheres.

2.1.4 Rheological Measurements

It should be emphasized that our laboratory also includes facilities for complete linear and nonlinear viscoelastic *rheological* measurements in shear flow, as well as phase-modulated birefringence measurements in steady or transient Couette flows. The former are a key to determining the relaxation time spectra for multi-mode constitutive models.

In addition, as suggested above, we may use the stress-optical relationship between the stress and refractive index tensor,

$$\underline{\underline{\sigma}} = C\underline{\underline{n}}$$

(1)

to deduce stress components from birefringence measurements in the extensional or extension-dominated flows that exist in the four- and two-roll mills. This relationship is valid provided the dominant polymer contribution to the stress is elastic in origin, and the degree of chain stretch is small enough that the end-to-end vector for chain segments displays Gaussian statistics so that a linear spring approximation is adequate. In practice, this limits applicability to flow regimes corresponding to fractional extensions of 0.4-0.5 of the maximum contour length. [18] Although , it is generally necessary to specify three rheological functions to completely specify the state of stress in a linear flow (e.g., the viscosity and two normal stress differences in a simple shear flow), it is convenient, for present purposes, to specify the rheological state in terms of the *generalized extensional viscosity*. For planar, linear motions of the type considered here, we can express the velocity gradient tensor in the form

$$\nabla \underline{u} = \dot{\gamma} \begin{pmatrix} 0 & 1 & 0 \\ \lambda & 0 & 0 \\ 0 & 0 & 0 \end{pmatrix}$$

(2)

where we use a coordinate system that is oriented at 45° relative to the principle axes of the rate-of-strain tensor. The class of extension-dominated flows corresponds to $0 < \lambda \le 1$. Simple shear is obtained for $\lambda = 0$. The corresponding rate-of-strain tensor is

$$\underline{\underline{E}} = \frac{\dot{\gamma}}{2} \begin{pmatrix} 0 & 1+\lambda & 0 \\ 1+\lambda & 0 & 0 \\ 0 & 0 & 0 \end{pmatrix}.$$

(3)

The *generalized extensional viscosity* is then defined as the ratio

$$\eta \equiv \frac{\sigma_{12}}{\dot{\gamma}(1+\lambda)} \quad .$$

(4)

For $\lambda=1$, it is the standard extensional viscosity for planar extensional (hyperbolic) flow. According to the stress optical law (1), σ_{12} can be obtained directly from birefringence measurements

$$\sigma_{12} = \frac{\Delta n}{2C} \sin 2\chi \qquad (5)$$

where C is the stress-optical coefficient, Δn is the birefringence and χ is the angle of the principle axis of the refractive index tensor relative to the x-axis fixed in the flow (i.e., relative to the bisector between the two rollers of a co-rotating two-roll mill). The generalized viscosity function is then obtained from equation (4) using measured values of $\dot{\gamma}$ and λ.

2.2 Framework for Theoretical Studies

Our recent theoretical studies have had three main thrusts: first, evaluation of polymer dynamics theories for the class of two-dimensional strong flows by direct comparison between model predictions and measured behavior; second, more general computational studies for Boger-like fluids, aimed at establishing *qualitative relationships* between flow behavior and the physical assumptions, mathematical approximations or parameter values in dumbbell-based constitutive theories; and third, the development of numerical methods to solve flow problems via the complete configuration-space distribution function rather than the corresponding closure-based constitutive theories.

When full flow field calculations are not available, the best that one can do to test models is to use *measured* flow data as *input*, and then compare predictions with data for birefringence or some other measurable quantity. [5,14] *Especially for entangled systems, there is currently a very significant gap between the most promising constitutive theories (e.g., extensions of the reptation-based theory of Doi-Edwards to accommodate segment stretch), [19] and the class of models that are currently being studied via viscoelastic flow computations (e.g., nonlinear dumbbell extensions of Oldroyd-B). [20]* One primary reason is that fluid mechanics calculations are only currently practical for molecularly-based constitutive models (except the linear dumbbell) by introducing *ad hoc* closure approximations to substitute calculations of the second moments of the chain configuration distribution function, for calculations of the distribution function itself.

For nonlinear dumbbell-based generalizations of Oldroyd B, the closure assumptions known as "pre-averaging," lead to well known constitutive models such as FENE-P, Chilcott-Rallison and others for the relatively dilute, but highly elastic liquids known as Boger fluids. Although it is clear that (single-mode versions of) dumbbell-based models *cannot* describe short-time or short-lengthscale chain dynamics, multi-bead models are too complex for simulations that involve dynamic changes in chain configuration from equilibrium to full extension. Even for the simple class of dumbbell-based models, however, there are many *unresolved* issues, both in understanding predicted flow behavior, and in trying to distinguish properties of the model that are "caused" by the closure approximations from those which emanate from the basic physical assumptions of the underlying molecular model.[†]

[†] Solutions for the complete distribution function, ψ, can be obtained via the *stochastic* method of Öttinger,[21] or via direct solution of the Schmolukowski (diffusion) equation for which we have recently developed a very efficient (and highly parallelizable) coordinate-free, Lagrangian technique known as smoothed particle hydrodynamics[22] However, to date, both of these techniques have been applied only for the idealization of *homogeneous* flows.

160

For reptation-based models no generally accepted closure approximations exist. Even the derivation of the BKZ model from the simplest version of the Doi-Edwards theory involves the largely discredited "independent-alignment" approximation. *Thus, though it is clearly desirable in the long run to carry out flow calculations based on models, such as the Doi-Edwards theory, which account explicitly for entanglements, this cannot be done at the moment.* The most recent efforts to simulate *flows* of concentrated (and entangled) solutions have, in fact, used either the Giesekus [23] or Bird-DeAguiar [24] generalizations of dumbbell models, [5] or the Phan-Thien and Tanner [25] modification of network theory, [7] with parameters *empirically* chosen using linear viscoelastic, and steady shear data. These approaches, while clearly not the long term answer, do *presently* make some sense if we take account of the fact that the models are already significantly more complex than the Maxwell/Oldroyd-B models that were studied numerically over the past decade, and if we remember that we are still at the stage of seeking even *qualitative* understanding of the flow behavior for the simplest viscoelastic fluid models.

2.2.1   Reptation-Based Constitutive Theories

The well known reptation-based constitutive theory of Doi and Edwards is currently believed by many workers in the field, [26] to provide the best opportunity for a realistic description of the nonlinear rheological behavior of highly entangled "linear chain" polymer solutions or melts. For the narrow MWD systems and the strong flows that we deal with, however, perhaps the most important deficiency of the original Doi-Edwards theory is its failure to deal explicitly with *segmental stretching*, which will occur whenever the "shear rate" exceeds the inverse Rouse time scale, τ_R^{-1}. An extension of the Doi-Edwards theory to incorporate this effect was only proposed relatively recently (cf. Pearson *et al.* 1991). [19] In shear flow, this model adds a qualitatively correct prediction of overshoot for N_1 for $\dot{\gamma} > \tau_R^{-1}$ and also modifies the high shear rate overshoot for the *shear* stress to better accord with data.[†] *The critical question that our studies are intended to address is whether the Doi-Edwards model, with or without segmental stretch, does an equally good job of capturing nonlinear behavior for the class of "strong," extension-dominated flows.*

2.2.2   Dumbbell-Based Constitutive Theories

Recent work on viscoelastic flow calculations has been dominated by the nonlinear, elastic dumbbell-based theories. These models attempt to represent the configuration of a polymer molecule in terms of a single vector, which can be oriented and stretched by a flow, and relaxes back toward equilibrium via an entropic spring, plus Brownian fluctuations in orientation and length. Although there is no *direct* account of chain-chain interactions (though see 23,24), and the physical picture is very simple, existing comparisons with birefringence and flow data, [28] as well as visual observation of the stretching of single strands of DNA in extensional flow, [11] suggest that the Chilcott-Rallison model yields qualitatively reasonable predictions, at least for the class of strongly elastic dilute solutions known as *Boger fluids,* and for steady or "slowly-varying" extension-dominated flows where the dominant polymer response is overall alignment and stretching. Recent numerical work [20] has focused largely on the qualitative behavior of this model in nonhomogeneous flows for various values of the model parameters, c, L and τ. [8,29]

[†] Recent work on the model to incorporate a weak chain stretch for *steady* shear at high $\dot{\gamma}$, also appears to improve the problem of too fast thinning of the shear viscosity with the shear rate.[27]

2.2.3 Numerical Methods

The development of numerical methods for the solution of nontrivial viscoelastic flow problems at finite Deborah numbers, De≥0(1), has occupied many researchers worldwide for almost twenty years, and this effort has been reflected in a series of nine international workshops, many with published proceedings, [20] over the same period. A modest number of groups, including our own, have finally reached a point where two-dimensional and axisymmetric problems, with or without geometric singularities like re-entrant corners, can be solved for many of the most common differential-type constitutive equations up to Deborah numbers of at least 0(10). Since our technique is already described in easily accessible publications, [29,30] we do not provide details here. It is, however, important to emphasize the impact of the very recent development of successful numerical techniques. Without these techniques, it had been virtually impossible to achieve any meaningful tests of proposed constitutive theories beyond steady or time-dependent homogeneous shear flows. The removal of this major roadblock to progress in understanding the behavior of different models in non-trivial flows is already having a profound effect on the rate of research progress.

3. RECENT RESULTS

The research activities described above have been ongoing for several years, and during this period we have worked on a significantly broader spectrum of problems than is discussed below. However, in view of space and time restrictions, we limit our listing of recent accomplishments to work that is *directly* related to the use of the two- and four-roll mill for evaluating polymer dynamics and constitutive model behavior in extension-dominated flows.

3.1 Dilute Solutions

There have been many prior studies of dilute polymer solutions both from our lab, [31] from Bristol, [32] and elsewhere. [33] Recently, however, two new components have been introduced: first, the ability to obtain full numerical fluid mechanics solutions based upon dumbbell models that are thought to be applicable for these fluids in the class of flows considered here; and second, the capability of making detailed velocity gradient measurements under transient flow conditions. Together, these provide a basis for quantitative evaluation of constitutive models for dilute, Boger-type fluids that was not previously possible.

We have recently carried out a new set of experimental measurements [28] for start-up flows in the two-roll mill using both ultra-dilute (40 ppm) polystyrene solutions where no measurable change occurs in the flow, and a Boger-like fluid corresponding to 520 ppm of 3.84 x 10^6 MW polystyrene in a viscous solvent consisting of TCP to which 10% by weight of polystyrene oligomer (MW=5000) was added. Results were then compared with predictions using the Chilcott-Rallison model, and the Chilcott-Rallison model with conformation-dependent friction added, [34] with model parameters c, L and τ selected to match the experimental liquids. Both full fluid mechanics calculations, and model predictions of birefringence based upon *measured* flow data were considered. In both cases, an empirical correction was made to account for the molecular weight distribution.

The result for ultradilute solutions is that the Chilcott-Rallison solution, with conformation-dependent bead friction in the weak form suggested by Lardon and Magda, [35] provides essentially a quantitative agreement with the measured transient birefringence. The birefringence shows a relative sharp onset to the steady state plateau at a critical strain, $\dot{\gamma}\sqrt{\lambda t}$, of approximately ten, and except for a small difference in the weak birefringence levels prior to this onset point, there is excellent agreement with model predictions. The main point of disagreement between predictions and observations is that the dumbbell rotates toward alignment with the outflow axis right from its initial equilibrium configuration, as opposed to the measured orientation angle which only begins to rotate at the onset of chain stretch. However, this early rotation has little effect on the measured Δn in any case.

A typical birefringence data set for the 520 ppm solution showing the magnitude Δn and orientation χ versus *measured* values of $\dot{\gamma}$ and λ is presented in Figure 1, together with *predictions* from the same dilute solution model with *the same material constants* using (a) *measured flow data*, and (b) obtained via full fluid mechanics calculations using the Chilcott-Rallison model with c = 0.1 and L^2 = 2500. Shown in Figure 2 is the predicted, and measured time-dependent values for $\dot{\gamma}$ and λ. Although 520 ppm is only approximately 0.1c*, there are now very significant differences between the theoretical model predictions for a dilute solution and the data. First, the specific birefringence, $\Delta n/c$, which should be independent of c for any specific set of flow conditions, is very much lower than for the dilute limit (~2x). Previous data at other concentrations showed a similar result with the fractional decrease in $\Delta n/c$ dependent on c/c*. [36] Previous authors [32] have suggested that these depressed values for Δn were due in some unexplained way to "entanglements." There is no question that hydrodynamic interactions between chains can be induced when there is sufficient stretching, even for relatively small c/c*. However, it is not clear to us at this time why these interactions should lead to reduced stretching (or less birefringence), or what exactly they may have to do with entanglements. [37]

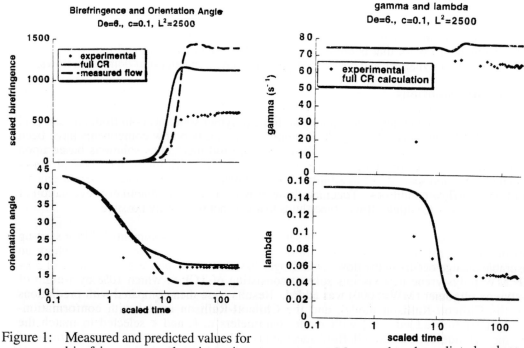

Figure 1: Measured and predicted values for birefringence and orientation angle.

Figure 2: Measured and predicted values for $\dot{\gamma}$ and λ (see eqn. 2).

The experimental and computed data for $\dot{\gamma}$ and λ both show a *modest* change in $\dot{\gamma}$ relative to Newtonian fluid values, but a much *stronger* decrease in λ. Hence, *qualitatively*, the two agree. At a more quantitative level, however, the flow changes predicted via the Chilcott-Rallison model are significantly stronger than the measured changes, perhaps consistent with the fact that the *predicted* birefringence (and thus, presumably, the elastic stress contribution) is also stronger. The major discrepancy in predicted and measured birefringence levels suggests that the Chilcott-Rallison model has a fundamental flaw for application to solutions that are not ultradilute. One possibility that we are currently exploring is that viscous stress contributions which have recently been predicted [38] and apparently measured [2] for similar fluids should be included in the constitutive model.

3.2   Entangled Solutions

From a technological point of view, the most interesting regime of polymeric fluid behavior is at high concentrations and/or molecular weights where the chain-chain topological interactions known as entanglements play a dominant role in determining their dynamics. When this occurs, the qualitative rheological behavior is fundamentally altered, and constitutive models of the bead-spring or dumbbell type would not generally be expected to give reasonable predictions. This is the regime of temporary entanglement network theory, [39] or the more recent reptation-based models that were discussed earlier. [19]

It is, perhaps, worthwhile to briefly discuss the expected qualitative differences in the molecular dynamics of unentangled and entangled solutions or melts, for this focuses on the reasons why they exhibit such distinctive rheological behavior. We concentrate here on extensional flows, or extension-dominated flows which are characterized by a single preferred alignment direction and exponential stretching of material lines in this direction. This is the class of motions experienced by polymer in the neighborhood of the stagnation point in the two- and four-roll mills.

For all polymeric liquids, for which the dominant contributions to stress are *elastic* in origin, the stress can be envisioned as arising from two distinct mechanisms: chain-alignment, and chain stretch (i.e., an increase of the length of statistical subunits of the chain). Typically, when chain stretching occurs, it is the dominant factor in determining the polymer contribution to stress, and yields an *increase* in the generalized extensional viscosity, η, with strain rate. In contrast to chain stretch, chain alignment typically leads to a *decrease* in η with increase of the strain rate. [15]

For dilute solutions, such as the Boger fluids considered in the preceding section, we have seen that chain alignment and stretch typically occur *simultaneously* in extension-dominated flows, i.e., the onset point for both is $|\nabla \underline{u}| \approx \tau_R^{-1}$ (the longest Rouse timescale in an ideal or good solvent), and they occur in an inseparable way over the same range of $|\nabla \underline{u}|$. Hence, the viscosity η in such fluids typically *increases* rapidly and monotonically with increase of strain rate.

For a highly entangled solution or melt, on the other hand, the diffusion process that leads to relaxation of flow-induced alignment is strongly inhibited and slowed so that it occurs on a timescale τ_D that is increased by a factor proportional to the molecular weight relative to the timescale for diffusion-induced retraction from a stretched state, which is still the Rouse timescale, τ_R. This means that there is a wide range of flow conditions *prior* to the onset of chain stretch,

$$\tau_D^{-1} \leq \dot{\gamma}\sqrt{\lambda} < \tau_R^{-1}$$

where an entangled polymer will exhibit strong viscoelasticity due to flow-induced alignment alone. Once chain stretching commences, for $\dot{\gamma}\sqrt{\lambda} \geq \tau_R^{-1}$, further changes in the stress are dominated by this process, and the generalized viscosity will increase monotonically with increase in the strain rate just as in dilute solutions.[†] Prior to this, however, the *entangled* fluid

---

[†] note that chain stretch is not predicted to occur at steady-state for the limit, $\lambda=0$, of simple shear flow for any of the current reptation-based models (nor in dumbbell models with constant isotropic bead friction). These models treat a polymer chain, from a hydrodynamics point of view, as an object of zero cross-

exhibits an extensive regime of nonlinear viscoelastic behavior dominated by flow-induced *alignment* (it is this regime that is described by the original Doi-Edwards constitutive theory). In this regime, the generalized viscosity function for all flows, $\lambda \geq 0$, is expected to *decrease* monotonically with increase of strain rate. Hence, the behavior of the generalized viscosity function, η, in strong, extension-dominated flows provides a fundamental basis to distinguish between dilute and entangled polymer solutions.

We have recently carried out experimental studies for high molecular weight, nearly monodisperse, solutions of polystyrene for both steady and transient flows in the co-rotating two-roll mill. [14] Earlier investigations of non-dilute solutions, typically included measurements only of the birefringence. [40] However, our recent studies include both birefringence and flow measurements. We have included three different combinations of molecular weight (2.89, 3.84 and 8.42 x 10^6) and concentration (7.6, 4.8, 3.3 wt%) that all correspond approximately to ten entanglements per chain (N). [41] We have also made similar measurements for a solution of the 8.42 x 10^6 MW sample at 2.2 wt% which corresponds approximately to five entanglements per chain. The three more concentrated solutions all fall within the regime identified by prior investigators, [42] as highly-entangled, semidilute solutions where reptation theory is expected to apply. We have confirmed this qualitatively by means of linear viscoelastic measurements which show a distinct plateau and other characteristics consistent with the expected ratio of relaxation times, $\tau_D/\tau_R = 3N \sim 30$. In any case, we cannot reach higher N values in the present flow systems using polystyrene solutions because higher concentrations would yield viscosities in excess of the several thousand Poise which our flow devices are designed to handle. We limit our observations below to the results for the 2.89 x 10^6 MW solution.

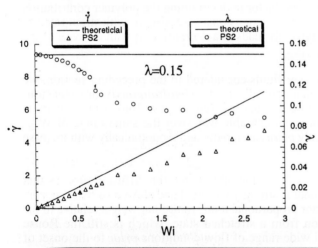

Figure 3:
Measured values for $\dot{\gamma}$ and λ for the two-roll mill configuration corresponding to a Newtonian value of $\lambda=0.15$. The solid lines are the values of $\dot{\gamma}$ and λ for a Newtonian fluid.

sectional dimension. Hence, as the chain becomes aligned in the flow direction for $\lambda=0$, it is at 45° to the principle strain rate directions, and there is no hydrodynamic force remaining to cause it to stretch. For $\lambda>0$, on the other hand, the chain aligns in the direction of the principle eigenvalue of $\nabla \underline{u}$, and in this direction the extension rate is $\dot{\gamma}\sqrt{\lambda}$, so that steady chain stretch occurs when $\dot{\gamma}\sqrt{\lambda} > \tau_R^{-1}$. Since steady chain stretch does not occur for $\lambda=0$, the shear viscosity η is predicted by the Doi-Edwards theory to *decrease* monotonically for *all* $\dot{\gamma}$. Recent experimental work has shown that these predictions for $\lambda=0$ are not quite correct, and "weak" chain stretching does actually occur at steady state for large enough shear rates. Modified models have recently been proposed which retain a very small residual cross-sectional dimension of a fully aligned chain.[27] This has virtually no effect on model predictions for $\lambda>0$, but does change the qualitative behavior for simple shear flow by allowing stretch at very high shear rates. One consequence, understandable from the preceding discussion, is that this weak stretching process is predicted to decrease the rate of shear-thinning, which is known to be excessive in the original Doi-Edwards theory.

3.1.1 Steady Flow Data

Velocity gradient and birefringence data were obtained at the stagnation point of the two-roll mill for four different configurations, which would correspond to flow types λ=0.02, 0.06, 0.15 and 0.20 for a Newtonian fluid. A typical set of velocity gradient data is shown in Figure 3 where we plot $\dot\gamma$ and λ as a function of Weissenberg number $\mathrm{Wi} = \dot\gamma\sqrt{\lambda}\tau_R$ (based on *Newtonian* values of $\dot\gamma_N$, λ_N and the equilibrium state relaxation time τ_R.) The solid lines show the expected behavior for a *Newtonian* fluid—i.e., $\dot\gamma$ increases *linearly* and λ is a *constant*. We see that the velocity gradient is strongly influenced by the viscoelastic nature of the polymer solution. Both the magnitude, $\dot\gamma$, and the flow type parameter λ are strongly decreased. The flow thus becomes less extensional and more shear-like in nature. The strength of the flow, either measured by the extension rate in the direction of steady state alignment, $\dot\gamma\sqrt{\lambda}$, or by the magnitude of the strain rate, $\dot\gamma/2(1+\lambda)$, decrease monotonically with increase of Wi. Thus, for any comparison between experiment and theoretical model predictions, and especially in using birefringence data to infer rheological properties, like the generalized viscosity, η, it is essential to incorporate *measured* flow data. Although the decrease in $\dot\gamma$ and λ at the stagnation point is qualitatively similar to that which occurs for dilute Boger-like fluids (both measured experimentally and predicted by dumbbell-based constitutive models for these systems), the flow modification is much more localized for the dilute fluids compared to the present entangled fluids.

3.1.2 Steady Birefringence Data

Steady state birefringence data for the 2.89×10^6 polystyrene solution is shown in Figure 4, in this case plotted as a function of the Weissenberg number based upon *measured* values of $\dot\gamma$ and λ, for λ=0.02, 0.06 and 0.15. Also shown are predictions from the extended version of the Doi-Edwards theory which includes segmental stretch, [19] based upon the measured values of $\dot\gamma$ and λ (recall that viscoelastic flow calculations are not presently possible for this model).

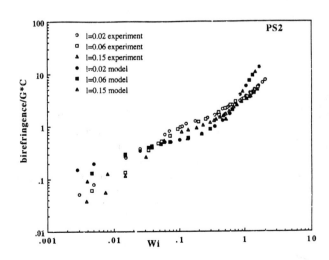

Figure 4:
Measured birefringence values compared with predictions from the extended Doi-Edwards theory (ref. 18,19) using measured flow data.

Focusing first on the experimental data, there are a number of important features. First, the data shows clear evidence of chain stretching. One indication is that birefringence levels are found for Wi > 0(1) which exceed the expected plateau values based upon literature values for C and the plateau modulus, G_N^0. A second indicator of segmental stretching is that the birefringence data for all values of λ collapses to a single curve when plotted versus $\dot{\gamma}\sqrt{\lambda}\tau_R$. This is predicted by the extended Doi-Edwards theory, [18] but is also expected on general grounds whenever there is significant stretch and alignment in the outflow direction. A second feature that is evident without direct comparison with model predictions, is that there is only a very weak plateau region. Linear viscoelastic data for the same fluid shows a much broader and more distinct plateau for G, consistent with the expected separation of timescales τ_D and τ_R by a factor of $3N \approx 30$. A simple interpretation of the birefringence data in Figure 4 is that the flow-induced alignment narrows the gap between τ_R and τ_D very dramatically. As polymer alignment is increased, the topological constraints that restrict relaxation of flow-induced alignment become less effective (in the language of reptation models this is equivalent to saying that the tube radius *dilates* by an amount that increases with increased alignment). Although current models of concentrated solutions of rod-like polymer molecules have included this effect, [43] it has not yet been incorporated into models for flexible chain systems (though a proposal for doing so has been suggested by Yavich [14]).

Comparing experimental data with model predictions, there are several additional points worth comment. First, the data for Wi < 0(1) does not correlate with $\dot{\gamma}\sqrt{\lambda}$ as it does for Wi > 0(1). Indeed, the data spread from model calculations is very similar to what is seen experimentally. However, it is obvious that the measured birefringence levels fall significantly below those predicted. We believe that this may be an additional sign of the flow-induced alignment effect on τ_D that was mentioned above. The minimum Wi based upon the equilibrium estimate of τ_D is only 0.3, and by this point significant chain alignment will have already occurred. The decrease of τ_D, or increase of tube radius, could account for the differences in Δn that are seen in Figure 4, though additional experimental and theoretical work is necessary to confirm this speculation. The other obvious discrepancy between theory and experiment is that the predicted rate of chain stretch is obviously much stronger than what is seen experimentally. This may reflect a problem with the current chain-stretching part of the extended Doi-Edwards model, [19] but it is important to remember that no account of alignment-induced tube dilation has been included in the current model calculations, and this will clearly reduce the predicted rate of chain stretching even with no other changes in the model.

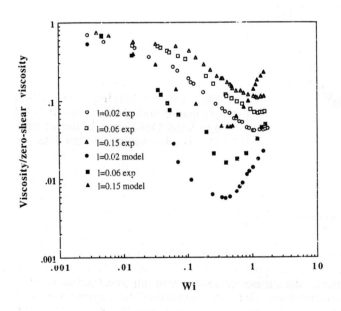

Figure 5:
Generalized extensional viscosity (eqn. 4) obtained via the stress-optical relationship (eqn. 1) from measured birefringence compared with predictions from the extended Doi-Edwards theory (ref. 18,19) using measured flow data.

3.1.3 Steady Generalized Extensional Viscosity

We have noted earlier that the measured birefringence values can be used in conjunction with the stress-optical formula to deduce corresponding rheological functions. Results for the generalized extensional viscosity function versus measured values of Wi are shown in Figure 5, together with calculated results for the extended Doi-Edwards theory [19] using the measured $\dot{\gamma}$ and λ. The three sets of curves correspond to three different configurations of the two-roll mill; if the fluids were Newtonian these would correspond to three constant values of λ. Since both $\dot{\gamma}$ and λ are functions of the flow conditions, each curve in Figure 5 represents the viscosity η along a trajectory of parameter space corresponding to increasing $\dot{\gamma}$ and decreasing λ (and this is also true of the model predictions since these are all based on the *measured* $\dot{\gamma}$ and λ values). A comprehensive picture of the predicted dependence of η on $\dot{\gamma}$ and λ is shown in Figure 6, which is reproduced from our recent modeling paper. [18] Several brief comments should be noted about these results (a more comprehensive discussion is contained in a recent paper [14]).

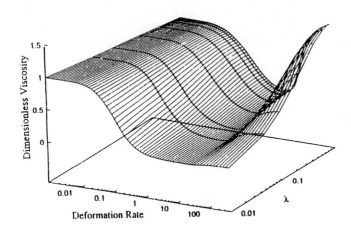

Figure 6: Predicted dependence of the generalized extensional viscosity on $\dot{\gamma}$ and λ (eqn. 2) for the extended Doi-Edwards theory (ref. 18,19).

First, the viscosity *decreases* sharply with the strain rate. This is strong evidence that the present solution, while corresponding to only ten entanglements per chain, is still dominated by entanglement effects. Second, at Wi $\approx 0(1)$, there is a very distinct change toward increase of η with strain rate. This is a clear signal of chain stretching, as explained earlier. Although the range of Wi values is too small to establish the ultimate rate of η increase, this is a consequence of the viscoelastic changes in the flow, which reduce both λ and $\dot{\gamma}$ from their "design," values and current experiments are being carried out to encompass larger values of Wi. Comparing with model predictions, we also see that the rate of decrease of η with Wi is smaller in the experiments and that there is a considerable offset in absolute values. The former effect is likely due to the flow-induced dilation of tube radius that was suggested earlier. It is well known that the original Doi-Edwards theory predicts excessively strong *shear*-thinning. The introduction of an alignment-induced tube dilation would likely cure this defect, both for shear flow and for the positive λ flows considered here. The offset in absolute values of η is a consequence of a sharp difference in the orientation angle χ between model predictions and experimental measurements (see eqn. 5). The experimental values of χ are typically smaller and this leads to predictions which exceed measured values for η. We do not currently have an explanation for the difference in orientation angle.

4. FUTURE RESEARCH

We have suggested above that the differences between predictions using the extended Doi-Edwards theory [19] and experimental measurements may be due to the neglect of flow-induced tube dilation as the polymer chains become increasingly aligned. However, we also recognize that ten entanglements per chain is quite marginal for applicability of reptation ideas, and this will be exacerbated if the number of entanglements per chain is reduced by tube dilation. Hence, to completely distinguish deficiencies of the model from its inapplicability due to small N, we are currently carrying out measurements for polyisoprene in the oligomer "squalene" where we can achieve equilibrium values of $N=0(40)$ without introducing excessive viscosities for use in our flow cells. At the same time, we are seeking to reach larger Weissenberg numbers in order to obtain information on chain stretching, and also extending our studies to include measurements in a newly constructed four-roll mill so that we obtain a purely extensional flow $(\lambda=1)$.

Acknowledgments: We thank the Polymers Program of the National Science Foundation for support of this work. The dilute solution measurements were made by Graham Harrison, and the corresponding numerical solutions were obtained by Johan Remmelgas. The work on entangled solutions comes mainly from the PhD thesis of Dmitry Yavich. We also thank Enrique Geffroy, Pushpendra Singh and David Mead for their contributions to this project.

References

1. Harlen, O. G., G. H. McKinley and S. Spielberg, in preparation (1996).

2. Tirtaatmadja, V. and T. Sridhar, *J. Rheol.* **39**, 1133 (1995).

   James, D. F. and T. Sridhar, *J. Rheol.* **39**, 713 (1995).

   Tirtaatmadja, V. and T. Sridhar, *J. Rheol.* **37**, 1081 (1993).

3. G. Astarita, *J. non-Newt. Fl. Mech.* **6**, 69 (1979)

   W. L. Olbricht, J. M. Rallison and L. G. Leal, *J. Non-Newtonian Fluid Mechanics* **10**, 291 (1982)

4. James, D. F. and Walters, K. in*Techniques in Rheological Measurements* edited by A. A. Collyer, Elsevier, London (1993).

   Keller, R. A., *J. non-Newt. Fl. Mech.* **42**, 49 (1992).

5. Quinzani, L. M., Armstrong, R. C. and Brown, R. A., *J. non-Newt. Fl. Mech.* **52**, 1 (1994).

   Quinzani, L. M., Armstrong, R. C. and Brown, R. A., *J. Rheol.* **39**, 1201 (1995).

6. McKinley, G. H., Armstrong, R. C. and Brown, R. A., *Phil. Trans. Roy. Soc. London* **A 344**, 265 (1993).

   Arigo, M. T., Rajagopalan, D., Shapley, N. and McKinley, G. H., *J. non-Newt. Fl. Mech*, **60**, 225 (1995).

7. H. P. W. Baaijens, Peters, G. W. M., Baaijens, F. P. T. and Meijer, H. E. H., *J. Rheol.* **39**, 1243 (1995).

   F. P. T. Baaijens, Baaijens, H. P. W., Peters, G. W. M. and Meijer, H. E. H., *J. Rheol.* **38**, 351 (1994).

8. J. V. Satrape and Crochet, M. J., *J. non-Newt. Fl. Mech.* **55**, 91 (1994).

9. S. P. Carrington, PhD Thesis, U. of Bristol (1995).

10. F. R. Cottrell, Merrill, E. Q. and Smith, K. A., *J. Polym. Sci. A-2*, **27**, 1415 (1969).

    R. C. Armstrong, Gupta, S. K. and Basaran, O., *Polym. En.g Sci.* **20**, 466 (1980).

    M. J. Manasveta and Hoaglund, D. A., *Macromolecules*, **24**, 3427 (1991).

11. R. G. Larson, Perkins, T. T., Smith, D. E. and Chu, S., preprint (1996)

    (see also *Science* **268**, 83 (1995))

12. J. T. Wang, Yavich, D. and Leal, L. G., *Phys. Fluids A* **6**, 3519 (1994).

13. G. E. Fuller, Rallison, J. M., Schmidt, R. L. and Leal, L. G., *J. Fl. Mech.* **100**, 555 (1980).

14. D. Yavich, PhD Thesis, UCSB (1995)

    D. Yavich, D. Mead and L. G. Leal, in preparation (1996)

    D. Yavich, D. Mead and L. G. Leal, in preparation (1996)

    Enrique Geffroy, PhD. Thesis, Calif. Inst. of Tech. (1990)

15. M. Doi and Edwards, S. F., *The Theory of Polymer Dynamics* Oxford Science Publications, Oxford (1986).

16. D. L. Davidson, Graessley, W. W. and Schowalter, W. R., *J. non-Newt. Fl. Mech.* **49**, 345 (1993).

17. M. J. Solomon and Müller, S. J., *J. non-Newt. Fl. Mech.* **62**, 81 (1996).

18. D. Mead, D. Yavich and L. G. Leal, D. Mead and L. G. Leal, *Rheological Acta*, **34**, 339-359 (1995).

    Rheological Acta, **34**, 360-383 (1995).

19. D. Pearson, E. A. Herbolzheimer, G. Marrucci and N. Grizutti, *J. Polym. Sci. Polym. Phys. Ed.* **29**, 1589 (1991).

20. B. Casewell, *J. non-Newt. Fl. Mech.* **62**, 99 (1996).

21. H. C. Öttinger, *Stochastic Processes in Polymeric Fluids*, Springer-Verlag, Berlin (1995).

22. A. Srinivasan, C. V. Chaubal, Egecioglu, O. and Leal, L. G., preprint (1996).

23. H. Giesekus, *J. non-Newt. Fl. Mech.* **11**, 69 (1982).

24. R. B. Bird, and DeAguiar, J. R., *J. non-Newt. Fl. Mech.* **13**, 149 (1983).

25. N. Phan-Thien and Tanner, R. I., *J. non-Newt. Fl. Mech.* **2**, 353)1977).

26. R. G. Larson, *Constitutive Equations for Polymer Melts and Solutions* Butterworths, Boston (1988).

27. N. A. Spenley and Cates, M.E., *Macromolecules* **27**, 3850 (1994).

    M. E. Cates, McLeish, T. C. B. and Marrucci, G., *Europhysics Letters* **21**, 451 (1993).

28. G. M. Harrison, Remmelgas, J. and Leal, L. G., preprint (1996). also presented in program on "Dynamics of Complex Fluds" at the Isaac Newton Institute, Cambridge UK, Feb. (1996).

29. P. Singh and L. G. Leal, *J. Rheol.* **38(3)**, 485-517 (1994).

30. P. Singh and Leal, L. G., *Theor. and Comp. Fluid Mechanics* **5**, 107-137 (1993).

31. G. G. Fuller and L. G. Leal, *Rheol. Acta* **19**, 580 (1980).

    P. N. Dunlap and L. G. Leal, *J. Non-Newtonian Fluid Mechanics* **23**, 5 (1987).

    P. N. Dunlap, C. H. Wang and L. G. Leal, *J. Polymer Sci., Part B: Polymer Physics* **25**, 2211 (1987).

32. A. Keller and J. A. Odell, *Colloid and Polymer Sci.,* **263**, 181-201 (1985).

33. A. Ouibrahim and D. H. Fruman, *J. non-Newt. Fl. Mech.* **7**, 315-331 (1980).

    D. F. James and J. H. Saringer, *J. Fluid Mechanics* **97**, 655-771 (1980).

34. G. G. Fuller and L. G. Leal, *J. Non-Newtonian Fluid Mechanics* **8**, 271 (1981).

    P. Singh and L. G. Leal, *J. Non-Newtonian Fluid Mechanics,* accepted (1996).

35. R. G. Larson and J. Magda, *Macromolecules* **22**, 3004-3010 (1989).

36. R. C.-Y. Ng and L. G. Leal, *J. Rheol.* **37**, 443-468 (1993).

37. R. C.-Y. Ng and L. G. Leal, *Rheol. Acta,* **32** 25-35 (1993).

38. J. M. Rallison, preprint (1996).

39. G. G. Fuller and L. G. Leal, *J. Polymer Sci.* (Polymer Physics Ed.) **19**, 531 (1981).

40. G. G. Fuller and L. G. Leal, *J. Polymer Sci.* (Polymer Physics Ed.) **19**, 557 (1981).

    Enrique Geffroy and L. G. Leal, *Journal of Polymer Science, Polymer Physics Ed.* **30**, 1329-1349 (1992).

41. K. Osaki, Y. Nishimura and M. Kurata, *Macromolecules* **18**, 1153-1157 (1985).

    K. Osaki, E. Takatori, Y. S. Tsunashima and M. Kurata, *Macromolecules* **20**, 525-529 (1987).

    K. Osaki, K. Nishizawa and M. Kurata, *Macromolecules* **15**, 1068-1071 (1982).

    K. Osaki, S. Kimura, K. Nishizawa and M. Kurata, *Macromolecules* **14**, 455-456 (1981).

42. W. W. Graessley, *Adv. Poly. Sci.* **47**, 67 (1982).

43. M. Doi, *Adv. in Coll. and Interface Sci.,* **17**, 233 (1982).

Theoretical and Applied Mechanics 1996
T. Tatsumi, E. Watanabe and T. Kambe (Editors)
© 1997 Elsevier Science B.V. All rights reserved.

Acoustics of Unstable Flows

A P Dowling

Department of Engineering, University of Cambridge,
Trumpington Street, Cambridge, CB2 1PZ, United Kingdom

A generalised energy equation is used to discuss self-excited flow oscillations within a unified framework. Four different types of flow instability are examined in detail: these are combustion instabilities, shock oscillations and edge and cavity tones. Theoretical results are compared with experimental data, both from laboratory rigs and from full-scale installations.

1. INTRODUCTION

Acoustic waves play an important role in the self-excited oscillations of many unstable flows. We discuss four different examples in which acoustic or hydrodynamic feedback leads to self-sustaining oscillations at discrete frequencies. These are combustion instabilities, shock oscillations, edge and cavity tones. They are chosen because they involve different forms of interaction between pressure waves and flow disturbances, highlighting in turn the rôle of unsteadiness in the rate of combustion, in entropy production and in vortex shedding from a sharp edge as driving mechanisms for instability. In these flows, there is a steady solution to the equations of fluid motion. The generalised energy equation enables us to determine under what conditions linear perturbations from this steady flow are unstable, and grow into self-sustaining finite amplitude oscillations.

Combustion oscillations can occur whenever combustion takes place within a resonator, essentially because unsteady combustion generates acoustic waves, while these acoustic waves act back to disturb the flame still further. If the phase relationship is suitable, the acoustic waves gain energy from their interaction with the unsteady combustion. When this energy gain exceeds that lost of reflection at the ends of the pipe, perturbations grow in amplitude. Rayleigh [1] developed criterion for the occurrence of thermoacoustic oscillations by investigating the energetics of the acoustic waves. These ideas were subsequently expressed in more mathematical terms by Chu [2]. Consideration of acoustic energy has also proved useful in understanding self-excited vortex flows [3 - 5], since it provides a convenient way of mapping the exchange of energy between the acoustical and vortical flow fields. Howe [6] developed a simple expression for the acoustical power production due to vorticity.

In Section 2 we derive a generalised acoustic energy equation, which encompasses both the Rayleigh source term due to unsteady heating and the Howe term due to vorticity. This energy equation also identifies the acceleration of density inhomogeneities or convected 'hot spots' as an additional source of acoustic energy. Thermoacoustic oscillations are discussed in detail in Section 3. Examples where the Rayleigh source term is the driving mechanism for flow instability are described. In Section 4, we consider flows destabilised by the acoustic field generated by convected entropy fluctuations. We demonstrate a thermoacoustic oscillation where, according to Rayleigh's criterion the unsteady combustion would be stable, but the flow is destabilised by the acoustic power generated by entropy inhomogeneities accelerating through a nozzle. A second example involves a supersonic underexpanded jet impinging on a bluff body. At certain flow conditions, the stand-off shock formed upstream of the body undergoes self-excited oscillations. These occur because shock motion changes the effective shock strength and leads to downstream entropy fluctuations. The deceleration of these entropy inhomogeneities in the stagnation region generates pressure waves which propagate upstream and sustain the shock oscillation. Finally, in Section 5, we use the same energy equation to discuss self-excited oscillations in some practical vortex flows.

2. A GENERALISED ACOUSTIC ENERGY EQUATION

Consider the generic flow illustrated in Figure 1. It may involve combustion and a separated flow region with vorticity. For such a flow, there is a steady solution to the equations of fluid motion and we still denote the 'steady-state' variables by an overbar. Any unsteadiness in the combustion or vorticity can generate acoustic waves, and we want to investigate when the 'steady-state' flow is unstable by seeing whether small perturbations to it grow or decay with time.

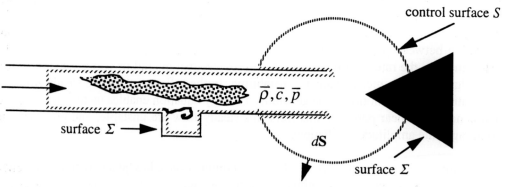

Figure 1 Combustion and vorticity within a control surface S

Decompose the particle velocity field into steady, solenoidal and irrotational elements by writing

$$\mathbf{u}(\mathbf{x},t) = \bar{\mathbf{u}}(\mathbf{x}) + \mathbf{u}_\omega(\mathbf{x},t) + \mathbf{u}_a(\mathbf{x},t), \tag{2.1}$$

where

$$\nabla \cdot \mathbf{u}_\omega = 0 , \qquad\qquad \nabla \times \mathbf{u}_\omega = \mathbf{w} - \bar{\mathbf{w}} \tag{2.2a}$$

$$\nabla \cdot \mathbf{u}_a = \nabla \cdot (\mathbf{u} - \bar{\mathbf{u}}), \qquad \nabla \times \mathbf{u}_a = \mathbf{0}. \tag{2.2b}$$

The normal components of $\bar{\mathbf{u}}$, $\mathbf{u}_\omega$ and $\mathbf{u}_a$ are to be zero on any rigid surfaces Σ:

$$\bar{\mathbf{u}} \cdot \mathbf{n} = \mathbf{u}_\omega \cdot \mathbf{n} = \mathbf{u}_a \cdot \mathbf{n} = 0 \qquad \text{on } \Sigma. \tag{2.3}$$

$\mathbf{u}_\omega(\mathbf{x}, t)$ is the velocity field due to the unsteady vorticity, $\mathbf{w} - \bar{\mathbf{w}}$, and its images in the surfaces Σ. We will refer to the irrotational velocity $\mathbf{u}_a(\mathbf{x},t)$ as 'acoustic', not making undue distinction between when the propagation distances are large in comparison with the acoustic wavelength so that compressibility is important, and when the distances are so short that the sound speed is effectively infinite and the feedback hydrodynamic. Since $\mathbf{u}_a(\mathbf{x},t)$ is irrotational, it can be expressed in terms of a scalar potential $\phi(\mathbf{x}, t)$:

$$\mathbf{u}_a = \nabla \phi. \tag{2.4}$$

Write the equation of mass conservation, $\dfrac{D\rho}{Dt}+\rho\nabla\cdot\mathbf{u}=0$, in the form:

$$\frac{1}{\bar{\rho}\bar{c}^2}\frac{\partial}{\partial t}\left(p-\bar{p}+\tfrac{1}{2}\bar{\rho}\left(u^2-\bar{u}^2\right)\right)+\frac{\mathbf{u}}{\bar{\rho}\bar{c}^2}\cdot\nabla(p-\bar{p})+\nabla\cdot\mathbf{u}_a$$

$$=-\frac{1}{\rho}\frac{D\rho}{Dt}-\nabla\cdot\bar{\mathbf{u}}+\frac{1}{\bar{\rho}\bar{c}^2}\frac{D}{Dt}(p-\bar{p})+\frac{\mathbf{u}}{\bar{c}^2}\cdot\frac{\partial\mathbf{u}}{\partial t}. \tag{2.5}$$

p is the pressure, ρ the density and c the speed of sound. The overbar denotes the steady-state flow.

The momentum equation,

$$\rho\frac{Du_i}{Dt}+\frac{\partial p}{\partial x_i}=\frac{\partial\tau_{ij}}{\partial x_j}, \tag{2.6}$$

can be rewritten as

$$\bar{\rho}\frac{\partial u_{ai}}{\partial t}+\frac{\partial}{\partial x_i}\left(p+\tfrac{1}{2}\bar{\rho}u^2\right)=-(\rho-\bar{\rho})\frac{Du_i}{Dt}-\bar{\rho}(\mathbf{w}\times\mathbf{u})+\frac{\partial\tau_{ij}}{\partial x_j}+\tfrac{1}{2}u^2\frac{d\bar{\rho}}{dx_i}-\bar{\rho}\frac{\partial u_{\omega i}}{\partial t}, \tag{2.7}$$

where τ_{ij} is the viscous stress tensor and we have made use of the vector identity: $\mathbf{u}\cdot\nabla\mathbf{u}=\mathbf{w}\times\mathbf{u}+\nabla\left(\tfrac{1}{2}u^2\right)$. After subtracting the steady-state flow from (2.7), we obtain

$$\bar{\rho}\frac{\partial u_{ai}}{\partial t}+\frac{\partial}{\partial x_i}\left(p-\bar{p}+\tfrac{1}{2}\bar{\rho}\left(u^2-\bar{u}^2\right)\right)=-(\rho-\bar{\rho})\frac{Du_i}{Dt}-\bar{\rho}(\mathbf{w}\times\mathbf{u}-\bar{\mathbf{w}}\times\bar{\mathbf{u}})+\frac{\partial}{\partial x_j}\left(\tau_{ij}-\bar{\tau}_{ij}\right)$$

$$+\tfrac{1}{2}\left(u^2-\bar{u}^2\right)\frac{d\bar{\rho}}{dx_i}-\bar{\rho}\frac{\partial u_{\omega i}}{\partial t}. \tag{2.8}$$

An acoustic energy equation follows from multiplying (2.5) by $\left(p-\bar{p}+\tfrac{1}{2}\bar{\rho}\left(u^2-\bar{u}^2\right)\right)$ and adding it to the product of (2.8) and $\mathbf{u}_a+\mathbf{u}\left(p-\bar{p}+\tfrac{1}{2}\bar{\rho}\left(u^2-\bar{u}^2\right)-\bar{\rho}\mathbf{u}\cdot\mathbf{u}_a\right)/\bar{\rho}\bar{c}^2$. After some algebra, the resulting equation can be expressed in the form:

$$\frac{\partial e}{\partial t}+\nabla\cdot\mathbf{I}=E_1+E_2+E_3+E_4+E_5+E_6, \tag{2.9}$$

where

$$e=\tfrac{1}{2}\bar{\rho}u_a^2-\frac{\bar{\rho}}{2\bar{c}^2}(\mathbf{u}\cdot\mathbf{u}_a)^2+\frac{1}{2\bar{\rho}\bar{c}^2}\left(p-\bar{p}+\tfrac{1}{2}\bar{\rho}\left(u^2-\bar{u}^2\right)\right)^2, \tag{2.10}$$

$$I_i=\left(p-\bar{p}+\tfrac{1}{2}\bar{\rho}\left(u^2-\bar{u}^2\right)\right)\left(u_{ai}+u_i\frac{(p-\bar{p})}{\bar{\rho}\bar{c}^2}\right)+\bar{\rho}\phi\frac{\partial u_{\omega i}}{\partial t}$$

$$-\left(\tau_{ji}-\bar{\tau}_{ji}\right)\left(u_{aj}+\frac{u_j}{\bar{\rho}\bar{c}^2}\left(p-\bar{p}+\tfrac{1}{2}\bar{\rho}\left(u^2-\bar{u}^2\right)-\bar{\rho}\mathbf{u}\cdot\mathbf{u}_a\right)\right) \tag{2.11}$$

and

$$E_1=\left(-\frac{1}{\rho}\frac{D\rho}{Dt}-\nabla\cdot\bar{\mathbf{u}}+\frac{1}{\bar{\rho}\bar{c}^2}\frac{D}{Dt}(p-\bar{p})+(p-\bar{p})\nabla\cdot\left(\frac{\mathbf{u}}{\bar{\rho}\bar{c}^2}\right)\right)\left(p-\bar{p}+\tfrac{1}{2}\bar{\rho}\left(u^2-\bar{u}^2\right)\right), \tag{2.12a}$$

$$E_2 = \nabla\bar{\rho} \cdot \left(\phi \frac{\partial \mathbf{u}_\omega}{\partial t} + \frac{1}{2}\left(u^2 - \bar{u}^2\right)\left(\mathbf{u}_a + \frac{\mathbf{u}}{\bar{\rho}\bar{c}^2}\left(p - \bar{p} + \frac{1}{2}\bar{\rho}\left(u^2 - \bar{u}^2\right) - \bar{\rho}\mathbf{u}\cdot\mathbf{u}_a\right)\right)\right), \tag{2.12b}$$

$$E_3 = -(\rho - \bar{\rho})\frac{D\mathbf{u}}{Dt}\cdot\left(\mathbf{u}_a + \frac{\mathbf{u}}{\bar{\rho}\bar{c}^2}\left(p - \bar{p} + \frac{1}{2}\bar{\rho}\left(u^2 - \bar{u}^2\right) - \bar{\rho}\mathbf{u}\cdot\mathbf{u}_a\right)\right), \tag{2.12c}$$

$$E_4 = -\bar{\rho}\left(\mathbf{w}\times\mathbf{u} - \bar{\mathbf{w}}\times\bar{\mathbf{u}}\right)\cdot\left(\mathbf{u}_a + \frac{\mathbf{u}}{\bar{\rho}\bar{c}^2}\left(p - \bar{p} + \frac{1}{2}\bar{\rho}\left(u^2 - \bar{u}^2\right) - \bar{\rho}\mathbf{u}\cdot\mathbf{u}_a\right)\right), \tag{2.12d}$$

$$E_5 = -\left(\tau_{ij} - \bar{\tau}_{ij}\right)\frac{\partial}{\partial x_j}\left(u_{ai} + \frac{u_i}{\bar{\rho}\bar{c}^2}\left(p - \bar{p} + \frac{1}{2}\bar{\rho}\left(u^2 - \bar{u}^2\right) - \bar{\rho}\mathbf{u}\cdot\mathbf{u}_a\right)\right), \tag{2.12e}$$

and

$$E_6 = \frac{\bar{\rho}\,\mathbf{u}\cdot\mathbf{u}_a}{\bar{c}^2}\frac{\partial}{\partial t}\left(\bar{\mathbf{u}}\cdot\mathbf{u}_\omega + \frac{1}{2}\left(u_\omega^2 - u_a^2\right)\right) - \left(\frac{1}{2}\left(u^2 - \bar{u}^2\right) - \mathbf{u}\cdot\mathbf{u}_a\right)\frac{\mathbf{u}}{\bar{c}^2}\cdot\nabla\left(p - \bar{p} + \frac{1}{2}\bar{\rho}\left(u^2 - \bar{u}^2\right)\right) \tag{2.12f}$$

Note that equation (2.9) is exact, but is not unique: there are many ways of combining the equations of mass and momentum conservation into an acoustic energy equation. However, this form is convenient for our purposes. For small amplitude perturbations from the steady-state, all the terms in (2.9) depend quadratically on the amplitude of the perturbations, as is appropriate for an energy equation. For a subsonic flow, e is positive definite. It only involves the acoustic kinetic energy, and excludes the kinetic energy due to the vortical flow. Similarly, it depends on an incompressible approximation to stagnation pressure and, in particular, is uninfluenced by entropy fluctuations. In an irrotational flow, e is equivalent to Morfey's energy density [7], and in the geometrical acoustics limit it reduces to the Blokhinstev form [8].

We can use equation (2.9) to investigate the change of acoustic energy within a volume V bounded by a control surface S. After integration over V it yields

$$\frac{\partial}{\partial t}\int_V e\,dV = \sum_{n=1}^{6}\int_V E_n\,dV - \int_S \mathbf{I}\cdot d\mathbf{S} \tag{2.13}$$

The term on the left-hand side of equation (2.13) is the rate of change of acoustic energy within V. Taking a short-time average shows that acoustic waves grow in magnitude if

$$\sum_{n=1}^{6}\tilde{E}_n\,dV > \int_S \tilde{\mathbf{I}}\cdot d\mathbf{S}, \tag{2.14}$$

where the tilde denotes an average over one period of the oscillation. $\mathbf{I}$ denotes the flux of acoustic energy through the surface S. Note that when S is a fixed impenetrable surface, the boundary conditions (2.3) ensure that the only non-zero term in $\mathbf{I}$ is due to viscous stress. The terms E_n describe exchange of energy between the acoustic field and the heat addition and vortical flow. These terms have a straightforward interpretation.

To understand the energy source term E_1, first note that the chain rule of differentiation shows that

$$\frac{D\rho}{Dt} = \left.\frac{\partial\rho}{\partial s}\right|_p \frac{Ds}{Dt} + \frac{1}{c^2}\frac{Dp}{Dt}. \tag{2.15}$$

The material derivative of the entropy s is substantial whenever combustion or heat transfer take place, and substitution for $D\rho/Dt$ in (2.12a) shows that

$$E_1 = \left(-\frac{Ds}{Dt}\frac{1}{\rho}\frac{\partial\rho}{\partial s}\Big|_p - \nabla\cdot\overline{\mathbf{u}} - \frac{1}{\rho c^2}\frac{Dp}{Dt} + \frac{1}{\bar{\rho}\bar{c}^2}\frac{D}{Dt}(p-\bar{p}) + (p-\bar{p})\nabla\cdot\left(\frac{\mathbf{u}}{\bar{\rho}\bar{c}^2}\right) \right)$$

$$\times \left(p-\bar{p}+\tfrac{1}{2}\bar{\rho}\left(u^2-\bar{u}^2\right) \right). \tag{2.16}$$

E_1 therefore describes the rate of gain of acoustic energy due to unsteady heat input and is the driving term in most thermoacoustics oscillations, which are considered further in Section 3. E_2 accounts for the effect of mean density gradients, and is also discussed in Section 3. E_3 describes the acoustic power generated by the acceleration of density inhomogeneities. We show in Section 4 that this can be considerable when there is a nozzle contraction downstream of a low Mach number combustion zone, and that it can play a major rôle in shock oscillations. E_4 is the acoustic power generated by an usteady vortical flow and is discussed in more detail in Section 5. E_5 is usually negative and accounts for the dissipation of energy by viscosity. For small amplitude perturbations, the terms in E_6 are multiplied by a factor $(\overline{\mathbf{u}}/\bar{c})^2$ and do not usually play an important rôle in low Mach number flows.

3. THERMOACOUSTIC OSCILLATIONS

In this section we concentrate on the rôle of unsteady heat addition as the driving mechanism for flow instability. To highlight its contribution, consider the flow to be irrotational and neglect viscous and heat conduction effects. Then

$$\rho T \frac{Ds}{Dt} = q(\mathbf{x},t), \tag{3.1}$$

where T denotes temperature and q is the rate of heat input/unit volume. For a perfect gas

$$\frac{\partial\rho}{\partial s}\Big|_p = -\frac{\rho}{c_p} = -\frac{\rho T(\gamma-1)}{c^2}, \tag{3.2}$$

c_p is the specific heat capacity at constant pressure and γ the ratio of specific heats. Equations (3.1) and (3.2) can be used to simplify E_1 in (2.16). E_2 and E_3 have simple forms for small amplitude disturbances of an irrotational flow. E_4 is identically zero, and E_6 vanishes to second order in the amplitude of the small disturbances. E_5 is zero when viscous effects are neglected. Hence, for small amplitude disturbances, the energy equation in (2.9) reduces to

$$\frac{\partial e}{\partial t} + \nabla\cdot\mathbf{I} = (p'+\bar{\rho}\overline{\mathbf{u}}\cdot\mathbf{u}_a)\left(\frac{(\gamma-1)q'}{\gamma\bar{p}} - \frac{(\gamma-1)^2\bar{q}p'}{\gamma^2\bar{p}^2} - \left(\frac{\mathbf{u}_a}{\gamma\bar{p}} + \frac{p'\overline{\mathbf{u}}}{\gamma^2\bar{p}^2} \right)\cdot\nabla\bar{p} \right)$$

$$+ \left(u_{ai} + \frac{\bar{u}_i p'}{\bar{\rho}\bar{c}^2} \right)\left(-\rho'\overline{\mathbf{u}}\cdot\nabla\bar{u}_i + \overline{\mathbf{u}}\cdot\mathbf{u}_a\frac{d\bar{\rho}}{dx_i} \right). \tag{3.3}$$

The prime denotes a perturbation from the 'steady-state' flow. Each term in this equation depends quadratically on these small perturbations. When there is no mean flow, the right-hand side of (3.3) reduces to the Rayleigh source term $(\gamma-1)p'q'/\gamma\bar{p}$: the component of unsteady heat input in phase with the pressure perturbation drives the acoustic waves to instability. A mean flow and mean heat input lead to additional source terms, which are in general complicated, but simplify when the steady state flow is one dimensional. Then the equations of conservation of mass, momentum and energy can be combined to relate the mean flow gradients to $\bar{q}$. This shows that [9]:

$$\frac{d\bar{u}}{dx} = -\frac{1}{\bar{\rho}\bar{u}}\frac{d\bar{p}}{dx} = -\frac{\bar{u}}{\bar{\rho}}\frac{d\bar{\rho}}{dx} = \frac{(\gamma-1)\bar{q}}{(1-\bar{M}^2)\gamma\bar{p}} \tag{3.4}$$

where $\bar{M} = \bar{u}/\bar{c}$ is the local mean Mach number. Substitution into (3.3) leads to

$$\frac{\partial e}{\partial t} + \frac{\partial I}{\partial x} = (p' + \bar{\rho}\bar{u}u_a)\frac{(\gamma-1)q'}{\gamma\bar{p}}$$

$$-\frac{(\gamma-1)\bar{q}}{(1-\bar{M}^2)}\left[(\gamma-1-\gamma\bar{M}^2)\left(\frac{p'^2}{\gamma^2\bar{p}^2} + \frac{\bar{M}p'u_a}{\gamma\bar{p}\bar{c}}\right) + (1-\bar{M}^2)\frac{u_a^2}{\bar{c}^2} + \frac{\bar{M}\rho'}{\bar{\rho}}\left(\frac{u_a}{\bar{c}} + \bar{M}\frac{p'}{\gamma\bar{p}}\right)\right] \tag{3.5}$$

in agreement with the energy equation given by Bloxsidge et al [9] for one-dimensional flow. Oscillations of a confined flame, burning in the wake of a bluff-body flame-holder, saturate when the flow reverses [10]. Hence in a limit cycle, u_a is of the same order as the mean velocity, $\bar{u}$. This leads to oscillations with $u_a/\bar{u}$ and possibly $q'/\bar{q}$ and $\rho'/\bar{\rho}$ of order unity, while $p'/\bar{p}$ and $u_a/\bar{c}$ are only $O(\bar{M})$. Therefore, for a low Mach number mean flow, the energy source term in (3.5) reduces to the simple Rayleigh source, $(\gamma-1)p'q'/\gamma\bar{p}$.

It is evident from (3.3) and (3.5) that knowledge of the phase relationship between the unsteady heat input and pressure is crucially important in determining the stability. The form of coupling between the unsteady heat input and the flow also affects the frequency of oscillation. This can be illustrated by considering a simple case with no mean flow [11]. Then linear pressure perturbations satisfy an inhomogenous wave equation:

$$\frac{1}{\bar{\rho}\bar{c}^2}\frac{\partial^2 p'}{\partial t^2} - \nabla\cdot\left(\frac{1}{\bar{\rho}}\nabla p'\right) = \frac{\gamma-1}{\bar{\rho}\bar{c}^2}\frac{\partial q'}{\partial t}. \tag{3.6}$$

The eigenfrequencies ω are determined by considering pressure modes proportional to $e^{i\omega t}$ and determining the values of ω which satisfy the boundary conditions. Real ω then gives the frequency of oscillation, while the sign of Imaginary ω determines whether oscillations grow or decay in time.

To demonstrate the influence of the coupling between q' and the flow on these eigenfrequencies, consider the simple one-dimensional problem illustrated in Figure 2. Heat addition is concentrated at a single plane $x = b$, i.e. $q'(\mathbf{x},t) = Q'(t)\delta(x-b)$. In $x<b$, the mean temperature is $\bar{T}_1$, the density $\bar{\rho}_1$, and the speed of sound $\bar{c}_1$, while in $x>b$, the mean fluid properties have values $\bar{T}_2$, $\bar{\rho}_2$ and $\bar{c}_2$. We take boundary conditions

$$\frac{\partial p'}{\partial x} = 0 \quad \text{on} \quad x = 0 \text{ and } p' = 0 \quad \text{on} \quad x = \ell. \tag{3.7}$$

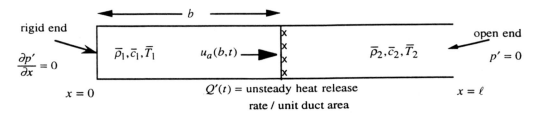

Figure 2 One-dimensional disturbances in a duct

Results for two idealised forms of unsteady heat input are shown in Figure 3. In Case I, $Q'(t) \equiv 0$, there is no unsteady heat input/unit volume. This is the limiting case when the reaction time of the heating process is much longer than the period of oscillation. Then, even though the fluid might enter the heating zone unsteadily, the rate of heat input is unable to adjust sufficiently rapidly and remains constant. In Case II, the instantaneous rate of heat input is taken to be directly proportional to the instantaneous mass flow rate into the heating zone, the heat input per unit mass being that required to raise the mean temperature from $\bar{T}_1$ to $\bar{T}_2$: i.e. $Q'(t) = c_p(\bar{T}_2 - \bar{T}_1)\bar{\rho}_1 u_a(b,t)$. Such a heat input would be appropriate when the reaction time is so much faster than the period of oscillation that the rate of heat input responds quasi-steadily to changes of flow rate into the heating zone. The method of solution is described in reference [11] and results are given in Figure 3. When $\bar{T}_2 = \bar{T}_1$, the lowest frequency of oscillation is that of the $\frac{1}{4}$-wave mode and $\omega = \frac{1}{2}\pi\bar{c}_1 / \ell$. As the temperature difference increases, the form of the unsteady heat release rate affects the frequency. Indeed, for $\bar{T}_2/\bar{T}_1 = 6$, there is nearly a 60 % difference between the frequency of oscillation for Case I and Case II.

Of course, it is well known that the unsteady heat input affects the frequency. Rayleigh commented that "the pitch is raised if heat be communicated to the air a quarter period before the phase of greatest condensation: and the pitch is lowered if the heat be communicated a quarter period after the phase of greatest condensation". Numerical calculations show that this shift in frequency can be significant, and Dowling [11] discusses how this effect is accounted for in some of the methods of solution in the literature.

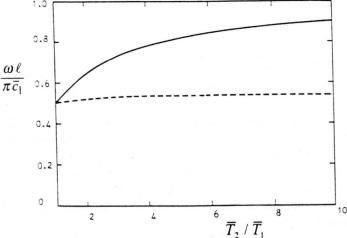

Figure 3 The lowest frequency of oscillation as a function of temperature ratio $\bar{T}_2/\bar{T}_1$ for $b = \ell / 2$. ———— Case I; ------------ Case II, (from reference [11]).

178

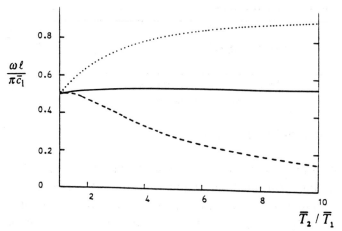

Figure 4   The lowest frequency of oscillation for Case II for $b = \ell/2$, —— Correct solution; ·········'Acoustic resonance frequency' and the value predicted by Green function technique;   ------ Frequency predicted by linearised Galerkin theory (from reference [11])

Figure 4 is taken from reference [11]. It compares the lowest frequencies of oscillation for Case II predicted by various approximate methods with the exact result. Many authors just assume that thermoacoustic oscillations occur at the 'acoustic resonance frequency' of a duct. By this, they invariably mean a frequency at which a solution of the homogenous wave equation (equation (3.6) with $q' \equiv 0$) satisfies the boundary conditions. This procedure clearly neglects any influence of the unsteady heat input. A 'Green function' technique used by Hedge *et al* [12] similarly neglects effects of the unsteady heating. The linearised Galerkin method of Culick [13] and others recognises that the form of the coupling between the unsteady heat input and the flow does affect the frequency, but linearises the change in frequency and mode shape. The form of the unsteady heat input in Case II vanishes identically for $\overline{T}_2 = \overline{T}_1$ and is infinitesimally small for small $\overline{T}_2 - \overline{T}_1$. Hence, the frequency predicted from the linearised Galerkin method and the true frequency agree exactly for $\overline{T}_2 = \overline{T}_1$. The two curves of frequency against temperature ratio also have the same slope at this point, but they diverge significantly for larger temperature ratios, as shown in Figure 4. By a temperature ratio of six, the Galerkin method predicts a frequency which is less than half the exact value.

The generic problem of a ducted flame burning in the wake of a bluff-body flame holder has been studied extensively [14] because it models the essential features of the afterburner of an aeroengine. Afterburners are notoriously susceptible to damaging, low frequency combustion oscillations. It is necessary to include the mean flow and mean heat release to correctly predict the onset fuel-air ratio, frequencies of oscillation and mode shapes [11,15]. Moreover, as we have just seen in Figure 3, the coupling between the unsteady heat release rate and the flow is crucially important. Bloxsidge *et al* [15] investigated this important relationship experimentally by exciting a stable flame at a range of frequencies and measuring the change in the rate of heat release at different axial positions in the duct. Instantaneous rates of heat release were found to lag the velocity at the flame-holder, $u_1(t)$, which led to a flame model of the form [10, 15]

$$\tau_1 \frac{dq'}{dt} + q' = \frac{\overline{q}(x)}{\overline{u}_1} u_1'\big(t - \tau_2(x)\big). \tag{3.8}$$

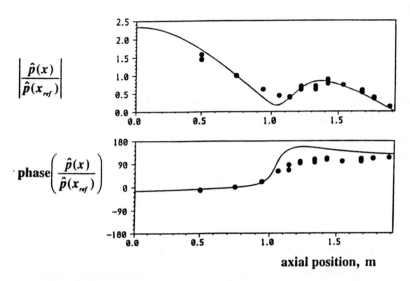

Figure 5 Comparison of theory ⎯⎯ and experiment ● for a low Mach number inlet flow, equivalence ratio, $\phi = 0.70$, $\overline{M}_1 = 0.08$, $\overline{T}_1 = 288$ K, frequency = 82 Hz (calculated), 77 ±1Hz (experimental), from reference [15].

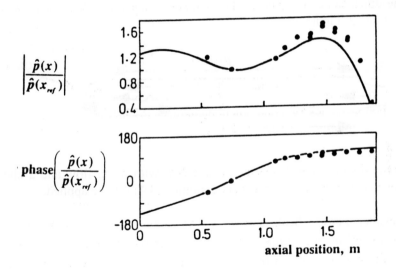

Figure 6 Comparison of theory ⎯⎯ and experiment ● for a higher Mach number inlet flow, equivalence ratio, $\phi = 0.66$, $\overline{M}_1 = 0.23$, $\overline{T}_1 = 540$ K, frequency = 161 Hz (calculated), 159 ±2Hz (experimental), from reference [16].

The addition of a nozzle at the downstream end of the duct was found both experimentally and theoretically to be highly destabilising. This is because the acceleration of hot spots through the nozzle is a source of additional sound which can propagate upstream and perturb the flame still further. The rôle of these convected entropy inhomogeneities as a source of acoustic energy is discussed in the next section.

180

4. CONVECTED ENTROPY INHOMOGENEITIES

In a low Mach number irrotational flow, E_3 in (12.2c) simplifies to

$$E_3 = -(\rho - \bar{\rho})\frac{D\mathbf{u}}{Dt} \cdot \mathbf{u}_a. \tag{4.1}$$

This describes the acoustic energy input due to the acceleration of a 'hot spot', whose density ρ differs from the mean density $\bar{\rho}$. For small amplitude disturbances, E_3 reduces to

$$E_3 = -\rho'\frac{D\bar{\mathbf{u}}}{Dt} \cdot \mathbf{u}_a, \tag{4.2}$$

an energy source term which is also displayed explicitly in (3.3). Averaging over one acoustic cycle shows that there is a net energy input from this term if

$$-\rho' u_{a_s} > 0, \tag{4.3}$$

where u_{a_s} is the component of the acoustic velocity, $\mathbf{u}_a$, in the direction of the mean flow acceleration. A convected density inhomogeneity is most destabilising when a hot spot, with negative ρ', is in phase with u_{a_s}.

As a first example of a flow which is destabilised by convected entropy inhomogeneties, consider one-dimensional thermoacoustic oscillations in a duct with mean flow and a choked exit, as illustrated in Figure 7.

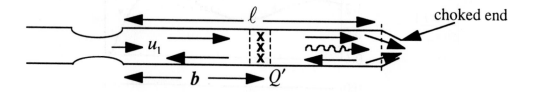

Figure 7 One dimensional disturbances in a duct with mean flow and a choked nozzle. # denotes the plane on which the boundary condition (4.4) is to be applied.

We will consider a case where, as in Case I of Section 3, there is a steady heat addition, which raises the mean temperature from $\bar{T}_1$ to $\bar{T}_2$ across the surface $x = b$, and no unsteady heat input/unit time, i.e. $Q'(t) \equiv 0$. For a short nozzle, the downstream boundary condition can be transferred from the exit to the constant area duct just upstream of the nozzle. This leads to

$$2\frac{u_a}{\bar{u}} + \frac{\rho'}{\bar{\rho}} = \frac{p'}{\bar{p}} \tag{4.4}$$

at the downstream end of the constant area duct [17].

The eigenfrequencies of the thermoacoustic oscillations in such a configuration can be determined by the method described by Dowling [11]. The only modification required is to change the fifth line of the matrix X in equation (3.15) of reference [11] to describe the choked end boundary condition in (4.4), instead of the open end condition considered in that paper. Results for a range of temperature ratios are shown in Figure 8.

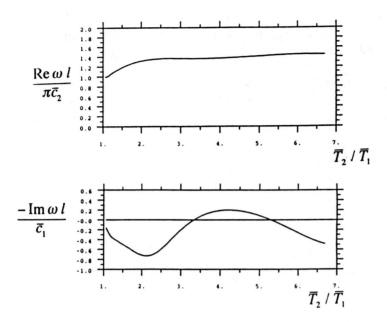

Figure 8   The fundamental frequency of oscillation for $b = \ell/2$, $\overline{M}_1 = 0.1$, and Case I, $Q'(t) \equiv 0$.

For $\overline{T}_2 = \overline{T}_1$ and low mean flow Mach numbers, the eigenfrequencies correspond to half-wavelength modes of the duct, $\omega \approx \pi \overline{c}_1 / \ell$. As the temperature ratio increases, the fundamental resonant frequency changes. In particular, Imaginary ω becomes negative for temperature ratios between 3.3 and 5.3. Then linear perturbations are unstable, growing exponentially with time. The instability is driven by the acoustic waves generated as density inhomogeneities accelerate through the nozzle. Indeed, the Rayleigh energy source term is identically zero for the particular case considered here, $Q'(t) \equiv 0$.

A downstream nozzle introduces the possibility of a low frequency, almost incompressible, 'bulk' mode. For a low Mach number mean flow, the period of oscillation of this mode depends on the convection time of density inhomogeneities from the combustion zone to the nozzle [18]. Its frequency is shown as a function of $\overline{M}_1$ in Figure 9, and this mode is also unstable for some mean flow parameters.

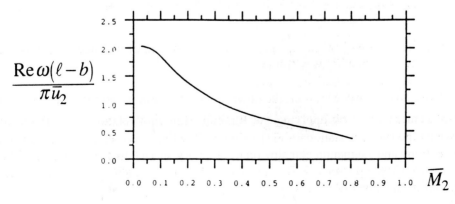

Figure 9   The frequency of the 'bulk' mode for $b = \ell/2$, $\overline{T}_2 = 6\overline{T}_1$, and Case I, $Q'(t) \equiv 0$.

The more realistic form of the instantaneous rate of heat release in (3.8) leads to theoretical predictions which are in good agreement with experimental results for flames in ducts terminated by a choked nozzle. See Macquisten & Dowling [16], Figure 11.

Figure 10 illustrates another flow which is destabilised by the acceleration/deceleration of density inhomogeneities. It consists of an underexpanded supersonic jet impinging upon a flat plate. Flows similar to this occur for VSTOL aircraft at take-off and for rockets at launch. The stand-off shock formed above the plate can be unstable for certain flow conditions, causing the shock to undergo self-excited discrete frequency oscillations (see for example the experimental results of Semiletenko *et al* [19]). We can use the energy equation (2.13) to investigate the driving mechanism for this instability.

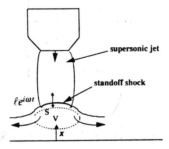

Figure 10 An underexpanded supersonic jet impinging on a flat plate

The important feedback between the shock and the plate occurs near the jet centre-line and there the flow is well approximated by a normal shock with a downstream irrotational flow [20, 21]. Pressure disturbances in the stagnation region perturb the shock. This in turn alters the shock strength, partly because the shock *velocity* alters the relative velocity between the shock and the oncoming gas, and partly because the shock *displacement* moves it to a new position with a different Mach number, $\overline{M}_1(x)$. The changing shock strength leads to downstream flow perturbations. We can see whether the direct energy input from shock is able to sustain oscillations in the stagnation volume V by evaluating the flux of acoustic energy, $-\int_S \tilde{\mathbf{I}} \cdot d\mathbf{S}$, across a stationary surface just downstream of the shock. For small amplitude perturbations of an irrotational inviscid flow, equation (2.11) shows that flux of acoustic energy into the stagnation region per unit shock area is given by

$$-I_x = -\left(p_2' + \overline{\rho}_2 \overline{u}_2 u_2'\right)\left(u_2' + \frac{\overline{u}_2 \, p_2'}{\overline{\rho}_2 \, \overline{c}_2^2}\right) \tag{4.5}$$

where the suffix 2 denotes the flow at a fixed position downstream of the shock and u is the velocity in the x-direction which is away from the plate.

When the shock oscillates at frequency ω, the downstream flow can be expressed in terms of the shock displacement $\ell e^{i\omega t}$ by applying the Rankine-Hugoniot relations in a frame of reference in which the shock is stationary [21]. Kuo & Dowling give explicit forms for the axial velocity, entropy and stagnation enthalpy on a stationary surface just downstream of the shock (see reference [21], equations (2.4), (A3) and (A4)). Using their expressions for the flow perturbations in equation (4.5) leads to

$$-\tilde{I}_x = -\left(A + B\omega^2\right)|\ell|^2 \tag{4.6}$$

where

$$A = -\frac{4\bar{\rho}_2\bar{c}_1^3\left(\bar{M}_1^2-1\right)^2}{\bar{M}_1^5(\gamma+1)^3}\frac{d\bar{M}_1}{dx}\left[\frac{d\bar{M}_1}{dx}+\frac{\alpha}{2\bar{c}_1}\bar{M}_1(\gamma+1)-\frac{d\bar{M}_1}{dx}\frac{\bar{M}_2\bar{c}_1}{\bar{M}_1\bar{c}_2}\frac{2\left(\bar{M}_1^2-1\right)^2}{(\gamma+1)}\right] \qquad (4.7)$$

and

$$B = \frac{2\bar{\rho}_2\bar{c}_1\left(\bar{M}_1^2-1\right)\left(2+(\gamma-1)\bar{M}_1^2\right)}{\bar{M}_1^5(\gamma+1)^3}\left[\bar{M}_1^2+1+\frac{\bar{M}_2\bar{c}_1\left(\bar{M}_1^2-1\right)}{\bar{M}_1\bar{c}_2(\gamma+1)}\left(2+(\gamma-1)\bar{M}_1^2\right)\right] \qquad (4.8)$$

The suffix 1 denotes the flow at a fixed position just upstream of the shock, $\bar{M}_1 = -\bar{u}_1/\bar{c}_1$ is positive and $\alpha = d\bar{u}_2/dx$ describes the mean stagnation flow between the shock and the plate.

When the shock is at a location at which it is statically stable, A is positive. It is clear from the definition of B in (4.8) that B is also positive. Hence, from the form of the intensity in (4.6), we must conclude that the flux of acoustic energy across a surface just downstream of the shock is *towards* the shock: the fluid in the stagnation region V does work on the shock, rather than receiving energy directly from it. The acoustic energy flux across this surface therefore tends to dampen any acoustic oscillations in the stagnation region. And yet self-sustaining oscillations occur! The explanation is that the oscillating shock also leads to entropy fluctuations, whose deceleration through the stagnation is a source of acoustic energy.

A full linear stability analysis involves solving an inhomogeneous wave equation in the stagnation region to determine the acoustic field generated by the decelerating entropy fluctuations. Application of appropriate boundary conditions on the shock and the plate then leads to predictions for the complex eigenfrequencies as functions of jet pressure ratio, Mach number and nozzle-to-plate distance, L. Such a stability analysis confirms our view of the instability mechanism. Mørch [20] neglects the effect of the entropy fluctuations and predicts only decaying disturbances. Inclusion of the source terms due to entropy fluctuations leads to unstable modes, whose frequencies agree well with experiment (see references [21] and [22]). The impinging jet in Figure 10 is also susceptible to another form of flow oscillation, involving the interaction of pressure waves with vortex shedding from the jet lip and this class of flow instability is discussed in the next section.

5. VORTEX FLOWS

Even a subsonic jet impinging on a flat plate can produce a discrete frequency tone and the frequency of this oscillation can be crudely estimated from the travel time of disturbances. Unsteady forcing at the jet lip forces the shedding of vortical structures. Let us suppose that these convect downstream with a speed U_c. The vorticity interacts with the plate to generate pressure waves which propagate to the lip with speed c, and trigger further vortices. The total travel time is therefore $L/U_c + L/c$, where L is the nozzle-to-plate distance. If any phase changes during the generation and receptivity of the sound can be neglected, then the total travel time should be an integer multiple of the period of oscillation, i.e.

$$\frac{L}{U_c}+\frac{L}{c}=\frac{N}{f}, \text{ where } N \text{ is an integer and } f \text{ denotes the frequency of oscillation in Hertz.} \qquad (5.1)$$

In spite of the gross simplifications in equation (5.1), experimental data often collapses onto such curves with surprising accuracy. Figure 11 illustrates the classical experimental results of Ho & Nosseir for a jet of Mach number $\overline{M} = 0.9$ [23]. The multiple stages predicted by (5.1) do indeed occur.

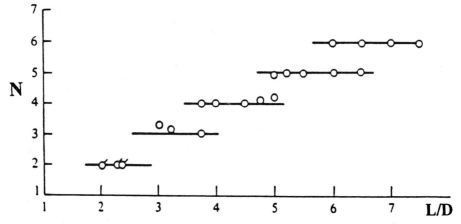

Figure 11 Ho & Nosseirs experimental data [23]: Variation of $N = fL\left(U_c^{-1} + c^{-1}\right)$ with L/D. $U_c = 0.63$ of mean jet velocity, c = speed of sound

The generation of discrete tones also occurs in high Reynolds number devices. One such phenomenon is known as 'flap hoot' and is heard when vortex shedding from the trailing edge of the flap shroud interacts with the leading edge of the flap to produce a self-sustaining oscillation (see Figure 12). As the flaps are deployed, the distance L increases and the frequency drops. Figure 12 shows data from an aircraft in flight. The frequencies of the intense sound generated collapse onto the line $N = fL\left(U_c^{-1} + c^{-1}\right)$ over a 3-fold increase in L.

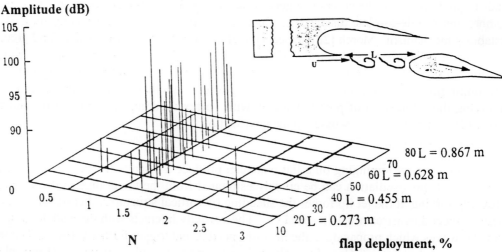

Figure 12   Flap hoot geometry & experimental data (BAe data analysed by Graham & Dowling)

Self-sustaining flow tones have been the subject of extensive study and reviews are given in [24] and [25]. Here we will go on to consider cases where the feedback is enhanced by acoustic resonances. The energy equation in (2.13) helps to understand these flows. For a low Mach number isentropic flow, the main energy source term is (see equation 2.12d)

$$E(t) = -\bar{\rho} \int_V (\mathbf{w} \times \mathbf{u} - \overline{\mathbf{w}} \times \overline{\mathbf{u}}) \cdot \mathbf{u}_a dV \qquad (5.2)$$

This energy source term was derived by Howe [6, 26] and Nelson *et al* [3] and has been used extensively to investigate resonant acoustic-vortex interactions, for example near Helmholtz resonators [3], and due to baffles and splitter plates in ducts [5,27]. We will concentrate on the geometry illustrated in Figure 13, investigated in detail by Bruggeman *et al* [4]. Vortex shedding occurs near the lip of a deep cavity or closed side branch. The flow oscillations can become particularly intense at the resonant frequencies, f, of the cavity, i.e. for $f = c/4L$, $3c/4L$ etc. Bruggeman *et al* note that in gas pipelines, with a total pressure of 60 bar, the pressure perturbations due to such an acoustic-vortex interaction can be of the order of 2.5 bar. They show how the energy equation (5.2) can be used to interpret experimental results and to deduce modifications that reduce the amplitude.

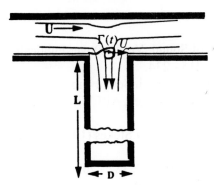

Figure 13 Vortex shedding near a closed side branch

For an acoustic field at resonance with frequency ω, we write

$$\mathbf{u}_a(\mathbf{x},t) = \mathbf{u}_0(\mathbf{x})\sin\omega t \qquad (5.3)$$

The streamlines of the velocity field $\mathbf{u}_0(\mathbf{x})$ have the form sketched in Figure 13. Vortex shedding is observed to be initiated when the acoustic velocity is first inwards into the side-branch, i.e. at time $t = 0$, $2\pi/\omega, \ldots$, for the time origin chosen in (5.3). Bruggeman *et al* find that the subsequent vortex motion and growth is dominated by the mean flow. For the first cycle they take

$$\mathbf{w}(\mathbf{x},t) = -\Gamma(t)\delta(x - U_c t)\delta(y)\mathbf{e}_z, \text{ with } \Gamma(t) = \tfrac{1}{2}U^2 t \text{ and } \mathbf{U}_c = 0.38U\mathbf{e}_x \qquad (5.4)$$

Substitution for $\mathbf{u}_a$, $\mathbf{w}$ and $\mathbf{u} = \mathbf{U}_c$ from (5.3) and (5.4) into (5.2) leads to

$$E(t) = \bar{\rho}\,\Gamma(t)U_c\,u_{oy}(\mathbf{U}_c t)\sin\omega t \qquad (5.5)$$

u_{0y} is negative(inwards into the side-branch). During the first half cycle, $0 < t < \pi/\omega$, $\sin\omega t$ is positive, and it is clear from (5.5) that $E(t)$ is negative, describing absorption of acoustic energy by the vortex interaction. Over the next half cycle, $E(t)$ is positive, and the acoustic waves gain energy. If the vortex has not completed its passage across the duct open by the time $2\pi/\omega$, the reversal of the acoustic field leads to a second period of energy absorption. Since the vortex takes a time D/U_c to cross the duct opening, the acoustic waves gain maximum power from their interaction with the vortex if $fD/U_c = N$, for some integer N. This energy argument shows how to attenuate damaging fluctuations: a sharp upstream corner increases the acoustic sink, while a rounded downstream decreases the acoustic energy source.

6. CONCLUSIONS

The generalised acoustic energy equation (2.13) shows conditions under which linear acoustic waves grow in amplitude. In particular, it identifies the energy source terms from thermoacoustic oscillations, from acoustic-vortex interactions and from accelerated 'hot spots', thereby providing physical insight into the acoustics of a range of unstable flows.

ACKNOWLEDGEMENTS

I am grateful to BAe Regional Aircraft Ltd for allowing me to use their unpublished data in Figure 11. The combustion instability work described in Section 3 has been carried out with the support of Rolls-Royce plc and the Defence Research Agency. Figure 10 is published with the permission of the Journal of Fluid Mechanics.

REFERENCES

[1]   JWS Rayleigh (1945), The Theory of Sound, Vol II, Dover Publications.
[2]   BT Chu (1964), Acta Mechanica,1, 215 - 234.
[3]   PA Nelson, NA Halliwell & PE Doak (1983), J. Sound Vib., 91, 375 - 402.
[4]   JC Bruggeman, A Hirschberg, MEH van Dongen, APJ Wijnands & J Gorter (1989), Trans ASME, J. Fluids Eng, 111, 484 - 491.
[5]   K Hourigan, MC Welsh, MC Thompson & AN Stokes (1990), J. Fluids & Structures, 4, 345-370.
[6]   MS Howe (1984), IMA J. of Applied Maths, 32, 187 - 209.
[7]   CL Morfey (1971), J. Sound Vib., 14, 159 - 170.
[8]   DI Blokhinstev (1946), Acoustics of a Nonhomogenous Moving Medium, translated in NACA 1399 (1956).
[9]   GJ Bloxsidge, AP Dowling, N Hooper & PJ Langhorne (1988), AIAA-J.,26, 783-790.
[10] AP Dowling (1996), AIAA-96-1749.
[11] AP Dowling (1995), J. Sound Vib., 180(4), 557 - 5810.
[12] UG Hedge, D Reuter, BT Zinn & BR Daniel (1987), AIAA-87-0216.
[13] FEC Culick (1988), AGARD-CP-450.
[14] PJ Langhorne (1988), Fluid Mechanics 193, 417-443.
[15] GJ Bloxsidge, AP Dowling & PJ Langhorne (1988), Fluid Mech, 193, 44 -473.
[16] MA Macquisten & AP Dowling (1993), Combustion and Flame, 94, 253-264.
[17] FE Marble & SM Candel (1977), J. Sound Vib. 55(2), 225-243.
[18] JJ Keller (1995), AIAA J. 33, 12, 2280-2287.
[19] BG Semiletenko, BN Sobkolov & VN Uskov (1974), Fluid Mech, 3, 90-95.
[20] KA Mørch (1964), J. Fluid Mech, 20, 141-159.
[21] CY Kuo & AP Dowling (1996), J. Fluid Mech, 315, 267-291.
[22] CY Kuo & AP Dowling (1996), AIAA 96-1747.
[23] CM Ho & NS Nosseir (1981), J. Fluid Mech, 105, 119-142.
[24] D Rockwell & E Naudascher (1979), Ann. Rev. Fluid Mech, 11, 67-94.
[25] WK Blake & A Powell (1986), The development of contemporary views of flow-tone generation, in Recent Advances in Aeroacoustics, edited by A Krothapalli & CA Smith, Springer-Verlag.
[26] MS Howe (1975), J. Fluid Mech, 71, 625-673.
[27] SAT Stoneman, K Hourigan, AN Stokes & MC Welsh (1988), J. Fluid Mech, 192, 455-484.

Theoretical and Applied Mechanics 1996
T. Tatsumi, E. Watanabe and T. Kambe (Editors)

The Acoustics of Ocean Waves

Michael J. Buckingham

Marine Physical Laboratory
Scripps Institution of Oceanography
University of California, San Diego
9500 Gilman Drive
La Jolla, CA 92093-0213, USA
and

Institute of Sound and Vibration Research
The University, Southampton SO17 1BJ, England

ABSTRACT

When a wave breaks, it creates a sub-surface population of bubbles, each of which "rings" for several milliseconds after the instant of formation. These wind-generated bubble sources are important contributors to the ambient noise field in the ocean. Although the noise is a stochastic phenomenon, it possesses a number of stable properties, as a result of which it can be inverted to provide useful information on a variety of ocean parameters and processes. Several ocean-noise inversions schemes are currently being developed, not only in connection with surface phenomena, including bubble production and diffusion at the sea surface, but also with the aim of determining the geo-acoustic parameters of the seabed: the bubble-generated noise contains information about both boundaries, the surface and the bottom. Moreover, the noise also acts as a form of natural acoustic illumination, analogous in some ways to daylight in the atmosphere, offering the potential for imaging objects in the ocean. An ambient noise imaging scheme is described, based on the ADONIS acoustic lens, and an example of an image obtained using only ambient noise illumination is presented. The possibility of obtaining images of the evolving source structure within a breaking wave is also explored.

1. INTRODUCTION

When a wave breaks on the ocean surface, a population of bubbles is created, forming a visible whitecap and a sub-surface plume which extends to a depth of several metres[1,2]. The whitecap and the plume are two-phase mixtures, consisting of air and seawater, with the air in the form of numerous bubbles. At the instant of closure, a bubble is not in equilibrium with its surroundings[3], either because it supports an excess pressure or because it has an initial radial velocity, or a combination of the two. To reach equilibrium, the bubble oscillates in a radial mode, thus acting as an acoustic monopole, a very efficient acoustic radiator. This sound producing mechanism was originally investigated in the laboratory by Minnaert[4], who appreciated that the sound of running water, whether it be from a tap, a babbling brook or a waterfall, originates largely in the radial resonances of newly formed bubbles.

The same is true with breaking waves. Each breaking event is acoustically energetic, and the source of the sound is principally oscillating bubbles. The acoustic signature spans a broad band of frequencies, from a few tens of hertz up to 100 kHz or thereabouts, reflecting the fact that the size distribution of the bubbles extends from tens of microns up to centimetres, satisfying an approximate power-law distribution of the form a^{-n}, where a is bubble radius and n lies somewhere between 3 and 6[5-7].

According to a simple linear model of bubble resonance, the pressure pulse radiated by a bubble immediately after formation takes the form of an exponentially decaying sinusoid. An expression for the resonance frequency, f_0, was first derived by Minnaert[4], who balanced the inertia of the liquid mass surrounding the bubble against the compressibility of the gas enclosed within the bubble. His result is the well-known expression

$$f_o = \frac{1}{2\pi a} \sqrt{\frac{3\gamma P}{\rho}} \; , \tag{1}$$

where γ is the ratio of the specific heats of the gas in the bubble, P is ambient pressure, and ρ is the density of the surrounding fluid. Surface tension affects the resonance frequency to some extent, more so for smaller bubbles[8], but is neglected in Eq. (1). For an air bubble just below the sea surface, where P is one atmosphere, it follows from Eq. (1) that the product of the resonance frequency and the bubble radius is given by

$$f_o a \approx 3.3 \; . \tag{2}$$

Thus, a bubble with a radius of 100 microns oscillates at a frequency of 33 kHz.

At least three mechanisms contribute to the exponential decay of the bubble oscillations: thermal losses associated with the polytropic compression and expansion of the enclosed gas, shear viscosity acting at the gas-liquid interface, and radiation damping arising from the spherical acoustic wave propagating away from the bubble[9]. The Q-factor arising from these loss mechanisms varies weakly with frequency, approximately as $f^{-1/4}$, taking values of 60 at 100 Hz and 10 at 100 kHz.

The exponential decay predicted by the linear theory is rarely observed in practice. Departures from the ideal form of damping may occur for a variety of reasons, including the fact that in natural bodies of water most bubbles are not "clean", but rapidly acquire a surface film[10] which could modify their behaviour at resonance. Medwin and Beaky[11] have identified several species of bubble-pulse signatures from extensive measurements made in a wave tank. The variability in pulse shape is exemplified in Fig. 1, which shows the pressure pulse produced when an air bubble was released from an underwater nozzle. The experiment was performed in a freshwater pool, with the nozzle located sufficiently far from the boundaries to avoid any interference from acoustic reflections. It is clear that the linear theory does not provide a satisfactory description of the observed bubble-pulse signature, with perhaps the most notable discrepancy being that the decay is often not monotonic. At present the reason for this is unknown.

As with individual bubbles in a tank, the oscillations of bubbles in the ocean are more complicated than simple theory would suggest. Non-linear effects may be significant[12] and collective oscillations[13, 14] of bubble clouds may also occur, producing sound at low frequencies (< 500 Hz). Although the details of all the sound-generation processes associated with breaking surface waves may not be fully understood, it is now well accepted that bubbles play a very important role as acoustic sources at the sea surface. Amongst other things, the bubble sound sources offer the prospect providing useful information on air-sea interactions[15], with the potential for yielding gas fluxes across the sea surface from simple, passive acoustic measurements. Before such benefits can be realized, however, improved models of gas diffusion[16-18] across the air-sea interface are required, since such models will form the basis of the acoustic inversions.

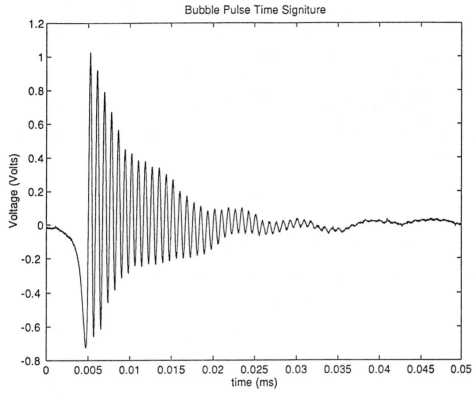

Figure 1. Example of a pressure pulse radiated by a bubble on being released from an underwater nozzle.

Although it is not yet possible to invert for the gas fluxes, other types of acoustic inversion based on sound generated naturally at the sea surface are not only feasible but are currently being implemented. Three such techniques are discussed in this article, one concerned with the void fraction profile through sea surface bubble layer, a second yielding the geo-acoustic parameters of the seabed, and finally an acoustic imaging technique that is analogous to optical photography using daylight. In the case of the latter, the acoustic system that has been developed to perform the imaging, a high resolution, multibeam sonar, can be directed at the sea surface to provide the spatial and spectral structure of the acoustic sources within an individual wave as it evolves through the breaking process. Thus, a system that was originally designed to exploit the sound from breaking waves, has now been turned on the waves themselves to yield fundamental information on the acoustics of wave breaking.

2. THE SEA-SURFACE BUBBLE LAYER

Newly created bubbles from wave-breaking act as acoustic sources, but become quiescent after a few milliseconds. They then become part of a background population of older bubbles that exists immediately beneath the sea surface. Within this surface layer, the volume fraction of air decreases with depth, showing a profile that depends on the details of the gas transport processes acting beneath the surface. The principal forces are turbulent diffusion, advecting the bubbles into the ocean , and buoyancy which opposes the downward trend. Dissolution is also a factor, but may be insignificant when the surface layer is supersaturated with gas, as is often the case in higher sea states. One effect of buoyancy is preferentially to remove larger bubbles, leaving the smaller microbubbles (of order 100 microns radius) to form the residual background layer.

The shape of the void fraction profile is governed by the turbulent diffusion coefficient, D. For the simplest case, in which D is independent of depth, the void fraction profile decays exponentially with increasing depth[16, 17]. An alternative possibility is that a constant stress layer exists beneath the surface, resembling the turbulent boundary layer found near a rigid wall, in which case the diffusion coefficient increases linearly with depth. Under these conditions, the solution of the diffusion equation for the void fraction as a function of depth, $\beta(z)$, is the so-called inverse-square profile[18]:

$$\beta = \frac{\beta_o}{(1 + bz)^2} \tag{3}$$

where b is a constant and β_0 is the void fraction at the sea surface.

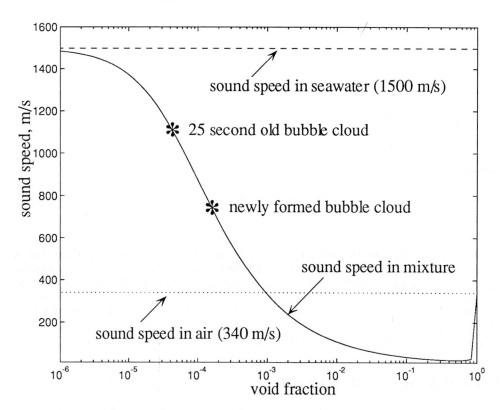

Figure 2.   Wood's curve showing the sound speed in a mixture of air and seawater.

According to Wood[19], the presence of air in water reduces the sound speed due to the increased compressibility and reduced density of the two-phase medium. Fig. 2 shows the sound speed as a function of void fraction, as computed from Wood's expression, and annotated to indicate typical void fraction values at different times during the evolution of the bubble plume beneath a breaking wave. Note the very rapid reduction in sound speed as the void fraction increases from 1 part in 10^5, and in particular that the sound speed falls below the sound speed in air when the void fraction is 1 part in 10^3, to a minimum of about 20 m/s at equal concentrations of air and water.

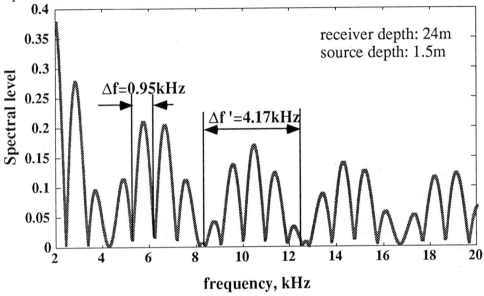

Figure 3.   Theoretical inverse-square spectrum showing bi-periodic structure.

It is evident from Wood's curve, taken in conjunction with Eq. (3), that one effect of the void fraction profile beneath the sea surface is to introduce a sound speed profile that increases with increasing depth. Such a profile is upward refracting, thus forming a surface acoustic waveguide along which sound may propagate very efficiently. The form of the sound speed profile corresponding to the void fraction profile in Eq. (3) will be referred to as the "inverse-square profile". This is a three-parameter curve, the implication being that if the three parameters could be determined from an acoustic inversion, then the void fraction as a function of depth could be inferred from an appropriate inversion.

About six years ago, Farmer and Vagle[2] reported an experiment in which they placed a hydrophone in the surface bubble layer and listened to the sound of breaking waves at a range of some tens of metres. Over a 20 kHz bandwidth, they observed spectra that showed fairly well defined structure, which they attributed to wave-guiding effects.

Buckingham[20] subsequently developed a theory of acoustic propagation in an inverse-square profile, which lead to spectra showing structure similar to those observed in the ocean experiments . Fig. 3, showing an example of a theoretical inverse-square spectrum, serves to illustrate the structure that is imposed by the upward refracting profile. Two periodicities are present, a rapid "carrier" with peaks of width Δf, which is controlled by the receiver depth, and a slower "modulation" of width $\Delta f'$, which is governed by the source depth. The profile parameters affect both spectral intervals. By measuring the values of Δf and $\Delta f'$ in the experimental spectra, and performing an inversion based on the inverse-square theory, profile parameters were obtained that agreed closely with those actually measured in the bubble layer using an independent technique (an upward-looking sonar). Estimates of the source depth, that is to say, the depth of the acoustically active bubbles, were also obtained from the inversions.

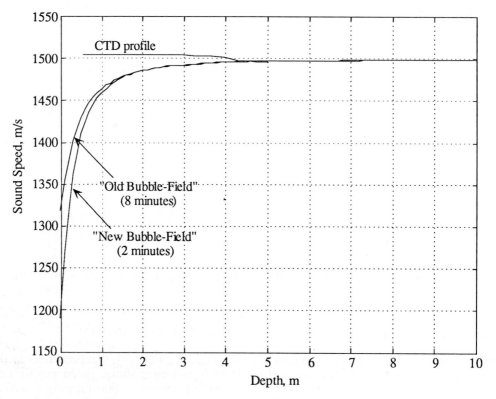

Figure 4.   Sound speed profiles in an evolving wake, from inverse-square inversions, and a CTD profile.

Encouraged by these early inversions, a more controlled experiment was conducted recently, in which a surface bubble layer was created artificially by a small motor boat in Saanich Inlet, British Columbia. Conditions were calm, with minimal surface disturbance from wind. A dedicated broadband source at a nominal depth of 1 m transmitted sound through the bubbly wake to a pair of vertically aligned hydrophones at a range of 10 m . Each set of measurements lasted about half an hour, corresponding roughly to the time that the wake remained visible. A series of inversions based on the inverse-square theory yielded the void fraction profile, and thence the sound speed profile, as the wake slowly dissipated. Fig. 4 shows examples of two sound speed profiles, as obtained from the acoustic inversions, at times of 2 and 8 minutes after the formation of the wake. The evolution back to the bubble-free profile is quite evident from these curves: the new wake shows a surface sound speed of 1220 m/s, which, six minutes later has increased to 1320 m/s. In the absence of the wake, the surface sound speed was close to 1500 m/s, as indicated by the CTD profile.

Of course, the CTD is insensitive to the presence of bubbles, yielding the same profile irrespective of whether the wake is present or not. Thus, CTD casts are incapable of providing information on the depth structure of the surface bubble layer, which is one reason why the acoustic inversion technique is of interest. There are still a number of questions to be resolved in connection with the inverse-square inversions, especially in the open ocean when naturally breaking waves provide the source of sound, but nevertheless the new technique offers the potential for obtaining useful information about the sea-surface bubble layer that would otherwise be difficult to acquire.

3. WAVE-GENERATED NOISE INVERSIONS FOR THE GEO-ACOUSTIC PARAMETERS OF THE SEABED

In shallow water, where the depth is less than about 200 m, the sound generated by surface waves undergoes multiple reflections from the bottom interface. As with sound from a dedicated source, these reflections modify the noise in a way which depends on the nature of the bottom, suggesting that, with an appropriate inversion technique, the noise in the water column could yield the geo-acoustic parameters of the seabed[21]. In this type of inversion, it is important to identify some feature of the noise field that is robust and repeatable if reliable results are to be obtained.

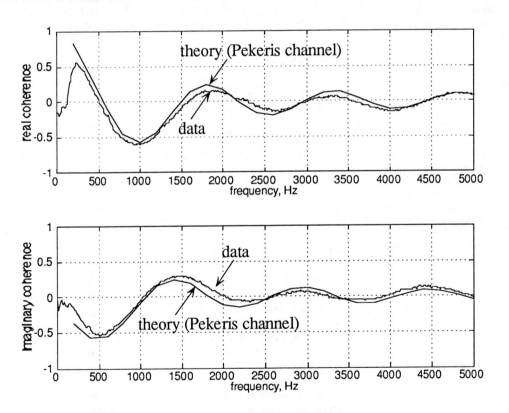

Figure 5. Real and imaginary parts of the noise coherence function. Data collected off Eureka, northern California, where the bottom is a saturated sediment.

Since a shallow-water channel acts as a geometrical waveguide, a pronounced vertical structure is imparted to the noise as it propagates along between the surface and bottom boundaries. In fact, the vertical directionality of the noise is largely determined by the seabed, and hence is identified as the feature that offers the best potential for returning the geo-acoustic parameters of the bottom. The obvious method of extracting the vertical direction-ality would be with a vertical line array: by steering the beam from the upward vertical, down through the horizontal to the downward vertical, the narrowband noise power as a function of polar angle could be determined. Alternatively, a pair of vertically aligned sensors could be used to determine the broadband coherence of the noise, which is related to the directionality through a Fourier transformation[22]. Thus, the directionality and the coherence contain similar information and, in principle, either could be inverted for the parameters of the seabed. In practice, the coherence technique has the advantage of simple instrumentation, since it relies on just two hydrophones rather than a fully populated array.

In June 1996, two vertically aligned hydrophones, separated by 0.996 m, were deployed in shallow water off Eureka, northern California, to record wind-generated noise data over a bandwidth of 20 kHz. The measured coherence between the noise fluctuations at the two phones is shown in Fig. 5, where it can be seen that a well-defined oscillatory structure is evident in both the real and imaginary parts of the coherence function, extending up to at least 5 kHz. Moreover, the data are closely matched by the theoretical curves shown in the figure, which were derived from a model of surface-generated ambient noise in a Pekeris channel[23], that is, an isovelocity water column overlying a fast, fluid basement. The model is exact in the sense that it includes the propagating modes, the continuous spectrum, and the head wave, although the latter is negligible under the conditions of the experiment. Input parameters were chosen that match the geo-acoustic parameters of the seabed at the measurement site, since these were known from independent surveys.

The agreement between theory and data over such a wide frequency band is really quite remarkable. It is not, however, an isolated example. Similar fits have been found between noise data taken in Hauraki Gulf, New Zealand, in January 1996 and the theoretically derived vertical coherence for the site, which has also been independently surveyed. Such a close match between theory and data is supportive of the idea that the vertical structure of shallow-water, surface-generated noise is governed by the bottom boundary in a way that is accurately predictable. Conversely, the vertical noise coherence appears to be suitable for inversion, to provide estimates of the geo-acoustic parameters of the seabed.

A shallow-water noise inversion scheme has been developed, in which the measured coherence is compared with the theoretically predicted coherence in a matched-field type of procedure[24]. At present, the principal parameters that are determined from the inversions are the compressional and shear speeds of the bottom, although density and compressional attenuation are candidates that are also being considered. To some extent, the effect of these parameters on the coherence separates out rather conveniently. For instance, the real part of the coherence is sensitive to the compressional speed but not the density, whereas the imaginary part is affected significantly by the density at low frequencies. At higher frequencies, the compressional attenuation[25] modifies the real part, through mode stripping, and also the imaginary part by introducing asymmetry into the distribution of the noise about the horizontal. From considerations such as these, it may in the future be possible to exploit the broadband nature of the noise inversion technique to determine a full set of geo-acoustic parameters for fluid and elastic seabeds.

4. IMAGING OBJECTS IN THE OCEAN WITH AMBIENT NOISE

The sound generated by breaking waves has application, not only for investigating the properties of the channel boundaries, but also for obtaining information about objects located in the body of the ocean. Ambient noise (acoustic daylight) may be regarded as a form of acoustic illumination, which in some ways is analogous to daylight in the atmosphere. Any object in the ocean is irradiated by the noise, which will be scattered or reflected as a result of the acoustic impedance mismatch presented by the body's boundaries. The scattered component of the noise represents an acoustic signature that is characteristic of the body itself. To extract the information contained in the signal, an acoustic lens is required to focus the scattered energy onto an image plane, the analogy being that of a conventional camera with an optical lens and a film-plane where the image is formed. In the case of the acoustic lens, the film is replaced with an array of hydrophones, which converts the intensity variation across the image into electronic signals that are manipulated through software to appear as a pictorial representation of the object space on a VDU.

To implement ambient noise imaging, a high resolution, multibeam receiver is required, to sample the intensity of the sound scattered from many points in the object space. Each beam generates a pixel in the final image, and the intensity of a pixel represents the strength of the acoustic scatterer in the corresponding beam. ADONIS (Acoustic Daylight Ocean Noise Imaging System), an acoustic lens that was designed to perform incoherent, ambient noise imaging[26], consists of a 3 m diameter spherical dish, faced with closed-cell neoprene rubber acting as an efficient reflector, with an array of 128 hydrophones in the focal surface (Fig. 6). From the geometry of the sensors in relation to the dish, this system produces 128 receive-only beams, each associated with a particular hydrophone in the array head. No phase-delay or time-delay beamforming is necessary, which has considerable benefit in terms of the computational power needed to produce the images.

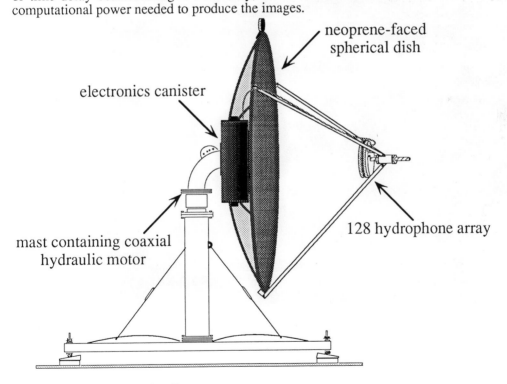

Figure 6.   The ADONIS reflector.

196

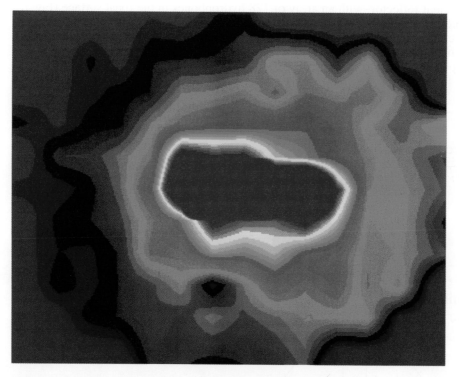

Figure 7. Horizontal bar target above and interpolated acoustic
daylight image below.

ADONIS operates over a decade of bandwidth, from 8 to 80 kHz. At the highest frequency the beam width is just under 1°, corresponding to an angular resolution of about 1 m at 60 m range. A spectral scan is performed on each of the 128 channels, which is the basis for generating coloured images, and the frame rate is 30 Hz, fast enough to produce fairly fluid movement. A hydraulic motor located in the mast supporting the dish provides a facility for scanning in azimuth, which is useful for generating moving images of static targets. All the data processing is performed in real time on a desktop computer, using specially developed imaging software. Ultimately, the data are transferred to CD ROM for permanent storage.

Theoretically, acoustic daylight imaging has been investigated analytically[27] and by computer simulation[28], with a view to establishing the acoustic contrast that may be expected from pixel to pixel. Both approaches show that the contrast depends on the degree of anisotropy exhibited by the noise, but a typical value is in the region of 2 dB. This is consistent with the results of the early experiments that were performed to determine the effect of a target on the received noise level[29]. More recently, ADONIS has been deployed off R/P ORB, which is essentially a large rectangular barge with a moonpool that makes it ideal for the placement of the 3 m dish on the seabed. The ORB experiments were conducted in near-coastal waters off Point Loma, southern California, where the water depth is about 7 m.

Several types of target were used in the ORB deployments, including geometrical shapes, some of which were formed from sheets of aluminium faced with closed cell neoprene rubber, whilst others were made of a variety of materials with different surfaces. Barrel targets were also imaged, either suspended in the water column or partially embedded in the seabed, which is a fine-silt sediment. Fig. 7 shows an example of an interpolated acoustic daylight image of a rectangular bar-like target, 1m high by 3 m wide, at a range of 40 m from the dish. As a reminder, perhaps it is in order to say that in obtaining this and our other acoustic daylight images, no dedicated sound source was used, and the targets themselves were not radiating.

5. IMAGING THE ACOUSTIC SOURCE STRUCTURE WITHIN A BREAKING WAVE

Acoustic daylight imaging exploits the ambient noise in the ocean, regardless of its origin. As it happens, the principal source of noise in the geographical region where the image in Fig. 7 was obtained was biological, in the form of snapping shrimp. Surface conditions were calm, and the shrimp were very energetic, transmitting very narrow pulses of sound with a duration of about 10 μsec, corresponding to a bandwidth of at least 100 kHz[30]. In deeper, rougher waters, wave-breaking could be the predominant contributor to the noise field, and images similar to that in Fig. 7 should be obtainable, although perhaps with a modified shadowing structure.

However, the acoustic structure of the waves themselves is of considerable interest in connection with bubble formation processes and the hydrodynamics of energy dissipation through breaking. ADONIS, with its high resolution, multibeam capability, is an ideal system for probing the internal acoustic structure of breaking waves. Visual images of the sources within the wave can be obtained, showing both the temporal and spectral distribution through a single breaking event. Sixteen frequencies are available, uniformly spaced on a logarithmic scale between 8 and 80 kHz, allowing the spectral and spatial evolution of the breaking process to be examined from frame-to-frame, that is to say, with a temporal resolution of 30 ms.

In April 1996, the ADONIS receiver was mounted on R/P FLIP and deployed at a site in the Pacific Ocean about 20 km west of San Diego. FLIP is a vessel with a unique facility for upending on station, in effect turning itself into an enormous spar buoy, thus providing a very

198

stable research platform. With FLIP in the upright position, the ADONIS array was located about 50 m beneath the sea surface, looking upward at a 50° slant angle. At the highest frequency, this arrangement provided a surface footprint of order 10 m by 10 m, within which each pixel corresponded to an area of about 1 m by 1m.

Many images of the internal structure of breaking waves were obtained in the experiment, which was conducted in fairly light seas (SS 1 to 2). In the preliminary images, the evolution of a wave as observed over a period of about 5 seconds is clearly visible, from which it is evident that the acoustic processes occurring during wave-breaking are extremely interesting, showing an unexpected degree of complexity in terms of temporal, spectral and spatial behaviour. It is hoped that the images may help in developing theoretical models of the wave breaking process.

6. CONCLUDING REMARKS

Breaking surface waves are an acoustically complicated phenomenon. However, the sounds that they generate can be used to advantage for probing the ocean, using a variety of techniques that are currently under development. Surface processes and bottom parameters can both be addressed using the wave-generated noise, and even imaging of objects in the water column is possible, by using the natural acoustic illumination in the ocean. It is interesting that some of the instrumentation that has been developed to exploit the noise, can now be turned on the sources themselves to provide a detailed view of the internal structure of the waves. Thus, the wheel has turned full circle: initially, the waves were treated simply as the source of the ambient sound; and now, the sound is the source of the information on the waves.

ACKNOWLEDGMENT

I am indebted to Drs. Grant Deane and John Potter, and to my graduate students, Chad Epifanio, Holly Burch, Nicholas Carbone, Thomas Hahn and Thomas Berger, whose efforts underlie much of the material described in the text. The research was supported by the Office of Naval Research under contract numbers N00014-93-1-0054, N00014-91-J-1118, and N00014-94-1-0431, for which I am grateful.

REFERENCES

1. E.C. Monahan and M. Lu, "Acoustically relevant bubble assemblages and their dependence on meteorological parameters," IEEE J. Ocean. Eng. **15**, 340-349 (1990).

2. D.M. Farmer and S. Vagle, "Waveguide propagation of ambient sound in the ocean-surface bubble layer," J. Acoust. Soc. Am. **86**, 1897-1908 (1989).

3. M.S. Longuet-Higgins, "Bubble noise mechanisms - a review," in *Natural Physical Sources of Underwater Sound* (Kluwer, Dordrecht, 1993), pp. 419-452.

4. M. Minnaert, "On musical air-bubbles and the sounds of running water," Philosophical Magazine **16**, 235-248 (1933).

5. B.R. Kerman, "Sea Surface Sound: Natural Mechanisms of Surface Generated Noise in the Ocean," in *NATO Advanced Science Institute Series C: Mathematical and Physical Sciences* (Kluwer, Dordrecht, 1988), pp. 639.

6. B.R. Kerman, "Natural Physical Sources of Underwater Sound: Sea Surface Sound (2)," in *Proceedings of the Conference on Natural Physical Sources of Underwater Sound Cambridge, U.K. July 3-6, 1990* (Kluwer, Dordrecht, 1993), pp. 750.

7. M.J. Buckingham and J.R. Potter, "Sea Surface Sound '94 Proceedings of the III International Meeting on Natural Physical Processes Related to Sea Surface Sound," in World Scientific, Singapore, 1995), pp. 494.

8. A.P. Dowling and J.E. Ffowcs Williams, "Sound and Sources of Sound," Ellis Horwood, Chichester, 1983), pp. 321.

9. C. Devin Jr., "Survey of thermal, radiation, and viscous damping of pulsating air bubbles in water," J. Acoust. Soc. Am. **31**, 1654-1667 (1959).

10. S.A. Thorpe, "On the clouds of bubbles formed by breaking wind-waves in deep water, and their role in air-sea gas transfer," Philosophical Transactions of the Royal Society **A 304**, 155-210 (1982).

11. H. Medwin and M.M. Beaky, "Bubble sources of the Knudsen sea noise spectra," J. Acoust. Soc. Am. **86**, 1124-1130 (1989).

12. M.S. Longuet-Higgins, "Nonlinear damping of bubble oscillations by resonant interaction," J. Acoust. Soc. Am. **91**, 1414-1422 (1992).

13. A. Prosperetti, "Bubble dynamics in oceanic ambient noise," in *Natural Mechanisms of Surface Generated Noise in the Ocean* (Kluwer, Dordrecht, 1988), pp. 151-171.

14. W.M. Carey and D. Browning, "Low frequency ocean ambient noise: measurement and theory," in *Natural Mechanisms of Surface Generated Noise in the Ocean* (Kluwer, Dordrecht, 1988), pp. 361-376.

15. W.K. Melville, "The role of surface-wave breaking in air-sea interaction," Annu. Rev. Fluid. Mech. **28**, 279-321 (1996).

16. S.A. Thorpe, "A model of the turbulent diffusion of bubbles below the sea surface," J. Phys. Oceangr. **14**, 841-854 (1984).

17. S.A. Thorpe, "On the determination of K_v in the near-surface ocean from acoustic measurements of bubbles," J. Phys. Oceangr. **14**, 855-863 (1984).

18. M.J. Buckingham, "Sound speed and void fraction profiles in the sea surface bubble layer," J. Appl. Acoustics **in press**, (1996).

19. A.B. Wood, "A Textbook of Sound," G. Bell and Sons Ltd., London, 1964), pp. 610.

20. M.J. Buckingham, "On acoustic transmission in ocean-surface waveguides," Phil. Trans. Roy. Soc. Lond. A **335**, 513-555 (1991).

21. M.J. Buckingham and S.A.S. Jones, "A new shallow-ocean technique for determining the critical angle of the seabed from the vertical directionality of the ambient noise in the water column," J. Acoust. Soc. Am. **81**, 938-946 (1987).

22. M.J. Buckingham, "A theoretical model of ambient noise in a low-loss, shallow water channel," J. Acoust. Soc. Am. **67**, 1186-1192 (1980).

23. C.L. Pekeris, "Theory of propagation of explosive sound in shallow water," in *Geological Society of America Memoir 27 Propagation of Sound in the Ocean* (Geological Society of America, New York, 1948), pp. 1-117.

24. M.J. Buckingham, G.B. Deane, and N.M. Carbone, "Determination of elastic sea floor parameters from shallow-water ambient noise," in *2nd European Conference on Underwater Acoustics* (European Commission, Lyngby, Denmark, 1994), pp. 19-25.

25. M.J. Buckingham, "Theory of acoustic attenuation, dispersion and pulse propagation in granular materials including marine sediments," J. Acoust. Soc. Am. **submitted**.

26. M.J. Buckingham, J.R. Potter, and C.L. Epifanio, "Seeing underwater with background noise," Scientific American **274**, 40-44 (1996).

27. M.J. Buckingham, "Theory of acoustic imaging in the ocean with ambient noise," Journal of Computational Acoustics **1**, 117-140 (1993).

28. J.R. Potter, "Acoustic imaging using ambient noise: Some theory and simulation results," J. Acoust. Soc. Am. **95**, 21-33 (1994).

29. M.J. Buckingham, B.V. Berkhout, and S.A.L. Glegg, "Imaging the ocean with ambient noise," Nature **356**, 327-329 (1992).

30. D.H. Cato and M.J. Bell, *Ultrasonic ambient noise in Australian shallow waters at frequencies up to 200 kHz 1992*, DSTO Materials Research Laboratory:

Theoretical and Applied Mechanics 1996
T. Tatsumi, E. Watanabe and T. Kambe (Editors)
© 1997 Elsevier Science B.V. All rights reserved.

Notions of Material Uniformity and Homogeneity

M. Epstein[a] and G.A. Maugin[b]

[a]Department of Mechanical Engineering, The University of Calgary, Calgary, Alberta T2N 1N4, Canada

[b]Laboratoire de Modélisation en Mécanique, Université Pierre et Marie Curie (Paris VI), Tour 66, 4-Place Jussieu, 75252 Paris, France

1. INTRODUCTION

We wish to start by expressing our gratitude to the organizers of the XIX-th ICTAM for inviting us to participate in this minisymposium. It is, also, most appropriate that a lecture on the theory of continuous distributions of inhomogeneities be presented in Japan, where almost half a century ago it found its modern origins through the work of Professor Kondo [1]. The ideas of Kondo, paralleled in Europe by the works of Nye [2], Bilby [3], Kröner [4], and others, were based, at least partially, on the recognition of the discontinuous nature of matter and, particularly, on the consideration of defective crystalline lattices. A heuristic passage to the limit as the crystalline characteristic lattice unit is allowed to approach zero, elicited the identification of differential-geometric measures of inhomogeneity. We shall refer to this metodology as the *structural approach*.

A typical illustration of the use of this approach is provided by the consideration of the *Burgers vector* of a dislocation (Figure 1). In a defective lattice, such as the one shown, a circuit consisting of, say, three interatomic steps successively applied in each of the four cardinal directions, leaves an opening in the East-West direction, which would vanish in a perfect lattice. The lack of closure **b** is known as the Burgers vector of the dislocation. An almost

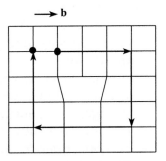

Figure 1. Burgers Circuit

202

identical picture emerges in the depiction of a purely differential concept: the *Cartan torsion* of a connection. This kind of analogy, in the hands of skillful researchers with a keen physical insight and a good dose of mathematical expertise, led to significant results permitting the characterization of the type and density of continuously distributed defects.

In contraposition to the structural approach, there stands the *continuum approach*, based exclusively on the a-priori definition of a material body as a *differentiable manifold*, and thus devoid of any interpretation at the atomic level. In fact, the material behaviour of a continuous body is completely characterized by one or more *constitutive functions*, such as

$$W = W(\mathbf{F}, \mathbf{X}) \tag{1}$$

representing, for example, the *strain-energy density* per unit volume of the body B in a given *reference configuration* as a function of the point $X \in B$ and of the *deformation gradient* $\mathbf{F}$ at that point (Figure 2). In a *simple hyperelastic body* one such scalar function is sufficient to fully describe the mechanical response of the body. To clarify the terminology in a manner useful for later use, we recall that the deformation gradient is a linear transformation that maps, say, an infinitesimal brick at X into its

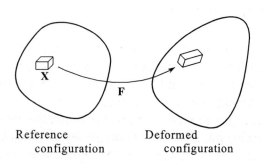

Reference configuration   Deformed configuration

Figure 2. The deformation gradient

deformed counterpart, namely, an infinitesimal parallelepiped. Obviously, as linear maps go, any differential volume about X can be used to describe the same deformation gradient tensor $\mathbf{F}$. In particular, there is no remnant of any preferred lattice directions at X. It would appear, therefore, that the continuum description, with its paucity of detail, would be unable to capture the geometric intricacy required for the description of distributed inhomogeneities. Surprisingly, nevertheless, it was shown by Noll [5] and Wang [6] that the constitutive equation alone, when properly analyzed, can provide a self-contained and rigorous geometric-differential description of continuous distributions of inhomogeneities, whereby the geometric constructs emerge naturally as a consequence of the formulation, rather than as ad-hoc assumptions. The fact that both approaches have led in some cases to identical or very similar geometrical results, raised expectations which turned out to be beyond the terms of reference of each theory. On the other hand, such coincidences are certainly not entirely accidental, and work remains to be done so as to shed light onto the zone of penumbra between the two domains. In this introductory presentation we shall limit ourselves to explore the basis of the continuum approach and some of its possible applications and generalizations.

2. MATERIAL UNIFORMITY AND HOMOGENEITY

Given a constitutive law of a body in the form (1) in some reference configuration, how can it be ascertained whether or not all its points X are made of the same material? Consider a pair of points, X_1 and X_2, as shown in Figure 3. It may be the case that they are both made of the same material, but that they happen to be differently deformed in the particular reference configuration chosen, so that the dependence of W on F is different at X_1 as compared with X_2. Nevertheless, it is clear that there exists a linear *transplant operation* $P_{1\,2}$ which will deform a brick at X_1 into an element around X_2 in such a way that the responses are exactly matched for any further deformation gradient F, namely:

$$W(F\ P_{1\,2},\ X_1) = J_{P_{1\,2}}\ W(F,\ X_2) \tag{2}$$

identically for all non-singular F. The determinant $J_{P_{1\,2}}$ of $P_{1\,2}$ has been included to correct for the change in volume of the elementary parallelepipeds. We conclude, then, that the answer to the original question is as follows: A body is *uniform*, i.e., all its points are made of the same material, if for each pair of points, X_1 and X_2, a linear map $P_{1\,2}$ can be found such that Eq. (2)

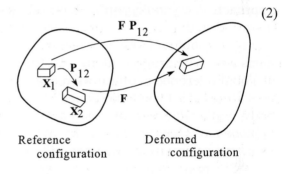

Figure 3. The transplant operation

is satisfied identically for all possible deformation gradients F. An equivalent way to characterize uniformity is by fixing one point, X_0 say, and comparing its response with all the rest. We can imagine, for clarity, that a typical elementary brick at X_0 has been removed from the body, and we may suggestively call this entity a *reference crystal*, although no real connection with the crystallographic lattice is implied. If we denote by $P(X)$ the transplant map from X_0 to the generic point X, the uniformity condition can be expressed as the existence of a field of linear maps $P(X)$ such that

$$W(F,\ X) = J_{P^{-1}(X)}\ W_0(F\ P(X)) \tag{3}$$

204

where W_0 represents the constitutive equation at $X = X_0$. We shall call $P(X)$ a *uniformity field* (Figure 4). A body is uniform if such a field exists. If, as we assume, $P(X)$ can be chosen smoothly in a neighbourhood of each point, the body is said to be *smoothly uniform*. A nice way to visualize a uniformity field is as follows: Let e_1, e_2, e_3 be a fixed basis in the reference crystal (Figure 5). Then the uniformity map $P(X)$ maps this basis into another basis $E_1(X)$, $E_2(X)$, $E_3(X)$ at each point X of the reference

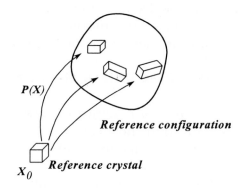

Figure 4. A uniformity field

configuration. These distorted bases constitute what may suggestively be called a *material or crystallographic basis*, provided one bears in mind that the terminology in no way implies that any use has been made of a lattice-based approach. A crystallographic basis shows how the different nighbourhoods are distorted with respect to each other in the given reference configuration. If a change of global reference configuration were to be effected, the crystallographic basis would be dragged along and the question can be asked: is there a change of global reference configuration that leads to a straightening out, an alignment as it were, of all the local bases? If such a change is possible we say that the body is *globally homogeneous* and we call the new reference configuration a *homogeneous configuration*. In a homogeneous reference configuration the constitutive equation becomes independent of position.

It may happen that the straightening out of the crystallographic basis can be accomplished in a neighbourhood of each point, rather than globally. That is, for each point a change of reference configuration can be found such that the crystallographic basis becomes a parallel field in a neighbourhood of that point. We then say that the body is *locally homogeneous*. A classical example of a locally homogeneous body is provided by an initially straight homogeneous

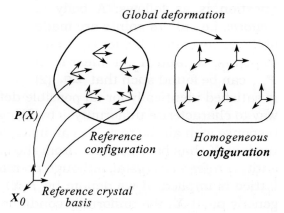

Figure 5. A "crystallographic" basis

rod whose free ends are welded together to form a ring.

3. CRITERION FOR LOCAL HOMOGENEITY

A smooth uniformity field $P(X)$, or its associated material basis $E_\alpha(X)$, ($\alpha = 1, 2, 3$), determines a *material parallelism*. Indeed, two vectors at different points, X and Y, of a neighbourhood are said to be *materially parallel* if they have the same components in their respective local material bases, $E_\alpha(X)$ and $E_\alpha(Y)$. Let $P_\alpha^I(X)$ represent the matrix of Cartesian components of the local material basis, viz.:

$$E_\alpha(X) = P_\alpha^I(X) f_I, \qquad \alpha, I = 1, 2, 3 \tag{4}$$

where f_I is the fixed Cartesian basis and where the summation convention for diagonally repeated indices is implied. Then, the components $V^\alpha(X)$ of a vector $V(X)$ in the material basis $E_\alpha(X)$ are related to their Cartesian counterparts $V^I(X)$ by:

$$V^I(X) = P_\alpha^I(X) V^\alpha(X) \tag{5}$$

One can now define [5] a *material connection* by means of the covariant derivative (semicolon) operation:

$$V_{;J}^I = V_{,J}^I + \Gamma_{KJ}^I V^K \tag{6}$$

where a comma stands for the ordinary partial derivative and where the *material Christoffel symbols* are given by:

$$\Gamma_{KJ}^I = -P_{\alpha,J}^I (P^{-1})_K^\alpha = P_\alpha^I (P^{-1})_{K,J}^\alpha \tag{7}$$

This covariant derivative is associated with the material parallelism by the property that it vanishes for all materially parallel vector fields, as can be verified directly by differentiating (5) when the components $V^\alpha(X)$ are constant and substituting the result in Eq. (6). The material Christoffel symbols are in general not symmetric in the lower indices, so that the *torsion tensor* with components:

$$\tau_{JK}^I = \Gamma_{JK}^I - \Gamma_{KJ}^I \tag{8}$$

does not vanish.

Consider now the question of local homogeneity: Is there a global deformation that renders the material bases Euclidean-wise parallel in a neighbourhood? This will be the case if, and only if, the material bases are adapted to some curvilinear coordinate system, $\phi^\alpha = \phi^\alpha(X^I)$, in the neighbourhood, that is, if:

$$(P^{-1})_K^\alpha = \frac{\partial \phi^\alpha}{\partial X^K} \tag{9}$$

For, if this is the case, we can clearly regard the functions $\phi^\alpha = \phi^\alpha(X^I)$ as the definition of a deformation which straightens the coordinate lines. The integrability conditions of (9), namely, the conditions for the existence of functions $\phi^\alpha = \phi^\alpha(X^I)$ with the property (9), are given by the equality of the mixed partial derivatives, viz.:

$$(P^{-1})_{K,J}^\alpha = (P^{-1})_{J,K}^\alpha \tag{10}$$

A comparison of (10) with (8) yields the homogeneity conditions:

$$\tau_{JK}^I = 0 \tag{11}$$

In other words, a body is locally homogeneous if the torsion tensor of a material connection vanishes identically neighbourhood-wise.

4. THE ROLE OF THE MATERIAL SYMMETRY GROUP

The local homogeneity criterion just described would be absolute, were it not for the fact that, in general, the uniformity field (for a fixed reference configuration and a fixed reference crystal) is not unique. The reason for this possible lack of uniqueness is to be found in the presence of *material symmetries* , namely, local deformations of the reference configuration that do not affect the material response at a point. More precisely, a linear transformation G_X is a material symmetry at X if

$$W(\boldsymbol{F}, \boldsymbol{X}) = W(\boldsymbol{F}\boldsymbol{G}_X, \boldsymbol{X}) \tag{12}$$

identically for all $\boldsymbol{F}$. It is usually assumed, on physical grounds, that all material symmetries are *unimodular* (i.e., of determinant +1). All material symmetries at $\boldsymbol{X}$ form a multiplicative group $\mathcal{G}_X$, as can be easily verified and understood. For a uniform material, these *symmetry groups* at different points cannot be independent of each other. In fact, since the mechanical properties at two points $\boldsymbol{X}$ and $\boldsymbol{Y}$ are the same modulo a transplant $\boldsymbol{P}_{XY}$, it follows that the symmetry groups at $\boldsymbol{X}$ and $\boldsymbol{Y}$ are *conjugate* (Figure 6) according to the formula:

$$\mathcal{G}_y = \boldsymbol{P}_{XY}\, \mathcal{G}_X\, \boldsymbol{P}_{XY}^{-1} \tag{13}$$

Conversely, if $\boldsymbol{P}_{XY}$ is a transplant map, so is any map $\boldsymbol{P}_{XY}{}'$ of the form

$$\boldsymbol{P}_{XY}{}' = \boldsymbol{G}_Y\, \boldsymbol{P}_{XY}\, \boldsymbol{G}_X \tag{14}$$

with $\boldsymbol{G}_X \in \mathcal{G}_X$ and $\boldsymbol{G}_Y \in \mathcal{G}_y$. Viceversa, if $\boldsymbol{P}_{XY}$ and $\boldsymbol{P}_{XY}{}'$ are two transplant maps between $\boldsymbol{X}$ and $\boldsymbol{Y}$, there exist $\boldsymbol{G}_X \in \mathcal{G}_X$ and $\boldsymbol{G}_Y \in \mathcal{G}_y$ such that (14) is satisfied. As a matter of fact, by virtue of the conjugation condition (13), for any given $\boldsymbol{P}_{XY}$ and $\boldsymbol{P}_{XY}{}'$ one can always find $\boldsymbol{G}_X{}' \in \mathcal{G}_X$ and $\boldsymbol{G}_Y{}' \in \mathcal{G}_y$ such that

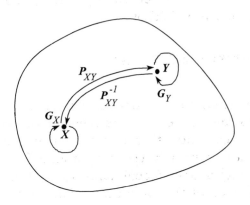

Figure 6. Conjugation of groups

$$\boldsymbol{P}_{XY}{}' = \boldsymbol{G}_Y{}'\, \boldsymbol{P}_{XY} \tag{15}$$

and

$$\boldsymbol{P}_{XY}{}' = \boldsymbol{P}_{XY}\, \boldsymbol{G}_X{}' \tag{16}$$

This means that the totality $\mathcal{P}_{XY}$ of transplant maps between X and Y can be obtained from any given $\boldsymbol{P}_{XY}$ as

$$\mathcal{P}_{XY} = \boldsymbol{P}_{XY}\,\mathcal{G}_X = \mathcal{G}_Y\boldsymbol{P}_{XY} \tag{17}$$

Choosing $\boldsymbol{X} = \boldsymbol{X}_0$, we may also write:

$$\mathcal{P}_Y = \boldsymbol{P}(Y)\,\mathcal{G}_o = \mathcal{G}_Y\boldsymbol{P}(Y) \tag{18}$$

using the notation of Eq. (3).

If the symmetry group of the material is discrete, this lack of uniqueness presents no problems since, by continuity, two uniformity fields $\boldsymbol{P(X)}$ and $\boldsymbol{P'(X)}$ must everywhere differ by a constant element $\boldsymbol{G}_0$ of $\mathcal{G}_o$:

$$\boldsymbol{P'(X)} = \boldsymbol{P(X)}\,\boldsymbol{G}_0 \tag{19}$$

Introducing this result in the definition of the material Christoffel symbols (7), we conclude that the material connection, and hence its torsion, is unique when the symmetry group is discrete.

If, however, the symmetry group is continuous, we may specify a new uniformity field by:

$$\boldsymbol{P'(X)} = \boldsymbol{P(X)}\,\boldsymbol{G}_0(\boldsymbol{X}) \tag{20}$$

where $\boldsymbol{G}_0(\boldsymbol{X})$ is a smoothly varying choice of the symmetry element. In this case, the new material Christoffel symbols will in general differ from the original ones. We conclude, therefore, that when the material symmetry group is continuous, the non-vanishing of the torsion of one material connection does not necessarily imply a lack of local homogeneity, but may simply be the result of an unhappy choice of the uniformity field. A simple example is provided by a perfectly homogeneous and isotropic solid lamina in a homogeneous configuration. If, after having chosen as a material basis a Cartesian one, we should apply a variable rotation at each point (as we are permitted by the rotational isotropy of the material) we would obtain another material basis field with non-vanishing torsion.

To better describe this situation, consider the following geometrical entity: at each point $\boldsymbol{X}$ of the body B (in a given reference configuration and for a given reference crystal) we attach the corresponding set $\mathcal{P}_X$ as per Eq. (18), or, what is the same, the set of all the corresponding material bases (Figure 7). This entity is technically known as a *fibre bundle* and the set $\mathcal{P}_X$ is called the *fibre at* $\boldsymbol{X}$. Each fibre is typified, according to (18), by the symmetry group of the

209

reference crystal, also called the
typical fibre. In other words, the
typical fibre represents the
available degree of freedom in each
fibre. A *section* of the fibre bundle
consists of the choice of one
element from each fibre, and is
thus the mathematical counterpart
of a crystallographic basis.

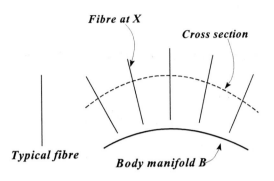

Figure 7. A fibre bundle

This peculiar fibre bundle
can be seen as a reduction of the
principal frame bundle, where *all*
possible bases are included in the
fibre, rather than just those bases
allowed by the symmetry group.
Such a reduction is known as a *G-structure*, and it can be shown [7] that the
physical notion of local homogeneity corresponds exactly to the mathematical
integrability of the G-structure.

5. THE ESHELBY TENSOR

So far we have only considered a fixed inhomogeneity pattern embedded,
as it were, in the body and characterized by a uniformity field of material bases
that can be chosen with a certain degree of freedom, as permitted by the
symmetries of the material. We now assume that this pattern is no longer fixed,
but is allowed to evolve in time. This means that, as time goes on, we permit the
material bases to attain formerly unreachable triads, according to the new
inhomogeneity pattern developed. In the terminology of fibre bundles, we are
now permitting the cross section to wander out of the restricted G-structure
and into the principal bundle of frames. It makes then physical sense to
calculate the derivative of the strain energy function with respect to P. When
this is done [8], starting from Eq. (3) and using the formula for the derivative
of a determinant, one obtains:

$$\frac{\partial W}{\partial P} = -(W\,I - F^T\,T)\,P^{T^{-1}} \tag{21}$$

where I is the identity and where

$$T = \frac{\partial W}{\partial F} \tag{22}$$

is the *first Piola-Kirchhoff* stress. The expression within brackets:

$$b = W I - F^T T = - \frac{\partial W}{\partial P} P^T \tag{23}$$

can be recognized as the *Eshelby tensor*, originally introduced by Eshelby [9] in 1951. Since the variable tensor P assumes the role of an internal variable, we see that, as originally intended by Eshelby, b has roughly the meaning of the change of elastic energy involved in moving the inhomogeneity. A similar result was anticipated in [10], though not within the framework of the theory of uniform materials.

If the body is in equilibrium, and in the absence of body forces, the Cartesian divergence of the Eshelby tensor is related to the material connection by:

$$b^J_{I,J} = b^J_M \Gamma^M_{JI} \tag{24}$$

so that it vanishes identically in a homogeneous reference configuration.

It is physically relevant to note [8] that if $P(X)$ undergoes a small change *within* the degree of freedom permitted by a continuous symmetry group, then the work done on this small change by the Eshelby tensor vanishes. The importance of this result can be understood by pointing out that if the Eshelby tensor is to be regarded as the natural driving force behind a change in the inhomogeneity pattern (i.e., the motion of dislocations), then it is crucial that it should be work-wise insensitive to a mere rate of change of crystallographic basis within the same inhomogeneity pattern.

6. RESTRICTIONS ON LAWS OF EVOLUTION

The last remark brings us to a more general question which is relevant, among other areas, to the theory of plasticity. Are there any formal restrictions to all possible laws of evolution [11]? For definiteness, let us consider an evolution law of the form:

$$\Phi(P,\ \dot{P},\ b)\ =\ 0 \tag{25}$$

where a superimposed dot denotes time derivative. Although this law is expressed in terms of one particular uniformity field $P(X)$, it is obvious that the intention is to represent a law of evolution of the inhomogeneity pattern itself. In the case of a discrete symmetry group, any two uniformity fields differ by a *constant* element of the symmetry group of the reference crystal, as per Eq. (19). In the pictorial language of fibre bundles, each uniformity field is represented by a section, and two sections which differ everywhere by a constant multiplicative element will appear as "parallel" (Figure 8). As time goes on, these parallel sections keep moving in parallel, after the fashion of synchronized swimmers. One should require, therefore, that the law of evolution be form-invariant under parallel changes of sections. In the case of a continuous symmetry group, one and the same inhomogeneity pattern can be represented by an infinity of sections which are not necessarily parallel, but we shall still require that under parallel changes of sections within the symmetry group the law of evolution must be form-invariant. We shall refer to this criterion as the *principle of G-covariance.*

It is not difficult to translate this verbal statement into mathematical terms. Let $P(X,\ t)$ and $Q(X,\ t)$ represent two sections (uniformity fields) that evolve in parallel. This means that there exists a fixed element G_0 of the symmetry group of the reference crystal such that:

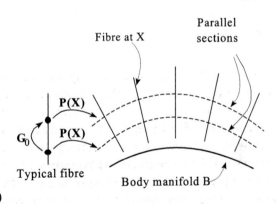

Figure 8. Parallel sections

$$Q(X,\ t)\ =\ P(X,\ t)\ G_0 \tag{26}$$

for all times t. Differentiating this equation with respect to time and eliminating G_0 we obtain

$$\dot{Q}\ Q^{-1}\ =\ \dot{P}\ P^{-1} \tag{27}$$

We conclude, therefore, that sections evolving in parallel have at all times the same *inhomogeneity velocity gradient,* defined as

$$L_P = \dot{P} \, P^{-1} \tag{28}$$

Conversely, let two evolving uniformity fields $P(X, t)$ and $Q(X, t)$ have at all times the same inhomogeneity velocity gradient, i.e.

$$L_P = L_Q \tag{29}$$

If at some paricular time t_0 they were parallel, that is , if

$$Q(X, t_0) = P(X, t_0) \, G_0 \tag{30}$$

then they will evolve in parallel at all times [11]. It follows that the property of having equal inhomogeneity velocity gradients completely characterizes sections evolving in parallel. Therefore, the principle of G-covariance is tantamount to the requirement that the uniformity field and its derivative enter the law of evolution through the inhomogeneity velocity gradient only, viz.:

$$\Phi(L_P \, , \, b) = 0 \tag{31}$$

We note that the inhomogeneity velocity is a tensor with both indices in the reference configuration, the same as the Eshelby tensor.

The second formal restriction will be called the *principle of actual evolution*, and it is designed specifically to cover a situation that may arise in the case of continuous symmetry groups. Indeed, when the symmetry group is continuous, it is possible to move smoothly from one uniformity field to another always staying within the same inhomogeneity pattern. In the terminology of frame bundles this means that, when the symmetry group is continuous, one can smoothly change the cross section within the same G-structure. If that is the case, however, the inhomogeneity pattern is not evolving at all, since all such sections represent the same distribution of inhomogeneities. This unwanted apparent evolution will happen if, and only if, the section at time t is related to the section at a fixed time t_0 by the equation:

$$P(X, t) = G_X(X, t) \, P(X, t_0) \tag{32}$$

where $G_X(X, t)$ is a time-dependent member of the symmetry group at X in the reference configuration. Differentiating (32) we obtain at time t_0

$$\dot{P} = \dot{G}_X P \tag{33}$$

which can be rewritten as

$$L_P = \dot{G}_X \tag{34}$$

Since by (32) $G_X(X, t_0)$ must be the identity, $\dot{G}_X$ is nothing but an element of the Lie algebra of the symmetry group at X, that is, an infinitesimal generator of the group. The principle of actual evolution can then be stated in the following terms:For a law of evolution to prescribe at all times an actual change of inhomogeneity pattern, the prescribed L_P must not belong to the Lie algebra of the instantaneous symmetry group $\mathcal{G}_X$.

A particular case of interest is that of a fully isotropic solid, for which the symmetry group of the reference crystal (assumed in a stress-free configuration) is the full orthogonal group. The elements of the associated Lie algebra are then the infinitesimal orthogonal transformations, namely, the skew-symmetric matrices. Since, by Eq. (13), the symmetry groups in the reference configuration are P-conjugates of the orthogonal group, so are the corresponding G-algebras, namely, an element of the G-algebra at a point in the reference configuration is given by

$$\Omega_X = P \, \Omega \, P^{-1} \tag{35}$$

where Ω is a skew symmetric matrix. The law of evolution must prescribe an inhomogeneity deformation gradient which can not be brought into this form.

7. ANELASTIC DISSIPATION

A complete thermodynamic formulation can be obtained [12,13] by including the absolute temperature θ and its referential gradient $\nabla\theta$ among the independent constitutive variables. The thermodynamic Eshelby tensor is defined in terms of the free energy Ψ per unit volume as:

$$b = \psi \, I - F^T \, T = -\frac{\partial\psi}{\partial P} \, P^T \tag{36}$$

The second law of thermodynamics, in the form of the Clausius-Duhem inequality, implies the following dissipation condition:

$$trace(\boldsymbol{b}^T \boldsymbol{L}_p) + \theta^{-1} \boldsymbol{q} \cdot \nabla\theta \leq 0 \qquad (37)$$

where $\boldsymbol{q}$ is the heat flux vector per unit area in the reference configuration. While the second term in the left-hand side of (37) represents the classical dissipation by heat conduction, the first term represents the dissipation associated with the motion of inhomogeneities, confirming once more that the driving force behind such motion is indeed the Eshelby tensor. This form of the second law of thermodynamics naturally imposes further restrictions in the possible form of an evolution equation. Thus, for example, in the absence of heat conduction, a proportional relation between the inhomogeneity velocity gradient tensor and the Eshelby tensor would imply a negative value for the constant of proportionality.

8. FURTHER EXTENSIONS OF THE THEORY

The foregoing theory can be extended to the case of electrified materials [14], as well as to second-grade elasticity and to Cosserat media [15 - 18]. In the latter case, the body itself being a fibre bundle, one expects the appearance of further geometric measures of inhomogeneity. Indeed, let the constitutive equation of a general Cosserat medium be given by a scalar function such as:

$$W = W(\boldsymbol{F}, \boldsymbol{K}, \nabla\boldsymbol{K}, \boldsymbol{X}) \qquad (38)$$

where, in addition to the deformation gradient $\boldsymbol{F}$ of the underlying body, we also have at each point a non-singular linear transformation $\boldsymbol{K}$ representing the homogeneous deformation of the microstructure. The uniformity condition involves in this case the existence of three fields of maps: $\boldsymbol{P}(\boldsymbol{X})$, with the same meaning as before; $\boldsymbol{Q}(\boldsymbol{X})$, governing the transplant associated with $\boldsymbol{K}$; and $\boldsymbol{R}(\boldsymbol{X})$, a third order tensor governing the transplant associated with the gradient of $\boldsymbol{K}$. The question of local homogeneity can be settled in terms of three connections rather than just one. Their Christoffel symbols are given by:

$$\Gamma^I_{JK} = (\boldsymbol{P}^{-1})^\alpha_{J,K} \, \boldsymbol{P}^I_\alpha$$

$$\Delta^I_{JK} = (\boldsymbol{Q}^{-1})^\alpha_{J,K} \, \boldsymbol{Q}^I_\alpha$$

$$\Lambda^I_{JK} = -R^{\ I}_{\alpha\beta} \, (\boldsymbol{Q}^{-1})^\alpha_J \, (\boldsymbol{P}^{-1})^\beta_K \qquad (39)$$

The local homogeneity conditions can be shown to be expressible in terms of the vanishing of the torsion of Γ, as before, and of the tensor difference $\sigma = \Delta - \Lambda$ of the other two connections, representing the *Cosserat inhomogeneity tensor* . The case of second-grade materials (materials with couple stresses) can be obtained from the foregoing theory by setting $K = F$ and ∇K symmetric in its lower indices.

ACKNOWLEDGEMENTS: The authors gratefully acknowledge the support of the Natural Sciences and Engineering Research Council (Canada) and of the Centre National de la Recherche Scientifique (France).

REFERENCES

1. K. Kondo (ed.), *Memoirs of the Unifying Study of the Basic Problems in Engineering Sciences by Means of Geometry*, Tokyo Gakujutsu Bunken Fukyu-Kai (1955)
2. J.F. Nye, Acta Metall., **1** (1953), 153.
3. B.A. Bilby, Progr. Solid Mechs., **1** (1960), 329.
4. E. Kröner, Arch. Rat. Mech. Anal., **4** (1960), 273.
5. W. Noll, Arch. Rat. Mech. Anal., **27** (1967), 1.
6. C.-C. Wang, Arch. Rat. Mech. Anal., **27** (1967), 33.
7. M. Elżanowski, M. Epstein and J. Śniatycki, J. Elasticity, **23** (1990), 167.
8. M. Epstein and G.A. Maugin, Acta Mechanica, **83** (1990), 127.
9. J.D. Eshelby, Phil. Trans. Royal Soc. London, **A-244** (1951), 87.
10. A. Golebiewska-Herrmann, Int. J. Solids Structures, **17** (1981), 1.
11. M. Epstein and G.A. Maugin, Acta Mechanica, **115** (1996), 119.
12. M. Epstein, *Nonlinear Thermomechanical Processes in Continua*, TUB Dokumentation, Berlin, **61** (1992), 147.
13. M. Epstein and G.A. Maugin, C.R. Acad. Sci. Paris, **320-II** (1995), 63.
14. G.A. Maugin and M. Epstein, Proc. Royal Soc.London, **A-433** (1991), 299.
15. M. deLeón and M. Epstein, Reports Math. Phys., **33** (1993), 419.
16. M. deLeón and M. Epstein, C.R. Acad. Sci. Paris, **319-I** (1994), 615.
17. M. Epstein and M. deLeón, Acta Mechanica, **114** (1996), 217.
18. M. Epstein and M. deLeón, J. Elasticity (in press).

Theoretical and Applied Mechanics 1996
T. Tatsumi, E. Watanabe and T. Kambe (Editors)
© 1997 Elsevier Science B.V. All rights reserved.

On micromechanics of inelastic and piezoelectric composites

G. J. Dvorak[a] and Y. Benveniste[b]

[a]Center for Composite Materials and Structures, and Department of Mechanical Engineering, Aeronautical Engineering and Mechanics, Rensselaer Polytechnic Institute, Troy, NY 12180-3590, U.S.A.

[b]Department of Solid Mechanics, Materials and Structures, Faculty of Engineering, Tel Aviv University, Ramat Aviv, Tel Aviv 69978, Israel.

The theory of uniform fields in elastic heterogeneous solids is summarized for multiphase and two-phase systems with arbitrary or fibrous microstructures. The results are used to derive certain exact connections for the elastic moduli as well as for the mechanical and transformation strain influence functions in the transformation field analysis method. The method is a general procedure for evaluation of internal fields and overall response caused by distributions of local eigenstrain. Applications of the method are shown in incremental analysis of inelastic composites and laminates. These topics are covered in Sections 1 and 2.

The third section of the paper provides a brief review of microstructure-independent exact connections for the effective moduli of piezoelectric composites. The basic method of derivation is the method of uniform fields and constitutes the unifying link with the preceeding sections of the paper. In two-phase fibrous systems, a field decoupling formalism allows the derivation of additional exact relations among the effective moduli. Composites with piezoelectric and piezomagnetic phases are also considered; these exhibit an effective magnetoelectric effect which is not present in the constituents. The section concludes with a brief discussion of phase-interchange connections in piezocomposites.

1. UNIFORM FIELDS IN HETEROGENEOUS SOLIDS

1.1 Background

Fibrous and particulate composite materials, and polycrystals, are typical examples of heterogeneous aggregates consisting of two or more bonded, elastically homogeneous phases. The local geometry of such materials is typically known only in terms of approximate phase shapes and their volume fractions. Therefore, in general, it is not possible to solve exactly for the local fields in the phases, even under homogeneous overall boundary conditions. However, exact solutions for local stresses and strains exist in many cases. They do not depend on the local geometry, in fact, such local fields are uniform or piecewise uniform in the aggregate volume. Some but not all uniform fields can be created by superimposing the mechanical fields caused by application of uniform overall strain or stress to the aggregate with the residual fields caused by certain internal eigenstrains or eigenstresses which are independent of the current mechanical loads. In fibrous systems, uniform fields exist also during loading along a certain proportional path. Here we show how such fields can be created in fibrous composites, in aggregates of arbitrary geometry, and in certain laminated plates. Applications of the uniform fields in derivation of many exact connections between overall and phase elastic constants, and between the mechanical and residual local fields will be discussed. For original work on the subject, see Dvorak (1983, 1986, 1990), Dvorak and Chen (1989), Benveniste and Dvorak (1990a, 1990b, 1992a) Bahei-El-Din (1992) and Dvorak and Benveniste (1992).

In what follows, the constitutive relations for the phases are written as,

$$\varepsilon_r(\mathbf{x}) = \mathbf{M}_r\sigma_r(\mathbf{x}) + \mu_r \quad \sigma_r(\mathbf{x}) = \mathbf{L}_r\varepsilon_r(\mathbf{x}) + \lambda_r \tag{1}$$

where $r = 1, 2, \ldots N$, $\mathbf{L}_r$ are the (6x6) phase stiffness matrices, $\mathbf{M}_r = \mathbf{L}_r^{-1}$ are phase compliances, μ_r are uniform phase eigenstrains, and $\lambda_r = -\mathbf{L}_r\mu_r$ are phase eigenstresses; the latter will be referred to jointly as *transformation fields*.

A representative volume V of a statistically homogeneous aggregate is selected such that it responds to both external loads and local transformation fields as any larger volume. Perfect bonding at all internal interfaces is assumed. Surface tractions in equilibrium with a uniform overall stress σ^0, or surface displacements compatible with a uniform overall strain ε^0 are prescribed at the surface of V. The overall constitutive relations of the aggregate are then written as,

$$\varepsilon = \mathbf{M}\sigma^0 + \mu \quad \sigma = \mathbf{L}\varepsilon^0 + \lambda \tag{2}$$

where $\mathbf{L}$ and $\mathbf{M} = \mathbf{L}^{-1}$ are overall elastic stiffness and compliance, μ and $\lambda = -\mathbf{L}\mu$ are the overall eigenstrains and eigenstresses caused by the local transformation fields.

The local and overall fields are connected by the relations,

$$\varepsilon_r(\mathbf{x}) = \mathbf{A}_r(\mathbf{x})\varepsilon^0 + \sum_s \mathbf{D}_{rs}(\mathbf{x})\mu_s \quad \sigma_r(\mathbf{x}) = \mathbf{B}_r(\mathbf{x})\sigma^0 + \sum_s \mathbf{F}_{rs}(\mathbf{x})\lambda_s \tag{3}$$

where $\mathbf{A}_r(\mathbf{x})$ and $\mathbf{B}_r(\mathbf{x})$ are mechanical influence functions (Hill 1963) and the $\mathbf{D}_{rs}(\mathbf{x})$ and $\mathbf{F}_{rs}(\mathbf{x})$ are the transformation influence functions; $r, s = 1, 2, \ldots N$. The mechanical influence functions can be found for a given phase geometry, but if that is known only in terms of phase volume fractions, approximate techniques such as the Mori-Tanaka (1973) or self-consistent methods (Hill 1965) are available for evaluation of their volume averages, the mechanical concentration factors $\mathbf{A}_r$ and $\mathbf{B}_r$. These are well-known results (Benveniste 1987, Walpole 1969), hence the $\mathbf{A}_r$ and $\mathbf{B}_r$ can be regarded as known quantities. The transformation influence functions are, of course, related to the Green's function of a certain comparison medium $\mathbf{L}^0$ (Dvorak and Benveniste 1992); however, more direct evaluations will be derived below.

Moreover, exact connections between local and overall averages must be respected (Hill 1964, Levin 1967),

$$\epsilon = \frac{1}{V} \sum_r \int_{V_r} \epsilon_r(\mathbf{x}) dV \quad \sigma = \frac{1}{V} \sum_r \int_{V_r} \sigma_r(\mathbf{x}) dV \tag{4}$$

$$\mu = \frac{1}{V} \sum_r \int_{V_r} \mathbf{B}_r^T(\mathbf{x}) \mu_r dV \quad \lambda = \frac{1}{V} \sum_r \int_{V_r} \mathbf{A}_r^T(\mathbf{x}) \lambda_r dV \tag{5}$$

Using (4) in (2) and (3) provides after some algebra the following expressions for the overall $\mathbf{L}$ and $\mathbf{M}$ in (2)

$$\mathbf{L} = \mathbf{L}_1 + \sum_{r=2}^N c_r(\mathbf{L}_r - \mathbf{L}_1)\mathbf{A}_r, \quad \mathbf{M} = \mathbf{M}_1 + \sum_{r=2}^N c_r(\mathbf{M}_r - \mathbf{M}_1)\mathbf{B}_r \tag{6}$$

where $r = 1$ denotes a matrix phase, if present, or any chosen reference phase, and the phase volume fractions $c_r = V_r/V$; $\Sigma c_r = 1$. According to (4), the concentration factors must satisfy the relations $\Sigma c_r \mathbf{A}_r = \Sigma c_r \mathbf{B}_r = \mathbf{I}$. Therefore, for two-phase solids, the concentration factors can be found in terms of the overall $\mathbf{L}$ or $\mathbf{M}$; this is useful if the latter are known, e.g., from experimental measurements on actual systems,

$$c_\alpha \mathbf{A}_\alpha = (\mathbf{L}_\alpha - \mathbf{L}_\beta)^{-1}(\mathbf{L} - \mathbf{L}_\beta) \quad c_\alpha \mathbf{B}_\alpha = (\mathbf{M}_\alpha - \mathbf{M}_\beta)^{-1}(\mathbf{M} - \mathbf{M}_\beta) \tag{7}$$

1.2 Uniform fields in two-phase fibrous composites

Consider a composite material consisting of two distinct, homogeneous elastic phases, perfectly bonded at interfaces generated by lines parallel to the x_1-axis of a fixed Cartesian coordinate system. Matrix-based composites reinforced by aligned cylindrical fibers of any cross section, and two-phase layered or lamellar materials are examples of such systems. The geometry of the microstructure in the transverse $x_2 x_3$-plane can be arbitrary, provided that, apart from certain exceptions, the aggregate is statistically homogeneous. This implies that the overall elastic symmetry of the aggregate is similar to that of a monoclinic crystal; no other restrictions on local or overall elastic symmetry need to be imposed.

The objective is to find a certain overall stress $\sigma^0 = \hat{\sigma}$ such that the overall strain (2) and the local strains (3) satisfy the uniform strain condition in V,

$$\hat{\epsilon}_\alpha(\mathbf{x}) = \hat{\epsilon}_\beta(\mathbf{x}) = \hat{\epsilon} \tag{8}$$

where $r = \alpha, \beta$ denote the two phases. The solution can be constructed by a decomposition procedure where the phases are first separated and subjected to certain as yet unknown transformation strains μ_r and surface tractions in equilibrium with a uniform local stress $\hat{\sigma}_r$. The local strains follow from (1) and (3), and are now uniform. The local stresses and phase eigenstrains need to be adjusted such that the local strains satisfy (8), while the phase surface tractions are in equilibrium. The traction continuity conditions that need to be satisfied at the phase interfaces imply that the phase stresses comply with the conditions,

220

$$\hat{\sigma}_1^\alpha \neq \hat{\sigma}_1^\beta, \quad \hat{\sigma}_j^\alpha = \hat{\sigma}_j^\beta \quad \text{for} \quad j = 2, 3, 4, 5, 6 \tag{9}$$

where x_1 is parallel to the interface generators, and the stress components are written in the usual contracted notation. The aggregate can now be reassembled such that the strains satisfy (8) and the local and overall stresses are related by,

$$\hat{\sigma}_1 = c_\alpha \hat{\sigma}_1^\alpha + c_\beta \hat{\sigma}_1^\beta, \quad c_\alpha + c_\beta = 1 \tag{10}$$

$$\hat{\sigma}_j = \hat{\sigma}_j^\alpha = \hat{\sigma}_j^\beta \quad \text{for} \quad j = 2, 3, 4, 5, 6 \tag{11}$$

The desired magnitude of $\hat{\sigma}$ is finally found by solving the following system of equations derived by using (11) and (12) in (1) and then in (8),

$$M_{i1}^\alpha \hat{\sigma}_1^\alpha - M_{i1}^\beta \hat{\sigma}_1^\beta + \sum_{j=2}^{6} \left(M_{ij}^\alpha - M_{ij}^\beta \right) \hat{\sigma}_j + \mu_i^\alpha - \mu_i^\beta = 0; \quad i = 1 - 6. \tag{12}$$

If this system can be solved for a given composite, then the aggregate can be reassembled and loaded by the overall stresses (10) and (11), and the uniform phase eigenstrains; the overall and local strains then follow from (2) and (1), and satisfy (8). Since the rank of the (7x7) coefficient matrix of the unknowns in (12) is not greater than six, at least one of the unknown stresses can be chosen as a parameter in the solution. Even if no local eigenstrains are prescribed, a nontrivial solution depending on an arbitrary constant may exist for the homogeneous system; this would create a uniform strain field during proportional loading. Also, for any applied overall stress, (12) provides the difference $\mu_\beta - \mu_\alpha$ that would make the local strains uniform and dependent on one of the axial normal stresses subject to (10).

For systems consisting of at most transversely isotropic phases, the general and homogeneous solutions have been found by Dvorak (1990). Of interest here is the homogeneous solution, for the case of zero phase eigenstrains. An overall axisymmetric stress state supports the uniform internal fields,

$$S_A/S_T = q\left[\left(c_\alpha l_\alpha + c_\beta l_\beta \right) \left(l_\alpha - l_\beta \right) - \left(c_\alpha n_\alpha + c_\beta n_\beta \right) \left(k_\alpha - k_\beta \right) \right] \tag{13}$$

and the overall strain components are proportional,

$$\hat{\varepsilon}_1/\hat{\varepsilon}_2 = \hat{\varepsilon}_1/\hat{\varepsilon}_3 = -2\left(k_\alpha - k_\beta \right)/\left(l_\alpha - l_\beta \right) \tag{14}$$

The shear stresses and strains vanish. In the above, we used the definitions,

$$S_A = \hat{\sigma}_1 \quad S_T = \left(\hat{\sigma}_2 + \hat{\sigma}_3 \right)/2$$
$$q^{-1} = \left(l_\alpha k_\beta - k_\alpha l_\beta \right) \quad k_r = \left(L_{22}^r + L_{23}^r \right)/2 \quad l_r = L_{12}^r \quad n_r = L_{11}^r \tag{15}$$

where k_r, l_r and n_r are Hill's moduli of transversely isotropic phases.

One application of the uniform fields is in finding connections between local and overall elastic constants. In particular, consider a monoclinic fibrous aggregate made of transversely isotropic phases, arranged such that the transverse x_2x_3-plane is the only one plane of elastic symmetry. Nominally, there are 13 nonzero overall stiffness coefficients. Apply the surface tractions (13) to create a uniform stress field and follow this by removing the corresponding overall strain, to return to zero overall stress. The loading sequence actually is,

$$\hat{\sigma}_0 - \mathbf{L}\hat{\varepsilon}_0 = 0 \tag{16}$$

where $\mathbf{L}$ is the overall stiffness. A substitution from (13) eventually provides a system of equations that can be reduced to the following four connections between the local and overall elastic moduli,

$$\frac{\frac{1}{2}(L_{12}+L_{13})-c_a l_a - c_b l_b}{L_{11}-c_a n_a - c_b n_b} = \frac{\frac{1}{2}(L_{22}+L_{23})-k_a}{L_{12}-l_a} = \frac{\frac{1}{2}(L_{23}+L_{33})-k_a}{L_{13}-l_a} = \frac{\frac{1}{2}(L_{26}+L_{36})}{L_{16}} = \frac{k_a - k_b}{l_a - l_b} \tag{17}$$

Therefore, there are only nine independent elastic moduli in this fiber system. If the aggregate is transversely isotropic, (17) reduce to Hill's (1964) universal connections between moduli, which were derived by decomposing the volume averages of local strain fields.

Extending the loading/unloading sequence (16) to the local uniform fields,

$$\hat{\varepsilon}_r^0 - \mathbf{A}_r(\mathbf{x})\hat{\varepsilon}_0 = 0, \qquad \hat{\sigma}_r^0 - \mathbf{B}_r(\mathbf{x})\hat{\sigma}_0 = 0 \tag{18}$$

one can recover after some algebra connections between the coefficients of the mechanical influence functions in a monoclinic system made of transversely isotropic phases. In particular, the strain influence functions $\mathbf{A}_r(\mathbf{x})$ in (3) are subject to the following six constraints,

$$\left.\begin{array}{l} 2(\Delta k/\Delta l)(A_{11}^r(\mathbf{x})-1)-A_{12}^r(\mathbf{x})-A_{13}^r(\mathbf{x})=0 \\ 2(\Delta k/\Delta l)A_{21}^r(\mathbf{x})-A_{22}^r(\mathbf{x})-A_{23}^r(\mathbf{x})+1=0 \\ 2(\Delta k/\Delta l)A_{31}^r(\mathbf{x})-A_{32}^r(\mathbf{x})-A_{33}^r(\mathbf{x})+1=0 \\ 2(\Delta k/\Delta l)A_{41}^r(\mathbf{x})-A_{42}^r(\mathbf{x})-A_{43}^r(\mathbf{x})=0 \\ 2(\Delta k/\Delta l)A_{51}^r(\mathbf{x})-A_{52}^r(\mathbf{x})-A_{53}^r(\mathbf{x})=0 \\ 2(\Delta k/\Delta l)A_{61}^r(\mathbf{x})-A_{62}^r(\mathbf{x})-A_{63}^r(\mathbf{x})=0 \end{array}\right\} \quad r=\alpha, \beta \tag{19}$$

where $\Delta k/\Delta l = (k_\alpha - k_\beta)/(l_\alpha - l_\beta)$. Similar relations exist for $\mathbf{B}_r(\mathbf{x})$ (Dvorak 1990). The remarkable feature of this result is its validity for any transverse geometry of the fibers, regardless of the actual shape, whereas the mechanical influence functions have been generally regarded as dependent on the shape, as well as on the local and overall moduli. Additional applications of the uniform fields, in piezoelectric fibrous systems, are discussed below.

1.3 Uniform fields in systems with arbitrary phase geometry

Consider again a representative volume V of a multiphase solid, such as a polycrystal, loaded by uniform overall stress. The local fields in the phases are, of course, not uniform. However, a uniform strain field $\varepsilon = \varepsilon_r$ can be created in V by introducing a uniform stress field into all phases together with piecewise uniform eigenstrains μ_r, to satisfy the relation

$$\varepsilon = \varepsilon_r = \mathbf{M}_1\sigma^0 + \mu_1 = \ldots = \mathbf{M}_p\sigma^0 + \mu_p = \mathbf{M}_q\sigma^0 + \mu_q = \ldots = \mathbf{M}_N\sigma^0 + \mu_N \quad (r = 1,2\ldots,N). \quad (20)$$

where the $\mathbf{M}_r$ denote the phase compliances. A similar procedure involving a uniform overall and local strain field and piecewise uniform eigenstresses, creates a uniform stress field in V. The magnitude of the phase eigenstrains required to satisfy (20) is found as,

$$\mu_r = \mu_q^0 - (\mathbf{M}_r - \mathbf{M}_q)\sigma^0, \quad \varepsilon = \varepsilon_r = \mathbf{M}_q\sigma^0 + \mu_q^0 \quad (21)$$

where r = q is any chosen phase where a given eigenstrain may be introduced.

Recall that the local strain field in a multiphase medium loaded by overall uniform strain ε^0 and piecewise uniform phase eigenstrains μ_r is described by (3_1), rewritten here as,

$$\varepsilon_S(\mathbf{x}) = \mathbf{A}_S(\mathbf{x})\varepsilon^0 + \sum_r \mathbf{D}_{sr}(\mathbf{x})\mu_r \quad (22)$$

Substituting from (21), one finds,

$$\sum_{r=1}^{N} \mathbf{D}_{sr}(\mathbf{x})(\mathbf{M}_r - \mathbf{M}_q) = -(\mathbf{I} - \mathbf{A}_s(\mathbf{x}))\mathbf{M}_q \quad (23)$$

If this connection is to hold for any $\mathbf{M}_q$, then in each local volume V_s,

$$\sum_{r=1}^{N} \mathbf{D}_{sr}(\mathbf{x}) = \mathbf{I} - \mathbf{A}_s(\mathbf{x}), \quad \sum_{r=1}^{N} \mathbf{D}_{sr}(\mathbf{x})\mathbf{M}_r = \mathbf{0} \quad (24)$$

The existence of the uniform strain field (21) thus reveals two connections between the transformation influence functions in a multiphase solid. Additional relation of this kind was found from the elastic reciprocal theorem by Dvorak and Benveniste (1992) as,

$$c_s\mathbf{D}_{sr}\mathbf{M}_r = c_s\mathbf{M}_s\mathbf{D}_{rs}^T \quad (25)$$

Moreover, the connections (5) and $\Sigma c_r \mathbf{A}_r = \mathbf{I}$, used in (3), provide another relation for the transformation concentration factors,

$$\sum_{r=1}^{N} c_r\mathbf{D}_{rs} = \mathbf{0} \quad (26)$$

It turns out that of the four equations (24) to (26), only (24_1) and either (24_2) or (25) are independent; (26) can be derived from the last two. Therefore, for the (NxN) transformation influence functions in an N-phase solid, there are (2xN) exact independent relations between these functions. The implication is that the transformation functions can be found in terms of their mechanical counterparts only in two-phase solids. The result follows readily from appropriate forms ($r = \alpha, \beta$) of the two relations (36) and from the identity that holds for any two-phase system $\left(\mathbf{L}_\alpha - \mathbf{L}_\beta\right)^{-1}\mathbf{L}_\alpha = -\mathbf{M}_\beta\left(\mathbf{M}_\alpha - \mathbf{M}_\beta\right)^{-1}$, as,

$$\mathbf{D}_{r\alpha}(\mathbf{x}) = \left(\mathbf{I} - \mathbf{A}_r(\mathbf{x})\right)\left(\mathbf{L}_\alpha - \mathbf{L}_\beta\right)^{-1}\mathbf{L}_\alpha, \quad \mathbf{D}_{r\beta}(\mathbf{x}) = -\left(\mathbf{I} - \mathbf{A}_r(\mathbf{x})\right)\left(\mathbf{L}_\alpha - \mathbf{L}_\beta\right)^{-1}\mathbf{L}_\beta \qquad (27)$$

An analogous result can be derived for the eigenstress influence functions as

$$\mathbf{F}_{r\alpha}(\mathbf{x}) = \left(\mathbf{I} - \mathbf{B}_r(\mathbf{x})\right)\left(\mathbf{M}_\alpha - \mathbf{M}_\beta\right)^{-1}\mathbf{M}_\alpha, \quad \mathbf{F}_{r\beta}(\mathbf{x}) = -\left(\mathbf{I} - \mathbf{B}_r(\mathbf{x})\right)\left(\mathbf{M}_\alpha - \mathbf{M}_\beta\right)^{-1}\mathbf{M}_\beta \qquad (28)$$

These results can be also found directly from a loading/unloading sequence similar to that leading to (16), but involving a uniform field with phase eigenstrains (Dvorak 1990). For system with more than two phases, exact evaluation of $\mathbf{D}_{rs}$ in terms of $\mathbf{A}_r$ and/or $\mathbf{A}_s$ is not possible, as additional information about the local geometry must be included. However, the above (2xN) exact connections between the transformation influence functions still provide useful insight, for example, Dvorak and Benveniste (1992, eqn. 52) show for a three-phase system of any geometry six relations between the self-induced and transmitted transformation influence functions, which suggest that only the self-induced $\mathbf{D}_{11}(\mathbf{x})$, $\mathbf{D}_{22}(\mathbf{x})$ and $\mathbf{D}_{33}(\mathbf{x})$ need to be determined to complete the set.

In the absence of detailed geometry description for a multiphase medium, recourse can be made to the self-consistent or Mori-Tanaka methods. In particular, Dvorak and Benveniste (1992), using abundantly various uniform field solutions, found that either method provides the following expressions for the transformation influence functions,

$$\mathbf{D}_{sr} = \left(\mathbf{I} - \mathbf{A}_s\right)\left(\mathbf{L}_s - \mathbf{L}\right)^{-1}\left(\delta_{sr}\mathbf{I} - c_r\mathbf{A}_r^\mathrm{T}\right)\mathbf{L}_r,$$
$$\mathbf{F}_{sr} = \left(\mathbf{I} - \mathbf{B}_s\right)\left(\mathbf{M}_s - \mathbf{M}\right)^{-1}\left(\delta_{sr}\mathbf{I} - c_r\mathbf{B}_r^\mathrm{T}\right)\mathbf{M}_r, \qquad (29)$$

with the understanding that the overall $\mathbf{L}$ and $\mathbf{M}$, as well as the mechanical concentration factors must all be estimated by the same method. The Kronecker symbol is used here, but no summation is implied by repeated subscripts.

Apart from the many insights into evaluation of and connections between the transformation influence functions, the uniform fields in multiphase media of any micro-geometry have specific applications in polycrystals. Certain polycrystals with columnar type crystals admit a *planar isotropic* uniform stress and strain fields. These fields result in exact connections for the components of the effective compliance of the polycrystal, Benveniste (1994a, 1996a). Some three-dimensional polycrystals, with possible grain boundary sliding, admit *isotropic* uniform stresss and strain fields. A consequence of these fields are exact relations for the effective thermal expansion tensor, Benveniste (1996b).

1.4 Uniform fields in laminate plates

Bahei-El-Din (1992) extended the results for fibrous systems in Section 1.2 to laminated composite plates loaded by in-plane stresses, equal transverse normal stresses on each plane surface, and uniform changes in temperature. Identical plies of different fiber orientation were considered, with isotropic matrix and transversely isotropic fibers. To create the uniform field, the plies are first separated, the thermal phase eigenstrains are introduced, and each ply is loaded by an axisymmetric stress state such that in superposition with the thermal fields, it creates an isotropic stress state in the matrix. The in-plane strains are also isotropic in each ply. The plate can be thus reassembled, the transverse stresses applied at the surface planes, and the resultants of axial stresses supported by the overall in-plane components. This decomposition allows to replace the effect of uniform thermal change by application of equivalent mechanical stresses. At the same time, since the matrix stresses are isotropic, the equivalent mechanical loads do not contribute to matrix plastic flow.

Uniform fields of this kind are useful in analysis of thermal hardening in elastic-plastic systems. In a plate that does not yet experience inelastic deformation, or is being unloaded, they define the rigid-body translation of the cluster of ply yield surfaces in the overall stress space. Since both in-plane and out-of-plane translation is present, the sections of the yield surface cluster undergo significant changes. The above procedure provides an elegant way of analyzing the yield surface rearrangement.

2. TRANSFORMATION FIELD ANALYSIS OF INELASTIC COMPOSITES

2.1 Formulation

The transformation field analysis is a new method for evaluation of overall response and local fields in composite materials and polycrystals. It can be used as long as the total strains remain small, and can be additively decomposed into elastic and inelastic components. Apart from physical admissibility, no other restrictions need to be placed on phase or overall elastic symmetry, inelastic constitutive relations of the phases which may differ from one phase to another, or local geometry.

The additive decomposition requirement suggests that the total phases stress or strains can be written as,

$$\sigma_r(\mathbf{x}) = \sigma_r^e(\mathbf{x}) + \sigma_r^{re}(\mathbf{x}), \quad \varepsilon_r(\mathbf{x}) = \varepsilon_r^e(\mathbf{x}) + \varepsilon_r^{in}(\mathbf{x}) \tag{30}$$

where $\varepsilon_r^{in}(\mathbf{x})$ is the inelastic strain that has accumulated along the loading path, and $\sigma_r^{re}(\mathbf{x})$ is the corresponding relaxation stress that would be caused at $\mathbf{x}$ by application of the $\varepsilon_r(\mathbf{x})$ during the deformation history; $\sigma_r^{re}(\mathbf{x}) = -\mathbf{L}_r \varepsilon_r^{in}(\mathbf{x})$, and $\varepsilon_r^{in}(\mathbf{x}) = -\mathbf{M}_r \sigma_r^{re}(\mathbf{x})$, where $\mathbf{L}_r$ and $\mathbf{M}_r$ are the elastic stiffness and compliance of the phase r. If thermal strains are also present, (1) can be recast as,

$$\sigma_r(\mathbf{x}) = \mathbf{L}_r \varepsilon_r(\mathbf{x}) + l_r \theta + \sigma_r^{re}(\mathbf{x}), \quad \varepsilon_r(\mathbf{x}) = \mathbf{M}_r \sigma_r(\mathbf{x}) + \mathbf{m}_r \theta + \varepsilon_r^{in}(\mathbf{x}) \tag{31}$$

where $\mathbf{m}_r$ is the (6x1) thermal strain vector listing the linear coefficients of thermal expansion of the phases, and $\mathbf{l}_r = -\mathbf{L}_r\mathbf{m}_r$ is the thermal stress vector; θ is a uniform change in temperature from that of the stress-free state. Analogous decompositions can be written for the overall stresses and strains. These results are readily reconciled with (1) and (2) if the local and overall transformation fields are defined as,

$$\lambda_r(\mathbf{x}) = \mathbf{l}_r\theta + \sigma_r^{re}(\mathbf{x}), \quad \mu_r(\mathbf{x}) = \mathbf{m}_r\theta + \varepsilon_r^{in}(\mathbf{x}) \tag{32}$$

$$\lambda = \mathbf{l}\,\theta + \sigma^{re}, \quad \mu = \mathbf{m}\theta + \varepsilon^{in} \tag{33}$$

A general form of the inelastic constitutive relations can be formally written as,

$$\sigma_r^{re} = \mathbf{g}(\varepsilon_r), \quad \varepsilon_r^{in} = \mathbf{f}(\sigma_r) \tag{34}$$

where the $\mathbf{f}$ and $\mathbf{g}$ functions reflect the influence of past deformation history; in some constitutive theories, such as plasticity, (34) need to be written for rates or increments.

Suppose now that the composite aggregate has been subdivided into certain subvolumes V_r, V_s, etc., such that each subvolume is a part of only one phase volume. In each subvolume, the actual elastic and inelastic fields (31) are approximated by piecewise constant functions. The governing equations for evaluation of local fields in such subdivided aggregates can be now written by substituting (34) into (32), converting from eigenstrains to eigenstresses and vice versa, and finally using the results in subvolume-averaged version of (3) to provide,

$$\varepsilon_s + \mathbf{D}_{ss}\mathbf{M}_s\mathbf{g}(\varepsilon_s) + \sum_{\substack{r=1 \\ r \neq s}}^{N} \mathbf{D}_{sr}\mathbf{M}_r\mathbf{g}(\varepsilon_r) = \mathbf{A}_s\varepsilon + \mathbf{a}_s\theta, \quad r,s = 1,2,\dots,N \tag{35}$$

$$\sigma_s + \mathbf{F}_{ss}\mathbf{L}_s\mathbf{f}(\sigma_s) + \sum_{\substack{r=1 \\ r \neq s}}^{N} \mathbf{F}_{sr}\mathbf{L}_r\mathbf{f}(\sigma_r) = \mathbf{B}_s\sigma + \mathbf{b}_s\theta, \quad r,s = 1,2,\dots,N \tag{36}$$

2.2 Elastic-plastic systems

As an example, consider a composite system with phases which, when loaded to a certain effective stress magnitude, may experience elastic-plastic deformation. The elastic response is given by (31). If certain yield condition is satisfied, then inelastic strains may accumulate incrementally, following a constitutive law of the form

$$d\sigma_r^{re} = \ell_r^P d\varepsilon_r + \ell_r^P d\theta, \quad d\varepsilon_r^{in} = m_r^P d\sigma_r + m_r^P d\theta \tag{37}$$

Following now the procedure leading to (35) and (36), one recovers the governing equations for the local fields in the form,

$$d\varepsilon_s + \sum_{r=1}^{N} \mathbf{D}_{sr}\mathbf{M}_r \ell_r^p d\varepsilon_r = \mathbf{A}_s d\varepsilon + \left(\mathbf{a}_s - \sum_{r=1}^{N} \mathbf{D}_{sr}\mathbf{M}_r \ell_r^p \right) d\theta \tag{38}$$

$$d\sigma_s + \sum_{r=1}^{N} \mathbf{F}_{sr}\mathbf{L}_r m_r^p d\sigma_r = \mathbf{B}_s d\sigma + \left(\mathbf{b}_s - \sum_{r=1}^{N} \mathbf{F}_{sr}\mathbf{L}_r m_r^p \right) d\theta \tag{39}$$

where the thermal concentration factors are defined as,

$$\mathbf{a}_s = \sum_{r=1}^{N} \mathbf{D}_{sr}\mathbf{m}_r, \quad \mathbf{b}_s = \sum_{r=1}^{N} \mathbf{F}_{sr}\ell_r \tag{40}$$

The solution of (38) or (39) is sought in the form,

$$d\varepsilon_r = \mathcal{A}_r d\varepsilon + a_r d\theta, \quad d\sigma_r = \mathcal{B}_r d\sigma + \ell_r d\theta \tag{41}$$

where the coefficients are represented by instantaneous mechanical and thermal concentration factor tensors.

If a fine subdivision into many subvolumes is used, then a numerical solution is indicated. However, a closed-form solution can be found for a crude subdivision of the microstructure, r, s = α, β,

$$\mathcal{A}_\alpha = \left[\mathbf{I} + \mathbf{D}_{\alpha\alpha}\mathbf{M}_\alpha \ell_\alpha^p - \left(c_\alpha / c_\beta \right)\mathbf{D}_{\alpha\beta}\mathbf{M}_\beta \ell_\beta^p \right]^{-1} \left[\mathbf{A}_\alpha - \left(1/c_\beta \right)\mathbf{D}_{\alpha\beta}\mathbf{M}_\beta \ell_\beta^p \right]$$

$$a_\alpha = \left[\mathbf{I} + \mathbf{D}_{\alpha\alpha}\mathbf{M}_\alpha \ell_\alpha^p - \left(c_\alpha / c_\beta \right)\mathbf{D}_{\alpha\beta}\mathbf{M}_\beta \ell_\beta^p \right]^{-1} \left[\mathbf{a}_\alpha - \mathbf{D}_{\alpha\alpha}\mathbf{M}_\alpha \ell_\alpha^p - \mathbf{D}_{\alpha\beta}\mathbf{M}_\beta \ell_\beta^p \right] \tag{42}$$

$$\mathcal{B}_\alpha = \left[\mathbf{I} + \mathbf{F}_{\alpha\alpha}\mathbf{L}_\alpha m_\alpha^p - \left(c_\alpha/c_\beta \right)\mathbf{F}_{\alpha\beta}\mathbf{L}_\beta m_\beta^p \right]^{-1} \left[\mathbf{B}_\alpha - \left(1/c_\beta \right)\mathbf{F}_{\alpha\beta}\mathbf{L}_\beta m_\beta^p \right]$$

$$\ell_\alpha = \left[\mathbf{I} + \mathbf{F}_{\alpha\alpha}\mathbf{L}_\alpha m_\alpha^p - \left(c_\alpha/c_\beta \right)\mathbf{F}_{\alpha\beta}\mathbf{L}_\beta m_\beta^p \right]^{-1} \left[\mathbf{b}_\alpha - \mathbf{F}_{\alpha\alpha}\mathbf{L}_\alpha m_\alpha^p - \mathbf{F}_{\alpha\beta}\mathbf{L}_\beta m_\beta^p \right] \tag{43}$$

and an exchange of subscripts furnishes the results for phase β. Note that apart from the instantaneous plastic stiffness and compliance from (37), the coefficients in (38) and (39) depend only on elastic moduli and phase (or subvolume) volume fractions. Therefore, these quantities are typically constant, except in applications involving large changes in temperature which may affect the moduli. Once these coefficient matrices have been evaluated, one can readily write the governing equations for the chosen constitutive relations of the phases. This has been illustrated for viscoelastic and viscoplastic systems by Dvorak et al. (1994). Applications of the transformation method to damage by interface decohesion are outlined in Dvorak et al. (1995).

3. EXACT CONNECTIONS FOR THE EFFECTIVE PROPERTIES OF PIEZOELECTRIC COMPOSITES

3.1. Background

Piezoelectric composites constitute an important branch of modern engineering materials. The use of piezocomposites in ultrasonic transducers is today well established, with several applications existing in underwater acoustics and medical imaging, see for example, Newnham *et al.* (1978), Smith (1989,1993), and Zhang *et al.*(1996).

The topic of exact connections between the effective moduli of composite media, overviewed and discussed in the article by Milton (1996) in this conference proceedings, gains a special significance in the field of piezocomposites in view of the numerous effective constants which describe these solids. The third section of the present paper is concerned with a brief review of the exact connections derived by the second author for the effective moduli of piezoelectric composites in the last years. The basic method of derivation is the "method of uniform fields" described in Section 1. For fibrous systems it is shown that a "field-decoupling method" by Milgrom and Shtrikman (1989) is applicable and provides additional connections between the effective properties. Finally, in some modes of deformation and electric fields use is made of certain "translation and rotation mappings" of the constitutive tensors similar to the transformations used by Milton (1988) ; these correspondence relations are used in establishing phase interchange connections in piezoelectric composites.

3.2. Constitutive laws

Let us consider a composite medium with piezoelectric phases obeying the following constitutive law

$$\sigma^{(r)} = \mathbf{L}_r \varepsilon^{(r)} - \mathbf{e}_r^T \mathbf{E}^{(r)} + \beta_r \theta_0 \quad , \quad \mathbf{D}^{(r)} = \mathbf{e}_r \varepsilon^{(r)} + \kappa_r \mathbf{E}^{(r)} + \mathbf{p}_r \theta_0 \quad , \tag{44}$$

where σ, ε, $\mathbf{E}$, $\mathbf{D}$ are the stress, strain tensors, and the electric field intensity and electric displacement vectors respectively. The elastic, piezoelectric and dielectric constants of the constituents are represented by the fourth, third, and second order tensors $\mathbf{L}_r, \mathbf{e}_r, \kappa_r$ respectively, and $(\mathbf{e}^T)_{ijk} = (\mathbf{e})_{kij}$. The thermal stress tensor and the pyroelectric vector are denoted by β_r and $\mathbf{p}_r$, and θ_0 is a temperature change from a reference temperature. The superscript or subscript "r" denotes that the relevant property belongs to phase "r" of the composite.

We consider a state of static equilibrium so that in the absence of body forces and electrical charges, there is:

$$\sigma_{ij,j} = 0 \quad , \quad D_{i,i} = 0. \tag{45}$$

The strain tensor and electric field vector are derived from the mechanical displacements u_i and electric potential ϕ through

$$\varepsilon_{ij} = (u_{i,j} + u_{j,i})/2 \quad , \quad E_i = -\phi_{,i} \quad . \tag{46}$$

The following boundary conditions prevail at the interfaces of two adjacent phases

$$u_i^{(r)} = u_i^{(s)} \quad , \quad \phi^{(r)} = \phi^{(s)} \quad , \quad \sigma_{ij}^{(r)} n_j = \sigma_{ij}^{(s)} n_j \quad , \quad D_i^{(r)} n_i = D_i^{(s)} n_i \quad , \tag{47}$$

where $\mathbf{n}$ denotes the normal to the interface.

Let now the piezoelectric composite be statistically homogeneous, and apply homogeneous boundary conditions in the form

$$\mathbf{u}(S) = \varepsilon^0 \mathbf{x} \quad , \quad \phi(S) = -\mathbf{E}^{(0)} \mathbf{x} \quad , \quad \theta(S) = \theta_0 \quad , \tag{48}$$

under which there is:

$$\overline{\varepsilon} = \varepsilon^0 \quad , \quad \overline{\mathbf{E}} = \mathbf{E}^0 \quad , \quad \theta = \theta_0 \quad , \tag{49}$$

where an overbar denotes a representative volume average. The effective behaviour of the composite is then given by:

$$\mathbf{L}\,\varepsilon^0 - \mathbf{e}^T\mathbf{E}^0 + \beta\theta_0 = \sum_{r=1}^{N} c_r [\mathbf{L}_r \overline{\varepsilon}^{(r)} - \mathbf{e}_r^T \overline{\mathbf{E}}^{(r)} + \beta_r \theta_0]$$

$$\mathbf{e}\varepsilon^0 + \kappa\mathbf{E}^0 + \mathbf{p}\theta_0 = \sum_{r=1}^{N} c_r [\mathbf{e}_r \overline{\varepsilon}^{(r)} + \kappa_r \overline{\mathbf{E}}^{(r)} + \mathbf{p}^{(r)}\theta_0] \quad , \tag{50}$$

where c_r denote the volume fractions of the phases and $(\overline{\chi}^{(r)})$ is the average of the quantity χ over the phase "r" in the representative volume. In the following sections we are interested in looking for possible microstructure- independent connections between some of the effective properties $\mathbf{L}, \mathbf{e}, \kappa, \beta$ and $\mathbf{p}$.

3.3 Two-phase systems with arbitrary phase geometry and general anisotropy.

We start by subjecting the two-phase composite to the boundary conditions

$$\mathbf{u}(S) = \hat{\varepsilon}\mathbf{x} \quad , \quad \phi(S) = -\hat{\mathbf{E}}\mathbf{x} \quad , \quad \theta(S) = \theta_0 \quad , \tag{51}$$

and look for a *specific* set $(\hat{\varepsilon}, \hat{\mathbf{E}}, \theta_0)$ under which the strain and electric fields within the composite aggregate are uniform

$$\varepsilon^{(r)}(\mathbf{x}) = \hat{\varepsilon} \quad , \quad \mathbf{E}^{(r)}(\mathbf{x}) = \hat{\mathbf{E}} \quad , \quad r = 1,2 . \tag{52}$$

If (52) prevails, this results in a mechanical displacement field and electric potential which trivially satisfy the interface continuity conditions $(47)_1$, and $(47)_2$ and produces stresses and electric displacements which equally fulfill (45). Thus the problem reduces to finding forms of $\hat{\varepsilon}$ and $\hat{\mathbf{E}}$ in terms of θ_0 which would insure the continuity of the tractions and of the normal component of the electric displacements in $(47)_3$ and $(47)_4$. Since we do not wish to involve the microstructure in the derivation, this is achieved by demanding :

$$\sigma_{ij}^{(1)} = \sigma_{ij}^{(2)} \quad , \quad D_i^{(1)} = D_i^{(2)} \quad , \tag{53}$$

which readily result in the following equations:

$$\mathbf{L}_1\hat{\varepsilon} - \mathbf{e}_1^T\hat{\mathbf{E}} + \beta_1\theta_0 = \mathbf{L}_2\hat{\varepsilon} - \mathbf{e}_2^T\hat{\mathbf{E}} + \beta_2\theta_0 \quad , \quad \mathbf{e}_1\hat{\varepsilon} + \kappa_1\hat{\mathbf{E}} + \mathbf{p}_1\theta_0 = \mathbf{e}_2\hat{\varepsilon} + \kappa_2\hat{\mathbf{E}} + \mathbf{p}_2\theta_0 \ . \tag{54}$$

The expressions for $\hat{\varepsilon}$ and $\hat{\mathbf{E}}$ in terms of θ_0 derivable from (54) completes the construction of the uniform field solution.

Let now the composite be subjected to (51) and let its effective properties be defined under (50). Making use of the fact that

$$\bar{\varepsilon}^{(1)} = \bar{\varepsilon}^{(2)} = \hat{\varepsilon} \quad , \quad \bar{\mathbf{E}}^{(1)} = \bar{\mathbf{E}}^{(2)} = \hat{\mathbf{E}} \quad , \tag{55}$$

and substituting the solution of $\hat{\varepsilon}$ and $\hat{\mathbf{E}}$ from (54) into (50) readily provides the sought expressions of β and $\mathbf{p}$ in terms of $\mathbf{L}, \mathbf{e}, \kappa$ and the constituent properties, and have been given in Benveniste (1993a, Part II, Equation (40)). Similar exact results were derived independently by Dunn (1993) by other means. In the special case of thermoelasticity these results reduce to those given before by Levin (1967), Cribb (1968), Schapery (1968), Rosen and Hashin (1970), and Laws (1973). It should finally be emphasized here that the uniform fields approach described in the present paper *goes beyond* providing connections between the effective properties ; as shown in the previous sections, it furnishes exact relations for *the pointwise local fields* induced by several loading configurations in the considered systems. Similar local relations exist in piezocomposites and have been given in Benveniste (1993a,b).

3.4 Fibrous systems with cylindrical microgeometry

We consider now an N-phase piezoelectric composite with a fibrous structure characterized by the fact that the phase boundaries are surfaces which can be generated by straight lines parallel to the x_3 axis (note that in Section 1.2 the phase boundaries were parallel to the x_1-axis). The transverse microgeometry in the x_1x_2 plane is thus invariant. Let us denote the stresses, strains, electric fields and electric displacements by:

$$\sigma = \begin{Bmatrix} \sigma_{11} \\ \sigma_{22} \\ \sigma_{33} \\ \sigma_{32} \\ \sigma_{13} \\ \sigma_{12} \end{Bmatrix}, \quad \varepsilon = \begin{Bmatrix} \varepsilon_{11} \\ \varepsilon_{22} \\ \varepsilon_{33} \\ 2\varepsilon_{23} \\ 2\varepsilon_{13} \\ 2\varepsilon_{12} \end{Bmatrix}, \quad \mathbf{E} = \begin{Bmatrix} E_1 \\ E_2 \\ E_3 \end{Bmatrix}, \quad \mathbf{D} = \begin{Bmatrix} D_1 \\ D_2 \\ D_3 \end{Bmatrix}. \tag{56}$$

The individual phases are transversely isotropic about the x_3-axis. Due to the transversely isotropic constituents and assumed cylindrical microgeometry, the most general admissible overall symmetry of the composite is that of a monoclinic crystal with a point symmetry of class 2. The constituent matrices for the phases and the composite are therefore given by (see Nye (1957), for example):

$$\mathbf{L}_r = \begin{bmatrix} k_r+(G_T)_r & k_r-(G_T)_r & l_r & 0 & 0 & 0 \\ k_r-(G_T)_r & k_r+(G_T)_r & l_r & 0 & 0 & 0 \\ l_r & l_r & n_r & 0 & 0 & 0 \\ 0 & 0 & 0 & (G_L)_r & 0 & 0 \\ 0 & 0 & 0 & 0 & (G_L)_r & 0 \\ 0 & 0 & 0 & 0 & 0 & (G_T)_r \end{bmatrix} , \mathbf{e}_r^T = \begin{bmatrix} 0 & 0 & (e_{31})_r \\ 0 & 0 & (e_{31})_r \\ 0 & 0 & (e_{33})_r \\ 0 & (e_{15})_r & 0 \\ (e_{15})_r & 0 & 0 \\ 0 & 0 & 0 \end{bmatrix} ,$$

$$\mathbf{\kappa}_r = \begin{bmatrix} (\kappa_{11})_r & 0 & 0 \\ 0 & (\kappa_{11})_r & 0 \\ 0 & 0 & (\kappa_{33})_r \end{bmatrix} , \mathbf{\beta}_r^T = \{(\beta_T)_r \quad (\beta_T)_r \quad (\beta_L)_r \quad 0 \quad 0 \quad 0\} , \mathbf{p}_r = \begin{Bmatrix} 0 \\ 0 \\ (p_L)_r \end{Bmatrix} \qquad (57)$$

and,

$$\mathbf{L} = \begin{bmatrix} L_{11} & L_{12} & L_{13} & 0 & 0 & L_{16} \\ L_{12} & L_{22} & L_{23} & 0 & 0 & L_{26} \\ L_{13} & L_{23} & L_{33} & 0 & 0 & L_{36} \\ 0 & 0 & 0 & L_{44} & L_{45} & 0 \\ 0 & 0 & 0 & L_{45} & L_{55} & 0 \\ L_{16} & L_{26} & L_{36} & 0 & 0 & L_{66} \end{bmatrix} , \mathbf{e}^T = \begin{bmatrix} 0 & 0 & e_{31} \\ 0 & 0 & e_{32} \\ 0 & 0 & e_{33} \\ e_{14} & e_{24} & 0 \\ e_{15} & e_{25} & 0 \\ 0 & 0 & e_{36} \end{bmatrix} , \mathbf{\kappa} = \begin{bmatrix} \kappa_{11} & \kappa_{12} & 0 \\ \kappa_{12} & \kappa_{22} & 0 \\ 0 & 0 & \kappa_{33} \end{bmatrix} , \mathbf{\beta} = \begin{Bmatrix} \beta_{11} \\ \beta_{22} \\ \beta_{33} \\ 0 \\ 0 \\ \beta_{12} \end{Bmatrix} \qquad (58)$$

with $\mathbf{p}$ having the same structure as that in (57).

Let us now subject the composite solid to the boundary conditions (48). The cylindrical microstructure together with the transverse isotropy of the constituents allow to decompose this loading and the corresponding solutions into two different problems:

Problem I

$$\begin{Bmatrix} u_1(S) \\ u_2(S) \\ u_3(S) \end{Bmatrix} = \begin{bmatrix} \varepsilon_{11}^0 & \varepsilon_{12}^0 & 0 \\ \varepsilon_{12}^0 & \varepsilon_{22}^0 & 0 \\ 0 & 0 & \varepsilon_{33}^0 \end{bmatrix} \begin{Bmatrix} x_1 \\ x_2 \\ x_3 \end{Bmatrix} , \quad \phi(S) = -E_3^0 x_3 , \quad \theta(S) = \theta_0 \qquad (59)$$

with the solution formally being represented as:

$$u_1 = u_1(x_1, x_2), \quad u_2 = u_1(x_1, x_2), \quad u_3 = \varepsilon_{33}^0 x_3, \quad \phi = -E_3^0 x_3, \quad \theta = \theta_0. \tag{60}$$

Problem II

$$\begin{Bmatrix} u_1(S) \\ u_2(S) \\ u_3(S) \end{Bmatrix} = \begin{bmatrix} 0 & 0 & (\varepsilon_{13})_0 \\ 0 & 0 & (\varepsilon_{23})_0 \\ (\varepsilon_{13})_0 & (\varepsilon_{23})_0 & 0 \end{bmatrix} \begin{Bmatrix} x_1 \\ x_2 \\ x_3 \end{Bmatrix}, \quad \phi(S) = -(E_1)_0 x_1 - (E_2)_0 x_2 \tag{61}$$

and the solution being in the form of:

$$u_1 = \varepsilon_{13}^0 x_3, \quad u_2 = \varepsilon_{23}^0 x_3, \quad u_3 = \varphi(x_1, x_2) - \varepsilon_{13}^0 x_1 - \varepsilon_{23}^0 x_2, \quad \phi = \phi(x_1, x_2). \tag{62}$$

It is noted that Problem I is a generalized plane strain problem coupled to longitudinal electric field, whereas Problem II is an anti-plane mechanical problem coupled to an in-plane electric field.

As in the subsection above we now look for a *specific* set $(\hat{\varepsilon}, \hat{E}, \theta_0)$, under which the strains and electric fields in the composite will be uniform. Proceeding along the same lines, it is observed that the fields $(\hat{\varepsilon}, \hat{E})$ need to be determined from a condition which will insure the continuity of the tractions and normal component of the electric displacement across the constituent interfaces of the multiphase composite. Due to the cylindrical geometry of these interfaces, this can be achieved by applying the conditions which were given in (53), but this time omitting the σ_{33} component for the stresses and D_3 component for the electric displacements.

For two-phase composites a possible set of uniform fields can be shown to be given by (Benveniste,(1993b)):

$$\begin{Bmatrix} \hat{\varepsilon}_{11} \\ \hat{\varepsilon}_{22} \\ \hat{\varepsilon}_{33} \\ 2\hat{\varepsilon}_{23} \\ 2\hat{\varepsilon}_{13} \\ 2\hat{\varepsilon}_{12} \end{Bmatrix} = \xi_1 \begin{Bmatrix} 1 \\ 1 \\ r_3 \\ 0 \\ 0 \\ 0 \end{Bmatrix} + \xi_2 \begin{Bmatrix} 0 \\ 0 \\ 1 \\ 0 \\ 0 \\ 0 \end{Bmatrix} + \eta_1 \begin{Bmatrix} 0 \\ 0 \\ \tilde{r}_3 \\ 0 \\ 0 \\ 0 \end{Bmatrix} \theta_0, \quad \begin{Bmatrix} \hat{E}_1 \\ \hat{E}_2 \\ \hat{E}_3 \end{Bmatrix} = \xi_2 \begin{Bmatrix} 0 \\ 0 \\ S_3 \end{Bmatrix}, \tag{63}$$

where ξ_1, ξ_2, and η_1 are arbitrary constants and r_3, $\tilde{r}_3$, and S_3 are defined as:

$$r_3 = -\frac{2(k_1 - k_2)}{l_1 - l_2}, \quad \tilde{r}_3 = -\frac{(\beta_T)_1 - (\beta_T)_2}{(l_1 - l_2)}, \quad S_3 = \frac{(l_1 - l_2)}{(e_{31})_1 - (e_{31})_2}. \tag{64}$$

Substitution of these uniform field solutions in the equation for the effective properties yields the following exact connections among the effective properties:

$$\frac{(L_{11}+L_{12})/2-\sum_{r=1}^{2}c_r k_r}{L_{13}-\sum_{r=1}^{2}c_r l_r}=\frac{(L_{12}+L_{22})/2-\sum_{r=1}^{2}c_r k_r}{L_{23}-\sum_{r=1}^{2}c_r l_r}=\frac{(L_{13}+L_{23})/2-\sum_{r=1}^{2}c_r l_r}{L_{33}-\sum_{r=1}^{2}c_r n_r}=$$

$$=[(L_{16}+L_{26})/(2L_{36})]=(k_1-k_2)/(l_1-l_2),$$

$$\frac{e_{31}-\sum_{r=1}^{2}c_r(e_{31})_r}{L_{13}-\sum_{r=1}^{2}c_r l_r}=\frac{e_{32}-\sum_{r=1}^{2}c_r(e_{31})_r}{L_{23}-\sum_{r=1}^{2}c_r l_r}=\frac{e_{33}-\sum_{r=1}^{2}c_r(e_{33})_r}{L_{33}-\sum_{r=1}^{2}c_r n_r}=\frac{\sum_{r=1}^{2}c_r(\kappa_{33})_r-\kappa_{33}}{e_{33}-\sum_{r=1}^{2}c_r(e_{33})_r}=$$

$$=e_{36}/L_{36}=[(e_{31})_1-(e_{31})_2]/(l_1-l_2), \tag{65}$$

$$\beta_{11}=[\sum_{r=1}^{2}c_r l_r-L_{13}]\tilde{r}_3+\sum_{r=1}^{2}c_r(\beta_T)_r \quad,\quad \beta_{22}=[\sum_{r=1}^{2}c_r l_r-L_{23}]\tilde{r}_3+\sum_{r=1}^{2}c_r(\beta_T)_r$$

$$\beta_{33}=[\sum_{r=1}^{2}c_r n_r-L_{33}]\tilde{r}_3+\sum_{r=1}^{2}c_r(\beta_L)_r \quad,\quad p_3=[\sum_{r=1}^{2}c_r(e_{33})_r-e_{33}]\tilde{r}_3+\sum_{r=1}^{2}c_r(p_L)_r \quad,$$

$$\beta_{12}=L_{36}[(\beta_T)_1-(\beta_T)_2]/(l_1-l_2). \tag{66}$$

It is seen that all the constants entering in the above relations are those which would appear in Problem I. Variants of the above connections can also be obtained by using alternative forms of the uniform fields to those given in (63), and can be found in Benveniste (1993b). It is emphasized however that any such relations are not independent from those in (65) and (66). Equations (65) show that there exists nine exact connections for the fourteen effective constants $L_{11}, L_{12}, L_{13}, L_{22}, L_{23}, L_{33}, L_{16}, L_{26}, L_{36}, e_{31}, e_{32}, e_{33}, e_{36}, \kappa_{33}$. Equation (66), on the other hand, shows that the effective thermal terms of the pyroelectric composite can be expressed in terms of the effective $\mathbf{L}, \mathbf{e}, $ and κ tensors and the constituent properties. Special forms of (65) and (66) applicable to fibrous systems with effective overall behaviour of the kind 2mm, 4mm, and 6mm have been studied in Benveniste and Dvorak (1992b), and Benveniste (1993b). By generalizing the analysis of Hill (1964) for purely elastic systems, Schulgasser (1991) has also derived the special forms of (65) for fibrous composites with overall transverse isotropy.

Results similar to those given in (65) and (66) can be derived for three-phase fibrous composites with cylindrical microgeometry. Such results which were again obtained by the uniform fields approach can be found in Benveniste (1993b,1994b). Chen (1993) has used an analysis similar to that of Hill (1964) in obtaining exact connections among the $\mathbf{L}, \mathbf{e}, $ and κ tensors in three-phase systems. Four-phase fibrous systems admit only exact relations of the form (66), and do not possess forms similar to (65), Benveniste (1993b).

It is finally mentioned that exact expressions for some of the effective moduli can be obtained in multiphase fibrous piezoelectric composites in which the constituent phases possess the same transverse modulus G_T; these results can be derived by following an approach of Hill (1964) in fibrous elastic systems, and have been given in Benveniste (1994b,1994c), and Chen (1993).

Let us now turn our attention to the effective constants $L_{44}, L_{45}, L_{55}, e_{15}, e_{25}, e_{14}, e_{24}, \kappa_{11}, \kappa_{12}$, and κ_{22} which appear in Problem II. In this case, the constitutive equations for the phases can be summarized as:

$$\left\{ \begin{array}{c} \sigma_{13} \\ D_1 \end{array} \right\}^{(r)} = \mathbf{C}_r \left\{ \begin{array}{c} \dfrac{\partial \varphi}{\partial x_1} \\ \dfrac{\partial \phi}{\partial x_1} \end{array} \right\}^{(r)} \quad , \quad \left\{ \begin{array}{c} \sigma_{23} \\ D_2 \end{array} \right\}^{(r)} = \mathbf{C}_r \left\{ \begin{array}{c} \dfrac{\partial \varphi}{\partial x_2} \\ \dfrac{\partial \phi}{\partial x_2} \end{array} \right\}^{(r)} \quad ; \quad \mathbf{C}_r = \left[\begin{array}{cc} L_{55} & e_{15} \\ e_{15} & -\kappa_{11} \end{array} \right]^{(r)} , \tag{67}$$

whereas the equilibrium equations for the stresses and electric displacements reduce to:

$$(L_{55})_r \nabla^2 \varphi^{(r)} + (e_{15})_r \nabla^2 \phi^{(r)} = 0 \quad , \quad (e_{15})_r \nabla^2 \varphi^{(r)} - (\kappa_{11})_r \nabla^2 \phi^{(r)} = 0 \quad . \tag{68}$$

The problem consists therefore of curl-free driving vector fields $(\nabla\varphi, \nabla\phi)$ and divergenceless fluxes $(\sigma, \mathbf{D})$, and the field decoupling procedure of Milgrom and Shtrikman (1989) for *two-phase media* readily applies. According to this procedure, a matrix $\mathbf{W}$ can always be found which diagonalizes the two symmetric matrices of the constituents $\mathbf{C}_r$ with $r = 1, 2$ by the transformation $\mathbf{W}\mathbf{C}_r\mathbf{W}^T$. Next, transformed new fields are defined in the form of

$$\left\{ \begin{array}{c} \tilde{\sigma}_{13} \\ \tilde{D}_1 \end{array} \right\}^{(r)} = \mathbf{W} \left\{ \begin{array}{c} \sigma_{13} \\ D_1 \end{array} \right\}^{(r)} \quad , \quad \left\{ \begin{array}{c} \tilde{\sigma}_{23} \\ \tilde{D}_2 \end{array} \right\}^{(r)} = \mathbf{W} \left\{ \begin{array}{c} \sigma_{23} \\ D_2 \end{array} \right\}^{(r)} ,$$

$$\left\{ \begin{array}{c} \dfrac{\partial \tilde{\varphi}}{\partial x_1} \\ \dfrac{\partial \tilde{\phi}}{\partial x_1} \end{array} \right\}^{(r)} = (\mathbf{W}^T)^{-1} \left\{ \begin{array}{c} \dfrac{\partial \varphi}{\partial x_1} \\ \dfrac{\partial \phi}{\partial x_1} \end{array} \right\}^{(r)} \quad , \quad \left\{ \begin{array}{c} \dfrac{\partial \tilde{\varphi}}{\partial x_2} \\ \dfrac{\partial \tilde{\phi}}{\partial x_2} \end{array} \right\}^{(r)} = (\mathbf{W}^T)^{-1} \left\{ \begin{array}{c} \dfrac{\partial \varphi}{\partial x_2} \\ \dfrac{\partial \phi}{\partial x_2} \end{array} \right\}^{(r)} . \tag{69}$$

It is readily verified that the constitutive equations decouple in the transformed fields, and the transformed stresses and electric displacements fulfill equilibrium conditions similar to (45). It thus follows the transformations (69) decouple the problem and the matrix $\mathbf{W}$ should diagonalize the effective tensor of the composite as well. For overall monoclinic symmetry of class 2, the effective constitutive equations for the fields in the present mode are

$$\left\{ \begin{array}{c} \overline{\sigma}_{13} \\ \overline{D}_1 \end{array} \right\} = \left[\begin{array}{cc} L_{55} & e_{15} \\ e_{15} & -\kappa_{11} \end{array} \right] \left\{ \begin{array}{c} \dfrac{\partial \overline{\varphi}}{\partial x_1} \\ \dfrac{\partial \overline{\phi}}{\partial x_1} \end{array} \right\} + \left[\begin{array}{cc} L_{54} & e_{25} \\ e_{14} & -\kappa_{12} \end{array} \right] \left\{ \begin{array}{c} \dfrac{\partial \overline{\varphi}}{\partial x_2} \\ \dfrac{\partial \overline{\phi}}{\partial x_2} \end{array} \right\} \quad ,$$

$$\left\{ \begin{array}{c} \overline{\sigma}_{23} \\ \overline{D}_2 \end{array} \right\} = \left[\begin{array}{cc} L_{54} & e_{14} \\ e_{25} & -\kappa_{12} \end{array} \right] \left\{ \begin{array}{c} \dfrac{\partial \overline{\varphi}}{\partial x_1} \\ \dfrac{\partial \overline{\phi}}{\partial x_1} \end{array} \right\} + \left[\begin{array}{cc} L_{44} & e_{24} \\ e_{24} & -\kappa_{22} \end{array} \right] \left\{ \begin{array}{c} \dfrac{\partial \overline{\varphi}}{\partial x_2} \\ \dfrac{\partial \overline{\phi}}{\partial x_2} \end{array} \right\} \quad . \tag{70}$$

Since the same matrix **W** diagonalizes not only the constituent matrices but the effective matrices as well, this imposes some restrictions among the components of these matrices. It can be shown that the exact connections in this case are, Benveniste (1994b):

$$\text{Det}\begin{bmatrix} L_{55} & e_{15} & -\kappa_{11} \\ (L_{55})_1 & (e_{15})_1 & -(\kappa_{11})_1 \\ (L_{55})_2 & (e_{15})_2 & -(\kappa_{11})_2 \end{bmatrix}=0 \ , \ \text{Det}\begin{bmatrix} L_{44} & e_{24} & -\kappa_{22} \\ (L_{55})_1 & (e_{15})_1 & -(\kappa_{11})_1 \\ (L_{55})_2 & (e_{15})_2 & -(\kappa_{11})_2 \end{bmatrix}=0 \quad ,$$

$$\text{Det}\begin{bmatrix} L_{54} & e_{14} & -\kappa_{12} \\ (L_{55})_1 & (e_{15})_1 & -(\kappa_{11})_1 \\ (L_{55})_2 & (e_{15})_2 & -(\kappa_{11})_2 \end{bmatrix}=0, \qquad e_{14}=e_{25}. \tag{71}$$

An important property of the connections in (71) is that they also imply

$$\text{Det}\begin{bmatrix} L_{55} & e_{15} & -\kappa_{11} \\ L_{44} & e_{24} & -\kappa_{22} \\ L_{54} & e_{14} & -\kappa_{12} \end{bmatrix}=0 \quad , \tag{72}$$

which is a relation which does not involve the constituent properties. Relations with such a property can also be obtained by manipulating some of the connections in (65):

$$\frac{L_{16}+L_{26}}{L_{36}}=\frac{L_{11}-L_{22}}{L_{13}-L_{23}} \quad , \quad \frac{e_{31}-e_{32}}{L_{13}-L_{23}}=\frac{e_{36}}{L_{36}}. \tag{73}$$

It is remarkable that the simple fact the fibrous material is two-phase and possesses an overall symmetry of the monoclinic system of class 2, imposes restrictions as (72) and (73) on its effective parameters.

3.5 Exact Connections for the Effective Properties of Composites with Piezoelectric and Piezomagnetic Phases

Composite systems with piezoelectric and piezomagnetic phases have been recently studied by Harshe et al. (1993), Avellaneda and Harshe (1994), Nan (1994), and Benveniste (1995a).These composites exhibit a magnetoelectric effect which is not present in the phases. In (Benveniste (1995a)) two-phase fibrous composite systems with transversely isotropic piezoelectric and piezomagnetic phases have been considered. The overall behaviour of the composite has also been assumed to be transversely as well. A development parallel to Section 3.4 results in microstructure -independent connections for the effective properties of the nature given in (65), (66) and (71). These connections have been implemented in a simple derivation of the effective moduli of the composite cylinder assemblage microgeometry of the considered composite aggregate.

3.6 Phase Interchange Relations in Piezoelectric Composites under Anti-plane Mechanical and In-Plane Electrical Fields

Exact connections between the effective properties of a two-phase piezoelectric solid and those of the medium in which the phases have been interchanged are a classic example of microstructure independent relations in composite media. In a pioneering study, Keller (1964) showed that such phase interchange connections for effective properties can be derived in the context of heat conduction in periodic two-dimensional composites. Among the works which discuss phase interchange relations in composite media, of particular interest is a study by Milton (1988) in which the existence of an isomorphism between any given two-dimensional conductivity problem and a whole family of associated conductivity problems has been established. This correspondence is established by means of a general transformation of the local conductivity tensor, and is based primarily on the observation that a two-dimensional divergence-free vector field when rotated locally by 90 degrees produces a curl-free field and vice versa. The idea of these rotation transformations appear in Dykhne (1971), except that he considered the case of scalar conductivites only. Recently, correspondence relations of the nature described by Milton (1988) have been shown to exist in certain heterogeneous piezoelectric solids in an anti-plane mode of deformation which is coupled to an in-plane electrical field, Benveniste (1995b). It is shown that a certain mapping of the local constitutive law of the composite into a different constitutive law results in the same mapping between the effective laws. We conclude this paper by stating some phase interchange relations in two-phase piezoelectric media which are a direct consequence of the correspondence relations established in that last work.

Let us consider a two-phase piezoelectric fibrous composite of the nature described in Section 3.4. Consider an anti-plane mode of deformation in this solid which is coupled to an in-plane electrical field induced by the loading (61). Let the phase constitutive laws be defined by $\mathbf{C}_r$ with $r = 1,2$ in (67). The composite possesses overall monoclinic symmetry of class 2, which is the most general possible anisotropy under its cylindrical fibrous microgeometry. Denote the matrices in equation (70), which describe this kind of anisotropy by:

$$\mathbf{C}_{11} = \begin{bmatrix} L_{55} & e_{15} \\ e_{15} & -\kappa_{11} \end{bmatrix}, \ \mathbf{C}_{12} = \begin{bmatrix} L_{54} & e_{25} \\ e_{14} & -\kappa_{12} \end{bmatrix}, \ \mathbf{C}_{21} = \begin{bmatrix} L_{54} & e_{14} \\ e_{25} & -\kappa_{12} \end{bmatrix}, \ \mathbf{C}_{22} = \begin{bmatrix} L_{44} & e_{24} \\ e_{24} & -\kappa_{22} \end{bmatrix} \quad (74)$$

Interchange now the phases and let the effective behaviour of the new medium be defined by matrices of the type in (74), denoted now by: $\mathbf{C}'_{11}, \mathbf{C}'_{12}, \mathbf{C}'_{21}, \mathbf{C}'_{22}$. The following phase interchange relations have be shown to exist between the original and new medium, Benveniste (1995b):

$$\mathbf{C}_1^{-1}\mathbf{C}'_{11}\mathbf{C}_2^{-1}\mathbf{C}_{22} - \mathbf{C}_1^{-1}\mathbf{C}'_{12}\mathbf{C}_2^{-1}\mathbf{C}_{12} = \mathbf{I} \ , \quad -\mathbf{C}_1^{-1}\mathbf{C}'_{11}\mathbf{C}_2^{-1}\mathbf{C}_{12} + \mathbf{C}_1^{-1}\mathbf{C}'_{12}\mathbf{C}_2^{-1}\mathbf{C}_{11} = \mathbf{0}$$

$$\mathbf{C}_1^{-1}\mathbf{C}'_{12}\mathbf{C}_2^{-1}\mathbf{C}_{22} - \mathbf{C}_1^{-1}\mathbf{C}'_{22}\mathbf{C}_2^{-1}\mathbf{C}_{12} = \mathbf{0} \ , \quad \mathbf{C}_1^{-1}\mathbf{C}'_{22}\mathbf{C}_2^{-1}\mathbf{C}_{11} - \mathbf{C}_1^{-1}\mathbf{C}'_{12}\mathbf{C}_2^{-1}\mathbf{C}_{12} = \mathbf{I} \quad (75)$$

If the medium is effectively transversely isotropic and also insensitive to phase interchange on the effective scale, then equations (75) determine uniquely its effective properties. Phase

236

interchange relationships in piezoelectric media have also been recently studied by Chen (1995).

Acknowledgment

The work of G. J. Dvorak was supported by the Office of Naval Research ; Dr. Yapa Rajapakse served as program monitor.

References

Avellaneda, M. and Harshe, G. 1994 *J. of Intell. Mater. Syst. and Struct.* **5**, 501-513.
Bahei-El-Din,Y.A. 1992 *Int. J. Plasticity* **8**,867-892
Benveniste, Y. 1987 *Mech of Mat.* **6** 147-157.
Benveniste,Y. 1993a *J. Appl. Mech.* **60**, 265-275.
Benveniste, Y. 1993b *Proc. R. Soc. Lond. A* **441**, 59-81.
Benveniste, Y. 1994a *Proc. R. Soc. Lond. A* **447,** 1-22.
Benveniste, Y. 1994b *J. Eng. Mat. and Tech.* **116**, 260-267.
Benveniste, Y. 1994c *Mech.of Mat* **18**, 183-193.
Benveniste, Y. 1995a *Phys. Rev. B* **51**, 16424-16427.
Benveniste, Y. 1995b *J. Mech. Phys. Solids* **43**, 553-571.
Benveniste, Y. 1996a *J. Mech. Phys. Solids* **44**, 137-153.
Benveniste, Y. 1996b *Int. J. Solids Struct.* to appear.
Benveniste, Y. and Dvorak, G.J. 1990a in *Micromechanics and Inhomogeneity ,* (T. Mura Anniversary Volume), Eds. G.J. Weng , M. Taya, and H. Abe, eds., Springer-Verlag, New-York, 65-81.
Benveniste, Y. and Dvorak, G.J. 1990b in *Inelastic Deformation of Composite Materials, IUTAM Symposium,* edited by G.J. Dvorak, 77-98.
Benveniste Y. and Dvorak, G.J. 1992a *J. Appl. Mech.* **59,** 1030-1032.
Benveniste,Y., and Dvorak., G.J. 1992b *J. Mech. Phys. Solids* **40**, 1295-1312.
Chen, T. 1993 *J. Mech. Phys. Solids* **41**, 1781-1794.
Chen, T. 1995 *J. Appl. Phys.* **78**, 7413-7415.
Cribb, J. L. 1968 *Nature* **220,** 576-577.
Dunn, M. L. 1993 *Proc. R. Soc. Lond. A* **441**, 549-557.
Dvorak,G.J. 1983 in *Mechanics of Composite Materials: Recent Advances*, Eds. Z. Hashin and C. Herakovich, Pergamon Press 73-91.
Dvorak, G.J. 1986 *J. Appl. Mech.* **53**, 737-743.
Dvorak, G.J. 1990 *Proc. R.. Soc. Lond. A* **431**, 89-110.
Dvorak, G.J. 1992 *Proc. R. Soc. Lond. A* **437,** 311-327.
Dvorak, G.J., Bahei-El-Din, Y.A. and Wafa, A.M. 1994 *Comp. Mech.* **14**, 201-228.
Dvorak, G.J. and Benveniste, Y. 1992 *Proc. R. Soc. Lond. A* **437**, 291-210.
Dvorak, G.J. and Chen.T. 1989 *J. Appl. Mech.* **56,** 418-422.
Dvorak, G.J., Senjoha, M and Srinivas, M., 1995 in *IUTAM Symposium on Micromechanics of Plasticity and Damage in Multiphase Materials,* Eds. A. Zaoui and M. Pineau, Kluver Academic Publishing
Dykhne, A. M. 1971 *Sov. Phys. JETP* **32**, 348-351.

Harshe, G., Dougherty, J.P., and Newnham, R.E. 1993 *Int. J. Appl. Electromagn. Mater.* **4**, 145-159.

Hill, R. 1963 *J. Mech. Phys. Solids* **11**, 357-372.

Hill, R. 1964 *J. Mech. Phys. Solids* **12**, 199-212.

Hill, R. 1965 *J. Mech. Phys. Solids* **13**, 213-222.

Keller, J. B. 1964 *J. Math. Phys.* **5**, 548-549.

Laws, N. 1973 *J. Mech. Phys. Solids* **21**, 9-17.

Levin, V. M. 1967 *Mekh. Tverd. Tela* **2** 88-94, English Translation: *Mech. of Solids* **11**, 58-61.

Milgrom, M. and Shtrikman, S. 1989 *Phys. Rev.* **40**, 5991-5994.

Milton, G.W. 1988 *Phys. Rev.B* **38**, 11296-11303.

Milton, G.W. 1996 in IUTAM Conference Proceedings, Kyoto.

Mori, T. and Tanaka, K. 1973 *Acta Metall.* **21**, 571-574.

Nan, C.W. 1994 *Phys.Rev. B* **50**, 6082-6088.

Newnham, R. E. Skinner, D.P. and Cross, L.E. 1978 *Mater. Res. Bull.* **13**, 525-536.

Nye, J.F. 1957 *Physical Properties of Crystals: Their Representation by Tensors and Matrices,* Clarendon Press, Oxford.

Rosen, B.W. and Hashin, Z. 1970 *Int. J. Engng.Sci.* **8**, 157-173.

Schapery, R. A. 1968 *J. Comp. Mat.* **2**, 380-404.

Schulgasser, K. 1992 *J. Mech. Phys. Solids* **40**, 473-479.

Smith, W. A., 1989, *1989 Ultrasonics Symposium IEEE,* 755-766.

Smith, W. A. 1993 *IEEE Trans. Ultrason. and Freq. Cont* **40**, 41-49.

Walpole, L. J. 1969 *J. Mech. Phys. Solids* **17**, 235-251.

Zhang, Q. M., Wang. H, Zhao, J., Fielding, J.T., Newnham, R.E., Cross, L.E. 1996 *IEEE Trans. Ultrason. and Freq. Cont.* **43**, 36-43.

Theoretical and Applied Mechanics 1996
T. Tatsumi, E. Watanabe and T. Kambe (Editors)
© 1997 Elsevier Science B.V. All rights reserved.

Fracture and damage of nonhomogeneous solids

V.Tamuzs

Institute of Polymer Mechanics LAS
Aizkraukles str. 23, Riga LV 1006 Latvia

1. INTRODUCTION

Nonhomogeneous solids with regard to fracture phenomenon roughly can be divided in three groups: 1) dissimilar bodies which contain small number of parts or inclusions with different properties but size of crack is comparable with the size of inhomogeneities; 2) solids with continuously changed properties as function of coordinates (functionaly graded materials); 3) solids containing large number of nonhomogeneities having the size significantly less than typical size of body and crack propagated through.

Large number of solutions has been obtained for different problems of dissimilar bodies and functionaly graded materials, but having no chance to review all directions of the named theme, we shall concentrate our attention on the fracture peculiarities of the third group heterogeneous materials only. Even having so restricted scope, it is impossible to cover all investigations of the problem and to make complete list of references. Therefore, mainly the main papers on themes will be mentioned, sending the reader for more complete references to specialized revues of different topics on the theme. Some emphasis will be put to Russian results which in some cases had priority, but unfortunately are not widely known. It should be noted that selection of some topics is influenced by scientific interests of the author.

The linear fracture mechanics presupposes the ideal scheme of fracture process: crack growth initiates and continues when energy release rate at virtual crack enlargement reaches the critical value typical for material and loading conditions, the crack trajectory under uniform stress state is straight, but separated parts after failure retain their virgin mechanical quality.

In reality, as a consequence of material inhomogeneities, the fracture process is different, and all features mentioned above are violated. The crack trajectory is not rectilinear, SIF necessary for crack steady state propagation significantly exceeds (according to R curves) the SIF value at which crack growth initiates, and (what is the most important feature) the structural changes in material are induced by stress before or simultaneously with the main crack propagation as this propagation is embarrassed by inhomogeneities. These structural changes, named as damage, are essential, and consequently it can be noted that the main difference between fracture of homogeneous and nonhomogeneous solids consists in damage phenomenon which is absent in the first case and distinctly revealed in the last case. Therefore two problems are significant for investigation of fracture process in heterogeneous solids: the damage accumulation regularities and the crack-damage interaction in the final stage of failure.

2. DAMAGE ACCUMULATION. HYSTORICAL REMARKS

The notion of material damage is of an early origin and is likely to be attributed to 1924 [1]. Linear summation of damage was given in [2]. An analysis of the volume fracture which initiates and gives rise to the actual failure process by scalar function of damage was elaborated by Kachanov [3] and Rabotnov [4]. Later the second order damage tensor was introduced as a measure of damage [5]. (More complate list of references see in [6]). The new characteristic of damage state by function on sphere, which takes into account the orientation of damage in space, was introduced in 1968 [7] (See also [8]). It was shown that such damage characteristic includes in particular cases the symmetrical tensors of different orders and can be applied for damage accumulation and durability analysis at complex stress state and complex loading.

It is remarkable that later the intensive development of damage mechanics shifted to western countries as Soviet scientists turned more their attention to the classical fracture mechanics. The corresponding references of latest progress in damage mechanics can be found in [9, 10].

The main disadvantage of continuum damage theory is that the introduced damage parameter $0 \leq D \leq 1$ has no clear physical sense. Therefore such theory is hardly applicable to fracture process of heterogeneous materials, where the accumulation of dispersed microcracks and microcavities can be observed directly by different physical means depending on the size of microcracks. In oriented polymers submicrocracks of size $10^{-6} \sim 10^{-5}$ can be detected by X-rays small angle scattering [8]. Microcracks in composites, ceramics and rocks can be observed by acoustic emission and visualization. In all cases the size of aroused defects correlates with typical size of structural elements - diameter of fibrils in oriented polymers [8], diameter of fibers in unidirectional composites, size of grains in metals and rocks.

Having the physical interpretation of damage as microcracks, it is possible to characterize the damage state quantitatively. Such quantitative assessment might be made through the parameter $\omega = Na^3$ which has been introduced by different authors for estimating the pennyshaped microcrack density in an isotropic material. Here, N is the number of cracks in a unit volume, and a is the crack radius. In the case of cracks with different radii, it is necessary to sum up their respective values; Kuksenko [8] has noted that the value $R = N^{-1/3}$ characterizes the mean distance between the microcrack centers and, if the typical linear size of a crack is $l \approx 2a$, then the coefficient $K = R/l = 1/2\omega^{-1/3}$ characterizes the ratio of the mean distance between the cracks to their size. In the theoretical calculations, it is convenient to introduce the assumption that a material consists of structural elements with linear size l, each of which can be broken, developing a crack of the same size. The damage is then characterized by a ratio of the broken elements n to their total number n_0.

The experimental data for the extreme level of damage of various materials are given in Tab.1 [11]. Nylon-6 and PMMA damage was determined by measurements based on small-angle X-ray diffraction [8]. The damage of rock-solid and polyester resin filled with PVC particles was estimated by direct crack calculation using a microscope [12, 13] after cutting the material. The damage of fibrous composites was determined by counting the fiber breaks which were revealed after burning out the matrix in the broken specimens. In all the investigated cases, the value of K changes from $5 \sim 2$ depending sensitively on the size of the specimen.

Table 1
Extreme level of damage [11]

Specimen Material	Type of Loading	a (μm)	$K=l^1N^{-1/3}$	$P=n/n_0$ 100% ($P \approx K^{-3}$ 100%)
Nylon-6	Tension	0.009	2.5	6
PMMA	Tension	0.17	3.5	2.3
Rock-Solid	Compression	200	5	0.8
PE+PVC	Low Cycle Fatigue	100-150	3.2	3
Epoxy Carbon $V_a=64\%$	Tension & Bending	10	3.5	2.3
Epoxy Boron $V_a=0.34$ Small Specimens	Tension		2.5	6
Epoxy Boron $V_a=0.34$ Small Specimens	Low Cycle Fatigue		2.1	10
Epoxy Boron $V_a=0.34$ Large Specimens	Tension		4.6	1

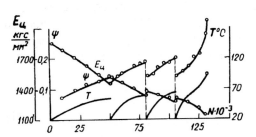

Fig. 1. Irreversible change of modulus E, energy dissipation $\Psi=\Delta W/W$ and selfheating temperature T during the cyclic loading of epoxy-glassfabric composite [16]. The effect of rest and cooling is shown.

Onset of microcracks causes a change of initial mechanical properties of material - reducing of modulus and increasing of damping properties. The first theoretical investigation has been done by Salganik 1973 [14] and next in [15]. One of the first experimental investigations is dated as far as in 1967 [16]. The irreversible change of modulus, damping and selfheating temperature of epoxy-glassfabric laminate composite during the cyclic loading was carefuly measured. (Fig. 1). Damage accumulation kinetics measured by temperature change was used to predict the individual cyclic durability of samples with good accuracy.

3. APPLICATION TO UNIDIRECTIONAL COMPOSITES

To substitute the real nonhomogeneous material by ideal model, which allows to carry out the calculations of statistical micro damage accumulation some hypothesis should be introduced. 1) Material is regarded as integral ensemble of structural elements. Each element can be in virgin or broken state due to onset of microcrack. The initial size of aroused microcracks is equal to size of structural element crossection. Process of crack enlargement is going by discrete steps. During each step the crack captivates new structural element and

242

increases in size by the size of element crossection (Fig. 2). 2) The strength of elements obeys the statistical strength distribution (usually the Weibull's distribution). Therefore for constructing of the model it is necessary to know:
1) typical size of elements
2) parameters of element strength distribution
3) stress concentration in elements surrounding the broken element or group of broken elements.

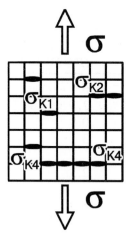

Fig. 2. Scheme of damage accumulation in heterogeneous material.

Unidirectional composite with brittle fibers and tough matrix can be mentioned as an example, closest to the ideal model. Investigation of fracture process of unidirectional composites includes many papers. The papers of Rosen [17] and Zweben [18] have to be refereed as the first. Then problem was specified by Harlow and Phoenix [19]. Batdorf [20] and Tamuzs [21] proposed the approximate procedures for describing the fracture process. Main idea in all papers is similar and consists in hypothesis that all composites is naturaly divided into structural elements containing single fiber of length 2δ, where δ is the zone around fiber break where stress in fiber is restored to nominal value. The δ is so called "ineffective length" of fiber in composite, depending upon fiber and matrix moduli, fiber radius, volume fraction, adhesion of matrix to fiber. Parameters of Weibull's strength distribution of element can be found by testing the reinforcement fibers, but stress concentration around the broken fiber or group of fibers has been investigated by many authors. Likely the first was the paper of Hedgepeth & Van Dyke [22]. The well elaborated procedure of fracture process calculation includes the finding of probability of appearance defects of different size. Usually the strength of fibers is determined by the Weibull's distribution

$$F(\sigma) = 1 - \exp\left[-\frac{l}{l_0}\left(\frac{\sigma}{\sigma_0}\right)^\beta\right]$$
(1)

and the probability of appearance of a single defect and groups of 2, 3 and more broken fibers is obtained. The final failure of specimen takes place when at least one defect of an arbitrarily large size appears. In order to simplify the over-stress coefficient calculation in the vicinity of the defect, it is assumed that the crack is penny-shaped, and only the fibers nearest to the defect are overstressed. The overstress along the fiber spreads on the ineffective length. The breakage calculation is reduced to the following approximate formulae:

$$P_1 = F(\sigma), \quad P_2 = P_1 \cdot P_{2,1}, \quad ..., \quad P_n = P_{n-1} \cdot P_{n,n-1}$$
(2)

where P_1 is the probability of a single break appearance. The function $F(\sigma)$ is determined by equation (1) for $l = \delta$. In formula (2), P_n $(n > 1)$ is the probability of appearance of a group consisting of n broken elements, $P_{2,1}$ is the probability of breaking of at least one element next to an already broken one, and $P_{n,n-1}$ is determined in a similar way. The probability $P_{2,1}$

is defined by the following formula:

$$P_{2,1} = 1 - \left[\frac{1 - F(k_1, \sigma)}{1 - F(\sigma)} \right]^{n_1} \qquad (3)$$

where k_1 is the overstress coefficient, $[1 - F(k_1, \sigma)]/[1 - F(\sigma)]$ is the conditional surviving probability of the overstressed element adjoining the single defect and n_1 is the number of elements subjected to the overstress k_1 and $P_{2,1}$ is the surviving probability of all the overstressed elements. The probability $P_{3,2}$ expressed by

$$P_{3,2} = 1 - \left[\frac{1 - F(k_2, \sigma)}{1 - F(k_1, \sigma)} \right]^{n_{21}} \left[\frac{1 - F(k_2, \sigma)}{1 - F(\sigma)} \right]^{n_2 - n_{21}} \qquad (4)$$

where n_2 is the number of elements surrounding the double defect and n_{21} is number of the adjoining elements which were subjected to overstress before the break of a second fiber.

The probability of developing at least one defect with size no smaller n in a specimen with N elements is determined by

$$W_n = 1 - \exp(-P_n N) . \qquad (5)$$

For sufficiently large n, all the curves for W_m $(m > n)$ merge. This limit curve determines the macrocrack appearance.

For epoxy carbon composite the probability curves Fig. 3a were calculated in [21]. The referenced approach reveals the size effect of strength and limit concentration of microcracks. Typical results are shown on Fig. 3b for carbon-epoxy unidirectional composite.

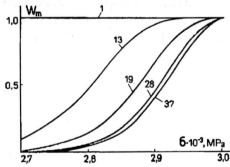

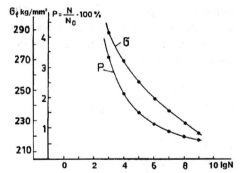

Fig. 3a. Probability curves of a different size defect appearance in unidirectional carbon-epoxy composite [21].

Fig. 3b. Ultimate fiber stress σ and limit damage level P of unidirectional epoxy carbon as function of sample size [21].

In reality, the fracture process of unidirectional composite on some stage differs from the ideal scheme, described above, because of low fracture toughness on shear mode along the fibers. For double branched crack at tensile loading (Fig. 4) it is clear that central part tends to be pulled out, and therefore the $K_{II} \neq 0$ occurs at the tips of branched crack. It is easy to see that for sufficiently long splitting size $L/a \gg 1$, the energy release rate hardly depends on

244

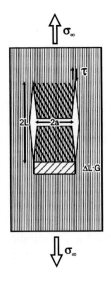

L (see paper of Dollar and Steif [23] for isotropic body) and

$$G_{II} = \frac{\sigma^2 a}{E_1} \; .$$

Estimating $\sigma = 1500$ MPa, $E = 1.5 \cdot 10^5$ MPa and $a = 0.1mm$, we get value of $G_{II} = 1.5$ kJ/m^2, which extends usual critical values for epoxy-carbon specimens ($G_{IIc} < 1000$ J/m^2) and is close to those for PEEK-carbon [24]. It means that fracture mode at defects of critical size as obtained above becomes unstable and can change from crack propagation by fiber breaking to sample splitting which cause brush-like fracture mode.

Fig. 4. Scheme of splitting of unidirectional composite at tensile loading. Unloaded region is marked shadowed. Energy release rate does not depend on the splitting length.

4. SOME EXAMPLES OF CONSTRAINED MULTI CRACKING OF HETEROGENEOUS SYSTEMS

4.1. Single fiber test

The limit density of microcracks will be higher in such heterogeneous systems where the aroused defects hardly influence a stress concentration in surrounding elements. The sample of single fiber embedded in tough matrix can be regarded as example of heterogeneous system with maximum allowable damageability, because each break of fiber is constrained by matrix and does not influence a stress state of fiber out of distance δ of ineffective length. Tensile tests of such samples were elaborated in early 90-ties to estimate the adhesion of fibers to matrix and the statistical strength distribution of fibers [25].

It follows from Weibull's distribution (2), that mean strength $<\sigma>$ of fiber is power function of size l

$$<\sigma> = \sigma_0 \left[\frac{l}{l_0} \right]^{-\frac{1}{\beta}} \Gamma \left(1 + \frac{1}{\beta} \right) \tag{6}$$

$$lg<\sigma> = -\frac{1}{\beta} \; lg \frac{l}{l_0} + B, \tag{7}$$

where

$$B = lg \left[\sigma_0 \Gamma \left(1 + \frac{1}{\beta} \right) \right].$$

So having enough fiber tests at different length, it is possible to find necessary Weibull's distribution parameters β and σ_0. It is easy to show that mean length of fiber fragmentation intervals $<l>$ versus applied fiber stress σ_f in single fiber test sample also follows analogical dependence [26]:

$$lg(<l>-\delta)=-\beta lg\left(\frac{\sigma_f}{\sigma_0}\right)+ lgl_0 \;; \tag{8}$$

where δ is ineffective length.

If $<l> \gg \delta$, the formula (8) simplifies

$$lg<l>=-\frac{1}{\beta}lg\sigma_f+C \;. \tag{9}$$

Therefore monitoring of fiber breaks and measurement of mean fragmentation length $<l>$ versus applied stress (strain) gives a valuable method to determine the fiber strength distribution parameters. This monitoring can be arranged by direct visualization [25] or by acoustic emission method [27]. The method possesses good accuracy and besides of statistical strength distribution offers information about fiber-matrix adhesion by measuring the mean fragmentation length in saturation state (Fig. 5). Analogical multicracking process is observed at tension of crossply composite [28], where the cracks in transverse ply accumulate till saturation (Fig. 6), and by monitoring this cracking process the strength distribution of ply in transverse direction to fibers can be obtained.

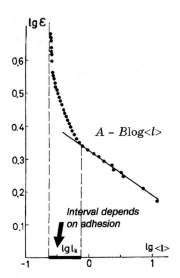

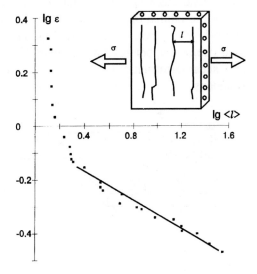

Fig. 5. Typical fiber fragmentation in single fiber test. Slope angle of straight line defines the Weibull's distribution shape parameter.

Fig. 6. Typical accumulation of transverse cracks in crossply composite.

Two possible generalizations of the method can be considered. Tensile test of unidirectional composite creates the multicracking of fibers, but the process stops by enlargement of multiple defects at relatively low stress value. It means that only the lower part of Weibull's distribution for fiber strength $\sigma/\sigma_0 \ll 1$ governs the fracture process. Therefore (1) changes to power form:

$$F(\sigma) \approx \frac{l}{l_0}\left(\frac{\sigma}{\sigma_0}\right)^\beta .$$
(10)

Monitoring of fiber breaks $n(\sigma)$ by acoustic emission, the Weibull's shape parameter β for fiber strength can be found by dependence

$$lgn = \beta\, lg\sigma + C .$$
(11)

The width of composite sample should be small enough to avoid the wrong acoustic signals, caused by splitting. It means that so called microplastic containing only the bundle of fibers should be used for the test.

4.2. Twodimensional multicracking.

The second exciting generalization of unidirectional cracking concerns a constrained twodimensional multicracking. The polygonal patterns of cracked brittle coatings can be observed on different scale levels-from microns in metallized thin films to meters in dried soil [29]. The attempt of describing the phenomenon, based on experience of unidirectional constrained cracking, was undertaken in [30].

A brittle lamina of thickness h_1 bonded to the substrate of thickness h_2 and influenced by thermal loading was considered (Fig. 7). For simplicity, the classical shear lag method was used (i.e. the upper ply has the modulus E, but the substrate with shear modulus G resists only shear deformation). Then equilibrium equations for coating ply turn to modified Lame equations

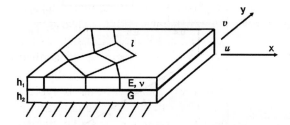

$$\frac{\partial}{\partial x}\left(\frac{\partial u}{\partial x} + \frac{\partial v}{\partial y}\right) + \frac{1-v}{1+v}\Delta u = \frac{u}{(1+v)\kappa^2}$$

$$\frac{\partial}{\partial y}\left(\frac{\partial v}{\partial y} + \frac{\partial u}{\partial x}\right) + \frac{1-v}{1+v}\Delta v = \frac{v}{(1+v)\kappa^2}$$
(12)

where

Fig. 7. Scheme of two dimensional constrained multi cracking.

$$\kappa = \left(\frac{E}{G}\frac{h_1 h_2}{1-v^2}\right)^{1/2}$$
(13)

is the only parameter determining the solution, and the boundary conditions on the crack lines l are

$$\sigma_{nn}\big|_l = -\sigma_0 = \frac{E\alpha\Delta T}{1-v} ; \qquad \sigma_{nt}\big|_l = 0 .$$

For long cracks ($l \gg \kappa$) asymptotic analysis reveals that SIF does not depend on crack size

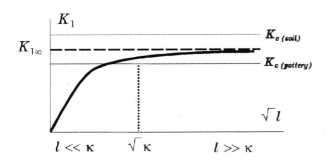

Fig. 8. Stress intensity coefficient versus crack size in brittle and tough coatings.

l and is determined at applied two dimensional stress σ (or ΔT) by

$$K_{1\infty} = \frac{E\alpha\Delta T}{\sqrt{2}} \sqrt{\frac{1+\nu}{1-\nu}} \sqrt{\kappa_1} . \qquad (14)$$

For very short cracks SIF K_I has the standard expression

$$K_1 = \sqrt{\pi}\,\sigma\sqrt{l} . \qquad (15)$$

Consequently, the functional dependence $K_I(\sqrt{l})$ can be illustrated by Fig. 8: a straight line at $l \ll \kappa$, $K_I = K_{I\infty}$ at $l \gg \kappa$, and some monotonously increasing graph in the transition zone.

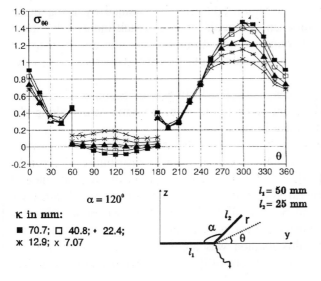

$\alpha = 120^0$

$l_1 = 50$ mm
$l_2 = 25$ mm

κ in mm:

■ 70.7; □ 40.8; ◆ 22.4;
✳ 12.9; ✕ 7.07

Fig. 9. Hoop stress distribution around a crack kink in two dimensional constrained cracking.

The crack behaviour depends essentially on the ratio $K_{I\infty}/K_c$ where K_c is the critical stress intensity coefficient for brittle lamina. If $K_{I\infty} > K_c$ the crack has some critical size l_c which becames unstable and propagates across all the material, or until an intersection with other preexisting crack occurs. Such type of cracking can be observed in very brittle coatings, e.g. in pottery coatings. If $K_{I\infty} < K_c$, the crack is always stable, and its enlargement can be caused only by change of the loading factor (increase of applied stress, decrease of temperature or moisture). Such cracking is typical for dried soil and, on the contrary to the "pottery mode," can be named the "desert mode."

For cracks of intermediate length and kinked or curvilined cracks, the numerical analysis is necessary. It revealed that around any crack kink point the hoop stress $\sigma_{\theta\theta}$ concentration in the outer region is observed while significant stress reduction in the inner region is revealed (Fig. 9). Such stress distribution leads to well known polygonal crack patterns [29].

5. INTERACTION OF MAIN CRACK WITH DAMAGE FIELD

Having the damage accumulation as essential stage of fracture process, the final failure of heterogeneous material nevertheless is caused mainly by macrocrack propagation through the body. Therefore the problem of crack interaction with damage field becomes fundamentally important.

It should be definitely noted that the problem consists of three subproblems: 1) creation of damage (or microcracks) in the vicinity of main crack; 2) interaction of crack with created or preexisted field of microcracks expressed in the change of SIF and 3) the propagation of macrocrack through the damaged region, including multiple kinking and blunting.

Large number of papers has been published on the topic during the last decade, but only the second part of the problem is elaborated sufficiently.

All solutions for the problem can be roughly divided in two groups: continuum approach and account of concrete microcracks' location. The idea of continuum approach started by papers [31, 32] consists in replacing of microcracked domain by continuous material with reduced stiffness. Certainly, if the crack tip is embedded into the inclusion with reduced modulus [33], the SIF will be reduced accordingly. But damage in the form of microvoids or microcracks influences the main crack not only by average softening of material but by local stress concentration around cavities too. This last influence is neglected in continuum approach, and as a result the shielding effect of damage located ahead of the crack tip is overestimated. Formaly, the solution of continuum approach is obtained by using the *J* integral through the damaged region, but each microvoid or microcrack induces the stress singularity or turns the domain in a multiconnected one. In both cases, the use of invariant *J* integral is incorrect and, its change because of local singularities at the microcrack tips can be calculated for each case.

Number of Kachanov's papers (references in [34]), dealing with account of concrete microcrack geometry are based on different assumptions (i.e. constant stress along the microcrack) and should be criticaly examined [35].

It should be mentioned however, that as early as in 1984 the assymptotic solution of two-dimensional problem for the system of finite main crack and arbitrary distributed microcracks under the tensile load was obtained [36]. The Muskhelishvili method of singular integral equations was used in the form of general system of N integral equations for N arbitrary located cracks, derived by Panasyuk and Savruk [37]. The solution was found using the expansion with respect to the small parameter $\lambda = a/a_0$ which equals to the ratio of the micro- to the macrocrack size. Subsequently, this method was applied to the crack-microvoids interaction [38], shear problem [39], thermoelastic one [40], and was extended by taking into account the crack closure [41].

In paper [36], the following expression was obtained for SIF at the tip of main crack of size a_0 in the presence of N arbitrary located microcracks, having equal size $a_k = a \big|_{k=1,...,N}$

$$K_I^\pm - K_{II}^\pm = K_0 \left[1 + \sum_{k=1}^{N} \lambda^2 F(u_k, \alpha_k) + \lambda^4 \Phi(u_k, \alpha_k) + O(\lambda^6) \right], \tag{16}$$

where $\lambda = a/a_0 \ll 1$, $K_0 = p\sqrt{a_0}$, is the SIF for a single crack, p - tension load applied at infinity perpendicular to the line of a main crack. The function $F(u_k, \alpha_k)$ is given by

$$F(u_k,\alpha_k)=\frac{1}{2}J_k\left(\frac{1}{(\bar{u}_k\mp1)\sqrt{u_k^2-1}}+\frac{e^{2i\alpha_k}}{(u_k\mp1)\sqrt{u_k^2-1}}\right)+$$

$$+\frac{1}{2}\bar{J}_k\left(1-\frac{e^{-2i\alpha_k}}{(\bar{u}_k\mp1)\sqrt{\bar{u}_k^2-1}}-(u_k-\bar{u}_k)\frac{2\bar{u}_k\pm1}{(\bar{u}_k\pm1)(\bar{u}_k^2-1)^{3/2}}e^{-2i\alpha_k}\right) \qquad (17)$$

where

$$J_k=e^{-2i\alpha_k}-1+e^{-2i\alpha_k}\frac{\bar{u}_k-u_k}{(\bar{u}_k^2-1)^{3/2}}+\frac{u_k}{\sqrt{u_k^2-1}}+\frac{\bar{u}_k}{\sqrt{\bar{u}_k^2-1}} \qquad (18)$$

$u_k = z_k/a_0$ - the nondimensional k microcrack midpoint coordinates in Cartesian coordinate system x, y (Fig. 10), $\bar{u}_k$ - conjugates coordinate, α_k - angle of the k-th microcrack to the x axis which is directed along the main crack. The upper sign refers to the right tip of the crack, the lower - to the left.

The principal result obtained in [36] consists in the proof that the term at λ^2 is composed by superposition of all microcrack influences. The mutual interaction of microcracks reveals only in terms at λ^4 and next. For collinear arrangement of microcracks the term Φ at λ^4 was derived in [42], and for rather close distances between microcracks it is not essential.

The general formula (2), (3) looks rather complicated, but it simplifies for concrete symmetrical arrangements of microcracks.

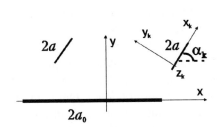

Fig. 10. Scheme of arbitrary located small cracks in the vicinity of the main crack.

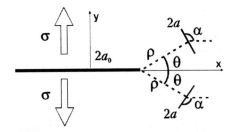

Fig. 11. Scheme of two symmetrical microcracks in the vicinity of crack tip.

For example, for two symmetrical microcracks located near the main crack tip, the following formula takes place:

$$K_I=K_0\left[1+\lambda^2\left(\frac{1}{\rho^2}F_1(\theta,\alpha)+\frac{1}{\rho^{3/2}}F_2(\theta,\alpha)+\frac{1}{\rho}F_3(\theta,\alpha)+\frac{1}{\rho^{1/2}}F_4(\theta,\alpha)\right)\right] \qquad (19)$$

where $\rho = r/a_0$ - nondimensional distance of microcrack midpoint from crack tip; θ and α

250

angles as noted in Fig. 11, but $F_i(\theta, \alpha)$ are defined by following formulae:

$$F_1 = \frac{1}{8}[8\cos(2\theta - 2\alpha) - 8\cos(4\theta - 2\alpha) + 8\cos2\theta + 11\cos\theta -$$

$$-6\cos(3\theta - 2\alpha) + 4\cos(\theta - 2\alpha) + 2\cos(\theta + 2\alpha) - 3\cos3\theta]$$

$$F_2 = \frac{\sqrt{2}}{4}\left[3\cos\left(\frac{7\theta}{2} - 2\alpha\right) - 3\cos\frac{7\theta}{2} - \cos\left(\frac{3\theta}{2} - 2\alpha\right) - \cos\frac{3\theta}{2} + 2\cos\left(\frac{3\theta}{2} + 2\alpha\right)\right]$$

$$F_3 = \frac{1}{32}\{-1 + 8\cos2\theta + 16\cos\theta + 9\cos4\theta + 8[\cos(\theta - 2\alpha) - \cos(3\theta - 2\alpha)] +$$

$$+2\cos2\alpha + 22\cos(2\theta - 2\alpha) - 6\cos(2\theta + 2\alpha) - 18\cos(4\theta - 2\alpha)\}$$

$$F_4 = -\frac{\sqrt{2}}{16}\left[\cos\left(\frac{\theta}{2} - 2\alpha\right) + 2\cos\left(\frac{\theta}{2} + 2\alpha\right) - \cos\frac{5\theta}{2} - 3\cos\frac{\theta}{2} + \cos\left(\frac{5\theta}{2} - 2\alpha\right)\right] \quad (20)$$

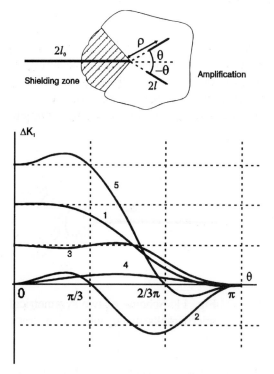

Fig. 12. Shielding and amplification zones for the case of two symmetrical microcracks at tension. Curve 1 ~ 4 correspond the terms at $1/\rho^2$, $1/\rho^{3/2}$, $1/\rho$, $1/\rho^{1/2}$ correspondingly; 5 - summary curve.

Calculations $\rho > \lambda$ reveal that for two radial symmetrical microcracks amplification to shielding changes at $\theta \approx 120°$ (Fig. 12), and the neutral angle hardly depends on the distance ρ. It is interesting to note that for seminfinite crack [43], the result, obtained by quite different method, fully coincides with $F_1(\theta, \alpha)$, but F_2, F_3, F_4 reveal the difference between solutions for finite and semi-finite crack.

In conclusion it can be noted that at present the problem of macrocrack interaction with preexisted field of microcracks for elastic material is rather clear. Many of published results overestimate the shielding effect of microcracks ahead of the main crack. Actualy it would be really strange if microcracks appearing ahead of the crack would resist the initiation of main crack growth. At the same time, the process of emerging microcracks consumes additional energy and therefore increases the fracture toughness. Besides there is a difference between main crack growth initiation and steady state propagation through the damaged region, which distinctly is revealed by experimentaly obtained R curves. These two problems are subject of future investigations.

REFERENCES

1. A.Palmgren, VDI-Z, Vol. 68, No. 14 (1924) pp. 339-341.

2. M.A.Miner, J. Appl. Mech., Vol. 12, No. 3 (1945) pp. 159-164.

3. L.M.Kachanov, Izv. akad. nauk SSSR, Otd. tekhn. nauk, No. 8 (1958) pp. 26-31.

4. Ju.N.Rabotnov, Voprosy prochnosti materialov i konstrukziy, Moscow (1959) pp. 5-7.

5. A.A.Il'ushin, Mekhanika tverdogo tela, No. 3 (1967) pp. 21-35.

6. H.Altenbach and H.Blumenauer, Neue Hütte, 34 Jahrgang, Heft 6 (1989).

7. V.Tamuzh and A.Lagzdinsh, Polymer Mechanics, No. 4 (1968) pp. 493-500.

8. V.S.Kuksenko and V.P.Tamuzs, Fracture Micromechanics of Polymer Materials, Martinus Nijhoff Publ. (1981) 311 p.

9. J.L.Chaboche, Continuum Damage Mechanics. J. Appl. Mech., Vol. 55 (1988) pp. 59-72.

10. D.Krajcinovic, J. Appl. Mech., Vol. 52 (1985) pp. 829-834.

11. V.Tamuzs, Some Peculiarities of Fracture in Heterogeneous Materials in "Fracture of Composite Materials", Ed. G.Sih and V.Tamuzs, Martinus Nijhoff Publ. (1982) pp. 131-138.

12. M.Ya.Mikel'son and L.Ya.Khokhbergs, Mechanics of Composite Materials, No. 1 (1980) pp. 26-33.

13. V.S.Kuksenko, D.I.Frolov and L.G.Orlov, Fracture of Composite Materials, Ed. G.C.Sih and V.P.Tamuzs, Sijthoff and Noordhoff (1979) pp. 25-33.

14. R.Salganik, Mekhanika tverdogo tela No. 4 (1973) pp. 149-158.

15. R.J.O'Connel and B.Budiansky, J. Geophys. Research, Vol. 79, No. 35 (1974) pp. 5412-5426.

16. P.P.Oldyrev and V.P.Tamuzh, Polymer Mechanics, No. 5 (1967) pp. 571-576.

17. B.W.Rosen, AIAA Journal, No. 2 (1967) pp. 1985-1994.

18. C.Zweben, AIAA J., No. 12 (1968) pp. 2325-2331.

19. P.Harlow and S.Phoenix, J. Composite Materials, Vol. 12 (1978) pp. 195-213.

20. S.Batdorf, Preprint of Weibull's Memorial Symp., Stockholm (1984).

21. V.P.Tamuzh, M.T.Azarova, V.M.Bondarenko, Yu.A.Gutans, Yu.G.Korabelnikov, P.E.Pikshe and O.F.Silujanov, Mechanics of composite materials, No. 1, (1982) pp. 27-33.

22. J.Hedgepeth and P.Van Dyke, J. Composite Materials, Vol. 1 (1967) pp. 294-309.

23. A.Dollar and P.S.Steif, ASME J. of Appl. Mech., Vol. 58 (1991) pp. 584-586.

24. P.Davies, W.Cantwell, H.Richard, C.Moulin and H.H.Kausch, Proc. ECC M-3 (1989) pp. 747-755.

25. H.D.Wagner and A.Eitan, Appl. Phys. Lett. 56 (20) (1990) pp. 1965-7.

26. V.V.Bolotin and V.Tamuzs, Mekhanika Kompozitnykh Materialov, No. 6 (1982) pp. 1107-10.

27. V.P.Tamuzhs, Yu.G.Korabelnikov, I.A.Rashkovan, A.A.Kārklin'sh, Y.A.Gorbatkina and T.Yu.Zakharova, Mech. Comp. Materials, 27 (4) (1991) pp. 413-18.

28. P.W.Manders, T.W.Chou, F.R.Jones and I.W.Roock, J. of Material Science, Vol. 18 (1983) pp. 2876-2889.

29. J.Walker, Scientific American, Vol. 255, No. 4 (1986) pp. 178-183.

252

30. V.Tamuzs, V.Beilin, R.Joffe and V.Valdmanis, Mech. Comp. Materials, Vol. 30, No. 6 (1994) pp. 529-539.

31. M.Ortiz, J. Appl. Mech., 54 (1987) pp.54-58.

32. J.W.Hutchinson, Acta Metall, 35 (1987) pp.1605-1619.

33. D.S.Steif, J. Appl. Mech., 54, (1987) pp.87-92.

34. M.Kachanov, Advances in Applied Mechanics, Vol. 30, (1994) pp. 259-445.

35. S.A.Meguid, P.E.Gaultier and S.X.Gong, Engineering Fracture Mechanics, 38, (1991) pp.451-465.

36. N.Romalis and V.Tamuzh, Mechanics of Composite Materials, 20 (1) (1984) pp. 35-43.

37. V.Panasyuk, M.Savruk and A.Datsyshin, Stress Distribution around Cracks in Plates and Shells, Naukova Dumka, Kiev (1976) (in Russian).

38. V.P.Tamuzh, N.B.Romalis, N.A.Dolotova and A.L.Polyakov, Mechanics of Composite Materials, 25, (1), (1989) pp. 88-95.

39. V.Tamuzs and V.Petrova, Physiko-Himicheskaya Mekhanika Materialov, No. 3 (1993) pp. 147-157 (in Russian).

40. V.Tamuzs, N.Romalis and V.Petrova, Theoretical and Applied Fracture Mechanics, 19 (1993) pp. 207-225.

41. V.Tamužs, V.Petrova and N.Romalis, Theoretical and Applied Fracture Mechanics, 21 (1994) pp. 207-218.

42. V.Petrova and V.Tamuzs, Prikladniye Zadachi Mehaniki Sploshnih Sred, Voronezhskiy Universitet, Voronezh, (1988) pp. 112-116 (in Russian).

43. S.X.Gong and H.Horii, J. Mech. Phys. Solids, Vol. 37 (1989) pp. 27-46.

Theoretical and Applied Mechanics 1996
T. Tatsumi, E. Watanabe and T. Kambe (Editors)

Mechanics of functionally gradient materials: Material tailoring on the micro- and macro-levels

Kikuaki TANAKA

Department of Aerospace Engineering, Tokyo Metropolitan Institute of Technology, Asahigaoka 6-6, J-191 Hino/Tokyo, Japan

A scheme of material tailoring in the functionally gradient materials (FGMs) is formulated under the mechanical and thermal multiobjectives with the help of the direct sensitivity analysis and the optimization technique associated with the heat conduction/thermal stress analysis by means of incremental FEM. The spatial distribution of the volume fraction of phases is optimally determined. A hollow cylinder in a ceramics-metal FGM is tailored successfully as a material with reduced thermal stresses conducting the least amount of heat.

1. INTRODUCTION

Composite materials have, when compared to the metallic materials, a potent characteristic to be easily designed so that they can exhibit a certain strength along a specified direction in a structure [e.g., 1,2]. This is a major reason why the composite materials are preferably used in the field of aerospace engineering, where the specific strength and the specific rigidity are the essential factors to be considered when selecting the materials. Aeroelastic tailoring is one of the most successful technologies based on and positively utilizing this advantage of the composite materials [e.g., 3,4]. A problem to be very carefully noted in the composite materials is the inevitable existence of the interfaces, where the different components contact mechanically or metallurgically/chemically. Since the material interface is a spot of the sharp change of the stress distribution and the stress concentration, it might very likely trigger the local plastification and subsequent initiation of the microcracking, which could, in the worst case, coalesce to be a fatal main crack propagating in the material [e.g., 5].

Since the material interface could be understood as a strong discontinuous change of the value of material parameters, the above-mentioned disadvantage would, if not disappear, decrease if the material is designed to exhibit a continuous, at least a step-wise, distribution of the value of material parameters. This material construction besides produce some additional functions which would be impossible to observe in the conventional composite materials or metallic materials. This is the idea of the functionally gradient materials (FGMs) [6-9]. Careful choice of the combination of components, together with an appropriate distribution of the fraction inside the material, realize, for example, such essentially contradicting functions as the thermal shielding and the reduction of thermal stresses. A gradient material composed of the ceramics (Al_2O_3, ZrO_2 or SiC, for example) on the high temperature side and the metallic materials such as Ti alloy on the low temperature side is one of the target FGMs. The R&D of FGMs are not limited in the mechanical field. The effective thermoelectric FGMs are recently under the development, in which the effective energy conversion materials are designed while still taking into account the reduction of the stresses [e.g., 10]

Tailoring in FGMs is usually carried out under the multiple objectives such as the reduction of the thermal stresses and the shielding of the heat flow. In many cases the objectives which are required to be fulfilled oppose each other from the physical point of view as in the above case. The material tailoring in the FGMs is, therefore, a methodology to find out a compromising solution which optimally satisfies all the objectives. It is performed by determining an optimal distribution of fraction of the phases composing the FGM. For this purpose the material parameters have firstly to be determined as the functions of the volume fraction of the phases. This is a topic in micromechanics [11,12] and one needs the data characterizing the microstructures in the FGM; the size and shape of the inclusion phase in a matrix, the orientation and spatial distribution of the inclusions in the matrix, and so on, meaning that the material tailoring is impossible to be carried out without enough knowledge of the microscopic status of the FGM.

In the present study the scheme of the material tailoring in FGMs is discussed within the mechanical and thermal fields [13,14]. The FGMs are modeled as a mixture composed of the element phases, the volume fraction of which characterizes an aspect of the microstructures in the FGMs. The fraction-dependence of the material parameters also represents the microscopic characteristics of the materials. The process of the thermoelastic material tailoring to reach an optimal solution is formulated by means of the incremental FEM heat conduction/thermoelastic stress analysis with the help of the direct sensitivity and optimization methods. The theory developed is applied to the mechanical and thermal material tailoring of a hollow cylinder subjected to an asymmetric thermal boundary condition.

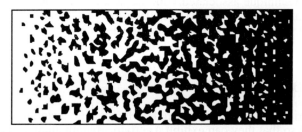

Figure 1. Micrograph in a FGM (Schematic) [15].

2. MATERIAL TAILORING OF FGMs

2.1. Mechanical modeling of FGMs

Figure 1 illustrates a schematic micrograph of cross-section in the FGM, which is composed of the metal phase (the white part in the figure) and the ceramics phase (the black part) [15]. The ceramics phase is on the higher temperature side to shield effectively the heat flow into the material system and to realize the high strength at the high temperature, whereas the metal phase on the lower temperature side is expected to increase the strength and toughness of the whole material system. The following points should be specially noted: The material element changes continuously from the metal phase to the ceramics phase without any macroscopic interface which might produce a critical stress jump. The microstructure of the material can be well modeled by the matrix-inclusions material system in both the metal-rich and the ceramics-rich regions. The size and shape of the inclusions are not uniform throughout the material. In the middle part the morphology is not so simple that a special consideration is required. The FGMs are also fabricated as a laminated composite; each layer, often the order of $10\,\mu$m or less in thickness, has a uniform composition of the component phases [16].

Some more detailed comments for the subsequent mechanical discussions: Suppose in the FGM a spherical control volume with a finite radius, and measure the volume fraction of the metal phase ξ; $(0 \leq \xi \leq 1)$ inside the control volume. If the value is attached to the center $\mathbf{x}$ of the control volume, a field of the volume fraction $\xi(\mathbf{x})$ is constructed. The field $\xi(\mathbf{x})$, which strongly depends on the size of the control volume, represents a macroscopic measure of the microstructure in the FGM. The fraction changes spatially from 0 to 1, which is different from the case of the composite materials with one value of ξ over the whole body. The situation is also different from the case of the materials in the process of transformation, in which the fraction changes not only spatially but also temporally [17]. When the FGM is composed of the

N phases, the N-1 volume fractions are necessary to specify a full status of the mixture of the phases [18]. In this paper, for simplicity, the FGMs composed of the two phases are discussed.

2.2. Material parameters

Mechanics of material tailoring in FGMs may be classified into the two phases; the macroscopic material tailoring and the microscopic material tailoring. In the macroscopic material tailoring the volume fraction of the phases should optimally be determined, under the multiobjective functions, by selecting a set of the design parameters $\mathbf{p}$ in

$$\xi = \xi(\mathbf{p}; \mathbf{x}) . \tag{1}$$

When evaluating the material parameters, the microstructures of the material are assumed *a priori*. If the material parameters are represented now by the elastic moduli tensor $\mathbf{E}$, it can be given at a material point $\mathbf{x}$ by

$$\mathbf{E} = \mathbf{E}(\xi; \mathbf{E}_m, \mathbf{E}_c; \dots) , \tag{2}$$

where $\mathbf{E}_m$ and $\mathbf{E}_c$ denote the elastic moduli tensor of the metal and ceramics phases, respectively. The spatial dependence of $\mathbf{E}$ inside the material comes through Eq.(1). Equation(2) states that a mixture of the metal and ceramics phases has the overall elastic moduli tensor $\mathbf{E}$. Determination of the overall or effective material parameters in the complex microscopic configurations is one of the main topics in micromechanics [11,12].

Considering the FGM as a matrix-inclusions material system, some aspect of the microstructures can be expressed by means of the size and shape of the inclusions. This is one of the ways of the microscopic material tailoring, in which the aspect ratio and the orientation of the inclusion, for example, can play the same role of the design parameters as the volume fraction of the inclusions ξ.

A direct description of the complicated microstructure, like the case in the middle part in Figure 1, needs the concept of topology [e.g., 15]. Statistical continuum mechanics, which studies the macroscopic and microscopic behavior of the materials characterized only statistically, can contribute to the microscopic material tailoring if it is coupled with the stochastic finite element method [19]. Percolation theory is also applied to evaluate the overall material parameters in FGMs [e.g., 20].

An effective estimation of the microstructures seems to be carried out by means of the fuzzy inference [21-22]. Suppose the FGM is composed of some types of the microstructures (such as the spherical particulate, the skeletal, the flake like inclusions, ...), which are assumed to be characterized by the membership function in fuzzy theory. The overall material property can be derived from the material properties of each microstructure and these membership functions. The method has successfully been applied to evaluate the material properties in the actual PSZ/SUS-FGM [23].

2.3. Objective functions

The material tailoring is carried out under the multiobjective functions. In the fabrication processes the mixture of the component phases are sprayed by means of the plasma torches onto the material surface to pile up a graded material. The FGM is heat-treated afterward. During these processes the FGM should not break, and just prior to the practical applications, a null or a specified residual stress distribution is expected inside the material. These are some examples of the objectives to be fulfilled in the *thermoplastic* material tailoring during these processes [7,23-25]. On the other hands, when the FGM is used in the practical situations, the mechanical and thermal criteria are imposed to the *thermoelastic* analyses from the view point of strength and toughness of materials.

The fracture distribution, and the microstructures when carrying out the microscopic material tailoring, should be determined to fulfill all these criteria *optimally*; meaning some conditions must be satisfied absolutely whereas others are expected to be fulfilled as much as possible. Posing and classifying the necessary and sufficient objective functions are a key to carry out an effective material tailoring. Some more details are explained in Sec.3.3, and the examples are shown in Chap.4.

3. THERMOELASTIC MATERIAL TAILORING

The procedures of macroscopic tailoring in FGMs are explained by means of FEM. For simplicity the FGM is assumed to be composed of the two heat conducting thermoelastic component materials. Both components are assumed to be mixed uniformly without any interface. The aim is to determine an optimal fraction distribution under which the thermomechanical state induced in the FGM never violate the specified criteria. The functional form of Eq.(2) is assumed to be given from the start since the microstructures are given *a priori*. The further specification of the problem will be discussed later in Chap.4.

3.1. Governing equation by means of FEM

The equilibrium equation and the equation of heat balance for the 3-D quasi-static thermoelastic problems are written as

$$\text{div } \boldsymbol{\sigma} + \rho \mathbf{f} = \mathbf{0}, \quad -\text{div } \mathbf{q} + Q = \rho c \dot{T} . \tag{3}$$

The symbols in this study have the following physical significance:

$\boldsymbol{\sigma}$:	stress tensor,	$\boldsymbol{\varepsilon}$:	strain tensor,
$\mathbf{u}$:	displacement vector,	T :	temperature,
ρ :	density,	$\mathbf{f}$:	body force vector,
$\mathbf{q}$:	heat flux vector,	Q :	heat production,
c :	specific heat,	$\mathbf{E}$:	elastic moduli tensor,
$\boldsymbol{\Theta}$:	thermoelastic tensor,	$\boldsymbol{\lambda}$:	thermal conductivity tensor.

The operators grad and div mean the gradient and the divergence with respect to the spatial coordinate $\mathbf{x}$, respectively. The superimposed dot means the time rate, whereas the quantity with Δ stands for the increment during the time interval Δt. The infinitesimal deformation is assumed.

The incremental form of the constitutive equations in thermomechanics and heat conduction are assumed;

$$\Delta \boldsymbol{\sigma} = \mathbf{E} : \Delta \boldsymbol{\varepsilon} + \boldsymbol{\Theta} \Delta T , \quad \Delta \mathbf{q} = -\boldsymbol{\lambda} \cdot \text{grad } \Delta T . \tag{4}$$

Equation(4)$_1$ should of course be replaced by the thermoplastic constitutive equation when the thermoplastic tailoring is carried out in the fabrication processes or the heat treatment procedures prior to the practical operation [7,23-25]. The strain increment $\Delta \boldsymbol{\varepsilon}$ is determined from the displacement increment $\Delta \mathbf{u}$ by

$$\Delta \boldsymbol{\varepsilon} = [\text{grad } \Delta \mathbf{u} + (\text{grad } \Delta \mathbf{u})^T]/2 . \tag{5}$$

The incremental form of the governing equation(3) can now be rewritten by

$$\text{div } (\mathbf{E} : \text{grad } \Delta \mathbf{u} + \boldsymbol{\Theta} \Delta T) + \rho \Delta \mathbf{f} = \mathbf{0} ,$$

$$-\text{div } (\boldsymbol{\lambda} \cdot \text{grad } \Delta T) + \Delta Q = \rho c \Delta \dot{T} . \tag{6}$$

Following the incremental FEM formulation, the time-dependent solution for the arbitrary thermomechanical boundary and initial conditions is given by solving the final global equation

$$\mathbf{K} \cdot \Delta \mathbf{X} = \Delta \mathbf{F} \tag{7}$$

successively for each time interval Δt. The solution ΔX is composed of the displacement increment Δu and the temperature increment ΔT of the nodes.

Since all material parameters in the governing equations (6) depend on the fraction ξ, the governing equation (7) can be reread as

$$K(\xi) \cdot \Delta X(\xi) = \Delta F(\xi) . \tag{8}$$

The fraction ξ can be regarded as a sensitivity variable from the material design point of view. Differentiation of Eq.(8) with respect to ξ leads the governing equation in the direct sensitivity analysis;

$$K \cdot \Delta X_\xi = \Delta F_\xi - K_\xi \cdot \Delta X , \tag{9}$$

where the suffix ξ means the differentiation with respect to ξ. The unknowns ΔX_ξ in this equation, the sensitivity increment, are composed of the displacement sensitivity increment Δu_ξ and the temperature sensitivity increment ΔT_ξ of the nodes. The stress sensitivity increment $\Delta \sigma_\xi$ can be determined by

$$\Delta \sigma_\xi = E : \text{grad } \Delta u_\xi + \Theta \Delta T_\xi + E_\xi : \text{grad } \Delta u + \Theta_\xi \Delta T , \tag{10}$$

which will be used in the optimization processes.

3.2. Material parameters

The fraction-dependence of the material parameters, Eq.(2), can be estimated by means of the Mori-Tanaka theory for the spherical inclusions [26-29]. In order to estimate a complicated topological structure around $\xi \approx 0.5$, like in Figure 1, the following three curves may be employed for the whole range of ξ:

$0 \leq \xi \leq 0.3$ and $0.7 \leq \xi \leq 1$: Mori-Tanaka theory
$0.3 \leq \xi \leq 0.7$: Interpolation between the above two curves by means of fuzzy inference [21-22]

The fuzzy inference procedures, used in the simulations in Chap.4, are explained below [14]: Let the value of a material parameter be $\Psi_1(\xi)$ for the composite material in which the SUS304 spherical inclusions are randomly dispersed in the PSZ matrix and the volume fraction of the inclusions is ξ. On the contrary, $\Psi_2(\xi)$ stands for the value of the material parameter in the composite material in which the PSZ spherical inclusions are randomly dispersed in the SUS304 matrix and the volume fraction of the inclusions is $1 - \xi$. Both $\Psi_1(\xi)$ and $\Psi_2(\xi)$ can be calculated according to Mori-Tanaka theory, which corresponds to the curves for the ranges $0 \leq \xi \leq 0.3$ and $0.7 \leq \xi \leq 1$, respectively. The value of the material parameter $\Psi(\xi)$ is now given in the side ranges of ξ ($0 \leq \xi \leq 1$) by

$$\Psi(\xi) = \begin{cases} \Psi_1(\xi) ; & 0 \leq \xi \leq 0.3 \\ \Psi_2(\xi) ; & 0.7 \leq \xi \leq 1 . \end{cases} \tag{11}$$

In the middle range $0.3 \leq \xi \leq 0.7$, $\Psi(\xi)$ is interpolated from $\Psi_1(\xi)$ and $\Psi_2(\xi)$ by means of the fuzzy inference. The membership functions

$$\mu_1(\xi) = \frac{2(\xi - \frac{1}{2})^3}{\delta^3} - \frac{3(\xi - \frac{1}{2})}{2\delta} + \frac{1}{2} , \quad \mu_2(\xi) = 1 - \mu_1(\xi) , \tag{12}$$

are first introduced for $\Psi_1(\xi)$ and $\Psi_2(\xi)$, respectively, where δ means the range of the fuzzy inference; $\delta = 0.4$ in this case. The value of the material parameter is now given by

$$\Psi(\xi) = \text{Gravity} (\Psi_1(\xi), \Psi_2(\xi), \mu_1, \mu_2 ; \xi) ; \quad 0.3 \leq \xi \leq 0.7 , \tag{13}$$

where the operator Gravity gives the coordinate of the center of gravity of the shaded figure formed by the two overlapped triangles with the coordinate of the center of gravity $\Psi_1(\xi)$ and $\Psi_2(\xi)$ and the grades $\mu_1(\xi)$ and $\mu_2(\xi)$ (cf. Figure 2).

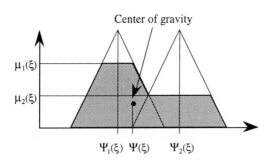

Figure 2. Fuzzy inference.

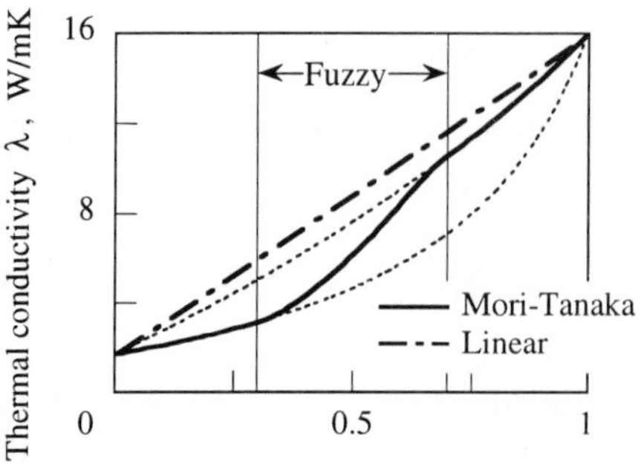

Figure 3. Fraction-dependence of thermal conductivity.

An example of the fraction-dependence of the material parameters is shown for the thermal conductivity in Figure 3 together with the linear approximation. The solid curve is used in the following simulations.

3.3. Design purpose

From the strength point of view the reduction of the stresses should be one of the main purposes of the material tailoring, but it does not always mean that a zero-stress distribution is the best solution. In some cases, when thinking of the strength and life time of the material itself, a certain non-zero stress distribution may be recommended. The reduction of the stresses should, therefore, be understood in the material tailoring to be a process to reach as close as possible a given reference stress distribution. From the practical point of view the reference stress distribution may be a broader allowable band which is settled from the strength and life time point of view. The stress is only required to be inside this allowable band. This process may be called the global reduction of stresses if it is compared with the conventional reduction of stresses, which pursues the reduction of the maximum value of the stress in the whole time-space domain.

In the ceramics-metal FGMs the full ceramic configuration often reduces mostly the thermoelastic stresses induced. The situation is, however, not acceptable from the toughness point of view. Some criterion to the toughness has, therefore, to be taken into account independently to the strength condition. The criterion can be expressed, like the case of the strength of materials, as an allowable band condition. The temperature field should also be constrained in some ways: Firstly the temperature-dependence of the strength in the component materials restricts the height of the temperature. Secondly some criterion to the heat flow should be taken into account when the FGM is expected to be used as a thermal barrier.

It should be noted that some criteria mentioned above are physically contradictory each other to be fulfilled. For example, the stress reduction is realized well by a flatter temperature field while from the better heat shielding point of view the temperature has to distribute in an appropriate way in the material from the higher temperature side to the lower temperature side. The material tailoring should be carried out so that the multi-purposes are fulfilled optimally. In some cases each criterion should be labeled by the priority ranking.

Summarizing, the material tailoring can be generally formulated as follows:

Minimize

$$F_s^J(\mathbf{p}) = \text{Max}\,[\,F_s^J(\boldsymbol{\sigma}(\mathbf{x},t),\, T(\mathbf{x},t),\, \xi(\mathbf{p};\mathbf{x}),\, \dots)\,]\,;\quad J = 1, 2, \dots, N \quad (14)$$

subject to constraints

$$F_h{}^I(\mathbf{p}) = \text{Max}\,[\,F_h{}^I(\boldsymbol{\sigma}\,(\mathbf{x},t),\,T(\mathbf{x},t),\,\xi(\mathbf{p};\mathbf{x}),\,...\,)\,] \leq 0\,;$$

$$I = 1,\,2,\,...\,,\,M\,.\qquad(15)$$

The constraints (15) are understood to be the hard objective which has to be absolutely fulfilled. The conditions (14) are, on the other hand, the soft objective which is required to be satisfied as much as possible. The actual cases will be explained later.

3.4. Optimization procedures

The multicriterial material tailoring can be performed by means of the direct sensitivity analysis and the multiobjective optimization technique [30-32], taking the following calculation procedures:

1. Choose an initial value $\mathbf{p}_0$ of the design parameter $\mathbf{p}$.
2. Carry out the heat conduction/thermal stress analyses with the fraction $\xi(\mathbf{p};\mathbf{x})$ for the whole time range.
3. When Eq.(15) is satisfied, seek an increment $\Delta\mathbf{p}$ of $\mathbf{p}$ to lower $F_s{}^J$ in Eq.(14). When Eq.(15) is not satisfied, seek an increment $\Delta\mathbf{p}$ of $\mathbf{p}$ to have a better estimation of $F_h{}^I$.
4. Find a new $\mathbf{p}$ by $\mathbf{p} = \mathbf{p}_0 + \Delta\mathbf{p}$ and go back to Step 2 with a new fraction distribution.

The optimization procedures give an ordered arrangement $(\xi_0,\,\xi_1,\,...\,)$ to reach finally an optimal solution. If the stress and the temperature fields at the J-th iteration process be $\boldsymbol{\sigma}_{J-1}(\mathbf{x},t;\,\xi_{J-1})$ and $T_{J-1}(\mathbf{x},t;\,\xi_{J-1})$, the change in volume fraction

$$\Delta\xi_J(\mathbf{x}) = \xi_J(\mathbf{x}) - \xi_{J-1}(\mathbf{x})\,,\qquad(16)$$

induces the new stress and temperature fields

$$\boldsymbol{\sigma}_J(\mathbf{x},t;\,\xi_J) = \boldsymbol{\sigma}_{J-1}(\mathbf{x},t;\,\xi_{J-1}) + \boldsymbol{\sigma}_{J-1\,\xi}\,\Delta\xi_J\,,$$

$$T_J(\mathbf{x},t;\,\xi_J) = T_{J-1}(\mathbf{x},t;\,\xi_{J-1}) + T_{J-1\,\xi}\,\Delta\xi_J\,,\qquad(17)$$

where

$$\boldsymbol{\sigma}_{J-1\,\xi} \equiv \left.\frac{\partial\boldsymbol{\sigma}}{\partial\xi}\right|_{\xi\,=\,\xi_{J-1}}\,,\qquad T_{J-1\,\xi} \equiv \left.\frac{\partial T}{\partial\xi}\right|_{\xi\,=\,\xi_{J-1}}\,,\qquad(18)$$

characterize the sensitivities with respect to the change in fraction $\Delta\xi_J$, which can be evaluated by means of Eqs.(9) and (10). The new values $\boldsymbol{\sigma}_J$ and T_J are used to check whether the choice $\Delta\xi_J$ gives a better solution with respect to the design criteria (14) and (15). The increments of fraction $\Delta\xi_J$ in Eq.(17) can be calculated, by means of $\Delta\mathbf{p}_J$ from

$$\Delta\xi_J = \frac{\partial\xi_J}{\partial\mathbf{p}_J}\cdot\Delta\mathbf{p}_J \equiv \Delta\xi_J(\Delta\mathbf{p}_J)\,,\qquad(19)$$

$$\Delta\mathbf{p}_J = A\,\varphi\,\Delta\mathbf{p}_{J-1}\,,$$

$$A \equiv \sum_K \{\,F^K(\boldsymbol{\sigma}\,(\mathbf{x},t),\,T(\mathbf{x},t),\,\xi(\mathbf{p};\mathbf{x}),\,...\,)\,/\,(F_0{}^K)\,\}^2\,,\qquad(20)$$

where $\Delta\mathbf{p}_J$ means the increment in design parameter estimated in the J-th iteration process. The function F^K represents $F_s{}^J$ or $F_h{}^I$ and $F_0{}^K$ stands for the prescribed reference value for $F_s{}^J$ or $F_h{}^I$ to work with the non-dimensionalized the terms. The acceleration factor φ is taken to be 1.25 in the later calculations.

4. ILLUSTRATIVE EXAMPLES

4.1. Problem to be solved

A hollow cylinder to be designed here, $2 r_0 = 100$ mm in inner diameter and $2 r_1 = 120$ mm in outer diameter, is made from a functionally gradient material composed of two phases; the ceramics phase (partially stabilized zirconia, PSZ) and the metal phase (stainless steel, SUS304) (cf. Figure 4). The volume fraction $\xi(r)$ of the SUS304 phase is assumed to changes only radially. The volume fraction of the PSZ phase is, therefore, denoted by $1-\xi(r)$.

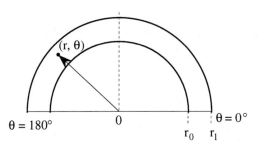

Figure 4. Geometry of cylinder.

The high temperature fluid, filled in the cylinder up to $\theta = 60°$ in the figure, runs at time $t = 0$ sec in the cylinder, which is surrounded with a low temperature coolant. The initial temperature of the cylinder is uniform and 300 K.

Neglecting the thermomechanical effect in the axial z-direction, the thermal and mechanical response of the cylinder can be discussed in a cross section, r-θ plane illustrated in Figure 4. The heat flows into and out of the cylinder by only means of the heat transfer. The unsteady heat conduction/thermal stress calculation starts just when the boundary condition is imposed to the cylinder.

The thermal boundary conditions are specified by the ambient temperature T_a and the coefficient of heat transfer h as follows:

On the inner surface in the fluid (r = 50 mm, $0° \le \theta \le 60°$);
   $T_a = 1600$ K , h = 700 W/m²K .

On the inner open surface (r = 50 mm, $60° \le \theta \le 180°$);
   $T_a = 1600 - 5 (\theta - 60)$ K , h = 100 W/m²K .

On the outer surface (r = 60 mm, $0° \le \theta \le 180°$);
   $T_a = 300$ K , h = 1200 W/m²K .

The second condition estimates the decrease in the heat inflow on the open surface by changing linearly the ambient temperature from the initial $T_a = 1600$ K just on the fluid surface ($\theta = 60°$) to $T_a = 1000$ K at the top ($\theta = 180°$).

Unsteady heat conduction/thermoelastic stress FEM analyses are carried out by means of the code MARC, with a mesh for a half domain ($\theta = 0° \sim 180°$) in Figure 4 discretized by the 80 equal size elements in the radial direction and 18 elements in the circumferential direction. The total number of elements and nodes is 1400 and 1539, respectively. The transient heat conduction analysis is first carried out to obtain the temperature field for a fraction distribution. The stress analysis is performed for the temperature distribution obtained. The critical stress in the present problem is always the circumferential stress σ_θ, which dominates among the stress components and is responsible for a possible failure or fracture.

Each phase is assumed to be isotropic both mechanically and thermally. The material parameters used in the actual thermoelastic material tailoring,

Young's modulus E ,   coefficient of thermal expansion α ,   Poisson's ratio ν ,
specific heat c ,   thermal conductivity λ ,   density ρ ,

are assumed to be temperature-independent. Their values are tabulated in Table 1 for both PSZ and SUS304 phases, and for Void used in Sec.4.3. The fraction-dependence of the material parameters is determined by means of the scheme explained in Sec.3.2.

Table 1 Material parameters.

	E GPa	α 1/K	ν	c kJ/kg K	λ W/mK	ρ kg/m³
PSZ	211	2.93×10^{-6}	0.31	0.467	1.67	5730
SUS304	193	14.87×10^{-6}	0.30	0.361	15.97	7930
Void	0	0	0	0.001	0.024	1.0

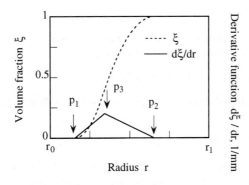

Figure 5. Identification of radial distribution of fraction.

The fraction ξ is expressed by means of the three design parameters $\mathbf{p} = (p_1, p_2, p_3)$ through a formula

$$\xi(\mathbf{p}; r) = \int_{r_0}^{r_1} (d\xi/dr)(r)\, dr \,, \tag{21}$$

where the function $d\xi/dr$ is given by the two linear lines shown in Figure 5 so that the area under $d\xi/dr$ is always equal to 1. The parameters p_1 and p_2, normalized to have a value in $[0, 1]$ between r_0 and r_1, stand for the end radius of the pure PSZ phase and the start radius of the pure SUS304 phase, respectively. The parameter p_3, normalized to have a value in $[0, 1]$ between $(r_1 - r_0)p_1 + r_0$ and $(r_1 - r_0)p_2 + r_0$, gives the radius of inflection of ξ-distribution.

The whole region is composed of the following phases:

pure PSZ phase: $r \leq (r_1 - r_0)p_1 + r_0$
FGM phase: $(r_1 - r_0)p_1 + r_0 \leq r \leq (r_1 - r_0)p_2 + r_0$
  PSZ-rich FGM phase: $(r_1 - r_0)p_1 + r_0 \leq r \leq (r_1 - r_0)(p_2 p_3 - p_1 p_3 + p_2) + r_0$
  SUS304-rich FGM phase: $(r_1 - r_0)(p_2 p_3 - p_1 p_3 + p_2) + r_0 \leq r \leq (r_1 - r_0)p_2 + r_0$
pure SUS304 phase: $r \geq (r_1 - r_0)p_2 + r_0$

4.2. Global stress reduction

The stress reduction is understood here in such a way that σ_θ should approach as close as possible to a given reference distribution $R(\xi(\mathbf{p}; r), t)$. An additional point to be noted in the

262

present material tailoring is that the circumferential stress distribution $\sigma_\theta(r, \theta, t)$ must always be globally reduced, meaning it must be inside an allowable band, which is settled through the consideration of strength and life time of the FGM itself. The band can be expressed by means of a limit tensile stress σ_{cr}^+ and the limit compressive stress σ_{cr}^- of a FGM characterized by ξ. These limit stresses must depend on the temperature T when thinking about the strong temperature-dependence of the material properties. They may also change in time due to some direct physical effect, which is not considered here. The band can be, therefore, given by the upper limit $\sigma_{cr}^+(\xi, T)$ and the lower limit $\sigma_{cr}^-(\xi, T)$.

The present material tailoring is now formulated as follows:

Minimize
$$F_2(\mathbf{p}) = \text{Max}[\,|\,\sigma_\theta(r, \theta, t) - R(\xi(\mathbf{p}; r), t)\,|\,]\,, \tag{22}$$

subject to constraints
$$F_1(\mathbf{p}) = \text{Max}[\,\sigma_\theta(r, \theta, t) - \sigma_{cr}^+\,,\ \sigma_{cr}^- - \sigma_\theta(r, \theta, t)\,] < 0\,,$$

$$\mathbf{P}(\mathbf{p}) \leq 0\,. \tag{23}$$

Equation(23)$_2$ represents a set of restrictions imposed on the design parameters $\mathbf{p}$.

The design parameters start from the initial value
$$\mathbf{p}_0 = (p_1\ p_2\ p_3) = (0.1\ 0.5\ 0.5)\,, \tag{24}$$

which gives the fraction distribution in Figure 10 labeled by $N = 1$ (Initial). The constraint $(23)_1$ on the parameters are given by
$$0 \leq p_1 \leq 0.3\,, \quad 0.3 \leq p_2 \leq 1\,, \quad 0 \leq p_3 \leq 1\,. \tag{25}$$

The preliminary calculations have revealed that the thicker PSZ phase generally reduces more the thermal stresses. Such a material tailoring is, however, not acceptable from the point of toughness in the FGM. Constraint$(25)_1$ is an indirect consideration to this issue.

Figure 6 illustrates the allowable band employed here with the linear upper and lower limit lines which depend only on the fraction, i.e., $\sigma_{cr}^+(\xi)$ and $\sigma_{cr}^-(\xi)$. High strength of PSZ phase in compression and of SUS304 phase in tension are taken into account. As Eq.(23) shows, the stress distribution σ_θ is required to be inside the band between these limit lines. A thin chain line stands for the reference distribution $R(\xi)$ determined as the middle line of the limit lines;

$$R(\xi) = (\sigma_{cr}^+ + \sigma_{cr}^-)/2\,, \tag{26}$$

which represents the safest stress distribution.

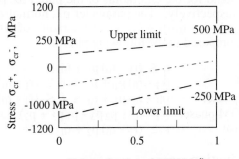

Figure 6. Band and reference distribution of stress.

Figure 7 shows the change in radial distribution of σ_θ at $\theta = 0°$ where maximum σ_θ is observed in most cases. The σ_θ distribution starts from the horizontal line and reaches the final steady state distribution drawn with the solid curve. The dotted lines represent the distribution at the intermediate time steps. The result reveals that the parts of the final steady distribution are outside the allowable band. It should first be noted that the allowable band is now highly nonlinear due to the nonlinear distribution of ξ determined from Eq.(25). The maximum deviation $F_1 = 208$ MPa occurs in the steady state at $t = 5000$ sec, labeled by $t = \infty$, and the radius $r = 51.3$ mm. After 5 runs of optimization procedure, an optimized solution given in Figure 8 is reached. The final deviation is $F_2 = 381$ MPa at (1700 sec/ 50.9 mm/ 180°). The optimal σ_θ distribution at $\theta = 0°$ is also plotted in the figure.

Although the purpose of this optimization after $N = 2$ is to minimize F_1, it also works to reduce the maximum value of σ_θ at each run, $\sigma_{\theta\,max}$, as shown in Figure 9. At the early stage of optimization $\sigma_{\theta\,max}$ has its maximum value in the final steady state whereas along with the progress of optimization the maximum value tends to appear at around 15 sec.

The change in the fraction distribution during optimization process is given in Figure 10 from the initial distribution, $N = 1$, to the final optimized distribution, $N = 5$ (Optimal). The dotted lines stands for the distributions at the intermediate steps. The result tells us that a PSZ-rich FGM phase distribution over the almost whole region is preferable.

When the reference distribution is selected as

$$R(\xi) = 0, \tag{27}$$

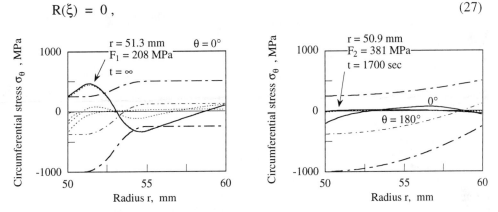

Figure 7. Radial distribution of stress ($N = 1$, Initial).

Figure 8. Optimal radial distribution of stress ($N = 5$).

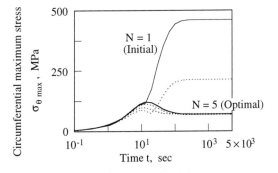

Figure 9. Change in maximum circumferential stress during optimization.

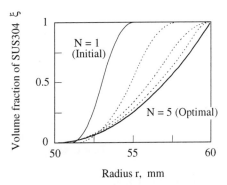

Figure 10. Change in fraction distribution during optimization.

the simulation gives a complete different solution although the band is the same as in Figure 6. Optimal solution is presented in Figure 11. The maximum value 356 MPa of F_2 is observed at (3.78 sec/ 50 mm/ 0°). This value is competitive with the value of F_1 at (5000 sec/ 50 mm/ 0°). Namely F_2 could be reduced in the subsequent optimization runs in the compensation for F_1 becoming positive, which is not acceptable. The final fraction distribution is given in Figure 12 with the solid curve. Almost no pure PSZ phase is also observed, but a large SUS304 phase is preferable in the present case.

The material tailoring could be performed by minimizing $\sigma_{\theta\,max}$ directly with no band constraint. The process to reach an optimal solution, after 13 runs of optimization procedure, is illustrated in Figure 13 as the time change in $\sigma_{\theta\,max}$. A sharp increase of $\sigma_{\theta\,max}$ at the middle time range in the initial run decreases rapidly with the progress of optimization. The value 90.0 MPa is competitive at both the points (8.7 sec/ 60 mm/ 0°) and (758 sec/ 53.2 mm/ 0°). The time change in σ_θ distribution is given in Figure 14. The arrows in the figure indicate the points of competition. The final optimized distribution of fraction is plotted in Figure 15 showing a PSZ-rich FGM phase in almost three quarters of the region.

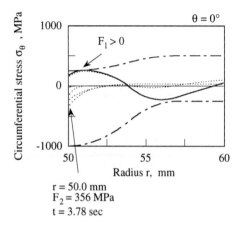

Figure 11. Radial distribution of stress (N = 4, Optimal).

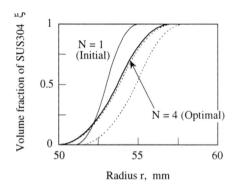

Figure 12. Change in distribution of fraction during optimization.

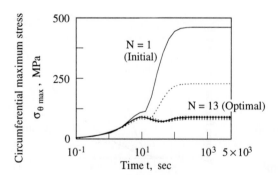

Figure 13. Change in maximum circumferential stress.

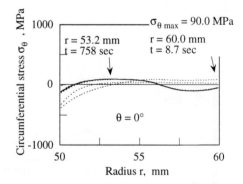

Figure 14. Radial distribution of stress (N = 13, Optimal).

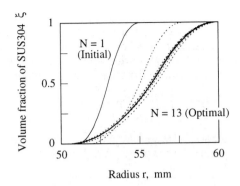

Figure 15. Change in distribution of fraction during optimization.

4.3. Thermal shielding

In most cases of applications the FGMs are expected to be a good thermal barrier which has enough strength in addition. A scheme of the material tailoring is discussed here under the design conditions of both the strength of materials and the thermal shielding [33]. The thermal shielding is expected by introducing the ill-heat conducting spherical voids inside the FGM which, however, inevitably lower the strength of the FGM. How to find out a point of compromise between the strength of materials and the thermal shielding is the point to be answered.

The volume fraction of void ξ_v is assumed to have the form, depending only on the radius,

$$\xi_v = p_4 \exp\{-(r^* - p_5)^2 / p_6\}, \qquad r^* = (r - r_0)/(r_1 - r_0), \qquad (28)$$

which is characterized by the three design parameters. The fraction of the SUS304 phase ξ distributes as before in Eq.(21).

The strength of the voided materials can be estimated by [34-35]

$$\sigma_{fr} = \sigma_0 \exp(-6\xi_v), \qquad (29)$$

where σ_0 stands for the strength of the material without voids, which is, if the linear mixture rule is employed for the fracture stresses σ_{fr}^{SUS} and σ_{fr}^{PSZ} of the PSZ and SUS304 phases, given in the PSZ-SUS304 FGM by

$$\sigma_0 = \frac{1 - \xi_v - \xi}{1 - \xi_v} \sigma_{fr}^{SUS} + \frac{\xi}{1 - \xi_v} \sigma_{fr}^{PSZ}. \qquad (30)$$

The following criteria can be employed as the criterion restricting the strength of FGM and the hard objective in Eq.(23), respectively:

$$f_S = \text{Max}(\sigma_d), \qquad \sigma_d = \sigma/\sigma_{fr}, \qquad (31)$$

$$F_1(\mathbf{p}) = \text{Max}(\sigma_d - 1) < 0. \qquad (32)$$

Equation(32) expresses that the FGM does not experience the fracture in the whole time-space. It should be noted that the stress means the circumferential stress σ_θ in the present problem.

In the thermal tailoring a non-dimensionalized factor

$$f_T = \frac{q}{q_{SUS}} \leq 1, \qquad q = \int h(T_a - T)\, dS, \qquad (33)$$

can be used as a criterion, where q stands for the heat flow out of the outer surface of cylinder, and T_a and T mean the temperatures of the outer coolant and at the outer surface of cylinder, respectively. The coefficient of heat transfer on the outer surface is denoted by h, and q_{SUS} stands for the value of q when the cylinder is made of SUS304 phase only.

The soft objective in Eq.(22) may be given as a linear combination of the strength criterion and the thermal criterion by

$$F_2(\mathbf{p}) = w f_S + (1-w) f_T \tag{34}$$

where w represents the weight of design. When $w=1$ only strength is considered, and $w=0$ means the thermal shielding be the main target of design. In the following material tailoring, the reference stress distribution is null, $R=0$.

The thermal boundary condition is simpler than before;

On the inner surface in the fluid (r = 50 mm, $0° \leq \theta \leq 60°$);
$T_a = 1600 \text{ K}$, h = 1000 W/m²K.

On the inner open surface (r = 50 mm, $60° \leq \theta \leq 180°$);
$T_a = 1600 \text{ K}$, h = 100 W/m²K.

On the outer surface (r = 60 mm, $0° \leq \theta \leq 180°$);
$T_a = 300 \text{ K}$, h = 1200 W/m²K.

The constraints for the 6 design parameters are given by

$$0 \leq p_1 \leq 0.5, \quad 0.5 \leq p_2 \leq 1, \quad 0.01 \leq p_3 \leq 0.99,$$
$$0 \leq p_4 \leq 0.5, \quad 0 \leq p_5 \leq 1, \quad 0.01 \leq p_6 \leq 10.0. \tag{35}$$

The strength is restricted by the allowable band shown in Figure 6.

The change in the circumferential stress σ_θ at $\theta = 0°$ is plotted for the case of $w=1$ (mechanical tailoring) in Figure 17 together with the allowable band in strength and the reference stress distribution. The unsteady solution at t = 2000 sec can be regarded as a steady solution. At the points (13.2 sec/ 50 mm/ 0°) and (100 sec/ 54.5 mm/ 0°) the hard objective is violated. Figure 18 shows the progress of tailoring during the optimization process. The voids are introduced only in the low stress region, otherwise the criterion (32) suffers a negative effect.

The temporal change in strength, measured by σ_d in Eq.(31), is plotted in Figure 19. Both optimal solutions for the thermal tailoring ($w=0$) and the mechanical tailoring ($w=1$) clear the criterion of the strength $\sigma_d < 1$. The mechanical tailoring gives the safer solution with respect to the strength of materials than the thermal tailoring. At the early stage of the mechanical tailoring, at around $t \approx 8$ sec, the compressive stress on the inner surface becomes critical. This state is sharply reduced during the optimization process by the increase of the SUS304 phase. In compensation for this the stress at around $r \approx 55$ mm becomes negative and continuously larger. This raises the σ_d - t curve again at around 50 sec and later. These two peaks becomes finally competitive in the iterations $N \geq 20$. Figure 20 sketches that as a result of thermal tailoring, at the stable state, almost 10% less amount of the heat flows through the cylinder.

5. CONCLUDING REMARKS

The thermoelastic material tailoring in FGMs developed with the help of the direct sensitivity analysis is shown to work well under the multiobjective optimization when solving the global reduction of the stress and the thermal shielding.

The evaluation of the material parameters, their fraction-dependence, is the key to get good solutions. The Mori-Tanaka theory based on Eshelby's inclusion theory is proved to be effective if coupled with the fuzzy inference. The optimal design of the microstructures in FGMs under some objective functions, the microscopic material tailoring, would be realized by the direct estimation of the material parameters by use of, for example, topology, fractal and

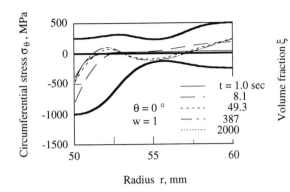

Figure 17. Change in circumferential stress.

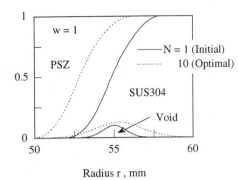

Figure 18. Change in fraction during optimization.

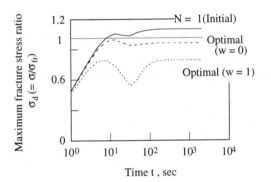

Figure 19. Time change in fracture stress ratio.

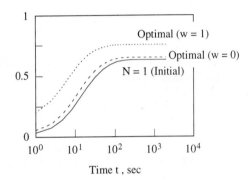

Figure 20. Time change in heat flow.

percolation theory. The macro/microscopic material tailoring should finally couple with the shape design, and even with the fabrication process design.

It should specially be noted that the different solutions are obtained in the material tailoring for the different boundary conditions even though the criteria are the same. One has, therefore, always to take strict notice of how much redundancy the material just tailored has, especially when the material tailoring is coupled with the optimal shape design.

Acknowledgment - The author wishes to thank Prof. V.F. Poterasu (Theoretical Mechanical Department, Technical University 'Gh. Asachi'/ Romania) and Prof. Y. Sugano (Department of Mechanical Engineering, Iwate University/ Japan) for their constructive criticism concerning an earlier version of the manuscript. Part of this work was financially supported by the Special Research Fund/ Tokyo Metropolitan Government.

REFERENCES

1. R.M.Jones, Mechanics of Composite Materials, MacGraw-Hill, New York, 1975.
2. K.K.Chawla, Composite Materials, Springer-Verlag, New York, 1987.
3. T.A.Weisshaar, J. Aircraft, **18** (1981) 669.

4. A.L.Librescu and J.Simovich, J. Aircraft, **25** (1988) 364.
5. G.C.Sih, G.F.Smith, I.H.Marshal and J.J.Wu, Composite Material Response, Elsevier Applied Science, London, 1988.
6. M.Yamanouchi, M.Koizumi, T.Hirai and I.Shiota, eds., Proc. 1st Int. Symp. on Functionally Gradient Materials, Sendai (1990).
7. K.Wakashima and H.Tsukamoto, ISIJ Int., **32** (1992) 883.
8. M.Koizumi, in J.B.Holt, M.Koizumi, T.Hirai and Z.A.Munir (eds.), Functionally Gradient Materials, The American Ceramic Society, Westerville (1993) 3.
9. B.H.Rabin and I.Shiota, MRS Bulletin, **XX** (1995) 14.
10. J.Teraki and T.Hirano, Proc. FGM'94 (1994) 133.
11. G.J.Weng, M.Taya, and H.Abe (eds.), Micromechanics and Inhomogeneity, Springer-Verlag, New York-Berlin-Heidelberg, 1990.
12. S.Nemat-Nasser and M.Hori, Micromechanics: Overall properties of heterogeneous materials, Elsevier Scientific Publishers B.V., Amsterdam, 1993.
13. K.Tanaka, Y.Tanaka, K.Enomoto, V.F.Poterasu and Y.Sugano, Computer Methods Appl. Mech. Engng, **106** (1993) 271.
14. K.Tanaka, H.Watanabe, V.F.Poterasu and Y.Sugano, Computer Methods Appl. Mech. Engng (1996) in press.
15. K.Muramatsu, A.Kwasaki, M.Taya and R.Watanabe, in M.Yamanouchi, M.Koizumi, T.Hirano and I.Shiota (eds.), Proc. 1st Int. Symp. Functionally Gradient Materials, Sendai (1990) 53.
16. S.Sampath, H.Herman, N.Shimoda and T.Saito, MRS Bulletin, **XX** (1995) 27.
17. K.Tanaka, D.Hasegawa, H.J.Bhöm and F.D.Fischer, Materials Sci. Research Int., **1** (1995) 23.
18. K.Wakashima and H.Tsukamoto, in M.Yamanouchi, M.Koizumi, T.Hirano and I.Shiota (eds.), Proc. 1st Int. Symp. Functionally Gradient Materials, Sendai (1990) 19.
19. V.F.Poterasu, K.Tanaka and Y.Sugano, Bull. TMIT, **8** (1994) 83.
20. T.Hirano, L.W.Whitlow and M.Miyajima, in J.B.Holt, M.Koizumi, T.Hirai and Z.A.Munir (eds.), Functionally Gradient Materials, The American Ceramic Society, Westerville (1993) 23.
21. E.Mamdani, Int. Man-Machine Studies, **8** (1976) 669.
22. T.Hirano, J.Teraki and T.Yamada, in M.Yamanouchi, M.Koizumi, T.Hirano and I.Shiota (eds.), Proc. 1st Int. Symp. on Functionally Gradient Materials, Sendai (1990) 5.
23. T.Hirano and K.Wakashima, MRS Bulletin, **XX** (1995) 40.
24. J.Teraki, T.Hirano and K.Wakashima, in J.B.Holt, M.Koizumi, T.Hirai and Z.A.Munir (eds.), Functionally Gradient Materials, The American Ceramic Society, Westerville (1993) 67.
25. M.J.Pindera, J.Aboudi and S.M.Arnold, in G.Z.Voyiadjis and J.W.Ju (eds.), Inelasticity and Micromechanics of Metal Matrix Composites, Elsevier Science B.V., London (1994) 273.
26. T. Mori and K. Tanaka, Acta Metall., **21** (1973) 571.
27. H.Hatta and M.Taya, J. Appl. Phys., **58** (1985) 2478.
28. Y.Benveniste, Mech. Materials, **6** (1987) 147.
29. K.Wakashima, H.Tsukamoto and B.H.Choi, Proc. Korea-Japan Metals Symp. on Composite Materials, Seoul/Korea (1988).
30. J.L.Cohon, Multiobjective Programming and Planning, Academic Press, New York, 1978.
31. M.Zeleny, Multiple Criteria Decision Making, McGraw-Hill, New York, 1982.
32. M.Matsumoto, J.Abe and M.Yoshiura, J. Mechanical Design, **115** (1993) 784.
33. K.Tanaka and H.Ikada, unpublished.
34. W.D.Kingery, H.K.Bowen and D.R.Uhlmann, Introduction to Ceramics, John Wiley & Sons, New York, 1976.
35. D.Liu, B.Lin and C.Fu, J. Ceram. Soc. Japan, **103** (1995) 878.

Theoretical and Applied Mechanics 1996
T. Tatsumi, E. Watanabe and T. Kambe (Editors)
© 1997 Elsevier Science B.V. All rights reserved.   269

Structural optimization of and with advanced materials*

Martin P. Bendsøe

Department of Mathematics, Technical University of Denmark
DK-2800 Lyngby, Denmark

Features of simultaneous design of overall structural performance together with design of the local material distribution itself can be clearly demonstrated by considering the minimum compliance problem. Here mathematical form permits a simple and clear distinction between these features. Emphasizing the material design aspect in linear elasticity, optimization of laminated plates through the laminate lay-up, the use of micro structures in generalized shape and topology design, and optimization by a free variation of the material tensor are discussed.

1. INTRODUCTION

Optimization of structures is traditionally performed through the variation of sizing and shape variables. With the appearance of composites and other advanced man-made materials it is natural to extend this variation to the material choice itself. Moreover, the study of generalized shape design and topology design leads one to consider design of micro structures, through the requirement for relaxation of initially ill-posed problems. The impetus to study micro structures and material design in a structural optimization framework has thus come from the increased industrial use of advanced materials as well as from developments within the scientific field itself.

From a physical perspective and for our understanding of nature it is extremely important to observe how optimization of structural performance automatically leads to structural hierarchy. The investigations by Cheng and Olhoff, 1981, and Lurie, Cherkaev and Fedorov, 1982, into the sizing optimization of solid elastic plates leading to formation of fields of thin ribs is a vivid illustration of this. A key point in the developments that follow is this concept of structural hierarchy. Macroscopic material parameters are varied at a global scale and the (optimal) realization of these material parameters are performed at a local, microscopic level.

The instructional nature of this presentation means that emphasis is given to introducing concepts and ideas in the framework of designing maximum stiffness structures in linear elasticity. As the developments will show, this can be a quite challenging problem in its own right, but it is of course only a beginning when addressing a concept as wide as the title of this paper indicates.

---

* This work was supported in part by the Danish Technical Research Council, through the Programme of Research on Computer Aided Design, and by the Danish Natural Sciences Research Council, grant # 9300720.

2. STIFFNESS OPTIMIZATION

In the following we will consider stiffness optimization of solids in plane elasticity, as this rather limited problem allows for a thorough analysis, giving insight and illustrating the basic concepts.*

2.1. Problem statement

For our developments it is convenient to express equilibrium in terms of the principle of minimum potential energy. The body we consider is a subset of a reference domain $\Omega \subset \mathbf{R}^2$ and the body has material properties described by the rigidity tensor E_{ijkl}, which may be varying over the body. For a (virtual) displacement u we denote by $\varepsilon_{ij}(u) = \frac{1}{2}\left(\frac{\partial u_i}{\partial x_j} + \frac{\partial u_j}{\partial x_i}\right)$ the linearized strains. Moreover, U is the space of kinematically admissible displacement fields, and p and t, are the body forces and the boundary tractions, respectively. The applied load and boundary conditions are defined on Ω (t is defined on $\Gamma_T \subset \Gamma \equiv \partial\Omega$). Using a standard tensor notation consistent with a Cartesian reference frame, equilibrium is then expressed as

$$\min_{u \in U} \ \Pi(u), \qquad \Pi(u) = \frac{1}{2}\int_{\Omega} E_{ijkl}(x)\varepsilon_{ij}(u)\varepsilon_{kl}(u)\,\mathrm{d}\Omega - \int_{\Omega} pu\,\mathrm{d}\Omega - \int_{\Gamma_T} tu\,\mathrm{d}s \tag{1}$$

The value of the potential energy at equilibrium equals minus one half of the compliance of the structure. For a structure of maximal stiffness we seek to minimize the compliance. This problem can thus be written as

$$\max_{\text{DESIGN}} \ \min_{u \in U} \ \left\{ \frac{1}{2}\int_{\Omega} E_{ijkl}(x)\varepsilon_{ij}(u)\varepsilon_{kl}(u)\,\mathrm{d}\Omega - \int_{\Omega} pu\,\mathrm{d}\Omega - \int_{\Gamma_T} tu\,\mathrm{d}s \right\} \tag{2}$$

and the physical meaning of this statement is decided through the specification of the dependence of the potential energy on design. For design of structure through material distribution and material choice, this dependence is given through the rigidity tensor E_{ijkl}.

3. DESIGN PARAMETRIZATION: THE RIGIDITY TENSOR

The exact physics of the problem statement (2) is determined through the definition and parametrization of the admissible rigidity tensor as it will appear in the potential energy. We are then able to cover a broad range of problems, with prominence here given to parametrizations which in some sense involves choice or design of the material itself.

3.1. Variable thickness sheets

For a variable thickness sheet, a material with properties E_{ijkl}^G is given and one seeks the thickness variation $h(x)$, $x \in \Omega$, which maximizes stiffness. For well-posedness, an isoperi-

---

* In the spirit of a tutorial presentation, the placing of the the developments in a broader historical perspective is postponed to section 7, which contains a brief literature survey.

metric constraint limiting the total amount of available material as well as upper and lower bounds on the thickness should be imposed. The set of rigidity tensors over which problem (2) is solved is then given as

$$E_{ijkl} = h(x) E_{ijkl}^G \quad ; \qquad \int_\Omega h(x) \, d\Omega \le V \quad ; \qquad 0 \le h_{\min} \le h \le h_{\max} < \infty \tag{3}$$

and the design variable is in reality the thickness function h. We only mention this case for comparison, as (2) with this design parametrization is a classical formulation of a sizing problem. Note that in an abstract sense the material properties in the form of the tensor E_{ijkl} are varied in this problem also, through the possibility of using the out-of-plane dimension to vary the properties linearly by the scaling parameter h. However, the basic properties of the tensor (directions of orthotropy etc.) remains unchanged.

3.2. Laminates

In a laminated plate the out-of-plane dimension is employed to tailor the material properties of the plate. This can be performed by varying ply thicknesses, fibre orientations and the stacking sequence of the laminate. Working with just one fixed ply material with in-plane material properties E_{ijkl}^P and keeping the thickness of the plate fixed at a constant thickness ($h = 1$ for convenience), the rotation formulas for fourth order tensors together with integration through the thickness gives an in-plane stiffness tensor of the plate of the form

$$E_{ijkl}(x) = E_{ijkl}^0 + \xi_1(x) E_{ijkl}^1 + \xi_2(x) E_{ijkl}^2 + \xi_3(x) E_{ijkl}^3 + \xi_4(x) E_{ijkl}^4 \tag{4}$$

In this expression E_{ijkl}^n, $n = 0,..,4$ are constant tensors given through E_{ijkl}^P, and the *lamination parameters* ξ_n, $n = 1,..,4$ are given by integrations through the thickness:

$$\xi_1 = \int_{-\frac12}^{\frac12} \cos 2\theta(z) \, dz, \quad \xi_2 = \int_{-\frac12}^{\frac12} \cos 4\theta(z) \, dz, \quad \xi_3 = \int_{-\frac12}^{\frac12} \sin 2\theta(z) \, dz, \quad \xi_4 = \int_{-\frac12}^{\frac12} \sin 4\theta(z) \, dz \tag{5}$$

as zero order moments of trigonometric functions. Here $\theta(z)$ is the in-plane rotation of the ply at the distance $z \in [-\frac12, \frac12]$ from the plate mid-plane.

It is readily seen that the set of lamination parameters constitute a convex set which is the convex hull of the curve $(\cos 2\theta, \cos 4\theta, \sin 2\theta, \cos 4\theta)$, $\theta \in [0, 2\pi]$ in $\mathbf{R}^4$. It can be shown that the set is given by the constraints

$$-1 \le \xi_2 \le 1, \quad \xi_1^2 + \xi_3^2 \le 1, \quad 2\xi_1^2(1 - \xi_2) + 2\xi_3^2(1 + \xi_2) + \xi_2^2 + \xi_4^2 - 4\xi_1\xi_3\xi_4 \le 1 \tag{6}$$

and that any lamination parameter satisfying these constraints can be realized by a lay-up consisting of not more than three plies at different angles. This realization is non-unique, leaving room for optimization with respect to other criteria over the set of stiffest designs.

3.3. Generalized shape design

In generalized shape design (also denoted topology design) one seeks the optimal distribution of a given volume of a fixed material within the spatial domain Ω. The shape is not parametrized by the boundaries of the body, but rather by the pointwise distribution of material, every point consisting of either material or void. Thus the set of admissible rigidity tensors becomes:

$$E_{ijkl} = 1_{\Omega^m} E_{ijkl}^I, \quad 1_{\Omega^m} = \begin{cases} 1 & \text{if } x \in \Omega^m \\ 0 & \text{if } x \in \Omega \setminus \Omega^m \end{cases}; \quad \int_\Omega 1_{\Omega^m} \, d\Omega = \text{Vol}(\Omega^m) \leq V \tag{7}$$

where Ω^m is the body of material points and where E_{ijkl}^I is the rigidity tensor for a given material (here *isotropic*).

It is now well-known that the design parametrization (7) leads to non-existence of solutions to problem (2). The reason for this is that in many cases stronger structures can be made by distributing the material partially over parts of the domain Ω, by constructing a structural hierarchy with a porous micro structure of void and the given material (as seen in bone etc.). As such structures clearly can arise as limits of designs given by the parametrization (7), existence of solutions require either a restriction of the design space (by disallowing structures with fine scale variations of shape and material) or by an extension of the design space to include any micro structures constructed from mixing the given material and void. In the spirit of this presentation we take the latter approach. We thus write the design parametrization as

$$E_{ijkl}(x) = E_{ijkl}(\text{composite of density } \rho(x)) \, ; \quad \int_\Omega \rho \, d\Omega \leq V \tag{8}$$

where ρ denotes the density in the composite of the given material. It is clear that working with this extension for the design problem will lead to the consideration of optimizing the micro structure (the generalized material) itself.

3.4. Free material design

For laminates and for the generalized shape design with micro structures we see that from a given material we generate a design parametrization which in essence constitutes an extended range of materials. Alternatively, one may say that the range of the rigidity tensors is extended by including a structural hierarchy at multiple scales.

Taking this extension of the admissible rigidity tensors to its fullest generalization one can consider design over all positive semi-definite constitutive tensors. The goal is to formulate a structural optimization problem in a form that encompasses the design of structural material in a broad sense, while also encompassing the provision of predicting the structural topologies and shapes associated with the optimum distribution of the optimized materials.

There is at first glance no natural cost function for this general design formulation. Instead, as we require that the optimal design solutions are independent of the choice of reference frame, we employ certain invariants of the stiffness tensor as the measure of cost. Additionally, to mimic thickness/density as a cost, we require the cost to be homogeneous of degree one. For exemplification, let us here then consider the Frobenius norm as cost measure, leading to a design parametrization

$$E \geq 0 \text{ in } \Omega; \quad \rho = \left[E_{ijkl} E_{ijkl} \right]^{\frac{1}{2}}; \quad \int_\Omega \rho \, d\Omega \leq V; \quad 0 \leq \rho_{\min} \leq \rho \leq \rho_{\max} < \infty \tag{9}$$

where we have included upper and lower bounds as well as an isoperimetric constraint on the density ρ of cost.

The information that we can obtain within this assumption of a locally unconstrained configuration of material gives insight into the nature of efficient local structures. For practical reasons it is important to understand how to design a particular local micro structure to the specific form of the elasticity tensors predicted. This is the subject of paragraph 6.3.

4. THE GLOBAL PROBLEM

4.1. Existence of solutions

One can prove existence of solutions to the minimum compliance problem (2) for each of the design parametrizations (3), (4+6), (8), and (9). Except for the case of generalized shape design the rigidity tensors depend linearly on the design variables (thickness, lamination parameters, the entries in E_{ijkl}). Moreover, the constraints are in all these cases convex, making it possible to invoke a saddle point argument for the problem (2).

For the generalized shape design problem the set (8) is constructed so as to include all limits of minimizing sequences of designs in (7). As the limiting process of computing effective moduli gives consistent convergence of compliance, we achieve existence of a solution with a performance which can be approximated arbitrarily well by a design in the set (7) (albeit with very small scale structure). For a more precise mathematical statement of this we refer to the literature.

4.2. Problem separation

In the generalized shape design problem as well as in the free material design a distinction can be made between the local (pointwise) use of material/resource and the global distribution of this, limited in total by the volume constraint. Moreover, in the lamination design problem with a pointwise variation of the lay-up, a similar local material design has to be carried out.

This separation into a local and global problem can be made precise by reformulating problem (2) in the form

$$
\max_{\substack{\rho_{min} \leq \rho(x) \leq \rho_{max} \\ \int_\Omega \rho \, d\Omega \leq V}} \quad \max_{\substack{E(x) \text{ for} \\ \text{density } \rho(x)}} \quad \min_{u \in U} \quad \Pi
\tag{10}
$$

Interchanging the inner maximization over local properties with the equilibrium minimization, using a saddle point argument, allows for a problem composition

$$
\max_{\substack{\rho_{min} \leq \rho(x) \leq \rho_{max} \\ \int_\Omega \rho \, d\Omega \leq V}} \quad \min_{u \in U} \quad \max_{\substack{E(x) \text{ for} \\ \text{density } \rho(x)}} \quad \Pi
\tag{11}
$$

We assume that the optimization of micro structure is pointwise so we can move the inner maximization under the integration over the domain. The final reformulation then becomes

$$
\max_{\substack{\rho_{min} \leq \rho(x) \leq \rho_{max} \\ \int_\Omega \rho \, d\Omega \leq V}} \quad \min_{u \in U} \quad \left\{ \int_\Omega \overline{W}(\rho, \varepsilon_{ij}(u)) \, d\Omega - \int_\Omega pu \, d\Omega - \int_{\Gamma_T} tu \, ds \right\}
\tag{12}
$$

where $\overline{W}(\rho, \varepsilon)$ denotes the pointwise optimal strain energy density expression

$$\overline{W} = \max_E \tfrac{1}{2} E_{ijkl} \, \varepsilon_{ij} \varepsilon_{kl} \tag{13}$$

which is the problem of generating the stiffest local structure, for a given strain (the stiffest material within the class of materials considered).

In case of multiple solutions to problem (13), the inner minimum potential energy problem of (12) can become non-smooth.

Analysis of problem (13), which we term the *local anisotropy* problem, will allow for insight into the properties of the optimal design. This is the focal point of the coming section.

5. THE LOCAL ANISOTROPY PROBLEM

5.1. Generalized shape design: energy bounds and the material science

In the case of generalized shape design with the inclusion of composites, problem (13) is in the form of the well-known problem in the theoretical materials science of generating bounds on the effective properties of mixings, here of two isotropic materials. There are several solutions to this problem, but for design purposes the so-called multiple scale (ranked) layered materials are particularly suited as analytical expressions exist for the effective properties of such materials. In our situation of a so-called single load problem, a rank-2 layering with two sets of orthogonal layerings at different scales suffices to obtain a solution to (13) (for multiple load problems (13) becomes a maximization of a sum of strain energies, leading to the need for rank-3 layerings).

For a rank-2 layering of material and void with primary layerings of density μ in the 2-direction and the secondary layer of density γ in direction 1, the effective material properties of the resulting material (which has a density $\rho = \mu + \gamma - \mu\gamma$) can easily be computed by recursive use of the formula for the effective properties of a single layered composite. The moduli are computed as the material is constructed, bottom up and the resulting material properties are:

$$E_{1111} = \frac{\gamma E}{\mu\gamma(1-v^2)+(1-\mu)}, \quad E_{1122} = \mu v E_{1111}, \quad E_{2222} = \mu E + \mu^2 v^2 E_{1111}, \quad E_{1212} = 0 \tag{14}$$

where E is Young's modulus and v is Poisson's ratio of the base material.

The optimal rank-2 material will have its layers directed along the directions of the principal strains $\varepsilon_I, \varepsilon_{II}$, so the apparent singularity $E_{1212} = 0$ is of no consequence. An analytical expression for $\overline{W}$ can be given and the result of the optimization (13) depends on the size of $\varepsilon_I, \varepsilon_{II}$ relative to the density ρ. The results are not very compact, so let us here restrict ourselves to the case where the optimal material micro structure has both layers at non-zero gauge. It turns out that in this case $\overline{W}$ is smooth. Moreover, the optimized energy functional corresponds to the energy of an equivalent linearly elastic *isotropic* material, $\overline{W} = \tfrac{1}{2}\overline{E}_{ijkl} \, \varepsilon_{ij}\varepsilon_{kl}$, with equivalent Young's modulus $\overline{E}$ and Poisson's ratio $\overline{v}$ given as

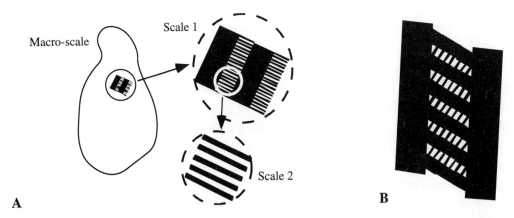

Fig. 1. Micro structures with layers at multiple scales **A:** The build-up of a second rank layered material, by successive layering of mutual orthogonal layers, resulting in an orthotropic material. **B:** A rank-3 layered material, by successive layering at three different scales. Here the layers are not orthogonal, possibly resulting in a generally anisotropic material.

$$\bar{E} = E / \left[(1-v)(2-\rho+\rho v)\right], \quad \bar{v} = (1-\rho+\rho v), \qquad \text{if } \varepsilon_I + \varepsilon_{II} < \rho(1-v)\varepsilon_I$$
$$\bar{E} = E / \left[(1+v)(2-\rho-\rho v)\right], \quad \bar{v} = -(1-\rho-\rho v), \qquad \text{if } \varepsilon_I - \varepsilon_{II} < \rho(1+v)\varepsilon_I \tag{15}$$

5.2. Laminates of extremal stiffness in a membrane state

For the problem of lay-up optimization of laminated plates, the local anisotropy problem (13) becomes a problem of minimizing a linear functional (in the lamination parameters) over a convex constraints set (given by (6)).

The lamination parameters can be defined by reference to any suitable frame. For our purposes it is convenient to use the frame of the principal strains $\varepsilon_I, \varepsilon_{II}$. Moreover, if we assume the ply material to be orthotropic, it is also convenient to measure the ply rotations relative to the frame of orthotropy (chosen so $E^P_{1111} \geq E^P_{2222}$), leading to $E^3_{ijkl} = 0$, $E^4_{ijkl} = 0$ in the expression (4). The local anisotropy problem can thus be expressed as

$$\max_{\xi_1,\xi_2} \tfrac{1}{2}\left(C_1\left(\varepsilon_I^2 + \varepsilon_{II}^2\right) + 2C_4\varepsilon_I\varepsilon_{II} + C_2\left(\varepsilon_I^2 - \varepsilon_{II}^2\right)\xi_1 + C_3\left(\varepsilon_I - \varepsilon_{II}\right)^2\xi_2\right) \tag{16}$$

subject to: $\;-1 \leq \xi_1 \leq 1, \;\; -1 \leq \xi_2 \leq 1, \;\; 2\xi_1^2(1-\xi_2)+\xi_2^2 \leq 1$

where the constraint set is the convex hull of the curve $(\cos 2\theta, \cos 4\theta)$, $\theta \in \mathbf{R}$, in $\mathbf{R}^2$, and

$$C_3 = \tfrac{1}{8}(E^P_{1111} + E^P_{2222} - 2E^P_{1122} - 4E^P_{1212}) \tag{17}$$
$$C_1 = \tfrac{1}{2}(E^P_{1111} + E^P_{2222}) - C_3, \qquad C_2 = \tfrac{1}{2}(E^P_{1111} - E^P_{2222}), \qquad C_4 = E^P_{1122} + C_3$$

By choice we have that $C_2 \geq 0$, so the solution will only depend on the sign of $\left(\varepsilon_I^2 - \varepsilon_{II}^2\right)$ and C_3. Whatever the sign of C_3, the optimal energy expression will be non-smooth.

If we have $C_3 \geq 0$ (we say that the ply material has low shear stiffness, cf. Pedersen, 1989), the optimal energy will be obtained from the lamination parameters $\xi_1 = 1$, $\xi_2 = 1$ if $\left(\varepsilon_I^2 - \varepsilon_{II}^2\right) \geq 0$ and $\xi_1 = -1$, $\xi_2 = 1$ if $\left(\varepsilon_I^2 - \varepsilon_{II}^2\right) \leq 0$. If $\left(\varepsilon_I^2 - \varepsilon_{II}^2\right) = 0$ (i.e. in terms of strains, uniform dilation or pure shear) the optimum can be obtained by any convex combination of these two lamination parameters, and elsewise the optimum is unique. As $\xi_1 = 1$, $\xi_2 = 1$ is realized by a single ply rotated zero degrees and $\xi_1 = -1$, $\xi_2 = 1$ is realized by a single ply rotated 90 degrees, we have that the optimal design for $\left(\varepsilon_I^2 - \varepsilon_{II}^2\right) \neq 0$ consists of a single ply rotated so that the stiffest direction is aligned with the numerically largest principal strain. However, if $\left(\varepsilon_I^2 - \varepsilon_{II}^2\right) = 0$, the optimal design can consist of a cross-ply consisting of two plies rotated zero or 90 degrees relative to the principal strain axes. The relative thicknesses of these two plies is in turn decided through the conditions of equilibrium (problem (12), without the optimization over density, as this is not relevant in this case).

If we consider the case $C_3 \leq 0$ (the ply material has high shear stiffness), a similar analysis shows that the optimal design is also here a single rotated ply, unless $\varepsilon_I = \varepsilon_{II}$, where again a (0,90) cross-ply is optimal. For the single ply situation (i.e., when $\varepsilon_I \neq \varepsilon_{II}$), the ply is rotated relative to the principal strains at an angle which is given by $\cos 2\gamma = -\frac{C_2}{4C_4} \frac{\varepsilon_I + \varepsilon_{II}}{\varepsilon_I - \varepsilon_{II}}$ or which is zero degrees if this does not make sense.

5.3. Extremal materials

For the free material design, the local anisotropy problem (13) is most easily solved by using a geometric interpretation. The dyadic 4-tensor $\varepsilon_{ij}\varepsilon_{kl}$ is given, and we seek the 4-tensor E_{ijkl} of length ρ which has the maximal projection on the given tensor $\varepsilon_{ij}\varepsilon_{kl}$. Thus the optimal E_{ijkl} is proportional to $\varepsilon_{ij}\varepsilon_{kl}$ and is uniquely given as

$$E_{ijkl} = \rho \frac{\varepsilon_{ij}\varepsilon_{kl}}{\varepsilon_{pq}\varepsilon_{pq}} \; ; \qquad E_{\text{matrix}} = \frac{\rho}{\varepsilon_I^2 + \varepsilon_{II}^2} \begin{pmatrix} \varepsilon_I^2 & \varepsilon_I \varepsilon_{II} & 0 \\ \varepsilon_I \varepsilon_{II} & \varepsilon_{II}^2 & 0 \\ 0 & 0 & 0 \end{pmatrix} \tag{18}$$

Here we have written the result in tensor notation as well as in a standard engineering matrix notation, with respect to the frame of the principal strains ε_I, ε_{II}. From the latter expression we see that the optimal rigidity tensor is orthotropic, with axes of orthotropy given by the principal strains (the numerically larger strain is aligned with the larger stiffness).

The reduced design problem (12) becomes in this case

$$\max_{\substack{\rho_{\min} \leq \rho(x) \leq \rho_{\max} \\ \int_\Omega \rho \, d\Omega \leq V}} \min_{u \in U} \left\{ \frac{1}{2} \int_\Omega \rho \, \varepsilon_{ij}(u) \varepsilon_{ij}(u) \, d\Omega - \int_\Omega p u \, d\Omega - \int_{\Gamma_T} t u \, ds \right\} \tag{19}$$

corresponding to a design problem of finding the maximum stiffness sheet made of an isotropic zero-Poisson-ratio material, with the density ρ playing the role of the thickness of the sheet (cf., paragraph 3.1). Note that the optimal tensor does not correspond to an isotropic zero-Poisson-ratio material, even though its corresponding energy functional does.

6. OPTIMAL STRUCTURES

The analysis of the local anisotropy problems leads to reduced optimization problems, where the local optimization of material properties is taken care of algebraically. This also gives considerable insight in the nature of the optimal designs. However, the full solution of the structural optimization problem as stated in (2) still requires considerable computational effort. There are several numerical very efficient methods available for this task, all based on a finite element discretization of the displacement field as well as of the design variables. The computations are a great challenge because of the large scale of the problems. Both the design variables and the displacements are distributed parameters, so we are treating a non-linear. optimization problem with a huge number of variables.

Most computational work employ algorithms based on successive FEM displacement analyses for fixed design, alternating with updates of the design variables using the so-called optimality criteria methods or mathematical programming techniques (sequential linear programming or sequential convex programming seems to be most popular, due to the large size of the problems). Approaches using algorithms for non-smooth optimization applied to various forms of the reduced problem (12) are also gaining in importance. Here large scale computations frequently hinge on some form of further reformulation, in many cases to smooth problems with many constraints, albeit dissimilar to the original problem (2) and with a structure suitable for the application of for example modern interior point methods.

When considering finite element discretization schemes for the design problems discussed here it is important to recognise that the problem at hand is inherently a saddle point problem, as evident from (2). This, of course, remains the underlying nature of the problem, independent of eventual reformulations. This has the effect that computational schemes in many cases can lead to designs which in some regions exhibit a material distribution in the form of checkerboard-like patterns. This is a purely numerical phenomenon and is not related to a formation of micro structure. Moreover it can be avoided by careful choice of discretization spaces.

6.1. Optimal distribution of optimal material

The local optimization problem for the free material design was in a sense very simple. Moreover, the resulting reduced optimization problem (19) also possess structure which in some sense makes this problem more tractable than the other situations considered here. First of all the reduced problem remains smooth, and moreover the density enters linearly in the optimized potential energy. This means that the problem in computational terms is similar to optimization of truss structures. This linearity also implies that a further interchange of maximization over density and minimization over displacements is possible. Solving for the density, we can reduce the optimization problem to the form of an equilibrium only problem with a non-smooth potential energy. With Λ denoting a Lagrange multiplier for the resource constraint, the reduced equilibrium-like problem is

$$\min_{\substack{u \in U \\ \Lambda \geq 0}} \left\{ \int_\Omega \max\left\{ \rho_{\min}\left[\tfrac{1}{2}\varepsilon_{ij}(u)\varepsilon_{ij}(u) - \Lambda \right], \rho_{\max}\left[\tfrac{1}{2}\varepsilon_{ij}(u)\varepsilon_{ij}(u) - \Lambda \right] \right\} d\Omega + \Lambda V - \int_\Omega pu\, d\Omega - \int_{\Gamma_T} tu\, ds \right\} \quad (20)$$

The optimal design is then given by the displacement and associated Lagrange multiplier at the optimum of (20) through the conditions of optimality for (19) which constitutes

278

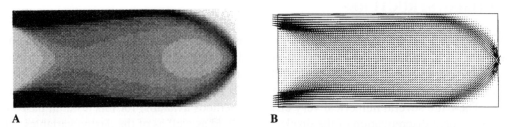

Fig. 2. Design of a medium aspect ratio cantilever using optimal materials (free material design). Single vertical load at mid right-hand point, supports at left hand vertical line. *(A)*: Distribution of resource, shown in a linear grey-scale, black illustrating maximum density. *(B)*: Directions and sizes of principal strains; directions correspond to direction of material axes.

$$\rho^*(x) = \rho_{\min} \text{ if } \left[\tfrac{1}{2}\varepsilon_{ij}(u^*)\varepsilon_{ij}(u^*)\right](x) < \Lambda^*, \quad \rho^*(x) = \rho_{\max} \text{ if } \left[\tfrac{1}{2}\varepsilon_{ij}(u^*)\varepsilon_{ij}(u^*)\right](x) > \Lambda^* \tag{21}$$

as well as the equation of equilibrium, as an equation which for given displacement determines the design in areas of specific energy satisfying $\left[\tfrac{1}{2}\varepsilon_{ij}(u^*)\varepsilon_{ij}(u^*)\right](x) = \Lambda^*$.

Note that the design can also be recovered from the Lagrange multipliers for a smooth reformulation of (20), written as

$$\min_{\substack{u \in U \\ \Lambda \geq 0 \\ \alpha}} \left\{ \int_\Omega \alpha \, d\Omega + \Lambda V - \int_\Omega p u \, d\Omega - \int_{\Gamma_T} t u \, ds \right\} \text{ subject to}: \begin{cases} \tfrac{1}{2}\varepsilon_{ij}(u)\varepsilon_{ij}(u) - \Lambda \leq \dfrac{\alpha}{\rho_{\min}} \\ \tfrac{1}{2}\varepsilon_{ij}(u)\varepsilon_{ij}(u) - \Lambda \leq \dfrac{\alpha}{\rho_{\max}} \end{cases} \tag{22}$$

This convex problem takes the form of the minimization of a linear functional, subject to quadratic constraints (which should be interpreted in their weak form). For a finite element based discretization of problem (2), the reformulations leading from (19) through (20) to (22) should be made consistent with the approximation spaces used for displacements and the material density.

6.2. Optimal two-ply laminate

For the optimal material problem the solution of the local anisotropy problem results in a smooth extremal potential energy functional. However, solving also for the optimal distribution produces a non-smooth problem, cf. eq. (20).

For the case of optimal laminates the solution of the local anisotropy problem in itself results in a non-smooth potential. Here the satisfaction of the equilibrium equation is required in order to determine the precise ratio of the layer thicknesses in the optimal cross-ply, i.e. global information is essential. For the extremal materials this information is not needed for the local problem, but the equilibrium information is also here vital for finding the in-plane material distribution.

For the laminated plate problem a smooth approximation to the reduced problem (12) (which in this case has no optimization over density) can be constructed by adding a term $\delta\|\xi\|^2$ to the objective function in (16) and the optimization problem can then be solved by alternate linear equilibrium analyses and the computation of solutions to the approximate (16). The penalty parameter δ is successively reduced during the optimization iterations,

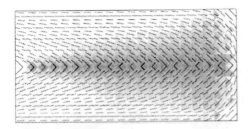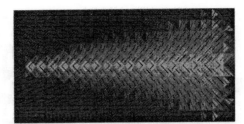

Fig. 3. Design of a medium aspect ratio cantilever as a two-ply laminate (membrane state), using lamination parameters. Load and supports as in figure 2. The thickness of the plies is shown in a linear grey-scale, black illustrating *zero thickness*, so that directions of material axes can be shown. Computational procedure realize the optimal lamination parameters by three plies, the third ply having zero thickness (single load case), (see text).

leading to a solution of the original problem. At the optimum, the optimal design is identified through an inverse algorithm which produces a laminate lay-up that realizes the optimal lamination parameters.

6.3. Designing material structures

The insight gained from the use of the full elasticity tensor as a design variable naturally leads one to consider what range of materials that can be built from a given material. One possible interpretation is to see this in the sense of constructing a structural hierarchy, with macroscopic properties determined through the design of material micro structures at a finer scale. From this viewpoint, the macroscopic design technique in the form of generalized shape design can be used to design micro structured media with prescribed material properties.

In the models described in the previous sections, locally optimal material properties are derived analytically. Moreover, for the laminated plate and the generalized shape design, a realization of the optimal properties through an optimal micro structure can be devised. This is performed to obtain the locally as well as globally optimal structure. When designing micro structure in a broader context, the local design is performed by computational means. For a single structural hierarchy consisting of a periodic repetition of a micro cell, generalized shape design can be employed to design the structural lay-out of the basis cell, for example in order to find a minimum mass micro structure which realizes certain specified effective material parameters, given through homogenization as generalized compliances.

Design of the micro structure itself takes the developments here 'full circle'. Finding an optimal structure can require the introduction of micro structure. Techniques developed for handling generalized shape design problems through a description of shape by a density of (composite) material has then been further extended to cover design of the micro structures themselves.

7. A BRIEF LITERATURE SURVEY

The various problem types described above have been the subject of intensive study over the last decades. The micro structural aspects, i.e., the presence of the material design in structural design problems has to a large extend been spurred by studies of the generalized

280

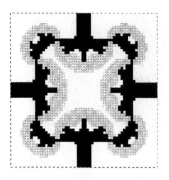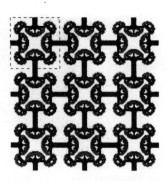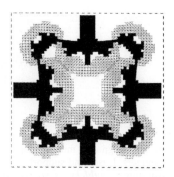

Fig. 4. Design of a micro structure with *negative* thermal expansion. Each base material has *positive* thermal expansion. White areas represents void, black regions material with low expansion and grey (hatched) regions material with high expansion. The main mechanism behind the negative resulting expansion is bi-material interfaces which make the cell contract when heated. The left-hand picture shows the design of the unit cell making up the periodic medium shown schematically in the middle. The right-hand picture shows (an exaggeration) of the deformation of the unit cell when the composite is heated. Sigmund and Torquato, 1996.

shape design problem and the so-called topology design problem (configuration design, layout design). Likewise, the use of advanced materials in production has given its significant stimulus to the inclusion of considerations for such materials in structural design studies. The presentation here attempts to combine these viewpoints through their mathematical and mechanical formulations and associated solution procedures, as seen from the viewpoint of the discipline of structural optimization.

7.1. Generalized shape design

The minimum compliance design problem is a classic in structural optimization, as its mathematical form allows for substantial insight. Referring to the developments in this paper, the variable thickness sheet problem was first considered by Rossow and Taylor, 1973, as a means to obtain information on the optimal shape, in a sense as a variant of the generalized shape design problem. The generalized shape design problem has been the subject of much study and it is difficult to give just credit to the many contributions which has now lead to our full understanding of this problem in its minimum compliance form (see for example Bendsøe, 1995). Here we will just point out suitable starting points for a foray in to this exciting subject which combines intricate mathematics and mechanics.

The generation of optimal bounds for the effective material properties composites is a closely related problem to the generalized shape design problem, as outlined in sections 4 and 5, and in many cases papers deal with both points of view. Examples of these theoretical developments can be found in Cherkaev and Kohn, 1996, Gibiansky and Cherkaev, 1987, Lurie and Cherkaev, 1986, Murat and Tartar, 1985, for problems in conduction and elasticity.

The development of computational schemes for generalized shape design based on the use of composites was first initiated by Bendsøe and Kikuchi, 1988, who named it the homogenization method for topology design, as homogenization theory plays a significant role. The discovery of the multiple scale layered materials as a possible realization of the stiffest micro structure gave rise to new types of computational schemes, as described by Jog, Haber, and Bendsøe, 1994, and Allaire and Kohn, 1993. The layered materials are not the only composites which accomplish the optimal bounds on stiffness, see for example Grabovsky and Kohn,

1995, and references therein. However, the layered materials do allow for a design parametrization in terms of the same parameters used above for the laminated plate (the ply angle is then here the layer direction), and this has been useful for theoretical studies as well as computational procedures, see for example Lipton, 1994a, and Díaz, Lipton, and Soto, 1995.

This presentation has largely ignored formulations of the generalized shape design problem that will provide solutions without micro structure, but it should be mentioned that a penalization of structural hierarchy is a way to achieve this, see for example Haber, Bendsøe and Jog, 1996.

When it comes to numerical schemes for the generalized shape design problem, much early work was based on the use of optimality criteria based methods, now supplemented by more standard mathematical programming techniques, see for example Thomsen, 1991, and Duysinx et al, 1995. Another aspect of the computational work is the formation of checkerboard patterns in the optimal material distributions and this numerical artifact has been studied in Bendsøe, Díaz and Kikuchi, 1993, Díaz and Sigmund, 1995, and Jog and Haber, 1996, where also guidelines for avoiding these spurious results can be found.

7.2. Laminates

Optimal structural design of laminated papers is traditionally handled by considering a design parametrization through ply angles, ply thicknesses and stacking sequence. This results in mixed integer-continuous variable optimization problems for which a number of various algorithms have been tried; see for example Fang and Springer, 1993, and Haftka, Gürdal and Kamat, 1992, for a survey and presentation of such techniques.

The lamination parameters are useful for problems where integral stiffness is the key mechanical property of the design optimization problem, and this design parametrization has since their introduction by Miki, 1982, been used by a number of authors (e.g., Fukanaga and Vanderplaats, 1991, Fukunaga and Sekine, 1992, Grenestedt and Gudmundson, 1993, Miki and Sugiyama, 1993).

For a full model of a laminated plate taking bending, membrane, and bending-membrane coupling into account twelve lamination parameters are needed (an algebraic characterization of this set is still an open problem). In Hammer et al, 1996, it is noted that the minimum compliance problem even for the general case only require a characterization of the set of membrane related lamination parameters (cf., (5)). By noticing that these parameters also enter in the formulas for the effective properties of layered materials, one can for laminates employ results from the field of optimal bounds on the characterization of the range of the lamination parameters (Avellaneda and Milton, 1989) as well as on how to realize given lamination parameters by an actual lay-up (Lipton, 1994b), linking design of laminates and generalized shape design in an intricate way.

7.3. Material design

The application of the full range of material tensors in optimum design as analyzed in Bendsøe et al, 1994, naturally lead to the work by Sigmund, 1994, 1995, on using topology design methods for solving the inverse problem of generating micro structures with prescribed properties. The computational work of Sigmund answered by the affirmative the question if any material tensor can be generated, a result also confirmed by the analytical work of Milton and Cherkaev, 1995 (see also Lakes, 1993, and Haslinger and Dvorak, 1995). This gives further mechanical relevance to the generalized material design problem. The discovery has furthermore (and perhaps more importantly) lead to a methodology for designing micro structures for which a range of macroscopic properties can be specified, see e.g., also Sigmund and Torquato, 1996. For laminates, the inverse problem for lamination parameters can be given analytical form (cf. remarks above), but also here a greater range of specified properties will require computational work (see Autio, Laitinen, and Pramila, 1992, for an example of this).

7.4. Extensions

It is evident that design of and with advanced materials require further mechanical considerations than those treated in this tutorial presentation. Examples are to consider generalized shape design and laminate design under strength constraints and an plastic deformation (see for example Hammer, 1995, Swan and Arora, 1996, Yuge and Kikuchi, 1995). On the level of the micro structure, optimization for resistance to micro buckling (Bendsøe and Triantafyllidis, 1990), to wear (Thomsen and Karihaloo, 1995) and to fracture (Wang and Karihaloo, 1995) are examples of extending the scope of the use of optimization techniques. Further examples of optimal structural design with and of advanced materials can be found in the conference proceedings by Pedersen, 1993, and by Olhoff and Rozvany, 1995.

8. CLOSURE

The purpose of this overview is to illustrate how design of structure, design with advanced materials and design of the materials themselves are closely interconnected subjects. Studies in this realm provides considerable insight into the form of structural efficiency, and this has lead to new ways in which we employ structural optimization techniques.

Acknowledgements

The author would like to thank co-authors and colleagues in Denmark and abroad for many fruitful discussions on the subject of this paper. Also, the generation of figures 3 and 4 by Ms. Velaja Hammer and Dr. Ole Sigmund, respectively, is gratefully acknowledged.

REFERENCES

Allaire G; Kohn RV (1993): Optimal Design for Minimum Weight and Compliance in Plane Stress using Extremal Microstructures. European J. Mech. A, 12, 839-878.

Ashby MF (1991): Materials and Shape. Acta Metall. Mater., 39, 1025-1039.

Avellaneda M; Milton GW (1989): Bounds on the Effective Elasticity Tensor Composites based on Two-point Correlation. In Hui D; Kozik TJ (eds.), Composite Material Technology, ASME, 89-93.

Autio M; Laitinen M; Pramila A (1992): Systematic Creation of Composite Structures with Prescribed Thermomechanical Properties. Comp. Eng., 3, 249-259.

Bendsøe MP (1995): Optimization of Structural Topology, Shape and Material. Springer Verlag, Berlin Heidelberg, 1995.

Bendsøe MP; Díaz AR; Kikuchi N (1993): Topology and Generalized Layout Optimization of Elastic Structures. In Bendsøe MP; Mota Soares CA (eds.), Topology Optimization of Structures, Kluwer Academic Press, Dordrecht, 159-206.

Bendsøe MP; Guedes JM; Haber RB; Pedersen P; Taylor JE (1994): An Analytical Model to Predict Optimal Material Properties in the Context of Optimal Structural Design. J. Applied Mech., 61, 930-937.

Bendsøe MP; Kikuchi N (1988): Generating Optimal Topologies in Structural Design Using a Homogenization Method. Comput. Meths. Appl. Mechs. Engrg. 71, 197-224.

Bendsøe MP; Triantafyllidis N (1990): Scale Effects in the Optimal Design of a Microstructured Medium against Buckling. Int. J. Solids Struct., 26, 725-741.

Cheng G; Olhoff N (1981): An Investigation Concerning Optimal Design of Solid Elastic Plates. Int. J. Solids Struct., 16, 305-323.

Cherkaev A; Kohn RV (1996): Topics in the Mathematical Modeling of Composite Materials. Birkhauser, in press.

Díaz AR; Lipton R; Soto CA (1995): A New Formulation of the Problem of Optimum Reinforcement of Reissner-Mindlin Plates. Comp. Meth. Appl. Mechs. Engrg., 123, 121-139.

Díaz A; Sigmund O (1995): Checkerboard Patterns in Layout Optimization. Structural Optimization, 10, 40-45.

Duysinx P; Zhang WH; Fleury C; Nguyen VH; Haubruge S (1995): A New Seperable Approximation Scheme for Topological Problems and Optimization Problems Characterized by a Large Number of Design Variables. loc.cit. Olhoff and Rozvany, 1995, 1-8.

Fang C; Springer GS (1993): Design of Composite Laminates by Monte Carlo Method. J. Comp. Mat., 27, 721-753.

Fukanaga H; Vanderplaats GN (1991): Stiffness Optimization of Orthotropic Laminated Composites Using Lamination Parameters. AIAA J, 29, 641-646.

Fukunaga H; Sekine H (1992): Stiffness Design Method of Symmetric Laminates Using Lamination Parameters. AIAA J., 30, 2791-2793.

Gibiansky LV; Cherkaev AV (1987): Microstructures of Composites of Extremal Rigidity and Exact Estimates of the Associated Energy Density. Edited translation, loc.cit. Cherkaev and Kohn, 1996.

Grabovsky Y; Kohn RV (1995): Microstructures Minimizing the Energy of a Two Phase Elastic Composite in Two Space Dimensions. J. Mech. Phys. Solids., 43, 933-947 (I: The Confocal Ellipse Construction), 949-972 (II: The Vigdergauz Microstructure).

Grenestedt JL; Gudmundson P (1993): Layup Optimization of Composite Material Structures loc. cit. Pedersen, 1993c, 311-336.

Haber RB; Bendsøe MP; Jog CS (1996): A new Approach to Variable-Topology Shape Design Using a Constraint on the Perimeter. Structural Optimization, 11, 1-12.

Haftka RT; Gürdal Z; Kamat MP (1992): Elements of Structural Optimization. 3rd edition, Kluwer Academic Publishers, Dordrecht.

Hammer VB (1995): Energy and Strenght Optimization of Laminates. loc.cit. Olhoff and Rozvany, 1995, 287-292.

Hammer VB; Bendsøe MP; Lipton R; Pedersen P (1996): Parametrization in Laminate Design for Optimal Compliance. Int. J. Solids Struct., to appear.

Haslinger J; Dvorak J (1995): Optimum Composite Material Design. Math. Modelling Num. Analysis, 29, 657-686.

Jog CS; Haber RB; Bendsøe MP (1994): Topology Design with Optimized, Self-Adaptive Materials. Int. J. Num. Meth. Engng., 37, 1323-1350.

Jog CS; Haber RB (1996): Stability of Finite Element Models for Distributed-Parameter Optimization and Topology Design. Comput. Meth. Appl. Mech. Engng., to appear.

Lakes R (1993): Materials with Structural Hierarchy. Nature, 361, 511-515.

Lipton R (1994a): On the Relaxation for Optimal Structural Compliance. JOTA, 81, 549-568.

Lipton R (1994b): On Optimal Reinforcement of Plates and Choice of Design Parameters. Control and Cybernetics, 23, 481- 493.

Lurie KA; Cherkaev AV (1986): The Effective Properties of Composites and Problems of Optimal Design of Constructions. Uspekhi Mekhaniki, 9, 1-81 (in Russian). Edited translation, loc.cit. Cherkaev and Kohn, 1996.

Lurie KA; Cherkaev AV; Fedorov AV (1982): Regularization of Optimal Design Problems for Bars and Plates. JOTA, 37, 499-522 (Part I), 37, 523-543 (Part II) and 42, 247-282 (Part III).

Miki M (1982): Material Design of Composite Laminates with Required In-Plane Elastic Properties. In Progress in Science and Engineering of Composites (Hayashi T; Kawata K; Umekawa S, eds), ICCM-IV, Tokyo, 1725-1731.

Miki M; Sugiyama Y (1993): Optimum Design of Laminated Composite Plates Using Lamination Parameters. AIAA J, 31, 921-922.

Milton GW; Cherkaev AV (1995): Which Elasticity Tensors are Realizable?. J. Engng. Mat. Techn., 117, 483-493.

Murat F; Tartar L (1985): Optimality Conditions and Homogenization. In Marino A et al (eds.), Nonlinear Variational Problems, Pitman, Boston, 1-8.

Olhoff N; Rozvany GIN, eds. (1995): WCSMO-1 - Proc. First World Congress of Structural and Multidisciplinary Optimization. Pergamon Press, Oxford.

Pedersen P (1989): On Optimal Orientation of Orthotropic Materials. Struct. Optim., 1, 101-106.

Pedersen P (ed.) (1993): Optimal Design with Advanced Materials. Elsevier, Amsterdam.

Rossow MP; Taylor JE (1973): A Finite Element Method for the Optimal Design of Variable Thickness Sheets. AIAA J. , 11, 1566-1569.

Sigmund O (1994): Materials with Prescribed Constitutive Parameters: An Inverse Homogenization Problem. Int. J. Solids Struct., 31, 2313-2329.

Sigmund O (1995): Tailoring Materials with Prescribed Elastic Properties. Mech. Mat., 20, 351-368.

Sigmund O; Torquato S (1996): Design of Materials with Extreme Thermal Expansion using a Three-Phase Topology Optimization Method. DCAMM Report No. 525, Danish Center for Applied Math. and Mech., Techn. Univ. of Denmark, DK.-2800 Lyngby, Denmark.

Swan, CC; Arora JS (1996): Topology Design of Material Layout in Structured Composites of High Stiffness and Strength. Preprint. Presented at 1995 NIST Workshop on Optimal Design for Materials and Structures, University of Utah, Salt Lake City, August, 1995.

Thomsen J (1991): Optimization of Composite Discs. Struct. Optim., 3, 89-98.

Thomsen NB; Karihaloo BL (1995): Optimum Microstrcuture of Transformation-Toughened Ceramics for Enhanced Wear Performance. J. Am. Ceram. Soc., 78, 3-8.

Wang J; Karihaloo BL (1995): Fracture Mechanics and Optimization - a Useful Tool for Fibre-Reinforced Composite Design. Composite Structures, 32, 453-466.

Yuge K; Kikuchi N (1995): Optimization of a Frame Structure Subjected to a Plastic Deformation. Structural Optimization, 10, 197-208.

Theoretical and Applied Mechanics 1996
T. Tatsumi, E. Watanabe and T. Kambe (Editors)
© 1997 Elsevier Science B.V. All rights reserved.

Minimization of Sound Transfer by Structural Optimization

L. Gaul[a], P. Kohmann[b], D. Wittekind[c]

[a]Institute A of Mechanics, University of Stuttgart, 70550 Stuttgart, Germany

[b]Dept. of Mech. Engg., University (FH) of Pforzheim, 75175 Pforzheim, Germany

[c]Thyssen Nordseewerke GmbH, Am Zungenkai, 26725 Emden, Germany

ABSTRACT

The main sound transfer paths between engines and supporting structures in general consist of structure borne sound through seatings, shaft coupling and exhaust system, coupled liquid and structure borne sound in flexible bellows, pipes and the airborne sound. The paper reports on measures for minimizing structure borne noise transmission from a diesel generator to the hull of a ship as well as from flanking paths of sound.

Resilient mounting systems are a common feature to minimize structure borne sound. For extreme isolation requirements double stage mounting systems are employed which consist of two levels of flexible mounts with an intermediate foundation. Losses of isolation efficiency are caused by resonances of the mounting system.

It is shown that design changes backed by experimental modal analysis and hybrid multibody calculations allow optimization of stiffness, inertial and damping force distribution in a way that isolation losses are avoided and the best theoretical performance of an associated idealized system can be achieved in the frequency range of interest.

Flanking paths of sound are generated by noise sources such as pumps and valves and transmitted in flexible bellows and pipes. Noise reduction in fluid filled pipes is improved by optimization of stiffness and damping properties in a new muffler design.

Experimental studies validate model simulation of the double stage mounting systems and muffler performance.

NOMENCLATURE

A	cross section of fluid	u_r	radial deformation of rubber rib
D_d	transmission loss	v	velocity
E_R	Youngs modulus of rubber material	V_0	original fluid volume
f	frequency	ΔV	change of fluid volume
j	$\sqrt{-1}$	Y_{il}	transfer mobility
K_f	bulk modulus of fluid	Z_∞	impedance of an infinitely long fluid column

286

K_{sub}	bulk modulus of substitute fluid	ζ_k	damping ratio of mode k
$\overline{K}$	bulk modulus of substitute fluid	η	loss factor
l	fluid column length in muffler	ν_R	Poissons ratio of rubber
m	mass of fluid column	ρ_f	density of fluid
p	sound pressure	ω	angular frequency
u_{ik}	displacement at location i of the mass normalized mode k	ω_{0k}	natural frequency of mode k

1. INTRODUCTION

Understanding of sound transfer between engines and supporting structures requires to analyze a number of paths. This is illustrated in *Figure 1* with the main paths from a resiliently mounted propulsion diesel engine to the adjacent ship structure [1].

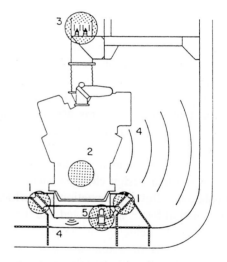

Structureborne sound paths via:
1. resilient support mounting, seatings,
2. flexible shaft coupling,
3. exhaust system, resilient mountings.

Airborne sound path:
4. air around the engine.

Coupled fluid/structureborne sound path:
5. flexible bellows, pipes.

Figure 1: Main sound transfer paths between a resiliently mounted engine and adjacent ship structure.

In the present paper the path types no. 1 and 5 are studied, i.e. the path via the resilient mounting system and the path via pipes.

Resilient mounting systems are a common feature on naval ships, research vessels and even merchant ships where the requirement for high comfort goes along with the contradictory requirement for light weight structures to maximize payload.

Double resilient mounting systems are employed when there is a demand for high structure borne noise attenuation. In addition to one elastic level of the single stage system the double resilient system contains another elastic level and an intermediate foundation between the two. The dynamic transfer behaviour of the intermediate foundation and the resilient mounts reduce the sound isolation performance when compared to an associated system with a rigid intermediate foundation and massless springs. Calculations based on a combined Multibody and Finite Element Approach (Hybrid MBS) show, that the intermediate foundation response significantly depends on the mass and stiffness distribution.

Experimental Modal Analysis (EMA) and HMBS calculations lead to an optimized design of the intermediate foundation. It offers advantages with respect to design compactness, tuning and application of damping material after installation and better isolation performance than conventional intermediate foundations with maximized stiffness [3, 4, 5, 6, 7].

After optimizing the sound spreading path of the resilient mounting system so called flanking paths become important. Particulary in shipborne applications amongst the most difficult one is the piping system attached to pumps.

For example centrifugal pumps are quite compact but deliver high water capacities. Consequently the piping system attached to in- and outlet is large compared to the pumps size and weight. Sources of hydraulical and structure borne noise such as pumps and valves generate sound predominately in the frequency range below 2 kHz. Common flexible hose inserts, so-called compensators, attached to the rigid or resilient piping system fail to reduce this kind of noise efficiently. This is why a new muffler is developped in the second part of the paper by stiffness and damping optimization. It combines the work principle of a passive reflexion absorber by a significant flexibility softening discontinuity with the work principle of a dissipating absorber [2], *Figure 2*.

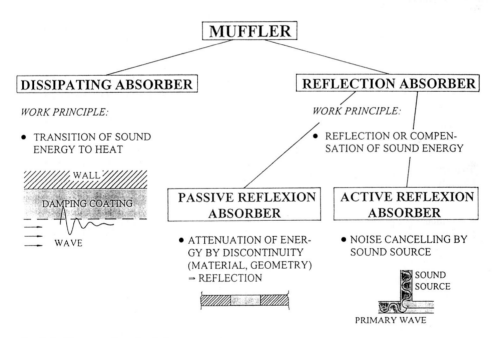

Figure 2: Muffler work principles.

2. TWO DOF MODEL OF DOUBLE STAGE MOUNTING SYSTEM

For comparing the force isolation gain of a double stage versus a single stage mounting system a two-degree-of-freedom-system has been chosen, *Figure 3*. Several transfer functions of the system are shown.

288

Theoretically the velocity level v_3 at the foundation, e.g. ship deck, can be predicted from the measured source level v_0 at the engine when multiplied with the calculated transfer impedance F_3 / v_0 and the measured foundation mobilty v_3 / F_3 according to *Figure 3*.

If the behaviour of the whole system is considered and compared to reality it must be taken into account that accelerations can be measured quite easily but forces cannot. What we are interested in is the true reduction of the force transmitted from the source to the foundation. The transmission loss of force amplitudes for the double stage 2 DOF mounting system $20 \log |F_0 / F_3|$ is compared with the transmission loss of an associated ($c_1 \to \infty$, $m_1 \to 0$) single stage mounting system $20 \log |F_0 / F_1|$. The gain of force transmission loss of the 2 DOF system is 24 dB/octave slope compared to 12 dB/octave slope of the SDOF system above the rigid body resonances.

However, neither the internal forces nor the force on the foundation can be measured very easily in practical systems. Besides, the three dimensional system has 12 rigid body DOF. To ensure optimum performance, all natural frequencies of the rigid body modes have to be lower than the second natural frequency in the vertical direction. The latter is determined by the weight of the masses and the spring stiffness and cannot be influenced beyond that. All others are adjustable by spring arrangement and by variations of the mass moments of inertia of the intermediate mass.

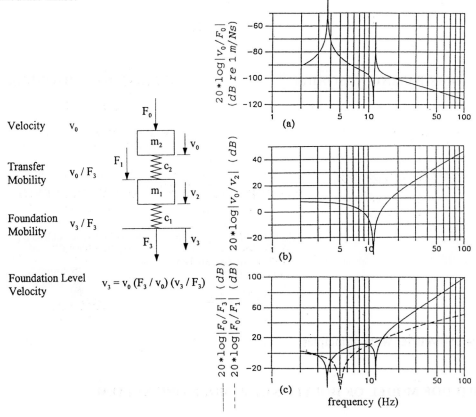

Figure 3: Two and one degree of freedom system: (a) Mobility of machine, (b) Attenuation of upper stage, (c) Attenuation of forces (— 2 DOF-system, --- 1 DOF-system).

All transfer functions taking account of the rigid body modes only have in common that, in the representation of *Figure 3* (level versus logarithm of frequency), they will merge into straight lines with sufficient distance above the highest rigid body mode. If the real system does not show this characteristic behaviour, this must be due to nonlinearities in the spring or other non-ideal performance of system parts.

One of the main reasons that ideal system behaviour is not observed in real installations is the intermediate mass not being rigid. This structure is usually designed as a steel frame. Special designs feature e.g. polymere concrete for the material. The influence of the springs on the dynamic behaviour of such a mass has been reported on in [8].

All designs have in common that their lowest-frequency mode shapes are bending modes. In the vicinity of the respective natural frequency, structureborne noise attenuation will be deteriorated. The questions arising are:

- what determines the degree of deterioration?
- how can negative effects be avoided?
- can attenuation be maximized over sufficiently large frequency ranges?

These questions have to be answered keeping in mind that no information about phase relationships of engine support motions are available.

2.1 EFFECT OF RESONANCES

The intermediate mass can be visualized as a flexible structure with defined points of excitation i (the attachment points of the springs of the upper level) and defined points of response l (the attachments of the springs of the lower level). Keeping the attenuation high means maximizing the transfer mobility between these points:

$$Y_{il}(\omega) = \sum_{k=1}^{n} \frac{j\,\omega\,u_{ik}u_{lk}}{\omega^2_{0k} - \omega^2 + 2j\,\omega\,\omega_{0k}\zeta_k}. \tag{1}$$

The value of this function at a given frequency is determined by the product of the modal components of the excitation and the response points.

If the weight of the intermediate mass is fixed only the stiffness/mass distribution can be varied. This optimization has to be done under severe space constraints e.g. in a submarine, where it is good practise to design the intermediate mass frame into a given space. By combining a rigid body model of the engine on upper foundation with a beam finite element model of the flexible intermediate foundation the principal behaviour of the system under dynamic load is investigated.

Figure 4 shows the attenuation of forces for a flexible intermediate foundation with homogeneous mass density compared to the attenuation of an associated system with rigid intermediate foundation of same mass. Despite the dips of attenuation at the bending resonances an averaged loss of 15 dB can be seen in between. This result can be derived analytically for the simple model of one finite beam element and consistent mass matrix [3]. The intermediate mass displacements at the spring attachment points have been calculated for rigid and flexible intermediate mass.

Their ratio is approximately $u(rigid)/u(flexible) = 1 - 0.83\,\mu\,l/(\mu\,l + 2\,m_z)$. It means that if no lumped masses are attached $m_z = 0$ only 17 % of the intermediate mass is acting. This causes the above mentioned loss of attenuation $20 \log \dfrac{1}{1 - 0.83} = 15.4\ dB$. Different stiffness/mass distributions significantly influence the attenuation of the system with equal weight.

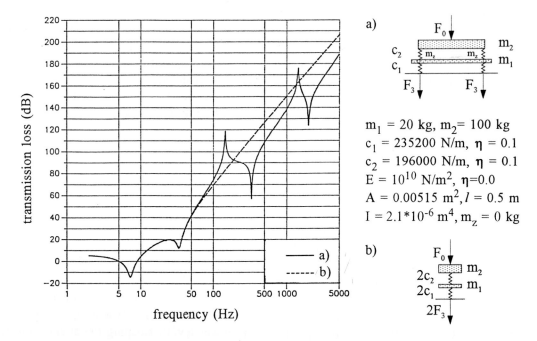

$m_1 = 20$ kg, $m_2 = 100$ kg
$c_1 = 235200$ N/m, $\eta = 0.1$
$c_2 = 196000$ N/m, $\eta = 0.1$
$E = 10^{10}$ N/m^2, $\eta = 0.0$
$A = 0.00515$ m^2, $l = 0.5$ m
$I = 2.1*10^{-6}$ m^4, $m_z = 0$ kg

Figure 4: Force attenuation of two systems with an intermediate mass of equal weight but different stiffness/mass distribution. Representation as in *Figure 3c*.

While resonances of the intermediate mass structure can not be avoided, the overall attenuation is affected differently. Damping of the intermediate mass would only influence the direct vicinity of the resonance. The absolute damping values of such systems are almost completely determined by the resilient elements - e.g. rubber springs - and other attachments [8, 9, 10].

2.2 OPTIMIZATION

It is obvious that three conditions have to be met if an eigenmode of the intermediate mass causes a reduction in attenuation:

a) the exciting frequency is close to a natural frequency of the intermediate mass
b) neither all excitation nor all response points are located at modal nodes
c) the excitation is at least partly in phase with the respective components of the mode.

The condition a) is necessary but not sufficient. From b) follows that attenuation is kept high if it can be ensured that the exciting points and/or the response points are located on or close to nodes and c) indicates a strong influence of the phase properties of the excitation.
Most emphasis for optimization will be put on b), because it can be calculated with good accuracy with the finite element method. The natural frequencies can be calculated with limited accuracy only. Approximate values are sufficient. A shift of a resonance is possible with simple structural modification, e. g. adding a mass at a mode maximum. It must only be

ensured that this potential is larger than the inaccuracy of the calculation. Because phase relationships of the source's supports are generally not known, c) cannot be adopted. In our calculations, we assume the machine to be rigid which is the worst case in most (not in all) applications.

A posible optimization philosophy is the variation of the stiffness/mass distribution. It leads to the attenuation depicted in *Figure 5* by several steps. Mass concentration at the attachment points of the lower level springs avoids the broad banded attenuation loss of *Figure 4*.

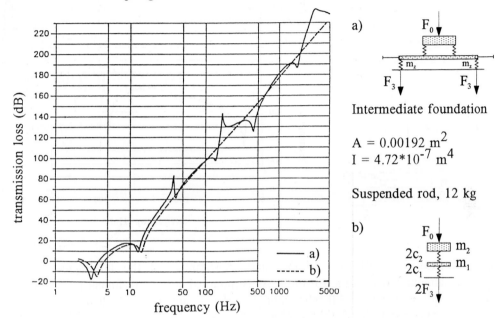

Figure 5: Force attenuation for lumped masses at lower level spring attachment points: (a) flexible intermediate foundation, (b) rigid intermediate foundation.

The drawback of the lower frequency of the first flexural mode can be circumvented by shifting the modes to the attachment points of the upper level springs. Instead of extending the size of the intermediate mass significantly this can be achieved by creating rotary inertia on properly damped suspensions. This optimized system comes close to the attenuation performance of the ideal rigid intermediate mass.

Figure 6 shows a structure which allowed variations of mass concentration locations. The representation of attenuation is according to *Figure 3b* indicated by the dotted line. It can clearly be seen that only one configuration yields good results in the frequency range shown. After verification by modal analysis and further finite element calculations the following rules were found for proper design of the intermediate mass:

- mass concentrations must be located at the edges of the structure
- almost all modes will show a node in the vicinity of the mass concentrations, consequently, the springs of the upper elastic level or the springs of the lower elastic level have to be mounted in the direct vicinity of the concentrations. At best, both are located close to the nodes
- the lowest vibration mode will be even lower in frequency but can easily be arranged in the minima of the excitation.

292

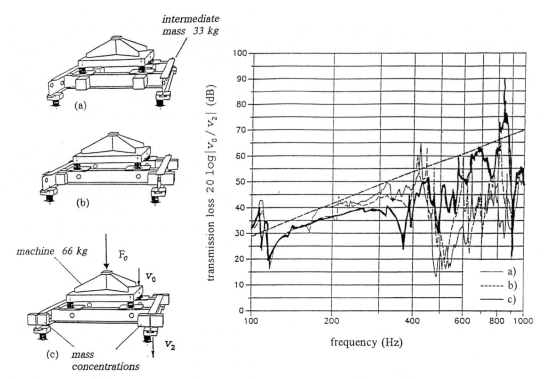

Figure 6: Measured influence of the mass distribution of the intermediate mass. The attenuation is highest if the mass is mounted at the attachment points of the lower springs (line c). All other mounting points are worse. Dotted smooth line is the ideal system according to *Figure 3b*.

2.3 FULL SIZE TESTS

The structure of *Figure 6* will not be applicable due to its bulky appearance in a system of true size. Further development of the principles lead to an even simpler structure shown in full scale in *Figure 7*. It consists of two transverse beams which serve as mass concentrations in total up to their first natural frequency in bending at around 600 Hz. The lower springs are attached to the ends of these beams. The machine is supported by the much weaker longitudinal beam. Although not necessary, damping has been applied to the logitudinal girder by glueing 4 hollow square shaped beams together with a damping material, which, however, was not optimized for this particular application. The 4 beams are welded to a flange which is screwed to the transverse beam. The total weight of the structure is 500 kg. It serves as an intermediate mass for a 1500 kg motor mounting.

To enable separation of intermediate mass and spring resonances the dynamic stiffness of the springs was measured on a test bed. The springs selected were stiffer than usually employed for such systems to ensure that structureborne noise transmission will prevail over airborne noise directly exciting the intermediate mass. For further comparison another intermediate mass with equal weight but designed to the conventional criteria was constructed. Although its stiffness was maximized the lowest bending mode was at 258 Hz, while that of the optimized mass was about 100 Hz. The exciting frequencies considered are 25 Hz harmonics. *Figure 8* compares the ideal system and real systems. The representation is as in *Figure 3b* with a linear frequency scale.

Figure 7: Optimized full size intermediate mass.

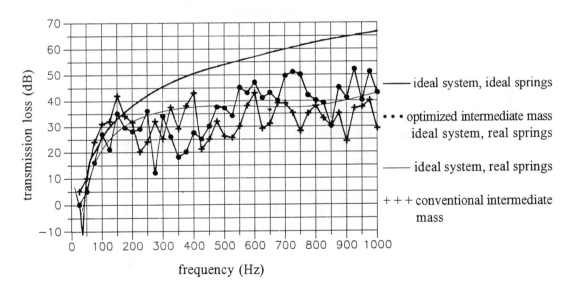

Figure 8: Attenuation of different intermediate mass configurations.

The classical intermediate mass is slightly better at low frequencies but much worse at higher. The optimized mass shows an attenuation minimum at 275 Hz which is due to a longitudinal resonance. Above 500 Hz the intermediate mass is even better than could be expected theoretically. Here, the machine is not rigid any more as assumed for the ideal system. The motion

294

of the supports will be uncorrelated and the transverse beam will be fully effective as a mass concentration.

3. MUFFLER WORK PRINCIPLES

Mufflers are a well-known measure to reduce noise in fluid filled pipes. Especially for naval ships and research vessels there are strong requirements for good noise control so as to be able to operate the ships in very quiet conditions [2, 3, 11]. Centrifugal pumps and valves generate predominately sound in the lower frequency range below 2 kHz. Since the wave length in water is comparatively long, noise control by sound dissipation is relatively inefficient. For this purpose, mufflers which insulate sound are normally used.

Figure 9 shows the transmission loss D_d versus frequency f for a representative selection of such mufflers. The transmission loss D_d is defined using the ratio of the sound pressure p_1 in front of the muffler to the sound pressure p_2 behind the muffler as follows,

$$D_d = 20 \log \left(\frac{p_1}{p_2} \right) . \tag{2}$$

As shown in *Figure 9*, the available mufflers either provide a limited transmission loss (*Figure 9a* and *Figure 9b*) or work in a small frequency range only (*Figure 9c* and *Figure 9d*) [12, 13]. In order to suppress noise propagation in the lower frequency range efficiently, it is therefore necessary to develop a new muffler. In addition to good transmisson loss, there are other important design requirements for the new muffler.

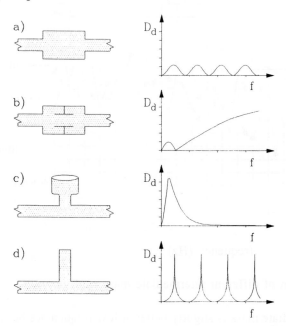

Figure 9: Transmission loss of commonly used mufflers. (a) single chamber muffler, b) double chamber muffler, c) rubber expansion joint, d) side-cut resonator).

Since the system pressure normally depends on the operating conditions of many different system parts, it is necessary that the effectiveness of the muffler be independent of system pressure. Because of severe space constraints on ships, the new muffler has to be compact as well. Finally, to facilitate retrospective installation on existing systems, and for good system efficiency generally, the flow resistance of the muffler must be as low as possible. In the paper at hand, a new muffler is optimized, which fulfils these requirements.

3.1 NEW MUFFLER DESIGN

Figure 10 shows the construction drawing of the new muffler. The muffler consists of a ribbed rubber hose with grooves in the circumferential direction. This rubber hose is vulcanized to steel flanges at both ends and embedded in a cylindrical steel pipe.

The acoustic behaviour of the muffler can be explained as follows. As a result of the elasticity of the ribbed rubber hose, the fluid column in the middle part of the muffler seems to be softer than the fluid columns in front or behind. This impedance jump generates reflections which lead to the desired attenuation. The effectiveness of the muffler is further improved by sound dissipation within the rubber material.

In order to calculate the transmission loss of the new muffler, an appropriate model has to be designed. For this purpose the fluid in the muffler can be modelled by three one-dimensional fluid columns, *Figure 11*.

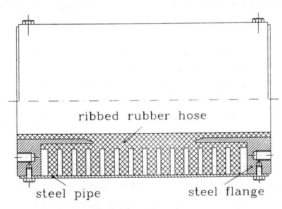

Figure 10: The new optimized muffler.

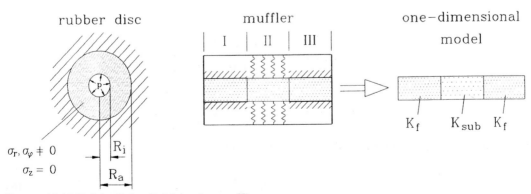

Figure 11: Model of the fluid in the muffler.

In the front and back parts of the muffler (*Figure 11*, part I and III) the two steel flanges prevent radial deformation of the rubber hose. These approximately rigid boundary conditions allow us to use the bulk modulus K_f of the fluid.

Only in the middle part of the muffler is the elasticity of the fluid affected by the softness of the ribbed rubber hose. In the model, this phenomenon can be taken into account by using a substitute fluid instead of the real fluid. The bulk modulus K_{sub} of this intended substitute fluid will be choosen so that the elastic properties of the one-dimensional fluid column will be equal to the elasticity of the real fluid in the rubber hose.

In order to account for sound dissipation in the rubber hose, this external damping effect can be modelled by material damping in the fluid. A common way to model this damping effect [14] is to use a complex modulus

$$\overline{K} = K_{sub} (1 + j\eta) . \tag{3}$$

In our case, the loss factor η is no longer the real material constant of the fluid. For this purpose the loss factor of the rubber material is used.

In order to get an idea of the effectiveness of the muffler, it is necessary to calculate the still unknown bulk modulus K_{sub}. This can be done by exploring the state of deformation of the rubber hose. Before we can start with an analytical investigation of the bulk modulus K_{sub} of the substitute fluid in the middle part of the muffler, it is necessary to make some simplifying assumptions as follows:

- the rubber hose in the middle part of the muffler will be uniformly stretched in radial direction by an internal pressure load (*Figure 12*),
- the deformation of the steel pipe which surrounds the rubber hose can be neglected (i.e. perfectly stiff pipe),
- the thin ribs of the rubber hose can be modelled as linear elastic discs, where the pressure load yields to plain stress (*Figure 12*),
- the inner cylindrical part of the rubber hose is perfectly stiff in radial direction, but perfectly flexible (i.e. zero stiffness) in circumferential direction,
- only the deformation of the rubber hose will be used to calculate the bulk modulus K_{sub}, the normally very low elasticity of the real fluid in the muffler will be neglected.

On the basis of these assumptions the radial deformation u_r of a rubber rib [15] is given by

$$u_r (r = R_i) = n p , \tag{4}$$

with

$$n = \frac{R_i (1 - v_R^2) \left(\left(\dfrac{R_a}{R_i} \right)^2 - 1 \right)}{E_R \left(1 + v_R + (1 - v_R) \left(\dfrac{R_a}{R_i} \right)^2 \right)} . \tag{5}$$

Using the material law of an ideal compressible fluid

$$p = - K_f \frac{\Delta V}{V_0} , \tag{6}$$

the required bulk modulus K_{sub} can be calculated

$$K_{sub} = \frac{(R_i + np)^3}{2R_i^2 n} \ .$$ (7)

So far, Eq.(7) allows to investigate the influence of the different construction parameters. This allows us to calculate the transmission loss and optimize the muffler design using the above model. However first we need to consider the behaviour of the muffler in a piping system. This will be presented in the next section.

3.2 A MODEL FOR TRANSMISSION LOSS PREDICTION

To get an idea of the effectiveness of the muffler in a piping system we will use the system shown in *Figure 12*. Normally, a muffler is installed nearby the noise source, in our case represented by a pump. The advantage of such a configuration is, that by reducing fluidborne noise directly at the source, minimal structureborne noise will be generated, and so optimal noise reduction can be obtained for the whole system.

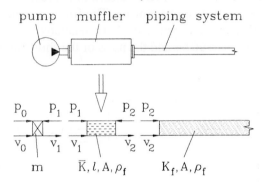

Figure 12: Model of muffler and piping system.

The following calculations are based on the one-dimensional model of this system, shown in the lower half of *Figure 12*.

The pump is modelled by a harmonic pressure p_0 and velocity v_0 simulation. The behaviour of the short fluid column between the pump and the acoustically active middle part of the muffler is represented by its mass. For the middle part of the muffler, a one-dimensional fluid column will be used. For this fluid a complex bulk modulus $\overline{K}$ (Eq.3) is taken. It is determined by the already known bulk modulus K_{sub} (Eq.7). The influence of the piping system downstream of the middle part of the muffler can be taken into account by an additional fluid - column. In this case we will assume an infinitely long fluid column, its input impedance Z_∞ is given by

$$Z_\infty = \frac{p_2}{v_2} = \sqrt{K_f \varrho_f} \ .$$ (8)

What we are interested in, finally, is the transmission loss of the muffler. This can be determined by calculating the sound pressure p_2 behind the muffler generated by a pressure stimulation p_0.

An efficient way of carrying out these calculations is the well known transmission matrix method. The idea of this method is to couple the transmission matrices of the system parts to a global transmission matrix for the whole system. In our case this global transmission matrix links the input stimulation p_0 and v_0 to the output pressure p_2 and velocity v_2. After taking Eq.8 into account, the desired transmission loss D_d can be calculated as follows

$$D_d = 20 \log \left| \cos \overline{\lambda}_d - \frac{\omega^2 m l}{\overline{K} A \overline{\lambda}_d} \sin \overline{\lambda}_d + j \frac{\dfrac{\omega m}{A} \cos \overline{\lambda}_d + \dfrac{\overline{K} \overline{\lambda}_d}{\omega l} \sin \overline{\lambda}_d}{Z_\infty} \right| ,$$

$$\overline{\lambda}_d = \omega l \sqrt{\frac{\varrho_f}{\overline{K}}} .$$

(9)

Figure 13 shows the transmission loss for the muffler calculated by Eq.9 versus stimulation frequency. In the undamped case ($\eta = 0$) resonance minima lead to discrete dips. These dips are caused by two different resonance phenomena. The first dip appears at 45 Hz. This is the eigenfrequency of the spring-mass system, which is generated by the stiffness of the fluid column in the middle part of the muffler and the mass of the short fluid column between muffler and pump.

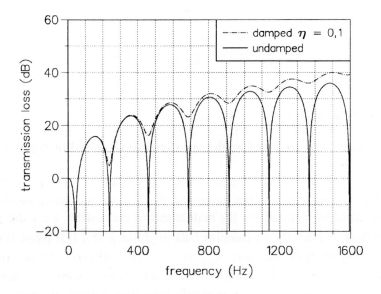

Figure 13: Calculated transmission loss of the muffler.

The following dips are caused by a second resonance phenomenon. If the excitation frequency is equal to the eigenfrequency of the fluid column in the middle part of the muffler, the sound pressure behind the muffler will become higher than the stimulation pressure and the transmission loss becomes negative. The frequency difference between two neighboring dips is always the same and is dependent on the construction parameters of the muffler.

A better transmission loss is predicted, if sound dissipation by the rubber hose is taken into account, *Figure 13*. As we can see, the dips at the eigenfrequencies of the fluid column in the muffler are no longer as deep as in the undamped case. Also the damping effect improves the effectiveness of the muffler in the upper frequency range. In order to determine the quality of this analytical approach, an experimental validation was carried out as presented in the next section.

3.3 EXPERIMENTAL VALIDATION

The experimental set-up used to measure the transmission loss of the new muffler is illustrated in *Figure 14*.

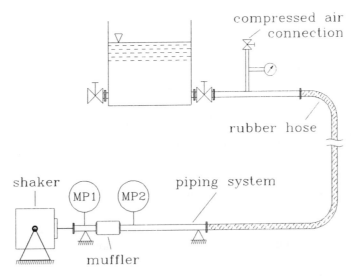

Figure 14: Experimental set-up to measure the transmission loss of the new muffler.

The sound was generated by a shaker connected to a thin membrane. This membrane was bolted to the face of a pipe in order to generate fluidborne noise. This noise then passed down a water filled piping system into which the muffler was integrated. In order to get a proper working section without reflections, a 15 meter long rubber hose was used to absorb the fluidborne noise at the end of the piping system.

The noise level was measured at two points (*Figure 14*, MP1 and MP2) by quartz pressure transducers. With these values the transmission loss of the muffler can be calculated. Compressed air was also used to facilitate different static pressures at the muffler.

Figure 15 shows the measured transmission loss of the new muffler at a system pressure of 2 bar. For the purpose of comparison, the second curve in this figure represents the previously calculated transmission loss (Eq.9).

We can see good overall agreement between the analytically calculated and experimentally measured transmission loss of the muffler. As expected, the transmission loss becomes higher with increasing frequency. Also, slight dips appear at the eigenfrequencies of the fluid column in the muffler. At the first of these eigenfrequencies (about 210 Hz), there is very good agreement between predicted and experimental results. At subsequent eigenfrequencies the agreement is not so good (e.g. 460 Hz predicted versus 550 Hz measured). This phenomenon can be explained by a stiffening effect of the rubber material which is also observed in context with rubber springs [10].

300

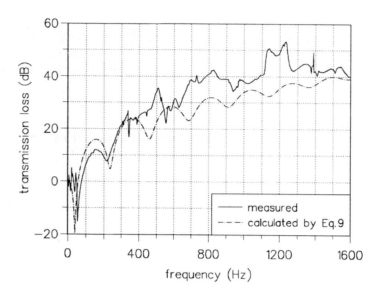

Figure 15: Comparison of the measured and calculated transmission loss of the new muffler.

As a result of the higher stiffness of the rubber material the bulk modulus of the fluid in the muffler will increase as well. Therefore the eigenfrequencies appear at higher frequencies and the distances between two neighboring eigenfrequencies will become continually larger.
The above presented analytical investigation is based on a constant bulk modulus for the substitute fluid and so does not take this phenomenon into account. Besides good transmission loss an additional aim of the new muffler design is for the transmission loss to be independent of system pressure. *Figure 16* illustrates that the new muffler fulfils this demand.

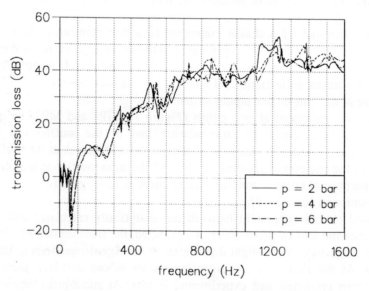

Figure 16: Measured transmission loss of the muffler for three different static system pressures.

The three curves represent the transmission loss at system pressures of p = 2 bar, 4 bar and 6 bar. Only in the lower frequency range up to 400 Hz is the transmission loss affected by system pressure. In the upper frequency range, the three curves are nearly the same.

4. CONCLUSION

The paper focusses on two examples of sound transfer reduction by structural optimization. The first takes the main path of structureborne sound between engine and supporting structure into account and the second the piping system as a flanking path. For extreme isolation requirements a double stage mounting system is employed for the engine. The application of simple analytical models combined with finite elements effectively serve as tools for improving the acoustic behaviour. For a test structure and a practical full size mounting system the theoretical optimum in structureborne noise attenuation could almost be achieved even with limited knowledge of the excitation properties.

For the attenuation of noise in a piping system a new muffler is presented. The performance of this muffler has been optimized according to the requirements of fluid filled pipes on ships. The new muffler is compact, has a low flow resistance and provides a high transmission loss largely independent of system pressure.

In order to optimize the design analytically, a mathematical model for the muffler has been introduced. This simple, one-dimensional model allows the transmission loss to be easily calculated.

The validity of the model has been demonstrated by measurements on a real muffler in a piping system, for which measured results compare very closely with those predicted by the model.

ACKNOWLEDGEMENT

Funds for this work provided by the Thyssen Nordseewerke GmbH Emden, Germany are gratefully acknowledged.

REFERENCES

[1]   Verheij, J. W.: Multipath Sound Transfer from Resiliently Mounted Shipboard Machinery, Institute of Applied Physics, TNO-TH, Delft, 1982.

[2]   Kohmann, P.; Gaul, L.: Noise Reduction in Fluid Filled Pipes on Ships by a New Muffler. Proc. 14th International Modal Analysis Conference, Dearborn, Michigan. SEM & Union College, Vol. II, 1996, pp. 1482-1487.

[3]   Wittekind, D.: Körperschalldämmung auf Schiffen durch doppelelastische Lagerungen. Thesis, University of the Federal Armed Forces Hamburg, Bericht aus dem Institut für Mechanik, 1992.

[4]   Wittekind, D.: Optimizing the Intermediate Mass of Double Resilient Mounting Systems, Proceedings of the Underwater Defence Technology Conference, London, 1992, pp. 181-186.

[5]   Wittekind, D.: Noise Reduction Measures of a Submarine Closed Cycle Diesel,

Proceedings of the International Conference on Submarine Systems, Stockholm,1992, pp. 265-271.

[6]   Wittekind, D.; Gaul, L.: Modal Analysis for Optimizing High Performance Resilient Mounting Systems on Ships. Proc. 11th International Modal Analysis Conference, Kissimmee, Florida, SEM&Union College, Vol.I, 1993, pp. 88-95.

[7]   Wittekind, D.; Gaul, L.: Optimization of High Performance Resilient Machinery Foundations on Ships. Kluwer Academic Publisher Dordrecht, Proc. IUTAM Symposium on Optimization of Mechanical Systems, Stuttgart, 1995, pp. 341-348.

[8]   Geissler, P.: Modal Tests under Different Boundary Conditions for a Polymere Concrete Machinery Foundation, 5th International Modal Analysis Conference, London, Vol. II, 1987, pp.1122-1128.

[9]   Gaul, L.; Bohlen, S.: Identification of Nonlinear Joint Models and Implementation in Discretized Structure Models, ASME New York, The Role of Damping in Vibration and Noise Control, 1987, pp. 213-219.

[10]   Gaul, L.; Chen, C.M.: Modeling of Viscoelastic Elastomer Mounts in Multibody Systems. Schiehlen, W. (ed.) Advanced Multibody System Dynamics, Kluwer Academic Publishers, Dordrecht,1993, pp. 257-276.

[11]   Kohmann, P.: Ein Beitrag zur Lärmminderung bei flüssigkeitsgefüllten Rohrleitungen auf Schiffen. Bericht aus dem Institut A für Mechanik der Universität Stuttgart, 1995.

[12]   Hoffmann, D.: Die Dämpfung von Flüssigkeitsschwingungen in Ölhydraulik-leitungen. VDI-Forschungsheft 575. Düsseldorf: VDI Verlag, 1976.

[13]   Rébel, J.: Systematische Übersicht über Dämpfungsmaßnahmen in Druckleitungen. In: Ölhydraulik und Pneumatik 20 (1976), Nr. 7.

[14]   Cremer, L.; Heckel, M.: Körperschall. Physikalische Grundlagen und technische Anwendungen. Berlin, Heidelberg, New York : Springer Verlag, 1982.

[15]   Szabó, I.: Höhere Technische Mechanik. 5. Auflage. Berlin, Heidelberg, New York: Springer Verlag, 1972.

Theoretical and Applied Mechanics 1996
T. Tatsumi, E. Watanabe and T. Kambe (Editors)
1997 Elsevier Science B.V.

OPTIMIZATION AND EXPERIMENTS - A SURVEY

Raphael T. Haftka
University of Florida
Gainesville, Florida 32611
and
Elaine P. Scott
Virginia Polytechnic Institute and State University
Blacksburg, Virginia 24061

Abstract

This paper is a review of optimization in relationship to experiments in four aspects. First, the use of optimization for designing experiments to provide the maximum amount of insight and information on the phenomena being investigated is discussed. Second, the use of experiments in the service of optimization is presented, with a particular interest in reducing the number of required experiments. Third, the use in numerical optimization of techniques developed for experimental optimization is discussed. Finally, the use of experimental validation of optimization is presented. Applications from a variety of different fields are provided for each of these aspects.

1. Introduction

Optimization can help a researcher investigating physical phenomena to design experiments that will shed maximum light on these phenomena. Conversely, experiments can be employed in the service of optimization when analytical models are not available or are not reliable enough. In that case the optimization can be conducted on the basis of experimental results, and we will call this practice experimental optimization. Because experiments are often expensive, in both of the above cases we usually try to accomplish the goal with a small number of experiments. This paper reviews both topics with special emphasis on the techniques employed to reduce the number of needed experiments.

The relation of experiments to optimization is also affected by the substantial experimental errors that are commonly present. Optimization may be used to design experiments so that in spite of the errors, they will yield the desired information. Conversely, methods of experimental optimization have to be chosen so that progress towards an optimum could be achieved in spite of the 'noise' associated with errors.

Section 2 of this paper deals with optimization in the service of experiments. Optimization can be employed to design efficient experiments, and optimization can be also employed to help extract data and physical insight from an experiment. The first category, commonly called design of experiments, is discussed in Section 2. The second category falls under the topic of system identification. We do not discuss optimization for system identification in this paper for two reasons. First, the topic has received much attention elsewhere. Second, with optimization used after the experiment, we lack the common thread, mentioned above, of the attempt to reduce the number of experiments. Additionally, in Section 2 we focus on papers that describe experiments rather than on papers that provide only analytical prescriptions on how to conduct them. This means that the large body of analytical papers on optimal excitation for system identification is only briefly touched upon.

Section 3 describes applications of experimental optimization, and Section 4 with the use in numerical optimization of techniques developed for experimental optimization., In the past few years there has been growing interest in such techniques as numerical simulations acquired some of the characteristics of laboratory or field experiments. Because our focus is on expensive experiments we deal very briefly in Sections 3 and 4 with situations where experiments are inexpensive. Process control is the archetypal case of inexpensive experiments because measurements are continually taken as part of the control. Optimization in such circumstances is usually called adaptive control, and employs methodologies which are quite different from those required when experiments are expensive.

Finally, Section 5 deals with experimental validation of experiments. There is a growing recognition of the importance of such validation, because the process of optimization is prone to take advantage of weaknesses of models, so that the modeling error at the `optimum' design may be much larger than the average modeling error in the entire design space.

2. Optimization of Experimental Design

Optimization can be used to improve the efficiency of experiments. In some cases it is used to improve the accuracy of parameters estimated from the experiments, and in others it is used to economize the experiments, either by reducing the number of experiments required or by reducing experimental costs. A third use is to design experiments that provide ideal conditions for distinguishing between competing models of physical phenomena.

The use of optimization to improve the accuracy for parameter estimation has been the focus of many studies. A survey of both qualitative and quantitative experimental design was conducted by Walter and Pronzato (1990). They define qualitative optimization as the selection of input and output parameters for the determination of the largest number of parameters. Quantitative experimental design is defined, however, as the minimization of uncertainty in the resulting parameter estimates; this is equivalent to improving the accuracy of the parameters, and therefore, it is of interest in this review.

Parameter estimation typically involves the determination of one or more parameters contained in the vector, β, which are inherent in a mathematical model of a measurable variable, η of some physical process. For example, η could be temperature and β could contain thermal conductivity and specific heat. The variable η is assumed to be a function of known independent variables, such as position and/or time. Thus, $\eta(\beta, x)$ contains values of η at specific values of the independent variables contained in x. The unknown values in β are typically found by minimizing a function which contains the sum of the squared differences between η and corresponding measured values of η at different values of x. Thus, for a simple function such as ordinary least squares, β are found by minimizing the function S_{OLS}, which is described as

$$S_{OLS} = (Y - \eta)^T (Y - \eta) \qquad (1)$$

where $\mathbf{Y}$ contains the measured values of η at different values of $\mathbf{x}$. The optimization of experimental design, therefore, involves the design of an experiment which provides the best information for the estimation of the parameters in β with the greatest accuracy.

To minimize the uncertainty of the parameter estimates, optimal experimental design theory involves first the establishment of an optimality criterion. Several optimality criteria have been proposed and are reviewed by Walter and Pronzato (1990) and also Kiefer (1974, 1975a). The optimality criteria are typically associated with the Fisher information matrix of the design (Kiefer, 1975a). The most commonly used criterion is called the D-optimal experimental design criterion, and it employs the inverse of the determinant of the Fisher information matrix. Thus, the D-optimality criterion is defined by

$$D = |\mathbf{M}|^{-1} \qquad (2)$$

where $\mathbf{M}$ is the Fisher information matrix,

$$\mathbf{M} = \mathbf{X}^{\mathrm{T}}\mathbf{X} \qquad (3)$$

and where $\mathbf{X} = \partial\eta/\partial\beta$ contains the sensitivity coefficients, which are the derivatives of the variable η, with respect to the vector β of unknown parameters. The reason for the wide use of the D-optimality criterion is that it is based on the asymptotic confidence regions for the maximum likelihood (ML) estimates. The ML parameter estimates are found by minimizing the ML sum of squares, S_{ML}, where S_{ML} can be described as

$$S_{ML} = (\mathbf{Y} - \eta)^{\mathrm{T}}\psi^{-1}(\mathbf{Y} - \eta) \qquad (4)$$

where $\mathbf{Y}$ and η are described above, and ψ is the covariance matrix of the observation errors. The confidence regions for ML estimates are ellipsoids. The effect of the D-optimality criterion is to minimize the volume of the these ellipsoids, thus also narrowing the confidence regions of the estimated parameters (Kiefer, 1981).

Proponents of the D-optimality criterion have also pointed out that this criterion is invariant under linear transformation of the estimated vector. In addition, this criterion tends to weight heavily on the parameters with the highest sensitivity (Kiefer, 1976b). For instance, in the design of experiments to estimate conductive and radiative properties of honeycomb core structures, Copenhaver (1996) pointed out that the use of the D-optimality criterion resulted in increased accuracy of the parameters with high sensitivity compared to those with low sensitivity.

Other optimality criteria include the A-, C-, E - and L-optimality criteria. These are all associated with the Fisher information matrix; the A-optimality criterion, for example, minimizes the trace of the Fisher information matrix, while the E-optimality criterion maximizes its minimum eigenvalue which acts to minimize the maximum diameter of the asymptotic confidence region ellipsoids. Pronzato and Walter (1989) defined another criterion, the V-criterion, which is similar to the D-optimality criterion, but prior lower and upper bounds are available for the noise associated with the measurements. Along the same lines, Lohmann et al. (1992) selected the minimization of the maximum scaled length of the confidence intervals as the optimal design criterion. Kalaba and Spingarn (1980) and Fredriksen (1996) also used similar techniques to demonstrate the optimization of input experimental parameters in order to maximize the sensitivity of experiments designed for the estimation of system parameters.

In the area of identification of the vibration properties of structures, there is another set of optimality criteria which are commonly used. Penny et al. (1994) reviewed some of the more popular criteria, including the Modal-Assurance criterion (Allemang and Brown, 1982) and the Effective-Independence criterion (Kammer, 1992). This latter method employs the Fisher information matrix, but it transforms it into the so-called idempotent matrix, $\mathbf{E}$,

$$\mathbf{E} = \mathbf{X}(\mathbf{X}^{\mathrm{T}}\mathbf{X})^{-1}\mathbf{X}^{\mathrm{T}} \qquad (5)$$

which has the property that its trace is equal to its rank. The Effective Independence criterion seeks to maximize the rank of $\mathbf{E}$. As the name implies, the criterion is tied to the independence of the measurements. Penny et al. (1994), however, suggest that a better measure of that independence may be the condition number of $\mathbf{X}^{\mathrm{T}}\mathbf{X}$, which may be reliably obtained by using the singular value decomposition.

Instead of using a general optimality criterion, the optimality criterion can also be customized for a specific experiment. For example, De Queiroz Orsini (1984) defined a suitable error function to optimize the resistance, inductance, and capacitance used in the experimental analysis of a resonant circuit. Also Furuya and Haftka (1993) based the optimization of actuator locations on large space structures on the set of selected members with the highest total fraction of elastic energy.

Once the optimality criterion is selected, a variety of different methods can be used to determine the optimal conditions. For example, in designing an experiment to estimate thermal properties of a material using the D-optimality criterion, the experimental parameters, such as the boundary conditions and geometry of the material (which can be adjusted by the experimentalist), are chosen to maximize the determinant shown in Eq. (2). Here, an analytical scheme could be utilized to determine the experimental parameters which satisfy the optimality criterion. This would involve the determination of the partial derivatives of the determinant of the Fisher information matrix with respect to each of the experimental parameters in question. Due to the complexity of the equations involved, this may be difficult to do analytically; in this case, numerical optimization schemes can be employed.

The simplest of the numerical optimization schemes to obtain a D-optimal design is sometimes referred to as a parametric study or an iterative method. In this case, the practical ranges of the experimental design variables are first determined. Then, the objective function is calculated for a limited number of points in this design space. These points are typically selected by incrementing each design variable uniformly, and then the design space is reduced based on these initial calculations. The revised design space is again segmented, using smaller increments, and the objective function is again determined for these points. If needed, the process is repeated to converge on an optimal design. Problems with this method are that it is often time intensive and tedious for even a very few design variables, and in addition, the process does not guarantee that the global optimum will be found.

Any number of more complex numerical schemes could also be used to maximize the optimality criterion. Recently, genetic algorithms have also been used to maximize the criterion. For example, a basic elitist genetic algorithm was used recently by Garcia and Scott (1996) in the design of experiments. Genetic algorithms, developed by Holland (1975), are based on the genetic rules and selection mechanisms of nature. Advantages of these methods are that they are easily programmed and they have shown with a high probability that the global optimum can be found (Krottmaier, 1993).

The D-optimality criterion has been applied to experimental design in many different areas. Taktak et al. (1993) used the D-optimality criterion in minimizing the area of the confidence region in the design of optimal experiments for the estimation of thermal properties of composite materials. Here, the experimental input parameters, such as sensor placement and the duration of an imposed heat flux, were optimized to provide estimates of thermal conductivity and heat capacity with the smallest confidence regions. The sensitivity coefficients used in the optimization procedure were calculated analytically using a one dimensional model for heat transfer through the composite, and the coefficients of the Fisher Information Matrix were determined by numerically integrating the $\mathbf{X}^T\mathbf{X}$ matrix over time. The optimal experimental parameters were determined using a parametric study, similar to that described previously. Moncman et al. (1995) also used the D-optimality criterion in the design of experiments used to estimate thermal properties of materials used in spacecraft structures. Here, the work by Taktak et al. was extended to include two dimensional heat flow and the estimation of both the in-plane and through the plane thermal properties. Analytical methods were used to determine the sensitivity coefficients in the optimization procedure, and again, the $\mathbf{X}^T\mathbf{X}$ matrix was numerically integrated over time. Again, due to the complexity of an analytical scheme to determine the optimal experimental input conditions, an iterative parametric approach, as described previously, was used to select the optimal design variables. In a related study, Copenhaver (1996) used a similar parametric study to optimize the duration of an imposed heat flux in the design of an experiment to simultaneously estimate three conductive and radiative properties of a honeycomb core sandwich structure. In this case, two different criteria were used in the optimization of the experiment. Optimal experimental designs were found using both the D-optimality criterion and a criterion based on scaled confidence intervals, similar to that proposed by Lohmann et al. (1992).

In applications related to system identification, optimal design-of-experiment theory has been used in the design of large space structures. Bayard et al. (1988) used the D-optimality criterion to determine the optimal sensor placement for identification experiments for large space structures, such as the Space Station. Here, the sensor placement problem was decoupled from the input design problem. Practical constraints on the number and location of the sensing instruments were implemented to define the limits of the possible input values. A gradient algorithm was employed to perform the optimization; however, due to a large number of local minima found through experimentation, a coarse grid with 100 initializing values was used to assure the global character of the solution. Large scale systems were also the focus of work by Koszalka (1987). The objective was to use the input information to simplify the design procedures. Two cases for the optimization process were considered: parallel structure, which means practically two parallel outputs or sets of measured data, and multi-stage structure, which means that the output of the first stage provides input for the second stage and results in the replacement of the global optimization problem by two sequential simpler ones. The D-optimality criterion was again used. Koszalkanotes that the result for the D-optimal design can be easily extended to A- and E- optimal designs.

In many biomedical tests, 'experiments' are often performed to provide samples for a particular medical test. These 'experiments' are not only expensive, but could result in excess disturbance or even harm to the patient. Thus, the reduction in the number of experiments (or samples) is not only important on a quantitative basis, but also on a qualitative basis. An example of a biomedical application is the determination of nitrogen washout in the function of the lung. This is an important test to provide information about the distribution of pulmonary ventilation. Lewis et al. (1982) found that the information from these tests can be maximized by providing an input breathing pattern with includes both 100% oxygen breaths combined with air at specific intervals. This was achieved using the D-optimality criterion to determine the optimal breathing pattern of inspired gas concentrations. A scheme was implemented to select the input patterns to be evaluated from over a billion possibilities by randomly selecting an input pattern and comparing that (through the D-optimality criterion) with the previous best pattern.

In other biomedical related studies, Cobelli and Ruggeri (1991) looked at the optimization of experiments for glucose tests in humans. Their objective was to reduce the number of experiments (samples) needed without affecting the accuracy of parameter estimation. Here, a stepwise approach was utilized. First, insulin was assumed to be a known input and the glucose concentration sampling schedule was optimized using a direct search procedure which employed the relaxation method to minimize the D-optimality criterion. Once this was determined, the insulin sampling schedule was determined by utilizing the output of a minimal model of insulin kinetics in the optimization procedure. The procedure resulted in reducing the sampling schedule from 33 to 14. The reduced sampling schedule was then validated with a new group of human experiments.

In another biomedical application, optimal design-of-experiment theory was utilized by Morris et al. (1988) in the design of experiments for estimating arterial wall transport parameters. In these tests, tracer experiments were conducted to determine the accumulation of low density lipoprotein in the arterial wall. The sensitivity coefficients associated with the parameter estimates were maximized to improve the accuracy of these transport parameters. A D-optimality criterion based analysis was utilized, and it was found that the results were moderately dependent on the initial estimates of the unknown parameters used to determine the sensitivity coefficients in the optimization procedure.

In a study related to fluid flow, Rose (1990) explored ways to minimize adverse measurement error effects through the optimization of experimental inputs in assessing the coupling effects in two-phase flow. Here, the dimensionless coefficients, $x_{i,j}$, of the sensitivity coefficient matrix, X, of the adverse errors in the measurement variables were minimized through the use of a least squares scheme as the optimization criterion. Thus, the optimal experimental inputs were chosen such that the sum of squares of the coefficients of the sensitivity coefficient matrix, $\sum_{i,j} x_{i,j}^2$, were minimized, resulting in an optimality condition of

$$\sum_{i,j} x_{i,j} dx_{i,j} = 0.$$

Peterson and Bayazitoglu (1991) introduced another important aspect of optimization of experiments, which is reducing the cost of the measurements. They used the cost as an objective function in the optimization of experiments involving fluid flow and heat transfer. They noted that measurements errors can be reduced by using more expensive instruments and fitted cost data with a polynomial function in terms of the uncertainty in the measurements. Then they used a gradient based optimization to minimize the cost subject to a constraint on the accuracy of the estimated quantity. That estimate was based on the root of the sum of the square of the errors induced by individual measurements, with these errors based on the sensitivity coefficients.

All of the above works focus on using sensitivity coefficients to reduce the errors in quantities estimated from experiments by using the D-optimality criterion or similar approaches. However, there are other aspects of experiment design that can benefit from optimization. In particular, optimization has been used to design "ideal" experiments, that is experiments where conditions are the best possible for testing the prediction of analytical models. For example, optimization can be used to design the ideal experiment for choosing between two competing models. Haftka and Kao (1990) suggested the use of optimization to evaluate differences between competing models for composite laminate failure, and van Wamelen et al. (1993) performed such an experiment design optimization. They designed a laminate stacking sequence and loading to maximize the ratio of the failure loads predicted by two competing criteria. Using genetic algorithms they arrived at a design with about 90 percent difference between the predicted failure loads. An experiment with this design confirmed one of the failure models.

A similar use of optimization to help in creating ideal experiments is reported by Albertini et al. (1990). They used numerical optimization to design the geometry of specimens which under torque develop two shear components. The objective functions used in the optimization were the uniformity of stresses in the specimens and the absence (or smallness) of other stress components. The resulting geometry resembled a small chalice and was accordingly dubbed "bichierino".

3. Experimental Optimization

In analytical optimization we have a set of design variables, and we evaluate the performance function and any constraints at a set of points in the space of design variables (called design points). Different optimization methods offer different strategies for selecting the sequence of design points that will lead to the optimum design. In experimental optimization we may use the same methods, except that the evaluation of the performance and the constraints is not done analytically, but instead by performing experiments.

Optimizing a product by conducting experiments has the advantage over analytical optimization because it does not require modeling the performance of the product. Rach (1990) cites the sensitivity of the optimum design to small imperfections as another reason for using tests instead of analysis. However, such sensitivity can also reduce the value of experimentally obtained optima (possibly not as much as analytical optima). Stuckman et al. (1990) provide another rationale for using experimental optimization. They discuss the design of a controller for a servo-motor control system for a robot. They point out that experimental optimization permits them to customize the optimum control design to the idiosyncrasies (due to manufacturing tolerances) of the robot, and to update the optimization as the properties of the robot change over time (for example, due to wear).

In most applications the cost of experiments is high, so that we employ experimental optimization mostly when we do not have reliable models of product performance. Occasionally we can conduct inexpensive experiments, and adaptive control problems are possibly the most common examples of inexpensive experimental optimization. In cases of process control, measurements of the process are continually taken for effecting feedback control, so that these measurements can be used to optimize the control procedure. For example, Semones and Lim (1989) describe a procedure to obtain on line the temperature and dilution rate that optimize the productivity of a continuous baker's yeast. They used the steepest descent method for the optimization. Because the focus of this paper is on situations where the experiments are expensive, the optimization during processes of adaptive control will not be further discussed here.

There are also borderline cases where experiments are only moderately expensive. Photoelastic determination of stresses may fall in this category. Durelli and his colleagues have used photoelasticity extensively to optimize the geometry of models for reducing stress concentrations (see Durelli et al., 1978, 1979, 1981, Durelli and Rajaiah, 1979a,b, 1981, Rajaiah and Durelli, 1981, Azarm et al., 1986, and Durelli and Ranganayakamma, 1988).

To reduce the cost of experimental optimization, we need to minimize the number of experiments required for the optimization. It may appear that this requirement is not different from the need to use efficient optimization methods when the cost of the analysis is high. However, the methods developed for efficient experimental optimization tend to be quite different from the methods developed for analytical optimization. It appears that the difficulty associated with obtaining derivatives from experiments, and the desirability of performing experiments in batches rather than singly are the main reasons for this difference.

In analytical optimization, derivatives of performance or constraint functions are available in many cases, and in many other cases they can be calculated reliably by repeated evaluation from finite difference formulae. This latter approach fails when numerical errors in the functions are very high. This case of high errors or noisy functions, as it is sometimes called, is the exception in analytical optimization, but it is the rule in experimental optimization. The performance measured in experiments has many sources of errors ranging from variability in manufacturing of the test specimens to performance measurement errors. These errors are often large enough to preclude the calculation of accurate derivatives. Additionally, without an analytical model, we may not even know whether the performance is a differentiable function of the design variables.

Because of the uncertainty about derivatives, experimental optimization usually avoids derivative-based methods. Such methods would require repeating experiments for small perturbations in design variables for the purpose of estimating derivatives, and the accuracy of the derivatives will be poor because of the noise in the function values. Adaptive control situations, where measurements are inexpensive, are the exception, in that large number of measurements may be taken to estimate derivatives in-spite of noise. One of the oldest methods for optimization without derivatives, the sequential simplex method of Spendly, Hext and Himsworth (1962) was developed with experimental optimization in mind, and it is still used. The method begins with experiments at points forming a simplex (a geometric figure which is the generalization of a triangle in two dimensions and a tetrahedron in three). The point with the poorest performance is discarded, and a new point is defined by reflecting the discarded point about the centroid of the simplex. The method also has provisions for shrinking the size of the simplex as we near the optimum design. The method is usually used with small number of design variables. For example, Khummongkol et al. (1992) applied it to the experimental optimization of activated carbon synthesis, with two design variables representing temperature and the weight ratio of NaCl (used as catalyst).

The sequential simplex method is a standard optimization method which is used also in analytical optimization when derivatives are not available. Usually, however, experimental optimization does not apply directly such standard optimization techniques. Instead, the performance in a region of design space is characterized by a number of experiments with substantial changes in the design variables. Then a simple function, such as a linear or quadratic polynomial, is fitted to the data obtained from the experiments. The simple function is usually called a response surface, and the approach is called the response-surface method (e.g., Box and Draper, 1987).

When we fit the coefficients of polynomials we need to solve a linear least-squares problem. The Fisher information matrix discussed in Section 2 then contains the derivatives of the polynomial with respect to its coefficients, that is, monomials. A lot of work has been done to find the best set of points that will minimize the variance of the polynomial due to noise in the measurements (e.g., Beck and Arnold, 1977, Myers and Montgomery, 1995). In particular, when the design space is a box or a sphere, there are well established sets of optimal points for conducting experiments in the design-of-experiments literature including the Box-Hunter design used, for example, by Villén et al. (1992).

Taguchi methods (e.g., Phadke, 1989), which offer a comprehensive approach to experimental optimization, also offer sets of points in what they call orthogonal arrays. The response surfaces generated by Taguchi methods are usually products of linear polynomials in each of the variables, so that higher order mixed terms are present, but the function is still linear in each design variable separately. For example, Mesenzhnik et al. (1988) describe the optimization of electrical cables, where the objective function is described by a polynomial in seven variables with linear terms and mixed quadratic and cubic terms.

When the design domain is not a box or a sphere, or the number of experiments that we can perform does not fit exactly one of the optimum designs available in the literature, we can use the optimality criteria described in Section 2, such as D-optimal design, to find an optimal set. Several software packages such as RS/Discover and EChip (see, e.g., Sell and Lepeniotis, 1992) are available that combine some of these optimality criteria with search methods for locating optimal or near optimal set of points.

While the search for the optimal set of points often employs the same optimality criteria as the search for optimal experimental conditions, the computational effort for experimental optimization is usually much higher. In problems of optimization of experiments, the number of experimental conditions to be optimized is usually small, so the search for an optimal set is usually a low-order optimization problem. However, in the case of experimental optimization, the number of unknown parameters is usually large. For example, if we have six design variables, and we attempt to fit the response as a quadratic polynomial, the number of unknown coefficients is 28, and we may need about 50 experiments to estimate these coefficients. Since each experiment is defined by six coordinates, finding the optimum set of points is an optimization problem in 300 variables.

Most algorithms for locating the optimum set of points do not tackle the continuous optimization problem. Rather they assume that a set of candidate points has been identified and seek the optimal sub-set that satisfies or nearly satisfies an optimality criterion, such as D-optimality. For example, in a six dimensional space, the analyst may start with four levels in each variable, to generate $4^6 = 4096$ points, and then search for the optimal subset of 50 of these 4096 points. One of the intuitively appealing approaches is to start with the full set of points, and successively discard points that contribute least to the determination of the unknown coefficients (e.g., Kammer, 1992). However this is often far from optimal. Instead, popular algorithms, such as the Detmax algorithm (Mitchell, 1974) use exchange strategies which try to replace points in the set with points outside the set. Genetic algorithms have also been used for this purpose, but they require more computational effort (e.g., Burgee et al., 1996).

The response surface fitted to the results of the experiments is used to perform local optimization. Such local optimization is called a design cycle. A second design cycle can then be performed around the optimum found by the first design cycle, and so on. Box (1969) developed this approach and called it Evolutionary Operation. This approach is quite similar to the sequential approximate optimization (see, for example, Haftka and Gürdal, 1992) employed for analytical optimization when the cost of analysis is very high. It is just that instead of the derivative-based approximations commonly used in analytical optimization, experimental optimization employs approximations based on sampling a small region in design space. This sampling region must be small enough to make the response surface locally valid, but large enough to avoid the noise problems previously mentioned. For example, Banerjee and Bhattacharyya (1993) used Evolutionary Operation for three variables, using ten experiments to obtain the direct effect of each variable plus all the interaction effects for each approximate optimization. They used this approach to optimize the concentration of three inducers to increase the production of the enzyme protease by microorganisms. Villén et al. (1992) used the response surface approach (quadratic polynomial in 3 variables) to optimize large volume sampling processes. They used the Box-Hunter experimental design to select design points. King and Buck (1991) used a response surface approach with 5 variables, and 18 experiments to optimize an etching process.

Because of the high cost associated with experimental optimization it is quite typical to conduct only a single design cycle of the approximate optimization. For example, Sandusta et al. (1987) used 12 experiments to establish a linear relationship between two parameters (grain size and type of filler) on the apparent density and ultimate compressive strength of corundum based thermal insulation. The best design was selected on the basis of these experiments. Broder et al. (1988) describe using 16 experiments to find the dependence of the performance of a coating on four variables that describe material fractions. Two experiments were carried out for each design variable, and the performance measured by a criterion called 'distinctness of image' (DOI) varied by up to 15 percent from one experiment to the next, while the difference between experiments with different parameters varied by up to 35 percent. The two experiments for each setting were averaged, and the results were used to obtain a good coating.

The single cycle approach can produce good improvement in existing designs when the optimization problem is simple. The existence of complex constraints or multiple objectives can frustrate this simple approach. For example, Gardner (1991) describes the use of Taguchi method for estimating the effects of seven variables (two levels each) on emission properties using an orthogonal array for sixteen experiments. After each of the properties was analyzed for the effects of the seven variables and their interactions, it was difficult to find a design that satisfied all the requirements.

4. Optimization of Numerical Models using Design-of-Experiment Techniques

In the past few years there has been a growing interest in using the techniques of experimental optimization for optimization based on numerical simulation, and in particular, the use of response surface techniques is becoming very popular. This popularity is driven by the fact that numerical simulations are fast acquiring most of the traits of laboratory experiments.

First, numerical simulations with models including hundreds of thousands of variables are very expensive and unwieldy. Developing such models is very time consuming, and they are often run on proprietary programs that are not easy to connect to an optimization routine. Consequently, like laboratory experiments, there is an advantage in running these numerical experiments in batches rather than a single point at a time when invoked by a typical sequential optimizer. The availability of a large number of work stations, which are often only sparsely used at nights or on weekends, also contributes to the advantage of performing numerical simulations in batches like laboratory experiments. Parallel computation capabilities are also easy to use when parallelization simply involves running the same simulation at many design points. Thus Balabanov et al. (1996) have used parallel computers to perform hundreds of analyses for the purpose of configuration optimization of a high speed civil transport.

Second, complex numerical simulations often acquire numerical noise from several sources (see Giunta et al. 1994). There is noise associated with the computation itself, such as round-off errors and errors associated with less than fully converged iterative processes. There are also errors associated with the process of discretization of physical models. Since the number of cells or grids is integer, small jumps in numerical predictions occur when an infinitesimal change in the design creates a unit change in the design discretization (see Narducci et al. 1995). Derivative-based methods, commonly used in analytical optimization, do not cope well with such numerical noise. On the other hand, response surface techniques filter the noise by using more experiments than coefficients in the response surface.

Response surface techniques provide several other advantages to designers using numerical simulations. First, the designer gets a view of the entire design space created by the response surface, while with traditional optimization techniques he gets to see only the points visited by the optimization procedure. Second, the response surfaces can be created by specialists well familiar with the complex software used for simulation instead of by designers who tend to be generalists and often lack the background to exercise these numerical simulation faultlessly. Finally, with the entire design domain being analyzed in a batch mode, it is easier to detect errors due to poor modeling techniques or plain mistakes, because the erroneous results stand out. In fact, the response surface can occasionally correct a small number of poor-accuracy analyses in the same way that it filters out noise (see Venter et al., 1996).

With all of these good properties, response surface methods, like most other experimental optimization techniques, are severely limited in the number of variables that they can handle. The number of coefficients of a quadratic polynomial, which is the most commonly used response surface, increases as the square of the number of design variables, and the number of analyses required to estimate these coefficients appears to increase even faster. For example, Kaufman et al. (1996) found that even about 1,000 numerical experiments did not appear to be sufficient for estimating the coefficients of a quadratic polynomial in 25 variables. For this reason, application of these techniques to design optimization has been limited to problems with a small number of design variables. One of the most popular applications appears to be aircraft and spacecraft system design (e.g., Chen et al., 1995, Engelund et al., 1993 Giunta et al., 1995, Lewis and Mistree, 1995, Stanley et al., 1992, and Tai et al. 1995)

Most of the work employing design-of-experiment techniques for optimization based on numerical simulations employed box-like design domain with standard point sets for the numerical experiments. For example, finite element structural analysis programs are often unwieldy and difficult to couple with optimization procedures. Mohr-Matuschek and Michaeli (1991) used a quadratic response surface approach with full factorial design for point selection for a single cycle optimization of snap-fit hinges analyzed by finite element models. Mason et al. (1994) used the same approach but with central composite design for point selection for the optimization of composite channel frames.

However, in most optimization problems there are constraints that result in odd-shaped design domains. The volume of the feasible design space is often a very small fraction of the volume of a box that contains it. Consequently, there is substantial incentive to limit the response surface approximation to the feasible design and in that case the choice of design points is typically based on optimality criteria such as D-optimal design (e.g., Kaufmann et al. 1996).

5. Validation of Optimization

When we optimize a system based on an analytical model of the response, we may drive the design beyond the region of applicability of the analytical model. This happens, for example, in structural optimization against buckling where the optimized structure can exhibit heightened imperfection sensitivity (Thompson and Supple, 1973, Mróz and Gawecki, 1975). For example, we expect optimization against buckling to cause the coalescence of two or more buckling loads, and such coalescence often produces imperfection sensitivity. The imperfection sensitivity, in turn, requires the application of more complex modeling for predicting the buckling load as a function of imperfection shape and amplitude. Geier et al. (1991) provide an interesting indication that optimization increases imperfection sensitivity. They optimized unstiffened composite cylindrical shells by varying ply angles, and then they reversed the optimization process to find the design that gives the lowest buckling load, which they call pessimum. They tested both optima and pessima and compared with analytical predictions. The experimental buckling loads for the optimal designs were up to 20 percent below the analytical results, while for the pessima, the experimental results were only between 1 and 9 percent below the analytical ones. Sun and Hansen (1988), who have performed an earlier validation of similar optimized shells, also found that optimal designs always buckled below the analytical prediction, with differences ranging from two to 21 percent.

If we anticipate such degradation of analytical models we can take preventive measures against it, such as adding constraints that prevent the design from moving outside the range of validity of the model. However, often the process takes us by surprise even if in retrospect we find that we could have anticipated the problem. Therefore, we may want to validate optimized designs to guard against this problem. The experiment may involve only the optimal design, to check that the optimization process has not taken us into a region of design space where our analysis method is not applicable. Instead, we can test also nearby designs, so as to check that errors in the analysis did not cause our design to become non-optimal, even if they are not large enough to cause substantial deterioration in performance. For example, Knight et al. (1992) confirmed with three tests the optimality of a finned heat sink design that was obtained with a single design variable. Adelman (1992) surveyed several studies that documented experimental validation of structural optimization. Some of these studies tested only the optimum design, while others tested also neighboring designs. Testing of optimized designs is particularly common in the design of control systems because of the relative ease of carrying out the experiments (e.g., Lee-Glauser et al., 1995, Kajiwara et al., 1996).

Most researchers document successes rather than failures, and for this reason, it is difficult to find papers which document the failure to validate optimization results. This may give the false impression that there is no need to validate optimization. However, occasionally such failures can be found in papers where the validation failure is only temporary or partial. For example, Haftka and Starnes (1988) showed a failure of plates with a hole with optimized local reinforcement near the hole. Eventually the failure was averted by removing rather than adding material. Messer et al. (1993) optimized parameters of an adaptive control law to reduce the control cost. They found that optimization with a small number of parameters was validated by experiments, while additional gains obtained by increasing the number of design variables could not be validated experimentally.

For probabilistic optimization another reason for the need for experimental validation is the uncertainty about the statistical data which is used to model variability in response. Ponslet et al. (1994) combined design of experiments with validation of optimization to demonstrate the utility of probabilistic optimization over deterministic optimization based on safety factors. As a demonstration problem they used the design of tuned dampers for a truss structure, where damper properties exhibit substantial variability. The objective was to obtain a design that satisfied best limits on dynamic response. Ponslet et al. searched circumstances where the difference between the deterministic optimum and the probabilistic optimum was the largest, and they then tested the two optimal designs. The experiments found that the probabilistic design failed in only one out of 29 experiments while the deterministic designed failed in six.

The need to validate optimization results is not confined to analytical optimization. Because experimental optimization is usually conducted on the basis of a simple response surface, there is a need to evaluate the accuracy of the response surface in the neighborhood of the predicted optimum. For example, Kaufmann and Stone (1987) used the Taguchi method to select process variables for optimizing the shear bond strength of a lap joint. They conducted a matrix of experiments using Taguchi orthogonal arrays (two-level partial factorial design) and they used additional experiments to validate the optimum. Likewise, Fox and

Lee (1990) used the Taguchi method for the optimization of metal injection molding and then performed confirmation experiments to measure the performance improvement with the optimized conditions. Similarly, Bicking (1992) used 22 factorial design to optimize temperature and pressure for extraction of aminehydrochloride, then checked the improvement by additional experiments.

Occasionally, validation is needed not only for the particular design that was obtained but as a demonstration that a new approach works. For example, Whitman (1991) proposed a method for designing experiments that improves upon the Taguchi arrays and demonstrated its efficiency for several problems including improving the strength of an outboard motor propeller by varying six parameters with 27 experiments. The quality of the procedure was checked by performing additional 18 experiments that agreed well with predictions based on the original 27.

6. Summary

In this paper, four aspects of the optimization of experiments are presented. First, analytical optimization can be used to design experiments to provide the maximum amount of information and insight on the phenomena being investigated. Optimization of experiments is usually based on analytical models of these phenomena. Experimental optimization, on the other hand, is used when a model of the system is not known and/or when high noise is present. Here, it is desired to reduce the number of experiments if the experiments are expensive. Additionally, because of the cost of setting up and scheduling difficulties, experiments are usually carried in batches rather than one at a time. Experimental optimization techniques have developed to cater to the need to filter out noise and work with a batch of experiments. The paper describes the response surface methodology that caters to these needs of experimental optimization. Furthermore, it describes a trend in optimization based on complex numerical simulation to use the same kind of response surface techniques. This is due to a growing similarity between laboratory experiments and complex numerical simulations. Finally, the validation of optimization through experiments was shown to be important, especially when the system is based on an analytical model which is not valid in the entire design space.

7. Acknowledgments

The work of the first author was sponsored by NASA grant NAG-1-1669, and the work of the second author was sponsored by NASA grants NAG-1-1507 and NCC-1-221.

References

Adelman, H. M., "Experimental Validation of the Utility of Structural Optimization," Structural Optimization, Vol. 5, 1992, pp. 3-11.

Albertini, C., Montagnani, M., Zyczkowski, M., and Laczek, S., "Optimal Design of a Specimen for Pure Double Shear Tests," Int. J. Mech. Sci., Vol. 32, No. 9, 1990, pp. 729-741.

Allemang, R. J. and Brown, D. L., "A Correlation Coefficient for Modal Vector Analysis," Proceedings of the 1st International Modal Analysis Conference (Orlando, FL), Union College, Schenectady, NY, 1982, pp. 110-116.

Azarm, S., Durelli, A. J., and Yu, W., "Shape Optimization of In-Plane Loaded Tall Beams," Proceedings of Modeling; Simulation the Annual Pittsburgh Conf., Vol 17, Part 1-2, 1986, pp. 715-719.

Balabanov, V., Kaufman, M., Giunta, A.A., Haftka, R. T., Grossman, B., Mason, W. H., and Watson, L. T., "Developing Customized Wing Weight Function by Structural Optimization on Parallel Computers," AIAA Paper 96-1336, Proceedings of the 37th AIAA/ASME/ASCE/AHS Structures, Structural Dynamics and Material Conference, Salt Lake City, UT, April 15-17, 1996, Part 1, pp. 113-125.

Banerjee, R., and Bhattacharyya, B. C., "Evolutionary Operation to Optimize Three-Dimensional Biological Experiments," Biotechnology and Bioengineering, John Wiley, Vol. 41, 1993, pp. 67-71.

Bayard, D. S., Hadaegh, F. Y., and Meldrum, D. R., "Optimal Experiment Design for Identification of Large Space Structures," Automatica, Vol. 24, No. 3, 1988, pp. 357-364.

Beck, J. V., and Arnold, K. J., Parameter Estimation in Engineering and Science, John Wiley, 1977.

Bicking, M. K. L., "A Simplified Experimental Design Approach to Optimization of SFE Conditions for Extraction of Amine Hydrochloride," J. of Chromatographic Sci., Vol. 30, No. 9, 1992, pp. 358-360.

Box, G. E. P., and Draper, N. R., Empirical Model-building and Response Surfaces, John Wiley, New York, 1987.

Box, G. E. P., Evolutionary Operation: a Statistical Method For Process Improvement, Wiley, 1969.

Broder, M., Kordomenos, P. I., and Thomson, D. M., "A Statistically Designed Experiment for the Study of a Silver Automotive Basecoat," J. of Coating Technology, Vol. 60, No. 766, 1988, pp. 27-32.

Burgee, S., Giunta, A. A., Narducci, R., Watson, L. T., Grossman, B. and Haftka, R. T., "A Coarse Grained Parallel Variable-Complexity Multidisciplinary Optimization Paradigm," The International Journal of Supercomputer Applications and High Performance Computing, (accepted for publication, 1996).

Chen, W., Allen, J. K., Mistree, F. and Tsui, K-L, "Integration of Response Surface Method with the Compromise DSP in Developing a General Robust Design Procedure," Advances in Design Automation, (Azarm, S., Dutta, D., Eschenauer, H., Gilmore, B. J., McCarthy, Yoshimura, M. Eds.), New York: ASME, 1995, pp. 485-492. ASME DE-Vol. 82-2.

Cobelli, C., and Ruggeri, A., "A Reduced Sampling Schedule for Estimating the Parameters of the Glucose Minimal Model from a Labeled IVGTT," IEEE Transactions on Biomedical Engineering, Vol. 38, No. 10, 1991, pp. 1023-1029.

Copenhaver, D. P., Thermal Characterization of Honeycomb Core Sandwich Structures, M. S. Thesis, Virginia Polytechnic Institute and State University, Blacksburg, VA 1996.

De Queiroz Orsini, L., "Parameter Estimation in a Simple Experiment: The Resonant Circuit Revisited," IEEE Transactions on Education, Vol. E-27, No. 1, 1984, pp. 47-49.

Durelli, A. J., Brown, K., and Yee, P., "Optimization of Geometric Discontinuities in Stress Fields," Exper. Mechanics, Vol. 28, No. 2, 1978, pp. 303-308.

Durelli, A. J., Erickson, M. and Rajaiah, K., "Optimum Shapes of Central Holes in Square Plates Subjected to Uniaxial Uniform Load," Int. J. Sol. and Struct., Vol. 17, 1981, pp. 787-793.

Durelli, A. J., and Rajaiah, K., "Optimum Hole Shapes in Finite Plates under Uniaxial Load," J. Appl. Mech., Vol. 46, Sept., 1979a, pp. 691-695.

Durelli, A. J., Rajaiah, K., "Optimized Inner Boundary Shapes in Circular Rings under Diametral Compression," Strain, Oct., 1979b, pp. 127-130.

Durelli, A. J., and Rajaiah, K., "Optimization of Inner and Outer Boundaries of Beam and Plates with Holes," J. Strain Analysis, Vol. 16, No. 4, 1981, pp. 211-216.

Durelli, A. J., Rajaiah, K., Hovanesian, J. D., and Hung, Y. Y., "General Method to Directly Design Stress-Wise Optimum Two-Dimensional Structures," Mech. Res. Comm., Vol. 6, No. 3, 1979, pp. 159-165.

Durelli A. J., and Ranganayakamma B., "Optimization Process in Tall Beams," Experimental Mechanics, Sept., 1988, pp. 259-265.

Engelund, W. C., Stanley, D. O., Lepsch, R. A., McMillin, M. M., and Unal, R., "Aerodynamic Configuration Design Using Response Surface Methodology Analysis," AIAA Paper 93-3967, August, 1993.

Fox, R. T., and D. Lee, "Optimization of Metal Injection Molding: Experimental Design," Int. J. of Powder Metallurgy, Vol. 26, No. 3, 1990, 223-243.

Frederiksen, P.S., ``Parameter Uncertainty and Design of Optimal Experiments for the Estimation of Elastic Constants," Danish Center for Applied Mathematics and Mechanics, Technical University of Denmark, Report No. 527, June 1996.

Furuya, H. and R. T. Haftka, "Genetic Algorithms for Placing Actuators on Space Structures," Proceedings of the Fifth International Conference on Genetic Algorithms, Urbana, IL, 1993, pp. 536-542.

Garcia, S., and E. P. Scott, "Use of Genetic Algorithms in Optimal Experimental Designs," Proceedings of the 2nd International Conference on Inverse Problems in Engineering/Theory and Practice, Le Croisic, France, June 9-14, 1996.

Gardner, T. P., "Optimization of Diesel Fuel Injection and Combustion System Parameters for Low Emissions Using Taguchi Methods," Fuels, controls and After treatment for Low Emission Engines, ICE, Vol. 15, ASME, 1991, pp. 99-106.

Geier, B., Klein, H., and Zimmerman, R., "Buckling Tests with Axially Compressed Unstiffened Cylindrical Shells Made from CFRP," in Buckling of Shell Structures, on Land, in the Sea and in the Air, (J. F. Julien, editor), Elsevier Applied Sci., London, 1991, pp. 498-507.

Giunta, A.A., Dudley, J. M., Narducci, R., Grossman, B., Haftka, R. T., Mason, W. H., and Watson, L. T., "Noisy Aerodynamic Response and Smooth Approximations in HSCT Design," Proceedings, 5th AIAA/USAF/NASA/ISSMO Symposium on Multidisciplinary Analysis and Optimization, Panama City, FL, Sept. 7-9, 1994, pp. 1117-1128.

Giunta, A.A., Narducci, R., Burgee, S., Grossman, B., Haftka, R. T., Mason, W. H., and Watson, L. T., "Variable Complexity Response Surface Aerodynamic Design of an HSCT Wing," Proceedings of the 13th AIAA Applied Aerodynamic Conference, pp. 994--1002, Paper No. 95-1896-CP, June 1995.

Haftka, R. T., and Gürdal, Z., Elements of Structural Optimization, 3rd Edition, Kluwer Publishers, 1992.

Haftka, R. T. and Kao, P.-J., "The Use of Optimization for Sharpening Differences between Models," ASME Winter Annual Meeting, Dallas Texas, 1990, November 25-30.

Haftka, R. T. and Starnes, J. H., "Stiffness Tailoring for Improved Compressive Strength of Composite Plates with Holes," AIAA J., Vol. 26, No. 1, 1988, pp. 72-77.

Holland, J. H., Adaptation of Natural and Artificial Systems, The University of Michigan Press, Ann Arbor, MI, 1975.

Kajiwara, I., Nagamatsu, A., Seki, K., and Kotani, V., "Integrated Optimum Design of Structure and Servosystem with Dynamic Compensator," AIAA Paper 96-1433, Proceedings of the 37th AIAA/ASME/ASCE/AHS Structures, Structural Dynamics and Material Conference, Salt Lake City, UT, April 15-17, 1996, Part 2, pp. 1005-1015.

Kalaba, R. E., and Spingarn, K., "Sensitivity of Parameter Estimates to Observations System Identification, and Optimal Inputs," Applied Mathematics and Computation 7, 1980, pp. 225-235.

Kammer, D. C., "Effect of Model Error on Sensor Placement for ON-Orbit Modal Identification and Correlation of Large Space Structures," Journal of Guidance, Control and Dynamics, Vol. 15, No. 2, 1992, pp. 334-341.

Kaufman, M., Balabanov, V., Burgee, S. L., Giunta, A.A., Grossman, B., Mason, W. H., Watson, L. T., and Haftka, R. T., "Variable-Complexity Response Surface Approximations for Wing Structural Weight in HSCT Design," Computational Mechanics, Vol. 18, No. 2, 1996, pp. 112-126.

Kaufmann, B. A., and Stone, D. H., "A Process Optimization Method for Lap Shear Bond Strength," Proceedings, 32nd Int. SAMPE Symposium, April 6--9, 1987, pp. 444-455.

Khummongkol, D., Charoenkool, A., and Pongkum, N., "Experimental Optimization of Activated Carbon Synthesis by the Simplex Search Method," Applied Energy, Vol. 41, No. 4, 1992, pp. 243-249.

Kiefer, J., "General Equivalence Theory for Optimum Designs (Approximate Theory)," The Annals of Statistics, Vol. 2, No. 5, 1974, pp. 849-879.

Kiefer, J., "Variation in Structure and Performance Under Change of Criterion," Biometrika, Vol. 62, No. 2, 1975a, pp. 277-288.

Kiefer, J., "Optimality Criteria for Designs," NSF report on Grant No. GP35816X, Dept. of Mathematics, Cornell University, Ithaca, New York, 1975b.

Kiefer, J., "The Interplay of Optimality and Combinatorics in Experimental Design," The Canadian Journal of Statistics, Vol. 9, No. 1, 1981, pp. 1-10.

King, D. L., and Buck, M. E., "Experimental Optimization of an Anisotropic Etching Process for Random Texturization of Silicon Solar Cells," Proceedings, Twenty Second IEEE Photovoltaic Specialists Conf., IEEE, Vol. 1. 1991, pp. 303-308.

Knight, R. W., Goodling, J. S., and Gross, B. E., "Optimal Thermal Design of Air Cooled Finned Heat Sinks - Experimental Verification," Proceedings of 1992 Inter Society Conf. on Thermal Phenomena, IEEE in Electronic Systems, 1992, pp. 206-212.

Koszalka, L., "Optimum Experiment Design for Identification of Large Scale System," IEEE Fifth Inter. Conf. on Systems Engineering, 1987, pp. 589-593.

Krottmaier, J., Optimizing Engineering Designs, McGraw-Hill Book Company, London, 1993.

Lee-Glauser, G., Juang, J-N., and Sulla, J. L., "Optimal Active Vibration Absorber: Design and Experimental Results," Journal of Vibration and Acoustics, Vol. 117, April 1995, pp. 165-171.

Lewis, K. and Mistree, F., "Designing Top-Level Aircraft Specifications: A Decision-Based Approach to a Multiobjective, Highly Constrained Problem," 36th AIAA/ASME/ASCE/AHS/ASC Structures, Structural Dynamics and Materials Conference, New Orleans, Louisiana, April 10-13, 1995, pp. 2393-2405. Paper No. AIAA-95-1431-CP.

Lewis, S. M., D'Argenio, D. Z., Bekey, G. A., and Mittman, C., "Optimal Inputs for Parameter Determination of Inert Gas Washout from the Lung," Respiration Physiology, Vol. 50, 1982, pp. 111-127.

Lohmann, T., Bock, H. G., and J. P. Schlöder, "Numerical Methods for Parameter Estimation and Optimal Experimental Design in Chemical Reaction Systems," Ind. Eng. Chem. Res., Vol. 31, 1992, pp. 54-57.

Mason, B. H., Haftka, R. T., and Johnson, E. R., "Analysis and Design of Composite Channel Frames," AIAA Paper 94-4364-CP, Proceedings, AIAA/USAF/NASA/ISSMO Symposium on Multi-disciplinary Analysis and Optimization, Panama City, FL, Sept. 7-9, 1994, Vol. 2, pp. 1023-1040.

Mesenzhnik, Ya. Z., Osyagin, A. A., Kuz'minov, S. V., and Kurganov, V. B., "Design of Experiment Investigation of Cables for the Oil and Gas Industry," Soviet Electrical Engineering, Vol. 59, No. 12, 1988, pp. 66-69.

Messer, S., Haftka, R. T., and Cudney, H. H., "The Cost of Model Reference Adaptive Control: Analysis, Experiments and Optimization," Proceedings, AIAA/ASME/ASCE/AHS/ASC 34th Structures, Structural Dynamics and Materials Conf., San Diego, CA, April 19-21, 1993, Part 6, pp. 3115-3125.

Mitchell, T. J., "An Algorithm for the Construction of D-Optimal Experimental Designs," Technometrics, Vol. 16, No. 2, pp. 203-220, May 1974.

Mohr-Matuschek, U., and Michaeli, W., "Statistical Experiment Design for the Optimization of Snap-Fit Hinges by Finite Element Analysis," Proceedings, Annual Technical Conf., Society of Plastic Engineers, Vol. 37, 1991, pp. 338-342.

Moncman, D. M., J. P. Hanak, D. P. Copenhaver, and E. P. Scott, "Optimal Experimental Designs for Estimating Thermal Properties," Proceedings of the 4th ASME/JSME Thermal Engineering Joint Conf., A.S.M.E., Maui, HI, 1995, pp. 461-168.

Morris, E. D., Saidel, G. M., and Chisolm III, G. M., "Optimal Design of Experiments for Estimating Arterial Wall Transport Parameters," IEEE Engineering in Medicine & Biological Society 10th Annual Inter. Conf., 1988, pp. 131-132.

Mróz, Z. and Gawecki, A., "Post-yield Behavior of Optimal Plastic Structures," Proceedings of IUTAM Symp. on Optimization in Structural Design, edited by Sawczuk, A. and Mróz, Z., Springer Verlag, Berlin, 1975, pp. 518-540.

Myers, R. H., and Montgomery, D. C., Response Surface Methodology: Process and Product Optimization Using Designed Experiments, John Wiley, New York, 1995.

Narducci, R., Grossman, B., and Haftka, R. T., "Sensitivity Algorithms for an Inverse Design Problems involving Shock Waves," Inverse Problems in Engineering, 2, pp. 49-83, 1995.

Penny, J. E. T., Friswell, M. I., and Garvey, S. D., "Automatic Choice of Measurement Locations for Dynamic Testing," AIAA Journal, Vol. 32, No. 2, 1994, pp. 407-414.

Peterson, J., and Y. Bayazitoglu, "Optimization of Cost Subject to Uncertainty Constraints in Experimental Fluid Flow and Heat Transfer," J. of Heat Transfer, Vol. 113, No. 3, 1991, pp. 314-320.

Phadke, M. S., Quality Engineering Using Robust Design, Prentice Hall, Englewood Cliffs, New Jersey, 1989.

Ponslet, E., Maglaras, G., Cudney, H. H., Nikolaidis, N. and Haftka, R. T., "Analytical and Experimental Comparison of Probabilistic and Deterministic Optimization," Proceedings, 5th AIAA/USAF/NASA/ISSMO Symposium on Multidisciplinary Analysis and Optimization, Panama City, FL, Sept. 7-9, 1994, pp. 544-559.

Pronzato, L. and E. Walter, "Experiment Design in a Bounded-error Context: Comparison with D-Optimality," Automatica, Vol. 25, No. 3, 1989, pp. 383-391.

Rach, V. A., "Optimization of Cylindrical Pressure Vessels with Respect to Weight," Mechanics of Composite Materials, Vol. 26, No. 3, 1990, pp. 366-370.

Rajaiah, K. and Durelli, A. J., "Optimization of Hole Shapes in Circular Cylindrical Shells under Axial Tension," Exper. Mechanics, May 1981, pp. 201-204.

Rose, W. D., "Optimizing Experimental Design for Coupled Porous Media Flow Studies," Exper. Thermal and Fluid Sci., Vol. 3, No. 6, 1990, pp. 613-622.

Sandusta, T. M., Kvasman, N. M., Pisareva, N. V., Gaodu, A. N., and Degtyareva, E. V., "Optimizing the Properties of Corundum-Based Thermal Insulation Products Using the Method of Experimental Planning," Refractories, Vol. 28, No. 5-6, 1987, pp. 310-313.

Sell, J. W., and Lepeniotis, S.S., "Thermoplastic Polymer Formulations: An Approach through Experimental Design," Advances in Polymer Technology, Vol. 11, No. 3, 1992, pp. 193-202.

Semones, G. B., and Lim, H. C., "Experimental Multivariable Adaptive Optimization of a Steady-State Cellular Productivity of a Continuous Baker Yeast Culture," Biotechnology and Bioengineering, Vol. 33, No. 1, 1989, pp. 16-25.

Spendley, W., Hext, G. R., and Himsworth, F. R., "Sequential Application of Simplex Designs in Optimisation and Evolutionary Operation," Technometrics, Vol. 4, No. 4, 1962, pp. 441-461.

Stanley, D., Unal, R.; and Joyner. R., "Application of Taguchi Methods to Dual Mixture Ratio Propulsion System Optimization for SSTO Vehicles," AIAA Paper 92-0213, *AIAA 30th Aerospace Sciences Meeting and Exhibit*, Reno, NV, January 1992.

Stuckman, B. E., Care, M. C., and Stuckman, P. L., "System Optimization using Experimental Evaluation of Design Performance," Eng. Opt., Vol. 16, 1990, pp. 275-290.

Sun, G., and Hansen, J. S., "Optimal Design of Laminated-Composite Circular-Cylindrical Shells Subjected to Combined Loads," Journal of Applied Mechanics, Vol. 55, March 1988, pp. 136-142.

Tai, J. C., Mavris, D. N., and Schrage, D. P., "Application of a Response Surface Method to the Design of Tipjet Driven Stopped Rotor/Wing Concepts," AIAA Paper 95-3965, *1st AIAA Aircraft Engineering, Technology, and Operations Congress*, Los Angeles, CA, September 19–21, 1995.

Taktak, R., J. V. Beck, E. P. Scott, "Optimal Experimental Design for Estimating Thermal Properties of Composite Materials," Int. J. of Heat and Mass Transfer, Vol. 36, No. 12, 1993, pp. 2977-2986.

Thompson, J. M. T., and Supple, J. W., "Erosion of Optimum Design by Compound Branching Phenomena," J. Mech. Phys. Solids, Vol. 21, 1973, pp. 135-144.

Van Wamelen A., Haftka, R. T., and Johnson, E. R., "Optimal Layups of Composite Specimens to Accentuate the Differences between Competing Failure Criteria," presented at the 8th Technical Conf. on Composite Materials, American Society of Composites, Cleveland, OH., October 19-21, 1993.

Venter, G., Haftka, R. T., and Starnes, J. H., Jr., "Construction of Response Surfaces for Design Optimization Application," AIAA Paper 96-4040, The AIAA/NASA/USAF/ISSMO Symposium on Multidisciplinary Analysis and Optimization, Bellevue, WA, Sept. 4-6, 1996.

Villén J., Señoráns, F. J., Herraiz, M., Reglero, G., and Tabera, J., "Experimental Design Optimization of Large Volume Sampling in a Programmed Temperature Vaporizer. Application in Food Analysis," J. of Chromatographic Sci., Vol. 30 No. 7, 1992, pp. 261-266.

Walter, E., and Pronzato, L., "Qualitative and Quantitative Experiment Design for Phenomenological Models - A Survey," Automatica, Vol. 26, No. 2, 1990, pp. 195-213.

Whitman, C. L., "Design of Experiments - Some Improvements on the Taguchi Method," Compaction, Quality Control; Training Advances in Powder Metallurgy, Metal Powder Industrial Federation, Vol. 1, 1991, pp. 287-298.

Theoretical and Applied Mechanics 1996
T. Tatsumi, E. Watanabe and T. Kambe (Editors)
© 1997 Elsevier Science B.V. All rights reserved.

Optimum Design of Metal Forming Based on Finite Element Analysis and Nonlinear Programming

Eiji Nakamachi

Department of Mechanical Engineering, Osaka Institute of Technology
5-16-1, Omiya, Asahiku, Osaka, 535, Japan

ABSTRACT

Recently, the metal forming simulation technology revealed great progress in the sense of practical application in the automotive, steel, electric/electronics and aerospace industries. The goal of metal forming simulation is to be embedded in the computational design system which consists of the analysis and synthesis modules. In the advanced computational system, the analysis module may adopt the nonlinear finite element (FE) analysis code, and the synthesis module may employ the mathematical programming code. This metal forming design system can detect the slackness of the deformed material and estimate the formability, and find the optimum condition for the forming operation. The bulk of this paper is devoted to provide the summary of theoretical bases for the explicit type FE analyses and nonlinear programming algorithm, and further the review of these applications to the metal forming analyses and design.

1. INTRODUCTION

The finite element (FE) modeling of metal forming process has been an active subject for more than two decades, because it includes the vital difficulties caused by large straining of material, unknown and changing contact boundaries, such as geometry and force, and further searching the optimum operation condition. A growing trend for the use of FE simulation and optimum design algorithm in the industrial metal forming operation and material fabrication increases the demand for robustness, accuracy, efficiency and performance. This paper provides the summary of theoretical bases for the explicit type nonlinear FE analyses, the advanced material modeling - crystalline plasticity theory -, and the optimization algorithm - "Sweeping Simplex Method" -, and its application to the sheet metal forming analyses and optimum design[1-9].

In 1970s, the theoretical bases for the nonlinear continuum mechanics[10-12], the elastic plastic material modeling[13-17] and nonlinear FE methodologies[18-32], have been established. In 1978,Wang and Budiansky have analyzed the hemi-spherical punch stretching using the static implicit (SI) elastic-plastic membrane FE method[33-34].

324

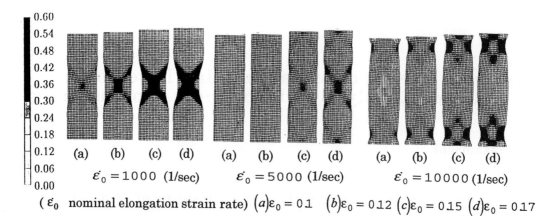

($\dot{\varepsilon}_0$ nominal elongation strain rate) $(a)\varepsilon_0 = 0.1$ $(b)\varepsilon_0 = 0.12$ $(c)\varepsilon_0 = 0.15$ $(d)\varepsilon_0 = 0.17$

Figure 1 Strain localization prediction by elastic/crystallin viscoplastic dynamic explicit FE analyses

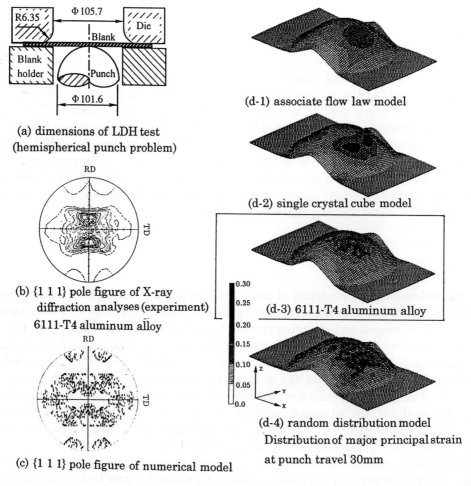

(a) dimensions of LDH test
(hemispherical punch problem)

(b) {1 1 1} pole figure of X-ray
diffraction analyses (experiment)
6111-T4 aluminum alloy

(c) {1 1 1} pole figure of numerical model

(d-1) associate flow law model

(d-2) single crystal cube model

(d-3) 6111-T4 aluminum alloy

(d-4) random distribution model
Distribution of major principal strain
at punch travel 30mm

Figure 2(a)
Studies of texture effect on the strain localization in LDH formability test

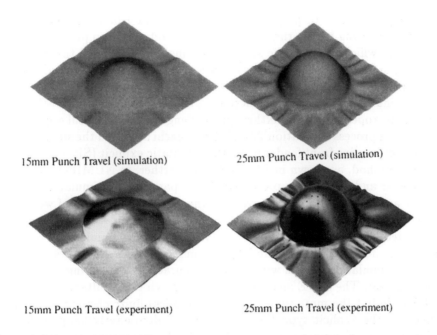

15mm Punch Travel (simulation)　　25mm Punch Travel (simulation)

15mm Punch Travel (experiment)　　25mm Punch Travel (experiment)

Figure 2(b)
Wrinkling deformation prediction　(DE FE simulation and experimental results)

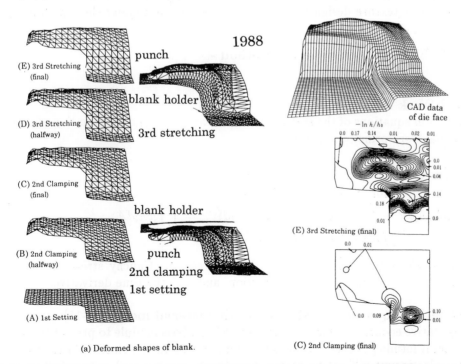

(a) Deformed shapes of blank.

Figure 3　Finite element simulation of HONDA frontfender forming (ROBUST -static explicit FE code)

After eight years blank of progress. In 1985 Toh and Kobayashi have analyzed the square-cup deep drawing by the rigid-plastic FE method, and also proposed the optimum design algorithm, which adopted the semi inverse method[35]. Until the middle of 1980s, the FE simulation has been recognized as the research tool to solve the academic problems, such like the plastic instability prediction as shown in Figure 1 and LDH formability test simulation - wrinkling and tearing predictions -, as shown in Figure 2(a)-(b)[36-38].

In 1988, there appeared the breakthrough challenge in the practical application for the industrial forming process simulation[28-32]. Nakamachi analyzed the single action stamping process of automotive front fender using the 3-D static explicit (SE) type elastic-plastic membrane FE method, as shown in Figure 3 [39]. Further, in NUMIFORM89 conference in 1989, Honecker and Mattiasson presented the oil-pan forming simulation results, which impressed the automotive engineers because of the detailed prediction of wrinkling during the forming process[40-41]. This DYNA-3D results attracted the engineers to develop the sophisticated FE code for the industrial forming process simulation. In 1st and 2nd NUMISHEET conferences, in 1991 and 1993, commercial DE FE codes have shown the panel stamping simulation results with enough accuracy by employing tens thousands finite elements[42-43]. The SI FE analyses also have shown the results with better accuracy than DE FE code but the more time consuming and less finite element number. Because of the efficiency, the explicit type FE codes are now dominantly adopted in the industries for the sheet metal forming analysis[44-45]. In the mean time, the optimum forming process designs were attempted by employing the inverse method and the mathematical programming method[46-49]. Because of higher order nonlinearity and almost impossible differentiation by the design parameter of the objective function, only the numerical computation is available for this optimum design system in conjunction with FE analyses. Finally, It is demonstrated that the two-stage sheet forming process optimization and the optimum crystalline texture design to fabricate the better formability sheet metal[9].

2.   ANALYSIS - explicit type FE method -

2.1. Virtual Work Equations

The weak form equations of the static equilibrium and the dynamic equation of motion yield the virtual work principles as follows;

$$\int_V (\dot{s}_{ij} + \sigma_{ij}\dot{u}_{i,j})\delta\dot{u}_{j,i}dV = \int_A \dot{p}_i\delta\dot{u}_i dA$$
$$\text{for static explicit FE method}$$
$$\int_V \rho\ddot{u}_i\delta\dot{u}_i dV + \int_V \nu\dot{u}_i\delta\dot{u}_i dV + \int_V \sigma_{ij}\delta\dot{u}_{i,j}dV = \int_A p_i\delta\dot{u}_i dA$$
$$\text{for dynamic explicit FE method.}$$

(1)

Here $\delta\dot{u}_i$ means the virtual velocity, s_{ij} Kirhhoff stress, σ_{ij} Cauchy stress, ρ the mass density, ν the viscosity coefficient, p_i the traction, and " $\cdot$ " the time derivative.

2.2. Constitutive Equations of elastic plastic material model [14-15]

The constitutive equation, which is utilized in FE analysis module to predict the strain during the metal forming process, are briefly summarized. In the elastic plastic material response, the plastic part, D_{ij}^p, of the strain rate, $D_{ij} = (\dot{u}_{i,j} + \dot{u}_{j,i})/2$, is specified through the classical plastic potential and the crystalline plasticity theories. Elastic response satisfies Hooke's law, which can be formulated as a linear relationship between the elastic strain rate, $D_{ij}^e = D_{ij} - D_{ij}^p$, and a suitable objective rate of Kirchhoff stress , $\hat{s}_{ij}$. The elastic-plastic constitutive equation can be expressed as

$$\hat{s}_{ij} = C_{ijkl}^e D_{kl}^e = C_{ijkl}^e (D_{kl} - D_{kl}^p) \equiv C_{ijkl}^{ep} D_{kl} \tag{2}$$

where C_{ijkl}^e is Hooke's elastic material coefficient tensor, C_{ijkl}^{ep} the elastic-plastic one, which will be discussed in the followings.

2.2.1. plastic potential model [14],[15] :
The plastic strain rate derived through the non-associated flow theories can be expressed as follow;

$$D_{ij}^p = \frac{\langle \cos \Theta \rangle}{H_t} \hat{\sigma}_{ij}' + b_N \frac{1}{h} n_{ij} n_{kl} \hat{\sigma}_{kl}'$$

$$\cos \Theta = \frac{n_{kl} \hat{\sigma}_{kl}'}{\sqrt{\hat{\sigma}_{kl}' \hat{\sigma}_{kl}'}}, \quad H_t = \frac{2}{3} E_s \frac{(1 + \rho_N \cos \Theta)}{1 + \rho_N}, \quad b_N = 1 - \frac{h \cos \Theta}{H_t} \tag{3}$$

where E_s means σ / ε , h is obtained from $1/h = 3/2 \cdot (1/E_t - 1/E)$, $E_t = d\sigma/d\varepsilon$, those can be determined by using uni-axial stress strain curve. $\hat{\sigma}_{ij}'$ means the objective rate of Cauchy stress deviator. ρ_N has the value $(-1 < \rho_N < +\infty)$: "-1" corresponds to the associated flow law (normality rule), " ∞ " the deformation theory. $n_{ij} = (\partial \Phi / \partial \sigma_{ij})/ (\| \partial \Phi / \partial \sigma_{ij} \|)$ means the unit normal to the yield surface $\Phi(\sigma_{ij})$ defined by the plastic potential function in the stress space.

2.2.2. crystalline plasticity model [7],[50-74]:
Crystalline materials can be characterized by their morphology and physical properties. The morphology characterizations include: (I) lattice structure, e.g. fcc,bcc,hpc... (II) texture (orientation distribution) and (c) grain shape, size and boundary.

For the polycrystal material modeling, the texture plays important role for the deformation induced anisotropy and formability of sheet. Grain shape and boundary are also important for the hardening and softening evolutions, caused by the interaction between crystal aggregation. According to the progress of measurement technology, those morphology properties at the meso scale can be measured by using Electron Microscope, X-ray Diffraction Analysis, Electromagnetic Acoustic Transducer and Atomic Force Microscope.

The physical characterization can be categorized as (I) hardening evolution caused by the single slip - self hardening - and the multi slip - interaction between slip systems - and (II) softening evolution caused by the thermal activation - temperature rising effect - and the meso-damage in the slip system. The hardening evolution was measured experimentally in case of the single crystal and it is extended to the case of polycrystal. Strain hardening in each slip system is an intrinsic property, it is caused by internal multiplication and interaction of dislocations in a grain. Other extrinsic hardening phenomena are caused by solution, second phase particles and grain boundaries. But, there still remain the problem to establish measurement technology at the meso scale to identify the parameter of the hardening and softening evolution equations.

The formulation of constitutive equation is reviewed briefly. The plastic strain is derived based on the single-crystal plasticity law.

$$D_{ij}^p = \sum_a P_{ij}^{(a)} \dot{\gamma}^{(a)}, \qquad 2P_{ij}^{(a)} = s_i^{(a)} m_j^{(a)} + s_j^{(a)} m_i^{(a)} \tag{4}$$

Here (a) designates the (a)th slip system with the slip direction $s_i^{(a)}$ and the normal $m_i^{(a)}$ to the slip plane. $\dot{\gamma}^{(a)}$ is the shear strain rate on the slip system (a) and the summation extends over the active slip systems. Peirce et al. proposed the rate dependent model to express the shear strain rate in slip system (a).

$$\dot{\gamma}^{(a)} = \dot{a}^{(a)} \left[\frac{\tau^{(a)}}{g^{(a)}} \right] \left[\left| \frac{\tau^{(a)}}{g^{(a)}} \right| \right]^{\frac{1}{m}-1} \tag{5}$$

Here $\tau^{(a)}$ means the resolved shear stress on the slip system (a). $g^{(a)}$ represents the reference shear stress, $\dot{a}^{(a)}$ the reference shear strain rate, and m the material rate sensitivity. The hardening evolution equation is introduced to define the evolution of $g^{(a)}$.

$$\dot{g}^{(a)} = \sum_{b=1}^{N} h_{ab} |\dot{\gamma}^{(b)}|, \quad h_{ab} = h(\gamma) q_{ab} \tag{6}$$

where N is the number of slip systems, and h_{ab} are the hardening moduli. A matrix q_{ab} is introduced to describe the self and latent hardening, proposed by Zhou et al.[69]. γ means the accumulated slip summation over all the slip systems.

Nakamachi and Dong proposed the over all hardening coefficient as follow[7];

$$h(\gamma) = \begin{cases} h_I, & \text{for stage I} \\ h_I sech^4 \left(\frac{\gamma - \gamma_{II}}{\Delta\gamma} \right) + h(\gamma - \gamma_{II}) \left[1 - sech^4 \left(\frac{\gamma - \gamma_{II}}{\Delta\gamma} \right) \right] & \text{for others,} \end{cases} \tag{7}$$

where h_I is the constant hardening modulus in stage I - single slip(easy gliding)-, γ_{II} the summation of slip during stage I, $\Delta\gamma$ the transition strain from stage I to stage II. $h(\gamma - \gamma_{II})$ means the hardening evolution at stage III and IV.

Taking into account the thermal softening caused by transformation from the plastic deformation energy to the heat energy, g_s is employed instead of g.

$$g_s = g(1 - \eta\theta), \quad \dot{\theta} = \frac{\sigma_{ij} D_{ij}^p}{(c\rho)} = \sigma_{ij} \sum_{(a)} \frac{P_{ij}^{(a)} \dot{\gamma}^{(a)}}{(c\rho)}. \tag{8}$$

Here, η is the thermal softening coefficient, θ the temperature rise, ρ is the mass density, and c the specific heat of the material.

Finally, the rate type constitutive equation is derived as;

$$\hat{s}_{ij} = C_{ijkl}^e D_{kl} - \sum_{a=1}^{N} (C_{ijkl}^e P_{kl}^{(a)} + W_{ik}^{(a)} \sigma_{kj} - \sigma_{ikj} W_{kj}^{(a)}) \dot{\gamma}^{(a)}$$
$$2W_{ij}^{(a)} = s_i^{(a)} m_j^{(a)} - s_j^{(a)} m_i^{(a)} \tag{9}$$

2.3. Contact Theory [1-8],[75-81]

The contact force is defined at the contact point between the material and the tool surfaces. The normal unit vector $\mathbf{n} = n_i \mathbf{e}_i$ to the contact tangential surface is defined as "mesh normal" or "tool normal," those are discussed by Keum et al.. The reference Cartesian coordinate system $\mathbf{e}_i - x_i$ and the contact surface coordinate $\mathbf{G}_i - \xi_i$ are introduced. $\mathbf{G}_3$ coincide with the normal vector $\mathbf{n}$. This $\mathbf{G}_i - \xi_i$ coordinate depends on the slide motion history. The coordinate ξ_1 corresponds to the sliding trace, and therefore $\mathbf{G}_1$ coincides with the slide direction vector.

2.3.1. Normal Contact Force

(A) DE FE model - penalty method -[8]: The normal component of contact force is determined by employing the penalty method or Lagrange multiplier method. The gap $\mathbf{g} = g_i\mathbf{e}_i$ between the FE node and the tool surface is only the required value to be calculated. Therefore the normal contact force in case of the penalty method is given by,

$$p_3 = \alpha \parallel \mathbf{g} \parallel \mathbf{n} \tag{10}$$

where α means the penalty factor.

(B) SE FE model - rate type formulation-[6]: The contact surface coordinate $\mathbf{G}_i - \xi_i$ is defined by using the relative sliding velocity vector $\dot{\mathbf{U}}(= \dot{U}_\alpha\mathbf{G}_\alpha)$ as follows;

$$\mathbf{G}_1 = \frac{\mathbf{n} \times \dot{\mathbf{U}} \times \mathbf{n}}{\parallel \mathbf{n} \times \dot{\mathbf{U}} \times \mathbf{n} \parallel}, \qquad \mathbf{G}_3 = \mathbf{n}, \qquad \mathbf{G}_2 = \mathbf{G}_3 \times \mathbf{G}_1; \mathbf{G}_i = X_i^j\mathbf{e}_j. \tag{11}$$

The Greek indices range over 1,2. Therefore, the normal component of the contact force can be related to the Cauchy stress tensor as follow;

$$p_3 = \sigma_{ij}n_in_j = \sigma_{ij}X_i^3X_j^3 \tag{12}$$

2.3.2. Tangential Contact Force [2],[3],[6]

(A) DE FE model - the total friction resistant force - :   The friction coefficient μ, the proportional ratio between the normal component of the contact force p_3 and the tangential component $\parallel \mathbf{p}_t \parallel$, has the state variables of material strainε, lubricant viscosityζ, surface roughness R, temperature θ, contact slide velocity $\parallel \dot{\mathbf{U}} \parallel$, and so on.

$$\parallel \mathbf{p}_t \parallel = \mu p_3, \quad \mu = \psi(\varepsilon, p_3, \zeta, R, \theta, \parallel \dot{\mathbf{U}} \parallel \ldots). \tag{13}$$

The unit vector of the slide direction $\mathbf{t}(= t_\beta\mathbf{G}_\beta)$ defined on the contact tangential plane is determined by the associated flow type friction law. This slide direction vector $\mathbf{t}$ can be obtained by introducing "friction yield surface," expressed as follow;

$$\mathbf{t} = \frac{\mathbf{L}}{\parallel \mathbf{L} \parallel}, \quad t_\beta = \frac{L_\beta}{\sqrt{L_\delta L_\delta}}, \quad L_\beta = \frac{\partial \phi}{\partial p_\beta}, \phi = a_1(p_1)^2 + a_2(p_2)^2 \tag{14}$$

Here, $a_1, a_2$are the friction anisotropy parameters. This means analogy to the associated flow law of the plastic potential theory. Therefore, the tangential friction force $\mathbf{p}_t$ is expressed as,

$$\mathbf{p}_t = \mu p_3\mathbf{t} = \mu p_3 t_\beta X_i^\beta\mathbf{e}_i \tag{15}$$

(B) SE FE model - the rate of contact force -[1],[6] : The rate of the base vector of the sliding trace coordinate $\dot{\mathbf{G}}_1$ is obtained as,

$$\dot{\mathbf{G}}_1 = {}_B\dot{\mathbf{G}}_1 + (\mathbf{t} - \mathbf{G}_1) = \dot{t}_1\mathbf{G}_1 + \dot{t}_2\mathbf{G}_2 + B_{\alpha 1}\dot{U}_\alpha\mathbf{G}_3 \tag{16}$$

where $B_{\alpha\beta}$ means the curvature tensor of tool surface ($\mathbf{b} = B_{\alpha\beta}\mathbf{e}_\alpha \otimes \mathbf{e}_\beta$). Introducing Nanson's equation and the Cauchy stress equation, the rate of the contact force $d\mathbf{f}$ defined on the contact surface segment da is derived as follows;

$$df = \dot{p}_k \mathbf{e}_k da$$
$$\dot{p}_k = (\dot{\sigma}_{ij} + \sigma_{ij}\dot{u}_{m,m} + \sigma_{im}\dot{u}_{j,m})n_i n_j(-\mu X_1^k + X_3^k)$$
$$+\sigma_{ij}n_j[-B_{\alpha 1}\dot{U}_\alpha X_i^1 - B_{\alpha 2}\dot{U}_\alpha X_i^2](-\mu X_1^k + X_3^k)$$
$$+\sigma_{ij}n_i n_j[-\mu(\dot{t}_1 X_1^k + \dot{t}_2 X_2^k + B_{\alpha 1}\dot{U}_\alpha X_3^k) - B_{\alpha 1}\dot{U}_\alpha X_1^k - B_{\alpha 2}\dot{U}_\alpha X_2^k)]. \tag{17}$$

2.4. FE Equations

(A) DE FE model - equation of motion - : Substituting the finite element descritization relationships into the virtual work principle, Equation(1), the motion equation of finite element for DE FE model is derived as,

$$\mathbf{M\ddot{u}} + \mathbf{C\dot{u}} = \mathbf{P} - \mathbf{F}. \tag{18}$$

Here $\mathbf{M}$ and $\mathbf{C}$ are the lumped mass and damping matrices, in which the real component value is multiplied by penalty number λ, respectively. $\mathbf{P}$ means the contact force, $\mathbf{F}$ the internal nodal force, $\mathbf{u}$ the node displacement. By using the central difference method, the displacement at the next step $t + \Delta t$ is given by;

$$\mathbf{u}^{t+\Delta t} = \left(\frac{\mathbf{M}}{\Delta t_2} + \frac{\mathbf{C}}{2\Delta t}\right)^{-1}\left(\mathbf{P} - \mathbf{F} + \frac{\mathbf{M}}{\Delta t^2}(2\mathbf{u}^t - \mathbf{u}^{t-\Delta t}) + \frac{\mathbf{C}}{2\Delta t}\mathbf{u}^{t-\Delta t}\right) \tag{19}$$

where Δt is the time integration step. For the stability of the time integration, Δt should be smaller than the smallest eigen period of the finite element divided by π.

(B) SE FE model - stiffness equation - : Next, Substituting the FE descritization relationships into the rate type virtual work equation (1), the incremental stiffness equation of SE FE model is obtained as follow;

$$\mathbf{K} \cdot \Delta\mathbf{u} = \Delta\mathbf{p}$$
$$(K_{jN}^{iM} + {}_\sigma K_{jN}^{iM} + {}_G K_{jN}^{iM} + {}_s R_{jN}^{iM})\Delta u_M^j = \Delta p_N^i \tag{20}$$

where K_{jN}^{iM} is the incremental stiffness matrix, ${}_\sigma K_{jN}^{iM}$ the initial stiffness matrix and ${}_G k_{jN}^{iM}$ the initial-rotation matrix. ${}_s R_{jN}^{iM}$ means the nonsymmetric incremental contact force matrix. $\Delta\mathbf{p}$ means the incremental force vector. The effective stress and strain are obtained by accumulating values in each incremental step. The small step increments of displacement, strain and stress are required to guarantee the accuracy, therefore we adopt so-called " r-minimum control scheme."

The total incremental displacement $\Delta\mathbf{u}$ of the FE node is given by the sum of the known rigid body motion, such as the punch travel increment $\Delta\mathbf{u}_p$ and an unknown additional incremental displacement, such as the sliding displacement $\Delta\mathbf{u}^*$. For non-contacting finite element node, $\Delta\mathbf{u}_p = 0$. Therefore FE stiffness equation can be written as follow;

$$\mathbf{K} \cdot \Delta\mathbf{u}^* = \Delta\mathbf{p} - \mathbf{K} \cdot \Delta\mathbf{u}_p. \tag{21}$$

3. OPTIMIZATION

Because of the difficulty of the differentiation of the objective function by the design parameter in this fabrication optimization, only the computational simulation algorithm is applicable, such like grid method, simplex method, genetic algorithm and neural network algorithm. In this paper the simplex method is selected because of its simplicity. "Sweeping Simplex Method" is newly proposed to find the global minimum condition in the design parameter space with high computational efficiency.

3.1. "Sweeping Simplex Method"

The conventional simplex method, which consists the operations of expansion, contraction, reflection and reduction, has the feature of no guarantee to find the global minimum of the objective function. Therefore, "Sweeping Simplex Method" is newly proposed, in which the whole feasible region is swept out. It means searching throughout the design parameter region. The parallel algorithm by starting the randomly selected simplex is adopted and the region of updated simplex is informed each other and therefore the overlapping of the region is avoided. Consequently, "simplex" sweeps out whole region and reach the global minimum point. The convergence criterion is given by,

$$\| \mathbf{x}_I - \mathbf{x}_J \| \leq \delta \tag{22}$$

where $\mathbf{x}_I$ means vertex of the candidate simplex, and $\mathbf{x}_J$ the simplex who has lowest value of the objective function. Accuracy of minimum point depends on the tolerance δ. "Sweeping Simplex" algorithm prevents the new simplex to invade the already swept out region. The total swept out region S can be defined by summation of regions where each simplex swept out.

$$S = \bigcup_{i=1}^{n} s_i$$
$$s_i = \{x \mid \| \mathbf{c} - \mathbf{x} \| \leq \rho R_{max} \} \tag{23}$$

Here $\mathbf{c}$ is the center of simplex, R_{max} the largest distance from $\mathbf{c}$ to $\mathbf{x}$, ρ the assigned coefficient. To judge the new simplex to get into the swept region, the following condition is employed.

$$\frac{m(S \cap s_m)}{m(s_m)} \leq \lambda \tag{24}$$

Here, m(s) means the region in the design parameter space, λ the threshold parameter.

3.2. Objective Function and Design Parameter

There expected huge number of design parameters in the industrial forming operations, such like material, lubricant, tool geometry and work processes. In this paper, two cases of optimization are demonstrated. First, introducing the experience based data base for the sheet metal forming, the geometry of the tool surface and punch height are selected as the design parameters. Second, the optimum texture design of the aluminum alloy sheet is discussed. Both cases adopted two objective functions as follow;

$$f_1 = min(| \varepsilon_t |), \qquad f_2 = \sqrt{\frac{1}{N} \sum_{i=1}^{N} (1 - \frac{h_i}{h_{ave}})^2} \tag{25}$$

where, f_1 means the minimum thickness strain of the deformed sheet, and f_2 the deviation from the uniform (average) thickness. In both cases, the optimization corresponds to search the global minimum point of objective function in the design parameter space.

4. APPLICATIONS

4.1. FE Benchmark tests

In this last decade, the interest in the sheet metal forming simulation by using FE analysis code is expanding while research in simulation of other fields of forming operations, e.g. forging, rolling, extrusion and casting, remains nearly constant, because of the enormous importance in the automotive, electrical and aeronautical industries. Until the end of 1980s, the research work has devoted the effort to firm the bases for the robust and efficient computation, which can be featured as the updated Lagrangian formulation, the plastic potential theory based elastic plastic material model, Lagrangian multiplier method for contact modeling. NUMIFORM(NUmerical Methods in Industrial FORMing Processes in 1989)[40],[44],[45] and NUMISHEET (NUMerical sImulation of 3-D SHEET metal forming processes in 1991,1993)[42],[43] have proposed Benchmark problems of sheet forming and revealed the progress of simulation technology in the sense of practical application. The first Benchmark problem (OSU) in NUMIFORM89 adopted very simple geometry tool problem, such like the hemi spherical punch stretching and drawing[40],[82]. Most of the participants employed the static implicit type integration scheme and simple material model such as rigid plastic or elastic plastic ones, in which the plastic potential theory - J2F, J2D, corner theory, void nucleation theory - was adopted. It concluded that the reliable results for the simple geometry tool and stretch dominant deformation, but uncertainty and convergence difficulty in case of large draw-in deformation. At the same conference, DE FE code analyzed successfully the large draw-in and wrinkling problems. DE FE code has shown the capability to predict such instability phenomena - wrinkling and localization-. During 1989 and 1991, the modification of finite element itself and the contact algorithm in conjunction with 3D tool description has been carried out. The achievement to overcome the convergence problem in SI FE method by using transient function has been appeared.

In 1991 NUMISHEET91(VDI) conference, 2nd Benchmark problem adopted the rather complex tool geometry, which is described by CAD surface data written in IGES standard format[42]. All SI,SE and DE FE results provided better agreement with the experiment than the first Benchmark problem, as shown in Figure 4. But still uncertainties of material and friction modelings were pointed out from the participants. The best agreement result was obtained by employing the free friction force and the isotropic hardening material models.

1993 NUMISHEET conference provided three Benchmark problems[43]; (I) square cup deep drawing, (II) front fender stamping and (III) 2D draw bending. From the comparison of results of (I)problem with the experiments -twenty seven FE simulation and ten experiment participants -, the availability of material and friction models has been investigated. The strain distribution results obtained by all SI, SE and DE FE analyses showed good agreement with the experiment. But it was pointed out that if there was no FLD curve obtained by the experiment,these FE codes could not predict the forming limit and formability. This result motivated the participants to develop more"physically true" material model based on the micro or meso scale constitutive law.

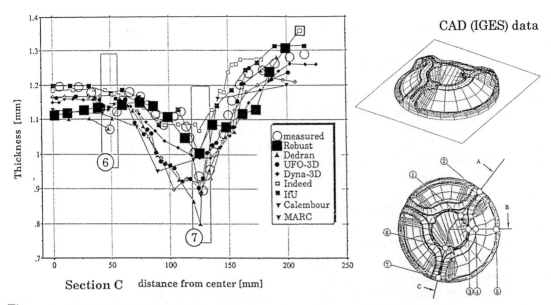

CAD (IGES) data

Figure 4   VDI NUMISHEET91 2nd Benchmark problem   (1991,[79])

Comparison of thickness strain with FE analyses and experimental measurement
(wheel house stamping)

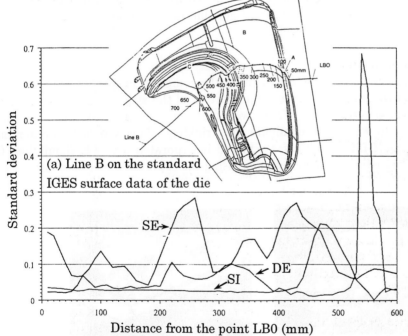

(a) Line B on the standard
IGES surface data of the die

Figure 5   NUMISHEET93 3rd Benchmark problem   (1993,[81])

Standard deviation of thickness strain on the Line B

(front fender stamping)

It suggests that the meso scale material modeling based on the crystalline plasticity is the future trend , which could be implemented in the FE analysis code, as shown in Figure 2(a). Further, both material and friction modelings required the comprehensive technology based on the micro scale experimental observation, which bridges the material science and mechanical engineering[6] ,[40-45]

In Benchmark problem (II), FE results - thirteen participants- were compared with the experiment. The standard deviation from the experimental results is shown in Figure 5. SI shows good accuracy - the iterative algorithm was improved very much-, but larger CPU time and the less finite element number. SE shows robustness, without the convergence difficulty, but poor accuracy and rather large CPU time. The less finite element number means poor prediction of the wrinkling mode and strain localization. DE FE results show good prediction of wrinkling because of huge number of finite element -tens thousands- and less CPU time compared with the others, because of no simultaneous equation solving procedure. But the little less accuracy of strain distributions than SI FE ones. There remained the problems, how to determine the damping parameter for the time scaling, and how to do mapping the high speed deformation to the quasi-static one.

In Benchmark problem (III), twenty FE simulations and eleven experiments were participated. All FE simulations showed very much scattered results. The material model is not sufficient to catch the strain and stress histories in the forming process. DE FE has the serious problem to overcome the infinity time scale difference.

In this year, 1996, 4th Benchmark problem in NUMISHEET96 will offer the opportunity to reveal the progress of time scaling algorithm in DE FE analysis and large scale descritization algorithm in SE and SI FE ones.

4.2. Virtual Manufacturing

The visualized inspection system by using the computer graphics(CG) scheme can detect the surface wrinkling and very small distortion of the wheel house inner as shown in Figure 6. "Virtual manufacturing(VM) " were performed to find the appropriate blank holding pressure loads, which can eliminate the wrinkling. This VM in Computer Integrated Manufacturing(CIM) system was achieved incorporated with FE simulation, CG detection and the heuristic optimum design by using the expertise experience database[6].

 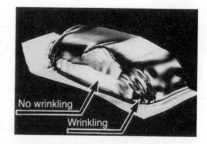

(1) low blank holding force   (2) appropriate holding force

Figure 6   Wheel House Forming Analyses -Highlight

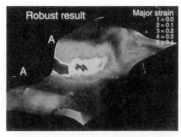
(a-1) SE FE simulation

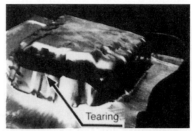
(b-1) Experimental verification

(Ⅰ) First Trial -tearing-

(a-2) SE FE simulation

(b-2) Experimental verification

(Ⅱ) Second Trial (open draw) -wrinkling-

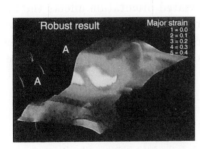
(a-3) SE FE simulation

(b-3) Actual production in the automotive plant

Figure 7　(Ⅲ) Final Trial -proper condition-
Virtual manufacturing of sheet forming in MITSUBISHI MOTOR Co. Plant
(a) FE simulation results of deformation and major strain distributions
(b) Experimental results of deformation

Next example demonstrates that after three virtual forming trials, the real production was carried out successfully in the stamping shop of MITSUBISHI MOTOR CO.[6]. Fig.8(a) and (b) show FE simulation and experimental results of deformations with major principal strain ; the first trial employed the virgin CAD tool data, the second one cut off one side of sheet to allow easy draw-in. The third (final) forming condition was obtained by investigating the previous simulation results and modifying the tool geometry. The real production was successfully carried out by employing the third virtual forming conditions as shown in Figure 7(a-3) and (b-3). The experiments at each step verify the FE simulation capability for the practical application.

4.3. Optimum Fabrication Design

The above mentioned VM requires the quantitative prediction of deformation behavior and estimation of fabrication performance. Especially, for the quantitative estimation of the performance of the forming operation, the mathematical programming approach is commended. In this paper, the fusion system, which consists SE and DE FE codes and "Sweeping Simplex Method," was applied to the optimization of two stage stamping operation and material fabrication. Figure 8 shows the optimization of the two stage hemisphere-prism punch stamping operation. The design parameter H1 and H2 mean punch travels of two cube cylindrical punches at the first stage forming operation. The objective function f_2 in Equation (25) was adopted and the global minimum point - H1=5.6mm and H2=3.0mm - in the design parameter space was found out by using "Sweeping Simplex Method." The good agreement with the experimental results was obtained.

Figure 9 shows the optimum texture design of aluminum alloy sheet material, by employing LDH formability test. Two deviation angles are selected as the design parameter, which mean the orientations of the crystal are uniformly distributed within the described value of angle deviation from the cube and copper textures. FE simulation results show how the combination of two texture with the deviated orientation affects the formability by adopting the objective function f_1 in Equation (25) . This texture design system can determine the optimum fabrication condition of aluminum alloy sheet. This optimum fabrication design system based on computational mechanics might innovate the classical one based on the human experience data base.

5. CONCLUSION

The virtual manufacturing technology, which consists the analysis and synthesis, is the future trend. The quantitative prediction of deformation behavior to achieve a vial and economical solution is the essence of approach to the optimum forming design. The elaboration simulation in this fusion system - the combination of FE analysis and mathematical programming - provides the extrapolated decision for the very complex problem in the industries, and places a responsibility on the engineers to use this power to establish the better forming operation system. The virtual reality and networking technologies and the super-parallel computing technology are also important. It suggested that the important role of this virtual manufacturing based on the highly integrated computational technology will be recognized more and more in the industrial hypersolution space.

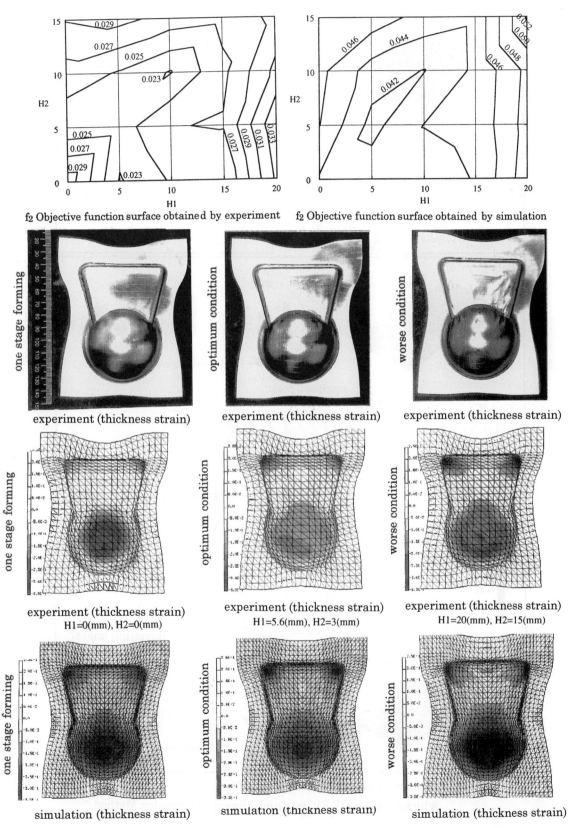

f2 Objective function surface obtained by experiment   f2 Objective function surface obtained by simulation

Figure 8 Optimization of two stage forming process of optimum condition

338

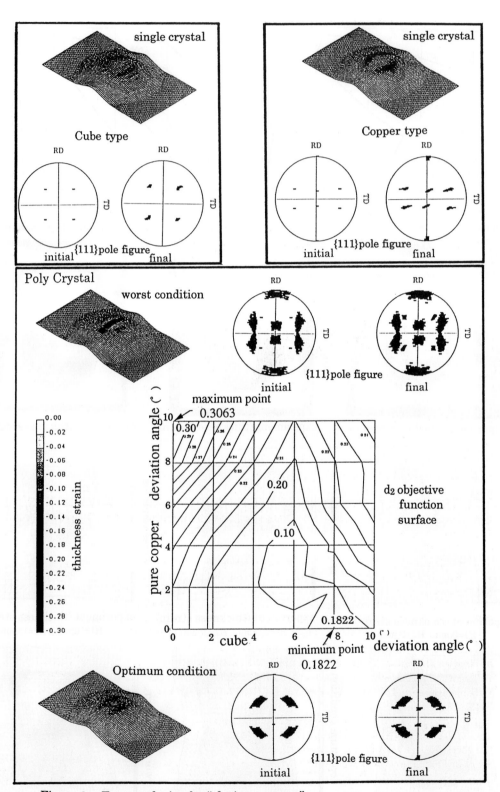

Figure 9 Texture design by " fusion system "
which consists DE FE code and " Sweeping Simplex Method."

References

1. E. Nakamachi, Int. J. Numer. Methods. Eng., 25 (1988) 283.
2. Y. T. Keum, E. Nakamachi, R. H. Wagoner and J. K. Lee,
   Int. J. Numer. Methods. Eng., 30 (1990) 1471.
3. E.Nakamachi and J.Komada, Trans. JSME, (in Japanese) 58-551(A)(1992)1228.
4. E. Nakamachi, J. Mater. Process. Technol., 50 (1995) 116.
5. T.Huo and E. Nakamachi, J. Mater. Process. Technol., 50 (1995) 180.
6. E. Nakamachi, Engineering Computations, 13-2/3/4(1996) 283.
7. E. Nakamachi and X. Dong, Engineering Computations, 13-2/3/4(1996) 308.
8. E. Nakamachi and T. Huo, Engineering Computations, 13-2/3/4(1996) 327.
9. T. Katayama, E. Nakamachi, T. Hasebe, T. Oguchi, and Y. Imaida,
   Proc. of 3rd World Congress on Computational Mechanics (Tokyo), (1994) 376.
10. Truesdel, W. Noll, The nonlinear fields theory, Springer Verlag,1962.
11. L.E. Malvern, Introduction to the Mechanics of a Continuum Medium, Prentice-Hall Inc., 1969.
12. A. E. Green and W. Zerna, Theoretical Elasticity, 2nd edition, Oxford, 1968.
13. R. Hill, The Mathematical Theory of Plasticity, Oxford, 1950.
14. M. Gotoh, Int. J. Solids Struct. 21-11(1985),1117.
15. M. Gotoh, Trans. JSME(A), (in Japanese), 55-518(1989),2080.
16. L.Barlat and J. Lian, Int. J. Plasticity, 5, (1989),51.
17.K.S.Havner, Finite Plastic Deformation of Crystalline Solids, Cambridge Univ. Press,1992.
18.O. C. Zienkiwicz, Finite Element Method, McGraw-Hill, London, 1977.
19. J.T. Oden, Finite Element of Non-linear Continua, McGraw-Hill, 1972
20. D. R. J. Owen and E. Hinton, Finite Elements in Plasticity: Theory and Practice,
    Pineridge Press Limited, 1980.
21. K.J. Bathe, Finite ElementProcedures in Engineering Analysis, Prentice-Hall,1982
22. G.W.Rowe, C.E.N.Sturgess, P.Hartley and I.Plllinger, Finite Element Plasticity and
    Metal Forming Analysis,Cambridge Univ. Press,1991.
23. Z-H. Zhong, Finite Element Procedures for Contact-Impact Problems,
    Oxford University Press, New York, 1993.
24. P.V.Marcal, Int. J. Mech. Sci. 7(1969) 229.
25. R.M. McMeeking and J.R.Rice, Int. J. Solids Struct., 11(1975)601.
26. R.D. Krieg and D.B. Krieg,Trans. ASME, J. Press. Vessel Technol., 101(1979) 226.
27. J.C. Nagtegaal, Comput. Methods. Appl. Mech. Eng., 33(1982) 469.
28. T. Belytschko, J. I. Lin and C. S. Tsay, Comput. Methods. Appl. Mech. Eng., 42 (1984) 225.
29. T. Belytschko, J. S. J. Ong and W. K. Liu, Comput. Methods. Appl. Mech. Eng.,
    44 (1984) 269.
30. K. J. Bathe and E. N. Dvorkin, Int. J. Numer. Methods. Eng., 21 (1985) 367.
31. M. Ortiz and E.P. Popov, Int. J. Numer. Methods. Eng., 21 (1985) 1561.
32. A. Matzenmiller and R.L. Taylor, Int. J. Numer. Methods. Eng., 37 (1994) 813.
33. N.M.Wang and B. Budiansky, Trans. ASME, J. Appl. Mech.,45(1978),73.
34. D.P.Koistinen and N.M.Wang, editors, Mechanics of Sheet Metal Forming, Plenum Press,1978.
35. C.H.Toh and S. Kobayashi,Int. J. Mach. Tool Des. Res. 25-1 (1985), 15.
36. S. Storen and J. R. Rice, J. Mech. Phys. Solids, 23 (1975) 421.
37. A. Needleman,Trans. ASME, J. Appl. Mech. 56 (1989) 1.
38. Y. Tomita, Appl. Mech. Rev., part 1, 47 (1994) 171.
39. E.Nakamachi and R.H.Wagoner, Proc. of the SAE Int. Congress and Exposition,
    No.880528, Detroit(1988)109.
40. E.G.Thompson,R.D.Wood, O.C.Zienkiewicz and A. Samuelsson, editors,
    Proc. of NUMIFORM 89, A.A.BALKEMA, 1989.
41. Honecker, A. and Mattiasson, K., reference [80] 457.
42. Proc. of FE-Simulation of 3-D Sheet Metal Forming Processes in Automotive Industry,
    VDI Verlag, 1991.
43. J.-L.Chenot, R.D.Wood and O.C.Zienkiewicz, editors, Proc. of NUMIFORM'92,

A.A.BALKEMA, 1992.

44. A.Makinouchi, E. Nakamachi, E.Onate and R.H.Wagoner, Proc. of NUMISHEET93, JSFSRG Press,RIKEN INSTITUTE Press, Saitama, Japan, 1993.
(J. Mater. Process. Technol., Special Volume 50, 1995)

45. S.-F.Shen and P.R.Dawson, editors, Proc. of NUMIFORM95, A.A.BALKEMA, 1995.

46. M.S. Bazaraa, H.D. Sherali and C.M. Shetty, Nonlinear Programming,
John Wiley and Sons, Inc., 1979

47. J.E.Jr. Dennis and R.B. Schnabel, Numerical Methods for Unconstrained Optimization and Nonlinear Equations, Prentice-Hall, Inc., 1983.

48. G.L. Nemhauser, Optimization; Handbooks in Operations Research and Management Science,Vol. 1, North-Holland, 1989.

49. N.P. Suh, The Principles of Design, Oxford University Press, 1990.

50. G.I. Taylor, J. Inst. Metals, 62 (1938) 307.

51. J. D. Eshelby, Proc. Roy. Soc. London, A 241 (1957) 376.

52. U.F. Kocks and T.J. Brown, Acta Metall., 14 (1966) 87.

53. J. W. Hutchinson, Proc. Roy. Soc. London, A 319 (1970) 247.

54. R. Hill and J. R. Rice, J. Mech. Phys. Solids, 20 (1972) 401.

55. R.W. Hertzberg, Deformation and Fracture Mechanics of Engineering Materials,
Wiley, New York, 1976.

56. R.J. Asaro, Acta Metall. 27, 445-453, 1979.

57. D. Peirce, R.J. Asaro and A. Needleman, Acta Metall. Mater., 30 (1982) 1087.

58. D. Peirce, R.J. Asaro and A. Needleman, Acta Metall. Mater., 31 (1983) 1951.

59. R.J. Asaro and A. Needleman, Acta Metall., 33 (1985) 923.

60. P.E. McHugh, R.J. Asaro and C.F. Shih, Acta Metall. Mater., 41 (1993) 1461.

61. T.Y. Wu, J.L. Bassani and C Laird, Proc. R. Soc. Lond. A, 435 (1991) 1.

62. J.L. Bassani and T.Y. Wu, Proc. R. Soc. Lond. A, 435 (1991) 21.

63. Q. Qiu and J.L. Bassani, J. Mech. Phys. Solids, 40 (1992) 813.

64. Q. Qiu and J.L. Bassani, J. Mech. Phys. Solids, 40 (1992) 835.

65. W. Yang and W.B. Lee, Mesoplasticity and its Applications, Springer-Verlag, 1993.

66. J.L. Bassani, Advances in Applied Mechanics, 30 (1994) 191.

67. A. S. Khan and S. Huang, Continuum Theory of Plasticity,
John Wiley & Sons, Inc., 1995.

68. A.M. Maniatty, P.R. Dawson and Y.-S. Lee, Int. J. Numer. Methods Eng., 35 (1992) 1565.

69. Y. Zhou, K.W. Neale and L.S. Toth, Int. J. Plasticity, 9 (1993) 961.

70. C.A. Bronkhorst, S.R. Kalidindi and L. Anand, Phyl. Trans. R. Lond., A 341 (1992) 443 .

71. A.J. Beaudoin, et al., Comput. Methods Appl. Mech. Engrg, 117 (1994) 49.

72. Y. Zhou and K.W. Neale, Int. J. Mech. Sci., 37 (1995) 1.

73. R. Becker and S. Panchanadeeswaran, Acta Metell., 43 (1995) 2701.

74. V. P. Smyshlyaev and N. A. Fleck, J. Mech. Phys. Solids, 44 (1996) 465.

75. N. Kikuchi and J. T. Oden, Contact Problems in Elasticity, Philadelphia, 1988.

76. T. Belytschko and J. I. Lin, Comput. Struct., 25 (1987) 95.

77. J. O. Hallquist, G. L. Goudreau and D. J. Benson, Comput. Methods. Appl. Mech.
Eng., 51, (1985) 107.

78. T. Belytschko and M. O. Neal, Int. J. Numer. Methods. Eng., 31 (1991) 547.

79. Y. Seguchi, A. Shindo and Y. Tomita, Sliding rule of friction in plastic forming of metal, Computational Methods in Nonlinear Mechanics, University of Texas at Austin, 683, 1974.

80. D.J. Benson, and J.O. Hallquist, Int. J. Numer. Methods Eng., 22 (1986) 723.

81. J.C. Simo, P. Wriggers, and R.L. Taylor, Comp. Meths. Appl. Mech. Engng., 50 (1985) 163.

82. J.K.Lee, R.H.Wagoner and E. Nakamachi, Proc. of Computational Mechanics'91, S.N.Atluri, D.E.Beskos, R.Jones and G.Yagawa, W.H.Wolfe Associates, Alpharetta, Georgia (1991)588.

Theoretical and Applied Mechanics 1996
T. Tatsumi, E. Watanabe and T. Kambe (Editors)
© 1997 Elsevier Science B.V. All rights reserved.

Quench stress and deformation: mathematical modelling and numerical simulation.

Sören Sjöström

Dept. TE, ABB STAL AB
S–612 82 FINSPÅNG, Sweden

and

Dept. of Mechanical Engineering, Linköping University
S–581 83 LINKÖPING, Sweden

ABSTRACT

The article deals with the modelling of heat treatment of steel by quenching. This process is modelled as a coupled thermal/metallurgical/mechanical process. A few application examples are shown.

1. FUNDAMENTALS OF THE MODEL

A first step in the setting up of a model of heat treatment of steel by quenching will be the insight that the process combines three simultaneous subproblems, namely
(1) a thermal problem (of heat conduction within the quenched component and heat transfer between quenchant and component),
(2) a metallurgical problem (of microstructure transformations), and
(3) a mechanical problem (of stress and deformation).
This is illustrated in Fig. 1 below, which shows the three corresponding problem classes.

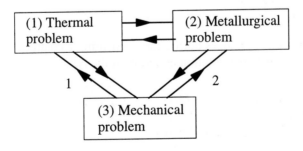

Fig. 1. Scheme of connections between the subprocesses involved.

Different variants of this modelling problem (using different degrees of refinement) have been published in, for instance, Refs. [1] through [16]. The only monography available on the subject is Ref. [17].

1.1. Thermal problem

For the solution of the heat conduction through the workpiece and its heat exchange with the environment, one uses the 1st law of thermodynamics. In a continuum description, we have,

$$\varrho \dot{e} = T : \dot{E} + r - \text{div} q, \tag{1}$$

where ϱ is the mass density, e is the internal energy per unit of mass, T is the Cauchy stress, E is the infinitesimal strain and div is the divergence operator. The rate of internal energy production r gives the phase transformation (latent) heat, *i.e.*

$$r = \mathcal{F}_r(\dot{z}), \tag{2}$$

while the heat flux vector q can be generally described within the workpiece by Foutrier's law

$$q = H(\Theta, z)\nabla\Theta, \tag{3}$$

and on the (cooled) outer surface by the boundary condition

$$q = \mathcal{F}_q(\Theta, t), \tag{4}$$

where $H(\Theta, z)$ is the heat conductivity, Θ is the temperature, t is time and ∇ is the gradient operator. In Eqs. (2) and (3), we have also used the microstructure description vector z, described in the next section

1.2. Metallurgical problem

At a particular instant, the internal microstructure of the material is defined by, for instance, a vector z, which may simply be described as a vector of phase volume fractions.

$$z = \{z_k\}_{k=1,\dots,K} \quad ; \quad \sum_{k=1}^{K} z_k = 1 \tag{5}$$

Due to the microstructure changes during the quenching, z changes continuously. For z, we must, therefore, establish an evolution equation of the type

$$\dot{z}(x, t) = \mathcal{F}_z'[\Theta(x, t), \dot{\Theta}(x, t), t] = \mathcal{F}_z(\Theta, \dot{\Theta}, z) \tag{6}$$

where x is the vector of Cartesian coordinates.

1.3. Mechanical problem

The basic equations of the mechanical (deformation/stress) part of the problem are the *equilibrium* (eq.), *compatibility* (comp.), and *constitutive* (const.) equations:

$$\text{div} T + \varrho b = 0 \qquad \qquad \text{(eq.)}, \tag{7}$$

$$E = \mathrm{sym}(\nabla u) = \tfrac{1}{2}(\nabla u + (\nabla u)^T) \qquad \text{(comp.)}, \qquad (8)$$

where

$$x = x_0 + u \qquad\qquad (9)$$

and

$$\dot{E} = \mathcal{F}_{\dot{E}}(T, \dot{T}, \Theta, \dot{z}; y, \Theta, z) \qquad \text{(const.)} \quad (10)$$

In Eqs. (7) through (10), b is the volume force per unit of mass, u is the displacement vector and x_0 is the material (*i.e.* Lagrangian) coordinate vector.

The stress state generated during the quenching is a self–equilibrating stress state. There are, therefore, good reasons to believe that the corresponding deformation will be limited, and an infinitesimal strain measure [Eq. (8)] will be sufficient.

2. DETAILS OF THE ABOVE MODELS

2.1. Thermal problem.

From Eqs.(1) through (3) we can, by usual arguments (developed in any standard textbook on heat transfer), derive the heat conduction equation

$$\mathrm{div}\, q = r + T:\dot{E}^p - (p - \Theta\frac{\partial p}{\partial \Theta})\cdot \dot{y} + \Theta\frac{\partial T}{\partial \Theta}:\dot{E}^e - \varrho c_v\dot{\Theta}, \qquad (11)$$

in which p is the thermodynamic force vector corresponding to the internal variable vector y.

For this particular application, the following function descriptions and simplifications are convenient:

$$r = \sum_{k=2}^{K} c^r{}_k \dot{z}_k = c^r \cdot \dot{z}, \qquad\qquad (12)$$

where c^r is a vector of constants $(c^r{}_k)_{k=2,\ldots,K}$ for the different possible phase transformations. Further,

$$q = H(\Theta, z)\nabla\Theta \quad \forall x \in \Omega \qquad\qquad (13)$$

$$q = n\mathcal{F}_{qs}(\Theta) \quad \forall x \in \partial\Omega \qquad\qquad (14)$$

where $\mathcal{F}_{qs}(\Theta)$ is a scalar–valued function of the temperature, describing the rate of heat transfer at the boundary, and n is the surface normal vector. The terms dependent on $\dot{E}^p$, $\dot{y}$ and $\dot{E}^e$, *i.e.*, the second, third and fourth terms of the right member of Eq. (11) can, further, be shown to be small for a typical heat treatment situation, in which there is no external load.

Eqs (11) through (14) lead to an equation in Θ, $\nabla\Theta$ and $\dot\Theta$,

$$\text{div } (H(\Theta,z)\nabla\Theta) = r - \varrho c_v \dot\Theta \qquad (15)$$

which forms the basis of a discretised (FE) model of the heat conduction part of the process.

Finally, we postulate that the temperature and microstructural composition dependence of H can be expressed as a linear mixture rule:

$$H(\Theta,z) = \sum_{k=1}^{K} z_k H_k(\Theta) \qquad (16)$$

where $H_k(\Theta)$ are the heat conductivities (as functions of Θ) of the different phases.

2.2. Metallurgical problem

The typical microstructural changes in steel during heat treatment by quenching are diffusional (e.g. pearlitic) or martensitic. It is therefore practical to separate the microstructure vector z into the three parts $z^a (= z_1)$ (fraction of the parent phase austenite), z^d [where $(z^d)_{k=2,...,K-1}$ are the fractions of diffusional transformation products], and $z^m (= z_K)$ (fraction of martensite):

$$z = \left\{ \begin{matrix} z^a \\ z^d \\ z^m \end{matrix} \right\} \qquad (17)$$

For the kth *diffusional* transformation product z_k^d one can, for instance, use the following classical evolution law by Johnson–Mehl–Avrami ([18] through [21]):

$$\dot z^d(x,t) = \mathcal{F}_z{}'(\Theta(x,t),\dot\Theta(x,t),t) = z_{,0}^a \left(1 - e^{-a_k t^{b_k}} \right), \qquad (18)$$

where $z_{,0}^a$ is the fraction of austenite at the beginning of the transformation, and where $a_k = a_k(\Theta)$ and $b_k = b_k(\Theta)$. $a_k(\Theta)$ and $b_k(\Theta)$ can, for instance, be taken from an IT diagram, which is a diagram that gives the time till, say, 10% and 90 % of the kth phase at the constant temperature Θ. Eq. (18) can, thus, be readily converted into the evolution–type equation

$$\dot z_k^d = \mathcal{F}_{z^d}(\Theta,\dot\Theta,z_k^d)$$

$$= -(z_k^d - z_{,0}^a)\left\{ -\frac{da_k}{d\Theta}\frac{1}{a_k}\ln(1 - \frac{z_k^d}{z_{,0}^a}) - b_k\frac{db_k}{d\Theta}[\ln(1 - \frac{z_k^d}{z_{,0}^a})]^{(b_k-1)/b_k}a^{1/b_k} \right\}\dot\Theta \qquad (19)$$

This basic model can be further refined by the introduction of an 'incubation' time before the start of the Johnson–Mehl–Avrami growth process. This technique was used in, for instance, Ref. [22].

For the *martensitic* transformation product z^m, the most frequent law used is the Koistinen–Marburger law [23]:

$$z^m = z^a_{,0}[1 - e^{-a(\Theta^m_s - \Theta)}],\tag{20}$$

in which Θ^m_s is the 'martensite start temperature', *i.e.* a threshold temperature, above which there is no martensitic transformation, and α is a characteristic constant. Eq. (20) can also be reformulated as an evolution law:

$$\dot{z}^m = \mathcal{F}_{z^m}(\dot{\Theta}, z^m) = -a(z^{a,0} - z^m)\dot{\Theta}\tag{21}$$

Common to diffusional as well as martensitic transformation is an influence even of the stress state on the respective evolution laws (see, for instance, [24] through [26]):

$$\dot{z} = \dot{z}(\Theta, \dot{\Theta}, z, T).\tag{22}$$

The knowledge of this influence is, however, still only fragmentary; further, the mechanisms behind this are of a micromechanical nature so that any modelling on a macro scale (for use in engineering computations) is not obvious even if these micro mechanisms should be sufficiently wellknown. For most of our work we have, therefore, chosen not to take this into account.

2.3. Mechanical problem

In the mechanical part of the problem, the constitutive description needs particular attention. Starting from the basic definition of Eq. (10), we split the strain rate into *elastic, thermal, plastic, viscoplastic* and *transformation* terms:

$$\begin{aligned}\dot{E} &= \mathcal{F}_E(T, \dot{T}, \dot{\Theta}, \dot{z}; y, \Theta, z)\\ &= E^e(\dot{T}; \Theta) + E^{th}(\dot{\Theta}; \Theta) + E^p(T, \dot{T}, \dot{\Theta}, \dot{z}; y, \Theta, z) + \dot{E}^v(T; z, \Theta) + E^{tr}(\dot{z}; T)\end{aligned}\tag{23}$$

Elastic strain rate $\dot{E}^e$ For the elastic strain, a linear equation can usually be justified, *i.e.*

$$E^e = \frac{1}{2\mu}T - \mathbf{1}\frac{\lambda}{2\mu(3\lambda + 2\mu)}\mathrm{tr}(T),\tag{24}$$

where λ and μ are the usual Lamé constants, which are each computed by linear fraction rules analogous with that of Eq. (16):

$$\mu = \mu(z, \Theta) = \sum_{k=1}^{K} z_k \mu_k(\Theta) \quad \text{and} \quad \lambda = \lambda(z, \Theta) = \sum_{k=1}^{K} z_k \lambda_k(\Theta)\tag{25}$$

Thermal strain rate $\dot{E}^{th}$ In our model, the thermal strain rate $\dot{E}^{tr}$ has been kept separated from the isothermal elastic strain rate $\dot{E}^e$, and again we postulate a linear fraction expression

$$E^{th} = \sum_{k=1}^{K} z_k \alpha_k(\Theta)\mathbf{1} \Rightarrow \dot{E}^{th} = E^{th}(\dot{\Theta}, \dot{z}, \Theta, z) = \ldots\tag{26}$$

Plastic strain rate E^p : Fundamental plasticity theory gives

$$\dot{E}^p = \dot{E}^p(T, \dot{T}, \Theta, \dot{z}, y, \Theta, z) = \dot{\lambda}^p \frac{\partial f^p}{\partial T} \tag{27}$$

where $\dot{\lambda}^p$ is the plastic multiplier and f^p is the yield function, following the usual complementarity principles:

$$f^p < 0 \Rightarrow \dot{\lambda}^p = 0 \tag{28}$$

$$f^p = 0 \text{ and } \dot{f}^p < 0 \Rightarrow \dot{\lambda}^p = 0 \tag{29}$$

$$f^p = 0 \text{ and } \dot{f}^p = 0 \Rightarrow \dot{\lambda}^p > 0 \tag{30}$$

Further, in Eq. (27), y is an internal variable, used for describing the plastic hardening. For linear kinematic hardening, the normal choice for y would be to use the plastic strain E^p. However, we now introduce as a postulate that the new phase forming by the simultaneous metallurgical process does not inherit the plastic hardening (as described by, for instance, the plastic strain) acquired during the previous austenitic state; but is, instead, plastically virgin. We therefore get a 'dilution' of the plastic strain and arrive at a new set V_k of internal variables (one for each microstructural constituent present):

$$\dot{V}_1 = \dot{E}^p \tag{31}$$

$$\dot{V}_k = \dot{E}^p - \frac{\dot{z}_k}{z_k} V_k \quad \forall k : 2 \leq k \leq K \tag{32}$$

where (as before) $k = 1$ stands for the parent phase (austenite). We therefore have the evolution law

$$\dot{E}^p = \dot{\lambda}^p(T, \dot{T}, \Theta, \dot{z}; V_k, \Theta, z) \frac{\partial f^p}{\partial T} \tag{33}$$

For the yield function f^p we use the von Mises definition:

$$f^p = [\tfrac{3}{2}(T' - T^b) : (T' - T^b)]^{1/2} - Y, \tag{34}$$

where T' and T^b are the stress deviator and the 'backstress', respectively, defined in the usual ways:

$$T' = T - \tfrac{1}{3} \text{tr}(T)I, \text{ and} \tag{35}$$

$$\dot{T}^b = \mathcal{F}_{T^b}(z, \dot{V}_k) = \sum_{k=1}^{K} c_k(\Theta) z_k \dot{V}_k. \tag{36}$$

Further, Y is the yield strength as measured in a uniaxial tensile test; for kinematic hardening we have, as usual,

$$Y = Y(\Theta, z), \tag{37}$$

where, again, a linear volume fraction rule has been postulated:

$$Y = \sum_{k=1}^{K} z_k Y_k(\Theta). \tag{38}$$

By the consistency condition $\dot{f}^p = 0$ during plastic flow, together with Eqs. (31) through (34), and Eqs. (36) and (38), we can now derive an expression for the plastic multiplier λ^p:

$$\dot{f}^p = \frac{\partial f^p}{\partial \boldsymbol{T}}\dot{\boldsymbol{T}} + \frac{\partial f^p}{\partial \boldsymbol{T}^b}\dot{\boldsymbol{T}}^b + \frac{\partial f^p}{\partial Y}\frac{\partial Y}{\partial \Theta}\dot{\Theta} + \frac{\partial f^p}{\partial Y}\frac{\partial Y}{\partial z} \cdot \dot{z} = 0 \tag{39}$$

$$\Rightarrow \lambda^p = \frac{-\frac{\partial f^p}{\partial \boldsymbol{T}} \cdot \dot{\boldsymbol{T}} + \frac{\partial f^p}{\partial \boldsymbol{T}^b} \cdot \sum_{k=2}^{K} c_k z_k V_k - \frac{\partial f^p}{\partial Y}\frac{\partial Y}{\partial \Theta}\dot{\Theta} - \frac{\partial f^p}{\partial Y}\frac{\partial Y}{\partial z} \cdot \dot{z}}{\frac{\partial f^p}{\partial \boldsymbol{T}^b} \cdot \frac{\partial f^p}{\partial \boldsymbol{T}} \sum_{k=1}^{K} c_k z_k} \tag{40}$$

Viscoplastic strain rate $\dot{\boldsymbol{E}}^v$: We postulate that for not–too–heavy steel sections no viscoplastic deformation ('creep') is likely to occur. This is, however, no to be considered as definite; we only postulate at this stage that the inelastic behaviour is dominated by instantaneous plastic rather than viscoplastic strain; hence $\boldsymbol{E}^v = 0$. For other materials, for instance aluminium, where similar processes have been analysed, we have, instead, found it necessary to use a viscoplastic model, which has been presented elsewhere.

Transformation strain $\boldsymbol{E}^{tr}$: For the modelling of steel quenching, there is experimental as well as theoretical evidence that one must go beyond the usual idea of an isotropic transformation strain by introducing also a deviatoric term. Though the transformation strain is in no way completely understood yet, there is at least a generally accepted frame for it, namely:

$$\dot{\boldsymbol{E}}^{tr}(\boldsymbol{T}, z, \dot{z}) = \dot{\boldsymbol{E}}^{tr,i}(\dot{z}) + \dot{\boldsymbol{E}}^{tr,d}(\boldsymbol{T}, z, \dot{z}) \tag{41}$$

For the isotropic transformation strain rate, we can assume the following equation:

$$\dot{\boldsymbol{E}}^{tr,i}(\dot{z}) = \sum_{k=2}^{K} \dot{z}_k E^{tr,i}_k \boldsymbol{1}. \tag{42}$$

The deviatoric term $\dot{\boldsymbol{E}}^{tr,d}(\boldsymbol{T}, z, \dot{z})$ of the transformation strain rate is a macro consequence of microincompatibilities between parent phase and transformation product ('Greenwood–Johnson mechanism', [27]) and of the microanisotropy of, for instance, the martensitic transformation ('Magee mechanism', [28]). $E^{tr,d}$ has been proved to be (at least partly) irreversible (*i.e.*, it is not recovered in a full transformation cycle austenite → martensite → austenite). Since this is a plasticity–like property, it has often been called transformation plastic tic strain. Reviews of theory, modelling and consequences of transformation plasticity are given in [29] and [30].

For diffusional transformation, different rate equations have been proposed, [31] through [39]. As an example, the one by Giusti *et al.* reads:

$$\dot{\boldsymbol{E}}^{tr,d}(T,\dot{z},z) = \sum_{k=2}^{K} c_k^{tr,T}(1 - z_k)\dot{z}_k T'. \tag{43}$$

In this equation, $c_k^{tr,T}$ is a constant.

For martensitic transformation, the interaction between micro–stress state and transformation kinetics is quite strong. A few attempts at a micromechanical modelling of the interaction between martensitic plate formation and local stresses are for instance, described in [40] through [43]. There is, however, at the moment no universally accepted model for the 'transformation plasticity' part of the transformation strain in the martensitic case, and we have so far chosen to use the same equation as for diffusional transformations.

3. NUMERICAL SIMULATION

It is now obvious that the different parts of the problem are closely interconnected (see Fig. 1). The inherent complexity, however, makes it desirable to find simplifications when preparing for a numerical simulation.

3.1. Semi–coupled model

Coupling No. 1 of Fig. 1 can, for instance, be identified as originating from the second, third and fourth terms of the right–hand member of Eq. (11). We can, however, on good reasons assume that there are no other internal variables present than the plastic strain $\boldsymbol{E}^p$, and further we know that the plastic strain rate $\dot{\boldsymbol{E}}^p$ must be low in this type of (externally unloaded) problems. We can therefore conclude that this coupling may well be left out without serious consequences.

Coupling No. 2 of Fig. 1 is described by Eq. (22). Referring to the previous discussion, we do not know very much about the influence of this coupling on the simulation; it is, however, obvious that if we postulate that even this coupling is of minor importance, we can set up a simplified 'semi–coupled' model, where the thermal/metallurgical and the mechanical solutions can be performed in a sequence instead of simultaneously as in the completely coupled model. This model (Fig. 2) will now be described in some detail, starting with the mechanical solution part.

3.1.1. Mechanical problem

The following *weak* formulation of the problem (see Fig. 3) can be set up:

$$\int_{\Omega} \boldsymbol{T}:\dot{\boldsymbol{E}}dV = \int_{\Omega} \boldsymbol{b} \cdot \boldsymbol{v}dV + \int_{\partial\Omega - \partial\Omega_v} \boldsymbol{t} \cdot \boldsymbol{v}dA$$

$$\forall \, \boldsymbol{v},\dot{\boldsymbol{E}} : \boldsymbol{v} \text{ satisfies the essential boundary conditions on } \partial\Omega_v \, , \, \dot{\boldsymbol{E}} = \text{sym}(\nabla\boldsymbol{v}) \tag{44}$$

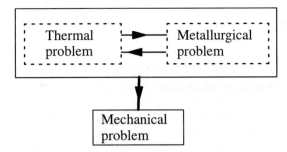

Fig. 2. Semi–coupled thermal/metallurgical/mechanical problem

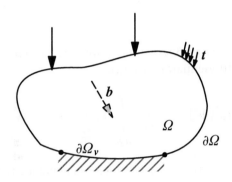

Fig 3. Basic definition of mechanical problem.

In Eq. (44), v is the velocity. After subdivision of Ω into E elements, each covering a domain Ω^e, and introduction of shape ('interpolation') functions $\Phi_n^e(x)$ in the usual finite-element way, defined on the elements Ω^e, and such that

$$
\begin{cases}
\Phi_n^e = 0 \, \text{for} \, x \notin \Omega^e \\
\\
0 \le \Phi_n^e \le 1 \, \text{for} \, x \in \Omega^e,
\end{cases}
\begin{cases}
\Phi_n^e = 0 \, \text{for} \, x = x_i, \, i \ne n \\
\Phi_n^e = 1 \, \text{for} \, x = x_n \\
\sum_{n=1}^{N_e} \Phi_n^e = 1 \quad \forall \, x \in \Omega^e
\end{cases}
, \tag{45}
$$

we have

$$
v(x) = \sum_{n=1}^{N_e} \Phi_n^e(x) v^e{}_n \tag{46}
$$

and

$$\dot{E}(x) = \sum_{n=1}^{N_e} (\mathrm{sym}(\nabla\Phi_n^e)v_n^e = \sum_{n=1}^{N_e} B_n^e(x)v_n^e, \tag{47}$$

where N_e is the total number of finite–element degrees of freedom of the eth element, and where v_n^e is the nth component of the element displacement vector v^e of the eth element.

The integrals in Eq (44) can now be formally calculated as sums over all elements Ω^e, and we get:

$$\sum_{e=1}^{E} \int_{\Omega^e} T : \sum_{n=1}^{N_e} B_n^e(x)v_n^e dV = \sum_{e=1}^{E} \int_{\Omega^e} b(x) \cdot \sum_{n=1}^{N_e} \Phi_n^e(x)v_n^e dV + \sum_{e=1}^{E} \int_{\partial\Omega^e} t(x) \cdot \sum_{n=1}^{N_e} \Phi_n^e(x)v_n^e dA \tag{48}$$

After converting to matrix/vector notations and replacing the *sum* operation by an *assembly* operation, and after time differentiation, we get:

$$\mathop{A}_{n=1}^{N} \int_{\Omega^e} \{\dot{T}\}^T[B]dV - \mathop{A}_{n=1}^{N} \int_{\Omega^e} \{\dot{b}\}^T[\Phi^e]dV - \mathop{A}_{n=1}^{N} \int_{\partial\Omega^e} \{\dot{t}\}^T[\Phi^e]dA = 0 \tag{49}$$

From Eqs.(23) through (26), (33), (34) and(40) we can conclude that there exists a linear mapping $\dot{T} \mapsto \dot{E}$, i.e.

$$\dot{E} = F^{tot} : \dot{T} \tag{50}$$

where $F^{tot} = F^{tot}(\Theta, \Theta, T, z, \dot{z}, V_k)$ is a 4th order tensor. Thus, under the assumption of sufficiently regular functions, we can write

$$\dot{T} = (F^{tot})^{-1} : \dot{E} \tag{51}$$

and can reformulate Eq. (49) as

$$\left\{ \mathop{A}_{n=1}^{N} \int_{\Omega^e} [B]^T[F^{tot}]^{-1}[B]dV \right\}\{v_n\} - \mathop{A}_{n=1}^{N} \int_{\Omega^e} [\Phi^e]^T\{\dot{b}\}dV - \mathop{A}_{n=1}^{N} \int_{\partial\Omega^e} [\Phi^e]^T\{\dot{t}\}dA = 0 \tag{52}$$

We now introduce the element tangent stiffness matrix $[K_v^e]$ and the element load rate vectors $\{\dot{f}_b^e\}$ and $\{\dot{f}_t^e\}$:

$$\left\{ \mathop{A}_{n=1}^{N} [K_v^e] \right\}\{v^n\} - \mathop{A}_{n=1}^{N} \{\dot{f}_b^e\} - \mathop{A}_{n=1}^{N} \{\dot{f}_t^e\} = 0 \tag{53}$$

which, by introduction of new notations, can be compactly written as

$$\begin{cases} [\mathbf{K_v}]\{\mathbf{v^n}\} - \{\mathbf{f_b}\} - \{\mathbf{f_t}\} = \{\mathbf{R}\} \\ \{\mathbf{R}\} = 0 \end{cases}.$$ (54)

For the *numerical solution of the mechanical problem*, we have used a time increment organisation, following a Newton–Raphson scheme for each increment Δt as shown in Fig. 4.

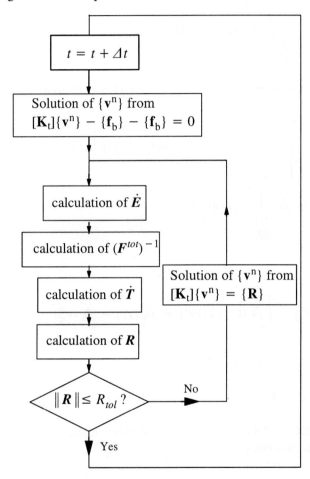

Fig 4. Newton–Raphson scheme

It must be noted that the correct calculation of $\dot{\mathbf{T}}$ (formally by Eq. (51)) is one of the main difficulties of the solution; for this we use what can mainly be characterised as a 'backward Euler' elastic predictor/plastic corrector scheme.

3.1.2. Thermal/metallurgical problem
The following weak formulation can be used:

$$\int_{\partial\Omega} \mathbf{q}(\mathbf{x}) \cdot \mathbf{n}(\mathbf{x})w(\mathbf{x})dA - \int_{\Omega} \mathbf{q}(\mathbf{x}) \cdot \nabla w(\mathbf{x})dV = \int_{\Omega} [r(\mathbf{x}) - \varrho c_v(\mathbf{x})\Theta(\mathbf{x})]w(\mathbf{x})dV \ \forall \, w(\mathbf{x}),$$ (55)

where $w(x)$ is an arbitrary weight function, which must be defined and well–behaving over the whole of Ω. Eq. (55) gives

$$\int_\Omega H(x)\nabla\Theta(x) \cdot \nabla w(x)dV + \int_\Omega \varrho c_v(x)\dot{\Theta}(x)w(x)dV$$

$$= \int_\Omega r(x)w(x)dV - \int_{\partial\Omega} q(x) \cdot n(x)w(x)dA \quad \forall \, w(x) \tag{56}$$

After manipulations analogous to those of the mechanical solution this gives:

$$\left\{ \underset{n=1}{\overset{N}{A}} \int_{\Omega_e} [\mathbf{B}]^T H[\mathbf{B}]dV \right\}\{\Theta^n\} + \left\{ \underset{n=1}{\overset{N}{A}} \int_{\Omega_e} \varrho c_v[\Phi^e]^T[\Phi^e]dV \right\}\{\dot{\Theta}^n\}$$

$$= \underset{n=1}{\overset{N}{A}} \int_{\Omega_e} r[\Phi^e]dV - \underset{n=1}{\overset{N}{A}} \int_{\partial\Omega_e} q \cdot n[\Phi^e]dA, \tag{57}$$

i.e.

$$\left\{ \underset{n=1}{\overset{N}{A}}[\mathbf{K}_\Theta^e] \right\}\{\Theta^n\} + \left\{ \underset{n=1}{\overset{N}{A}}[\mathbf{C}_\Theta^e] \right\}\{\dot{\Theta}^n\} = \underset{n=1}{\overset{N}{A}}\{\mathbf{r}_r^e\} - \underset{n=1}{\overset{N}{A}}\{\mathbf{r}_q^e\}, \tag{58}$$

which after the assembly process can be written

$$[\mathbf{K}_\Theta]\{\Theta_n\} + [\mathbf{C}_\Theta]\{\dot{\Theta}_n\} = \{\mathbf{r}_r\} - \{\mathbf{r}_q\} \tag{59}$$

For the numerical implementation of this, a standard implicit finite difference time incrementation technique has been used.

Note that the *metallurgical* process simulation part is hidden in the calculation of $r(x)$ (*i.e.* in FE terms in the calculation of $\{\mathbf{r}_r\}$), according to Eqs. (12), (19) and (21), and in the dependence of H on z [Eq. (16)].

3.2. A note on the coupled model

For a coupled treatment of the process, one must solve the two main equation systems of Eqs. (59) and (54) simultaneously. It should also be remembered that in the most complete formulation of the problem, the heat conduction problem is more complex than has been shown here [for instance, by the inclusion of the stress tensor T in the phase evolution rate equation (22)]. Therefore, a full theory will lead to the following final system of equations:

$$\begin{bmatrix} [\mathbf{K}_{vv}] & [\mathbf{K}_{v\Theta}] \\ [\mathbf{K}_{\Theta v}] & [\mathbf{K}_{\Theta\Theta}] \end{bmatrix} \left\{ \begin{array}{c} \{v^n\} \\ \{\Theta^n\} \end{array} \right\} + \left\{ \begin{array}{c} [\mathbf{C}_{v\Theta}]\{\dot{\Theta}^n\} \\ [\mathbf{C}_{\Theta\Theta}]\{\dot{\Theta}^n\} \end{array} \right\} = \left\{ \begin{array}{c} \{\mathbf{f}_v^n\} \\ \{\mathbf{f}_\Theta^n\} \end{array} \right\} \tag{60}$$

Among the problems with this formulation is that the system matrix symmetry is lost, so that some well–established numerical solution methods will fail.

An early version of such a complete solution, made in a simplified way, is presented in [22] and [26].

4 APPLICATIONS

In the Linköping group, we have implemented the model described in two different FE environments:
- The complete coupled model (though formally somewhat different from that of Eq. (60)) has been implemented in an in–house FE code (QUEST), written for the simplified geometry of an infinitely long cylinder. This program has served as a test bench during the early development of the mathematical/physical modelling.
- The semi–coupled strategy has been implemented in the commercial FE code ABAQUS by including the metallurgical and mechanical models as user–written subroutines.

In addition, there also exist commercially available FE packages that include (most of) the model described. The most well–known ought to be
- HEARTS (CRC Research Institute, Osaka, Japan) and
- SYSTUS/SYSWELD (FRAMASOFT, Paris, France).

4.1. First example: quenching of a long, axisymmetrically cooled cylinder of a carburised case–hardening steel

As an example of the performance possible, the quenching of a long, axisymmetrically cooled cylinder of a carburised case–hardening steel is shown. The case depth was 0.6 mm, and the quenching was in a medium–grade oil. The stress state history near the outer (quenched) surface is shown in Fig. 5, and the profile of residual stress is plotted against the radius in Fig. 6. The simulation has been performed with different degrees of refinement, and the numbers on the curves correspond to the data given in Table 1.

Table 1.
Data for simulations presented in Figs. 5 and 6.

Run No.	Semi–coupled/ coupled analysis?	Type of plastic hardening	E^p or V_k as internal variable?	Transformation plasticity?
21	semi–coupled	isotropic	E^p	no
22	semi–coupled	kinematic	E^p	no
23	semi–coupled	kinematic	V_k	no
25	coupled	kinematic	V_k	yes
24	semi–coupled	kinematic	V_k	yes

It can be seen that only after the introduction of transformation plasticity do the simulation results come near the experimental results.

4.2. Second example: the deformation of a case–hardened transmission wheel

As a second example, in Fig. 7 the axisymmetric half–cross section of a heavy truck transmission wheel is shown. The wheel has been 'horizontally' submerged in the quench bath

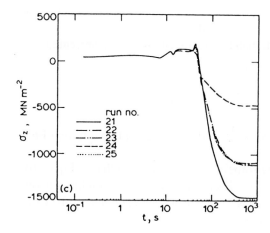

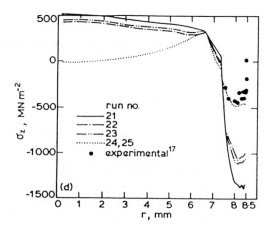

Fig. 5. Carburised 17 mm diameter cylinder, quenched in oil. Axial stress σ_z at the surface as a function of time t.

Fig. 6. Carburised 17 mm diameter cylinder, quenched in oil. Axial stress σ_z at room temperature as a function of radius r. Experimental values from [44]

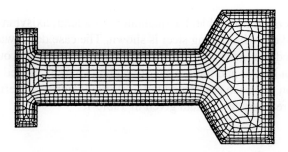

Fig. 7. Geometry and FE mesh of cross section of a heavy truck transmission wheel. Hub at left, cog base at right (cogs not modelled).

(which was an oil). Due to the flanges, the buoyancy flow of the quenchant will be much less efficient at the bottom face of the wheel than at the top (gas bubbles from the film boiling stage at the beginning of the process can even become trapped at the bottom face for most of the quenching process). This leads to a generally lower q at the bottom face, which, in turn, gives a tendency for bowl–shaped deformation of the wheel. The material is, again, a carburised case–hardening steel, and the deformation pattern given by the simulation (in this case by SYSWELD) is shown in Fig. 8, together with the results from measurements at the 0°, 90°, 180° and 270° cross sections. One further model refinement was introduced in this case by taking into account the fact that during a certain time interval after the initiation of the

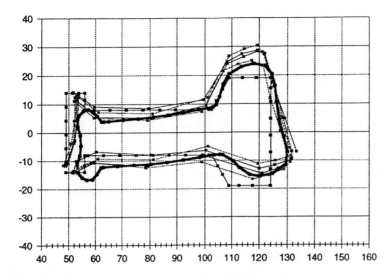

Fig. 8. Distortion of the heavy truck transmission wheel.
  Medium thickness solid line — original contour
  Bold solid line   — simulated distorted geometry
  Thin lines   — measured distorted geometries at the 0°, 90°, 180°
                                 and 270° cross sections.
  (Displacement magnification factor 100)

martensite formation (*i.e.*, while it is still rather hot) there is a certain amount of decomposition of the martensite into finely dispersed carbides.

This simulation is reported in [45].

5. CONCLUSIONS

The model described has been used in a number of different situations. Among the experience gathered we can list the following:
– The amount of data needed for characterisation of the behaviour of the material during the process is a major difficulty in the practical use.
– Even considering the rapid progress in the performance of computers, the complexity of the model still poses a limit on its use for large–scale 3D analyses (although we have, in fact, conducted a number of smaller 3–D test cases).
– In cases where we have been able to compare the calculated stress results with, for instance, X–ray diffraction measurements, we have found good agreement in case hardening simulations, whereas for through–hardening the results have not been as unambiguously good.

Latest reports from applications, however, point at a possible breakthrough in the design of heat–treated components for high–cycle fatigue use by the modelling described above.

REFERENCES

1. T Inoue, K Tanaka: *Int. J. Mech. Sci.* **17**(1975), 361–367
2. T Inoue, B Raniecki: *J. Mech. Phys. Sol.* **26**(1978), 187–
3. A Burnett, J Padovan: *J Thermal Stresses*, **2**(1979), 251–63.
4. B Hildenwall: Prediction of the residual stresses created during quenching, Linköping Studies in Science and Technology, Diss. No. 39, Linköping, Sweden, 1979.
5. T Inoue, S Nagaki, T Kishino, M Monkawa: *Ingenieur–Archiv* **50**(1981), 315–327.
6. F G Rammerstorfer, D F Fischer, W Mitter, K–J Bathe, M D Snyder: *Computers & Structures* **13**(1981), 771–779.
7. S Sjöström: The calculation of quench stresses in steel, Linköping Studies in Science and Technology, Diss. No. 84, Linköping, Sweden, 1982.
8. M Melander: A computational and experimental investigation of induction and laser hardening, Linköping Studies in Science and Technology, Diss. No. 124, Linköping, Sweden, 1985.
9. S Sjöström: *Mat. Sci. Techn.* **1**(1985), 823–
10. T Inoue, Z Wang: *Mat. Sci. Techn.* **1**(1985), 845–850.
11. A–M Habraken: Contribution à la modélisation du formage des métaux par la méthode des elements finis. Thèse Docteur en Sciences Appliquées, Université de Liège, Liège, Belgium, 1989.
12. N Järvstråt: Two–dimensional calculation of quench stresses in steel, Linköping Studies in Science and Technology, Thesis No. 245, Linköping, Sweden, 1990.
13. S Sjöström: The calculation of residual stress. In Hutchings, Krawitz (eds.): Measurement of Residual and Applied Stress using Neutron Diffraction, Kluwer Academic Publishers, Dordrecht, Netherlands, 1992. (Proceedings of the NATO Advanced Research Workshop on Measurement of Residual and Applied Stress using Neutron Diffraction, Oxford, U.K., 18th–22nd March, 1991.)
14. A–M Habraken, M Bordouxhe: *Eur. J. Mech. A/Solids*, **11**(1992), 381–402
15. T Inoue, D–Y Ju: *Int. J. Plasticity* **8**(1992), pp. 161–83
16. N Järvstråt, S Sjöström: Current status of TRAST, a material model subroutine system for the calculation of quench stresses in steel. Proc. ABAQUS Users' Conference, Aachen, Germany, 23rd–25th June, 1993, 273–288.
17. A J Fletcher: Thermal stress and strain generation in heta treatment, Elsevier, London, U.K., 1989.
18. W A Johnson , R F Mehl: *Trans. AIME*, **135**(1939), 416–58
19. M Avrami: *J Chem. Physics*, **7**(1939), 1103–
20. M Avrami: *J Chem. Physics*, **8**(1940), 212–
21. M Avrami: *J Chem. Physics*, **9**(1941), 177–
22. S Denis, S Sjöström, A Simon: *Met. Trans. A 18A*(1987), 1203–1212
23. D P Koistinen, R E Marburger: *Acta Metall* **7**(1959), 59–
24. E Gautier: Transformations perlitique et martensitique sous contrainte de traction dans les aciers, Thèse Docterur, L'institut National Poytechnique de Lorraine, Nancy, France, 1985
25. S Denis, E Gautier, A Simon, G Beck: *Mat Sci Techn* **1**(1985), 805–14
26. S Denis, E Gautier, S Sjöström, A Simon: *Acta metall.* **35**(1987), 1621–1632
27. G W Greenwood, R H Johnson: *Proc. Roy. Soc.*, **A 283**(1965), 403.
28. L. Magee: Transformation kinetics, microplasticity and aging of martensite in Fe–31Ni, PhD thesis, Carnegie–Mellon University, Pittsburgh, U.S.A., 1966

29. W Mitter: Umwandlungsplastizität und ihre Berücksichtigung bei der Berechnung von Eigenspannungen, Gebrüder Borntraeger, Berlin–Stuttgart, BRD, 1987.

30. F.D. Fischer, Q–P. Sun, K. Tanaka: *Appl. Mech. Rev.* **49**(1996), 317–364

31. Y Desalos, J Giusti, F Gunsberg: Déformations et contraintes lors du traitement thermique de pièces en acier, Irsid, Re902, Saint–Germain–en–Laie, 1982

32. J–B Leblond, G Mottet, J–C Devaux: *J. Mech. Phys. Solids*, **34**(1986), 395–409.

33. J–B Leblond, G Mottet, J–C Devaux: *J. Mech. Phys. Solids*, **34**(1986), 411–432.

34. J–B Leblond, J Devaux, J–C Devaux: *Int. J. Plasticity*, **5**(1989), 551–572.

35. J–B Leblond: *Int. J. Plasticity*, **5**(1989), 573–591.

36. F D Fischer: *Acta Metall. Mater.*, **38**(1990), 1535–1546.

37. J–F Ganghoffer, Modélisation micromécanique et calcul par éléments finis du comportement des materiaux en cours de transformations de phases: Cas des transformations perlitique et martensitique d'alliages ferreux, Diss., I.N.P.L., Nancy, France, 1992

38. J–F Ganghoffer, S Denis, E Gautier, A Simon, S Sjöström: *Eur. J. Mech. A/Solids,* **12**(1993), 21–32.

39. S. Sjöström, J–F. Ganghoffer, S. Denis, E. Gautier, A. Simon: *Eur. J. Mech. A/Solids,* **13**(1994), 803–817.

40. J–F Ganghoffer, S Denis, E Gautier, A Simon, K Simonsson, S Sjöström: *Journal de Physique IV,* **1**(1991), pp

41. J–F Ganghoffer, S Denis, E Gautier, A Simon, K Simonsson, S Sjöström: *Journal de Physique IV,* **1**(1991), C4/89–94.

42. K. Simonsson: Micomechanical FE simulations of the plastic behaviour of steels undergoing martensitic transformation, PhD thesis, Linköping University, Linköping, Sweden, 1994.

43. F. Marketz: Computational micromechanics study on martensitic transformation, PhD thesis, University for mining and metallurgy, Leoben, Austria, 1994.

44. M Knuuttila: Computer controlled residual stress analysis and its application to carburized steel, Linköping Studies in Science and Technology, Diss. No. 81, Linköping, Sweden, 1981.

45. T. Thors: Thermomechanical calculation of quench distortion with application to case hardening of steel, Thesis No. 566, Linköping, Sweden, 1996.

Theoretical and Applied Mechanics 1996
T. Tatsumi, E. Watanabe and T. Kambe (Editors)
© 1997 Elsevier Science B.V. All rights reserved.

359

Modeling large deformation and failure in manufacturing processes

D. J. Bammann[a], M. L. Chiesa[a] and G. C. Johnson[b]

[a]Sandia National Laboratories, Livermore, California 94551-0969
[b]Department of Mechanical Engineering, University of California, Berkeley, California 94720

A state variable model for describing the finite-deformation behavior of metals is proposed. A multiplicative decomposition of the deformation gradient into elastic, volumetric plastic (damage), and deviatoric plastic parts is considered. With respect to the natural or stress-free configuration defined by this decomposition, the thermodynamics of the state variable theory is investigated. This model incorporates strain rate and temperature sensitivity, as well as damage, through a yield surface approach in which the state variables follow a hardening minus-recovery format. The microstructural underpinnings of the model are presented along with a detailed description of the effects which each model parameter has on the predicted response. Issues associated with the implementation of this model into finite element codes are also discussed. The model has been applied in finite element simulations of several manufacturing processes. These include the prediction of anisotropy in hydroforming applications, metal cutting, and high temperature applications such as forging and welding. In each case, experiments are well described by predictions based on this model.

1. INTRODUCTION

Many manufacturing processes involving metals, such as forging, welding, sheet metal forming, cutting, etc., are associated with large deformations and large temperature excursions. As such, the material models developed to describe these processes must be capable of capturing these effects. The present paper describes a state variable model for the finite inelastic deformation of metals which incorporates the effects of temperature and strain-rate sensitivity, as well as the presence of damage.

In many forming processes, deformation takes place at elevated temperature, but the structure formed may be used at any temperature, from room temperature up to the forming temperature. This clearly necessitates the use of a model which accounts for the effects of temperature on the material response over the full range of temperatures and which accounts for the recovery which typically occurs while the material is at elevated temperature. Similarly, strain rate effects often play an important role in determining the final quality of a formed piece. Any model which attempts to capture the essence of a forming process which involves high-rate deformation must include rate effects in the model description. Finally, large deformation processes can lead to failure through the growth and coalescence of voids in the material.

Much research and modeling effort has been conducted in this area and a recent review of the status of this field can be found in the proceedings of the NUMIFORM conference held in 1995 [1]. Presented here is a model for the large-deformation plastic behavior of metal which includes all of these effects. Motivated by the microstructural processes which are responsible for the plastic deformation, this continuum-based model introduces a yield surface whose evolution is governed by the critical parameters of temperature, strain rate, and damage state. The model uses a hardening minus recovery format for the evolution of the internal (or state) variables which describe the size, shape, and location of the yield surface. In addition, an isotropic damage parameter is introduced, with evolution based upon the growth of spherical voids in an elastic matrix material.

This paper presents an overview of this model, and provides a detailed description of how the various constitutive parameters affect the predicted response in a series of uniaxial loading tests at different temperatures and strain rates. A variety of examples comparing the capabilities of this model with experimental data are also presented. These include modeling the failure of a series of notched-tensile specimens of 6061-T6 aluminum, the penetration of an aluminum disk by a projectile impacting at different velocities, the spall formation in aluminum plates, the anisotropic forming of a thin sheet and the HERF forging of a high pressure gas reservoir.

2. KINEMATICS

In the model used in this work, the total deformation gradient F is multiplicatively decomposed into terms which account for the elastic, plastic, damage, and thermal portions of the motion. Thus, we write

$$F = F_e F_p F_d F_\theta \qquad (1)$$

where F_e and F_p represent the elastic and (deviatoric) plastic portions, respectively, of the deformation gradient, F_d represents the (volumetric) deformation associated with any damage which may accompany the overall deformation, and F_θ represents the deformation associated with thermal expansion. In writing F in this way, we can think of a sequence of steps leading to the final state of the material, starting with a change in temperature leading to F_θ, then the evolution of damage leading to F_d. This motion is then followed by the plastic deformation F_p, which can be thought of as occurring at constant temperature and damage state, and finally the elastic deformation F_e.

Associated with the various parts of the deformation gradient are a series of different material configurations. Of particular interest for this work are the initial configuration prior to the deformation, the current configuration reached after the total deformation is completed, and the intermediate configuration reached by "unloading" through F_e^{-1} from the current configuration at current levels of temperature, damage, and plastic deformation. In identifying F_p and F_d separately, we are associating the volumetric portion of the inelastic deformation as "damage," while the deviatoric part of the inelastic deformation is referred to as the "plastic" deformation.

For simplicity, we take the damage and the thermal expansion to be isotropic. The damage can then be characterized by the void volume fraction ϕ, so that we can write

$$F_d = \frac{1}{(1-\phi)^{1/3}} 1, \qquad (2)$$

where 1 is the identity tensor. For most materials of interest, the deformation associated with thermal expansion is small, so that the thermal portion of the deformation gradient can be given in terms of the linear coefficient of thermal expansion β and the temperature change $\Delta\theta$ as

$$F_\theta = F_\theta 1 \cong (1 + \beta\Delta\theta) 1. \qquad (3)$$

The isotropic expressions in Eqs. (2) and (3) allow the total deformation gradient to be written as

$$F = \frac{F_\theta}{(1-\phi)^{1/3}} F_e F_p \ . \tag{4}$$

We identify with an overbar those quantities which are measured relative to the intermediate configuration associated with "unloading" from the current configuration through F_e^{-1}. The total strain $\overline{E}$ referred to this intermediate configuration may be written

$$\overline{E} = \frac{(1-\phi)^{2/3}}{F_\theta^2} F_p^{-T} E F_p^{-1} = \overline{E}_e + \overline{E}_{pd\theta} \ , \tag{5}$$

where

$$\overline{E}_e = \tfrac{1}{2}[F_e^T F_e - I], \qquad \overline{E}_{pd\theta} = \tfrac{1}{2}[I - \frac{(1-\phi)^{2/3}}{F_\theta^2} F_p^{-T} F_p^{-1}]. \tag{6}$$

The velocity gradient associated with this decomposition of F may be written as the sum of terms associated with each part of the deformation,

$$L = \dot{F} F^{-1} = L_e + L_p + L_d + L_\theta, \tag{7}$$

where

$$L_e = \dot{F}_e F_e^{-1}, \quad L_p = F_e \dot{F}_p F_p^{-1} F_e^{-1} = F_e \overline{L}_p F_e^{-1} \ , \quad L_d = \frac{\dot{\phi}}{3(1-\phi)} I, \quad L_\theta = \frac{\dot{F}_\theta}{F_\theta} I \ . \tag{8}$$

In this description, we have introduced the "plastic" velocity gradient defined in the intermediate configuration,

$$\overline{L}_p = \dot{F}_p F_p^{-1}. \tag{9}$$

Differentiating Eq.(6), and using Eq. (8)$_1$ gives

$$\dot{\overline{E}} = F_e^T D_e F_e \tag{10}$$

where D_e is the symmetric part of L_e.

The remainder of this paper will focus on problems for which the elastic strain is infinitesimally small. To quantify this, we write the elastic portion of the deformation gradient as

$$F_e = I + H \ , \qquad \sup \| H \| = \varepsilon \ll 1. \tag{11}$$

As in linear elasticity, the distinction between the current configuration and the reference configuration (in this case, the intermediate configuration) vanishes, and for purposes of the stress rate - strain rate relations,

$$\dot{\overline{E}}_e \approx D_e \ . \tag{12}$$

3. THERMODYNAMICS

The thermodynamics which we employ utilize the internal state variable theory of Coleman and Gurtin [2]. We utilize the natural configuration as the appropriate frame in which to formulate the thermodynamics. In this frame we assume that the free energy ψ, is a function of the elastic strain $\overline{E}_e$, defined with respect to this frame, the temperature, temperature gradient and deformation like internal state variables which will be motivated from dislocations in cell structures, cell interiors and voids, $\overline{A}$, $\overline{K}$, and $\overline{\Phi}$, respectively. Hence we assume,

$$\psi = \hat{\psi}(\overline{E}_e, \overline{K}, \overline{A}, \overline{\Phi}, \theta, \nabla\theta). \tag{13}$$

Then the Clausius -Duhem inequality with respect to the intermediate configuration can be written,

$$\dot{\psi} + \eta\dot{\theta} + \frac{1}{\overline{\rho}}\overline{\sigma}^T \cdot \overline{L} - \frac{\overline{q} \cdot grad\theta}{\overline{\rho}\theta} \geq 0, \tag{14}$$

where, η is the entropy, $\overline{q}$ the heat flux and $\overline{\rho}$ the density with respect to the intermediate configuration. Then substituting equation (13) into equation (14),

$$\left(\overline{\sigma}^T - \overline{\rho}\frac{\partial\psi}{\partial\overline{E}_e}\right) - \overline{\rho}\left(\eta + \frac{\partial\psi}{\partial\theta}\right) - \overline{\rho}\frac{\partial\psi}{\partial\,grad\,\theta}\cdot\overline{grad\,\theta}$$

$$+\overline{\sigma}^T \cdot \left(F_e\overline{l}_p F_e^{-1}\right) - q \cdot \frac{grad\,\theta}{\theta} - \overline{\rho}\frac{\partial\psi}{\partial\overline{A}}\cdot\dot{\overline{A}} - \overline{\rho}\frac{\partial\psi}{\partial\overline{K}}\cdot\dot{\overline{K}} \tag{15}$$

$$-\overline{\rho}\frac{\partial\psi}{\partial\overline{\Phi}}\cdot\dot{\overline{\Phi}} \geq 0$$

It then follows from classic arguments [3] that

$$\frac{\partial\psi}{\partial\,grad\,\theta} = 0$$

$$\eta = -\frac{\partial\psi}{\partial\overline{E}_e} \tag{16}$$

$$\sigma = \overline{\rho}F_e\frac{\partial\psi}{\partial\overline{E}_e}F_e^T$$

And defining α, κ, and Γ as the conjugate thermodynamic stresses

$$\alpha = \overline{\rho}F_e\frac{\partial\psi}{\partial\overline{A}}F_e^T$$

$$\kappa = \overline{\rho}F_e\frac{\partial\psi}{\partial\overline{K}}F_e^T \tag{17}$$

$$\Gamma = \overline{\rho}F_e\frac{\partial\psi}{\partial\overline{\Phi}}F_e^T$$

Evolution equations must now be written for the kinematic variables $\overline{A}$, $\overline{K}$, and $\overline{\Phi}$ motivated from micromechanics and related to he stresses α, κ, and Γ through equation (17). These stresses are the variables which appear in the flow rule of our constitutive model. Details of the relationships and the physical motivation of the evolution equations will be given in a later paper.

4. MODEL DESCRIPTION

Let us now define a measure of stress $\overline{\sigma}$ in the intermediate configuration, expressed in terms of the Cauchy stress as

$$\overline{\sigma} = J_e F_e^{-1} \sigma F_e^{-T}, \qquad (18)$$

where $J_e = \det F_e$. The constitutive equation relating the stress to the elastic strain in this configuration has the usual linear form,

$$\overline{\sigma} = K(\phi)\operatorname{tr}(\overline{E}_e)1 + 2\mu(\phi)\overline{E}_e', \qquad (19)$$

where $K(\phi)$ and $\mu(\phi)$ are the damage dependent bulk and shear moduli, respectively, and $\overline{E}_e'$ is the deviatoric part of $\overline{E}_e$. The forms of the damage dependence of the elastic moduli is the familiar form valid for dilute concentration of voids,

$$\mu(\phi) = \mu_0 \left[1 - 5\frac{3\kappa_0 + 4\mu_0}{9\kappa_0 + 8\mu_0}\phi \right] \qquad (20)$$

$$\kappa(\phi) = \kappa_0 \left[1 - \frac{3\kappa_0 + 4\mu_0}{4\mu_0}\phi \right] \qquad (21)$$

Taking the time derivative of Eq. (14), and linearizing in H allows us to express the constitutive relation in terms of a convective derivative of the Cauchy stress which involves the skew-symmetric part of the elastic velocity gradient tensor,

$$\overset{\circ}{\sigma} = \dot{\sigma} - W_e\sigma + \sigma W_e \approx \lambda(\phi)\operatorname{tr}(D_e)1 + 2\mu(\phi)D_e - \frac{\dot{\phi}}{1-\phi}\sigma. \qquad (22)$$

The model for the inelastic part of the motion is a strain rate and temperature dependent elastic-plastic model which uses internal variables which are introduced with respect to the unloaded configuration to describe the state of the material[4,5,6,7]. We begin with Hooke's law as defined in equation (22). From Eq. (7), we compute the elastic stretching and spin as

$$D_e = D - D_p - D_d - D_\theta, \qquad W_e = W - W_p. \qquad (23)$$

Motivated from dislocation mechanics, the plastic flow is assumed to depend upon the net deviatoric stress acting at a point. This is comprised of the deviatoric Cauchy stress as the outside driving force, plus/minus any internal stress fields which would tend to act upon the dislocation. From the thermodynamics, we choose the net stress to consist of the sum of the deviatoric Cauchy stress σ', the two dislocation related stresses α and κ, and the stress field associated with the presence of voids, Γ. Hence, the net stress ξ, is assumed to have the form

$$\xi = \sigma' - \alpha - \kappa + \Gamma. \tag{24}$$

In order to endow these tensors with certain characteristics which are desirable in describing macroscopic plastic flow, we make specific assumptions about the directional properties of the internal stresses. The tensor α is related to dislocations on slip systems in cell structure interiors. The tensor κ is related to dislocation substructure such as cell walls and averaged over several grains, it is assumed that this would lead to isotropic hardening (a lack of a preferred orientation dependence). Therefore, (for simplicity neglect the presence of voids) assume that κ is in the direction of the net stress

$$\kappa = \kappa \frac{\sigma' - \alpha}{|\sigma' - \alpha|}. \tag{25}$$

Then the net stress in the absence of voids ζ, is given as

$$\zeta = (\sigma' - \alpha)\left(1 - \frac{\kappa}{|\sigma' - \alpha|}\right). \tag{26}$$

We now assume that the presence of the voids tends to concentrate the net stress, analogous to a hole in a plate. Consistent with the kinematics we restrict our study to an isotropic distribution of spherical voids, therefore

$$\Gamma = \phi \zeta. \tag{27}$$

From equations (24), (25), (26) and (27) the magnitude of the net stress (effective stress) $|\xi|$, is given as

$$|\xi| = \xi \cdot \xi = (|\sigma' - \alpha| - \kappa)(1 + \phi). \tag{28}$$

The plastic flow rule is chosen to have a strong nonlinear dependence upon the deviatoric stress $|\xi|$, hence we choose the form

$$D_p = \left\langle \begin{array}{ll} 0 & \text{if } \beta < 0 \\ f(\theta)\sinh\left[\frac{\{|\sigma' - \alpha| - \kappa - Y(\theta)\}(1 + \phi)}{V(\theta)}\right]\frac{\sigma' - \alpha}{|\sigma' - \alpha|} & \text{if } \beta \geq 0 \end{array} \right., \tag{29}$$

where,

$$\beta = |\sigma' - \alpha| - \kappa - Y(\theta) \tag{30}$$

where $f(\theta)$ and $V(\theta)$ describe a rate dependence of the yield stress as a function of temperature θ. The damage ϕ tends to concentrate the effective stress in the flow rule, and for dilute concentrations of voids takes the more familiar form,

$$D_p = f(\theta)\sinh\left\{\frac{\{|\sigma' - \alpha| - \kappa - Y(\theta)\}}{(1-\phi)V(\theta)}\right\}\frac{\sigma' - \alpha}{|\sigma' - \alpha|}. \tag{31}$$

In order for the tensor variable α and the scalar variable κ to describe the deformed state of the material, evolution equations must be developed for both variables. The evolution of both state variables is cast into a hardening minus recovery format. Both dynamic and thermal recovery terms are included. The dynamic recovery is motivated from dislocation cross slip that operates on the same time scale as dislocation glide. For this reason, no additional rate dependence results from this recovery term. The thermal recovery term is related to the diffusional process of vacancy assisted climb. Because this process operates on a much slower time scale, a strong rate dependence is predicted at higher temperatures where this term becomes dominant. The evolution for these variables is defined by

$$\dot{\kappa} = H(\theta)|D_p| - \{R_s(\theta) + R_d(\theta)|D_p|\}\kappa^2 , \tag{32}$$

$$\dot{\alpha} - W_e\alpha + \alpha W_e = h(\theta)D_p - \{r_s(\theta) + r_d(\theta)|D_p|\}|\alpha|\alpha . \tag{33}$$

The evolution of the damage variable is based upon the solution of Cocks and Ashby for the growth of an array of spherical voids in a creeping material [ref]. In the context of this model, the evolution equation for the damage ϕ, takes the form

$$\dot{\phi} = \left\{\frac{1}{(1-\phi)^n} - (1-\phi)\right\}|D_p|\sinh\left[\frac{\mathrm{tr}\,\sigma}{|\xi|}\right]. \tag{34}$$

The tensor variable, α, represents a short transient and results in a smoother "knee" in the transition from elastic to elastic-plastic response in a uniaxial stress-strain curve. More importantly, this variable controls the unloading response and is critical in welding or quenching problems during the cooling cycle of the problem. It is termed a "short transient" in that it hardens rapidly and then saturates to a constant steady state value over a very short period of time during a monotonic loading at constant temperature and strain rate. This saturation value is maintained until the rate, temperature or loading path changes and the process repeats. This variable is responsible for the apparent material softening upon reverse loading termed the Bauschinger effect. The importance of the variable α in the prediction of localization in large deformation problems was detailed in [8].

The scalar variable, κ, is an isotropic hardening variable that predicts no change in flow stress upon reverse loading. This variable captures long transients and is responsible for the prediction of continued hardening at large strains. Unlike α, once steady state has been reached under constant conditions, this variable is not affected by a change in loading, though it is still affected by changes in temperature.

While the system of equations (31) - (34) appear quite complex, much insight can be gained about each particular model parameter for a special set of loading conditions. Consider the case of uniaxial stress at constant true strain rate and temperature along with the viscoplastic assumption that at large strains the total strain rate and the plastic strain rate are essentially the same. For this case, equations (32) and (33) can be integrated analytically to yield,

$$\alpha = \sqrt{\frac{h\dot{\varepsilon}}{r_d\dot{\varepsilon} + r_s}}\tanh\left\{\sqrt{\frac{h(r_d\dot{\varepsilon} + r_s)}{\dot{\varepsilon}}}\varepsilon\right\} \tag{35}$$

$$\kappa = \sqrt{\frac{H\dot{\varepsilon}}{R_d\dot{\varepsilon}+R_s}} \tanh\left\{\sqrt{\frac{H\dot{\varepsilon}(R_d\dot{\varepsilon}+R_s)}{\dot{\varepsilon}}}\ \varepsilon\right\} \tag{36}$$

Then by taking the magnitude of each side of equation for the flow rule and inverting, we can write the flow stress as,

$$\sigma = \alpha + \kappa + Y(\theta) + V(\theta)\sinh^{-1}\left\{\frac{|\dot{\varepsilon}|}{f(\theta)}\right\} \tag{37}$$

where, σ and ε are the true stress and true strain in uniaxial tension or compression. Using equations (35) - (37) we can now investigate the effects of each of the parameters on the predicted material response in a tension or compression test. Notice that the state variables saturate as a function of the hyperbolic tangent of the strain for this special case. From equation (37) we see that the model predicts that the stress will reach a steady state or saturation value as well. This value is given by,

$$\sigma_{sat} = \sqrt{\frac{h\dot{\varepsilon}}{r_d\dot{\varepsilon}+r_s}} + \sqrt{\frac{H\dot{\varepsilon}}{R_d\dot{\varepsilon}+R_s}} + Y(\theta) + V(\theta)\sinh^{-1}\left\{\frac{|\dot{\varepsilon}|}{f(\theta)}\right\}. \tag{38}$$

The strain at which this saturation occurs is proportional to

$$\varepsilon_{sat} \propto \sqrt{\frac{\dot{\varepsilon}}{H(R_d\dot{\varepsilon}+R_s)}}, \tag{38}$$

since κ is the long transient and therefore would determine the strain at which the stress would saturate for monotonic loading conditions.

We now can begin to address the physical meaning of these parameters in a uniaxial compression test. We will show that this model is capable of capturing very complex rate and temperature history dependent material response, but in problems where this level of sophistication is not required, the model reduces to a rate and temperature independent model of bilinear hardening. Consider the uniaxial true stress-true strain response of 304L stainless steel at a strain rate of $10^{-2}\,[s^{-1}]$ as depicted by the open circles in figures 1 and 2. By choosing all of the model parameters to be zero with the exception of H and Y, a bilinear fit of the data is achieved. In this case we illustrate two fits. In the case of a small strain problem, the initial elastic-plastic hardening slope is very high and the prediction of the model would not be very accurate after a strain of 0.1 (10%). The other fit, a large offset or back extrapolated fit, would be more appropriate for large strain problems in which initial yield information is not important. For problems requiring a more accurate representation of the stress response over the entire strain range, we include the dynamic recovery parameter, R_d. (See Fig. 2.) In this more accurate fit of the data, H is the initial small strain hardening slope from the previous example and R_d decreases the slope of the stress response to the shown saturated value.

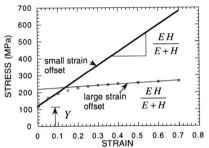

Figure 1- Two parameter fit of 304L SS data at 800C for both large and small strain offset definitions of yield. For two parameters, the model reduces to rate independent bilinear hardening.

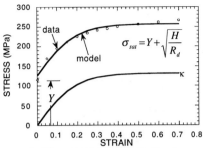

Figure 2- Three parameter fit for the same compression curve. In this case the model accurately captures both the hardening and recovery through the isotropic hardening variable κ.

The model with these parameters, would predict no effect of a change in load path. In particular, for the case of reverse loading, the material would yield at the same magnitude of stress level in the reverse direction, or at a stress level of about 250 [MPa] for a load reversal at a strain of 0.50. Many materials, however, exhibit a Bauschinger effect, an apparent softening upon load reversal. If this is an important aspect of the problem under consideration, the parameters associated with the short transient, a, must be included. The hardening parameter h, and the dynamic recovery parameter, r_d, play the same role in the saturation of a, as do the corresponding variables in the case of κ. Figure 3 demonstrates how the initial yield can be made up of contributions from both a and Y, and the large strain hardening is still described by the evolution of κ. In this case, reverse yield at a strain of 0.5 would be about 200 [MPa]. The importance of including a short transient in the description of large deformation problems involving significant load path changes will be illustrated in a later comparison between model prediction and experimental data.

We now consider the inclusion of strain rate and strain rate history effects. Figure 4 depicts an additional compression test at a higher rate of $10\,[s^{-1}]$. Notice that there is an initial rate dependence at yield and additional rate dependence which increases as a function of strain. The rate dependence of yield is described by the inverse hyperbolic sine, while the strain dependent rate effect results from a change in recovery mechanism and is therefore described by the static or thermal recovery.

Finally, in figure 5 we see that part of the rate dependence of the yield could be attributed to static or thermal recovery of the short transient. As before, there is little difference in these monotonic compression tests, but a significant difference in predictions would result for a change in load path.

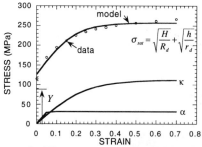

Figure 3 - Five parameter fit of 304l SS compression curve including the short transient α. This fit will more accurately capture material response during changes in load path direction.

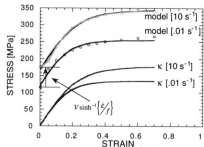

Figure 4 - Six parameter fit of 304L SS compression data with only the long transient κ but including the effects of rate dependence of yield through the parameters V and f. The strain dependent rate effect is captured by the static recovery parameter R_s.

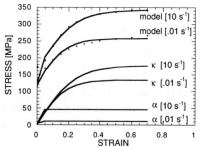

Figure 5 - Nine parameter fit of 304L SS compression data at two strain rates. This fit captures the initial rate dependence of the yield through the static recovery of the transient α.

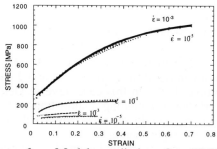

Figure 6 - Model prediction for 304L stainless steel tension tests or 304L stainless steel is depicted in Figure 1.

The model prediction for various strain rates and temperatures is shown in figure 6. The response is dominated by dynamic recovery at lower temperatures where the rate dependence is weak. The effects of thermal recovery become significant at higher temperatures where the rate dependence is strong. In Figure 7 the model prediction is compared with uniaxial compression tests that involve a significant change in temperature. Two specimens were loaded, one at 20C to a strain of nearly 0.5 and the other at 800C to a strain of 0.23. This specimen was unloaded and quenched rapidly. Once the specimen had cooled, it was reloaded at 20C. If the stress was a unique function of temperature and plastic strain, the flow stress upon reload would have been identical to the 20C specimen at that strain. As shown in Figure 7 [9], this is not the case. Rather, there is a strong temperature history effect upon reload that is adequately captured by the state variable model.

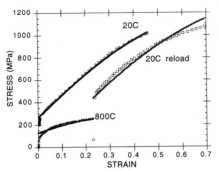 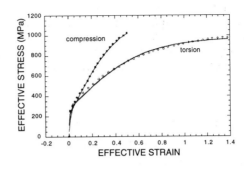

Figure 7 - Model prediction of compression tests and compression reload demonstrating temperature and history effects. The 800C test was quenched after being strained to 23% and reloaded at room temperature. The temperature history effect is demonstrated by the reload curve being much softer than the 20C curve.

Figure 8 - Model comparison with data for compression and torsion of 304L SS. The inclusion of the parameter η results in the model accurately tracking the evolving anisotropy typified by the differences in hardening between torsion and compression.

At this point, we consider the modification of the evolution of the variable κ, by introducing a parameter which quantifies the effect of directional changes in the load path. We introduce a scalar parameter η, which is defined as the inner product of the tensor, α, and the direction of plastic flow. Hence, equation (32) is rewritten,

$$\dot{\kappa} = \left(H_1(\theta) + H_2(\theta)\eta\right)\left|D_p\right| - \left\{R_s(\theta) + R_d(\theta)\left|D_p\right|\right\}\kappa^2, \tag{39}$$

where,

$$\eta = \frac{\xi \cdot \alpha}{|\xi||\alpha|}. \tag{40}$$

A form of this parameter was first introduced by Krieg and Key [10] in the hardening of the tensor variable to try to capture the shape of the uniaxial tensile stress in a load reversal test, while we choose to utilize it in the scalar variable to capture large strain evolving anisotropy. The ability of this parameter to track evolving anisotropy is shown in figure 8. Due to textural effects at the microstructural level, a significant difference in observed in the hardening response of 304L SS in compression and torsion, the hardening in torsion being much less than in compression. This effect is adequately simulated in the phenomenological model by the inclusion of the parameter η in the evolution of the isotropic hardening parameter κ. The inclusion of this parameter also results in a more accurate prediction of the apparent softening which occurs after large changes in load paths. This is best illustrated by considering the experimental work on the large deformation of aluminum 1100 by Armstrong et. al.[11]. Figure 8 depicts the uniaxial compression of this material to a true axial strain of 6. This was achieved by loading the material until just prior to barreling, re machining, then reloading. In the same figure is the multiaxial curve which is much softer than the uniaxial one. This curve is the result of loading a cube of material sequentially on the three orthogonal sides. Each loading resulted in a strain of 0.0075 and the sequence was continued until the same effective strain of 6.0 was achieved. Due to the continual large changes in loading path the material hardened at a much lower rate than in the case of uniaxial deformation. In the same figure, the model predictions are shown. The parameters were determined to fit this response, and with the addition of the coupling term the model is quite capable of capturing this type of response. A more accurate prediction of the early portion of the curve could possibly be achieved by assuming the tensor variable to be a much shorter transient, but this will be attempted when more reverse loading data is available to precisely determine the length of the transient.

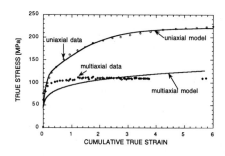
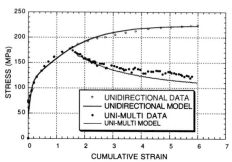

Figure 9 - Comparison of the model and data for aluminum 1100 for the case of uniaxial compression (unidirectional) and sequential loading of a cube on the three orthogonal sides (multidirectional). Data from Armstrong et. al. [11]

Figure 10 - In this figure the cube was initially loaded unidirectionally and then switched to multidirectional loading. This resulted in the stress tending to soften towards the multidirectional curve and was captured by the model.

5. IMPLEMENTATION

The model has been implemented into and tested in the explicit finite element codes, DYNA2D [12], DYNA3D [13], PRONTO2D [14], PRONTO3D [15], and the conjugate gradient code JAC3D [16]. The implementation is based upon the formulation of a numerical flow rule at each time step derived from a numerical consistency condition. When implementing a damage or temperature change associated with a model it is convenient to introduce an operative split. That is the damage, temperature and velocity gradient is assumed constant while updating the stress. The damage and temperature are then updated using the new values of stress, velocity gradient and state variables at the end of the time step. This results in a simpler and more efficient implementation and is consistent with the operator split between the constitutive equations and the equilibrium equations. For explicit codes, in which the time step is extremely small, this operator split is almost always insignificant. For implicit codes, in which the time step may be several orders of magnitude larger, inaccuracies may arise and subincrementing may be necessary.

For maximum efficiency the radial return method as proposed by Krieg[17] is employed to integrate the stresses. The implementation discussed in detail in[18] was used in the explicit codes and used without problems for most applications. In this implementation the plastic strain rate magnitude in the recovery and yield functions was replaced by the total strain rate magnitude. In solving the radial return equations, this resulted in a simple linear equation for the plastic strain increment. For most all applications this is a reasonable assumption. For extremely low strain rate applications, such as creep or relaxation a more complex implementation was used. This implementation resulted in a nonlinear equation for the plastic strain increment which was solved by an iterative process. The user must decide for his application whether the increased computation costs of the more accurate integration is required.

In the DYNA implementation the damage is limited to a maximum value of 0.99 to prevent numerical problems that arise when the porosity reaches 1. At this value, the elastic moduli and yield strength are one percent of their original values. In PRONTO the death option is used to completely remove the element when the damage reaches 0.99, After an element fails, the applied load is redistributed to surrounding elements, which then subsequently fail. In this manner, a diffuse crack (one element wide) is then propagated through the structure. It should be noted that only the first element failure is truly modeled correctly, since the stress concentration due to the failed element (and formation of a sharp crack) is not precisely modeled. Implementation of a dynamic rezoner would be required to model this effect. However, in most dynamic cases, reasonable results are obtained with the present formulation if a sufficiently fine mesh is used.

6. APPLICATIONS

6.1 Aluminum Notched Tensile Tests

The plasticity parameters for Aluminum 6061-T6 were determined using compression and tension tests. The fit of the model to the data is shown in Figure 9. The damage parameters are determined using the axisymmetric notched tensile specimens shown in Figure 10. The determination of these parameters requires a trail-and-error finite element analysis since a closed form solution does not exist for the three dimensional nonlinear boundary value problem. The notch specimens are used since the radius of curvature of the notch strongly influences the evolution of damage, due to the significant tensile pressure that develops in the specimen. The material parameters are typically determined from one of the notch tests and the remaining three are then used as verification. The progressive failure of a notched specimen is shown in Figure 11. Two elements appear to fail simultaneously in the axial direction, since only half the specimen was modeled. For the same reason, any formation of a cup-cone failure surface was precluded. A comparison of the predicted strain to failure with the test data is shown in Figure 12. This strain is the elongation over a one inch gauge length at first observed failure. The model accurately predicted the strain to failure over the entire range of radii tested.

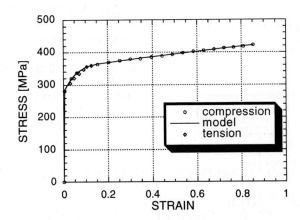

Figure 9. Comparison of Model with Aluminum 6061-T6 Compression Data

372

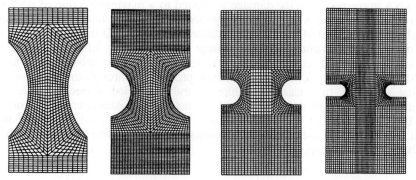

Figure 10. Notch Geometries Used For Failure Validation

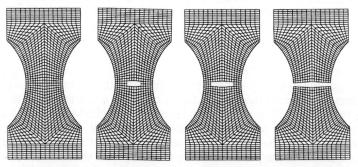

Figure 11. Progressive Failure of a Notch Tensile Specimen

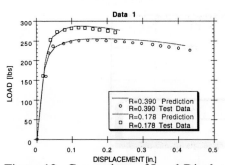

ELONGATION AT FAILURE

Radius	Test	Analysis
0.390	0.043	0.044
0.156	0.021	0.023
0.078	0.014	0.015
0.039	0.011	0.013

Figure 12. Comparison of Load-Displacement and Strain to Failure to Notch Specimens

6.2 Aluminum Disk Penetration Tests

Experiments were performed [18,19] in which 5.72 cm diameter, 3.2 mm thick, Aluminum 6061-T6 disks were impacted with a hardened steel rod. The disks were held in a manner that simulated free boundary conditions. The plasticity and damage parameters were the same as those used in section 7.1. Initial failure, in the form of visible cracks on the rear surface of the disk, occurred at an impact velocity between 79 and 84 m/s. The finite element model predicted failure would initiate between 84 and 89 m/s. In the experiments, complete perforation was obtained between 92 and 106 m/c. The analysis indicated that a velocity between 102 and 107 m/s would be required. Figure 13 shows the comparison of the model with the experiment for an impact velocity of 107 m/s. For this velocity complete perforation was seen. Figure 14 shows the comparison of test and calculation for an impact velocity of 91 m/s. At this velocity a crack is observed on the rear surface but has not propagated through the thickness. Figure 15 shows the sectioned test specimen which shows a crack through approximately 25 percent of the thickness which was also predicted by the finite element model.

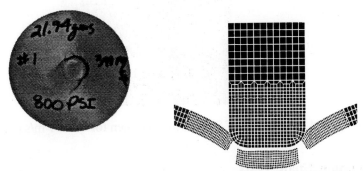

Figure 13. Comparison of Prediction and Experiment for V = 107 m/s

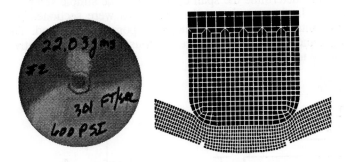

Figure 14. Comparison of Prediction and Experiment for V = 91 m/s

374

Figure 15. Etched Micrograph of Sectioned Specimen for V = 91 m/s (courtesy of M. Stout [20])

6.3 Aluminum Spall Tests

The plate impact tests of [21] were used to validate the spall prediction capability of the model. The experiment was a medium velocity plate on plate impact testing of Aluminum 6061-T6. Again, the same constants were used as in the previous two sections. Several impact velocities were used to determine the spall conditions. The simple setup for the experiments is shown in Figure 16. In the experiment the specimens were sectioned and then the resulting damage was then quantified according to amount of porosity observed. The results, except for complete separation and no observed damage, are thus somewhat qualitative. Figure 17 shows the comparison of the tests and predictions.

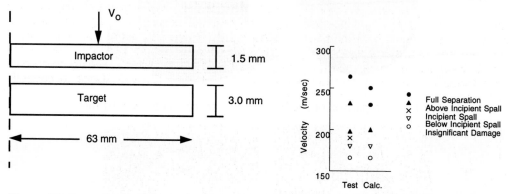

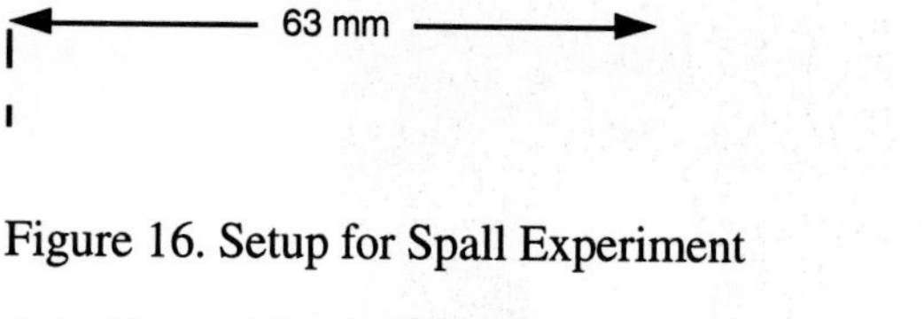

Figure 16. Setup for Spall Experiment

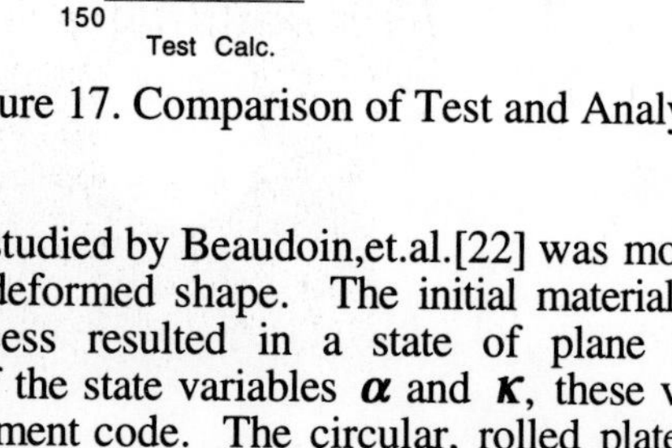

Figure 17. Comparison of Test and Analysis

6.4 Sheet Metal Forming

The hydroforming of an aluminum cup initially studied by Beaudoin,et.al.[22] was modeled to determine effects of inital material state on final deformed shape. The initial material state was modeled by assuming that the rolling process resulted in a state of plane strain compression. Having estimated the initial values of the state variables α and κ, these values were then input as initial conditions in the finite element code. The circular, rolled plate was forming into a hemispherical cup as shown in Figure 18. The analysis predicted an ovalling of the part as shown in Figure 19. The actual tested part [22] showed earring in addition to the ovalling. Current work includes further improvements to the model to capture this effect.

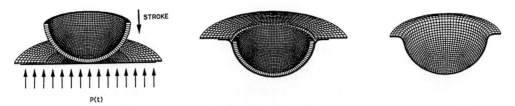

Figure 18. Modeling the Hydroforming of a Hemispherical Cup

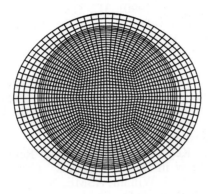

Figure 19. Predicted Ovalling of the Hemispherical Cup

6.5 HERF Forging

The model was used to predict the deformation and final material state of a closed die forging at high temperature and high rate (HERF). The high rate application required the inclusion of adiabatic heat generation from plastic work to account for large increases in temperature. The model was instrumental in assisting die design to result in optimum residual material state (i.e. material strength and grain flow direction for this application). The history dependent model has also been applied to multi-stage forgings where it is necessary to account for the deformation history during each stage and the recovery between stages to correctly predict the final material hardness distribution.

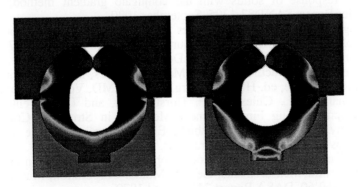

Figure 20. Modeling a HERF Forging With and Without Friction

REFERENCES

1. Shen and Dawson, eds., Simulation of Materials Processing: Theory, Methods and Applications, Balkema, Rotterdam, ISBN 90 5410 5534, 1995.

2. B.D. Coleman and M.E. Gurtin, "Thermodynamics with Internal State Variables," _J. Chem. Phys._, 47 (1968), pp. 597-611.

3. O.W. Dillon and J. Kratochvil, "Thermodynamics of an Elastic-Plastic Materials as a Theory with Internal State Variables," _J. Apll. Phys._, 4 (1969), pp. 41-54.

4. D.J. Bammann, "An Internal Variable Model of Viscoplasticity," _Int. J. Eng. Sci._, 22 (1984), pp. 1041-1053.

5. D.J. Bammann, "Modeling Temperature and Strain Rate Dependent Large Deformations of Metals," _Appl. Mech. Rev._, 1 (1990), 312-318.

6. D.J. Bammann and G.C. Johnson, "On the Kinematics of Finite Deformation Plasticity," _Acta Mechanica_, 70 (1987), pp. 1-13

7. D.J. Bammann and E.C. Aifantis, "A Model for Finite Deformation Plasticity," _Acta Mechanica_, 69 (1987), pp. 97-117.

8. D.J. Bammann and P.R. Dawson, "Modeling the Initial State of a material and It's Effect on Further Deformation," Material Parameter Estimation for Modern Constitutive Equations, eds. L.A. Bertram, F.B. Brown, A.D. Freed, ASME NY,ASME-AMD 168 (1993), pp.13-20.

9. W.A . Kawahara, Unpublished data.

10. R.D. Krieg and S.W. Key, "On the Accurate Representation of Large Strain Nonproportional Plastic Flow in Ductile Materials," _Constitutive Equations: Macro and Computational Aspects_, ed. K. J. Willam, ASME, Newyork, (1984), pp. 41-52.

11. B.E. Armstrong, J.E. Hockett, O.D. Sherby, 1982, "Large Strain Multidirectional Deformation of 1100 Aluminum at 300K," J. Mech. Phys. Solids, 1/2 (1982), pp. 37-58.

12. J.O Hallquist, User's manual for DYNA2D - an explicit two-dimensional finite element code with interactive rezoning and graphical display, Lawrence Livermore National Laboratory Report UCID-18756, Rev. 3, March 1988.

13. J.O. Hallquist, DYNA3D User's manual (nonlinear dynamics analysis of structures in 3D), Lawrence Livermore National Laboratory Report UCID-19592, Rev. 4, April 1988.

14. L.M. Taylor and D.P. Flanagan, PRONTO2D - a two-dimensional transient solid dynamics program, Sandia National Laboratories Report SAND86-0594, March 1987.

15. L.M. Taylor and D.P. Flanagan, PRONTO3D - a three dimensional transient solid dynamics program, Sandia National Laboratories Report SAND87-1912, March 1989.

16. J.H. Biffle, JAC3D - a three dimensional finite element computer program for nonlinear quasi-static response of solids with the conjugate gradient method, Sandia National Laboratories Report SAND87-1305, September 1990.

17. R.D. Krieg and D.B. Krieg, Accuracies of numerical solution methods for the elastic-perfectly plastic model, ASME, J. Pressure Vessel Tech. Vol.99 (1977), pp. 510-515.

18. D.J. Bammann, M.L. Chiesa, A. McDonald, W.A. Kawahara, J.J. Dike and V.D. Revelli, Prediction of Ductile Failure in Metal Structures, in Failure Criteria and Analysis in Dynamic Response, ed. H.E. Lindberg, ASME AMD, Vol. 107, Nov. 1990, pp. 7-12.

19. D.J. Bammann, M.L. Chiesa, M.F. Horstemeyer, and L.E. Weingarten, Failure in Ductile Materials Using Finite Element Methods, in Structural Crashworthiness and Failure, ed by N. Jones and T. Wierzbicki, Elsevier Applied Science, 1993, pp. 1-53.

20. M. Stout, Los alamos National Laboratories, unpublished results

21. W.M. Isbell and D.R. Christman, Shock Propagation and Fracture in 6061-T6 Aluminum from Wave Profile Measurements, General Motors Corporation, Materials and Structures Report MSL-69-60, DASA Report 2419, April 1970.

22. A.J. Beaudoin, P.R. Dawson, K.K. Mathur, U.F. Kocks and D.A. Korzekwa, "Application of Polycrystal Plasticity to Sheet Forming," Comp. Meth. Appl. Mech. and Engr., Vol 117, 1994, pp.49-70.

SECTIONAL LECTURES

Theoretical and Applied Mechanics 1996
T. Tatsumi, E. Watanabe and T. Kambe (Editors)

379

SOME DEVELOPMENT OF STRUCTURAL TOPOLOGY OPTIMIZATION

Gengdong Cheng

State Key Laboratory of Structural Analysis of Industrial Equipment
Dalian University of Technology, Dalian, 116024, CHINA,P.R.

ABSTRACT

The present paper reviews some development in the area of structural optimization with particular emphasis on one of the most challenging problems: structural topology optimization. We first revisit the previous study on the minimum compliance design problem of thin solid elastic plate with weight constraint. From its mesh dependent solutions we observed that optimization for thin isotropic plate leads to a plate with micro-structure, which is essentially a plate of macro-orthotropic materials. To obtain the optimum design we expanded the design space and regularized the problem formulation. The implication of the solution approach and numerical results is discussed in the light of the latest development in continuum topology optimization: homogenization method and artificial density method. These two approaches featured an integrated optimization of structural size, topology and material. They have opened a new way to continuum topology optimization and aroused the surging interest in application of structural optimization in industries as design tools.

We next review the recent development of topology optimization of discrete structure. With the new achievement in this direction, it is now possible to design optimum topology of large scale trusses for minimum compliance under weight constraints. Furthermore, the recent study showed that singular optima in topology optimization with stress constraints originates from discontinuities of constraint functions and leads to jelly-fish like feasible domain. Accordingly, structural topology optimization may be catalogued into two types: these with singular optimum and those without singular optimum. An ε-relaxed approach proposed recently transforms topology optimization problems of both types into structural size optimization ones by modifying the feasible domains.

1. INTRODUCTION

The history of structural optimization can be traced back to as early as 1904 when Michell [22] developed minimum weight truss design theory. Since the advent of modern electronic computer in 1960's, structural optimization with the aid of finite element method and operation research soon became an important branch of applied mechanics. According to the type of design variables structural optimization problem is normally classified into structural size, shape, topology and type optimization. Optimum design of truss structures with the bar areas being design variable is a typical structural size optimization problem. Many efficient numerical approaches and softwares were developed for solving them. From the late 1970's, parametrization and description of structural shape, sensitivity analysis with respect to shape design variables adds new dimension to research on structural optimization. Emergence of computer aided design techniques makes research on structural shape optimization become an urgent need and fruitful endeavor. Important progress has been seen in this area. However, the technical transfer from academic research of structural optimization to practical engineering design seems unsatisfactory. It is partly because of the fact that once the structural type and topology is fixed many constraints from manufacture, aesthetics and technology prevent the designer from making any further significant improvement. It looks very important to integrate the structural optimization idea in the primitive design stage.

By optimum topology of continuum such as membrane, plate or solid we mean the optimum connectivity of continuum structure. In other words, the goal is to determine whether voids are needed or how many voids are needed inside the continuum for the optimum performance of the continuum. By optimum topology of discrete structures such as truss or frame we mean the optimal connection between the set of given nodes including the supporting nodes and loading nodes. Early research activities on structural topology optimization were mainly placed on discrete structures. The optimum topology design provides very useful information for practical engineers in their early design stage. Furthermore, once the designers accept design tool for optimum structural topology, they are also better prepared for structural size and shape optimization. Therefore, systematic approach to structural topology optimum design is very important and attractive for engineers.

However, structural topology optimization has been considered as one of the most challenging areas in this field. Early achievement in this area includes Michell trusses [22] and the ground structure approach [15]. There were also many interesting numerical approaches [14,24,25,29,30,32]. Optimum layout theory developed since 1970's by to large extent Prager and Rozvany [27] provided solutions of a large class of interesting problems.

Research on optimum design of thin solid elastic plate [7] leads to a novel idea of introduction of micro-structures into topology optimization of continuum structures [2]. Based on micro-structure concept, the latest development in continuum topology optimization, i.e., the homogenization method and artificial density method featured an integrated optimization of structural size, topology and material. Their successful implementation pushed forward the application of continuum topology optimization in industries as design tools.

For the simplest possible problem, that is, the problem of finding the minimum compliance truss design for a given amount of material, important progress has been made recently [3] by Bendsoe and others. With their research effort, it is now possible to solve large scale truss topology design problems for minimum structural compliance or other global structural performance index. In many instances of the primitive design stage engineers need merely to know the structural global behaviors such as structural compliance and/or vibration frequency. Because of that, development by Bendsoe and others has opened the door to the practical application of structural optimization.

Instead of the great success of solving topology optimization problem aiming at global performance, structural topology optimization under local constraints such as stress constraints and local buckling constraints remains a challenging problem. Special difficulty caused by singular optima was observed and studied by Sved and Ginos [29], Kirsch [17,18] and others. Cheng and Jiang [8,9] pointed out that singular optimum (local or global) originated from discontinuities of constraint functions, the difference between topology optimization problems with local constraints and global constraints is significant. Accordingly, we cataloged structural topology optimization into two classes: these with singular optimum and those without singular optimum and proposed an ε-relaxed approach in order to transform topology optimization into structural size optimization problem [12]. The transformed problems can be solved by the existing algorithms.

2. OPTIMIZATION OF THIN, SOLID ELASTIC PLATES

Let us first revisit a well known problem, i.e., problem of optimum thickness distribution of thin, solid elastic plates for minimum structural compliance under the given static external load [7]. For the given structural material volume, material properties and boundary conditions, the plate optimization problem is formulated as follows:

To Find $\quad h(x,y) \quad\quad\quad (x,y) \in \Omega$

Min. $\quad \iint_{\Omega} p(x,y)w(x,y)dxdy$ $\hfill$ (1a)

St. $\quad V = \iint_{\Omega} h(x,y)dxdy$ $\hfill$ (1b)

$\quad\quad L(h(x,y),w(x,y)) = p(x,y), \quad\quad (x,y) \in \Omega$ $\hfill$ (1c)

$\quad\quad B(h(x,y),w(x,y)) = 0, \quad\quad\quad (x,y) \in \overline{\Omega} \setminus \Omega$ $\hfill$ (1d)

$\quad\quad h_{min} \le h(x,y) \le h_{max} , \quad\quad\quad (x,y) \in \Omega$ $\hfill$ (1e)

in which Ω is the domain occupied by the mid-plane of the plate, design variable $h(x,y)$ is the plate thickness distribution, and $w(x,y)$ is the deflection under the external load $p(x,y)$. The objective in (1a) is the structural compliance which to some extent is a measure inverse-proportional to structural stiffness. The set of constraints states that the total plate material volume is V (1b), deflection $w(x,y)$ should satisfy the differential equation (1c) and the boundary conditions (1d); the design variable, plate thickness $h(x,y)$, varies between the upper bound h_{max} and lower bound h_{min}as specified in (1e).

Based on formulation (1a-1e), earlier study showed that the optimal design of rectangular and circular plates is a smooth thickness distribution. However, these smooth optimal solutions are plagued by contradicting engineering experiences---plate stiffened by ribs is superior to plate with smooth thickness distribution. In fact, because of their better structural performance, plates with reinforced ribs and stiffeners are widely used in shipbuilding and civil engineering. The paradox was first resolved by the author and Olhoff in 1979. Using the formulation based on variation calculus, we derived the necessary condition for optimum plate designs. More importantly, using an efficient numerical algorithm based on the optimality criterion approach and finite element discretization, we showed that the final design of thin, solid elastic plate for minimum compliance is mesh dependent. Fig.1 shows numerical results for a square plate clamped along four edges with mesh 20*20 in Fig.1a and 40*40 mesh in Fig.1b. As the mesh becomes finer and finer, more and more thickness jumps between h_{min} and h_{max} appear in the final solution, resembling the rib-stiffened plate in engineering structure. Though the sequence of structural compliance converges with finer meshes, the sequence of plate thickness distribution does not converge. This leads to an important

382

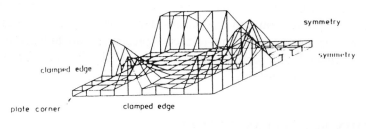

(a)

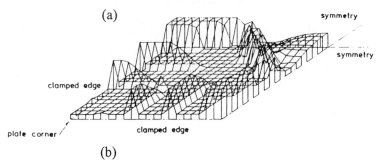

(b)

Fig.1 Quarter of a clamped square plate, $h_{max}/h_{min}=6$, $h_u/h_{min}=2$; Optimal design obtained with a) 20*20 mesh; b) 40*40 mesh; h_u is thickness of uniform plate with material volume V.

discovery that optimum thickness distribution is not a smooth one. Instead, the solution can be a stiffened plate, or even a plate stiffened with infinite number of stiffeners under some conditions.

From the above, we recognized that the formulation (1a-1e) prevents us from obtaining the optimum solution. To find the global optimum designs, the design space of plates with only smooth thickness distribution must be expanded to include possible designs with infinite number of infinitely thin stiffeners. We achieved such a formulation by adding the density of stiffeners as the design variables and therefore regularizing the problem. Thus the global optimum solution is described by both the optimal plate thickness and the optimum density distribution of integrated stiffeners (Fig.2). For the added design variable, zero and unit stiffener density corresponds to the minimum thickness h_{min} and thickness $h(x,y)$, respectively. Areas with density between 0 and 1 are the part of the stiffened plates. For annular plate, the minimum compliance design problem under the material volume constraint can be regularized as,

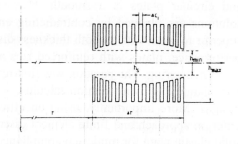

Fig.2. Annular plate stiffened with infinite thin stiffeners.

To Find $h(r), \mu(r)$

Min. $\int_\Omega p(r)w(r)rdr$ (2a)

St. $\int_\Omega [h(r)+\mu(r)(h_{max}-h(r))]rdr = \dfrac{V}{2\pi}$ (2b)

$h_{min} \le h(r) \le h_{max}$ $r \in \Omega$ (2c)

$$0 \le \mu(r) \le 1 \qquad r \in \Omega \qquad (2d)$$
$$L(h(r), \mu(r), w(r)) = p(r) \qquad r \in \Omega \qquad (2e)$$
$$B(h(r), \mu(r), w(r)) = 0, r = R_i, i = 0,1 \quad (2f)$$
$$\Omega = \Omega\{r | R_0 \le r \le R_1\} \qquad (2g)$$

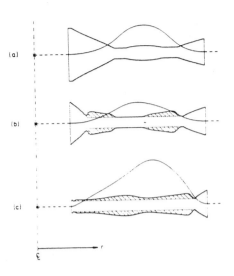

Here, (r, θ) are the polar coordinates with the origin at the plate center. The thickness distribution $h(r)$ of the solid part of the plate and the density $\mu(r)$ of stiffeners that run along the circumferential direction are assumed to be independent of θ. The plate deformation is easily determined as $w(r)*\sin(n\theta)$ under the given external load $p(r)*\sin(n\theta)$; R_0 and R_1 is the inner and outer radius, respectively. The effective stiffness of the plate can be evaluated based on principle of energy equivalence and continuity of plate deflection and slope. For annular plate, it is an explicit function of the plate thickness $h(r)$ for the solid parts and the density $\mu(r)$ of stiffeners. We can then discretize the plate of known effective stiffness by finite element and optimize the thickness $h(r)$ and density $\mu(r)$ by the optimality criterion method.

Fig.3 Annular plates of minimum compliance subjected to loads p=const*sin(nθ). h_u/h_{min} =1.6579, $h_{max}/h_{min}=5$, $R_0=0.2$, $R_1=1.0$, Plate edges are clamped; a) n=0; b) n=2; c) n=4;

Numerical results with finer and finer meshes showed successful convergence of not only the structural compliance, but also the thickness distribution and the density of stiffeners, which are not achievable without the regularization. As shown in Fig.3, the optimum design for different load and boundary conditions may be a solid smooth plate, a solid plate with finite number of stiffeners, or a plate with infinite number of infinitely thin stiffeners. Later the above results were confirmed by more accurate finite element simulations and even analytic methods. Fig.3c represents the most interesting solution. Though isotropic material is assumed in the problem formulation (2), minimization of plate compliance introduces micro-structure into the plate material, here in the form of infinite number of infinitely thin stiffeners. As a result, the plate material become macro-orthotropic.

The above results for thin, solid elastic plates have several implications:

1) It questions the existence of solutions for certain type of structural optimization problem. In this case, the conventional formulation of thin, solid elastic plate does not ensure a global optimum solution. Cheng and Olhoff [7] succeeded to construct a closed design space of the plate optimization problem simply based on intuitive insight. Many research work in connection with existence of optimum solution, G-closure and restriction of the design space were seen in literature (Kohn [20], Bendsoe [1], Lurie [21] and others).

2) Optimization of structural compliance of thin, solid elastic plates of isotropic material leads to plates with orthotropic materials, therefore bridging the gap between material optimization and structural optimization. It is the first example where material selection and structural optimization can be performed in a unified approach. Further research

includes concurrent design of material and structure, material micro structure design, prediction of effective properties for material with micro-structure. It is also related to optimum design of composite material.

3) Optimization of structural compliance of solid elastic plates introduces infinitely thin stiffeners in the problem formulation naturally. By optimizing stiffener density, we completed the selection of structural topology, i.e., plate with stiffeners or without stiffeners. It inspires a new approach, i.e., micro-structure concept to topology optimization.

3. MICRO-STRUCTURE CONCEPTS FOR TOPOLOGY OPTIMIZATION OF CONTINUUM STRUCTURES

3.1 Micro-structure concepts

A basic assumption of continuum mechanics is that material consisting of atoms, molecular or/and crystals is macro-homogeneous media. As a basis of elasticity, plasticity, structural mechanics and structural optimization, this assumption has been justified by scientific and engineering practice during the past several hundred years. Topology optimization of continuum structure aims at determing the connectivity of the continuum, or determing the number of holes and positions of the holes for the optimum performance of a structure. Taking a plane continuum as an example, a straightforward formulation is given as follows

To Find $N, \Omega_1, \Omega_2, ..., \Omega_N$

$$\text{Min.} \quad l(u) = \int_\Omega p(x,y)u(x,y)d\Omega + \int_\Gamma t(x,y)u(x,y)d\Gamma \tag{3a}$$

$$\text{St.} \quad a(u,v) = l(v), \qquad \text{for all } v \in V \tag{3b}$$

$$a(u,v) = \int_\Omega E_{ijkl}\varepsilon(u)\varepsilon(v)d\Omega \tag{3c}$$

$$\Omega = \Omega_0 \backslash (\Omega_1 \cup \Omega_2 \cup ... \cup \Omega_N) \tag{3d}$$

$$\int_\Omega d\Omega = S \tag{3e}$$

Here, the design variables are the number of holes N and the domain of each hole $\Omega_1, \Omega_2, ..., \Omega_N$. Ω_0 is the ground design domain possibly occupied by the continuum. Ω is the domain actually occupied by the continuum. The objective function in equation (3a), structural compliance $l(u)$, is calculated by applying body forces $p(x,y)$ on the actual domain and the surface forces $t(x,y)$ on the actual boundary. The body forces, surface forces and mid plane of membrane are coplanar. The equilibrium equations (3b) are expressed in a weak form with V representing the set of all kinematically admissible displacements. Since the material of plane continuum is assumed to be isotropic and homogeneous, the elasticity module tensor E_{ijkl} only depends on Young's module E and Possion ratio μ. The last equation (3e) describes the constraint on the total area of the continuum.

Though equations (3a-3e) describe the topology optimization problem of plane continuum, they are not operational. It is not possible to construct an algorithm to deal with the formulation. An alternative formulation of the problem is given in terms of an indicator function $\rho(x,y)$,

To Find $\rho(x,y)$ $(x,y) \in \Omega$

Min. $l(u) = \int_{\Omega_0} p(x,y)u(x,y)\rho(x,y)d\Omega_0 + \int_{\Gamma} t(x,y)u(x,y)\rho(x,y)d\Gamma$ (4a)

St. $a(u,v) = l(v),$ For all $v \in V$ (4b)

$a(u,v) = \int_{\Omega_0} E_{ijkl}\varepsilon_{ij}(x,y)\varepsilon_{kl}(x,y)\rho(x,y)d\Omega_0$ (4c)

$\rho(x,y) = (0,1)$ (4d)

$\int_{\Omega_0} \rho(x,y)d\Omega_0 = S$ (4e)

Here, the design variable $\rho(x,y)$ takes value of 1 or 0. If the obstacle in the formulation (3a-3e) is the parametrization of voids, by using the indicator function we render equations (4) an infinite dimensional 0-1 programming. Even with today's technology and high computational cost, it is unsolvable.

The approach to thin solid elastic plate optimization by Cheng and Olhoff [7] is instructive for constructing a rigorous, yet efficient formulation for topology optimization. It has been shown that introducing infinitely thin stiffeners and including the stiffener density as design variables regularize the solid plate problem. The design variable, i.e., the density variation represents a whole spectrum of topology changes: from plate with no stiffeners to plate with infinitely many stiffeners and plays the role of parameters which describe topology of the plates. In the past ten years, this idea is extended and very much enriched to solve topology optimization of continuum structures by Bendsoe, Kikuchi and others [2,4]. By introducing artificial micro-structure in the continuum they constructed an imaginary porous material. Parameters of the micro-structure such as the density of the porous material were chosen as design variables to satisfy the macro-scale design objective. Optimum connectivity of the continuum structure is generated as the result of structural size optimization and the latter can be solved very efficiently.

To apply the micro-structure concept to continuum optimization, a number of problems must be studied. For example, how to select a proper micro-structure since the final solution must be dependent on the type of micro-structure; how to establish the relation between the parameters of micro-structure and the material properties such as effective elasticities, mass density or so; how to avoid porous material in the final solution since we are interested in a clear structural topology and the micro-structure is rather artificial, in the final solution it is better to have merely solid and void. For more information, reader is referred to Ref.[4].

3.2 Homogenization Method

In 1988, Bendsoe and Kikuchi [2] proposed the above mentioned micro structure concept for the topology optimization of continuum structure. In their approach, they assume that material at any macro point of the given ground design domain consists of infinitely small, periodically distributed cells. Parameters of the cells, or the micro-structure are chosen as design variables and hopefully represent structural topology. Taking the advantage of periodicity, they applied the homogenization method, which was developed by Bensoussan, Lions, Papanicolaous [5] in 1978 based on the perturbation theory. It established the relation between the parameters of the cell and the macroscopic properties such as effective properties E_{ijkl} and mass density of porous material. The approach is known as homogenization method for topology optimization of continuum structures [2].

For plane continuum, one could consider that the micro-structure is a periodically distributed square cells with square hole, or circular hole or even more complicated shape. For simplicity, let us consider square cell with square hole, see Fig.4. Linear relative size $b(x,y)$ of the square hole with respect to the square cell is chosen as design variable. Area with $b(x,y)=1$ is void, area with $b(x,y)=0$ is solid and area with $0<b(x,y)<1$ is porous material. The ratio of the solid part and the cell area is $(1-b^2(x,y))$. The topology optimization problem (3) can be reformulated as

To Find $b(x,y)$, $\quad (x,y) \in \Omega_0$ (5a)

Min. $l(u) = \int_{\Omega_0} p(x,y)u(x,y)d\Omega_0 + \int_{\Gamma} t(x,y)u(x,y)d\Gamma$ (5b)

St. $a(u,v) = l(v)$, For all $v \in V$ (5c)

$a(u,v) = \int_{\Omega_0} E_{ijkl}(b(x,y))\varepsilon_{ij}(x,y)\varepsilon_{kl}(x,y)d\Omega_0$ (5d)

$0 \le b(x,y) \le 1$ (5e)

$\int_{\Omega_0}(1-b^2(x,y))d\Omega_0 = S$ (5f)

It is important to mention the difference of the formulation (4) and (5). In the formulation (4) of plane continuum topology optimization design variables are of discrete nature and its solution allows for only solid and void. In the relaxed formulation (5) design variables are parameters of micro-structure and its solution allows for porous material due to the fact that deign variable is a continuous design variable and may take any value between 0 and 1. In case that only very small part of continuum in the resulting design has intermediate hole size $0<b(x,y)<1$, it is reasonable to assume that the solution of (5) is a good approximation of (4). The portion of porous material in the solution of (5) very much depends upon the relation between effective elastic properties and material density. This in turn depends upon the shape of the cell. Numerical experience has shown that with the square cell and square hole in most of cases the solution of (5) has only small portion of the design domain of $0<b(x,y)<1$. Most of the area has either $b(x,y)=1$ (void) or $b(x,y)=0$ (solid). In addition to proper selection of the unit cell, we can also impose certain constraints as penalty to avoid artificial porous material.

Another aspect to determine the selection of unit cell is mathematical. If we need a G-closure formulation, rank-2 material is recommended based on rigorous mathematics study [4].

Optimization problem (5) can be solved by optimality criterion approach together with finite element discretization. The idea can be easily extended to three dimensional solid, plate bending and shell problems. Problems under multiple loading cases and/or with constraints or objective of other global measure of structural performance such as structural vibration frequency, buckling load can be dealt with too. Fig.4 shows one of the applications.

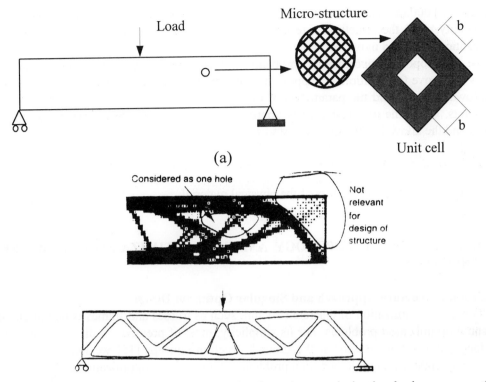

Fig.4 Typical unit cell and numerical example of topology optimization by homogenezation method.

3.3 Artificial Density Method

Instead of using homogenization method to establish a relation between effective elasticities properties, macro-mass density and parameters of micro-structure, one could use variable thickness sheet model for plane continuum [26] or ad-hoc artificial density approach [11,23]. In the later approach, we introduced an artificial relative material density $0 \le \rho(x,y) \le 1$ without referring any specific micro-structure. The density of physical material is $\rho(x,y)\rho_0$. Area of $\rho(x,y)=0$ represents void, $\rho(x,y)=1$ represents solid, whilst $0<\rho(x,y)<1$ represents porous material. Furthermore we assumed a nonlinear relation between the artificial density and structural constitutive matrix. For example, the Young's modulus E is assumed to be the form,

$$E = E_0 \rho^n(x,y) \qquad n > 1 \qquad (6)$$

where E_0 is the Young's modulus of the physical material. If n<1 the resulting design will have large area of porous material. Large n is good for concentration of material and thus gives a clear topology. By taking $\rho(x,y)$ as design variable, we can formulate the plane continuum topology optimization in a similar way with the formulation (5). Because of the simplicity, the problem can be solved by optimality criterion approach.

The homogenization method and other methods have been applied to solve two or three dimensional continuum structure optimization for minimum compliance and/or maximum natural vibration frequency. The approach is very efficient and the number of design variables may reach 10,000 or even more, see Fig.4. This makes it possible to solve practical problem in high quality. And the structural size optimization, topology optimization and the material optimization is unified in one formulation.

To treat topology optimization problem with constraints on local quantities, we may modify the above formulation and approach. For example, we may derive the relation between the allowable stress and the parameters of the cell in order to deal with problem with stress constraints. Or in the artificial density approach, we assume relation between the artificial density and the allowable stress as follows:

$$\sigma_{allow} = \sigma^0_{allow} \rho^m(x,y) \qquad m \geq 1 \tag{7}$$

where σ^0_{allow} is the allowable stress of the physical material.

4. INTEGRATION OF TOPOLOGY AND SIZE OPTIMIZATION OF DISCRETE STRUCTURES.

4.1 Ground Structure Approach and Singular Optimum Design

Topology optimization of discrete structure such as trusses and frames is one of classical structural optimization problems. For its simplicity, most of previous studies dealt with truss topology optimization. Based on the ground structure concept [15] we may formulate the minimum weight truss topology design problem in terms of size optimization as follows,

To Find $A_1, A_2, ..., A_N$

Min. $W = \sum_{i=1}^{N} \rho_i A_i L_i$ (8a)

St. $g_j(\mathbf{A}, \mathbf{U}) \leq 0,$ $j = 1, 2, ..., M$ (8b)

$0 \leq A_i \leq A_{i,max},$ $i = 1, 2, ..., N$ (8c)

Here A_i, L_i, ρ_i are the cross sectional area, length and specific weight of the ith bar, respectively. N is the total number of bars in the ground structure. The zero lower bound on cross sectional area represents possible elimination of bars, therefor change of the structural topology. The set of constraints $g_i(\mathbf{A}, \mathbf{U}) \leq 0$ may include constraints on structural compliance, fundamental vibration frequency, buckling load, nodal displacements and bar stresses. The advantage of the above formulation is that the topology optimization problem is in the form of size optimization and so very easy to deal with. Of course, the zero lower bound of cross sectional area results in difficulty of singular structural stiffness matrix. In addition, the number of bars N, i.e., the number of design variables is huge and the optimization problem (8) is very large because the ground structure may encompass many bars possibly connecting the all structural nodes, support nodes and loading nodes.

The ground structure approach and the formulation (8) has been frequently used as a basis of constructing various numerical algorithms, in which mathematical programming techniques are employed to remove unnecessary members during the process of optimization. In their study, Dorn and Greenberg [15] applied linear programming method to solve (8). Dobbs and Felton [14] used member cross sectional properties as topology design variables and applied a steepest descent-alternate mode algorithm to find the optimal topologies of trusses subjected to stress constraints and multiple loading cases. Their approach prohibits a removed member from re-entering the design. To avoid this pitfall and achieve globle optimal design, Sheu and Schmit [30] applied a branch and bound method to optimize the topology of trusses under multiple loading conditions with both stress and displacement constraints. In Ref.[25], in addition to the member areas, joint displacements were considered simultaneously as design variables for truss topology optimization problems. Branch and bound search technique was also employed in this work. Kirsch [18] suggested a heuristic two-stage design approach for topology optimization of discrete structures. For topology optimization of large scale ground structures subject to stress, local buckling, as well as displacement constraints, DCOC method was developed by Zhou and Rozvany [32].

Singular optima in structural topology optimization was first shown by Sved and Ginos [29]. They investigated a three-bar truss subjected to three loading cases and pointed out that a global optima could be obtained only by removing one of the members from the ground structure and, in effect, violating the stress constraint for that member. Problems where optimal topology is singular have been studied by Kirsch [17,18,19]. Kirsch stated that "in case of singular solutions, it might be difficult or even impossible to arrive at a true optima by numerical search algorithms".

Illustrating by truss topology optimization problems subject to stress constraints, Cheng and Jiang [9] demonstrated that the discontinuity of the bar's stress constraint function when its cross sectional area takes zero value is the essential cause of the existence of singular optima. They also showed that the singular optima is not an isolated feasible design, rather they are just end points of the line segment attached to the feasible domain. The whole feasible domain is of jelly-fish like shape. Because of that, there is an essential difference between topology and sizing optimization and it is very important to establish a rational formulation which can unify this two problems within the same framework if one wants to apply the sizing optimization techniques to solve the topology optimization problems. In Ref.[8] , Cheng discussed the shape of feasible domain for truss topology design when Euler buckling and compliance constraints are imposed. To this end, the authors introduced some basic definitions such as topology design variable and its critical value, the topology-dependent behavior constraints which are directly associated with topology design variables and the limiting values of these constraint functions [13]. They also presented the conditions for existence of singular optima and the classification of various structural topology optimization problems. Rozvany [28] defined the singular optima as an optima in whose vicinity the feasible domain is restricted to a k-dimension space with $k < n$, here n is the dimension of the design variable space.

4.2 ε-Relaxed Approach and Integration of Structural Size and Topology Optimization

Based on the continuity of constraint functions we devided truss topology optimization problems into two types. The first type is the one with continuous constraint function or without singular optimum, whilst the second is that with discontinuous constraint function or with singular optimum. Most of practical topology optimization problem is of mixed type.

For the first type, for example, minimum weight topology design problem with only constraint on structural compliance, or its dual formulation, one conventional approach is to approximate (8) by the following optimization problem (9).

To Find $A_1, A_2, ..., A_N$

Min. $W = \sum_{i=1}^{N} \rho_i A_i L_i$ (9a)

St. $g_j(\mathbf{A}, \mathbf{U}) \leq 0, \qquad j = 1, 2, ..., M$ (9b)

      $0 < \varepsilon \leq A_i \leq A_i^{max}, \qquad i = 1, 2..., N$ (9c)

In (9) instead of zero a small positive quantity ε is specified for the side constraints on topology design variables. By doing so, numerical problems caused by zero bar area are avoided and (9) is an ordinary structural size optimization problem. Its optimum design and the corresponding objective value, denoted by $A^{opt}(\varepsilon)$ and $W^{opt}(\varepsilon)$ respectively, depends on the small quantity ε. Based on the continuity of constraint functions it can be proven that they converge to the optimum design and its corresponding objective of (8) as the small quantity ε tends toward to zero. The solution of (9) is a good approximation of (8). In this way, the first type of problem can be solved by structural size optimization algorithm.

To solve topology optimization problem of the first type more efficiently, much research work [3] appeared recently in parallel with the study on continuum topology optimization. Bendsoe and others studied the mathematical structures of different formulations and pointed out a number of prominent features of the problem. For example, the problem may be reformulated as an unconstrained, convex and non smooth problem in the displacement variable only. By taking advantage of these properties a number of numerical methods were developed and tested. To illustrate the capability of those methods, one numerical example of large scale truss structure design is quoted here, see Fig.5. Similar approach can be applied to solve large scale truss topology optimization problems with constraints on structural buckling load, fundamental vibration frequency or other global structural behavior.

Structural topology optimization problems with bar stress constraints belong to the second type of problem. They behave very differently from the first type. To be complete we present the problem as follows,

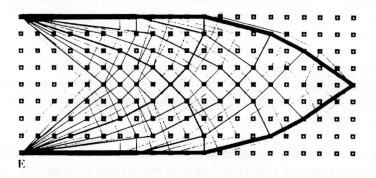

Fig.5 Example of minimum compliance design of truss for transmitting a single vertical force to vertical line of supports. 10940 potential bars, 21 by 9 nodes.(Bendsor, Ben-Tal and Zowe, 1993)

To Find $A_1, A_2, ..., A_N$

$$\text{Min.} \quad W = \sum_{i=1}^{N} \rho_i A_i L_i \tag{10a}$$

$$\text{St.} \quad \sigma_{jk}(\mathbf{A},\mathbf{U}) - \sigma_j^{max} \le 0, \qquad j = 1,2,...,N; \; k = 1,2,...K \tag{10b}$$

$$\sigma_j^{min} - \sigma_{jk}(\mathbf{A},\mathbf{U}) \le 0, \qquad j = 1,2,...,N; \; k = 1,2,...K \tag{10c}$$

$$0 \le A_i \le A_i^{max}, \qquad i = 1,2...,N \tag{10d}$$

Here σ_j^{max}, σ_j^{min} are the allowable tensile and compressive stresses of the jth bar, respectively. The truss structure being optimized is subjected to multiple loading cases, K is the number of loading cases. For problems with single loading case and no upper bound on design variables, if we neglect the compatibility conditions the above problem can be transformed into an Linear Programming one and solved by Simplex Method very efficiently. Fortunately, its optimum design can be statically determinate structure for which the neglecting of compatibility conditions can be justified. For problems under multiple loading cases, the solution becomes much more complicated.

Because of the discontinuity of stress function at zero cross sectional area, the feasible domain of (10) includes degenerated sub domains and has a jelly-fish like shape. Its optimum design may be singular, i.,e., one located at the degenerated feasible sub domain. If we apply an existing numerical optimization algorithm to formulation (10), for most of them the iteration will not converge to the optima , if it is singular, no matter how we select the initial designs. In particular, if we apply optimization numerical algorithm which is based on Kuhn-Tucker condition, iterations will converge to neither global optima nor local optima, but a design point which satisfies the Kuhn-Tucker condition and violates the constraints qualification [12]. On the other hand, if we approximate (10) in a similar way with (9), we would cut off the degenerated sub domain and miss the optima, if it is singular.

In Ref.[10], we replace the stress constraints by the constraints on bar internal force and establish a consistent formulation (11)

To Find $A_1, A_2,..., A_N$

$$\text{Min.} \quad W = \sum_{i=1}^{N} \rho_i A_i L_i \tag{11a}$$

$$A_j \sigma_{jk}(\mathbf{A},\mathbf{U}) - A_j \sigma_j^{max} \le 0, \qquad j = 1,2,...,N; \; k = 1,2,...K \tag{11b}$$

$$A_j \sigma_j^{min} - A_j \sigma_{jk}(\mathbf{A},\mathbf{U}) \le 0, \qquad j = 1,2,...,N; \; k = 1,2,...K \tag{11c}$$

$$0 \le A_i \le A_i^{max}, \qquad i = 1,2...,N \tag{11d}$$

The above formulation conceals the discontinuity of stress function, but does not change the shape of feasible domain. The optimum design of (11) may be still singular. To avoid the difficulty of singular optimum, we proposed a new approach [12], in which we further relaxed the constraints on internal forces and approximated the formulation (11) as,

To Find $A_1, A_2,..., A_N$

$$\text{Min.} \quad W = \sum_{i=1}^{N} \rho_i A_i L_i \tag{12a}$$

$$\text{St.} \quad A_j \sigma_{jk}(\mathbf{A},\mathbf{U}) - A_j \sigma_j^{max} \le \varepsilon, \qquad j = 1,2,...,N; \; k = 1,2,...K \tag{12b}$$

$$A_j \sigma_j^{min} - A_j \sigma_{jk}(\mathbf{A},\mathbf{U}) \le \varepsilon, \qquad j = 1,2,...,N; \; k = 1,2,...K \tag{12c}$$

392

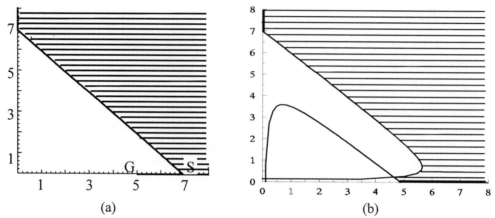

(a)　　　　　　　　　　　　(b)

Fig.6 Typical feasible domain of (12) for a) $\varepsilon = 0$ and b) $\varepsilon = 0.5$

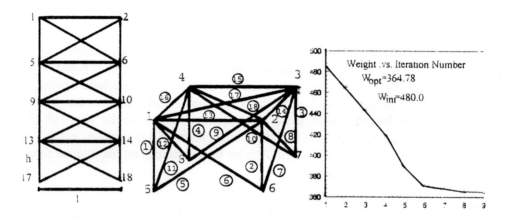

Fig.7 Numerical examples solved by the ε-relaxed approach, 72 bar truss, 16 design variable group. Stress and displacement constraints, optimial design(0.19, 0.53, 0.47, 0.64, 0.53, 0.52, 0,0, 1.32, 0.52, 0, 0, 1.90, 0.52, 0, 0) is singular.

$$\varepsilon^2 \leq A_i \leq A_i^{max}, \qquad i = 1,2..., N \tag{12d}$$
$$0 < \varepsilon << 1$$

We denote the optimum design and its objective of (12) as $A^{opt}(\varepsilon)$ and $W^{opt}(\varepsilon)$. Based on the mapping theory from the set of points to the set of domains [16], it can be proved that $A^{opt}(\varepsilon)$ and $W^{opt}(\varepsilon)$ converge to the corresponding optimum design and the objective of (8) when $\varepsilon \rightarrow 0$. It can be also shown that for formulation (12a-12d) singular optima does not exist and in the vicinity of degenerated feasible sub domain of formulation (8) the measure of the feasible domain is no longer zero. Fig.6 shows an example of modification of feasible domain by the ε-Relaxed approach. Therefore, it provides the opportunity for numerical search algorithms to arrive at the optima which is located at the degenerated sub domain of (8). With the formulation (12) we have solved a number of truss design examples by structural size optimization algorithm. Some of them are shown in Fig.7.

It is interesting to note that the formulation (12) actually includes the formulation (9) as a special case and can be used to deal with topology optimization problems of first type, second type and the mixed type. In other words, it provides an unified approach to integration of structural size optimization and topology optimization.

5. CONCLUDING REMARKS

The recent achievement in structural topology optimization is not only important for its own progress, but also for a wider application of structural optimization in engineering design.

Though important progresses have been seen in the area of structural topology optimization during the last decade, many difficult problems remain open. For most of topology optimization with constraints other than global structural quantities, the problem is non convex and we are still lacking of efficient method to reach the global optimum.

Acknowledgment

The support of National Natural Science Foundation in China under Research Project No.19572023 is gratefully acknowledged.

REFERENCES

1. M.P. Bendsoe, Existence proofs for a class of plate optimization problems, Lecture Notes in Control and Information Sciences, Vol.59, Springer Verlag, Berlin, 1984, pp773-779.
2. M.P. Bendsoe; N. Kikuchi, Generating optimal topologies in structural design using a homogenization method, *Comp. Meth. Appl.Mechs.Engrg.*,71 (1988) 197-224.
3. M.P. Bendsoe, A. Ben-Tal, J. Zowe, Optimization Methods for Truss Geometry and Topology Design, Structural Optimization, 7 (1994) 141-159.
4. M.P. Bendsoe, Optimization of Structural Topology, Shape, and Material, Springer-Verlag, Berlin Heidelberg New York, 1995.
5. A. Bensoussan, J-L Lions; G. Papanicolaou, Asymptotic Analysis for Periodic Structures, North-Holland, Amsterdam, 1978.
6. M. Bremicker, M. Chirehdast, N. Kikuchi, P. Papalambros, Integrated Topology and Shape Optimization in Structural Design. Mech. Struct. Mach., 19 (1992) 551-587.
7. K.T. Cheng(G.D. Cheng), N. Olhoff, An investigation concerning optimal design of solid elastic plates, *Int. J. Solid Struct.*,17 (1981) 305-323.
8. G.D. Cheng, Some Aspects of Truss Topology Optimization, Struct.Optim., 10 (1995) 173-179
9. G.D. Cheng and Z.Jiang, Study on Topology Optimization with Stress Constraints, Eng. Optim., 20 (1992) 129-148.
10. G.D. Cheng, and Z. Jiang, Numerical performances of two topology optimization of truss structures, Acta Mechanica Sinica, 6 (1994).

11. G.D. Cheng and D.X. Zhang, Topology optimization of planar continuum under stress constraints, Jour Dalian Univ. of Tech., 35 (1995)(1) 1-9. (in Chinese)

12. G.D. Cheng, X. Guo, ε-Relaxed Approach in Structural Topology Optimization, accepted by Struct. Optim.

13. G.D. Cheng, X. Guo, A Note on Jelly-fish Like Feasible Domain in Structural Optimization, accepted by Eng. Optim.

14. M. Dobbs, L. Felton, Optimization of Truss Geometry. J. Struct, Div. 95 (1968), No.ST-10, ASCE, 2105-2118.

15. W. Dorn, R. Gomory, and M. Greenberg, Automatic Design of Optimal Structures, J de Mecanique, 3 (1964) 25-52.

16. W.W. Hogan, Point-to-set maps in mathematical programming. SIAM Review 15 (1973) 591-603.

17. U. Kirsch, Optimal Topologies of Structures, Appl. Mech. Rev., 42 (1989)(8).

18. U. Kirsch, Fundamental Properties of Optimal Topologies, In: M.P. Bendsoe; C.A. Mota Soares,(eds.) Topology optimization of structures, Dordrecht: Kluwer, 1993, pp3-18.

19. U. Kirsch, On Singular Topologies in Optimum Structural Design, Struct Optim,2 (1990) 39-45.

20. R.V. Kohn, G. Strang, Structural Design Optimization, Homogenization and Relaxation of Variational Problems, Lecture Notes in Physics, Vol.154, Springer, Berlin, 1982, pp131-147.

21. K.A. Lurie, Direct relaxation of Optimal Layout Problems for Plates, Jour. Optim. Theory and Appl., 80 (1994)(1) 93-116.

22. A.G.M. Michell, The limits of economy of material in frame structures, Phil.Mag. 8 (1904) 589-597.

23. H.P. Mlejnek, Some aspects of the genesis of structures, Struc Optim, 5 (1992)(1-2) 64-69.

24. P. Pedersen, On the Optimal Layout of Multi-Purpose Trusses. *Comp and Struct* 2 (1972) 695-712.

25. U.T. Ringertz, A Branch and Bound Algorithm for Topology Optimization of Truss Structures, Eng. Opt., 10 (1986) 111-124.

26. M.P. Rossow, J.E. Taylor, A Finite Element Method for the Optimal Design of Variable Thickness Sheets, AIAA J., 11 (1973) 1566-1569.

27. G.I.N. Rozvany, Lay-out theory for grid-type structures, In: Bendsoe,M.P.,; Mota Soares,C.A.,(eds.) Topology optimization of structures, Dordrecht: Kluwer, 1993, pp251-272.

28. G.I.N. Rozvany, T. Birker, On Singular Topologies in Exact Layout Optimization, Struct. Opt., 8 (1994) 228-235.

29. G. Sved, and Z. Ginos, Structural Optimization under Multiple Loading, Int.J. Mech. Sci. (1968)

30. C. Sheu, L.A. Schmit, Minimum Weight Design of Elastic Redundant Truss Under Multiple Static Loading Condition, J. AIAA, 10 (1972) 155-162.

31. O. Sigund, Design of Material Structures using Topology Optimization, DCAMM special Report No.S69, Tech. Univ. Denmark, 1995.

32. M. Zhou, and G.I.N. Rozvany, DCOC: an optimality criteria method for large systems, Part I:theory, Struc Optim, 5 (1992)(1-2) 12-29.

Theoretical and Applied Mechanics 1996
T. Tatsumi, E. Watanabe and T. Kambe (Editors)
© 1997 Elsevier Science B.V. All rights reserved.

Kolmogorov–like behavior and inertial range vortex dynamics in turbulent compressible and MHD flows

Annick Pouquet,
CNRS URA 1362, OCA, BP 4229, 06304 Nice Cedex 4, France
pouquet@obs-nice.fr

Compressible flows are found in a variety of contexts, as in the combustion chamber of diesel motors, or in the interstellar medium where stars are born; such flows are made up of vortices, shocks, dense cold clouds and current sheets in the magnetized case interacting with(in) a turbulent background. If models can be devised in the subsonic case, the supersonic regime is best approached from the experimental side, in the laboratory as well as in numerical experiments.

A series of numerical simulations – both decay and forced runs, and in a variety of physical contexts – using various techniques, will be outlined, with resolutions of up to 1024^3 grid points. The prevalence of filamentary structures such as vortex and flux tubes, and their influence on the dynamics are described, in particular in the context of intermittency. Vorticity and magnetic current production is also discussed. This review ends with a succinct description of the energetics of the interstellar medium and the molecular clouds within, when self–gravity, magnetic fields and ionization effects such as ambipolar drift are all included.

1. Introduction

Compressible flows are found in a variety of contexts, from the combustion chamber of diesel motors to the interstellar medium of the galaxy where stars are formed. In the latter case, such flows are supersonic, strongly magnetized and self–gravitating. The basic ingredients of a theoretical description are thus vortices, shocks and rarefaction waves, together with magnetic flux tubes and current sheets, all such structures being embedded in a turbulent background. Various models have been developed for subsonic flows so as to take into account departures from the incompressible case in terms *e.g.* of the (small) Mach number, and the domain of acoustics in particular dealing with refined analysis of the Lighthill formula for far–field emission of noise is a vigorous branch of research (see Crighton [29], this Volume), both in the theoretical context and with industrial applications *e.g.* to aircrafts and now fast trains. The supersonic regime, however, is mostly studied with the help of laboratory experiments (such as in the case of supersonic jets, or of the collision of a vortex and a shock wave [115]), and numerical simulations.

In that light, a series of results stemming from several numerical experiments using different numerical techniques of integration in one, two and three space dimensions will be outlined, with resolutions of up to (but for a brief moment) 1024^3 points on a uniform grid. The complexity of the models diminishes as the dimension of space increases; for example, in one space dimension (but with all vector components included), magnetic fields, self–gravity and the ambipolar term due to the drift of charged particles with respect to neutral ones can all be included [7], and allow for a detailed parametric study and analysis of the structures that develop. After a short reminder of the various numerical techniques employed, both decay and forced runs with a variety of forcing according to the problem at hand will be described. A full model of the interstellar medium would require the inclusion of numerous effects – depending in particular on what scale the focus is on, from the global galactic scale of the kiloparsec (where 1 pc $\sim$ 3.2 light years), to that of giant molecular clouds (50 pc) to molecular clouds (1 pc), to cold dense cores of 0.01 pc, at the limit of present–day resolution with very long–based interferometry – and should in particular include some simplified form of radiative transfer. But the purpose here is rather to show that behind the apparent complexity and variety of such flows, a few general laws can be drawn, when stressing the common effects such as those due to intermittency, *i.e.* the scarcity of intense small–scale features, and the effect of such scarce features on both the temporal and spatial statistics; in other words, it is stressed that appropriate modifications of the Kolmogorov–like phenomenology can lead to some understanding of the global dynamical features of these diverse flows.

2. Basic equations

2.1. The models

We begin by writing a complete set of equations for a magnetized compressible fluid:

$$\frac{\partial \rho}{\partial t} + \nabla \cdot (\rho \mathbf{u}) = 0 \tag{1}$$

$$\frac{\partial \mathbf{u}}{\partial t} + \mathbf{u} \cdot \nabla \mathbf{u} = -\frac{\nabla P}{\rho} - \left(\frac{J}{M_a}\right)^2 \nabla \phi + \mathcal{L} + \mathcal{D}_N(\mathbf{u}) + \mathcal{D}_2(\mathbf{u}) + \mathbf{F} \tag{2}$$

$$\frac{\partial e}{\partial t} + \mathbf{u} \cdot \nabla e = -(\gamma - 1)e\nabla \cdot \mathbf{u} + \kappa \frac{\nabla^2 e}{\rho} + \mathcal{S}_E \tag{3}$$

$$\frac{\partial \mathbf{B}}{\partial t} = \nabla \times (\mathbf{u} \times \mathbf{B}) + \mathcal{H} + \mathcal{A} + \mathcal{D}_N(\mathbf{B}) + \lambda \nabla^2 \mathbf{B} \tag{4}$$

$$\nabla^2 \phi = \rho - 1 \tag{5}$$

$$P = (\gamma - 1)\rho e \ , \tag{6}$$

with

$$\mathcal{L} = \frac{1}{\rho}(\nabla \times \mathbf{B}) \times \mathbf{B} \tag{7}$$

the Lorentz force and

$$\mathcal{D}_2(\mathbf{u}) = \nu(\nabla^2 \mathbf{u} + \frac{1}{3}\nabla \nabla \cdot \mathbf{u}) \tag{8}$$

the standard dissipative term due to the viscosity ν. Here, the $\mathcal{D}_N(\mathbf{u})$ and $\mathcal{D}_N(\mathbf{B})$ terms represent either explicit or implicit numerical dissipation for both the velocity $\mathbf{u}$ and the magnetic field $\mathbf{B}$ (in fact, the induction) to allow the computation to be run at a high effective Reynolds number, *i.e.* in such a way that the large scales undergo little or no dissipation; this means that the overall dissipation is concentrated in the smallest resolved scales. For example, in the model of the interstellar medium of [88], the choice is made of a hyperviscosity namely $\mathcal{D}_N(*) = -\nu_8 \nabla^8(*)$ (see [53] for a discussion of hyperviscosity in the context of three–dimensional incompressible fluids). In the induction equation, the terms $\mathcal{A}$ and $\mathcal{H}$ represent respectively the ambipolar drift and the Hall current; such terms are present in the generalized form of Ohm's law for a weakly ionized (ambipolar drift) or partially ionized (Hall current) medium, and are given further below in §6. A perfect gas law has been chosen linking pressure P, density ρ and internal energy e. Source terms are also included, namely $\mathbf{F} = \mathbf{F}_c + \mathbf{F}_s$ with *a priori* on the one hand a compressible component $\mathbf{F}_c$ (with $\nabla \times \mathbf{F}_c = \mathbf{0}$), and on the other hand a solenoidal $\mathbf{F}_s$ (with $\nabla \cdot \mathbf{F}_s = 0$) component in the momentum equation to mimic respectively supernovae–like expansions or ionization winds emanating from young stellar objects and observed commonly in molecular clouds, and on the other hand galactic shear. Finally, in the energy equation $\mathcal{S}_E$ represents an as yet unspecified model for both heating and cooling, due *e.g.* to showering of cosmic rays or ultra–violet ionization (see *e.g.* [30]).

The parameters appearing in the equations are J, the gravitational number, with ϕ the gravitational potential, M_a the Mach number, and the kinetic and magnetic Reynolds numbers

$$R^V = U_0 L_0 / \nu \tag{9}$$

and similarly $R^M = U_0 L_0 / \lambda$ – where U_0 and L_0 are the characteristic velocity and length scale of the flow; the dissipative coefficients are κ the thermal diffusivity, ν the viscosity and λ the magnetic diffusivity. Included in direct numerical simulations (DNS), they may be omitted in modelization of turbulent flows, or put at a lower value than that used in DNS for the same flow at the same resolution, a standard procedure (they provide in fact a natural Gaussian filter at small scale).

2.2. A glimpse through various numerical techniques

A variety of numerical codes that have been developed over the years by several teams have been employed in order to obtain the results described here. On the one hand, one can use the direct numerical simulation approach in which dissipative terms are included explicitly. It has the advantage that it can be directly compared with experiments, but with the disadvantage that the kinetic and magnetic Reynolds numbers remain, in dimension three, substantially lower than in experimental set–ups or for geo–astrophysical flows. Of course lowering the space dimensionality – while keeping all three vector components – alleviate this problem linked to insufficient computing power.

On the other hand, one can develop "*sparse*" methods in which only part of the modes are kept. In fact, Fourier codes do just that since only commensurable wavenumbers are kept in such methods (see however [124]). Another example of a sparse method is one in which symmetries of the basic equations and of initial conditions (as well as the stirring mechanism, as in the case of the generation of magnetic fields) are enforced throughout, leading to substantial memory and CPU savings; examples are the Taylor–Green vortex [19] [20] and the Kida flow ([59] [89] and references therein). In both cases, singularity of the Euler (inviscid) flow have been looked for, but the question remains whether in the generic case when such symmetries are broken (and in fact they are, see [85]), a singular behavior still prevails. Other sparse methods consist in suppressing modes in a more or less *ad hoc* fashion, for example by a decimation in the dissipation regime [77], or otherwise [49] [117].

An altogether different approach is to discard the dissipative terms in the primitive equations, and introduce numerical dissipation instead, either relying on LES, or Large Eddy Simulation whereby transport coefficients derived from closure methods [69] are employed – or in a more geometrical form, whereby the steepening and sometimes the non–oscillating character (or essentially non–oscillating, ENO) of the shocks is enforced. Such methods have been developed in the context of compressible flows where they arise naturally, making use in particular of the existence of the Rankine–Hugoniot relationships. No equivalent law is known for incompressible flows; but one could consider modelizing small–scale incompressible fluids with a juxtaposition of Burgers (?) vortex filaments and the dissipative structures that enshroud them and which are probably filamentary as well [20] [103]. Alternatively, one can make use of detailed asymptotic solutions, as those available for slender cores, in order to obtain accurate solutions [65]. Two remarks are in order, here: all problems are related; and these methods complement each other. Indeed, direct numerical simulations and experiments (plus observations in geo–astrophysics) are essential to guide the theoretician towards consideration of the prevalent structures that arise in such flows; a theoretical understanding of the basic structures in interaction within complex turbulent flows is needed *per se*, but not solely since such models can help in improving numerical modeling; and numerical modeling in turn can help understand complex flows for which many – and thus, cheap – computations are needed. In this light, a keen exploration of the dynamics of vortex tubes and sheets, of their statistical importance and of their relation to statistical laws such as those developed by Kolmogorov [66] [67] [68] is a central theme of research today as exemplified in several instances in this conference. A better cartography of turbulent structures will indeed stem in part from a careful and detailed exploration of parameter space with a variety of experimental and numerical techniques.

3. Turbulent compressible flows

3.1. Introduction

Compressible flows may not be viewed simply as a superposition of shocks, but rather as a superposition of shocks and vortices interacting together and with a weaker possibly structureless turbulent background. Early works on turbulent flows concentrated on one–dimensional models, in particular on the Burgers equation. The dynamics of nonlinear waves using the reductive perturbation expansion method was further extended to the two–dimensional case [116] where it was shown that the four groups of nonlinear waves belonging to different families of characteristics behaved independently of each other. But the extension to three dimensions, with the introduction of the dynamical stretching of vorticity is mostly investigated numerically. As an example, the results of a series of three–dimensional computations of compressible turbulence performed with the PPM algorithm [126] using periodic boundary conditions is now outlined; initial conditions are random and centered in the large scales ($2\pi/L_0 \sim 1$), the initial ratio of compressible to solenoidal kinetic energy is $\sim 10\%$, the flow follows a γ–law with the adiabatic index $\gamma = 1.4$, and the initial *rms* Mach number is unity [94]. Runs on uniform grids of 64^3 to 1024^3 points have been performed. The computation on the largest grid is done for 0.2 eddy turnover times or acoustic time τ_{ac}, from $t = 1.80$ up to approximately the maximum of the enstrophy for that flow [96]. Flows are analyzed both with the raw data and with filtered data using Favre averaging and a Gaussian filter in Fourier space. The ratio of the integral scale to the Taylor scale for the 512^3 run is ~ 7.7 at the time of the peak in enstrophy (comparable to that in [54]) but, contrary to direct numerical simulations, large scales undergo only a negligible amount of dissipation. A second series of runs of compressible shear flows with average Mach numbers of order (but slightly below) unity were performed on grids ranging from 128^3 to 512^3 points, and results are comparable [95] in that the same type of structures emerge [1]; the driving is a unidirectional shear wave at $k_0 = 1$. In contrast with [61], we chose – in order to maintain a constant temperature on average, and thus a constant Mach number – to add a cooling law to the energy equation with a Stefan–like σT^4 dependence appropriate for optically thick media. In these stirred flows, on average, the kinetic energy is $E_K \sim 0.7$ and the heat energy $E_H \sim 1.5$; the *rms* Mach number is ~ 0.8, with density contrasts ρ_{max}/ρ_{min} of up to 30 and density fluctuations $\delta\rho/\rho_0$ of order unity.

3.2. The Kolmogorov range

One of the fundamental prediction for incompressible flows concerns the energy spectrum in the inertial range which should follow a $-5/3$ law. Corrections to that spectrum taking into account interactions with acoustic waves in the limit of small Mach numbers have been evaluated leading in the limit of $M_a \to 1$, to a k^{-2} spectrum [81]. From a numerical standpoint, the evaluation of the power–law is rendered difficult by the fact that large scales are dominated by energy–containing eddies, and small scales by dissipative eddies (both numerical or physical), so that the inertial range in a decaying flow is limited in extent to at best a decade in wavenumber. Such a limitation can be alleviated when dealing with stirred flows for which time–averaging smooths out fluctuations (see [61]). Nevertheless, there are clear indications that for the parameter range explored, the $-5/3$ power–law is present in the compressible decay runs [2]. The inertial range is followed by a shallower spectrum $E(k) \sim k^{-1}$. This latter spectrum is attributed to the prevalence of vortex filaments at small scales: such structures, in solid body rotation at the core, do not dissipate any energy so that there is an accumulation of energy which marks the end of the inertial range; this accumulation has also been observed [104] in the experimental data of Gagne [42] [3], as well as in second–order closure computations (see [72]). The longitudinal spectrum follows an approximate Kolmogorov law as well, as opposed to a k^{-2} law that is expected on the basis that the compressible part of the flow consists mainly of

---

[1]This forcing is purely solenoidal; the resulting ratio of compressible to vortical energies is ~ 10 %.

[2]In two dimensions, another power–law obtains for the solenoidal component of the energy spectrum [93], linked to an inverse transfer of energy to the large scales as in the incompressible case.

[3]See however [1] where a recent discussion of various plausible fitting laws is given.

velocity jumps. This shallower law may be due to the fact that, in equation (11) below, the dominant term is that due to the vorticity, and the (scalar) longitudinal component of the velocity simply adjusts to the dynamics of the transverse vortical components.

3.3. Vorticity production

Small–scale phenomena can be best understood by looking at the dynamical evolution of the velocity gradient matrix $\partial_i u_j$; here we concentrate on the equations for the potential vorticity $\omega_{\mathrm{p}} \equiv \omega/\rho = (\nabla \times \mathbf{u})/\rho$ and for $\nabla \cdot \mathbf{u}$; in the barotropic case, $P \sim \rho^\gamma$ they read:

$$\frac{\partial \omega_{\mathrm{p}}}{\partial t} + \mathbf{u} \cdot \nabla \omega_{\mathrm{p}} = \omega_{\mathrm{p}} \cdot \nabla \mathbf{u} \quad , \tag{10}$$

$$\frac{\partial \nabla \cdot \mathbf{u}}{\partial t} + \mathbf{u} \cdot \nabla[\nabla \cdot \mathbf{u}] = -\nabla^2 \frac{u^2}{2} + \omega^2 + \mathbf{u} \cdot \nabla^2 \mathbf{u} - \frac{\gamma}{\gamma - 1} \nabla^2 \rho^{\gamma-1} \quad , \tag{11}$$

where the dissipation and forcing terms have been omitted. In the full thermodynamical case, a term $\mathcal{B}/\rho$ must be added to (10), with $\mathcal{B}$ the baroclinic term defined as

$$\mathcal{B} = \frac{\nabla p \times \nabla \rho}{\rho^2} \quad . \tag{12}$$

In that latter case, and for initial conditions dominated by the solenoidal part of the velocity, computations have shown that, except at early time, the vortex stretching term dominates the baroclinic term, by roughly a factor twenty [95].

In the case of incompressible conducting flows, vorticity production also occurs through the Lorentz force; one has [98]:

$$D_t \omega = \mathbf{B} \cdot \nabla \mathbf{j} + \omega \cdot \nabla \mathbf{u} - \mathbf{j} \cdot \nabla \mathbf{B} \tag{13}$$

and

$$D_t \mathbf{j} = \mathbf{B} \cdot \nabla \omega + \mathbf{j} \cdot \nabla \mathbf{u} - \omega \cdot \nabla \mathbf{B} - 2 \sum_m \nabla u_m \times \nabla B_m \tag{14}$$

(see also [63]). The last two terms on the *rhs* of the equation for the current can also be written as $-\nabla(\mathbf{u} \cdot \nabla) \times \mathbf{B} + \nabla(\mathbf{B} \cdot \nabla) \times \mathbf{u}$. In terms of the Elsässer variables $\mathbf{z}^\pm = \mathbf{u} \pm \mathbf{B} = \nabla \times \psi^\pm$, $\omega^\pm = \omega \pm \mathbf{j}$, $\eta^\pm = \nabla \times \omega^\pm$ together with the notation $D_t^\pm = \partial_t + \mathbf{z}^\pm \cdot \nabla$, one has

$$D_t^- \mathbf{z}^+ = -\nabla P \tag{15}$$

and

$$D_t^- \omega^+ = \omega^+ \cdot \nabla \mathbf{z}^- + \sum_m \nabla z_m^+ \times \nabla z_m^- \quad , \tag{16}$$

$$D_t^- \eta^+ = -\nabla \mathcal{P}_2 - \eta^- \cdot \nabla \mathbf{z}^+ - \mathbf{z}^- \cdot \nabla \eta^+ + 2(\sum_{j,l}(\partial_j z_l^-)\partial_l \partial_j) \, \mathbf{z}^+ \quad , \tag{17}$$

where $\mathcal{P}_2 = -\nabla^2 P$; the equations for the $-$ variables are deduced by $\pm$ symmetry. Equation (17) can also be written as

$$D_t^- \eta^+ = -\nabla \mathcal{P}_2 + \eta^+ \cdot \nabla \mathbf{z}^- + \omega^+ \cdot \nabla \omega^- - \omega^- \cdot \nabla \omega^+ + 2 \sum_m \nabla \omega_m^+ \times \nabla z_m + \sum_m \nabla \times (\nabla z_m^+ \times \nabla z_m^-) \tag{18}$$

where the last two terms, once developed, also read $\sum_m [\,(\nabla z_m^- \cdot \nabla)\nabla z_m^+ - (\nabla z_m^+ \cdot \nabla)\nabla z_m^- + \eta_m^+ \nabla z_m^- - \eta_m^- \nabla z_m^+\,]$. The two–dimensional version obtains easily from the above (see also [15]), with A_M the magnetic potential and ψ the stream function:

$$\overset{(2D)}{\sum_m} \nabla z_m^+ \times \nabla z_m^- = 2\sum_m \nabla \partial_m A_M \times \nabla \partial_m \psi = \sum_m \nabla \partial_m \psi^+ \times \nabla \partial_m \psi^- \;, \qquad (19)$$

and

$$D_t^- \eta^+ = \eta^\pm \cdot \nabla \mathbf{z}^\mp \pm \mathcal{S} \qquad (20)$$

with

$$\mathcal{S} = \nabla \times \left(\sum_m \nabla z_m^+ \times \nabla z_m^-\right) \qquad (21)$$

where $\psi^\pm = \psi \pm A_M$ [4].

For non–conducting barotropic fluids, a cursory examination of equations (10, 11) shows that potential vorticity will grow only where strong velocity gradients occur, whereas the divergence of the velocity field can grow through many mechanisms.

Whether such vorticity production happens in a singular way or not in an incompressible flow is still strongly debated. The numerical work of Kerr [55] (see also [56] [57]) tends to show, but at moderate resolution, a tendency for the maximum of the vorticity to grow in a singular way as indicated in [8] [64]. More recently, computations at higher resolution using the highly symmetrical flow of Kida tends to confirm this singular behavior [89], a tendency also indicated analytically in the model developed in [13] [14] and linked to the sign of the fourth–order derivative of the pressure. Several models can indeed be written that lead to singular behavior in a finite time for an incompressible Euler flow (the compressible case, obviously, forming shocks rapidly), when *a priori* unjustified but simplifying assumptions are being made [12] [71] [120] [121] [110]. Whether the *generic* behavior of a highly turbulent flow is singular remains unclear. In particular, the role of structures and their interaction has to be asserted; for example, it was shown in [28] that the amplitude and the curvature of vortex filaments may be antinomic (the stronger, the straighter); and furthermore, the dynamics of the symmetric part of the velocity gradient matrix has not received sufficient attention (see however [57]).

3.4. The equivalent barotropic gas

Integrating in dimension three the equations given in §1 for an unmagnetized compressible fluid without forcing ($\mathbf{B} = 0$, $\mathbf{F_s} = \mathbf{F_c} = \mathcal{S_E} = 0$), it was noted in [60] (see also [62]) that such a fluid behaved on average as a barotropic gas $P \sim \rho^\gamma$ with $\gamma = c_p/c_v$ as given for that gas. These authors used a pseudo–spectral method on a grid of 64^3 points with the standard dissipative terms included; similar scatter plots obtain using the Euler equations with the PPM algorithm, and up to resolutions of 1024^3 points [95]. This result of course is related to the previously mentioned fact that the baroclinic term $\mathcal{B}$ is weak compared to the stretching term. The flow organizes locally as a barotropic gas, with in the (P, ρ) scatter plot a thickness related to entropy fluctuations in the vicinity of shocks.

---

[4]The 2D fluid version recovers with $\mathbf{z}^+ = \mathbf{z}^-$, see also [17].

When including self–gravity, rotation and magnetic fields, and more importantly heating and cooling as in [30], the same type of relationship obtains between pressure and density namely $P \sim \rho^{\gamma_e}$, but now with an *effective* polytropic index $\gamma_e \sim 0.3$ as shown in [88] [118] [119] (see also [35] for an analysis in the linear regime).

4. The prevalence of vortex tubes and their dynamics

The prevalence of vortex tubes for incompressible fluids is now well documented, since the early observations of Patterson (unpublished) and Siggia [109], further confirmed by higher resolution runs [103] [122] [54] [123] and experiments [33]. As is well known, the early evolution of a blob of vorticity embedded in a large–scale shear is to form vorticity sheets, as shown by both statistical dynamical arguments [12] and simple models (see *e.g.* [18]) developed over the years. This sheet is unstable to Kelvin–Helmholtz type of instabilities [73] [84] (see also [10]) or to a self–focusing instability [87] and leads to the development of numerous strong vortex filaments, as is also the case when the fluid is compressible (see §3). Vortex filaments clearly prevail when the imaging of the flow is made in such a way that thresholding is in amplitude: strong vortex structures are filamentary. However, careful analysis of the data has shown that at lower amplitudes, many sheets cohabit with such filaments. Recall that the rolling–up of these sheets into stretched spiral vortices leads to a Kolmogorov spectrum, as exemplified by Lundgren [74] and further developed in [100] and numerically in [76] [5]. Spiraling vortex sheets are also found in compressible flows [95] at intermediate values of the vorticity amplitude that fill most of the volume of the fluid (the intense vortex filaments represent only a few percent of the enstrophy budget).

Local dissipation $\sigma_{ij}\sigma_{ij}$ where σ_{ij} is the strain tensor is also filamentary [103] – although less markedly so – these fatter tubes surrounding the vortex filaments. The filamentary structure of dissipative events is at the core of the successful model of intermittency of She and Leveque [105] discussed briefly in the next Section.

When using the localized induction approximation – a simplified form of the Biot–Savart law, one can show that solitons propagate along vortex filaments [51] [58]; there is now a renewed interest in these issues, in part because the introduction of both the curvature and the torsion of the vortex line also leads to soliton equations (see [101], and [102] for an introduction).

An evolution equation for the curvature of a line or a surface is given in [92]; the similar equation for the torsion of vortex lines is found in [28], and for magnetic field lines in [22] (see also [99]), hence for the Frenet–Serret unit vectors $\hat{t}, \hat{n}, \hat{b}$ associated with such lines (with $\hat{t}$ the unit vector tangent to a vortex line, $\hat{n}$ the normal in the direction of the center of curvature, and $\hat{b}$ the bi–normal as usual). These equations have not been exploited fully as yet. For example, from the inviscid equations for the vorticity vector in the pure fluid incompressible case, one derives two equations for the vorticity magnitude w and for its tangent unit vector $\hat{\omega}$, namely (see also [27]):

$$D_T w = \bar{\alpha} w \tag{22}$$

$$D_T \hat{\omega} = -\bar{\alpha}\hat{\omega} + S_{ij}\hat{\omega}_j \ . \tag{23}$$

---

[5] An interpretation in terms of cascade is given in [47].

with $\bar{\alpha} = \hat{t} \cdot S \cdot \hat{t}$ where S is the symmetric part of the velocity gradient matrix [28]. From the above equations, one sees that when the vorticity grows in amplitude substantially, its direction remains constant, and vice–versa. In MHD, it is the full $\partial_i u_j$ matrix which acts since there is obviously no special relationship between $\mathbf{B}$ and $\mathbf{u}$, unlike vorticity; furthermore, the curl of the Lorentz force must be added to the above equations [22] (see also [99]), and for compressible flows one must take into account both the $\omega \nabla \cdot \mathbf{u}$ term and the baroclinic term [99]. Such structures provide good fit to numerical data, as shown in [80] where a large Reynolds number asymptotic theory for a vortex tube in a uniform non axi–symmetric strain is performed, with predictions for contours of energy dissipation agreeing well with numerical simulations; regions of large dissipation do *not* overlap with regions of large enstrophy (see [4] for MHD). In several of these cases, the modifications due to compressible effects have not been taken into account.

One can write exact solutions for the incompressible MHD equations which are the analogue to their fluid counterparts, such as for example the Burgers vortex [79] [99] [11]. In incompressible MHD, both the vorticity and the magnetic field are stretched in identical ways by the velocity gradients, so that one expects for short times an identical temporal and spatial behavior. Indeed, flux tubes are observed [44] for both the kinematical dynamo (*i.e.* when the velocity field is given, the initial seed magnetic field being too weak to react through the Lorentz force) and in the dynamical case [43] [23]. Note however that in a decay run with on average equal kinetic and magnetic energies, the current and vorticity in the MHD case seem rather to organize in sheets [90] (see also [78] for the dynamo regime in the presence of flux tubes). This observation may be related to the problem of intermittency, as indicated in the next Section, but may also be due to the low resolution (up to 180^3 grid points in [90]) whereby these sheets cannot destabilize.

5. Intermittency

Small–scale structures of turbulent flows are thus a combination of sheets, filaments and spirals. They have several effects on the dynamics. Filaments of vorticity provide no dissipation and thus may be the structure responsible for the bottleneck effect at the end of the inertial range; this corresponds, in the language of modes and statistics, to the fact that in the dissipation range the energy spectrum decreases rapidly and thus an overshoot of the inertial range is expected. From a statistical point of view, these small–scale structures, at least the most intense ones, are scarce. Is there a statistical effect due to their scarcity ? Numerous models (mostly phenomenological in fact) have been written to describe such an effect (see [41] for a recent introduction).

A simple diagnostic of intermittency comes from the observation that probability density functions of velocity differences at small scales depart from a Gaussian law, a result that should be expected since, as shown already by Betchov [12] the skewness – *i.e.* the normalized third order moment of the velocity field, which is equal to zero for a Gaussian – is non–zero and negative because of the (cubic) energy transfer rate to small scales. Wings of the *pdf* are either exponential or stretched exponentials, strong events not being as rare as for a normal distribution; however, these *pdf* do not appear to have power–law behavior. Such exponential wings are rather common; they are observed as well in MHD flows, in compressible flows, in the interstellar medium [38] [40], and in the data stemming from the analysis of foreign exchange markets [46] ! Strong events may in fact be an essential part of the dynamics of turbulent flows.

Longitudinal structure functions of order p, at scale ℓ of a field $\mathbf{v}$ are defined as $\delta v_\ell^p = <[v(x+\ell) - v(x)]^p >$, where v is the component of the vector field in consideration along the x direction; these functions may be assumed to behave in a range of scales (the inertial range) as a power–law, namely $\delta v_\ell^p \sim \ell^{\zeta_p}$. Dimensional scaling gives $\zeta_p = p/3$ for incompressible neutral three–dimensional fluids [66]. Corrections to such a linear law are expected [68], and indeed experimental and numerical data (see [2] for a recent assessment) – as well as geophysical data [24] – indicate a strong departure from the $p/3$ law, although the limit as the Reynolds number tends to infinity is not known. That the departure from the $p/3$ law be itself in the form of a power law or of a more complex functional is not known either and is difficult to assess experimentally, because of huge error bars that arise in the computation as one goes to high orders in p. Indeed, high–order structure functions put more weight on strong, thus rare, events leading to a need for very large data base (as of today, up to 10^{10} points have been used, giving reliable exponents up to $p \sim 10$).

The Kolmogorov relationship for $p = 3$ gives $\zeta_3 \equiv 1$, a result not of a phenomenological nature but in fact stemming [67] from the conservation of energy, [6] once isotropy, homogeneity, incompressibility and stationarity have been assumed (the restriction to isotropic flows can be relaxed, see [82] [41]).

Recently, a model has attracted a lot of attention because of its very good agreement with experimental data [105] (see also [34] for a probabilistic interpretation). It has been generalized to MHD as well [48] [91] in the framework of the slowing down of energy transfer to small scales because of Alfvén waves [70]. [7]

Many models of a very different nature as to what are the basic hypotheses can fit the data equally well (see [86][83] [16]), given the uncertainty in the data at high order where the models differ. However, a new method using the steepest descent algorithm has been devised recently [112] and allows for a better determination of the critical exponents; it is done in the framework of the multifractal analysis which relates ζ_p to the codimension of the fractal sets on which dissipative structures set in (for an extension of the multifractal formalism to compressible flows, see [106] [107] [108]).

In compressible flows, longitudinal and transverse velocity structure functions have been computed for the forced uni–directional shear flows mentioned in §3. In the inertial range, they both behave in a similar manner [97], and the ζ_p exponents are comparable to the incompressible case, with a strong departure from a linear law at high p and with $\zeta_3 \sim 1$.

As a final note, we remark that the intermittency of dissipative structures, which in a compressible flow lead to intense hot spots, may have dynamical consequences in the interstellar medium: in [39], it is shown using the wind tunnel data of [42], that within such

---

[6]An equivalent relationship for the third–order cross–correlation between a passive scalar such as temperature and velocity fluctuations can be found in [127]; its extension to the two–dimensional MHD incompressible case where the magnetic potential is a scalar (but non passive, although this does not in fact play a role) is straightforward [25].

[7]In this framework, we now have $\zeta_4^{(b)} = 1$ (see [26]), a relationship which does not appear to be of the same fundamental type as the equivalent Kolmogorov law for $\zeta_3^{(u)}$, and that obtains with the (possibly unrealistic assumption that the velocity and the magnetic field are uncorrelated.

hot spots, excitation lines of observed molecules can be obtained, which would otherwise not be explained due to the low temperature of dense cores (of the order of 10K). Similar effects of intermittency can be expected in combustion as well, with spatially spotty reaction events.

6. The interstellar medium

When the Hall term

$$\mathcal{H} = \nabla \times [\frac{1}{\rho\xi}\mathbf{j} \times \mathbf{B}] \tag{24}$$

is included in the induction equation (where now ρ is the total density of ions + neutrals, and where $\mathbf{u}$ is the velocity of the center of mass and ξ is the ionization of the medium), dispersive effects come in, that compete with nonlinear steepening. In the one–dimensional approximation, and with proper scaling, one arrives at the DNLS (for Derivative NonLinear Schrödinger) equation (see the recent review by Spangler [111]), except when the gas and magnetic pressures are equal, in which case another analysis must be performed as first shown in Hada [50]; the resulting equations in the case of equipartition contain nonlinearities that make them similar to the Burgers equation and in fact, without the dispersive terms, had been proposed by Thomas [113] [114] as a 1D model of MHD (see also [128]) [8]. These equations provide a testing ground for exploring the complex physics that goes on in the interstellar medium.

On the other hand, for a low ionization fraction (typically in the interstellar medium at the scale of dense cores, the ratio of ion to neutral densities is of the order of 10^{-6}), the Hall term in the generalized Ohm's law is negligible compared to the ambipolar drift term which reads

$$\mathcal{A} = a(\rho)\nabla \times [\frac{1}{\rho}(\mathbf{j} \times \mathbf{B}) \times \mathbf{B}] \ , \tag{25}$$

where the density function $a(\rho)$ depends on what is the main mechanism responsible for the ionization [75]. As opposed to a static gas as envisaged by Jeans, the turbulence which creates clumps also helps produce substantially more density contrast with nested sub–structures [7], a well–observed characteristic of molecular clouds. The inclusion of the ambipolar drift term makes collapse more efficient [7]; this may be related to the fact that the magnetic field tends to become force–free, as shown in [21].

All these avenues are being explored but the 1D approximation obviously is insufficient when dealing with structures which may be unstable to transverse perturbations. With, e.g. ISO and HST, great progress can be expected in the years to come.

7. Conclusion

In three–dimensional flows, the dominant structure of the inertial range is a vorticity network. Vorticity production occurs both through vortex stretching and the baroclinic term, the latter being significantly weaker in all cases explored up to date, including for a flow driven by a large–scale shear wave. Filaments are the prevalent form of vorticity in

---

[8]Recent numerical studies of one–dimensional MHD clouds are performed in [45] [7] .

the inertial range, as well as in the far end of the spectrum, as shown in numerous works. Energy transfer in the inertial range seem to occur mostly through bending or kinking [9]. These instabilities are impeded at small scales where the vorticity organizes into rigid vortex tubes. This different topology of the flow leads to an accumulation of energy in the early dissipation range, the level of which is associated with a time–lag effect that almost completely disappears in driven flows because they may be providing a natural mixing. This qualitative image of the dynamics of such a flow remains to be quantified. Flux tubes also prevail in MHD flows [44] [23]. One should recall as well the recent observations of filamentation in the Orion nebula [125] (see also [6] for a discussion).

The two–dimensional incompressible case on the other hand appears different, including for conducting fluids, because of the presence of an infinite number of invariants in the non dissipative limit; these invariants influence the dynamical evolution of the flow, but this has been little explored beyond the first two – energy and enstrophy (and three in MHD), except in the analysis of Eyink [37] where it is shown that these conservation laws provide constraints for the ζ_p exponents of structure functions when intermittency is taken into account.

The flows described here have a wide variety and complexity. However, the same basic ingredients can be found. For example, the alignment of the vorticity with the *second* eigenvector of the symmetrized velocity gradient matrix first found by Kerr [55], is now also found in compressible flows and MHD flows [23]; of course, the explanation – that a strong vortex creates a Biot–Savart swirling flow that dominates its own environment – holds for all three cases. Another example is that of intermittency in the sense that there is a departure from a linear law for the power–law exponents of the structure functions as one goes to high order, and in the sense that strong sparse structures are embedded in a weak noisy background, together with both temporal and spatial self–similarity à la Kolmogorov, the two possibly related because of what has been coined self–organized criticality or SOC [5]. Indeed, front steepening is simply described by the one–dimensional Burgers equation, which is also one of the simplest model of SOC, when one takes into account the conservation and symmetries needed for describing overlapping avalanches [52] (see also [31] in the context of fusion plasmas).

ACKNOWLEDGEMENTS

Some of the research described in this review has been done in collaboration with D. Balsara, T. Passot, H. Politano, D. Porter, E. Vazquez–Semadeni and P. Woodward. May they all be thanked here. Partial support from a variety of sources is acknowledged, among which GdR–CNRS "Mécanique des Fluides Géo–Astrophysiques" (SDU) and "Mécanique des Fluides Numériques" (SPI), and from the European Cooperative Network "Numerical Simulations of Nonlinear Magnetohydrodynamics" (ERBCHRXCT930410).

---

[9]Such vortex interactions may prevent singularities to occur in the incompressible Euler case as mentioned in [32] §7.8 in the context of a discussion of the Beale–Kato–Majda theorem of [8].

REFERENCES

1. Antonia, R., Shafi, H. & Zhu, Y. (1996) *Phys. Fluids* **8**, 2196.
2. Arneodo, A., Baudet, C., Belin, F., Benzi, R., Castaing, B., Chabaud, B., Chavarria, R., Ciliberto, S., Camussi, R., Chilla, F., Dubrulle, B., Gagne, Y., Hebral, B., Herweijer, J., Marchand, M., Maurer, M., Muzy, J.F., Naert, A., Noullez, A., Peinke, J., Roux, F., Tabeling, P., van de Water, W. & Willaime, H. (1996) *Europhys. Lett.* **34**, 411.
3. Ashurst, W., Kerstein, A., Kerr, R. & Gibson, R. (1987) *Phys. Fluids* **30**, 2343.
4. Bajer K. (1995) in "Small–scale structures in fluids and MHD", M. Meneguzzi, A. Pouquet & P.L. Sulem Editors, Notes in Physics **462** p. 255, Springer–Verlag.
5. Bak, P., Tang, C. & Wiesenfeld, K. (1987) *Phys. Rev. Lett.* **59**, 381.
6. Bally, J. (1996) *Nature* **382**, 114.
7. Balsara, D. & Pouquet, A. (1996) "Turbulent Flows within self–gravitating magnetized molecular clouds", preprint.
8. Beale J., Kato T. & Majda A. (1989) *Comm. Math. Phys.* **94**, 61.
9. Benzi, R., Ciliberto, S., Tripicciona, R., Baudet, C., Massaioli, F. & Succi, S. (1993) *Phys. Rev. E* **48**, R29.
10. Beronov, K. & Kida, S. (1996) "Linear two–dimensional stability of a Burgers vortex layer", Preprint, Kyoto University.
11. Beronov, K. & Beronova, V. (1996) "Linear solution models of dissipative MHD structures driven by large–scale fields", Preprint, Kyoto University.
12. Betchov R. (1956) *J. Fluid Mech.* **1**, 467.
13. Bhattacharjee, A. & Wang X. (1992) *Phys. Rev. Lett.* **69**, 2196.
14. Bhattacharjee, A., Ng C.S. & Wang X. (1996) *Phys. Rev. E*, to appear.
15. Biskamp, D. (1993) *Phys. Fluids B* **5**, 3893.
16. Boratav, O. (1996) "On recent intermittency models of turbulence", preprint, University of California at Irvine.
17. Brachet, M., Meneguzzi, M., Politano, H. & Sulem P. (1988) *J. Fluid Mech.*, **194**, 333.
18. Brachet, M.E., Meneguzzi, M., Vincent, A., Politano, H. & Sulem, P.L. (1992) *Phys. Fluids A* **4**, 2845.
19. Brachet, M.E., Meiron, D., Orszag, S., Nickel, B., Morf, R. & Frisch, U. (1983) *J. Fluid Mech.* **130**, 411.
20. Brachet, M.E. (1990) *C.R. Acad. Sci. II*, **311**, 775.
21. Brandenburg A. & E. Zweibel (1994) *Astrophys. Lett.* **427**, L91.
22. Brandenburg A., Procaccia I. & Segel D. (1995) *Phys. of Plasmas*, **2**, 1148.
23. Brandenburg, A., Jennings, R., Nordlund, A., Rieutord, M., Stein, R. & Tuominen, I. (1996) "Magnetic Structures in a dynamo simulation", *J. Fluid Mech.*, to appear.
24. Burlaga, L. (1991) *J. Geophys. Res.* **96**, 5847.
25. Caillol, P., Politano, H. & Pouquet, A. (1996) "The inverse cascade of magnetic potential and its scaling laws", Preprint Observatoire de la Côte d'Azur.
26. Carbone, V. (1994) *Phys. Rev. E* **50**, R671.
27. Constantin P. (1994) *SIAM Rev.* **36**, 73.
28. Constantin P., Procaccia I. & Segel D. (1995) *Phys. Rev. E* **51**, 3207.

408

29. Crighton, D.G. (1996), this Volume.

30. Dalgarno, A. & McCray (1972) *ARA&A* **10**, 375.

31. Diamond, P. H. & Hahm, T.S. (1995) *Phys. Plasmas* **2**, 3640.

32. Doering, C. & Gibbon, J. (1995) **Applied Analysis ot the Navier–Stokes Equations**, CTAM, Cambridge University Press.

33. Douady, S., Couder, Y. & Brachet M.E. (1991) *Phys. Rev. Lett.* **67**, 983.

34. Dubrulle, B. (1994) *Phys. Rev. Lett.* **73**, 959.

35. Elmegreen, B. G. (1991) *Astrophys. J.* **378**, 139.

36. Elmegreen, B. G. (1994) *Astrophys. J.* **433**, 39.

37. Eyink G. (1995) *Phys. Rev. Lett.* **74**, 3800.

38. Falgarone, E., D.C. Lis, T.G. Philips, D. Porter, Pouquet, A. & P. Woodward (1994) *Astrophys. J.*, **436**, 728.

39. Falgarone, E. & J. Puget (1995) *Astron. Astrophys.*, **293**, 840.

40. Falgarone, E. (1995) in "Small–scale structures in fluids and MHD", M. Meneguzzi, A. Pouquet & P.L. Sulem Editors, Notes in Physics **462** p. 377, Springer–Verlag.

41. Frisch, U. (1995) **Turbulence: The legacy of Kolmogorov**, Cambridge University Press.

42. Gagne, Y. (1987), Thèse, Université de Grenoble.

43. Galanti, B., Sulem, P.L. & Pouquet, A. (1992) *G. A. F D.* **66** 183.

44. Galloway, D.J. & Frisch U. (1986) *Geophys. Astrophys. Fluid Dyn.* **36**, 53.

45. Gammie, C. & Ostriker, E. (1996), *Astrophys. J.*, to appear.

46. Ghashgghaie, S., Breymann, W., Peinke, J., Talkner, P. & Dodge, Y. (1996) *Nature* **381**, 767.

47. Gilbert, A. (1993) *Phys. Fluids A* **11**, 2831.

48. Grauer R., J. Krug & C. Marliani (1994) *Phys. Letters A*, **195**, 335.

49. Grossman S. & D. Lohse (1991) *Phys. Rev. Lett.* **67**, 445.

50. Hada, T. (1993) *Geophys. Res. Lett.* **20**, 2415.

51. Hasimoto, H. (1972) *J. Fluid Mech.* **51**, 477.

52. Hwa, T. & Kardar, M. (1992) *Phys. Rev. A* **45**, 7002.

53. Jimenez J. (1994) *J. Fluid Mech.* **279**, 169.

54. Jimenez J., Wray A., Saffman P.G. & Rogallo R. (1993) *J. Fluid Mech.* **255**, 65.

55. Kerr, R. (1985) *J. Fluid Mech.* **153**, 31.

56. Kerr, R. (1993) *Phys. Fluids* **A5**, 1725.

57. Kerr, R. (1995) Lecture Notes in Physics **462**, M. Meneguzzi, A. Pouquet & P.L. Sulem Editors p. 17, Springer–Verlag.

58. Kida S. (1981) *J. Fluid Mech.* **112**, 397.

59. Kida S. (1985) *J. Phys. Soc. Japan* **54**, 2132.

60. Kida, S. & Orszag, S.A. (1990a) *J. Scientific Comp.* **5**, 1.

61. Kida, S. & Orszag, S.A. (1990b) *J. Scientific Comp.* **5**, 85.

62. Kida, S. & Orszag, S.A. (1992) *J. Scientific Comp.* **7**, 1.

63. Kinney, R., Tajima, T., McWilliams J. & Petviashvili N. (1993) *Phys. Plasmas* **1**, 260.

64. Klein, R., Majda, A. & McLaughlin R. (1992) *Phys. Fluids* **A4**, 2271.

65. Klein, R., Knio, O. & Ting, Lu (1996) *Phys. Fluids* **8**, 2415.

66. Kolmogorov, A. (1941a) *Dokl. Akad. Nauk SSSR* **30**, 299.

67. Kolmogorov A. (1941b) *Dokl. Akad. Nauk SSSR* **31**, 538.

68. Kolmogorov A. (1962) *J. Fluid Mech.* **13**, 82.

69. Kraichnan, R.H. (1959) *J. Fluid Mech.* **5**, 497.

70. Kraichnan, R.H. (1965) *Phys. Fluids* **8**, 1385.

71. Léorat, J. (1975) Thèse, Université Paris VII.

72. Lesieur M. (1990) **Turbulence in Fluids**, Second Edition, Kluwer.

73. Lin S. & G. Corcos (1984) *J. Fluid Mech.* **141**, 139.

74. Lundgren, T. (1982) *Phys. Fluids* **25**, 2193.

75. McKee, C. (1989) *Astrophys. J.* **345**, 782.

76. Mansour, & Lundgren, T (1996) *J. Fluid Mech.*, to appear.

77. Meneguzzi, M., Politano, H., Pouquet, A. & M. Zolver (1996) *J. Comp. Phys.*, **123**, 32.

78. Miller, R., Mashayek, F., Adumitroaie, V. & Givi, P. (1996) *Phys. Plasmas* **3**, 3304.

79. Moffatt, H.K. (1978) **Magnetic Field Generation in Electrically Conducting Fluids**, Cambridge University Press.

80. Moffatt, H.K., Kida, S. & Ohkitani, K. (1994) *J. Fluid Mech.* **259**, 241.

81. Moiseev, S., Petviashvili, V., Tur, A. & Yanovski, V. (1981) *Physica D* **2**, 218.

82. Monin, A.S. & Yaglom, A.S. (1975) **Statistical Fluid Mechanics: Mechanics of Turbulence**, Volume 2, The MIT Press, Cambridge.

83. Nelkin, M. (1995) *Phys. Rev. E* **52**, R4610.

84. Neu, J.C. (1984) *J. Fluid Mech.*, **143**, 253.

85. Nore, C., Brachet, C., Politano, H. & Pouquet, A. (1996), *Plasma Phys. Lett..*

86. Novikov, E.A. (1994) *Phys. Rev. E* **50**, R3303.

87. Passot, T., Politano, H., Sulem, P.L., Angillela J. R. & Meneguzzi, M. (1995) *J. Fluid Mech.* **282**, 313.

88. Passot, T., Vazquez, E. & Pouquet, A. (1995) *Astrophys. J.*, **441**, 702.

89. Pelz, R. & Boratav O. (1995) in Lecture Notes in Physics **462**, M. Meneguzzi, A. Pouquet & P.L. Sulem Editors p. 25, Springer–Verlag.

90. Politano, H., Pouquet, A. & Sulem, P.L. (1995) *Phys. Plasmas* **2**, 2931; see also *Lect. Notes Phys.* **462** p. 281, Springer–Verlag.

91. Politano, H. & Pouquet, A. (1995) *Phys. Rev. E* **52**, 6361.

92. Pope, S. (1988) *J. Engng. Sci.* **26**, 445.

93. Porter D., Pouquet, A. & Woodward, P. (1992) *Theor. Comp. Fluid Dyn.* **4**, 13.

94. Porter D., Pouquet, A. & Woodward, P. (1994) *Phys. Fluids A* **6**, 2133.

95. Porter, D., Pouquet, A. & Woodward, P. (1995) in "Small–scale structures in fluids and MHD", M. Meneguzzi, A. Pouquet & P.L. Sulem Editors, Notes in Physics **462** p. 51, Springer–Verlag.

96. Porter, D., Woodward, P. & Pouquet, A. (1996) submitted to *Phys.Fluids*.

97. Porter, D., Pouquet, A. & Woodward, P. (1996), in preparation.

98. Pouquet, A. (1994) in Proceedings "Plasma Trends in Astrophysics", V. Stefan Ed., Springer–Verlag.

99. Pouquet, A. (1996) in Proceedings European School of Astrophysics, C. Chiuderi & G. Einaudi Editors, *Lecture Notes in Physics* **468** p. 163, Springer–Verlag.

100. Pullin D. & P. Saffman (1992) *Phys. Fluids A* **5**, 126.

101. Ricca, R. (1994) *J. Fluid Mech.* **273**, 241.

410

102.Ricca, R. (1996) in "Proceedings of the School on Vortex and Flux Tubes", http//www.obs-nice.fr, H. Politano & A. Pouquet Eds, EPOCA (Electronic Publication of the Observatory of the Côte d'Azur), **1**, in press.

103.She, Z.S., Jackson, E. & Orszag S.A. (1990) *Nature* **344**, 226.

104.She, Z.S. & Jackson (1993) *Phys. Fluids A* **5**, 1526.

105.She, Z.S. & E. Lévêque (1994) *Phys. Rev. Lett.* **72**, 336.

106.Shivamoggi, B.K. (1992) *Phys. Lett. A* **166**, 243.

107.Shivamoggi, B.K. (1995) *Ann. Phys. New-York* **243**, 169.

108.Shivamoggi, B.K. (1995) *Ann. Phys. New-York* **243**, 177.

109.Siggia, E. (1981) *J. Fluid Mech.* **107**, 375.

110.Soria, J., Sondegaard, R., Cantwell, B., Chong, M. & Perry S. (1994) *Phys. Fluids* **6**, 871.

111.Spangler, S. (1994) in *Proc. of Kyoto Conf. on Chaos and Nonlinearity in Space Plasmas*, Tokru Hada Ed., World Scientific (Singapore).

112.Tcheou, J.M. & Brachet, M.E. (1996) "Multifractal scaling of probability density", preprint, Ecole Normale Supérieure, Paris.

113.Thomas, J. (1968) *Phys. Fluids* **11**, 1245.

114.Thomas, J. (1970) *Phys. Fluids* **13**, 1877.

115.Tokugawa, N., Takayama, F. & Kambe, T. (1996) *"Wave Scattering by the interaction of a vortex ring with a shock wave"*, XIXth ICTAM Book of Abstracts p. 539, Kyoto, August 1996.

116.Tokunaga, H. & Tatsumi, T. (1975) *J. Phys. Soc. Japan* **38**, 1167.

117.Vazquez, E. & Scalo, J. (1992) *Phys. Rev. Lett.* **68**, 2921.

118.Vazquez, E., Passot, T. & Pouquet, A. (1995) *Astrophys. J.*, **455**, 447.

119.Vazquez, E., Passot, T. & Pouquet, A. (1996) *Astrophys. J.*, December 20.

120.Vieillefosse P. (1982) *J. Phys.* **43**, 837.

121.Vieillefosse P. (1984) *Physica A* **125**, 150.

122.Vincent, A. & Meneguzzi, M. (1991) *J. Fluid Mech.* **225**, 1.

123.Vincent, A. & Meneguzzi, M. (1994) *J. Fluid Mech.*, **258**, 245.

124.Wirth, A. (1996), submitted to *J. Comp. Phys.*.

125.Wiseman, J. & Ho, P. (1996) *Nature* **382**, 139.

126.Woodward P. (1986) *"Numerical Methods for Astrophysicists,"* in *Astrophysical Radiation Hydrodynamics*, K.-H. Winkler & M. L. Norman eds., p. 245, Reidel.

127.Yaglom,A. (1949) *Dokl. Acad. Nauk SSSR* **69**, 743.

128.Yanase, A. (1996) preprint University of Okayama, to appear, *Phys. Fluids*.

Theoretical and Applied Mechanics 1996
T. Tatsumi, E. Watanabe and T. Kambe (Editors)

Essential structure of the damage mechanics theories

D. Krajcinovic

Mechanical and Aerospace Engineering, Arizona State University, Tempe AZ 85287, USA

1. INTRODUCTION

Microstructure of a typical engineering material is characterized by chemical, topological and geometrical disorder. The disorder can be either quenched (frozen in the microstructure) or annealed (if the energy imparted to the system fluctuates randomly). The damage, related to the existence of many defects (microcracks and microvoids) of different sizes and irregular shapes randomly distributed over the volume, is an important class of a quenched, topological disorder. The extent and nature of the microstructural disorder may have a profound influence on the macroscopic response and failure mode of a damaged structure subjected to the external stimuli. The principal objective of this study is to emphasize the effect of the microstructural damage, in the form of a diffuse distribution of microcracks, on the macroscopic response and failure of a structural system.

As a result of their efficiency the continuum models are an overwhelming choice in design analyses. However for a continuum models to be rational it must be consistent with the energy dissipating processes and mechanisms on the microscopic and atomic scales. The locations, sizes and shapes of microcracks are random parameters and the damage evolution is a stochastic process. Hence, it is by no means certain that the effect of microcracks on the response of a macroscopic system always admits a deterministic description by a finite number of smooth, continuous analytical functions. It is also not certain that a continuum representations of stress and strain fields, which ignore random local fluctuations in the vicinity of defects, provide a sufficiently reliable basis for the estimates of the system macro response and macro failure threshold. It is, therefore, important to define the conditions which must be satisfied for the application of continuum models. Finally, a discretization of a system into sub-systems (finite elements) should satisfy certain conditions to preserve the objectivity.

2. MODEL SCALE

The definition of a solid body depends on the observation scale. On the *atomic* scale a system is defined by an ensemble of a sufficiently large number of interacting discrete particles. These ensembles are depicted in the form of lattices formed by lumped masses (nodes) interconnected by bonds (links). A lattice is defined by the geometry, topology and the selected potential of bonds linking the particles. The location of a particle is specified by the position vector $\mathbf{r}_i$. Damage is related to the density and pattern of ruptured bonds between particles and the attendant deformation discontinuities within the ensemble. The material parameters of the pristine assembly can be determined from the lattice morphology and potentials (which, in a simplest case, define the relation between the attractive and repulsive forces and the change of the distance separating the two linked particles) and are characteristic of the considered material [1]. The statistics and evolution of the disorder (weak bonds, etc.) and damage (ruptured bonds) is an important part of the configuration. The quenched disorder can be often inferred from the band-width of the distribution of forces in links of the damaged atomic lattice. A large difference in link forces is typical of stress concentrations in severely damaged systems and often portend its impending failure. The determination of the effective properties of a damaged (locally disconnected) ensemble of particles is, in principle, always possible but in many cases requires a non-trivial computational effort.

On the *microscopic* scale a damaged solid is approximated by a continuous matrix inundated by a large number of microcracks defined as internal surfaces that support a local discontinuity of the deformation field. The adjective micro refers to the lengths that are commensurate to the intrinsic microstructural length L_m such as the grain size. Thus, a microcrack sees the surrounding matter as being anisotropic and heterogeneous. To render the analyses possible the cracks geometry is almost always simplified. Even in the simplest case the configurational space, which defines the geometry of an ensemble of N planar and penny-shaped microcracks of radius a distributed within a volume V, consists of $6N$ random scalars $\{r_i, a_i, \phi_i, \theta_i\}$ where ϕ and θ are the two angles that define the orientation of the normal to the crack surface.

From the continuum viewpoint a body is on the *macroscopic* scale "a given portion of matter ... treated as a collection of elements, called material particles, which at any given instant can be placed in a one-to-one correspondence with the points of a closed region of three-dimensional Euclidean space" [1]. In a pristine (damage-free) state these material particles fill the entire space without any gaps other than the macroscopic perforations. The current configuration of a continuous body weakened by a diffuse ensemble of microcracks is on the macro scale defined by the current position of each material particle (point) with respect to some reference (stress and damage free) configuration, spatial distribution of kinematic variables such as elastic strains (deformation gradients) and damage (internal) parameter(s).

Using the principles of determinism, local action, material objectivity and material invariance the constitutive relation for the macroscopic stresses are formulated in the form of tensor valued functionals of deformation gradients, temperature, position and time. The constitutive model is in a majority of cases simplified by assuming that the future response depends only on the current configuration (and not on the details of the process during which this configuration was created) and that the stress in a material point depends only on the kinematic variables in the same point. The ensuing simple and local material of grade one is a trade-off between accuracy and efficiency which may occasionally lead to an inadequate approximation of the actual material with a disordered microstructure. This is especially true with respect to the locality which is, by its very nature, contradictory to any description of a body with microstructure and statistical homogeneity which is almost never questioned.

3. HOMOGENIZATION

The adopted premise that the microstructural texture and disorder have a decisive effect on the macroscopic response may lead to a conclusion that a very fine resolution length ℓ is always needed to capture the physics of the process. Even though the physics of the process is conceptually simple at the atomic level the design efficiency can be secured only by the application of continuum models, i.e. at the expense of the resolution length. The tolerance level associated with the trade-off between the rigor with which the microstructural morphology is described by a model and its efficiency in design application can be established only by homogenization, i.e. formulation of relations between the analytical models on different scales with vastly different resolution lengths ℓ.

The essential aspects of the transition from mechanical models formulated on the atomic scale to those on the microscopic and ultimately macroscopic scale, referred to as the homogenization or coarse graining, are seldom carefully scrutinized. In the process of homogenization individual atoms are replaced by a continuous matrix and billions of ruptured atomic bonds by microcracks which are eventually described by a damage parameter in the form of a tensorial function of position and time. The enormous gain in computational efficiency and drastic reduction of configurational spaces needed to describe the structure of the material and the accumulated damage in each material point of the continuum at each instant of time is achieved at the expense of the loss of physical clarity and resolution length coarsening. The physical simplicity of Newton's laws, which define the change of the position and momentum of a particle in the molecular dynamics computations, is largely obfuscated by the introduction of many inferred, useful but artificial, continuum concepts. These phenomenological concepts are further related by constitutive laws which introduce "material" parameters that often cannot be uniquely determined by experiments. The statistical nature of the process and local fluctuations of considered random variables (i.e. higher statistical moments of their distribution) are lost in the course of the homogenization as a result of volume averaging and elimination of details considered unimportant on the basis of small samples, incomplete data or casual inferences. Criteria which must be satisfied to render a homogenization process objective should be based on the characteristic lengths which define the statistics of the disorder on all three scales and quantify the conditions which ensure the statistical homogeneity.

The transition from the atomic scale to the microscopic scale involves mapping of the average properties of a uniform ensemble of particles on the material point of a continuous matrix. This transition is objective if the ensemble of N atoms is uniform. The one-particle phase-space densities must add up to the (most probable) Maxwell-Boltzmann distribution, i.e. the ensemble entropy should tend to the entropy of the corresponding micro-canonical ensemble. Effective properties of a smaller ensemble of particles are size dependent (extensive).

A single grain of a typical polycrystalline ceramics consists of at least a billion atoms. Simulations on ensembles of this size exceed the capacity of current computational devices (especially in ionic crystals). To render the determination of material properties tractable they are computed for the matrix by averaging over a large enough sample of grains independently of the grain size defects. In the process the correlation between two random fields (cracks and grains) is neglected. All grain boundary (interfacial) cracks are approximated by cracks within a homogeneous medium. The potential influence of grain boundaries on the pattern of crack growth is ignored. The defects smaller than the grain and the effect of grain boundaries and triple joints on the materials strength are introduced through adjustable parameters.

The simplest description of a set of many microcracks in a volume V rests on the assumption that their exact locations within the volume and their correlations with the grain boundaries are not important. The size of a planar, elliptic microcrack is defined by $a = (2A^2 / \pi P)^{1/3}$, where A is the crack surface area and P the length of the crack perimeter. In the case of a penny-shaped crack a is its radius. The variation of the density of an ensemble of elliptic cracks with the orientations θ and ϕ of its bedding planes is often presented in the form of a histogram (rosette) of densities $w(a; \theta, \phi)$ in the planes of different orientations (θ, ϕ). If the orientations and sizes of penny-shaped microcracks are not correlated the density function of microcracks, defined as the number of cracks with radius a within the range $(a, a + da)$ and orientations (θ, ϕ) within the range $(\theta + d\theta, \phi + d\phi)$, per considered volume V, is [2, 3]

$$w(a; \theta, \phi) = \vartheta(a)\rho(\theta, \phi) \tag{1}$$

where

$$N = \int_{a^-}^{a^+} \vartheta(a)da \quad and \quad \int_o^{2\pi} d\theta \int_{-\pi/2}^{\pi/2} \rho(\theta, \phi)\cos\phi \, d\theta \, d\phi = 4\pi \tag{2}$$

The superscripts (+) and (-) define the upper and lower limits of the random variable distribution. The microcrack density function can be finally written in the form

$$w(a, \theta, \phi) = N\langle a^3 \rangle f(\theta^{+,-}, \phi^{+,-}) \quad where \quad N\langle a^3 \rangle = \int_{a^-}^{a^+} a^3 \vartheta(a)da \tag{3}$$

that generalizes the representation suggested for an isotropic distribution of orientations by Budiansky and O'Connell [4]. The relation between (3) and the results of stereological analysis of cuts through a volume are discussed in [4]. The representation of the microstructural disorder by (1) and (3.a) forms the foundation of the effective continuum and mean field models (EC&FM) and is widely held to be applicable to the dilute concentrations of non-interacting microcracks. It also implies that the macroscopic response is invariant to the changes of crack sizes and numbers for which the expression (3.b) is fixed.

The transition between a description based on $6N$ random scalars $\{r_i, a_i, \phi_i, \theta_i\}$ and the microcrack density (3) is possible only if the material within the volume V is statistically homogeneous with respect to the considered property of the volume. A material within the volume V is statistically homogeneous if the statistical properties of the considered fields are independent of the spatial locations of heterogeneities and defects. Hence, the one-point distribution and expectation of the microcrack distribution do not depend on their locations. The property of statistical homogeneity is not violated by the direct interaction of cracks if the two-point moment of this distribution depend only on the distance between two points $|\mathbf{x}_m - \mathbf{x}_n|$ but not on the actual location of microcracks (translational invariance). Subject to these limitations the considered distribution is ergodic.

As a result of the translational invariance each microcrack in a statistically homogeneous volume is subjected to the average (long range) stress. The effect of the local fluctuations attributable to the crack-to-crack interaction on the considered effective property can be neglected. The effective compliance attributable to the microcracks can be determined by superimposing the contributions of all individual microcracks. The compliance attributable to an ensemble of N active penny-shaped microcracks within volume V of an effective elastic matrix (defined by the effective shear modulus $\tilde{\mu}$ and Poisson's ratio $\tilde{\nu}$) is

$$\mathbf{S}^* = \frac{8}{3}\frac{1-\tilde{\nu}}{\tilde{\mu}(2-\tilde{\nu})} \int_{a^-}^{a^+} a^3 \vartheta(a)\,da \int_{\theta^-}^{\theta^+}\int_{\phi^-}^{\phi^+} \mathbf{F}(\theta,\phi)\rho(\theta,\phi)\cos\phi\,d\theta\,d\phi =$$

$$= \frac{8}{3\tilde{\mu}}\frac{1-\tilde{\nu}}{2-\tilde{\nu}} N\langle a^3\rangle \Phi(\theta^{+,-}, \phi^{+,-}) \tag{4}$$

where $(a^- \le a \le a^+, \theta^- \le \theta \le \theta^+, \phi^- \le \phi \le \phi^+)$[3]. The components of the fourth order tensor $\mathbf{F}$ are constants which depend only on the microcrack orientation that can be determined solving for the compliance attributable to a single penny-shaped crack embedded within an isotropic, homogeneous, elastic matrix [2, 3, 5]. The components of the fourth order tensor Φ are determined from the double integral in (4). The expressions (3) and (4) can be readily modified to consider elliptical cracks [4] and correlated orientations and sizes of cracks [6]. However, the derivation of the expression for the effective compliances which includes the microcrack interaction (assuming that the property of the statistical homogeneity was not violated) requires a substantial computational effort [7].

The effective parameters (4) of the solid are affected only by the microcracks which support a local discontinuity of deformation across the interface separating two mating surfaces. These microcracks are referred to as being *active*. All other microcracks, termed *passive*, do not count in the number density N defined by (2.a). The status of each microcrack depends on the local stress field. Microcracks subjected to: (a) tensile stresses in the direction of its normal or (b) shear stresses large enough to overcome the slip resistance (due to the friction or inter-locks) are typically active. Change of the microcrack status from active to passive and vice versa, during a non-proportional loading, may cause discontinuities in the effective parameters. Hence, a proper record of the deformation history, that accounts for all active and passive microcracks, may cause serious computational problems at large defect concentrations. In each instant of the deformation process, during which the direct microcrack interaction plays a significant role, it is necessary to guess the status of each microcrack before making any attempt to derive the effective properties of the considered volume of material. If the guess has not been substantiated by the result it becomes necessary to make new guess until one of them is substantiated by the result.

The mapping between the volume of a piecewise continuous material on micro scale and the corresponding material point of the effective continuum on the macroscopic scale makes sense only if the considered effective property is an intensive (size-independent) parameter of the volume. The smallest volume which satisfies this condition is the representative volume element (RVE) [2, 3, 8, 9, 10]. The considered properties of the continuum are unique and consistent with the micro-scale morphology only if they are averaged over a volume that is not smaller than RVE. Hence, the RVE is the smallest volume which is statistically homogeneous and which provides for a statistically homogeneous response when subjected to a tractions or temperatures imparted over the external surfaces of the RVE. The actual location of each defect within volume is then irrelevant.

This definition can be quantified by the comparison of several characteristic lengths. The minimal set of these lengths includes: (a) the specimen linear size L, (b) the characteristic microstructural length (grain size, etc.) L_m, (c) the range ξ over which the microcracks are correlated, and (d) the linear size L_{rve} of the RVE. Two microcracks are correlated if their direct interaction has a measurable effect on the threshold and pattern of their growth. A set of correlated defects forms a defect cluster. Hence, the size of the largest cluster is approximately equal to ξ. There are at lest three different types of clusters. (a) A macrocrack is a chain of concatenated microcracks. Its geometry may be irregular on the micro scale but the number of broken atomic bonds scales as L^{d-1} where d is the dimensionality of the problem. This type of a cluster is typically formed by the stress driven propagation of a pre-existing crack along the surfaces of inferior toughness (grain boundaries, etc.). The elastic energy release rate is concentrated within a very small volume near the macrocrack tip. (b) A percolation cluster has a ramified (gossamer-like) geometry which is also self-similar. The number of broken bond scales as L^D where the universal fractal exponent D is 91/48 and 2.53 for $d = 2$ and 3, respectively [3]. This cluster is formed exclusively by the random nucleation of microcracks in a specimen subjected to the triaxial compression. Hence, the elastic energy release rate does not play any role in the growth of this cluster. (c) A fault (shear band) consists of many roughly parallel(not intersecting) cracks. This type of a cluster is formed in weakly confined specimen subjected to the shortening in the axial direction. The fault growth is attributable to the direct interaction of cracks (cooperative effect). The elastic energy release rate is concentrated along the long side of the cluster perimeter. The volume of the fault cannot be defined directly and rigorously since it depends on the selected criterion of the interaction. The range over which the defects are correlated is proportional to its size and is referred to as the correlation length ξ. The failure caused by the growth of the largest defect cluster can be classified as a short to long range correlation transition.

The material and the response of the RVE subjected to a mean field (volume averaged) stress σ^o are statistically homogeneous (independent on the exact position of defects and heterogeneities within the RVE) only if the inequalities

$$L_m << L_{rve}, \qquad \xi \le L_{rve} \le L \quad and \quad \left|\frac{\partial \sigma^o}{\partial x}\right| L_{rve} << |\sigma^o| \qquad (5)$$

are satisfied. The discretization is objective (mesh-insensitive) only if the sub-systems are homogeneous (larger than the RVE), i.e. if a defect in a sub-system is not correlated with the defects belonging to the adjacent sub-systems.

If the three conditions (5) are satisfied the homogenization from the micro to macro scale is possible. A parameter of the RVE is simply mapped on the corresponding material point of the effective continuum. The size of the RVE depends on the parameter and may not exist at all in the case of extensive properties (rupture strength of a brittle material). The RVE size depends on the applied stress field: (a) directly through the gradient of the applied or mean stress field (5.c) and (b) through the effect of the sign of normal stress on the displacement field across the mating faces of the crack. The first of two conditions casts serious doubts on whether the EC&FM may be applied to determine the effective parameters of the process zone (wrapped around the tip of a macrocrack) in which the stress gradients are large. In the case (b) a microcrack may become passive if the stress normal to its bedding plane is compressive (and the shear stress negligible) or if the friction coefficient is large enough (as a result of the interlocks of the two mating surfaces of the crack) to eliminate the mode II and III crack deformation.

If the conditions (5) are satisfied the considered distribution of microcracks can be approximated by the density function (3). The effective compliance at a material point of the continuum is equal to the same effective property (4) of the corresponding RVE. In the limit of infinitesimal strains the macroscopic stresses of a simple, elastic and continuous material map on the macroscopic strains by

$$\bar{\varepsilon} = \bar{S} : \bar{\sigma} \tag{6}$$

where the bar above symbol implies averaging over the RVE. The fourth order compliance tensor $\bar{S}$ admits, in the context of linear elastic fracture mechanics, the additive form $\bar{S} = S + S^*$, where S and S^* are the overall compliance, compliance of the pristine matrix and compliance (4) attributable to the active microcracks, respectively. Thus, from (6) it follows

$$\bar{\varepsilon} = (S + S^*) : \bar{\sigma} \quad \text{or} \quad \bar{\varepsilon} = \bar{S} : \bar{\sigma} \tag{7}$$

4. DAMAGE PARAMETER

Damage parameter is primarily a continuum concept which implies that the influence of microcracks located within a RVE sized neighborhood of a material point on the macroscopic response can be described by a tensor characterized by continuous and smooth temporal and spatial gradients. The description of the damage depends on the observation scale. On the *atomic* scale the damage in a homogeneous ensemble of particles can be related to the distribution of the fraction of broken bonds between adjacent particles. A simple ratio between the number of broken bonds and the number of bonds in the pristine lattice is unlikely to be useful since the contribution of a bond to the transfer of momentum depends on its orientation. Removal of a bond carrying a small part of the applied traction will, therefore, affect the effective properties of the lattice much less than the rupture of a bond that carries a much larger fraction of external load. On the basis of the extensive lattice simulations reported in [12] it was shown in [11] that the lattice effective stiffness is the only size-independent measure of the accumulated damage. If a sub-system satisfies conditions (5) its effective stiffness does not depend on the details (higher statistical moments) of the microcrack distribution, local correlations within the system, variations of defect sizes, shapes and numbers for which the parameter (3.b) is fixed and other details that are typically not replicated in tests on "identical" specimens subjected to "identical" loading.

Microstructural analyses based on the EC&FM [2, 3, 4] suggest that the volume averaged product (3.b) is a proper measure of the isotropic distribution of the micro-mechanical damage in a brittle solid. The rosette histogram is an often used geometrical image of the microcrack distribution (3.a) within the volume. Each bin of this histogram is defined by an axial vector $\rho(\mathbf{n}_i)$ (where ρ_i is the microcrack density in the plane with normal $\mathbf{n}_i$ [13]). However, the manipulations of a large number of axial vectors does not lead to elegant and efficient algorithms. Instead it is preferable to approximate an arbitrary micro-crack distribution by a tensor parameter. The expressions for the fabric tensors were derived by Kanatani [14] who expanded an arbitrary microcrack density $\rho(\mathbf{n})$ into a Fourier-type infinite series of families of Laplace spherical harmonics. This method was explored in the context of damage mechanics in [15] and related to the particular microcrack distributions in [16]. The analytical expressions derived in [16] indicate that geometrically: (a) the scalar approximation of microcrack distributions cannot describe in a satisfactory manner any distribution that is not isotropic, (b) the second order tensor is useful only for the orthotropic microcrack distributions and (c) the fourth order tensor representation can describe an arbitrary anisotropic distribution of microcracks and provides a closer approximation of the arbitrary distribution than the second order tensor. The increase of the model complexity with the tensor order is very slight. A reasonable good fit of narrow-band microcrack distribution requires higher order tensor approximations.

The representations of microcrack distributions by the rosette histograms and fabric tensors implies that the orientation of their bedding planes is the only important aspect of their distribution. The exact position of the defect within the volume is assumed to have no effect on the momentum transfer. This assumption, analogous to one on which the EC&FM are based, is satisfied in the limit of the dilute concentration of microcracks and is consistent with expression (4) for the effective compliance. The required accuracy with which the geometry of a microcrack distribution is approximated by the fabric tensors is related to whether: (a) the geometrical details of the microcrack distribution are important per se and (b) the definitive conclusions can be made on the basis of the mean field methods that are valid only in the limit of the dilute concentration defects. Moreover, the representations of the type of expression (1), which reduces to the modified Budiansky, O'Connell parameter (3) imply that the RVE stiffness and response are invariant to the variations of microcrack sizes and number densities for which the volume averaged product (3.b) is fixed. It is by no means certain that all of these assumptions will always hold and, by inference, that (3) will always provide a rational basis for the selection of the damage measure in the case of brittle deformation.

The exact role of fabric tensors in damage mechanics hinges on the importance of accurate approximation of a particular microcrack distribution by tensors. The exact distribution of microcracks is never known since even the roughest estimates require a Sisyphean experimental effort, inordinate time and excessive cost. Moreover, the details of this distribution change from one material point (corresponding RVE) of a statistically homogeneous volume subjected to a homogeneous state of stress to the other and from one specimen (physical realization) to the other. Only parameters that can be readily identified and measured in laboratory and in situ tests [17, 18] are, in fact, the effective stiffness and compliance of a specimen. Fortunately, the effective compliance are, at least for a dilute concentration of microcracks, related by (4) to the microcrack density (1) and the ubiquitous Budiansky, O'Connell damage parameter (3.b).

To estimate the effect of errors introduced by the truncation of an infinite series of fabric tensors to a single tensor of zeroth to sixth order Krajcinovic and Mastilovic [19] compared the effective compliances computed from (4) using: (a) the exact delta function representation for a planar and a monoclinic distribution and (b) the corresponding fabric tensor approximations. The conclusion was that: (a) the scalar representation leads to the unacceptably large errors, (b) the second order tensor approximation provides in all considered cases good approximations, and (c) the fourth order tensor replicates the exact expressions of the effective compliance in the limit of the dilute concentrations of microcracks. These conclusions are in concert with those listed in [20]. The major deficiency of the second order tensor, which reduces anisotropic to orthotropic material, is numerically not significant.

All of these conclusions were based on the mean field methods which are valid only in the limit of dilute concentration of defects. The comparisons of effective compliances computed from an exact and tensorial approximation of microcrack distribution was extended in [19] to the analyses of results for the largest possible microcrack concentrations using the continuum percolation theory. The conclusions based on the comparisons of the effective compliances computed using the exact solution and tensor representations for the case of microcracks uniformly distributed within a pencil of angles $\beta^- \leq \theta \leq \beta^+$ are: (a) that none of the considered tensor approximations (from the zeroth to sixth order) is able to match the exact solution for the effective compliance when all micro-cracks are parallel (delta function case) $\beta^- = \beta^+$ and (b) that the accuracy was directly proportional to the order of the tensor approximation. Except for a very narrow range of angles $\beta^- = \beta^+ \geq 10^o$ the fourth and sixth order tensor provide almost identical expressions for the effective compliance which are by less then 10% underestimate the exact solution. The errors in estimating the effective compliance from the second order fabric tensor representation of the microcrack distribution exceeds by a factor of two or more the errors induced by the fourth order tensor representation of damage.

Based on these analyses it appears that the effective stiffness or compliance in a material point is a proper, and perhaps the only proper choice for the damage parameter on the macroscopic scale. This representation provides a meaningful measure of the effect that the microcracks have on the transfer of momentum in the volume tributary to a material point. This conclusion is supported by the facts that the effective stiffness of the material: (a) is easily identified and measured, (b) is size-independent (allowing for a direct relation between the test data measured on a scaled down specimens and the in situ collected data on full size structures), (c) is not sensitive to the details of the microcrack distributions which cannot be replicated from one specimen to the other and (4) exhibits universal behavior in the vicinity of the percolation threshold. Experimental determination of the effective compliance, assuming homogeneity, are robust since they depend on the likely events (averages). The effective compliance of a specimen has a clear physical meaning which is not limited to the domain of applicability of EC&FM. Moreover, this selection of the damage parameter is consistent with the process of homogenization during which an untold number of ruptured bonds is represented by the microcrack distribution (1) within RVE which is, in turn, replaced by few components of the compliance tensor (4) in the material point of the continuum. The adjective universality implies herein the independence of effective stiffness on the local fluctuations and correlations between defects. This property is manifested in the robust measurements and acceptable scatter of test data on the specimen scale. The effective stiffness can be measured even when the specimen is not statistically homogeneous. However, these measurements are size dependent and the determined property is, therefore, extensive.

From a purely pragmatic (and computational) viewpoint it also seems more reasonable to select the effective stiffness as a damage parameter than to introduce an additional, artificial measure of damage (which cannot be clearly and uniquely identified and measured) to compute that very same effective stiffness. This manipulation is time consuming, superfluous, non-physical and leads to an unnecessary ambiguity which cannot be supported on the purely rational grounds.

5. THERMODYNAMIC STATE

On the *atomic* scale the thermodynamic state of an ensemble of particles is defined by the positions of all particles and the potentials (which define the attractive and repulsive forces in particle links as a function of the distance between the particles) for the considered material [21]. The damage is related to the geometry and density of missing links in the lattice.

On the *microscopic* scale the state of the material within a RVE is defined by the volume averaged stresses, strains and the microcrack distribution defined by expressions (1) to (3).

The thermodynamic state in a material point x of a *macroscopic,* continuous body in equilibrium is defined by the specific internal energy $u(x)$, elastic strain $\bar{\varepsilon}(x)$ (averaged over the volume of RVE) and a set of internal variables ζ_i. The selection of these internal variables is not unique and the entropy production (Clausius - Duhem) inequality seems to be the only non-negotiable requirement that a particular choice must satisfy. As a result of the ill defined criteria for the selection of internal variables a veritable cornucopia of models, often based on clever but physically dubious artifices and less than compelling arguments, found their way into the archival literature causing considerable confusion. The sage advice in Kestin and Bataille [22] that "a gain in simplicity is purely formal, unless we can identify one among the multitude of possible parametrizations that is based on elements with a clear physical meaning" is often overlooked in a frantic search for a simple model and a needlessly close fit of a test curve (single physical realization of what is often a process that exhibits a large scatter of test data) relating two ad hoc paired macroscopic variables.

If the material is either statistically homogeneous (dilute crack concentration) or statistically self-similar (largest crack concentration - percolation threshold) the properties of the RVE can be computed, for a given distribution of microcracks (1), from (4). The determination of effective properties in the case of microcrack concentrations large enough to render the effect of direct crack interaction on the effective properties substantial represents a major problem. The statistics, including n-point correlation functions between points, crack perimeters and surfaces [23], that is needed to estimate of the effect of interactions of many defects on the effective properties is almost as a rule not available. Even if these functions were available an analytical formulation of this problem requires an a priori classification of all existing microcracks into active and passive sets and a string of radical simplifications of defect geometry and matrix structure. Final estimates of stress and strain fields require solutions of large systems of coupled integral [24] or algebraic [25] equations. Analytical models become all but impossible in three-dimensional cases and whenever the original material is not isotropic and homogeneous. Two-dimensional numerical simulations, which are either performed on a small sample [26] or are based on the cell method [27], may lead to the conclusions of arguable merit. The evolution of the correlation length is poorly approximated in both cases. The major role of the random geometry on the effective parameters, clearly exposed in [28, 29], is manifested by the multifractal nature of the stress field [11, 12] and the dominant role of the largest defect cluster (stress concentrations). The effect of the finite element mesh geometry on the breakdown threshold was addressed in [30]. In view of the problem complexity it is not surprising that the determination of effective properties in the non-local regime, characterized by a substantial effect of the direct interaction of defects on the effective parameters, has as yet to be addressed comprehensively.

6. CHANGE OF THE THERMODYNAMIC STATE

The determination of the change of the thermodynamic state (including the change of the position and momenta of all particles and damage accumulation in form of ruptured links) on the *atomic* scale is within the framework of the molecular dynamics method based directly on Newton's second law and the selected inter-atomic potential. The change of the state is often presented in the phase space defined by the position and momentum of each particles. The initial equilibrium, concentration of different defects, attendant stress concentrations and the degree of order is proportional to $\exp(-\Delta h / 2kT)$ where h is the enthalpy, k Boltzmann parameter and T absolute temperature. The computational algorithm is relatively straight-forward and the magnitude of the system size depends only on the capacity of the available computer and analyst patience.

The determination of the change of state on the microscopic scale requires the knowledge of the mode and rate of the change of damage accumulation. The estimate of the damage evolution requires data regarding the conditions (stress and temperature fields) which must be satisfied for new cracks to nucleate in and existing cracks to propagate through a disordered microstructure. Estimates of prospective nucleation sites and nucleation thresholds require knowledge of the spatial distributions of the existing stress concentrations (hot spots) and regions of inadequate toughness (weak links).

Modeling of the microcrack growth through a solid with a heterogeneous microstructure is a much more difficult problem. The Griffith's condition

$$G(\sigma,T;x) \geq R(x) \quad and \quad \frac{\partial G}{\partial a} > \frac{\partial R}{\partial a} \qquad (8)$$

where G and R are the elastic energy release rate and thermodynamic force that resists the crack growth, must be satisfied for the onset of unstable crack growth. The estimates of G and R for each microcracks requires a detailed knowledge of the locations of random local fluctuations of stresses and solid texture on the scale of the material characteristic length. This is especially the case when the damage evolution is controlled by the interaction induced stress fluctuations. Despite conceptual insights and a handful of clever models [31, 32] based on the simplified microstructural and defect geometry, analytical models which incorporate the microstructural disorder related to the statistics of the energy barriers, and their influence on the microcrack propagation, is still an item on the wish list. Results of the simulations on lattices, reviewed in [3], seem to confirm that the sign of the normal stresses and the band-width of the rupture strength of constituent phases (reinforcing fibers or particles) have a decisive influence on the deformation mode and intrinsic brittle to "ductile" (actually quasi-brittle) transition.

The deformation mode of a microcracked solid depends on the outcome of the competition between the microcrack nucleation and microcrack growth. In general, a narrow band-width of the microstructural toughness distribution (micro-homogeneous or damage sensitive solids $R \approx const.$) and long range tensile principal stresses promote the damage evolution (*brittle* response) modes that are controlled by the propagation of existing cracks. In contrast, the damage evolution in damage tolerant or micro-heterogeneous materials (wide band width of toughness distribution $R \neq const.$) and the absence of the long range tensile stresses is attributed to the nucleation of new defects which are immediately trapped (*quasi-brittle* response). The response mode also depends on the strain rates and temperature. Naturally, the precise estimates of the band-widths of the distribution of R-curves (i.e. $R(x)$) and the magnitude of the externally applied stress at which the response mode and failure type crosses over from brittle to quasi-brittle cannot be easily derived from an incomplete statistics of microstructural toughnesses. Existing data [3], that are obtained by numerical simulations, provide only a few qualitative guidelines to a very complex stochastic deformation process developing on a randomly evolving microstructure.

Most of the current continuum models for the deformation processes which are characterized by the irreversible rearrangements of the microstructure are based on the thermodynamics with internal variables [3], [8], [33-38]. Within this formalism a non-equilibrium thermodynamic process, which consists of a sequence of non-equilibrium states, is approximated by a sequence of accompanying constrained equilibrated states. This approximation is valid when a non-equilibrium state is "close" to its accompanying constrained state. The two states are defined as being "close" if the Deborah number is small, i.e. if the time interval needed for a system in a non-equilibrium state to relax to its accompanying state is short compared to the external characteristic time [37]. Each constrained state is uniquely defined by the set of functions $\{u(x), \bar{\varepsilon}(x), \zeta(x)\}$ (where $\bar{\varepsilon}$ is the macroscopic strain tensor and u the internal energy density). The change of state is defined by the increments of $\{\delta u(x), \delta\bar{\varepsilon}(x), \delta\zeta(x)\}$ in each material point x at all times t of interest. The set of thermodynamic fluxes (or rates of scalar internal parameters) $\delta\zeta(x)$ defines the irreversible (energy dissipating) modes of the dominant microstructural rearrangements.

The objective of a continuum analysis is to determine the density, motion and temperature fields by solving the equations of the balance of mass, momentum and internal energy at all x and for all t. A continuum model must, in the process, supplied by the constitutive relations between the fluxes $\{\delta\bar{\varepsilon}, \delta\zeta\}$ and their conjugate forces $\{\bar{\sigma}, f\}$. The principles of material indifference and entropy production are used to single out the class of admitted constitutive equations. In the limit of a dilute concentration of defects the effective compliance in a material point x of the (effective) continuum can be computed using the EC&FM in the form of the expression (4).

Within the restriction of infinitesimal strains and absence of plastic flow the relation mapping the rates of macroscopic stresses on macroscopic strains is from (6, 7)

$$\dot{\overline{\varepsilon}} = \overline{S} : \dot{\overline{\sigma}} + \dot{\overline{\varepsilon}}^* \quad where \quad \dot{\overline{\varepsilon}}^* = \dot{S}^* : \overline{\sigma} \tag{8}$$

where $\overline{S} = S + S^*$ is the overall effective compliance and $\dot{\overline{\varepsilon}}^i$ the inelastic strain rate. The inelastic strain rate can be uniquely determined from (8.b) if the compliance rate is known. The opposite is not true since the number of unknowns exceeds the number of available equations. This is an important issue related to the selection of the damage variable since the compliance rate is needed at each increment of the externally applied stimuli or time to update the overall effective compliance at each material point x of the continuous solid.

To establish a relation between the processes on the micro- and macroscopic scale consider the increment of the Gibbs' free energy density which corresponds to the transition between two neighboring constrained equilibrated states (corresponding to two different arrangements of the microstructure [38])

$$\delta \psi = \varepsilon_{ij} : \delta \sigma_{ij} + \frac{1}{V^o} \sum_{\alpha} f_\alpha \delta \zeta_\alpha + \eta \delta T \tag{9}$$

where η is the entropy and V^o the volume of the material in a reference state defined by the temperature T and absence of mechanical tractions at the volume boundaries. The thermodynamic force f_α, conjugate to the thermodynamic flux $\delta \zeta$, is defined by

$$f_\alpha = V^o \frac{\partial \psi(\sigma, T, \zeta)}{\partial \zeta_\alpha} \tag{10}$$

The inelastic change of the Gibbs' energy density associated with the progressive growth of an ensemble of α microcracks (brittle deformation process) is [32]

$$V^o(\delta^i \psi) = \sum_{\alpha} f_\alpha \delta \zeta_\alpha = \oint_{L_\alpha} [G(\sigma, \zeta) - R(\ell)] \delta a(\ell) \, d\ell \tag{11}$$

where the integral is taken along the entire length L of the crack perimeter, while G and R are the energy release rate and thermodynamic force which resists increase of the crack surface (cohesive energy). The integral in (11) is taken along the perimeters of all active cracks. The distance $\delta a(\ell)$ through which the crack front advances may vary along its perimeter. To establish the connection with the micromechanical models the thermodynamic flux ζ_i can be related to the increase of the microcrack area [3] and, consequently, to the Budiansky, O'Connell parameter (3) and the effective compliance (4). The corresponding conjugate thermodynamic force f_i is proportional to the energy release rate G integrated over the crack area and divided by the length of the crack perimeter.

The rigorous, analytical description of damage evolution in real materials with disordered microstructure, based on the nucleation and growth of all cracks within RVE, is not a feasible goal. Reliable estimates of the damage evolution, that combines nucleation of new ($\dot{N} > 0$) and growth of existing ($\dot{a} > 0$) microcracks, require a detailed knowledge of spatial distributions of energy barriers needed to trap the growing cracks and stress concentrations needed to nucleate new cracks. Microcracks can also grow as a consequence of the cooperative effect resulting from the direct interaction of closely spaced defects. An estimate of the influence of the direct crack interactions on the microcrack growth requires data needed to construct the n-point correlation functions [23] which define the statistics of relative positions and orientations of closely spaced cracks. In general, these data are not available. As a result the existing processes of the formulation of crack nucleation and growth criteria needed for the determination of realistic damage evolution patterns and rates are somewhat arbitrary.

Traditionally, the constitutive equations, which define the thermodynamic fluxes as a function of thermodynamic forces and material parameters, are almost always formulated in the terms of inelastic potentials and inelastic functions. The existence of a potential is by no means ensured in each deformation process and for each disordered microstructure. In the "local dependence" approximation [32, 33, 38] the form of the functional dependence of a thermodynamic flux

$$\dot{\zeta}_k = \chi(f_k, T, \zeta_j) \tag{12}$$

on the affinities, temperature and internal variables is based on the assumption that the growth of a crack depends on the stresses indirectly through its own affinity and that the presence of other defects is felt only through their effect on the effective (volume averaged) material parameters. In the absence of frictional effects [39] the condition (12) is satisfied whenever the material is locally statistically homogeneous, i.e. when the inequalities (5) hold. The exception is the case when the homogeneity is not violated by a limited interactions of few closely spaced cracks. The rate of the inelastic strain on the macroscopic scale is co-directional with the outward normal to the macroscopic potential Ω defined in the space of affinities (macroscopic stresses) [38]

$$\dot{\bar{\varepsilon}}^i = \frac{\partial \Omega(\bar{\sigma}, T, \zeta)}{\partial \bar{\sigma}} \tag{13}$$

The macroscopic inelastic potential is defined from the micro-potentials by

$$\Omega(\bar{\sigma}, T, \zeta) = \frac{1}{V^o} \int_o^{f(\bar{\sigma}, T, \zeta)} \dot{\zeta}_k(\mathbf{f}, T, \zeta_k) df_k \tag{14}$$

If the individual fluxes are related to the change of the microcrack surface each micro-potential can be related to the Griffith's condition $G(\bar{\sigma}, \bar{\mathbf{S}}) - R = 0$, where R is the resistance to the crack growth which is related to the microstructural fabric. The energy release rate G may depend only on the applied (macro) stress and effective compliance (damage density). The damage potential is in this case a convex hyper-surfaces in the space of affinities which is represented by the inner envelope of the micro-potentials [40, 41]. The potential (14), derived for a time-dependent process, can be in a time-independent approximation interpreted as being a damage surface viewed "as a singular clustering of constant flow potentials" [38]. Furthermore, the rate of the inelastic compliance rate is derived in [41] in the form analogous to (13) as

$$\dot{\bar{\mathbf{S}}} = \dot{\mathbf{S}}^* = \frac{\partial \Omega(\Gamma, T, \mathbf{S}^*)}{\partial \Gamma} \tag{15}$$

where $\Gamma_{ijmn} = \bar{\sigma}_{ij}\bar{\sigma}_{mn}/2$ is the thermodynamic force conjugate to the change of the effective compliance.

The damage potential Ω is dual, i.e. the inelastic rates of strain and effective compliance have identical functional dependence on the corresponding affinities σ and Γ. Hence, the potential $\Omega(\bar{\sigma}, T, \bar{\varepsilon}^i)$ for $\dot{\bar{\varepsilon}}^i$ is identical in form to the potential $\Omega(\Gamma, T, \mathbf{S}^*)$ for $\dot{\mathbf{S}}^*$. Therefore, the formulation of the potential Ω will, in associative models, require only modest modifications of the already available damage surfaces for rock and concrete [41 - 43] and even "yield" surfaces based on the Mohr-Coulomb or Drucker-Prager limit surfaces [40, 44].

The formulation of a continuum model in the case of long range tensile stresses can be rather simple. As long as the cracks grow in the cleavage mode the associative model represents a viable alternative since the frictional resistance in the absence of compressive stresses is not an issue. Situation becomes more complex in the non-associative case which can be expected almost always when all three principal stresses which define the state of stress are compressive [39, 43, 44]. In compression, the relative frictional sliding of crack mating faces invalidates the "local dependence" assumption (9) [39, 40, 44] rendering the associative model (15) inadequate. The traditional modes (isotropic and kinematic) of the damage surface kinematics will in all probability be useless in compression since only the crack area in planes perpendicular to principal tensile stresses will increase. Hence, the kinematics of the damage surface will be dominated by the distortion (vertex formation). Despite occasional successes in inferring damage surfaces by phenomenological arguments and observations it seems fair to conclude that the available test data are at the moment inadequate to fully utilize the potential of the existing continuum damage models in the engineering design.

The "local dependence" approximation (13) is not satisfied in the presence of the direct inter-action of closely spaced microcracks (i.e. at more substantial defect concentrations). Hence, the existence of the macroscopic inelastic potential is in this case not as yet proven. Moreover, the kinematics of the damage surface would in this case be very complicated since the micro-potentials are interdependent. At this point it is difficult to ascertain and estimate the error of several approximate schemes in the case when the "local dependence" approximation becomes questionable.

While the number of possible deformation modes may be bewildering few general trends make the recognition of the dominant deformation mode somewhat less arbitrary. In general, during the hardening phase of the volume averaged (macroscopic) stress - strain curve the growth of damage is stable. This is true when the damage evolution process is dominated by the microcrack nucleation or when the crack growth in damage tolerant materials is stable. During the hardening phase of the deformation the material is (with few exceptions) statistically homogenous on a small enough scale. Since the stress waves cannot propagate through a homogeneous softening solid [45], the specimen state that belongs to the softening regime cannot be statistically homogeneous. The onset of softening is, therefore, an unmistakable signal that either the cluster size(s) or the damage concentration is large enough to violate the inequalities (5). The loss of homogeneity in a damage tolerant material (with a strongly heterogeneous microstructure) can occur in both tension and compression [11, 12]. The deformation on the specimen scale is non-local since it strongly depends on the direct interaction of microcrack or microcrack clusters. The stress distribution is multifractal and the stress concentrations may be orders of magnitude larger than the volume average. The macroscopic response depends on the extreme statistics (largest microcrack clusters, smallest distance between adjacent microcracks, etc.) rather than on the averages.

Softening may be attributable to the: (a) propagation of a single macrocrack, (b) localization (formation of a shear band) or (c) percolation transition. The slope of the softening segment of the force-displacement curve decreases in that sequence. The apparent cause of softening depends on the state of stress and can be ascertained by the acoustic emission test. During the softening phase of the deformation the correlation length (size of the largest defect cluster) ξ increases. In the case of damage tolerant materials subjected to tensile tractions several large clusters will grow until one of them reaches its critical length. The statistical nature of the non-local effects, and the attendant multifractal distribution of cluster sizes and stress concen-trations, is reflected in the large scatter of the test data. Alternatively, a material with a reasonable homogeneous microstructure which is subjected to compressive tractions may soften either as a result of the incipient percolation of defects (at a strong confinement) or due to the imminent localization (at a moderate confinement). All three cases mentioned above are characterized by the loss of statistical homogeneity of the material and its response.

7. SUMMARY AND CONCLUSIONS

This brief overview of the damage mechanics in general and the application limits of the corresponding continuum model may be succinctly summarized as follows. The continuum models are applicable in the case of the dilute concentrations of microcracks and when all microcracks are of similar sizes, i.e. whenever the material is on a rather small scale statistically homogeneous. The microcrack distribution can in this case be represented by its orientation weighted density (3), and its effect on the local and global macroscopic response (damage) by the current effective stiffness or compliance (4). The details of the microcrack distribution and direct interaction of microcracks on the macroscopic response is slight. The ensuing continuum model may be either associative (in the case of long range tensile stresses) or non-associative (in solids subjected to the long range compressive stresses). The ensuing models are formally similar to those familiar from the plasticity theory. The damage can be measured by the change of the effective stiffness tensor and the inelastic potential can be often inferred, if not derived, from the Griffith's condition for the propagation of a single microcrack.

The damage evolution is precisely defined as an energy dissipating mode of a microstructural rearrangement during which the connectivity of the material is impaired through formation of internal surfaces and attendant local discontinuities in the deformation field. In contrast to the plastic deformation the total number of atomic bonds is reduced during the damage evolution. The residual strain accumulated during damage processes such as rocks and concrete is related to the frictional slips and interlocks making the modeling of unloading particularly difficult.

The range of the validity of continuum damage models depends on the sign of principal stresses and the heterogeneity of the microstructure. Damage evolution driven by the long range tensile stresses is, in general, of the cleavage type. The preferentially oriented cracks will grow much faster and rather soon invalidate the assumption that only the average crack size matters. The damage representation based on the parameter (3), claiming invariance with respect to the crack sizes and numbers, will in this case cease to provide a realistic estimate of the influence of microcracks on the system response. In materials with a homogeneous microstructure this transition from a regime during which the macroscopic response is almost equally influenced by many small microcracks (damage evolution) to a single crack dominated regime (fracture) may occur before any significant damage takes place. In materials with a heterogeneous microstructure (fiber or particle reinforced composites) a significant accumulation of damage may occur before the single crack propagation becomes the dominant feature of the deformation. Quite obviously, a damage mechanics model is a physically supported alternative only in the latter case.

The damage mechanics models is more often than not a proper choice when the long range tensile stresses are absent. The splitting of an unconfined rock and concrete specimen subjected to the uniaxial compression may be the only exception to this conclusion. In all other cases (laterally confined specimens) the damage evolution will govern the rate of the inelastic deformation and the material will remain statistically homogeneous during a large part of the deformation process. A carefully formulated continuum damage model is in these cases the only physically reasonable alternative.

At a first glance the limitations of the application of damage mechanics to the cases characterized by the absence of long range tensile stress field may appear to be indicative of its limited design utility. However, it should be kept in mind that the so-called "brittle" materials, such as concrete, rocks, ceramics, silicone, human bones, resins, etc., are either designed to be used in structures which minimize the tensile stresses or are reinforced by tougher materials (reinforced concrete, fiber reinforced composites, etc.) to make them damage tolerant by hindering the crack propagation. There are, of course, other circumstances in which the application of damage mechanics is the only rational strategy. Large velocity impacts, penetration of projectiles through many different materials, temperature induced microcracking of rocks in crustal condition, tertiary creep phase, intergranular corrosion of polycrystalline metals, crushing (comminution) of different materials, shin splits and bone breaks, curing of resins, etc. are just some of the important problems in which the application of any other theory would not make any sense.

Analytical modeling of brittle deformation processes is obviously more complex and richer in details than the modeling of ductile deformation processes. The strong dependence of the damage evolution on the sign of normal stresses is the most obvious but not the only difference between the brittle and ductile deformations. The fact that the properties of the system change on all scales and the dependence of the failure mode on the extreme statistics of defect distribution are two other idiosyncrasies of brittle deformation that must be taken into account. Material properties may change in a discontinuous manner during a non-proportional loading whenever the long range stresses change their sign. The microcrack growth, being one of the two basic damage evolution modes, may under propitious conditions cause an abrupt brittle failure at a rather modest accumulation of damage. Hence, more often than not, it will be necessary to formulate limit surface(s) in addition to the damage functions and potentials. This is, of course, an absolute necessity in the structures subjected to the long range tensile stresses. Several possible failure modes, including both intrinsic and extrinsic types, may cause further problems. A more substantial role of statistical modeling is one of the prerequisites of further development of damage mechanics. It seems that the research based solely on the mean field and classical continuum modeling is perilously close to the range of diminishing return. The insistence for better and more complex estimates of the thermodynamic forces driving the damage evolution must be complemented by the considerations of the thermodynamic forces that resist the damage evolution. Therefore, it seems safe to predict that the development of the damage mechanics (including formation of a reasonable comprehensive databank of experimental data) will take a much more intensive effort and a much more longer period of time than was anticipated during its period of adolescence only a couple of decades ago.

ACKNOWLEDGMENT

The author gratefully acknowledged the financial support in the form of a grant from the U.S. Department of Energy, Office of Basic Energy Sciences, Division of Engineering and Geosciences to the Arizona State University which made the research on which this study is based possible.

REFERENCES

1. J.H. Weiner, Statistical Mechanics of Elasticity, Willey & Sons, New York, NY, 1983.
2. S. Nemat-Nasser and M. Hori, Micromechanics: Overall Properties of Heterogeneous Materials, North-Holland, Elsevier Science, P.B.V., Amsterdam, The Netherlands, 1990.
3. D. Krajcinovic, Damage Mechanics, North-Holland, Elsevier Science P.B.V., Amsterdam, The Netherlands, 1996.
4. B. Budiansky and R.J. O'Connell, Int. J. Solids Structures, 12 (1976) 81.
5. V.A. Lubarda and D. Krajcinovic, Int. J. Damage Mechanics, 3 (1994) 38.
6. D. Krajcinovic and D. Fanella, Eng. Fracture Mech., 25 (1986) 585.
7. J.-W. Ju and K.H. Tseng, Int. J. Damage Mechanics, 3 (1995) 23.
8. P. Germain, Q.S. Ngyuen and P. Suquet, J. Appl. Mech., 50 (1983) 1010.
9. R. Hill, J. Mech. Phys. Solids, 11 (1963) 357.
10. M. Ostoja-Starzewski, Appl. Mech. Rev., 47 (1994) S221.
11. D. Krajcinovic and M. Basista, J. Phys. I, 1 (1991) 241.
12. A. Hansen, S. Roux and H.J. Herrmann, J. Phys. France, 50, (1989) 733.
13. R. Ilankamban and D. Krajcinovic, Int. J. Solids Structures, 23 (1987) 1521.
14. K. Kanatani, Int. J. Engng. Sci., 22 (1984) 149.
15. E.T. Onat and F.A. Leckie, J. Appl. Mech., 55 (1988) 1.
16. V.A. Lubarda and D. Krajcinovic, Int. J. Solids Structures, 30 (1993) 2859.
17. B. Audoin and S. Baste, J. Appl. Mech., 61 (1994) 309.

426

18. A.K. Pandey and M. Biswas, J. Sound and Vibr., 169 (1994) 3.
19. D. Krajcinovic and S. Mastilovic, Mech. Mater., 21 (1995) 217.
20. M. Kachanov, I. Tsukrov and B. Shafiro, in: M. Ostoja-Starzewski and I. Jasiuk (eds.), Micromechanics of Random Media, ASME Book No. AMR 139 (1994) S151.
21. D.H. Allen and D.J. Tildesley, Computer Simulations of Liquids, Clarendon Press, Oxford, UK, 1994.
22. J. Kestin and J. Bataille, in: J.W. Provan (ed.), Continuum Models of Discrete Systems, Univ. of Waterloo Press, Waterloo, Canada, (1978) 39.
23. S. Torquato, Appl. Mech. Rev., 44 (1991) 37.
24. I.A. Kunin, Elastic Media with Microstructure II, Springer-Verlag, Berlin, 1983.
25. G.J. Rodin, Int. J. Solids Structures, 30 (1993) 1849.
26. M. Kachanov, in: J. Hutchinson and T. Wu (eds.), Advances in Applied Mechanics, 29 (1993), 259.
27. A.R. Day, K.A. Snyder, E.J. Garboczi and M.F. Thorpe, J. Mech. Phys. Solids, 40, (1992) 1031.
28. T. Nakamura and S. Suresh, Acta Metall. Mater., 41 (1993) 1665.
29. H.J. Herrmann, D. Stauffer and S. Roux, Europhys. Lett., 3 (1987) 265.
30. S.J.D. Cox and L. Patterson, Comm. in Num. Meth. in Engng., 10 (1994) 413.
31. H. Horii and S. Nemat-Nasser, Phil. Trans. Royal Soc. London, 319 (1986) 337.
32. J.R. Rice, in: A.S. Argon (ed.), Constitutive Equations in Plasticity, MIT Press, Cambridge, MA, (1975) 23.
33. J. Kestin and J.R. Rice, in: E.B. Stuart, B. Gal and A.J. Brainard (eds.), A Critical Review of Thermodynamics, Mono-Book Corp., Baltimore, MD, (1970) 275.
34. J. Lemaitre and J.L. Chaboche, Mecanique des Materiaux Solides, Dunod, Paris, France, 1982.
35. D. Krajcinovic and J. Lemaitre (eds.), Continuum Damage Mechanics, Springer-Verlag, Wien, Austria, 1987.
36. G.A. Maugin, The Thermodynamics of Plasticity and Fracture, Cambridge University Press, Cambridge, UK, 1992.
37. J. Kestin, Int. J. Solids Structures, 29 (1992) 1827.
38. J.R. Rice, J. Mech. Phys. Solids, 19 (1971) 433.
39. M. Basista and D. Gross, (preprint), (1996).
40. J. Rudnicki and J.R. Rice, J. Mech. Phys. Solids, 23 (1975) 371.
41. V.A. Lubarda and D. Krajcinovic, Int. J. Plasticity, 11 (1995) 763.
42. D.J. Holcomb and L.S. Costin, J. Appl. Mech., 53 (1986) 536.
43. M.F. Ashby and C.G. Sammis, PAGEOPH, 133 (1990) 489.
44. S. Nemat-Nasser and M. Obata, J. Appl. Mech., 55 (1988) 24.
45. H. Kitagawa, JSME Int. J., 30 (1987) 1361.

Theoretical and Applied Mechanics 1996
T. Tatsumi, E. Watanabe and T. Kambe (Editors)

Fluid-Structure Interactions between Axial Flows and Slender Structures

Michael P. Païdoussis

Department of Mechanical Engineering, McGill University, 817 Sherbrooke Street W., Montreal, Québec, H3A 2K6 Canada

This lecture deals with certain aspects of the dynamics of slender structures interacting with axial flow; specifically, pipes and shells conveying fluid and cylinders in axial flow. The dynamics of such systems display many paradoxes and unexpected features, of much wider interest in applied mechanics; some of these are explored in this lecture.[†] (i) The critical flow velocity for flutter of cantilevered pipes conveying fluid, when plotted versus fluid/total mass ratio, undergoes a number of S-shaped "jumps" which are either separatrices or backbones of distinct behaviour, e.g. defining whether damping, nonlinearities and mass additions stabilize or destabilize the system, and whether the Hopf bifurcation is sub- or supercritical. Why? (ii) Energy considerations suggest that a cantilever aspirating, rather than discharging, fluid loses stability at arbitrarily small flow. Does it? (iii) 'Standard' CFD-type analyses of the dynamics of cylinders and shells interacting with flow may easily lead to the wrong conclusions on stability because of inappropriate specification of boundary and end conditions. (iv) Very long or tapered cylinders in axial flow develop very thick boundary layers *vis-à-vis* the radius; how does one deal with that? (v) Articulated systems in axial flow, differing in only two physical parameters, develop chaotic oscillations with increasing flow via three different routes.

1. INTRODUCTION

Fluid-structure interaction is a fascinating field. At its most interesting, it deals with the discovery and/or elucidation of a new phenomenon or of new dynamical behaviour of a structure interacting with flow. Examples range from the notorious Takoma Narrows bridge wind-induced collapse, to shell-type 'ovalling' oscillations of thin-walled metal chimneys in high wind, to 'fluidelastic instability' of heat-exchanger tube banks, and to chaotic oscillations of pipes conveying fluid.

Some of the work in this area is applications-oriented, i.e. it is fuelled by real engineering problems; e.g. wind-induced bridge oscillations, vibration of hydraulic structures [1] or heat-exchanger vibrations [2, 3]. Some of it is simply applications-*related*; e.g., most of the definitive research on the ovalling oscillation of chimneys was conducted in order to discover the underlying mechanism [4, 5, 6, 7], even though practical means for preventing the oscillations were already known. Finally, some of this work is purely

---

[†]The support of NSERC of Canada and FCAR of Québec is gratefully acknowledged.

curiosity-driven; e.g. the research on nonlinear and chaotic dynamics of pipes conveying fluid [8]. Of course, curiosity-driven research is the backbone of progress in Applied Mechanics and requires no apology, even though some of the purest such research may find unexpected applications five or fifteen years after it is done [9].

This lecture is confined to problems associated with relatively slender structures (pipes, cylinders, shells) with the fluid flow nominally axially disposed thereto, internally or externally. The axial-flow slender-structure combination implies the absence, or at most limited presence, of flow separation, which renders much easier both analytical modelling and interpretation of observed behaviour. Besides, cross-flow-induced phenomena associated with single or multiple bluff bodies (e.g., vortex shedding, galloping, fluidelastic instability) tend to receive the lion's share of attention in lectures, books and papers. Yet, axial-flow slender-structure interactions are just as fascinating; this is one of the reasons which has propelled this author into writing a book on the subject [10], wherein the interested reader will eventually find a much fuller account of the subject. Indeed, this lecture treats only some of the paradoxes in this area of fluid-structure interactions, focussing on some that are of wider interest in applied mechanics.

2. SENSITIVE DEPENDENCE OF DYNAMICAL BEHAVIOUR OF CANTILEVERED PIPES CONVEYING FLUID ON THE MASS RATIO

It is well known that a cantilevered pipe conveying fluid is a nonconservative system, in which energy may be transferred from the pipe to the fluid or *vice versa*, depending on the flow velocity, U. For sufficiently high U, the system losses stability by single-mode flutter and executes limit-cycle oscillation. In the simplest possible situation, for a pipe of mass per unit length m and flexural rigidity EI, conveying fluid of mass per unit length M and velocity U, the linearized equation of motion is

$$EI\frac{\partial^4 w}{\partial x^4} + MU^2\frac{\partial^2 w}{\partial x^2} + 2MU\frac{\partial^2 w}{\partial x \partial t} + (M+m)\frac{\partial^2 w}{\partial t^2} = 0 \, , \tag{1}$$

in which $w(x,t)$ is the lateral deflection, x the axial coordinate and t time, and dissipation has been ignored. When this equation is rendered nondimensional, it is found that its dynamics is governed by the following two dimensionless parameters

$$\beta = M/(M+m) \qquad \text{and} \qquad u = (M/EI)^{1/2}\, LU \, , \tag{2}$$

respectively the mass ratio (fluid/total mass per unit length; $0 < \beta < 1$) and the dimensionless flow velocity.

As shown in Figure 1, the critical value of u for flutter, u_c, depends on β in a rather peculiar fashion, displaying a number of S-shaped segments, at $\beta = \beta_S \simeq 0.30$, 0.67 and 0.90. If an experiment were possible in which β was increased gradually and then for each value of β the system was made to cross u_c, one would expect to see a jump up in u_c around these sensitive values of β, β_S, and a similar jump down for decreasing β.

An attempt was made to explain these 'jumps' in terms of different modes becoming unstable on either side of the β_S, but it was not wholly successful [11]. Thus, a change from the second to the third mode was found to occur across the first β_S; but, for instance,

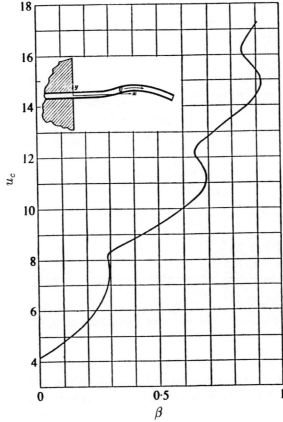

Figure 1: The critical value of u for flutter, u_c, versus β [12].

when gravity is present (vertical system) and $\gamma = [(M + m)L^3/EI]g = 10$, it is the same mode (the second) that becomes unstable across the jump.

With internal dissipation in the pipe material taken into account, the S-shaped segments are smoothed, and this agrees with experimental measurements of u_c [12, 13]. However, another interesting result emerges, as shown in Figure 2: dissipation stabilizes the system before the jump and destabilizes it after it — whether this is modelled as a hysteretic (structural) damping, as in Figure 2, or by a viscoelastic model. [The viscous damping, representing frictional losses with the surrounding ambient fluid has a different effect, at least for this value of the coefficient κ]. It is, of course, well known that weak damping may destabilize nonconservative systems [15]–[19]; of interest here is that the effect of damping is different on either side of the first β_S.

Another manifestation of peculiar things happening around these S-shaped curves was uncovered by Hill & Swanson [20], when a lumped mass is added at different locations along the pipe. Interestingly, for a mass M_a added at mid-point and $M_a/[(m+M)L] = 0.2$, destabilization occurs only for $\beta > 0.3$ approximately, whereas for $\beta < 0.3$ the effect is stabilizing — the separatrix being once again the first β_S.

The *nonlinear* dynamics of pipes conveying fluid has been studied extensively over the past twenty years and, in the course of these studies, a number of other peculiarities

430

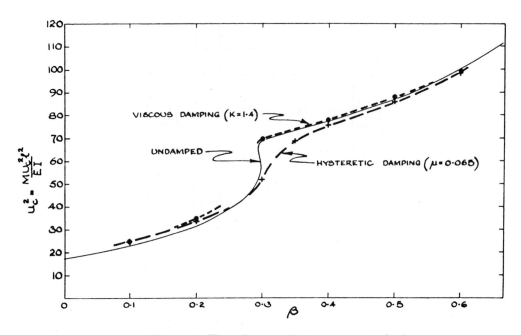

Figure 2: The effect of dissipation on u_c [14].

have been found to be associated with the S-curves. Rousselet & Herrmann [21] studied the effect of 'nonlinearities' (including some linear effects) on stability and found that, for some ranges of β the effect is stabilizing, whereas for others it is destabilizing, as shown on the left side of Table 1. The latter range ($0.21 < \beta < 0.42$) is centred about the first β_S. Lundgren *et al.* [22] studied the stability of the system with an inclined nozzle fitted at the free end of the pipe, both for motions in the plane of inclination of the nozzle ('in-plane') and perpendicular to it ('out-of-plane'). The results are shown on the right side of Table 1. It is seen that the in-plane oscillation regions straddle the β_S values. Moreover, for the range of β in which calculations were conducted by Rousselet & Herrmann, the degree of correspondence between these two different sets of phenomena is remarkable.

Table 1: Effect of 'nonlinearities' [21] and plane of flutter for a pipe fitted with an inclined end-nozzle [22] for different ranges of β

β	Effect of 'nonlinearities'	β	Plane of flutter
0.02 – 0.21	Stabilizing	0.00 – 0.23	Out-of-plane
0.21 – 0.42	Destabilizing	0.23 – 0.42	In-plane
0.42 – 0.66	Stabilizing	0.42 - 0.63	Out-of-plane
		0.63 – 0.70	In-plane
> 0.66	Calculations not done	0.70 – 0.85	Out-of-plane
		0.85 – 0.97	In-plane

A sophisticated analysis of the planar motions of the system (without the inclined nozzle) in the neighbourhood of the Hopf bifurcation leading to flutter was made by Bajaj *et al.* [23], using a variant of the theory in which the upstream pressure is considered to be constant. Thus, a flow equation becomes necessary, and a parameter $\alpha = f\,L/\sqrt{A}$ enters the problem, f being a friction factor, L the pipe length and A its internal cross-sectional area. As shown in Figure 3, for sufficiently small α (short pipes), stability is lost by a subcritical Hopf bifurcation, implying an unstable limit cycle (and beyond it, likely, a stable one); on the other hand, near the sensitive values of β, i.e. β_S, stability is lost via a supercritical Hopf bifurcation, irrespective of the value of α.

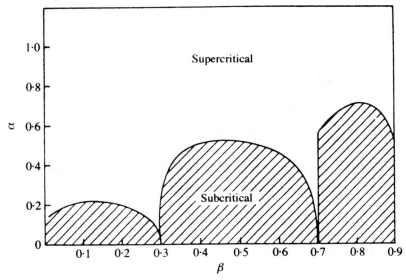

Figure 3: The different types of Hopf bifurcation depending on β and α [23].

This analysis was extended to 3-D motions [24, 25]. In this case, in addition to planar motions, which may be subcritical or supercritical, rotary motions of the pipe are also possible. Once more, however, the different kinds of motion are neatly separated by $\beta_S \simeq 0.3$ and 0.7, similarly to Figure 3.

No one knows for certain the underlying physical reasons for the pivotal role the β_S play in so many features of the dynamics of the system. However, it may all be related to the first finding: that, depending on β, dissipation can stabilize or destabilize the system; i.e., the energy transfer on either side of β_S may be different. A clue is provided by Benjamin's [17] work, related to the stability of compliant surfaces subjected to flow. In one of the three classes of instability possible, class A, dissipation is destabilizing. To explain, the phenomenon in simple terms, a one-degree-of-freedom mechanical system was considered, $m\ddot{q} + c\dot{q} + kq = Q$, where the generalized force $Q = M\ddot{q} + C\dot{q} + Kq$ is associated with the fluid flow. It is shown that for $m < M$, $k < K$ and $c > C$, oscillation can arise only if the system is allowed to do work against the external forces providing excitation; then, the effect of dissipation is always destabilizing. Unfortunately, as recognized by Benjamin, this system is not physical since $-M$ is the added mass and hence $M < 0$. If, however, a travelling wave is considered along an infinitely long compliant surface, the same dynamical behaviour may be shown to arise: for specific

432

ranges of a parameter depending on m, M, k and K, dissipation can be either stabilizing or destabilizing, similarly to the pipe system at hand.

Research into uncovering the mechanism making the dynamical behaviour so sensitively dependent on the β_S is actively being pursued.

3. FLUTTER IN ASPIRATING CANTILEVERS

The work done by the fluid forces over a cycle of oscillation of period T, is [26]

$$\Delta W = -MU \int_0^T \left[\left(\frac{\partial w}{\partial t} \right)_L^2 + U \left(\frac{\partial w}{\partial t} \right)_L \left(\frac{\partial w}{\partial x} \right)_L \right] \mathrm{d}t \,, \tag{3}$$

where $(\partial w/\partial t)_L$ and $(\partial w/\partial x)_L$ are, respectively, the lateral velocity and slope of the free end. Thus, for small enough U, the effect of the flow is to stabilize the system (Coriolis stabilization), since $[(\partial w/\partial t)_L]^2$ dominates and $\Delta W < 0$. For sufficiently high U, however, if, over most of the cycle, $(\partial w/\partial x)_L$ and $(\partial w/\partial t)_L$ have opposite signs, then $\Delta W > 0$ and the pipe gains energy from the flow, leading to amplified motions and flutter.

It was in the course of conducting research on aspirating pipes in the 1960s that the author realized that if the sign of U is reversed in (3), then the opposite dynamical behaviour would be predicted. In particular, the system would be unstable at infinitesimal flow velocities if dissipation is neglected; if it is taken into account, then U_c would merely be very small! Experiments were conducted but they were inconclusive since, as the aspiration was intensified to achieve instability, the loss of pressure in the pipe caused its shell-mode collapse near the support.

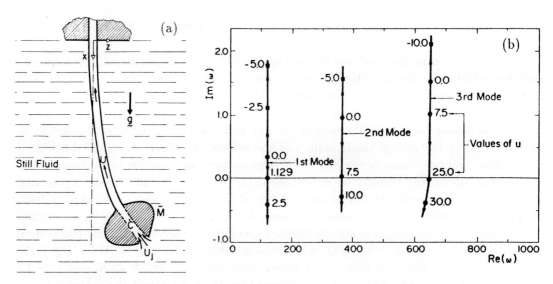

Figure 4: (a) The idealized system for ocean mining; (b) typical Argand diagram of the eigenfrequencies; note that, contrary to the sign convention in the text, here upward (sucking) flow corresponds to $u > 0$, while $u < 0$ represents downward flow [27].

The matter was taken up again theoretically, in conjunction with the stability of pipes used in Ocean Mining, for example to 'vacuum' up manganese nodules from the bottom of the sea [27]. The analytical results confirmed the foregoing: with zero damping, mirror-image eigenfrequency Argand diagrams are produced for $U > 0$ and $U < 0$, and opposite stability behaviour. The results with damping are only slightly different, as shown in Figure 4. Ever since, it was tacitly assumed that this dynamical behaviour is correct, and several papers were subsequently published with similar results, e.g. [28, 29].

A renewed attempt was made in the 1980s to obtain the predicted behaviour experimentally. This time the entire pipe, hung vertically, was immersed in water in a steel tank; water was supplied at the top of the tank, and was forced up the hanging pipe and out of the vessel. To achieve higher flows, compressed air was supplied at the top of the tank. In several experiments with different free-end configurations, the pipe remained unnervingly stable. The experiment was discontinued when, with ever increasing air-pressure, the rubber hose leading the water to the drain burst free of its clamp, spraying water all over the lab. At that point, the author was certain that something was wrong with *the theory*; for one thing, the flow into the pipe is not exactly tangential, replicating in reverse the outpouring jet in the case of $U > 0$.

It was while visiting Cambridge in 1995 and upon recounting this paradoxial behaviour to Dr D.J. Maull, that he recalled reading something similar in Richard Feynman's biography [30]. It turns out that in 1939 or 40, Feynman's and most other physicists' tea-time conversation at Princeton and the Institute for Advanced Study was dominated by this problem: if a simple S-shaped lawn sprinkler were made to suck up water instead of spewing it out, would it rotate backwards or in the same way as for normal operation? Feynman could apparently argue convincingly either way. (This problem was tied to the issue of reversibility of atomic processes!).

Eventually, Feynman decided to do an experiment, which was remarkably similar to the author's. He immersed the lawn sprinkler into a glass jar filled with water, with an outlet connected to the sprinkler and a compressed air supply to force the water into the sprinkler and out. With increasing pressure and flow, the sprinkler refused to budge, up to the point where the glass jar exploded, spraying water all over. The result was that Feynman was banished from the laboratory henceforth.

Clearly the flow field is entirely different in 'forward' and 'reverse' flow through the sprinkler. Besides, since the fluid has no net circulation going into the sprinkler, which being free does not impart any to the fluid going through it, no rotation can arise — by elementary control-volume considerations. The same is true for the aspirating pipe. There is no *a priori* reason why the dynamics with $U > 0$ and $U < 0$ should be mutually reversible, nor that the fluid boundary conditions associated with equation (1) at $x = L$ would be the same in the two cases. Indeed, in one case we have a jet (not a source flow) and in the other a sink. Hence, it is clear that aspirating pipes cannot aspire to flutter!

4. SENSITIVITY OF ANNULAR-FLOW-INDUCED INSTABILITIES TO BOUNDARY-CONDITION SPECIFICATION

Cylinders subjected to annular flows are notoriously prone to fluidelastic instabilities [2, 31]. Early work by Miller [32] was followed by more systematic work by several

434

researchers, e.g. [33]–[41].

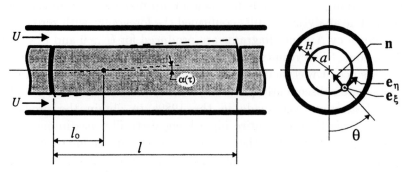

Figure 5: Geometry of the annular flow problem.

The system may be simplified into that shown in Figure 5, involving 'rocking motions' of a rigid centrebody about a hinge; alternatively, lateral translational motions may be considered. The system of Figure 5 was first analyzed for rocking motions [34] by means of potential flow theory. The boundary conditions were applied at the mean position of the centrebody; also, effectively, no conditions were specified for the transition in the flow between the moving centrebody and the upstream and downstream immobile extensions, i.e. across the 'end-gap'. It was found that the system develops flutter (by the negative fluid-dynamic damping becoming larger than the positive mechanical damping), if the hinge-point is farther downstream than $\frac{1}{2}l$. This is not surprising, since the motion-related perturbation pressure acting on the centrebody may be expressed as

$$p = -\rho a \left[(x - l_0) \ddot{\alpha} + 2 U \dot{\alpha} \right] G(h) \cos \theta \qquad (4)$$

[31], where ρ is the fluid density, a the centrebody radius, $G(h)$ is a function of $h = H/a$, and H, l_0, α and θ are defined in the figure. The first term in the square brackets is an added-mass term, while the second is a Coriolis term which controls the instability. Using the similarity with internal flow, it is clear that when the body rotates about the hinge, the Coriolis-related forces will all act in the same direction, opposing motion on the part of the body downstream of the hinge, while aiding it on the upstream portion; hence the critical location of the hinge is at $l_0 = \frac{1}{2} l$. For nonuniform flow passages, this is modified, but the basic mechanism is the same; in particular, a divergent flow passage has a destabilizing effect, while a convergent one is stabilizing, in agreement with Miller's [32] original findings.

It is important to note the absence of a stiffness term in (4) — i.e. a term proportional to α — for uniform flow; hence the impossibility of a divergence (static, or buckling, instability) to occur according to this model. When viscous effects are taken into account, albeit approximately [35], stiffness terms do arise and they can be negative (for $l_0 > \frac{1}{2} l$). Hence, in principle, loss of stability by divergence is also possible, although for all the cases considered in Ref. [35] the dynamics continued being dominated by flutter.

Two developments occurred subsequently, forcing a closer look at these results. The first was the development of CFD codes, e.g. [36]–[42], which necessitated the specification of exactly how the perturbation pressure evolves from the vibrating to the nonvibrating part of the system across the end-gap. The CFD techniques involved a fixed-grid computational domain based on a finite-difference formulation with primitive variables, and real-time discretization of the Navier-Stokes equations for unsteady incompressible laminar flows based on a three-time-level implicit scheme, utilizing a pseudo-time integration with artificial compressibility. The second development was an experimental program in which, instead of the centrebody being free to move, it was fixed, while equivalently a portion of the outer containment pipe could move. In this case, it is not possible to have the moving and immobile portions of the pipe unconnected, since then the fluid could leak out of the system! Thus, at least a membranous connection has to exist. Also, in the CFD codes, continuity from the last node on the moving portion of the system to the immobile portion imposes a similar type of connection. This means that one should consider the system domain to be not $x \in [0, l]$ but $x \in [-\varepsilon, l + \varepsilon]$, $\varepsilon << l$, and that at these two limits the displacement of the system is zero. Then, by analogy to the pipe problem, this system is wholly conservative if analyzed by potential flow theory; thus, the added damping (fluid damping analogous to added mass, in this case the net effect of the Coriolis forces) is zero, while the added stiffness is not, as shown in Figure 6. The system can therefore lose stability only by divergence!

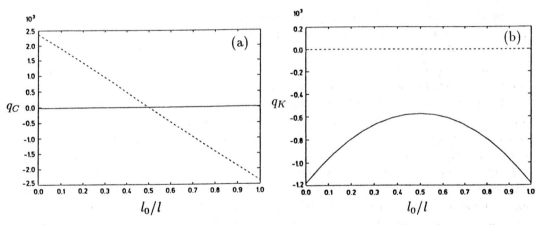

Figure 6: Variation with l_0/l of (a) fluid added damping and (b) fluid added stiffness for rocking motion, according to potential flow theory: – – – , ignoring end effects; —— , accounting for them [43].

Of course, once viscous effects are considered, the system becomes nonconservative and hence may lose stability either by divergence or flutter [31].

These results are summarized in Table 2. The case of rocking motions and potential flow was studied extensively by de Langre et al. [43]. It is noted that the two cases of viscous flow, although leading to the same qualitative result (divergence or flutter in both cases), the added damping when the ends are presumed fixed is quite different from the other case, since the effect of the Coriolis forces in the first case is null.

It should also be noted that if the centrebody is nonuniform, or more generally the annular passage is nonuniform, stiffness terms may arise even in potential flow analyses, since the flow velocity is not uniform across the gap. Furthermore, in more refined CFD models in which the boundary conditions are applied on the moving boundary of the cylinder rather than at its mean position, e.g. [44], then again stiffness terms may arise, even with potential flow — though they would be small for small-amplitude motions.

Table 2: The effect of end-conditions on stability of a uniform rigid flexibly-mounted centrebody in a uniform annular flow, executing rocking or lateral translational motions

Motion	Flow model	End effects	Added stiffness	Added damping	Possible instability
Rocking	Potential	Ignored	Zero	Non-zero	Flutter
	Potential	Accounted	Non-zero	Zero	Divergence
	Viscous	Ignored	Non-zero	Non-zero	Div./Fl.
	Viscous	Accounted	Non-zero	Non-zero	Div./Fl.
Lateral	Potential	Ignored	Zero	Zero	—
	Potential	Accounted	Non-zero	Zero	Divergence
	Viscous	Accounted	Non-zero	Non-zero	Div./Fl.

Another instance of sensitivity to boundary conditions was encountered in the stability analysis of cantilevered cylindrical shells subjected to annular flow. The system was first analyzed by analytical techniques taking viscous flow effects into account approximately [45], and flutter instabilities were predicted in various shell modes and confirmed experimentally with reasonably good agreement [46]. It was then decided to adapt the aforementioned CFD technique to the analysis of this problem [47]. It was found that, with the no-slip boundary condition applied at the mean postion of the wall which coincides with the boundary of the fixed computational mesh, the CFD solution was completely devoid of any unsteady fluid-dynamic effects. In fact, the *steady* solution is then reproduced; hence, only divergence can be predicted, and this solely due to steady-flow pressurization effects! The crux of the matter is that the fixed numerical grid does not coincide with the continuously deforming wall. Therefore, the no-slip condition on this grid boundary is not $\mathbf{U}_w = \dot{\mathbf{x}}_w$, where $\mathbf{U}_w$ is the fluid velocity at the wall and $\dot{\mathbf{x}}_w$ the wall velocity. By a Taylor series expansion, the no-slip condition on the grid boundary is

$$\mathbf{U}_0 = \dot{\mathbf{x}}_w - \mathbf{x}_w \cdot (\nabla \mathbf{U})_0 , \qquad (5)$$

where the subscript 0 denotes the mean position of the shell (the boundary of the mesh) and $\mathbf{x}_w$ is the wall displacement from the mean position. This is the condition used in Ref. [40]. A variant of this was used for the more complex shell problem, in which, at the beginning of each iteration within each physical time step, the flow velocities at all the grid points initially in contact with the motionless shell are updated [47, §3.3]; clearly, some of these points now cyclically come in contact with the mean axial flow. In this way the unsteady viscous forces are recovered. Both divergence and flutter can now be predicted, in fact leading to better agreement with experiment [46, 47] than the analytical method could give.

Therefore, apparently small refinements or additional specifications in the boundary and end conditions in this sort of problem can have profound effects on its predicted dynamical behaviour. None of the final forms of the analyses discussed in the foregoing are totally *wrong*, though some are more appropriate than others for particular physical systems, and occasionally not for those intended in the first instance!

5. TAPERED CYLINDERS WITH THICK BOUNDARY LAYERS

It is well known that slender cylinders in external axial flow are subject to fluidelastic instabilities, namely divergence (buckling), followed at higher flows by flutter [48]–[50]. The equation of motion, again in its simplest form, is remarkably similar to (1):

$$EI\frac{\partial^4 w}{\partial x^4} + M U^2\frac{\partial^2 w}{\partial x^2} + 2 M U\frac{\partial^2 w}{\partial x\partial t} + F_N - \left\{\int_x^L F_L \,\mathrm{d}x\right\}\frac{\partial^2 w}{\partial x^2} + (m + M)\frac{\partial^2 w}{\partial t^2} = 0\,,\quad (6)$$

where F_N and F_L are the frictional forces per unit length in the normal and longitudinal directions and $M = \rho A$ is the virtual or 'added' mass per unit length.

It is also known that, in the case of cantilevered cylinders, these instabilities materialize provided the free end of the cylinder is fairly well streamlined. Indeed, in the course of such an experiment, a streamlined end glued-on to the cylinder came off while the cylinder was fluttering; whereupon the cylinder came instantly to rest — as recorded on film.

It was consequently presumed that a conical cylinder, in which the 'streamlined end' would in effect extend over the whole length, would be at least as liable to instability as a uniform one with a tapered end. It was, therefore, a big surprise when it was found that the conical cylinder remained stable — to the maximum attainable flow, several times higher than that causing instability in uniform cylinders [51]–[53].

A possible explanation of the observed behaviour is that for the conical system there exists a much thicker boundary layer than for a uniform cylinder. However, attempts to suck it off through small holes on the surface of the cylinder and then out through an internal flow-passage were unsuccessful, as the internal flow began to affect the dynamics. Hence, an analytical explanation was sought, as follows. The inviscid forces, obtained by slender-body theory, which control the dynamics of a uniform cylinder, give a lateral force per unit length

$$\mathcal{L} = -\left[(\partial/\partial t) + U\,(\partial/\partial x)\right]\left\{\rho A\left[(\partial w/\partial t) + U\,(\partial w/\partial x)\right]\right\}\,,\quad (7)$$

which may be thought of as being made up of a cross-flow due to lateral motions $\partial w/\partial t$, and another component due to the mean flow around the inclined cylinder at an angle $\sim\partial w/\partial x$. For the case of a thick boundary layer, this may be modified as follows. For the lateral flow due to $\partial w/\partial t$ the boundary layer is neglected, while for the inclined mean flow its presence is felt, thus generating a modified form of (7), namely $\mathcal{L}' = -\left[(\partial/\partial t) + U\,(\partial/\partial x)\right]\left\{\rho A\,(\partial w/\partial t) + \rho A^*U\,(\partial w/\partial x)\right\}$, where A^* is the cross-sectional area, augmented by the boundary-layer thickness. Moreover, towards the free end of the cylinder, the boundary-layer thickness increases sufficiently for the boundary layer to become insensitive to motion of the cylinder and, more importantly, for the cylinder to become effectively 'insulated' from the mean axial flow beyond. This amounts to the cylinder feeling a progressively smaller flow velocity with increasing x, namely a flow velocity U^*, which may very approximately be taken as $U^* = U(A/A^*)$, thereby giving

$$\mathcal{L}^* = -[(\partial/\partial t) + U^*(\partial/\partial x)]\{\rho A[(\partial w/\partial t) + U(\partial w/\partial x)]\}.\tag{8}$$

With this modified lift, a new theoretical model was constructed, and the theoretical results agree with experimental observations reasonably well, as shown in Figure 7. For a sufficiently slender conically tapered cylinder ($\varepsilon = L/D(0)$ large, where $D(0)$ is the value of D at the fixed end), no instabilities are obtained either theoretically or experimentally.

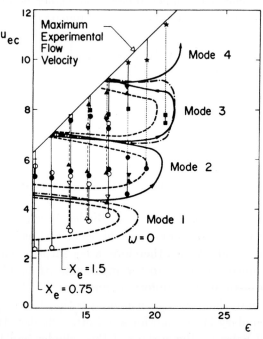

Figure 7: Critical flow velocity u_{ec} versus slenderness $\varepsilon = L/D(0)$ when the thick boundary layer on the conical cylinder is taken into account; lines: theory; points interconnected by vertical lines: experimental ranges of instability [53].

6. CHAOS IN ARTICULATED SYSTEMS IN ANNULAR FLOW

Recent work has shown that pipes conveying fluid can develop chaotic oscillations, when the basic system is 'perturbed' by motion-limiting restraints [54, 55], or by the addition of a magnetic attractor [56] or an end-mass [57] — see Refs [8] and [58]. This type of work was recently extended to systems subjected to external flow, and displayed even more varied dynamical behaviour [59].

The system considered here is a three-segment articulated cylinder consisting of three rigid cylinders interconnected by flexible joints, supported upstream and free downstream, centrally located within a cylindrical conduit and subjected to annular flow. For sufficiently high flow velocities the system becomes subject to flutter and to impacting with the external conduit, modelled in an idealized fashion as a trilinear or cubic spring. Eventually, for high enough dimensionless flow velocity u, the oscillation becomes chaotic. It was found that by changing only two physical parameters, namely the width of the annulus and the streamlining of the downstream end, *three different routes* to chaos are followed [60]. Specifically, these two parameters are: $h = D_h/D$, the hydraulic diameter divided by the cylinder diameter; and f, a parameter tending to 1 for a perfectly streamlined end and 0 for a blunt end.

The first case considered ($h = 0.1$, $f = 0.4$) develops chaotic oscillations via the classical period-doubling route, as shown in Figure 8(a). The system loses stability by a pitchfork bifurcation, leading to two new equilibrium points (S.F.B. = stable fixed points) on either side of the trivial ones. Then, Hopf bifurcations lead to limit-cycle (L.C.) motions, which eventually through a sequence of period-doublings lead to chaos. As an example, Figure 8(b) shows period-4 motion; $\phi_i(\tau)$ is the angular displacement of the ith articulation. The Feigenbaum number was found to be Fei = 4.34, not too far away from the classical Fei = 4.667 [61].

For the second case considered ($h = 0.5$, $f = 0.4$) chaos arises via intermittency of type III [62]. Figure 8(c) shows the characteristic unsteadiness in the maximum amplitude, associated with the alternating periods of 'quiet vibration' and more violent 'bursts'.

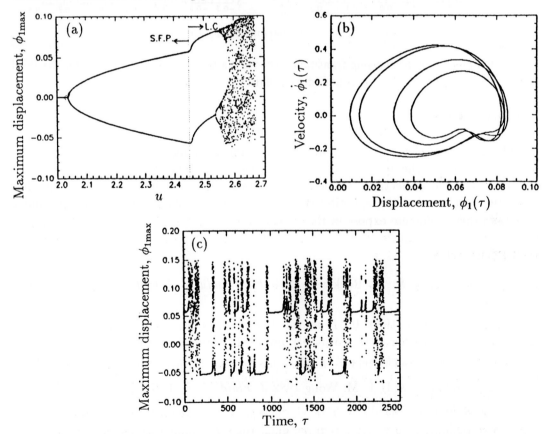

Figure 8: (a) Bifurcation diagram showing transition to chaos via period-doubling bifurcations and (b) phase-plane diagram showing period-4 motion, for the first articulated system studied ($h = 0.1, f = 0.4$); (c) maximum displacement versus time for chaos via the intermittency route, for the second system ($h = 0.5, f = 0.4$) [50].

For another set of parameters ($h = 0.5$, $f = 0$) chaotic motion develops via the Ruelle & Takens quasiperiodic route. The transition from quasiperiodicity ($u = 6.25$) to chaos ($u = 6.35$) is quite clear in the Poincaré maps of $\dot{\phi}_2(\tau)$ versus $\phi_2(\tau)$ when $\phi_1(\tau) = 0$, Figure 9(a,b). At $u = 6.25$, two base frequencies are observed in the PSD, f_1 and f_2, while all other peaks are related thereto via $f = nf_1 + mf_2$, with n and m integers.

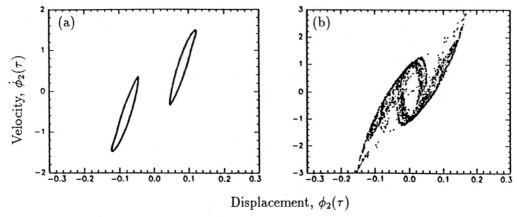

Figure 9: Poincaré maps showing transition to chaos via the quasiperiodic route, for the third system ($h = 0.5, f = 0$): (a) quasiperiodic oscillation; (b) chaotic oscillation [50].

7. CONCLUSION

The brevity of this paper does not permit even touching upon many other, equally interesting facets of the dynamics of slender bodies interacting with axial flows, nor could the topics covered in Sections 2-6 be fully explored. However, it is hoped that this has been sufficient to whet the appetite; the interested reader should, in due course, find a fuller and more complete exposé in Ref. [10].

REFERENCES

1. E. Naudascher & D. Rockwell 1994 *Flow Induced Vibration*, A.A. Balkema, Rotterdam.

2 M.P. Païdoussis 1980 In *Practical Experiments with Flow-Induced Vibrations* (eds E. Naudascher & D. Rockwell), pp. 1-81, Springer, Berlin.

3. M.K. Au-Yang (ed.) 1993 *Technology for the '90s*, ASME, New York.

4. M.P. Païdoussis & D.T.-M. Wong 1982 *J. Fluid Mech.* **115**, 411-426.

5. M.P. Païdoussis, S.J. Price & H.-C. Suen 1982 *J. Sound Vib.* **83**, 533-553.

6. M.P. Païdoussis, S.J. Price & H.-C. Suen 1982 *J. Sound Vib.* **83**, 555-572.

7. M.P. Païdoussis, S.J. Price & S.-Y. Ang 1988 *J. Fluids Struct.* **2**, 95-112.

8. M.P. Païdoussis & G.X. Li 1993 *J. Fluids Struct.* **7**, 137-204.

9. M.P. Païdoussis 1993 *ASME J. Press. Vessel Tech.* **115**, 2-14.

10. M.P. Païdoussis 1997/8 *Fluid-Structure Interactions: Slender Bodies and Axial Flow*, Academic Press, London.

11. M.P. Païdoussis 1969 Mech. Eng. Res. Lab. Report No. 69-3, McGill U.

12. R.W. Gregory & M.P. Païdoussis 1966 *Proc. Roy. Soc. (London)* A**293**, 512-527.

13. R.W. Gregory & M.P. Païdoussis 1966 *Proc. Roy. Soc. (London)* A**293**, 528-542.

14. M.P. Païdoussis 1963 *Oscillations of Liquid-Filled Flexible Tubes*, Ph.D. Thesis, U. of Cambridge.

15. H. Ziegler 1952 *Principles of Structural Stability*, Blaidsdell, Waltham, Mass.

16. T.B. Benjamin 1960 *J. Fluid Mech.* **9**, 513-532.

17. T.B. Benjamin 1963 *J. Fluid Mech.* **16**, 436-450.

18. S. Nemat-Nasser, S.N. Prasad & G. Herrmann 1966 *AIAA J.* **4**, 1276-1280.

19. V.V. Bolotin & N.I. Shinzher 1969 *Int. J. Solid Struct.* **5**, 965-989.

20. J.L. Hill & C.P. Swanson 1970 *J. Appl. Mech.* **37**, 494-497.

21. J. Rousselet & G. Herrmann 1981 *J. Appl. Mech.* **48**, 943-947.

22. T.S. Lundgren, P.R. Sethna & A.K. Bajaj 1979 *J. Sound Vib.* **64**, 553-571.

23. A.K. Bajaj, P.R. Sethna & T.S. Lundgren 1980 *SIAM J. Appl. Math.* **39**, 213-230.

24. A.K. Bajaj & P.R. Sethna 1984 *SIAM J. Appl. Math.* **44**, 270-286.

25. A.K. Bajaj & P.R. Sethna 1991 *J. Fluids Struct.* **5**, 651-679.

26. T.B. Benjamin 1961 *Proc. Roy. Soc. (London)* A**261**, 457-486.

27. M.P. Païdoussis & T.P. Luu 1985 *ASME J. Energy Resources Tech.* **107**, 250-255.

28. J.H. Sällström & B.A. Åkesson 1990 *J. Fluids Struct.* **4**, 561-582.

29. M. Kangaspuoskari, J. Laukkanen & A. Pramila 1993 *J. Fluids Struct.* **7**, 707-715.

30. J. Gleick 1972 *Genius*, Pantheon Books, New York.

31. M.P. Païdoussis 1987 *Appl. Mech. Rev.* **40**, 163-175.

32. D.R. Miller 1970 Argonne National Lab. Report ANL-83-43, Argonne, Il, U.S.A.

33. D.E. Hobson 1982 *Proc. BNES 3rd Int'l Conf. Vibration in Nucl. Plant*, pp. 440-463.

34. D. Mateescu & M.P. Païdoussis 1985 *ASME J. Fluids Eng'g* **107**, 421-427.

35. D. Mateescu & M.P. Païdoussis 1987 *J. Fluids Struct.* **1**, 197-215.

36. D. Mateescu, M.P. Païdoussis & F. Bélanger 1988 *J. Fluids Struct.* **2**, 615-628.

37. D. Mateescu, M.P. Païdoussis & F. Bélanger 1994 *J. Fluids Struct.* **8**, 489-507.

38. D. Mateescu, M.P. Païdoussis & F. Bélanger 1994 *J. Fluids Struct.* **8**, 509-527.

39. M.P. Païdoussis, D. Mateescu & W.-G. Sim 1990 *J. Appl. Mech.* **57**, 232-240.

40. F. Bélanger, E. de Langre, F. Axisa, M.P. Païdoussis & D. Mateescu 1994 *J. Fluids Struct.* **8**, 747-770.

41. F. Bélanger, M.P. Païdoussis & E. de Langre 1995 *AIAA J.* **33**, 752-755.

442

42. G. Porcher & E. de Langre 1996 In *Flow-Induced Vibration - 1996*, (eds M.P. Païdoussis *et al.*) PVP-Vol. 328, pp. 199-208, ASME, New York.

43. E. de Langre, F. Bélanger & G. Porcher 1992 In *Axial and Annular Flow-Induced Vibration and Instabilities* (eds M.P. Païdoussis & M.K. Au-Yang), PVP-Vol. 244, ASME, New York.

44. D. Mateescu, A. Mekanik & M.P. Païdoussis 1996 *J. Fluids Struct.* **10**, 57-77.

45. M.P. Païdoussis, A.K. Misra & V.B. Nguyen 1991 *J. Fluids Struct.* **5**, 127-164.

46. V.B. Nguyen, M.P. Païdoussis & A.K. Misra 1993 *J. Fluids Struct.* **7**, 913-930.

47. V.B. Nguyen, M.P. Païdoussis & A.K. Misra 1994 *J. Sound Vib.* **176**, 105-125.

48. M.P. Païdoussis 1966 *J. Fluid Mech.* **26**, 717-736.

49. M.P. Païdoussis 1966 *J. Fluid Mech.* **26**, 737-751.

50. M.P. Païdoussis 1973 *J. Sound Vib.* **29**, 365-385.

51. M.J. Hannoyer & M.P. Païdoussis 1978 *ASME J. Mech. Design* **100**, 328-336.

52. M.J. Hannoyer & M.P. Païdoussis 1979 *J. Appl. Mech.* **46**, 45-51.

53. M.J. Hannoyer & M.P. Païdoussis 1979 *J. Appl. Mech.* **46**, 52-57.

54. M.P. Païdoussis & F.C. Moon 1988 *J. Fluids Struct.* **2**, 567-591.

55. M.P. Païdoussis & C. Semler 1993 *Nonlin. Dynamics* **4**, 655-670.

56. D.M. Tang & E.H. Dowell 1988 *J. Fluids Struct.* **2**, 263-283.

57. G.S. Copeland & F.C. Moon 1992 *J. Fluids Struct.* **6**, 705-718.

58. M.P. Païdoussis 1994 *Proc. 1st Int'l Conf. on Flow Interaction* (eds N.W. Ko *et al.*), pp. 40-53, Hong Kong.

59. M.P. Païdoussis & R.M. Botez 1993 *J. Fluids Struct.* **7**, 719-750.

60. M.P. Païdoussis & R.M. Botez 1995 *Nonlin. Dynamics* **7**, 429-450.

61. F.C. Moon 1992 *Chaotic and Fractal Dynamics*, John Wiley, New York.

62. P. Bergé, Y. Pomeau & C. Vidal 1984 *Order Within Chaos*, John Wiley, New York.

Theoretical and Applied Mechanics 1996
T. Tatsumi, E. Watanabe and T. Kambe (Editors)

Composites: A Myriad of Microstructure Independent Relations

Graeme W. Milton[a]*

[a]Department of Mathematics, The University of Utah,
Salt Lake City, Utah 84112, U.S.A.

Typically, the elastic properties of composite materials are strongly microstructure dependent. So it comes as a pleasant surprise to come across exact formulae for (or linking) effective moduli that are universally valid no matter how complicated the microstructure. Such exact formulae provide useful benchmarks for testing numerical and actual experimental data, and for evaluating the merit of various approximation schemes. This paper presents a sampling of results in the field.

1. INTRODUCTION

The word myriad has its origin as the Greek word for 10,000. It would be a real challenge to present that many microstructure independent relations amongst the effective properties of composites, especially in a short paper. Instead this article presents an appetizer of results pertaining mostly to elasticity and thermoelasticity. The sampling is sufficiently diverse to encompass the main ideas used to generate microstructure independent relations in many different contexts, including thermoelectricity and piezoelectricity, where a host of microstructure independent results have been obtained: see [1–5] and references therein.

2. UNIFORM FIELDS

Consider a bimetal strip. When the temperature is raised the strip bends. This is due to the difference in thermal expansion of the two metals. Now consider the bimetal strip immersed in water. As the pressure in the water is increased the strip bends due to the difference in bulk moduli of the two metals. Now one can imagine applying just the right combination of temperature increase and water pressure increase or decrease so both phases expand at exactly the same rate and there is no distortion. Of course this same argument applies not just to bimetal strips but to any geometric configuration of two isotropic phases and in particular to a two-phase composite.

Suppose the two-phases are isotropic so that a block of phase 1 or phase 2 immersed in a fluid heat bath at temperature T and pressure p expands or contracts isotropically as T and p are varied. Let $\rho_1(T,p)$ and $\rho_2(T,p)$ denote the mass density of phase 1 or

*The author is grateful to Yury Grabovsky for many helpful comments on the manuscript and thanks him, Johan Helsing, and Alexander Movchan for permission to discuss joint work prior to publication. The support of the National Science Foundation through grants DMS-9501025 and DMS-9402763 is gratefully acknowledged.

phase 2 relative to some base temperature T_0 and base pressure p_0: thus $1/\rho_1(T,p)$ and $1/\rho_2(T,p)$ measure the relative change in the volume of each phase as the temperature and pressure changes from (T_0, p_0) to (T,p). According to this definition we have

$$\rho_1(T_0, p_0) = \rho_2(T_0, p_0) = 1. \tag{1}$$

So the two surfaces $\rho_1(T,p)$ and $\rho_2(T,p)$ intersect at $(T,p) = (T_0, p_0)$. Unless the surfaces are tangent at this point they will intersect along a trajectory passing through (T_0, p_0). Along this trajectory $(T(h), p(h))$ parameterized by h both phases expand or contract at an equal rate.

Now suppose a composite is manufactured at the base temperature T_0 and pressure p_0 with no internal residual stress. When this composite is placed in the heat bath at temperature T and pressure p there is no reason to suppose the composite will expand or contract isotropically as T and p are varied. Indeed by considering the example of the bimetal strip it is clear that internal shear stresses and warping can occur. However along the trajectory $(T(h), p(h))$ the composite will expand isotropically and its density relative to its density at the base temperature and pressure will be

$$\rho_*(T(h), p(h)) = \rho_1(T(h), p(h)) = \rho_2(T(h), p(h)) \quad \forall h. \tag{2}$$

Rewriting this relation as $1/\rho_*(T(h), p(h)) = 1/\rho_1(T(h), p(h)) = 1/\rho_2(T(h), p(h))$ and differentiating with respect to h, gives

$$3\alpha_* \frac{dT(h)}{dh} - \frac{1}{\kappa_*}\frac{dp(h)}{dh} = 3\alpha_1 \frac{dT(h)}{dh} - \frac{1}{\kappa_1}\frac{dp(h)}{dh} = 3\alpha_2 \frac{dT(h)}{dh} - \frac{1}{\kappa_2}\frac{dp(h)}{dh}, \tag{3}$$

where

$$\kappa_a(T,p) = -\left\{\frac{\partial(1/\rho_a)}{\partial p}\right\}^{-1}, \quad \alpha_a(T,p) = \frac{1}{3}\frac{\partial(1/\rho_a)}{\partial T}, \quad a = 1, 2 \text{ or } *, \tag{4}$$

are the tangent bulk moduli and thermal expansion constants of the phases and composite along the trajectory. Provided the trajectory has been suitably parameterized dp/dh and dT/dh will not be both zero. Hence the determinant of the system of equations (3) must vanish which gives the well-known relation

$$\alpha_* = \frac{\alpha_1(1/\kappa_* - 1/\kappa_2) - \alpha_2(1/\kappa_* - 1/\kappa_1)}{1/\kappa_1 - 1/\kappa_2}, \tag{5}$$

between effective bulk moduli and effective thermal expansion coefficients due to Levin [6]. Thus (2) is a non-linear generalization of Levin's formula. This simple observation is joint but unpublished work with J. Berryman presented in Pittsburgh in 1994 at the SIAM Meeting on Mathematics and Computation in the Materials Sciences.

There is another viewpoint which sheds light on Levin's work. Let us begin with a result that applies to elasticity, and not just to thermoelasticity. Suppose the elasticity tensor field of a composite has the property that there exist symmetric matrices v and w with $C(x)v = w$ for all x. Then the uniform strain field which equals v everywhere and the uniform stress field which equals w everywhere are solutions of the elasticity equations. Consequently the effective tensor C_* must satisfy $C_*v = w$. (This remark was made to me by A. Cherkaev, although in the context of thermal expansion it dates back to work of Cribb [7]; see also Dvorak [8] who extended the idea.) In an isotropic polycrystal where the pure crystal has cubic symmetry this condition is satisfied, with $v = I$, and the result implies Hill's microstructure independent formula [9] for the effective bulk modulus of such a polycrystal.

A corollary is that if in a two phase composite the tensor $C_1 - C_2$ is singular with

$$(C_1 - C_2)v = 0 \quad \text{then} \quad (C_* - C_2)v = 0. \tag{6}$$

Now consider the equations of thermoelasticity. These take the form

$$\begin{pmatrix} \epsilon(x) \\ \varsigma(x) \end{pmatrix} = \begin{pmatrix} S(x) & \alpha(x) \\ \alpha(x)^T & c_p(x)/T_0 \end{pmatrix} \begin{pmatrix} \sigma(x) \\ \theta \end{pmatrix} \quad \text{with} \quad \nabla \cdot \sigma = 0, \; \epsilon = [\nabla u + (\nabla u)^T]/2, \tag{7}$$

where $\theta = T - T_0$ is the change in temperature T measured from some constant base temperature T_0, $\epsilon(x)$ and $\sigma(x)$ are the strain and stress fields, $\varsigma(x)$ is the local increase in entropy per unit volume over the entropy of the state where $\sigma = \theta = 0$, $S(x)$ is the compliance tensor, $\alpha(x)$ is the tensor of thermal expansion and $c_p(x)$ is specific heat per unit volume at constant stress. Macroscopically the average fields satisfy

$$\begin{pmatrix} \langle \epsilon \rangle \\ \langle \varsigma \rangle \end{pmatrix} = \begin{pmatrix} S_* & \alpha_* \\ \alpha_*^T & c_{*p}/T_0 \end{pmatrix} \begin{pmatrix} \langle \sigma \rangle \\ \theta \end{pmatrix}, \tag{8}$$

and this serves to define the effective compliance tensor S_*, the effective tensor of thermal expansion S_*, and the effective constant of specific heat at constant stress c_{*p}.

The important observation is that because the entropy field $\varsigma(x)$ is not subject to any differential constraints, we can ignore it completely when computing the effective compliance tensor and effective thermal expansion tensor. In other words it suffices to work with the reduced set of equations

$$\epsilon(x) = M(x) \begin{pmatrix} \sigma(x) \\ \theta \end{pmatrix} \quad \text{with} \quad \nabla \cdot \sigma = 0, \; \epsilon = [\nabla u + (\nabla u)^T]/2, \tag{9}$$

where for three-dimensional thermoelasticity $M(x) = (S(x) \; \alpha(x))$ is representable as a 6×7 matrix in an appropriate basis. Now in a two phase composite the matrix $M_1 - M_2$, like any matrix which has more columns than rows, is necessarily singular, i.e. there necessarily exists a v such that $(M_1 - M_2)v = 0$. Consequently the effective tensor M_* must satisfy $(M_* - M_2)v = 0$. When the phases are isotropic this reduces to Levin's formula (5). When the phases are anisotropic (but each with constant orientation) it reduces to the formula of Rosen and Hashin [10]. Having obtained particular solutions for the stress and strain field with a non-zero value of θ it is an easy matter to determine the corresponding entropy field $\varsigma(x)$ and thereby obtain a formula for the effective constant of specific heat c_{*p} in terms of the effective elasticity tensor [10].

The same reasoning can be applied to obtain an exact expression for effective constant of specific heat and effective thermal expansion tensor in terms of the effective compliance tensor for polycrystaline materials constructed from a single crystal [11,12]. The compliance tensor S_0 and thermal expansion tensor α_0 of the single crystal must be such that $S_0 I$ and α_0 are both uniaxial with a common axis of symmetry. This is ensured if the crystal has hexagonal, tetragonal or trigonal symmetry. The contact between crystals need not necessarily be ideal; slippage along grain boundaries is allowed [13]. When the constitutive relation involves more than one constant field in addition to θ, such as humidity causing expansion due to moisture absorption, then uniform field arguments yield exact relations even when $S_0 I$ and α_0 are biaxial, provided they share the same three principal axes [14].

There is also a direct mathematical correspondence between the equations of poroelasticity and those of thermoelasticity (the fluid pressure plays the role of the temperature). Consequently these exact microstructure independent relations extend to effective poroelastic moduli of two phase media [15,16].

Uniform field arguments are also important three dimensional elastic two phase media when the microstructure and stress and strain fields are independent of the x_3 coordinate. The constitutive relation can be expressed in the form

$$\begin{pmatrix} g(x) \\ \sigma_{33}(x) \end{pmatrix} = \begin{pmatrix} L(x) & \alpha(x) \\ \alpha(x)^T & C_{3333}(x) \end{pmatrix} \begin{pmatrix} h(x) \\ \epsilon_{33} \end{pmatrix}, \tag{10}$$

where

$$L = \begin{pmatrix} C_{1111} & C_{1122} & \sqrt{2}C_{1112} & \sqrt{2}C_{1123} & -\sqrt{2}C_{1113} \\ C_{1122} & C_{2222} & \sqrt{2}C_{2212} & \sqrt{2}C_{2223} & -\sqrt{2}C_{2213} \\ \sqrt{2}C_{1112} & \sqrt{2}C_{2212} & 2C_{1212} & 2C_{2312} & -2C_{1312} \\ \sqrt{2}C_{1123} & \sqrt{2}C_{2223} & 2C_{2312} & 2C_{2323} & -2C_{2313} \\ -\sqrt{2}C_{1113} & -\sqrt{2}C_{2213} & -2C_{1312} & -2C_{2313} & 2C_{1313} \end{pmatrix}, \tag{11}$$

and

$$\alpha = \begin{pmatrix} C_{1133} \\ C_{2233} \\ \sqrt{2}C_{3312} \\ \sqrt{2}C_{3323} \\ -\sqrt{2}C_{3313} \end{pmatrix}, \quad g = \begin{pmatrix} \sigma_{11} \\ \sigma_{22} \\ \sqrt{2}\sigma_{12} \\ E_1' \\ E_2' \end{pmatrix}, \quad h = \begin{pmatrix} \epsilon_{11} \\ \epsilon_{22} \\ \sqrt{2}\epsilon_{12} \\ D_1' \\ D_2' \end{pmatrix}, \tag{12}$$

in which $E_1'(x_1, x_2)$, $E_2'(x_1, x_2)$, $D_1'(x_1, x_2)$, and $D_2'(x_1, x_2)$ are components of the vector fields

$$E' = \begin{pmatrix} E_1' \\ E_2' \end{pmatrix} = \begin{pmatrix} \sqrt{2}\sigma_{23} \\ -\sqrt{2}\sigma_{13} \end{pmatrix}, \quad D' = \begin{pmatrix} D_1' \\ D_2' \end{pmatrix} = \begin{pmatrix} \sqrt{2}\epsilon_{23} \\ -\sqrt{2}\epsilon_{13} \end{pmatrix}. \tag{13}$$

Here the $C_{ijk\ell}$ are the cartesian components of the elasticity tensor field $C(x_1, x_2)$. Let us also introduce two-dimensional stress and strain fields

$$\sigma' = \begin{pmatrix} \sigma_{11} & \sigma_{12} \\ \sigma_{12} & \sigma_{22} \end{pmatrix}, \quad \epsilon' = \begin{pmatrix} \epsilon_{11} & \epsilon_{12} \\ \epsilon_{12} & \epsilon_{22} \end{pmatrix}. \tag{14}$$

Now the three dimensional displacement field is necessarily of the form $u(x) = v(x_1, x_2) + x_3 w$ where w is constant, and this together with the constraints on the three dimensional stress field $\sigma(x)$ implies that

$$\nabla \times E' = 0, \quad \nabla \cdot D' = 0, \quad \nabla \cdot \sigma' = 0, \quad \epsilon = [\nabla u' + (\nabla u')^T]/2, \tag{15}$$

and that ϵ_{33} [like θ in the thermoelastic problem (7)] is constant, where $u' = (v_1, v_2)$ is a two dimensional displacement field. Thus if we interpret E' and D' as two-dimensional electric and electric displacement fields, then the equation $g = Lh$ can be regarded as a two-dimensional piezoelectric equation incorporating a positive definite symmetric tensor $L(x)$, which will have an associated effective tensor L_*. Since the field $\sigma_{33}(x_1, x_2)$ is not subject to any differential constraints we can drop it from the equation (10). By applying the uniform field argument we thereby obtain expressions for the components of the three-dimensional effective elasticity tensor C_* in terms of the components of the two-dimensional effective piezoelectric tensor L_*, assuming it is a two-phase medium. If the medium has more than two phases then, by setting $\epsilon_{33} = 0$, 15 of the 21 components of C_* can be determined from the elements of L_*.

When the elasticity tensors C_1 and C_2 of the two phases are both invariant under the reflection transformation $x_3 \rightarrow -x_3$ then the two-dimensional piezelectric problem decouples into a planar elastic problem and a two-dimensional dielectric problem (the antiplane elastic problem). In particular if both phases are elastically isotropic and the composite is transversely isotropic, then the relation between C_* and L_* reduces to Hill's formulae [17] for C_* in terms of the effective bulk and shear moduli of the planar elastic problem and the effective axial shear modulus of the antiplane elastic problem.

3. TRANSLATING BY A NULL–LAGRANGIAN

The translation discussed below originates in the work of Lurie and Cherkaev [18] on the plate equation. Its existence accounts [19] for certain invariance properties of the stress field discovered by Dundurs [20]. More general stress invariance properties, under a linear rather than a constant shift of the compliance tensor, have recently been discovered by Dundurs and Markenscoff [21]. The following analysis is largely based on the papers of Cherkaev, Lurie and Milton [22] and Thorpe and Jasiuk [19].

In a two-dimensional, simply-connected, possibly inhomogeneous, elastic body with no body forces present the components of the stress field σ can be expressed in terms of a potential ϕ, known as the Airy stress function, through the equations $\sigma_{11} = \phi_{,22}$, $\sigma_{12} = -\phi_{,12}$ and $\sigma_{22} = \phi_{,11}$. Here as elsewhere we use a comma in a subscript to denote differentiation with respect to the indices that follow the comma: thus, for example, $\phi_{,22} = \partial^2 \phi / \partial x_2^2$. The relation between the stress and Airy stress function can be expressed in the equivalent form

$$\sigma = \mathcal{R} \nabla \nabla \phi \quad \text{where} \quad \nabla \nabla \phi = \begin{pmatrix} \phi_{,11} & \phi_{,12} \\ \phi_{,12} & \phi_{,22} \end{pmatrix}, \tag{16}$$

and $\mathcal{R}$ is the fourth order tensor with cartesian elements

$$\mathcal{R}_{ijk\ell} = \delta_{ij}\delta_{k\ell} - (\delta_{ik}\delta_{j\ell} + \delta_{i\ell}\delta_{jk})/2, \tag{17}$$

whose action in two-dimensions is to rotate a matrix by 90°. Now the key point is to recognize that $\nabla \nabla \phi$ satisfies the same differential constraints as a strain field: it derives from the "displacement field" $\nabla \phi$. In other words, the stress field rotated at each point by 90° produces a strain field. We use this observation to rewrite any solution of the two-dimensional elasticity equations

$$\epsilon = \mathcal{S}\sigma, \quad \epsilon = [\nabla u + (\nabla u)^T]/2, \quad \nabla \cdot \sigma = 0, \tag{18}$$

in the equivalent form

$$\epsilon' = \boldsymbol{S}'\sigma, \quad \epsilon' = [\nabla \boldsymbol{u}' + (\nabla \boldsymbol{u}')^T]/2, \quad \nabla \cdot \sigma = 0, \tag{19}$$

where

$$\boldsymbol{S}' = \boldsymbol{S} + t\boldsymbol{R}, \qquad \boldsymbol{u}' = \boldsymbol{u} + \nabla\phi. \tag{20}$$

Evidently if the strain and stress fields ϵ and σ solve the elasticity equations in a medium with compliance tensor $\boldsymbol{S}$ then the strain and stress fields ϵ' and σ solve the elasticity equations in a translated medium with compliance tensor $\boldsymbol{S}'$. The basic Euler-Lagrange equations for the Airy stress function are the same in both media: hence the name null-Lagrangian. General characterizations of null-Lagrangians, or quasicontinuous functionals have been given by Ball, Currie and Olver [23] and by Murat [24].

When the medium under consideration is a composite, we have from (19) that

$$\langle \epsilon' \rangle = \langle \boldsymbol{S}\sigma \rangle + t\boldsymbol{R}\langle \sigma \rangle = (\boldsymbol{S}_* + t\boldsymbol{R})\langle \sigma \rangle. \tag{21}$$

Since it is this linear relation which defines the effective tensor $\boldsymbol{S}'_*$ of the translated medium we deduce that

$$\boldsymbol{S}'_* = \boldsymbol{S}_* + t\boldsymbol{R}. \tag{22}$$

Thus the effective tensor undergoes precisely the same translation as the local tensor.

For example, consider a locally isotropic planar elastic material which is macroscopically elastically isotropic. The local compliance tensor $\boldsymbol{S}(\boldsymbol{x})$ and effective compliance tensor $\boldsymbol{S}_*$ have elements

$$\begin{aligned} S_{ijk\ell}(\boldsymbol{x}) &= (\delta_{ik}\delta_{j\ell} + \delta_{i\ell}\delta_{jk})/2E(\boldsymbol{x}) - [\delta_{ij}\delta_{k\ell} - (\delta_{ik}\delta_{j\ell} + \delta_{i\ell}\delta_{jk})/2]\nu(\boldsymbol{x})/E(\boldsymbol{x}), \\ S^*_{ijk\ell}(\boldsymbol{x}) &= (\delta_{ik}\delta_{j\ell} + \delta_{i\ell}\delta_{jk})/2E_* - [\delta_{ij}\delta_{k\ell} - (\delta_{ik}\delta_{j\ell} + \delta_{i\ell}\delta_{jk})/2]\nu_*/E_*, \end{aligned} \tag{23}$$

where $E(\boldsymbol{x})$ and E_* are the local and effective in plane Young's modulus and $\nu(\boldsymbol{x})$ and ν_* are the local and effective in plane Poisson's ratio. It follows from (17) and (20) that under translation these moduli transform to

$$E'(\boldsymbol{x}) = E(\boldsymbol{x}), \qquad \nu'(\boldsymbol{x}) = \nu(\boldsymbol{x}) - tE(\boldsymbol{x}), \tag{24}$$

i.e. the Young's modulus remains unchanged, while the ratio of the Poisson's ratio to Young's modulus is shifted uniformly by $-t$. Under this translation the result (22) implies that the effective Young's modulus E_* and Poisson's ratio ν_* transform in a similar fashion,

$$E'_* = E_*, \qquad \nu'_* = \nu_* - tE_*. \tag{25}$$

A nice application of this result is to a metal plate with constant moduli E and ν that has a statistically isotropic distribution of holes punched into it. Under the translation (24) the holes remain holes (since the holes effectively correspond to a material with zero Young's modulus) while the Young's modulus E of the metal is unchanged, and its Poisson's ratio is shifted from ν to $\nu - tE$. By dimensional analysis it is apparent that the ratio, E_*/E can only depend on ν and on the geometry. But (25) implies this ratio remains invariant as t and hence ν varies. We conclude that E_*/E only depends on the geometry, and is not influenced by ν. This result was observed numerically by Day, Snyder, Garboczi and Thorpe [25] and subsequently proved by Cherkaev, Lurie and Milton [22]. The extent to which it holds in three dimensions was explored by Christensen [26].

Moreover when there are so many holes that the plate is about to fall apart, then E_* is close to zero, and (25) implies that ν_* is also independent of ν in this limit [25,19]. This is a striking result: no matter what the geometry of the configuration happens to be (so long as it is just about to fall apart) the effective Poisson's ratio takes a universal value which is independent of both the Young's modulus and the Poisson's ratio of the plate.

Translations are also useful for deriving microstructure independent results in the context of three-dimensional elasticity. The following is an extension of a two-dimensional argument [18,22,19] used to rederive Hill's result [17] for the effective bulk modulus of a locally isotropic planar elastic medium with constant shear modulus. [22,19].

First consider the rather extreme example of a locally isotropic three-dimensional medium with a compliance tensor $\boldsymbol{S}(\boldsymbol{x})$ with cartesian elements

$$S_{ijk\ell}(\boldsymbol{x}) = \delta_{ij}\delta_{k\ell}/9\kappa(\boldsymbol{x}). \tag{26}$$

This material has infinite shear modulus and finite bulk modulus $\kappa(\boldsymbol{x})$, i.e. at each point its Poisson's ratio is -1. (Although seemingly unphysical, materials with Poisson's ratio arbitrarily close to -1 can in fact be constructed: see [27,28] and references therein.) The deformation of the material is conformal since any change of angles corresponds to shear. In three dimensions the only conformal mappings are inversion in a sphere and uniform dilation, and since we are looking for periodic solutions the first can be ruled out. Therefore the strain $\boldsymbol{\epsilon}(\boldsymbol{x})$ must equal $\alpha\boldsymbol{I}$ where α is constant. The three dimensional stress field $\boldsymbol{\sigma}(\boldsymbol{x})$ being symmetric and divergence-free derives from a 3×3 symmetric matrix valued potential $\boldsymbol{\phi}(\boldsymbol{x})$:

$$\boldsymbol{\sigma}(\boldsymbol{x}) = \nabla \times (\nabla \times \boldsymbol{\phi}(\boldsymbol{x}))^T, \tag{27}$$

Substituting this in the constitutive equation $\boldsymbol{S}\boldsymbol{\sigma} = \boldsymbol{\epsilon} = \alpha\boldsymbol{I}$ gives an equation for the matrix potential $\boldsymbol{\phi}$:

$$\text{Tr}[\nabla \times (\nabla \times \boldsymbol{\phi}(\boldsymbol{x}))^T] = 9\kappa(\boldsymbol{x})\alpha. \tag{28}$$

Since this is a single equation it seems plausible to look for solutions of the form

$$\boldsymbol{\phi}(\boldsymbol{x}) = \phi(\boldsymbol{x})\boldsymbol{I} \quad \text{with} \quad \phi(\boldsymbol{x}) = \phi_0(\boldsymbol{x}) + (\boldsymbol{x} \cdot \boldsymbol{x})f, \tag{29}$$

in which the scalar function $\phi_0(\boldsymbol{x})$ is periodic and f is a constant determining the average value of the stress. The associated stress field is

$$\boldsymbol{\sigma} = \boldsymbol{I}\Delta\phi - \nabla\nabla\phi, \tag{30}$$

which when substituted in the constitutive law gives the equation

$$2\Delta\phi_0 = 9\kappa(\boldsymbol{x})\alpha - 6f, \tag{31}$$

for the periodic potential ϕ_0. This has a solution if and only if the average value of the right hand side is zero, which thereby determines the value of f and the associated average value of the stress:

$$f = 3\langle\kappa(\boldsymbol{x})\rangle\alpha/2, \quad \langle\boldsymbol{\sigma}\rangle = 2f\boldsymbol{I} = 3\langle\kappa(\boldsymbol{x})\rangle\alpha\boldsymbol{I} = 3\langle\kappa(\boldsymbol{x})\rangle\langle\epsilon\rangle. \tag{32}$$

Clearly the effective bulk modulus of this composite is microstructure independent and equal to

$$\kappa_* = \langle\kappa(\boldsymbol{x})\rangle. \tag{33}$$

Now let $\boldsymbol{\mathcal{T}}$ denote the fourth order tensor with cartesian elements

$$\mathcal{T}_{ijk\ell} = \delta_{ij}\delta_{k\ell}/2 - (\delta_{ik}\delta_{j\ell} + \delta_{i\ell}\delta_{jk})/2. \tag{34}$$

This tensor has the important property that for stresses of the form (30) the field $\boldsymbol{\mathcal{T}}\boldsymbol{\sigma} = \nabla\nabla\phi$ satisfies the same differential constraints as a strain: it derives from the "displacement field" $\nabla\phi$. The translated medium $\boldsymbol{S} + t\boldsymbol{\mathcal{T}}$ now will have three dimensional inverse bulk and shear moduli

$$1/\kappa'(\boldsymbol{x}) = 1/\kappa(\boldsymbol{x}) + 3t/2, \quad 1/\mu' = -2t. \tag{35}$$

By applying the same sort of analysis as before it follows that inverse effective bulk and shear moduli of the translated medium equal

$$1/\kappa'_* = 1/\kappa_* + 3t/2 = 1/\langle\kappa(\boldsymbol{x})\rangle + 3t/2, \quad 1/\mu'_* = -2t. \tag{36}$$

By combining these formulae we see that a locally isotropic elastic medium with constant shear modulus μ' and bulk modulus $\kappa'(\boldsymbol{x})$ has effective shear and bulk moduli μ'_* and κ'_* given by

$$\mu'_* = \mu, \quad \frac{1}{4/\kappa'_* + 3/\mu'} = \left\langle\frac{1}{4/\kappa'(\boldsymbol{x}) + 3/\mu'}\right\rangle \tag{37}$$

which is Hill's result [17].

4. DUALITY FOR ANTIPLANE ELASTICITY

The duality relations for antiplane elasticity have their origins in the work of Keller [29] and Dykhne [30]. Mendelson [31], upon whose work the following treatment is based, extended their analysis to arbitrary inhomogeneous anisotropic planar media.

Let $u(x_1, x_2)$ be the vertical displacement in a state of anti-plane shear and let $\varphi(x_1, x_2)$ be the shear stress potential. The equations of antiplane elasticity take the form

$$\begin{pmatrix} \sigma_{31} \\ \sigma_{32} \end{pmatrix} = \boldsymbol{m}(\boldsymbol{x})\begin{pmatrix} 2\epsilon_{31} \\ 2\epsilon_{32} \end{pmatrix}, \quad \begin{pmatrix} \sigma_{31} \\ \sigma_{32} \end{pmatrix} = \begin{pmatrix} \varphi_{,2} \\ -\varphi_{,1} \end{pmatrix}, \quad \begin{pmatrix} 2\epsilon_{31} \\ 2\epsilon_{32} \end{pmatrix} = \begin{pmatrix} u_{,1} \\ u_{,2} \end{pmatrix}, \tag{38}$$

where $\boldsymbol{m}(x_1, x_2)$ is a symmetric 2×2 anti-plane shear elasticity matrix. Now let us introduce a new vertical displacement $u'(x_1, x_2) = \varphi(x_1, x_2)$ and a new shear stress potential

$\varphi'(x_1, x_2) = -u(x_1, x_2)$. The associated stress and strain field components satisfy the relations

$$\begin{pmatrix} \sigma'_{31} \\ \sigma'_{32} \end{pmatrix} = \begin{pmatrix} \varphi'_{,2} \\ -\varphi'_{,1} \end{pmatrix} = \boldsymbol{R}_\perp \begin{pmatrix} 2\epsilon_{31} \\ 2\epsilon_{32} \end{pmatrix}, \quad \begin{pmatrix} 2\epsilon'_{31} \\ 2\epsilon'_{32} \end{pmatrix} = \begin{pmatrix} u'_{,1} \\ u'_{,2} \end{pmatrix} = \boldsymbol{R}_\perp \begin{pmatrix} \sigma_{31} \\ \sigma_{32} \end{pmatrix}, \tag{39}$$

where

$$\boldsymbol{R}_\perp = \begin{pmatrix} 0 & 1 \\ -1 & 0 \end{pmatrix} \tag{40}$$

is the matrix for a 90° rotation. In two dimensions a curl free vector field when rotated pointwise by 90° produces a divergence free vector field and vice-versa. This key fact explains why the new stress and strain fields given by (39) satisfy the required differential constraints. These fields are linked through the constitutive relation

$$\begin{pmatrix} \sigma'_{31} \\ \sigma'_{32} \end{pmatrix} = \boldsymbol{m}' \begin{pmatrix} 2\epsilon'_{31} \\ 2\epsilon'_{32} \end{pmatrix}, \tag{41}$$

in which

$$\boldsymbol{m}'(\boldsymbol{x}) = [\boldsymbol{R}_\perp^T \boldsymbol{m}(\boldsymbol{x}) \boldsymbol{R}_\perp]^{-1} = \boldsymbol{m}(\boldsymbol{x})/\det[\boldsymbol{m}(\boldsymbol{x})]. \tag{42}$$

In other words these potentials solve the antiplane shear problem in a dual medium with anti-plane shear elasticity matrix $\boldsymbol{m}'(\boldsymbol{x})$.

By taking averages of the fields we deduce that the dual medium has effective shear matrix

$$\boldsymbol{m}'_* = [\boldsymbol{R}_\perp^T \boldsymbol{m}_* \boldsymbol{R}_\perp]^{-1} = \boldsymbol{m}_*/\det[\boldsymbol{m}_*], \tag{43}$$

where $\boldsymbol{m}_*$ is the effective shear matrix of the original medium. Thus duality relations link the effective tensors of two different media. If the medium is a composite of two isotropic phases, then $\boldsymbol{m}(\boldsymbol{x})$ and $\boldsymbol{m}'(\boldsymbol{x})$ take the form

$$\boldsymbol{m}(\boldsymbol{x}) = \chi(\boldsymbol{x})\mu_1 \boldsymbol{I} + (1 - \chi(\boldsymbol{x}))\mu_2 \boldsymbol{I}, \quad \boldsymbol{m}'(\boldsymbol{x}) = [\chi(\boldsymbol{x})\mu_2 \boldsymbol{I} + (1 - \chi(\boldsymbol{x}))\mu_1 \boldsymbol{I}]/\mu_1\mu_2, \tag{44}$$

where μ_1 and μ_2 are the shear moduli of the two phases and $\chi(\boldsymbol{x})$ is the characteristic function representing the microstructure of phase 1 (taking the value 1 when $\boldsymbol{x}$ is in phase 1 and zero otherwise.) So, the phase interchanged medium is obtained from the dual medium by multiplying $\boldsymbol{m}'(\boldsymbol{x})$ by the factor $\mu_1\mu_2$ and therefore its effective tensor is obtained by multiplying $\boldsymbol{m}'_*$ by the same factor $\mu_1\mu_2$. If we consider the effective tensor $\boldsymbol{m}_*$ as a function $\boldsymbol{m}_*(\mu_1, \mu_2)$ of the two-phases then (43) implies

$$\boldsymbol{m}_*(\mu_2, \mu_1) = \mu_1\mu_2 \boldsymbol{m}_*(\mu_1, \mu_2)/\det[\boldsymbol{m}_*(\mu_1, \mu_2)]. \tag{45}$$

It may happen that the geometry is phase interchange invariant like a checkerboard. Then $\boldsymbol{m}_*(\mu_2, \mu_1) = \boldsymbol{m}_*(\mu_1, \mu_2)$ and it follows from the above equation that $\det[\boldsymbol{m}_*(\mu_1, \mu_2)] = \mu_1\mu_2$. In particular if the effective shear matrix is isotropic then we have $\boldsymbol{m}_* = \mu_* \boldsymbol{I}$ where $\mu_* = \sqrt{\mu_1\mu_2}$ [30]. By this procedure we have obtained an exact expression for the effective antiplane shear modulus μ_* of the composite without solving for the fields directly.

5. DUALITY FOR PLANAR ELASTICITY

The duality relations for incompressible planar elastic media discussed here are due to Berdichevski [32]. The extension of the duality relations to compressible planar elastic media with a constant bulk modulus and to certain other anisotropic planar media is due to Helsing, Milton and Movchan [33].

Consider a planar elastic medium which is incompressible at each point. Since $\nabla \cdot \boldsymbol{u} = 0$ there exists a potential $\psi(\boldsymbol{x})$ such that $u_1 = \psi_{,2}$ and $u_2 = -\psi_{,1}$. Let us introduce the matrices

$$\boldsymbol{a}_1 = \frac{1}{\sqrt{2}} \begin{pmatrix} 1 & 0 \\ 0 & 1 \end{pmatrix}, \quad \boldsymbol{a}_2 = \frac{1}{\sqrt{2}} \begin{pmatrix} 1 & 0 \\ 0 & -1 \end{pmatrix}, \quad \boldsymbol{a}_3 = \frac{1}{\sqrt{2}} \begin{pmatrix} 0 & 1 \\ 1 & 0 \end{pmatrix}, \tag{46}$$

as a basis on the space of 2×2 symmetric matrices. The two-dimensional stress and strain fields $\boldsymbol{\sigma}(\boldsymbol{x})$ and $\boldsymbol{\epsilon}(\boldsymbol{x})$ can be expanded in this basis,

$$\boldsymbol{\sigma}(\boldsymbol{x}) = \sigma_1(\boldsymbol{x})\boldsymbol{a}_1 + \sigma_2(\boldsymbol{x})\boldsymbol{a}_2 + \sigma_3(\boldsymbol{x})\boldsymbol{a}_3, \quad \boldsymbol{\epsilon}(\boldsymbol{x}) = \epsilon_2(\boldsymbol{x})\boldsymbol{a}_2 + \epsilon_3(\boldsymbol{x})\boldsymbol{a}_3, \tag{47}$$

where the coefficients satisfy the equations

$$\begin{pmatrix} \epsilon_2 \\ \epsilon_3 \end{pmatrix} = \boldsymbol{S}(\boldsymbol{x}) \begin{pmatrix} \sigma_2 \\ \sigma_3 \end{pmatrix}, \quad \begin{pmatrix} \epsilon_2 \\ \epsilon_3 \end{pmatrix} = \frac{1}{\sqrt{2}} \begin{pmatrix} 2\psi_{,12} \\ \psi_{,22} - \psi_{,11} \end{pmatrix}, \quad \begin{pmatrix} \sigma_2 \\ \sigma_3 \end{pmatrix} = \frac{1}{\sqrt{2}} \begin{pmatrix} \phi_{,22} - \phi_{,11} \\ -2\phi_{,12} \end{pmatrix}, \tag{48}$$

and $\sigma_1 = (\phi_{,11} + \phi_{,22})/\sqrt{2}$. Here the 2×2 matrix $\boldsymbol{S}(\boldsymbol{x})$ represents the non-singular part of the compliance tensor in this basis and $\phi(\boldsymbol{x})$ is the Airy stress function. We now introduce dual potentials $\phi'(x_1, x_2) = \psi(x_1, x_2)$ and $\psi'(x_1, x_2) = -\phi(x_1, x_2)$ and the associated stress and strain field components

$$\begin{pmatrix} \epsilon'_2 \\ \epsilon'_3 \end{pmatrix} = \frac{1}{\sqrt{2}} \begin{pmatrix} 2\psi'_{,12} \\ \psi'_{,22} - \psi'_{,11} \end{pmatrix} = \boldsymbol{R}_\perp \begin{pmatrix} \sigma_2 \\ \sigma_3 \end{pmatrix}, \quad \begin{pmatrix} \sigma'_2 \\ \sigma'_3 \end{pmatrix} = \frac{1}{\sqrt{2}} \begin{pmatrix} \phi'_{,22} - \phi'_{,11} \\ -2\phi'_{,12} \end{pmatrix} = \boldsymbol{R}_\perp \begin{pmatrix} \epsilon_2 \\ \epsilon_3 \end{pmatrix}, \tag{49}$$

and $\sigma'_1 = (\phi'_{,11} + \phi'_{,22})/\sqrt{2}$. These field components satisfy the constitutive relation

$$\begin{pmatrix} \epsilon'_2 \\ \epsilon'_3 \end{pmatrix} = \boldsymbol{S}'(\boldsymbol{x}) \begin{pmatrix} \sigma'_2 \\ \sigma'_3 \end{pmatrix} \quad \text{where} \quad \boldsymbol{S}'(\boldsymbol{x}) = [\boldsymbol{R}_\perp^T \boldsymbol{S}(\boldsymbol{x}) \boldsymbol{R}_\perp]^{-1} = \boldsymbol{S}(\boldsymbol{x})/\det[\boldsymbol{S}(\boldsymbol{x})]. \tag{50}$$

In other words the dual potentials solve the planar elasticity equations in an incompressible medium with $\boldsymbol{S}'(\boldsymbol{x})$ being the non-singular part of the compliance tensor in the basis (46). By taking averages of the fields we deduce that the non-singular part of the effective compliance tensor for the dual medium is

$$\boldsymbol{S}'_* = [\boldsymbol{R}_\perp^T \boldsymbol{S}_* \boldsymbol{R}_\perp]^{-1} = \boldsymbol{S}_*/\det[\boldsymbol{S}_*], \tag{51}$$

in which $\boldsymbol{S}_*$ is the non-singular part of the compliance tensor of the original medium in the basis (46).

As an example, consider an isotropic incompressible planar elastic composite of two isotropic phases. Then the matrices $\boldsymbol{S}(\boldsymbol{x})$ and $\boldsymbol{S}_*$ take the form

$$\boldsymbol{S}(\boldsymbol{x}) = \chi(\boldsymbol{x})\boldsymbol{I}/(2\mu_1) + (1 - \chi(\boldsymbol{x}))\boldsymbol{I}/(2\mu_2), \quad \boldsymbol{S}_* = \boldsymbol{I}/(2\mu_*), \tag{52}$$

where μ_1, μ_2 and μ_* are the shear moduli of the phases and composite, while $\chi(\boldsymbol{x})$ is the characteristic function representing the microstructure of phase 1 (taking the value 1 when $\boldsymbol{x}$ is in phase 1 and zero otherwise.) Then, by direct analogy with the relation (45) for antiplane shear, we see that the effective shear modulus μ_* as a function $\mu_*(\mu_1, \mu_2)$ of the shear moduli of the phases satisfies the phase interchange relation $\mu_*(\mu_1, \mu_2)\mu_*(\mu_2, \mu_1) = \mu_1\mu_2$. By translation we can extend this result to two phase planar elastic composites that have compressible phases sharing a common bulk modulus κ. Then the phase interchange relation takes the form

$$E_*(E_1, E_2)E_*(E_2, E_1) = E_1E_2, \tag{53}$$

where $E_*(E_1, E_2)$ is the effective in plane Young's modulus expressed as a function of the in-plane Young's moduli E_1 and E_2 of the two phases. In particular if the composite is phase interchange invariant, like a two-dimensional checkerboard, then (53) implies its effective in plane Young's modulus is $\sqrt{E_1E_2}$. If the bulk modulus is not the same in both phases then Gibiansky and Torquato [34] have shown that the effective elastic moduli of the composite and phase interchanged material are linked by inequalities which reduce to the relation (53) when the bulk moduli are equal.

A related example is that of an isotropic two-dimensional polycrystal of incompressible crystals. The individual crystals, being incompressible, necessarily have square symmetry and are characterized by two shear moduli $\mu^{(1)}$ and $\mu^{(2)}$. The duality result (51) implies that the effective shear modulus μ_* of the polycrystal is given by the formula $\mu_* = \sqrt{\mu^{(1)}\mu^{(2)}}$ of Lurie and Cherkaev [18]. Using translations they generalized this result to two-dimensional polycrystals, comprised of compressible grains with square symmetry and found that the effective shear modulus μ_* of the polycrystal is given by

$$\mu_* = \frac{\kappa}{-1 + \sqrt{(\kappa + \mu^{(2)})(\kappa + \mu^{(1)})/(\mu^{(1)}\mu^{(2)})}}, \tag{54}$$

where κ, $\mu^{(1)}$ and $\mu^{(2)}$ are the planar bulk and two shear moduli of the crystal.

More generally, planar elastic duality transformations can be applied whenever there exists a matrix $\boldsymbol{v}$ and constant t such that $(\boldsymbol{S}(\boldsymbol{x}) - t\boldsymbol{R})\boldsymbol{v} = 0$ for all $\boldsymbol{x}$ [33]. High accuracy numerical results for the effective compliance tensor of periodic media comprised of two orthotropic phases confirm the predictions of the theory.

6. LINKING ANTIPLANE AND PLANAR ELASTICITY PROBLEMS

Given that duality relations hold for both antiplane and planar elasticity, one might wonder if these problems are are linked in some way. Such a link would be a surprise because antiplane problems involve a second order shear matrix, whereas planar elastic problems involve a fourth order elasticity tensor. A formal similarity between incompressible elasticity and antiplane elasticity is known [35] but this does not provide a correspondence between the fields solving the planar and antiplane problems. Here we establish a direct correspondence. The ensuing analysis is based on the papers of Milton and Movchan [36,37] and Helsing, Milton and Movchan [33].

The constitutive relation in a simply connected, planar, locally orthotropic medium, with the axes of orthotropy aligned with the coordinate axes takes the form

$$
\begin{pmatrix} u_{1,1} \\ u_{2,2} \\ (u_{1,2} + u_{2,1})/\sqrt{2} \end{pmatrix} = S \begin{pmatrix} \sigma_{11} \\ \sigma_{22} \\ \sqrt{2}\sigma_{21} \end{pmatrix}, \quad S = \begin{pmatrix} s_1 & s_2 & 0 \\ s_2 & s_4 & 0 \\ 0 & 0 & s_6 \end{pmatrix},
\tag{55}
$$

and the equilibrium constraint $\nabla \cdot \sigma = 0$ implies there exist stress potentials $\phi_1(\boldsymbol{x})$ and $\phi_2(\boldsymbol{x})$ such that

$$
\begin{pmatrix} \sigma_{11} & \sigma_{12} \\ \sigma_{21} & \sigma_{22} \end{pmatrix} = \begin{pmatrix} \phi_{1,2} & \phi_{2,2} \\ -\phi_{1,1} & -\phi_{2,1} \end{pmatrix}.
\tag{56}
$$

Let us substitute these expressions back into the constitutive law and into the relation $\sigma_{12} = \sigma_{21}$, implied by symmetry of the stress field. Manipulating the resulting four equations so the terms involving derivatives with respect to x_2 appear on the left while terms involving derivatives with respect to x_1 appear on the right gives an equivalent form of the elasticity equations

$$
\boldsymbol{\eta}_{,2} = \boldsymbol{N}\boldsymbol{\eta}_{,1}
\tag{57}
$$

introduced by Ingebrigtsen and Tonning [38], where

$$
\boldsymbol{\eta} = \begin{pmatrix} u_1 \\ u_2 \\ \phi_1 \\ \phi_2 \end{pmatrix}, \quad \boldsymbol{N} = \begin{pmatrix} 0 & -1 & -s_6 & 0 \\ s_2/s_1 & 0 & 0 & s_2^2/s_1 - s_4 \\ 1/s_1 & 0 & 0 & s_2/s_1 \\ 0 & 0 & -1 & 0 \end{pmatrix}.
\tag{58}
$$

The matrix $\boldsymbol{N}(\boldsymbol{x})$ is known as the fundamental elasticity matrix. The associated effective fundamental elasticity matrix $\boldsymbol{N}_*$ governs the relation between the average fields,

$$
\langle \boldsymbol{\eta}_{,2} \rangle = \boldsymbol{N}_* \langle \boldsymbol{\eta}_{,1} \rangle,
\tag{59}
$$

and is related to the effective compliance matrix $\boldsymbol{S}_*$ in the same way that the fundamental elasticity matrix $\boldsymbol{N}(\boldsymbol{x})$ is related to the local compliance matrix $\boldsymbol{S}(\boldsymbol{x})$.

Now notice that the equations can be rewritten in the equivalent form

$$
\boldsymbol{\eta}'_{,2} = \boldsymbol{N}'\boldsymbol{\eta}'_{,1}, \quad \langle \boldsymbol{\eta}'_{,2} \rangle = \boldsymbol{N}'_* \langle \boldsymbol{\eta}'_{,1} \rangle,
\tag{60}
$$

where

$$
\boldsymbol{\eta}'(\boldsymbol{x}) = \boldsymbol{K}\boldsymbol{\eta}(\boldsymbol{x}), \quad \boldsymbol{N}'(\boldsymbol{x}) = \boldsymbol{K}^{-1}\boldsymbol{N}(\boldsymbol{x})\boldsymbol{K}, \quad \boldsymbol{N}'_* = \boldsymbol{K}^{-1}\boldsymbol{N}_*\boldsymbol{K},
\tag{61}
$$

and $\boldsymbol{K}$ is an arbitrary constant, non-singular 4×4 matrix. In other words, when the fundamental matrix field $\boldsymbol{N}(\boldsymbol{x})$ undergoes a constant similarity transformation then the effective fundamental matrix $\boldsymbol{N}_*$ undergoes the same similarity transformation. The translation of the compliance tensor discussed in section 3 corresponds a particular similarity transformation of the fundamental matrix, as do the duality transformations for antiplanar and planar elasticity. (For antiplane elasticity the associated fundamental matrix is a 2×2 matrix). Other duality transformations of the fundamental matrix form of the equations have been analysed by Nemat-Nasser and Ni [39].

Similar mappings between equivalent sets of equations, obtained by taking linear combinations of potentials and fluxes separately, have been applied to coupled field problems by Straley [1] and Milgrom and Shtrikman [2] among others. For media with isotropic phases they use these mappings to transform to a diagonal form of the equations where no couplings are present, and thereby obtain exact relations between the effective thermoelectric moduli in a two-phase medium. By mixing the potentials and flux potentials one obtains a more general class of equivalence transformations for two-dimensional coupled field problems (see Milton [40], Benveniste [5] and references therein). Under these the tensor entering the constitutive law undergoes a fractional linear transformation. Such transformations are equivalent to the similarity transformations of the fundamental matrices considered here. Working with the fundamental matrices has the advantage that the transformation takes a simpler form and is therefore easier to analyze.

For simplicity, let us suppose the moduli are such that for all $\boldsymbol{x}$

$$\Delta(\boldsymbol{x}) = (s_2(\boldsymbol{x}) + s_6(\boldsymbol{x}))^2 - s_1(\boldsymbol{x})s_4(\boldsymbol{x}) > 0. \tag{62}$$

Then the eigenvalues of the $\boldsymbol{N}(\boldsymbol{x})$ at each point $\boldsymbol{x}$ are

$$\lambda_1 = -\lambda_2 = -i\alpha_1, \quad \lambda_3 = -\lambda_4 = -i\alpha_2, \tag{63}$$

where $\alpha_1(\boldsymbol{x})$ and $\alpha_2(\boldsymbol{x})$ are the two real positive roots of the polynomial

$$s_1(\boldsymbol{x})\alpha^4 - 2(s_2(\boldsymbol{x}) + s_6(\boldsymbol{x}))\alpha^2 + s_4(\boldsymbol{x}) = 0. \tag{64}$$

The corresponding eigenvectors are

$$\boldsymbol{v}_1 = \begin{pmatrix} -p_1 \\ i\alpha_1 p_2 \\ i\alpha_1 \\ 1 \end{pmatrix}, \quad \boldsymbol{v}_2 = \begin{pmatrix} -p_1 \\ -i\alpha_1 p_2 \\ -i\alpha_1 \\ 1 \end{pmatrix}, \quad \boldsymbol{v}_3^{(j)} = \begin{pmatrix} -p_2 \\ i\alpha_2 p_1 \\ i\alpha_2 \\ 1 \end{pmatrix}, \quad \boldsymbol{v}_4^{(j)} = \begin{pmatrix} -p_2 \\ -i\alpha_2 p_1 \\ -i\alpha_2 \\ 1 \end{pmatrix}, \tag{65}$$

in which

$$p_1(\boldsymbol{x}) = -s_6(\boldsymbol{x}) + \sqrt{\Delta(\boldsymbol{x})}/2, \qquad p_2(\boldsymbol{x}) = -s_6(\boldsymbol{x}) - \sqrt{\Delta(\boldsymbol{x})}/2. \tag{66}$$

Now suppose p_1 and p_2 do not depend on $\boldsymbol{x}$. (This holds if and only if s_6 and Δ are both independent of $\boldsymbol{x}$.) Then $\boldsymbol{v}_1$ and $\boldsymbol{v}_2$ will span a two-dimensional space that does not depend on $\boldsymbol{x}$, and $\boldsymbol{v}_3$ and $\boldsymbol{v}_4$ will span a two-dimensional space that does not depend on $\boldsymbol{x}$. Thus with an appropriate choice of $\boldsymbol{K}$ the matrix $\boldsymbol{N}'(\boldsymbol{x})$ will be block diagonal. Specifically, the choice

$$\boldsymbol{K} = \begin{pmatrix} -p_1 & 0 & 0 & -p_2 \\ 0 & p_2 & p_1 & 0 \\ 0 & 1 & 1 & 0 \\ 1 & 0 & 0 & -1 \end{pmatrix} \quad \text{gives} \quad \boldsymbol{N}' = \begin{pmatrix} 0 & -1 & 0 & 0 \\ \alpha_1^2 & 0 & 0 & 0 \\ 0 & 0 & 0 & \alpha_2^2 \\ 0 & 0 & -1 & 0 \end{pmatrix}. \tag{67}$$

As a consequence the equation $\boldsymbol{\eta}'_{,2} = \boldsymbol{N}'\boldsymbol{\eta}'_{,1}$ decouples into a pair of equations that can be expressed in the form

$$\begin{pmatrix} \eta'_{2,2} \\ -\eta'_{2,1} \end{pmatrix} = \boldsymbol{m}_1 \begin{pmatrix} \eta'_{1,1} \\ \eta'_{1,2} \end{pmatrix}, \quad \begin{pmatrix} \eta'_{3,2} \\ -\eta'_{3,1} \end{pmatrix} = \boldsymbol{m}_2 \begin{pmatrix} \eta'_{4,1} \\ \eta'_{4,2} \end{pmatrix}, \tag{68}$$

where $\boldsymbol{m}_1(\boldsymbol{x})$ and $\boldsymbol{m}_2(\boldsymbol{x})$ are the 2×2 matrix valued fields

$$\boldsymbol{m}_1 = \begin{pmatrix} \alpha_1^2 & 0 \\ 0 & 1 \end{pmatrix}, \quad \boldsymbol{m}_2 = \begin{pmatrix} \alpha_2^2 & 0 \\ 0 & 1 \end{pmatrix}. \tag{69}$$

These can be regarded as equations of antiplane elasticity in two different inhomogeneous anisotropic media with $\boldsymbol{m}_1(\boldsymbol{x})$ and $\boldsymbol{m}_2(\boldsymbol{x})$ being the antiplane shear matrix fields of these media. In other words, when s_6 is constant and Δ is constant and positive, the original planar elasticity equations can be reduced to a pair of uncoupled antiplane elasticity equations. The uniform field argument implies that when s_6 is constant the effective compliance matrix $\boldsymbol{S}_*$ is necessarily orthotropic with its axes aligned with the co-ordinate axes having $s_{*6} = s_6$. From the effective antiplane shear matrices

$$\boldsymbol{m}_{*1} = \begin{pmatrix} \alpha_{*1}^2 & 0 \\ 0 & 1 \end{pmatrix}, \quad \boldsymbol{m}_{*2} = \begin{pmatrix} \alpha_{*2}^2 & 0 \\ 0 & 1 \end{pmatrix}, \tag{70}$$

associated with $\boldsymbol{m}_1(\boldsymbol{x})$ and $\boldsymbol{m}_2(\boldsymbol{x})$ we can compute the remaining elements s_{*1} s_{*2} and s_{*4} of the effective compliance matrix $\boldsymbol{S}_*$ associated with $\boldsymbol{S}(\boldsymbol{x})$ by solving the three equations

$$(s_{*2} + s_{*6})^2 - s_{*1}s_{*4} = \Delta, \quad s_{*1}\alpha_{*j}^4 - 2(s_{*2} + s_{*6})\alpha_{*j}^2 + s_{*4} = 0, \quad j = 1, 2. \tag{71}$$

This correspondence between the moduli of the effective antiplane shear matrices and the moduli of the effective compliance tensor has been verified numerically [33]. When s_6 is constant and Δ is constant and negative there is still a correspondence with antiplane elasticity: the original planar elasticity equations can then be reduced to a single viscoelastic antiplane problem, with a complex shear matrix field $\boldsymbol{m}(\boldsymbol{x})$.

7. A QUESTION OF PERCOLATION

The following is joint work with Yury Grabovsky. For more details, and for references to the relevant results from partial differential equation theory, see [41].

When we think of percolation it is usually in the context of current flow, where the current may be electrical, thermal or fluid flow. Perhaps one is considering a composite of two isotropic phases where phase 1 is permeable to current while phase 2 is impermeable to it. Or perhaps one is considering a polycrystal where each individual crystalline grain only allows the current to flow in certain directions within that crystal. At a percolation transition the rank of the effective conductivity or permeability tensor changes. For isotropic three-dimensional composites of two isotropic phases. the transition is from a tensor of rank 0, when phase 2 blocks all current flow, to a tensor of rank 3, when paths of phase 1 form a connected labyrinth of infinite extent. In the context of elasticity we can study the analogous percolation question: given that the local compliance tensor (or local elasticity tensor) is singular in some parts or in all of the material, is the rank of the effective compliance tensor (or effective elasticity tensor) dependent on the microstructure?

Curiously, the rank of the effective compliance tensor is independent of the microstructure in a planar elastic material where the local compliance tensor is rank 1, of the form

$$\boldsymbol{S}(\boldsymbol{x}) = \boldsymbol{s}(\boldsymbol{x}) \otimes \boldsymbol{s}(\boldsymbol{x}), \tag{72}$$

where the 2×2 matrix valued field $s(x)$ is positive definite for all x. Specifically, the effective elasticity tensor takes exactly the same rank 1 form,

$$\mathcal{S}_* = s_* \otimes s_*, \tag{73}$$

where s_* is a 2×2 positive definite matrix. The assumption of positive definiteness of $s(x)$ is necessary: it follows from the work Bhattacharya and Kohn [42] (see sections 5.2 and 5.3) that the existance of "percolating" stress or strain fields can be microstructure dependent when $s(x)$ is not positive definite.

This result has a surprising corollary. If we take a two-dimensional planar elastic polycrystal constructed from a crystal with a positive definite compliance tensor $\mathcal{S}_0$ such that the translated tensor $\mathcal{S}_0' = \mathcal{S}_0 - t\mathcal{R}$ is rank 1 of the form $s_0' \otimes s_0'$ for some value of t, then necessarily the effective compliance tensor $\mathcal{S}_*$ of the polycrystal must be such that the translated effective tensor $\mathcal{S}_*' = \mathcal{S}_* - t\mathcal{R}$ is rank 1 of the form $s_*' \otimes s_*'$. (The positive definiteness of the tensor $\mathcal{S}_0$ ensures the positive definiteness of the matrix s_0'.) In particular, if the polycrystal is elastically isotropic then its two-dimensional shear modulus is microstructure independent and equal to $1/(2t)$. The bulk modulus, by contrast, is microstructure dependent. This result was first derived [43] from the optimal bounds on the bulk and shear moduli of two-dimensional planar elastic polycrystals constructed from an orthotropic crystal.

A related result is that if a planar elasticity tensor field $\mathcal{C}(x)$ is rank 2 with a positive definite matrix in its null-space for all x, then the associated effective elasticity tensor $\mathcal{C}_*$ is rank 2 with a positive definite matrix in its null-space.

The proof of (73) is technical and rests upon certain results from elliptic partial differential equation theory. Only those readers interested in getting a rough idea of the steps involved should read the remainder of this section as it is rather condensed.

If we apply an average stress field such that the resulting strain field in the material is non-zero then this "percolating" strain field $\epsilon(x)$ must be of the form

$$\epsilon(x) = \alpha(x)s(x), \tag{74}$$

for some scalar field $\alpha(x)$. The infinitesimal strain compatibility condition that $\nabla \cdot (\nabla \cdot \mathcal{R}\epsilon(x)) = 0$ (which ensures that $\epsilon(x)$ derives from some displacement field) requires that $\alpha(x)$ satisfies the second order elliptic partial differential equation,

$$\nabla \cdot (\nabla \cdot (\alpha(x)\hat{s}(x)) = 0 \quad \text{where} \quad \hat{s}(x) = \mathcal{R}s(x). \tag{75}$$

If we impose the normalization constraint that $\langle \alpha \rangle = 1$, then it is known from partial differential equation theory that a solution to the above equation for $\alpha(x)$ exists and is unique. [For example, when $s(x) = \beta(x)I$ where $\beta(x) > 0$ for all x, as in (26) but in two dimensions, the solution is $\alpha(x) = \langle 1/\beta \rangle / \beta(x)$]. From the solution for $\alpha(x)$ we determine the average strain

$$\langle \epsilon(x) \rangle = \langle \alpha(x)s(x) \rangle = \alpha_* s_*(x), \tag{76}$$

where α_* is some constant. Thus up to a proportionality constant, we can determine the matrix $s_*(x)$ from the solution for $\alpha(x)$. The uniqueness of the solution for $\alpha(x)$ is what guarantees that $\mathcal{S}_*$ is at most rank 1, of the form (73). It is known that the field $\alpha(x)$ is

458

positive everywhere and consequently s_* is either a positive definite or negative definite matrix, depending on the sign of α_*. Since we are free to change the signs of s_* and α_* we can take s_* to be positive definite.

The associated stress field $\boldsymbol{\sigma}(\boldsymbol{x})$ must be such that

$$\text{Tr}(s(\boldsymbol{x})\boldsymbol{\sigma}(\boldsymbol{x})) = \alpha(\boldsymbol{x}). \tag{77}$$

By substituting the relation (16) into this we obtain another second order elliptic partial differential equation,

$$\text{Tr}(\hat{s}(\boldsymbol{x})\nabla\nabla\phi) = \alpha(\boldsymbol{x}), \tag{78}$$

this time for the Airy stress function $\phi(\boldsymbol{x})$, which can be taken to have the form

$$\phi(\boldsymbol{x}) = \phi_0(\boldsymbol{x}) + \boldsymbol{x} \cdot \boldsymbol{F}\boldsymbol{x}, \tag{79}$$

where $\phi_0(\boldsymbol{x})$ is periodic and the constant matrix $\boldsymbol{F}$ is determined by the average value of the stress field: $\langle\boldsymbol{\sigma}\rangle = 2\boldsymbol{\mathcal{R}}\boldsymbol{F}$. Thus the periodic function $\phi_0(\boldsymbol{x})$ satisfies

$$\text{Tr}(\hat{s}(\boldsymbol{x})\nabla\nabla\phi_0) = \alpha(\boldsymbol{x}) - 2\text{Tr}(\hat{s}(\boldsymbol{x})\boldsymbol{F}). \tag{80}$$

From elliptic partial differential equation theory it is known that (80) has a unique solution for $\phi_0(\boldsymbol{x})$ if and only if the right hand side is orthogonal to $\alpha(\boldsymbol{x})$, i.e. if and only if

$$0 = \langle\alpha^2 - 2\text{Tr}(\alpha\hat{s}\boldsymbol{F})\rangle = \langle\alpha^2\rangle - \alpha_*\text{Tr}(s_*\langle\boldsymbol{\sigma}\rangle), \tag{81}$$

where we have used (76). [When $s(\boldsymbol{x}) = \beta(\boldsymbol{x})\boldsymbol{I}$ and $\alpha(\boldsymbol{x}) = \langle 1/\beta\rangle/\beta(\boldsymbol{x})$ equation (80) becomes $\Delta\phi_0 = \langle 1/\beta\rangle/\beta^2 - 2\text{Tr}(\boldsymbol{F})$ and because $\Delta\phi_0$ has zero average value so must $\langle 1/\beta\rangle/\beta^2 - 2\text{Tr}(\boldsymbol{F})$ which accounts, in this case, for the condition (81).]

From (76) and the effective constitutive law we have $\text{Tr}(s_*\langle\boldsymbol{\sigma}\rangle) = \alpha_*$ which with (81) gives $\alpha_* = \langle\alpha^2\rangle^{1/2}$. Thus, we can determine the effective compliance tensor completely by solving (75) for $\alpha(\boldsymbol{x})$ and using (76) and the identity $\alpha_* = \langle\alpha^2\rangle^{1/2}$ to determine s_*.

Incidentally, percolation type questions often arise in the context of finding optimal microgeometries that attain bounds on effective moduli. For example, the task of finding three-dimensional polycrystalline microstructures that have the lowest possible effective conductivity is equivalent [44,45] to the task of finding non-trivial periodic rotation fields $\boldsymbol{R}(\boldsymbol{x})$ (satisfying $\boldsymbol{R}(\boldsymbol{x})^T\boldsymbol{R}(\boldsymbol{x}) = \boldsymbol{I}$) such that the equation

$$\nabla\boldsymbol{u}(\boldsymbol{x}) = \alpha(\boldsymbol{x})\boldsymbol{R}(\boldsymbol{x})^T\boldsymbol{A}\boldsymbol{R}(\boldsymbol{x}) \tag{82}$$

has a solution for the vector potential $\boldsymbol{u}(\boldsymbol{x})$ for some choice of scalar field $\alpha(\boldsymbol{x})$, where $\boldsymbol{A}$ is a given positive definite 3×3 diagonal matrix. Notice the similarity of (74) and (82).

REFERENCES

1. J.P. Straley, J. Phys. D.: Appl. Phys. 14 (1981) 2101-2105.
2. M. Milgrom and S. Shtrikman, Physical Review A 40 (1989) 1568-1575.
3. K. Schulgasser, J. Mech. Phys. Solids 40 (1992) 473-479.
4. Y. Benveniste and G. Dvorak, J. Mech. Phys. Solids 40 (1992) 1295-1312.
5. Y. Benveniste, J. Mech. Phys. Solids 43 (1995) 553-571.

6. V.M. Levin, Mech. Solids 2 (1967) 58-61.

7. J.L. Cribb, Nature 220 (1968) 576-577.

8. G. Dvorak, Proc. Roy. Soc. Lond. A 431 (1990) 89-110.

9. R. Hill, Proc. Phys. Soc. Lond. A 65 (1952) 349-354.

10. B.W. Rosen and Z. Hashin, Int. J. Engng. Sci. 8 (1970) 157–173.

11. Z. Hashin, J. Mech. Phys. Solids 32 (1984) 149-157.

12. K. Schulgasser, J. Mech. Phys. Solids 35 (1987) 35-42.

13. Y. Benveniste, Int. J. Solids Struct., to appear.

14. K. Schulgasser, J. Mat. Sci. Lett. 8 (1989) 228-229.

15. J.G. Berryman and G.W. Milton, Geophysics 56 (1991) 1950-1960.

16. A. Norris, J. Appl. Phys. 71 (1992) 1138-1141.

17. R. Hill, J. Mech. Phys. Solids 12 (1964) 199-212.

18. K.A. Lurie and A.V. Cherkaev, J. Opt. Theor. Appl. 42 (1984) 305-316.

19. M.F. Thorpe and I. Jasiuk, Proc. R. Soc. Lond. A 438 (1992) 531-544.

20. J. Dundurs, Journal of Composite Materials 1 (1967) 310-322.

21. J. Dundurs and X. Markenscoff, Proc. Roy. Soc. Lond. A 443 (1993) 289-300.

22. A. Cherkaev, K. Lurie and G.W. Milton, Proc. Roy. Soc. Lond. A 438 (1992) 519-529.

23. J. Ball, J.C. Currie and P.J. Olver, J. Funct. Anal. 41 (1981) 135-174.

24. F. Murat, Ann. Scuola Norm. Sup. Pisa Cl. Sci. 8 (1981) 69-102.

25. A.R. Day, K.A. Snyder, E.J. Garboczi and M.F. Thorpe, J. Mech. Phys. Solids 40 (1992) 1031-1051.

26. R.M. Christensen, Proc. Roy. Soc. Lond. A 440 (1993) 461-473.

27. R.S. Lakes, Science 235 (1987) 1038-1040.

28. G. W. Milton, J. Mech. Phys. Solids 40 (1992) 1105-1137.

29. J.B. Keller, J. Math. Phys. 5 (1964) 548-549.

30. A.M.Dykhne, Zh. Eksp. Teor. Fiz. 59 (1970) 110-115. [Soviet Physics JETP 32 (1971) 63-65.]

31. K.S. Mendelson, J. Appl. Phys. 46 (1975) 4740-4741.

32. V.L. Berdichevski, Variational principles in mechanics of continuum media, Nauka, Moscow, 1983.

33. J. Helsing, G.W. Milton and A.B. Movchan, J. Mech. Phys. Solids, to appear.

34. L.V. Gibiansky and S. Torquato, Int. J. Eng. Sci., to appear.

35. G. Francfort, Proc. Royal Soc. Edinburgh 120A, (1992) 25-46.

36. G.W. Milton and A.B. Movchan, Proc. R. Soc. Lond. A 450 (1995) 293-317.

37. G.W. Milton and A.B. Movchan, European Journal of Mechanics A-Solids, submitted.

38. K.A. Ingebrigtsen and A. Tonning, Phys. Rev. 184 (1969) 942-951.

39. S. Nemat-Nasser and L. Ni, Int. J. Solids Struct. 32(1995) 467-472.

40. G.W. Milton, Phys. Rev. B 38 (1988) 11296-11303.

41. Y. Grabovsky and G.W. Milton, Proc. Roy. Soc. Edinburgh, submitted.

42. K. Bhattacharya and R.V. Kohn, Arch. Rat. Mech. Anal., submitted.

43. M. Avellaneda, A.V. Cherkaev, L.V. Gibiansky, G.W. Milton and M. Rudelson, J. Mech. Phys. Solids 44 (1996) 1179-1218.

44. M. Avellaneda, A.V. Cherkaev, K.A. Lurie and G.W. Milton, J. Appl. Phys. 63 (1988) 4989-5003.

45. V. Nesi and G.W. Milton, J. Mech. Phys. Solids 39 (1991) 525-542.

Theoretical and Applied Mechanics 1996
T. Tatsumi, E. Watanabe and T. Kambe (Editors)
© 1997 Elsevier Science B.V. All rights reserved.

461

Plasticity: Inelastic Flow of Heterogeneous Solids at Finite Strains and Rotations

Sia Nemat-Nasser

Center of Excellence for Advanced Materials
Department of Applied Mechanics and Engineering Sciences
University of California, San Diego, La Jolla, CA 92093-0416

The theoretical basis of rate- and temperature-dependent finite-deformation plasticity is examined at the dislocation scale, leading to specific constitutive models for both bcc and fcc crystals. Using the results obtained through some novel experimental techniques, and for illustration, constitutive parameters are obtained for some commercially pure tantalum. Then constitutive relations are developed for single crystals (mesoscale), based on exact kinematics of crystollographic slip due to the dislocation motion, accompanied by elastic lattice distortion. The transition from the single (mesoscale) to the polycrystal (macroscale) response requires homogenization and averaging procedures. Hence, some fundamental averaging theorems applicable to finite strains and rotations, including the generalization of Eshelby's and the double-inclusion results to finite deformations, are outlined. The emphasis of the paper is to illustrate the essential roles that the heterogeneities at various scales, from nano to macro, play in defining the properties of metals.

1. INTRODUCTION

Classical rate-independent plasticity is a well-defined mathematical model developed in the theoretical and applied mechanics community, based on the notion of the convex yield surface in either stress- or strain-space, within which the material response is assumed to be elastic, and on which the response may be elastoplastic. The physical motivation stems from the uniaxial stress-strain relation for metals which displays a sharp yield point and little strain-rate dependence.[1] Aided by powerful modern computational tools, the formalism has been of great service to technology, especially in the areas of metalforming, wire drawing, high-speed machining, and crashworthy design, just to mention a few applications.

The resistance of most metals to inelastic flow is rate-dependent, although some, like mild steel, are not very sensitive to rate effects, while others, like certain aluminums, are mildly rate-sensitive, and others, such as tantalum-tungsten (Ta-W) alloys, can be highly rate-sensitive, as is shown in Figures 1a,b,c. The classical plasticity theory has been modified to account for the rate dependence. Furthermore, the effect of temperature has been integrated into most of the rate-dependent models.[2] Whether rate-sensitive or rate-

---

[1] See, Hill (1956), Naghdi (1960), Hill and Rice (1972, 1973), Havner (1982, 1992), Nemat-Nasser (1983, 1992), and references cited therein.

462

insensitive, the inelastic flow of metals is a manifestation of their *heterogeneity* and their *discrete* and *heterogeneous deformation pattern at the microscale.* This heterogeneity and heterogeneous deformation pattern are also instrumental for the strengthening and workhardening of metals. Dislocations, interstitial and substitutional atoms, and alloying elements provide the necessary microstructural inhomogeneities that profoundly affect the overall material strength and response. Modern plasticity, therefore, combines microstructural observations with controlled *recovery* experiments, over a broad range of strains, strain rates, and temperatures, to develop physically-based theories which capture the essential microstructural features that underlie the basic metal plasticity. The theory is then embedded in computational codes through effective algorithms for simulation and design purposes.

This presentation provides a brief account of the theoretical basis of rate- and temperature-dependent finite-deformation plasticity, at the dislocation (*i.e.,* nano) scale, leading to specific constitutive models for both bcc and fcc crystals. Using the results obtained through some novel experimental techniques, constitutive parameters are obtained for some commercially pure tantalum, as an illustration. Constitutive relations are developed for single crystals (mesoscale), based on kinematics of crystollographic slip, accompanied by elastic lattice distortion.

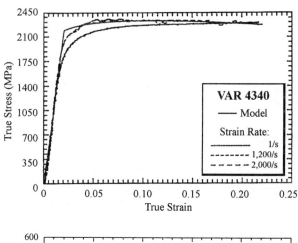

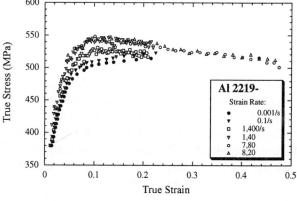

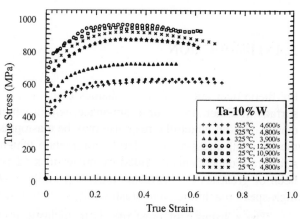

Figure 1. Stress-strain relations for: (a) 4340 steel, (b) 2219-T87 aluminum, and (c) Ta-10%W.

---

[2] See, Rice (1971), Senseny, Duffy, and Hawley (1978), Perzyna (1980, 1984), Klepaczko and Chiem (1986), Rashid *et al.* (1992), and references cited therein.

The transition from the single (mesoscale) to the polycrystal (macroscale) response requires homogenization and averaging procedures. Hence, some fundamental averaging theorems applicable to finite strains and rotations, including the generalization of Eshelby's and the double-inclusion results to finite deformations, are outlined. The emphasis of the paper is to illustrate the essential roles that the heterogeneities at various scales, from nano to macro, play in defining the properties of metals. Also, the interplay among physically-based models, recovery experiments, rigorous mathematical formulation of continuum crystal plasticity, and effective computational simulations, is illustrated and emphasized.

2. MICROSTRUCTURAL INHOMOGENEITIES

Plastic flow in the range of temperatures and stain rates where diffusion and creep are not dominant, occurs basically by the motion of dislocations. The flow stress,[3] τ, therefore, is essentially defined by the resistance to the dislocation motion due to the presence of *microstructural inhomogeneities*. The motion of dislocations is opposed by both *short-range* and *long-range obstacles*. The short-range barriers may be overcome by *thermal activation*, whereas the resistance due to long-range obstacles is essentially independent of the temperature and strain rate, *i.e.*, it is *athermal*. The short-range barriers may include the lattice resistance (generally called the *Peierls stress*), point defects such as vacancies and self-interstitials, other dislocations which intersect the slip planes, alloying elements, and solute atoms (interstitials and substitutional). The long-range barriers may include grain boundaries, farfield forests of dislocations, and other microstructural inhomogeneities with farfield influence. Once a barrier is crossed, the dislocation moves to the next barrier under the action of the net shear stress $\tau_n = \tau - \tau_a - \tau_D$, where τ_a is the stress due to the farfield barriers (*athermal*), and τ_D is the *drag resistance*. The stress difference, $\hat{\tau} - \tau - \tau_a = \hat{\tau} - \tau^*$, must be overcome by the thermal vibrations at each barrier, where $\hat{\tau}$ is the barrier's threshold stress, above which the dislocation crosses the barrier without thermal assistance. For low dislocation velocities, the effect of τ_D may be insignificant, but at high strain rates, it may play a prominent role. If the net shear stress, τ_n, is positive, dislocations accelerate from obstacle to obstacle, and may even overshoot one or several obstacles before they stop and thermal activation resumes. For simplicity, it is often assumed that $\tau_n = 0$, so that $\tau_D = \tau^* = \tau - \tau_a$ between obstacles.

2.1. A Simple Model

A simple but effective model may be used to quantify the flow stress of metals at various strain rates, where both thermal activation and viscous drag affect the motion of dislocations.[4] Here, the model is used to motivate the derivation of general constitutive expressions which are then viewed in a phenomenological context.

As a starting point, it is assumed that the average time required for dislocations to move from one barrier to the next, consists of two parts, a waiting period, t_w, and a running period, t_r. The waiting period is defined by

---

[3] We use τ for stress and γ for the conjugate strain. These stand for axial stress and strain in uniaxial cases, for the resolved shear stress and strain in the case of crystals, and for the effective stress and strain in phenomenological models.

[4] See, *e.g.*, Kocks *et al.* (1975), Clifton (1983), Regazzoni *et al.* (1987), and references cited therein.

$$t_w = \omega_0^{-1} \exp\{\Delta G / kT\}, \tag{2.1}$$

where ΔG is the required activation energy, k is the Boltzman constant, T is the absolute temperature, and ω_0 is the attempt frequency which, among other factors, depends on the structure and composition of the crystal, as well as on the core structure of the dislocation; ω_0 is in the range of 10^{11} to $10^{13} s^{-1}$. The running time may be expressed as

$$t_r = d / \bar{v}_r, \quad \bar{v}_r = \tau^* b / D, \tag{2.2a,b}$$

where b is the magnitude of the Burgers vector, D is the *drag* coefficient,[5] and $\bar{v}_r$ is the average running velocity of the dislocations. Hence, the *average dislocation velocity* is estimated by

$$\bar{v} = \frac{d + \lambda}{t_w + t_r} = \frac{(1 + \lambda')d}{t_w + t_r}, \quad \lambda' = \lambda / d, \tag{2.3a-c}$$

where d is the average travel distance between barriers and λ is the average barrier width, usually assumed to be much smaller than d, *i.e.*, $\lambda \ll d$; this assumption is not used in what follows. Between the barriers, the frictional stress, $\tau_D = D\bar{v}_r / b$, must be overcome by the driving stress, $\tau^* = \tau - \tau_a$, so that $\tau_D \leq \tau^*$.

The effective plastic strain rate, $\dot{\gamma}$, may be written in terms of the density of the mobile dislocations, ρ_m, their average velocity, $\bar{v}$, and the magnitude of the Burgers vector, b, as

$$\dot{\gamma} = b \rho_m \bar{v} = \frac{\dot{\gamma}_r (1 + \lambda') \tau^*}{\tau_D^0} \left\{ 1 + \frac{\tau^*}{\tau_D^0} \exp\{\Delta G / kT\} \right\}^{-1}, \tag{2.4}$$

where $\dot{\gamma}_r = b \rho_m \omega_0 d$ is the reference strain rate, and $\tau_D^0 = \omega_0 D d / b$ is the reference drag stress. For the activation energy, ΔG, consider,[6]

$$\Delta G = G_0 \left\{ 1 - \left[\frac{\tau^*}{\hat{\tau}} \right]^p \right\}^q, \quad \tau^* = \tau^*(\gamma, \dot{\gamma}, T), \quad \hat{\tau} = \tau^*(\gamma, \dot{\gamma}, 0), \tag{2.5a,b}$$

where $0 < p \leq 1$ and $1 \leq q \leq 2$. When $\tau^* = \tau - \tau_a > \hat{\tau}$, then the resistance to the flow is solely due to the athermal, τ_a, and frictional, τ_D, parts of the process, so that $\Delta G \approx 0$. In this case, only viscous drag and the stress field due to farfield inhomogeneities resist the motion of dislocations. Then, each dislocation vibrational "attempt" is successful, and $t_w = \omega_0^{-1} \ll t_r$.

The athermal part, τ_a, is generally viewed to depend on the density of the dislocations, as well as on the grain size and other microstructural inhomogeneities. Here, for illustration, it will be assumed that τ_a is a nondecreasing function of the effective plastic strain, γ, and a nonincreasing function of the average grain size, d_G, writing

$$\tau_a = f(\gamma, d_G), \tag{2.6}$$

where, for illustration, it may be assumed that $\tau_a \approx \tau_a^0(1 + a\gamma)^n + k_0 d_G^{-1/2}$, where τ_a^0, a, and k_0 are material parameters; the linear dependence of τ_a on $d_G^{-1/2}$ is called the Hall-Petch effect.

---

[5] In crystals with strong Peierls resistance, the drag may be associated with the lattice friction. Then the drag coefficient D in (2.2b) may be a function of the average running velocity of the dislocations, $\bar{v}_r$. When the drag is due to the interaction of the moving dislocation with phonons and, at low temperatures, with electrons, D may be assumed to be a constant; see Kocks *et al.* (1975).

[6] The parameters p and q define the shape of the energy barrier. Ono (1968) suggests that p = 2 / 3 and q = 2 may be assumed for most cases.

2.2. Application to bcc Metals

At strain rates less than $10^5 s^{-1}$, the drag resistance may be neglected. Then from $\dot{\gamma} = b\,\rho_m\,d/t_w$, (2.11), (2.15), and since $\tau = \tau^* + \tau_a$, it follows that

$$\tau(\gamma, \dot{\gamma}, T) = \hat{\tau}\left\{1 - \left[-\frac{kT}{G_0} \ln\frac{\dot{\gamma}}{\dot{\gamma}_r}\right]^{1/q}\right\}^{1/p} + \tau_a \quad \text{for } T \le T_c, \qquad T_c = -\frac{G_0}{k}\left\{\ln\frac{\dot{\gamma}}{\dot{\gamma}_r}\right\}^{-1}. \tag{2.7a,b}$$

Here, T_c is the critical temperature, above which only long-range barriers resist plastic flow; hence $\tau = \tau_a$ for $T > T_c$.

Recently, Nemat-Nasser and Isaacs (1996) have shown that (2.7) can be used to represent the flow stress of commercially pure tantalum over temperatures from liquid nitrogen (77K) to over 1000K, and strain rates from quasi-static to over $10^4 s^{-1}$. Assuming that Peierls' barrier (lattice resistance) is the only obstacle to the dislocation motion, $G_0 = 1 eV/atom$, $q = 2$, $p = 2/3$, and $\dot{\gamma}_r \approx 5.46 \times 10^8 s^{-1}$ fit all available experimental data. These authors first identify that at a strain rate of $5000 s^{-1}$, for which they produce considerable experimental data, $T_c \approx 1000K$, and that above this critical temperature, $\tau_a = \tau_a^0 \gamma^{1/5}$ fits all these data. This leaves $\hat{\tau}$ and τ_a^0 as the only remaining parameters. Figure 2 shows experimental and theoretical results; see the next section for a brief account of the experimental techniques, and Nemat-Nasser and Isaacs (1996) for additional comments and comparisons.[7]

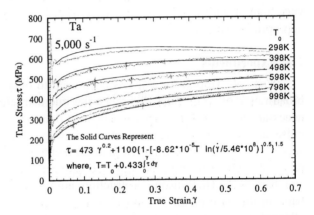

Figure 2. Comparison of adiabatic flow stress for Ta at a 5,000s⁻¹ stain rate with model predictions.

2.3. Application to fcc Metals

For most fcc metals, the dislocations which intersect the slip planes are the main barriers to the plastic flow. In this case, the force, $\hat{\tau}\,b$, associated with the threshold stress $\hat{\tau}$, measured per unit dislocation length, depends on the barrier spacing along the moving dislocation line, and hence on the dislocation density. This spacing, l, may be estimated by $l \approx \rho_c^{-1/2}$, where ρ_c is the dislocation density which increases with increasing plastic strain, γ.

Hence, assume that $l = l_0/\hat{f}(\gamma)$, $\hat{f}(0) = 1$, with $\hat{f}$ a nondecreasing function. For example, $\hat{f}(\gamma) = (1 + a_0\gamma)^{n_0}$ or $\hat{f}(\gamma) = (1 + a_0\gamma^{n_0})$, with $a_0, n_0 > 0$, may be assumed, where a_0 and n_0 are constitutive parameters. Since the stress $\hat{\tau}$ required to cross a barrier without thermal assistance is $\hat{\tau} = G_0/(b\lambda l)$, where λ is the average width of the barrier, it may be assumed that $\hat{\tau}$ for the dislocation barriers relates to the plastic strain by

$$\hat{\tau} = \hat{\tau}_0 \hat{f}(\gamma), \quad \hat{f}(0) = 1, \quad \hat{f}'(\gamma) \ge 0, \quad \hat{\tau}_0 = \frac{G_0}{b\lambda l_0}. \tag{2.8a-d}$$

---

[7] Here, τ and γ represent the uniaxial stress and strain, respectively.

The final expression for the flow stress is now obtained by substituting into (2.7a) for τ_a from (2.6) and for $\hat{\tau}$ from (2.8a). Note that the reference strain rate, $\dot{\gamma}_r$, is now also a function of γ, since $d \approx l \approx \rho_c^{-1/2}$, and hence $d = d_0/\hat{f}_1(\gamma)$, where function $\hat{f}_1$ has the same properties as $\hat{f}$, defined in (2.8). Therefore,

$$\dot{\gamma}_r = \dot{\gamma}_r^0/\hat{f}_1(\gamma), \qquad \dot{\gamma}_r^0 = b\,\rho_m\,\omega_0\,d_0\,. \tag{2.9a,b}$$

This expression must be used in (2.7a) for $\dot{\gamma}_r$, *when the intersecting forests of dislocations are the main barriers to the motion of the mobile dislocations.* One then obtains

$$\tau(\gamma, \dot{\gamma}, T) = \hat{\tau}_d^0\left\{1 - \left[-\frac{kT}{G_0}\left[ln\frac{\dot{\gamma}}{\dot{\gamma}_r^0} + ln\,\hat{f}_1(\gamma)\right]\right]^{1/q}\right\}^{1/p}\hat{f}(\gamma) + f(\gamma, d_G) \text{ for } T \le T_c, \tag{2.10}$$

where T_c now depends on the strain, γ; substitute for $\dot{\gamma}_r$ from (2.9) into (2.7b). Constitutive relations (2.7) for bcc metals, and (2.9) for fcc metals must be modified according to (2.4) at very high strain rates. In such a case, substitute for ΔG from (2.5) into (2.4), and use the appropriate expression for $\tau^* = \tau - \tau_a$.

2.4. Experiments

The split Hopkinson bar has been used to obtain the plastic response of metals, over a broad range of strain rates. To relate the microstructural changes to the loading history, it is necessary to control the experiments such that the sample is subjected to only a given stress pulse and then is recovered, without having been subjected to any additional loads. To accomplish this, the classical Hopkinson technique has been modified such that all the stress pulses that emanate from the interface between the sample and the incident and the transmission bars of the Hopkinson construction, are trapped, once they reach the free ends of the two bars. In the case of the compression test, it is only necessary that no compression pulse is allowed to return and reload the sample. The details for the tension, compression, and compression followed by tension, *recovery experiments* are given in Nemat-Nasser, Isaacs, and Starrett (1991). The mechanism that is used to trap the incoming tension pulse, has also been used to change (increase or decrease by about a third) the *strain rate* during the compression loading at high strain rates; see, Nemat-Nasser *et al.* (1993). A similar change in strain rate can be accomplished in UCSD's tension recovery Hopkinson technique. Figure 3 is a sketch of UCSD's recovery compression split Hopkinson bar.

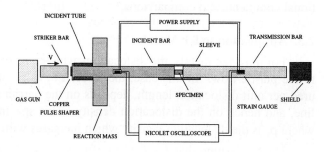

Figure 3. UCSD's recovery compression split Hopkinson bar: the sample is subjected to a single pulse and recovered without any additional loading.

For high-temperature, high strain-rate experiments, the recovery Hopkinson system has been supplemented by the addition of a furnace capable of heating the sample to 1,000°C. In this design, the furnace heats the sample to the required temperature, while both the incident and the transmission bars are kept at suitably low temperatures, outside of the heating portion of the furnace. Once the desired temperature is attained, the transmission and incident bars are brought into contact with the specimens, just microseconds before the stress pulse reaches the end of the incident bar and loads the sample. This is accomplished as follows: The specimen is kept at the center of the furnace by means of the thermocouples that actually measure the temperature of the sample. Then, the same gas gun that propels the striker bar toward the incident bar, also activates two *bar movers* which are attached to the transmission bar and move this bar, bringing it into contact with the sample and then moving the sample into contact with the incident bar. Figure 4 shows the bar movers in relation to the Hopkinson construction. Using this novel technique, it is possible to develop isothermal flow stresses even of refractory metals which, at high strain rates, produce considerable heating due to the high-strain rate plastic deformation.

To obtain an isothermal flow stress at a high strain rate, the sample is heated to the required temperature in the furnace attached to the recovery Hopkinson bar, and then loaded incrementally. After the application of each load increment, the sample is unloaded without being subjected to any additional stress pulses. The sample is then allowed to return to the furnace temperature before the application of the next strain increment. Since the unloading, the cooling of the sample to its initial temperature, and the reloading, may affect the microstructure and hence the thermomechanical properties of the material, it is necessary to check these in each case. To this end, a sample which has been loaded, unloaded, and cooled to its

initial temperature, may then be reheated to the temperature that it had just prior to its unloading. If there are no substantial changes in the microstructure that affect the flow-stress properties, then the flow stress, upon loading, should follow the previous stress-strain curve. For a tantalum-10% tungsten (Ta-10%W) alloy, this is illustrated in Figure 5.

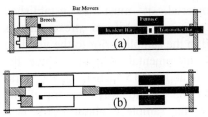

Figure 4. UCSD's recovery compression Hopkinson bar for high strain-rate testing at high temperatures of up to 1,000°C; (a) before, and (b) during a test.

In Figure 5, the dashed curve is obtained by loading the sample at about a 5,700s^{-1} strain rate, to a true strain of about 75%. This curve represents the (essentially) adiabatic, true stress-strain relation for the material. The solid curve (1) is obtained by taking another sample of the same material, loading it to about 23% strain at the same strain rate, and then unloading it. The fact that curve (1) follows closely the dashed curve is an indication of the good quality of the test and the fact that the two samples represent the same material with essentially the same initial conditions.

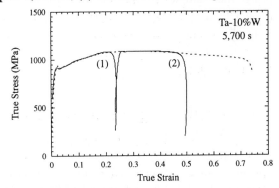

Figure 5. The true stress-true strain relations for Ta-10%W at a 5,700s^{-1} strain rate; the dashed curve and curve 1 are adiabatic, with an initial temperature of 25°C; curve 2 represents a continuation of curve 1 after preheating the specimen to 135°C.

Now, to check if any recovery has occurred due to unloading and cooling, we have calculated the total plastic work and, assuming that essentially all of it has been used to heat the sample, have estimated that the sample temperature just prior to unloading must have been about 110°C more than its initial room temperature. Thus, the sample which had then attained room temperature after unloading, was first heated to about 135°C in the furnace, and loaded at the same strain rate. This has produced the second solid curve, marked (2), shown in Figure 5. This also follows closely the dashed curve. Two important results follow from these results: (1) if there was any recovery, it did not affect the flow stress noticeably; and (2) essentially all the plastic work had been converted into heat.

Figure 6 gives the true stress-true strain relations for tantalum-10% tungsten tested at an initial 25°C (room temperature), at strain rates of 10^{-3}, 1, and 5,700s^{-1}. The curves marked (4), (5), and (6), are obtained by incremental straining, in the manner discussed above. Curve (3) is the adiabatic stress-strain relation at 5,700s^{-1}. Curve (7), obtained by connecting the initial yield points of curves (4), (5), and (6), represents the isothermal flow stress at room temperature at a 5,700s^{-1} strain rate. Note that curves (1), (2), and (7) are almost parallel. Their differences, therefore, are essentially due to the differences in their strain rates.

Before accepting curve (7) as the high strain-rate isothermal flow stress of this material, it must be established whether or not a uniform stress state is attained in the sample, prior to the yielding in each incremental straining. Since the sample is about 0.15 inch long, the duration for the elastic wave to travel the length of the sample is about 1.2μs. It is easy to show that the stress state in the sample is uniform after 5ms. The time required by the applied stress pulse to bring the sample to the yield state is greater than 10μs. Hence, a uniform stress state in the sample is attained long before the material begins to yield at each incremental loading.

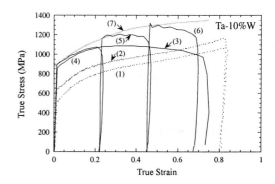

Figure 6. The true stress-true strain relations for Ta-10%W at 25°C; curves 1, 2, and 7 are isothermal relations at strain rates of 10s^{-3} 1, and 5,700s^{-1}; curve 5 is adiabatic at a 5,700s^{-1} strain rate.

3. MESOSTRUCTURAL INHOMOGENEITIES

At the scale of microns, most metals consist of aggregates of single crystals with various orientations and sizes. The aggregate properties are defined by the collective properties of the crystals. Single crystals sustain plastic deformation by the motion of dislocations. Dislocations tend to move on closed-packed crystallographic planes and in closed-packed directions which constitute the crystal's slip systems. For each crystal structure, there is a well-defined set of slip systems which may be activated under suitable loading conditions. For example, room temperature plastic deformation in face-centered cubic (fcc) metals (*e.g.*, Al, Cu, Ni) is by dislocation motion on the {111} crystallographic planes and in the <110> directions, where {111} and <110> are the standard Miller indices. For the fcc crystals, the {111}<110> are the *primary slip systems*. In certain cases, slip on these primary systems may be accompanied by slip on *secondary slip systems*. For example, at high temperatures (*e.g.*, above 450°C), aluminum slips also on the {100} and other planes. The slip direction is always the most closely packed <110> family. In body-centered (bcc) crystals, {110}, {112}, and {123} are all often viewed as the primary slip planes, depending on the test temperature, with <111> being the slip directions. The primary slip planes in the hexagonal closed-packed (hcp) crystals, are the family of the basal planes {0001}, as well as the pyramidal, {10$\bar{1}$1}, and prismatic, {11$\bar{2}$0}, planes.

The inelastic deformation by crystallographic slip, leaves the lattice structure unaffected. *The lattice distortion is due to elastic deformation only.* The molecules are viewed to flow through the lattice in this picture of crystal plasticity. This flow consists of *simple shearing* due to the dislocation motion associated with a finite number of active slip systems, resulting in both *plastic deformation and rigid-body rotation of the material relative to the lattice.*

3.1. Decomposition of Deformation and its Rate

Consider a finitely deformed crystal with initial particle positions denoted by $\mathbf{X}$ and the current ones by $\mathbf{x}$, respectively. The deformation gradient, $\mathbf{F} = (\partial\mathbf{x}/\partial\mathbf{X})^{\mathrm{T}}$, is decomposed as

$$\mathbf{F} = \mathbf{F}^*\mathbf{F}^{\mathrm{p}} = \mathbf{V}^{\mathrm{e}}\mathbf{R}^*\mathbf{F}^{\mathrm{p}}, \quad \mathbf{F}^* = \mathbf{V}^{\mathrm{e}}\mathbf{R}^*. \tag{3.1a-c}$$

In general, neither $\mathbf{F}^*$ nor $\mathbf{F}^{\mathrm{p}}$ may be compatible, *i.e.*, they may *not* be gradients of some smooth displacement fields. Their product, $\mathbf{F}$, is compatible. Figure 7 shows the initial, C_0, the final C, and two intermediate configuraions, $C_{\mathbf{F}^{\mathrm{p}}}$ and C_{R}, of the crystal, as well as the associated deformation gradients. Here, $\mathbf{F}^{\mathrm{p}}$ denotes the deformation gradient corresponding to the plastic flow of matter through the lattice by slip-induced simple shearing, whereas $\mathbf{V}^{\mathrm{e}}$ is the pure elastic deformation, and $\mathbf{R}^*$ is the rigid rotation of the lattice.

The velocity gradient, $l \equiv \dot{\mathbf{F}}\,\mathbf{F}^{-1}$, is decomposed as

$$l = l^* + l^{\mathrm{p}} = (\mathbf{d}^* + \mathbf{w}^*) + (\mathbf{d}^{\mathrm{p}} + \mathbf{w}^{\mathrm{p}}), \quad \mathbf{d}^* = \mathbf{d}^{\mathrm{e}} + \frac{1}{2}(\mathbf{V}^{\mathrm{e}}\boldsymbol{\Omega}^*\mathbf{V}^{\mathrm{e}-1} - \mathbf{V}^{\mathrm{e}-1}\boldsymbol{\Omega}^*\mathbf{V}^{\mathrm{e}}),$$

$$\mathbf{w}^* = \mathbf{w}^{\mathrm{e}} + \frac{1}{2}(\mathbf{V}^{\mathrm{e}}\boldsymbol{\Omega}^*\mathbf{V}^{\mathrm{e}-1} + \mathbf{V}^{\mathrm{e}-1}\boldsymbol{\Omega}^*\mathbf{V}^{\mathrm{e}}), \tag{3.2a-d}$$

where $l^{\mathrm{p}} = \mathbf{F}^*\tilde{l}^{\mathrm{p}}\mathbf{F}^{*-1}$, $\tilde{l}^{\mathrm{p}} = \dot{\mathbf{F}}^{\mathrm{p}}\mathbf{F}^{\mathrm{p}-1}$, $l^{\mathrm{e}} = \dot{\mathbf{V}}^{\mathrm{e}}\mathbf{V}^{\mathrm{e}-1}$, and $\boldsymbol{\Omega}^* = \dot{\mathbf{R}}^*\mathbf{R}^{*\mathrm{T}}$. The velocity gradient l^{p} measures the rate of plastic distortion with reference to the current configuration, whereas $\tilde{l}^{\mathrm{p}}$ is with respect to the elastically relaxed configuration, $C_{\mathbf{F}^{\mathrm{p}}}$, in which the lattice is still in its initial unrotated orientation.

3.2. Plastic Distortion

Denote the slip direction and slip normal of a typical slip system in the *undeformed configuration,* by $\mathbf{s}_0^{\alpha}$ and $\mathbf{n}_0^{\alpha}$, respectively. As stated before, *the plastic deformation leaves the lattice structure unaffected.* Hence, the slip direction and slip normal in the intermediate configuration, $C_{\mathbf{F}^{\mathrm{p}}}$, are the same as those in the initial undeformed configuration, C_0. The plastic part of the velocity gradient due to crystallographic slip may be expressed by

$$\tilde{l}^{\mathrm{p}} = \tilde{\mathbf{d}}^{\mathrm{p}} + \tilde{\mathbf{w}}^{\mathrm{p}} = \sum_{\alpha=1}^{n}\dot{\gamma}^{\alpha}\mathbf{s}_0^{\alpha}\otimes\mathbf{n}_0^{\alpha}, \quad \tilde{\mathbf{d}}^{\mathrm{p}} = \sum_{\alpha=1}^{n}\dot{\gamma}^{\alpha}\mathbf{p}_0^{\alpha}, \quad \tilde{\mathbf{w}}^{\mathrm{p}} = \sum_{\alpha=1}^{n}\dot{\gamma}^{\alpha}\mathbf{r}_0^{\alpha}, \tag{3.3a,b}$$

where $\dot{\gamma}^{\alpha}$ is the slip rate of the α'th slip system measured relative to the undeformed lattice, n is the total number of active slip systems in the single crystal, $\mathbf{p}_0^{\alpha} \equiv (\mathbf{s}_0^{\alpha}\otimes\mathbf{n}_0^{\alpha} + \mathbf{n}_0^{\alpha}\otimes\mathbf{s}_0^{\alpha})/2$, and $\mathbf{r}_0^{\alpha} \equiv (\mathbf{s}_0^{\alpha}\otimes\mathbf{n}_0^{\alpha} - \mathbf{n}_0^{\alpha}\otimes\mathbf{s}_0^{\alpha})/2$. When measured relative to the current configuration, these expressions become

470

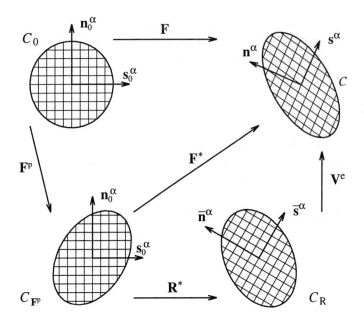

Figure 7. Decomposition of deformation gradient $\mathbf{F}$ into plastic deformation $\mathbf{F}^p$, lattice rotation $\mathbf{R}^*$, and lattice distortion (elastic deformation) $\mathbf{V}^e$, $\mathbf{F}^* = \mathbf{V}^e \mathbf{R}^*$ and $\mathbf{F} = \mathbf{F}^* \mathbf{F}^p$.

$$l^p = \mathbf{d}^p + \mathbf{w}^p = \sum_{\alpha=1}^{n} \dot{\gamma}^\alpha \mathbf{s}^\alpha \otimes \mathbf{n}^\alpha = \sum_{\alpha=1}^{n} \dot{\gamma}^\alpha \mathbf{p}^\alpha + \sum_{\alpha=1}^{n} \dot{\gamma}^\alpha \mathbf{r}^\alpha, \quad \mathbf{s}^\alpha = \mathbf{F}^* \mathbf{s}_0^\alpha, \quad \mathbf{n}^\alpha = \mathbf{F}^{*-T} \mathbf{n}_0^\alpha, \quad (3.4\text{a,b})$$

where $\mathbf{p}^\alpha \equiv (\mathbf{s}^\alpha \otimes \mathbf{n}^\alpha + \mathbf{n}^\alpha \otimes \mathbf{s}^\alpha)/2$, and $\mathbf{r}^\alpha \equiv (\mathbf{s}^\alpha \otimes \mathbf{n}^\alpha - \mathbf{n}^\alpha \otimes \mathbf{s}^\alpha)/2$.

3.3. Crystal Elasticity

The *elastic* distortion of the lattice may be measured by the left-stretch tensor, $\mathbf{V}^e$, relative to configuration C_R, in which the undistorted lattice is rotated by $\mathbf{R}^*$ from its initial orientation. The strain measure can be arbitrary. A possible choice is the Lagrangian (elastic) strain $\overline{\mathbf{E}}^e = \frac{1}{2}(\mathbf{V}^{e2} - 1)$, with an associated stress $\overline{\mathbf{S}}^P$, relating through the elastic potential, $\overline{\phi}$, by

$$\overline{\mathbf{S}}^P = \mathbf{V}^{e-1} \boldsymbol{\tau} \mathbf{V}^{e-1} = \frac{\partial \overline{\phi}}{\partial \overline{\mathbf{E}}^e}, \qquad (3.5\text{a,b})$$

where $\boldsymbol{\tau}$ is the Kirchhoff stress. The corresponding elasticity tensor, $\overline{L}$, measured relative to C_R, then becomes

$$\overline{L} = \frac{\partial^2 \overline{\phi}}{\partial \overline{\mathbf{E}}^e \, \partial \overline{\mathbf{E}}^e}, \qquad (3.6)$$

Define the objective (convected with $\mathbf{F}^*$) rate of change of the Kirchhoff stress as

$$\overset{\nabla}{\boldsymbol{\tau}}{}^* \equiv \mathbf{F}^* (\mathbf{F}^{*-1}\boldsymbol{\tau}\,\mathbf{F}^{*-T})^{\boldsymbol{\cdot}}\mathbf{F}^T = \dot{\boldsymbol{\tau}} - \boldsymbol{l}^*\boldsymbol{\tau} - \boldsymbol{\tau}\,\boldsymbol{l}^{*T}, \tag{3.7a,b}$$

and note the following exact constitutive relation:

$$\overset{\nabla}{\boldsymbol{\tau}}{}^* \equiv \mathcal{L}:\mathbf{d}^* = \mathcal{L}:(\mathbf{d}-\mathbf{d}^p), \tag{3.7c,d}$$

where $\mathcal{L}$ is defined in terms of $\bar{\mathcal{L}}$ by

$$\mathcal{L}_{ijkl} = V_{ia}^e\, V_{jb}^e\, V_{kc}^e\, V_{ld}^e\, \bar{\mathcal{L}}_{abcd}, \quad i,\,j,\,k,\,l,\,a,\,b,\,c,\,d = 1,\,2,\,3\,. \tag{3.7e}$$

In terms of the stress rate convected with the *total deformation gradient*, obtain

$$\overset{\nabla}{\boldsymbol{\tau}} \equiv \mathbf{F}\,(\mathbf{F}^{-1}\boldsymbol{\tau}\,\mathbf{F}^{-T})\boldsymbol{\cdot}\mathbf{F}^T = \overset{\nabla}{\boldsymbol{\tau}}{}^* - \boldsymbol{l}^p\boldsymbol{\tau} - \boldsymbol{\tau}\,\boldsymbol{l}^{pT} = \mathcal{L}:\mathbf{d} - \sum_{\alpha=1}^{n} \dot{\gamma}^\alpha \boldsymbol{\lambda}^\alpha,$$

$$\boldsymbol{\lambda}^\alpha \equiv \mathcal{L}:\mathbf{p}^\alpha + (\mathbf{s}^\alpha \otimes \mathbf{n}^\alpha \boldsymbol{\tau} + \boldsymbol{\tau}\,\mathbf{n}^\alpha \otimes \mathbf{s}^\alpha)\,. \tag{3.8a-d}$$

The inverse constitutive relation is

$$\mathbf{d} = \mathcal{M}:\overset{\nabla}{\boldsymbol{\tau}} + \sum_{\alpha=1}^{n} \dot{\gamma}^\alpha \boldsymbol{\mu}^\alpha, \quad \boldsymbol{\mu}^\alpha \equiv \mathcal{M}:\boldsymbol{\lambda}^\alpha = \mathbf{p}^\alpha + \mathcal{M}:(\mathbf{s}^\alpha \otimes \mathbf{n}^\alpha \boldsymbol{\tau} + \boldsymbol{\tau}\,\mathbf{n}^\alpha \otimes \mathbf{s}^\alpha), \tag{3.9a-c}$$

where $\mathcal{M} = \mathcal{L}^{-1}$ is the current elastic compliance tensor. In terms of the Jaumann rate of the Kirchhoff stress, $\overset{\circ}{\boldsymbol{\tau}} = \dot{\boldsymbol{\tau}} - \mathbf{w}\boldsymbol{\tau} + \boldsymbol{\tau}\mathbf{w}$, obtain

$$\overset{\circ}{\boldsymbol{\tau}} = \mathcal{L}':\mathbf{d} - \sum_{\alpha=1}^{n} \dot{\gamma}^\alpha \boldsymbol{\lambda}^\alpha, \quad \mathcal{L}'_{ijkl} = \mathcal{L}_{ijkl} + \frac{1}{2}\,(\delta_{il}\,\tau_{jk} + \delta_{jl}\,\tau_{ik} + \delta_{ik}\,\tau_{jl} + \delta_{jk}\,\tau_{il})\,. \tag{3.10a,b}$$

As shown by Hill and Rice (1972), the difference $\overset{\circ}{\boldsymbol{\tau}} - \mathcal{L}':\mathbf{d} = -\sum_{\alpha=1}^{n} \dot{\gamma}^\alpha \boldsymbol{\lambda}^\alpha$ is independent of the chosen strain measure or the reference state. This is illustrated by (3.8) and (3.10). In the first case, $\overset{\nabla}{\boldsymbol{\tau}}$ corresponds to the Lagrangian strain, while in the latter case, $\overset{\circ}{\boldsymbol{\tau}}$ relates to the log-arithmic strain. As is seen, the elasticity tensor, and hence $\boldsymbol{\mu}^\alpha$, depend on the chosen strain and the reference state.

The rate of change of the resolved Kirchhoff shear stress is given by

$$\dot{\tau}^\alpha = \boldsymbol{\lambda}^\alpha:\mathbf{d}^* = \boldsymbol{\lambda}^\alpha:\mathcal{M}:\overset{\nabla}{\boldsymbol{\tau}}{}^* = \boldsymbol{\mu}^\alpha:\overset{\nabla}{\boldsymbol{\tau}}{}^*, \quad \tau^\alpha = \boldsymbol{\tau}:\mathbf{p}^\alpha\,. \tag{3.11a,b}$$

3.4. Slip Rates

In the classical rate-independent approach to crystal plasticity, the slip rates, $\dot{\gamma}^\alpha$, are related to the rate of change of the resolved shear stress, $\dot{\tau}^\alpha$. As noted before, the deformation of metals generally is rate dependent. Empirical relations, such as a power law, have been used to express $\dot{\gamma}^\alpha$ in terms of the driving shear stress. It is, however, more appropriate to employ the dislocation-based physical model of Section 2, since it directly applies to slip-induced crystal flow.

To this end use (2.4) to express $\dot{\gamma}^\alpha$ in terms of the *corresponding* resolved shear stress, $\tau^{*\alpha} = \tau^\alpha - \tau_a^\alpha = \mathbf{p}^\alpha:(\boldsymbol{\tau}-\boldsymbol{\tau}_a)$, where $\boldsymbol{\tau}$ is the Kirchhoff stress, and $\boldsymbol{\tau}_a$ is the athermal stress field. For certain bcc single crystals with Peierls' stress as the most dominant resistance to the dislocation motion, the athermal stress may be due to the forest of dislocations and the material impurities, as well as the overall crystal boundaries. On the other hand, if the dislocations are the essential barriers, only the boundaries and other crystals may cause the athermal resistance. In either case, τ_a^α should be very small for an isolated crystal. For the sake of completeness, the athermal stress is included in what follows. This requires a clear definition of the overall plastic strain, γ, in terms of the slip rates, $\dot{\gamma}^\alpha$. Since γ is used to represent the overall activity of the mobile dislocations, it may be reasonable to define it as

$$\gamma = \int_0^t \dot{\gamma}(\eta)\,d\eta, \quad \dot{\gamma} = \sum_{\alpha=1}^n \{2(\dot{\gamma}^\alpha\mathbf{p}^\alpha):(\dot{\gamma}^\alpha\mathbf{p}^\alpha)\}^{1/2} = \sum_{\alpha=1}^n \dot{\gamma}^\alpha, \quad \dot{\gamma}^\alpha \geq 0. \tag{3.12a,b}$$

Now with ΔG^α given by (2.5a,b) in terms of the corresponding shear stress, $\tau^{*\alpha}$, the threshold stress, $\hat{\tau}$, and the plastic strain, γ, rewrite (2.4) as

$$\dot{\gamma}^\alpha = \dot{\gamma}_r^\alpha \frac{(1+\lambda')\tau^{*\alpha}}{\tau_D^0}\left\{1 + \frac{\tau^{*\alpha}}{\tau_D^0}\exp\{\Delta G^\alpha/kT\}\right\}^{-1} \equiv g(\tau^\alpha, \gamma, T). \tag{3.13a,b}$$

Note that $\hat{\tau}$ is constant for Peierls' barrier, and depends on the plastic strain γ for the dislocation barrier. In this latter case, $\hat{\tau}$ is given by (2.8). Hence, in general, $\dot{\gamma}^\alpha$ depends on the plastic strain, γ, as well as on the resolved shear stress, τ^α, and the temperature, T. When non-Schmid effects are also included, then $\dot{\gamma}^\alpha$ may depend on all stress components.

3.5. Constitutive Relations

For averaging purposes, the nominal stress and the deformation gradient, as well as their rates, are best suited when finite deformations and rotations are involved; see Hill (1972), and Havner (1982). The reference state, however, can be arbitrary. Denote the nominal stress rate, referred to the current configuration, by $\dot{\mathbf{n}}$, and note that

$$\dot{\mathbf{n}} = \overset{\triangledown}{\mathbf{\tau}} + \mathbf{\tau}\mathbf{l}^T. \tag{3.14}$$

The constitutive relation (3.8) now becomes

$$\dot{\mathbf{n}} = \mathcal{L}:\mathbf{d} + \mathbf{\tau}\mathbf{l}^T - \sum_{\alpha=1}^n \dot{\gamma}^\alpha\mathbf{\lambda}^\alpha, \tag{3.15a}$$

where $\mathbf{\lambda}^\alpha$ is defined by (3.8d). It is convenient to express this equation as

$$\dot{\mathbf{n}} = \mathcal{F}':\mathbf{l} - \sum_{\alpha=1}^n \dot{\gamma}^\alpha\mathbf{\lambda}^\alpha, \tag{3.15b}$$

where $\mathcal{F}'$ is the associated instantaneous modulus tensor with components $\mathcal{F}'_{ijkl}$, which is nonsymmetric with respect to the exchange of the indices i and j, and k and l, but not ij and kl.

For incremental calculations, it may be effective to modify $\dot{\gamma}^\alpha$ in (3.15) to account for the change in the slip rate over the corresponding time increment, Δt; Pierce *et al.* (1985). For this, let

$$\Delta\gamma^\alpha = [(1-\theta)\dot{\gamma}^\alpha(t) + \theta\dot{\gamma}^\alpha(t+\Delta t)]\Delta t, \quad 0 \leq \theta \leq 1.$$

$$\dot{\gamma}^\alpha(t+\Delta t) \approx \dot{\gamma}^\alpha(t) + g_\alpha\Delta\tau^\alpha + g_\gamma\Delta\gamma^\alpha, \quad g_\alpha \equiv \partial g/\partial\tau^\alpha, \quad g_\gamma \equiv \partial g/(\partial\gamma), \tag{3.16a-d}$$

where g is defined by (3.13). From these and (3.11), obtain the following results:

$$\Delta\gamma^\beta = \sum_{\alpha=1}^n M^{\beta\alpha}[\dot{\gamma}^\alpha(t) + \theta\,\Delta t\,g_\gamma\mathbf{\lambda}^\alpha:\mathbf{d}]\Delta t,$$

$$[M^{\beta\alpha}] = [\delta^{\alpha\beta}(1-\theta\,\Delta t\,g_\gamma) + \theta\,\Delta t\,g_\alpha\mathbf{\lambda}^\alpha:\mathbf{p}^\beta]^{-1}. \tag{3.16e,f}$$

The final constitutive relation now is

$$\dot{\mathbf{n}} = \mathcal{F}:\mathbf{l} - \mathbf{s}, \quad \mathbf{s} \equiv \sum_{\alpha,\beta=1}^n M^{\alpha\beta}\mathbf{\lambda}^\alpha\dot{\gamma}^\beta,$$

$$\mathcal{F} = \mathcal{F}' - \sum_{\alpha, \beta = 1}^{n} \{\theta \, \Delta t \, g_\beta M^{\beta\alpha} \boldsymbol{\lambda}^\beta \otimes \boldsymbol{\lambda}^\alpha\} . \qquad (3.17\text{a-c})$$

Here s is viewed as the *relaxation* stress produced by rate-dependent plastic flow; Nemat-Nasser and Obata (1986).

4. FROM MESOSCALE TO MACROSCALE: A SELF-CONSISTENT HOMOGENIZATION

Transition from the mesoscale (crystal) to the macroscale (polycrystal) requires a homogenization *model*. Here, consider the self-consistent method of Hill (1965), as generalized by Iwakuma and Nemat-Nasser (1984) and Nemat-Nasser and Obata (1986), for application to finite deformation problems in elastoplasticity. In this section we give an outline of some of the essential steps, and leave the details and related issues to another occasion. For the choice of the kinematical and dynamical variables, it turns out that the deformation gradient, $\mathbf{F}$, its rate, $\dot{\mathbf{F}}$, and the nominal stress, $\mathbf{S}^N$, and its rate, $\dot{\mathbf{S}}^N$, are suitable deformation and stress measures for the purpose of averaging.[8] Their unweighted volume averages are completely defined by the surface data (whether uniform or not), and for either the uniform traction or the linear displacement boundary data, they lead to many useful relations which can be employed to effectively characterize the overall aggregate response. For example, for any boundary data,

$$< \mathbf{F} \mathbf{S}^N > - < \mathbf{F} > < \mathbf{S}^N > = \frac{1}{V} \int_{\partial V} (\mathbf{x} - < \mathbf{F} > \mathbf{X}) \otimes \{\mathbf{N} \cdot (\mathbf{S}^N - < \mathbf{S}^N >)\} \, dA , \qquad (4.1)$$

where $< ... >$ denotes the volume average of the corresponding quantity. Identity (4.1), due to Hill (1984), remains valid if $\mathbf{F}$ or $\mathbf{S}^N$ (or both) are replaced by their rates, $\dot{\mathbf{F}}$ and $\dot{\mathbf{S}}^N$; from (4.1) other identities can be obtained, but they are not discussed here.

The most essential element in a systematic homogenization is a concentration tensor which yields the local (*i.e.*, at the crystal level) field variables in terms of the overall quantities.

4.1. Concentration Tensor

It is common to assume that the stress and deformation fields are uniform within each grain of a suitable RVE (representative volume element) of a polycrystal of volume V and boundary ∂V. Consider such an RVE, subjected to suitable uniform boundary data. The stress and deformation measures are, in general, nonuniform within the RVE. Thus, even though these measures are assumed to be uniform over each grain, they change from grain to grain.

A fundamental issue is to calculate the local quantities, *i.e.*, the quantities within a typical grain, in terms of suitable uniform data prescribed on the boundary of the RVE. In the self-consistent method of Hill, this is done by considering an ellipsoidal inclusion (which represent a typical crystal) embedded in a *homogeneous matrix which has the overall properties of the polycrystal*, and which is subjected to the overall uniform data at infinity. In this manner the local quantities are computed and averaged over relevant orientations and sizes of the crystals to obtain the overall properties. For finite deformation problems, this is

---

[8] Here an arbitrary reference state is assumed. When the current state is the reference one, the nominal stress and its rate are denoted by n and $\dot{n}$, respectively, with l standing for the crystal's velocity gradient.

accomplished incrementally, using appropriate variables which, as pointed out before, are the *nominal stress* and the *velocity gradient*, measured relative to any suitable reference state.

Here, however, we shall directly consider the current configuration as the reference one and use the nominal stress rate, $\dot{\boldsymbol{n}}$, the relaxation stress rate, $\boldsymbol{s}$, and the velocity gradient, $\boldsymbol{l}$, as our basic variables, where now these lower case letters refer to the crystal's field quantities. For the crystal and the uniform matrix, the constitutive relations, respectively are

$$\dot{\boldsymbol{n}} = \mathcal{F}^{\Omega} : \boldsymbol{l} - \boldsymbol{s} , \quad \dot{\mathbf{N}} = \mathcal{F} : \mathbf{L} - \mathbf{S} . \tag{4.2a,b}$$

If the overall modulus tensor $\mathcal{F}$ is such that the operator $\mathcal{F}_{\text{risj}} \partial^2(...) / (\partial x_r \partial x_s)$, is positive-definite, then it can be shown that (see Nemat-Nasser and Obata, 1986)

$$\boldsymbol{l} = \mathcal{A}^{\Omega} : \mathbf{L} + \mathcal{A}^{\Omega} : \boldsymbol{\mathcal{P}}^{\Omega} : (\mathbf{S} - \boldsymbol{s}) , \quad \mathcal{A}^{\Omega} = \{ \mathbf{1}^4 - \boldsymbol{\mathcal{P}}^{\Omega} : (\mathcal{F} - \mathcal{F}^{\Omega}) \}^{-1} , \tag{4.3a,b}$$

where the fourth-order tensor $\boldsymbol{\mathcal{P}}^{\Omega}$ is computed by integrating the Green function of the operator $\mathcal{F}_{\text{risj}} \partial^2(...) / (\partial x_r \partial x_s)$ over the ellipsoidal region Ω. The procedure is to find the solution of

$$\mathcal{F}_{\text{risj}} G^{\infty}_{jm,rs}(\mathbf{x}, \mathbf{y}) + \delta(\mathbf{x} - \mathbf{y}) \delta_{im} = 0 \quad \mathbf{x} \text{ in } V^{\infty}, \quad G^{\infty}_{jm}(\mathbf{x}, \mathbf{y}) \to 0 \quad \text{as } |\mathbf{x}| \to \infty, \tag{4.4a,b}$$

where $\delta(\mathbf{x} - \mathbf{y})$ is the delta function. Then to form the following tensor:

$$\Gamma^{\infty}_{\text{irjs}}(\mathbf{y} - \mathbf{x}) \equiv -\frac{1}{2} \{ G^{\infty}_{ij,rs}(\mathbf{y} - \mathbf{x}) + G^{\infty}_{ji,sr}(\mathbf{y} - \mathbf{x}) \} , \tag{4.5a}$$

where comma followed by the subscripts r and s denotes differentiation with respect to y_r and y_s or x_r and x_s; note that the tensor field $\Gamma^{\infty}_{\text{irjs}}$ has the following symmetries:

$$\Gamma^{\infty}_{\text{irjs}} = \Gamma^{\infty}_{\text{jris}} = \Gamma^{\infty}_{\text{jsir}} = \Gamma^{\infty}_{\text{isjr}} . \tag{4.5b}$$

Hence, while $\mathcal{F}_{\text{irjs}}$, in general, may have, at most, 45 distinct components, $\Gamma^{\infty}_{\text{irjs}}$ has, at most, only 36 distinct components. Finally, one obtains

$$\boldsymbol{\mathcal{P}}^{\Omega} = \int_{\Omega} \Gamma^{\infty}(\mathbf{y} - \mathbf{x}) \, dV_y \tag{4.6}$$

which, for ellipsoidal Ω, is a constant tensor; Iwakuma and Nemat-Nasser (1984).

4.2. Generalized Eshelby's Tensor

Expression (4.3a) gives the local velocity gradient, $\boldsymbol{l}$, in terms of the overall data, $\mathbf{L}$ and $(\mathbf{S} - \boldsymbol{s})$, using the concentration tensor, $\mathcal{A}^{\Omega}$. This tensor is expressed in terms of $\boldsymbol{\mathcal{P}}^{\Omega}$ which can be employed to define a generalized Eshelby tensor, $\boldsymbol{\mathcal{S}}^{\Omega}$, for large-deformation elastoplasticity problems.

Indeed, if an ellipsoidal region, Ω, within a homogeneous, finitely (but homogeneously) deformed, infinitely extended, elastoplastic solid, undergoes a transformation which, if free from the constraint imposed by the surrounding uniform materials of instantaneous effective modulus tensor $\mathcal{F}$, would have a constant transformation velocity gradient equal to $\dot{\mathbf{F}}^*$, then, in the presence of the surrounding constraint, the velocity gradient in Ω would be uniform and given by

$$\dot{\mathbf{F}} = \boldsymbol{\mathcal{S}}^{\Omega} : \mathbf{F}^* , \quad \boldsymbol{\mathcal{S}}^{\Omega} = \mathcal{F} : \boldsymbol{\mathcal{P}}^{\Omega} . \tag{4.7a,b}$$

Now, consider an arbitrary finite region W in an unbounded domain V^∞ of pseudo-modulus tensor $\mathcal{F}$. The region W may consist of several disconnected subregions, say, W_α, $\alpha = 1, 2, ..., n$, of arbitrary shape. Suppose that *an arbitrary transformation velocity gradient field*, $\dot{\mathbf{F}}^*(\mathbf{x})$, is distributed in W. Let Ω be an arbitrary ellipsoidal domain in V^∞, such that W is totally contained within Ω; see Figure 8. Then, the *average* velocity gradient and the corresponding *average* nominal stress rate, taken over domain Ω, are completely determined by the generalized Eshelby tensor for Ω.

Indeed, denoting the average value of $\dot{\mathbf{F}}^*(\mathbf{x})$ over W by $< \dot{\mathbf{F}}^* >_W$, and those of $\dot{\mathbf{F}}(\mathbf{x})$ and $\dot{\mathbf{S}}^N$ over Ω by $< \dot{\mathbf{F}} >_\Omega$ and $< \dot{\mathbf{S}}^N >_\Omega$, respectively, obtain

$$< \dot{\mathbf{F}} >_\Omega = \frac{W}{\Omega} \, \mathcal{S}^\Omega : < \dot{\mathbf{F}}^* >_W,$$

$$< \dot{\mathbf{S}}^N >_\Omega = \frac{W}{\Omega} \, \mathcal{F} : (\mathcal{S}^\Omega - \mathbf{1}^4) : < \dot{\mathbf{F}}^* >_W, \qquad \text{(4.8a,b)}$$

where W and Ω stand for their volumes.

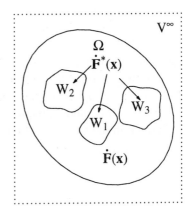

Figure 8. An ellipsoidal Ω in an unbounded uniform V^∞, contains $W = W_1 + W_2 + W_3$; arbitrary transformation velocity gradients $\dot{\mathbf{F}}^*(\mathbf{x})$ are distributed within W.

The remarkable *exact* results (4.8a,b) are valid for any finite W, containing any arbitrary (integrable) transformation velocity gradient, $\dot{\mathbf{F}}^*$. They can be used to homogenize a heterogeneous solid and to obtain the corresponding overall pseudo-moduli, as was discussed before.

4.3. Double Inclusion Problem

As an application of the general results (4.8a,b), consider an ellipsoid Ω_1 within another ellipsoid Ω_2, embedded in a uniform unbounded domain V^∞ of pseudo-modulus tensor $\mathcal{F}$; see Figure 9.

Let Ω_1 contain an arbitrary region W, in which an arbitrary transformation velocity gradient $\dot{\mathbf{F}}^*(\mathbf{x})$ is distributed. It can now be shown that the average value of the velocity gradient over the annulus $\Omega_2 - \Omega_1$, is given by

$$< \dot{\mathbf{F}} >_{\Omega_2 - \Omega_1} = \frac{W}{\Omega_2 - \Omega_1} \{ \mathcal{S}(\Omega_2) - \mathcal{S}(\Omega_1) \} : < \dot{\mathbf{F}}^* >_W. \qquad \text{(4.9)}$$

Here, $\mathcal{S}(\Omega_1)$ and $\mathcal{S}(\Omega_2)$ are the generalized Eshelby tensors corresponding to the ellipsoidal domains Ω_1 and Ω_2, respectively. The average stress rate in $\Omega_2 - \Omega_1$ is given by

$$< \dot{\mathbf{S}}^N >_{\Omega_2 - \Omega_1} = \frac{W}{\Omega_2 - \Omega_1} \, \mathcal{F} : \{ \mathcal{S}(\Omega_2) - \mathcal{S}(\Omega_1) \} : < \dot{\mathbf{F}}^* >_W. \qquad \text{(4.10)}$$

These are the finite-deformation versions of the so-called Tanaka-Mori result (Tanaka and Mori, 1972; see also, Nemat-Nasser and Hori, 1993, Section 11).

476

The generalized Eshelby tensor depends on both the shape and the orientation of the ellipsoid. However, only the ratios of the axes enter the components of this tensor in a coordinate system coincident with the directions of the principal axes of the ellipsoid. Therefore, if the corresponding principal axes of Ω_1 and Ω_2 have common ratios and directions, and since both are embedded in the same unbounded domain, then

$$S(\Omega_2) = S(\Omega_1), \quad <\dot{\mathbf{F}}>_{\Omega_2-\Omega_1} = \mathbf{0}, \quad <\dot{\mathbf{S}}^N>_{\Omega_2-\Omega_1} = \mathbf{0},$$

(4.11a-c)

i.e., the average velocity gradient (and, hence, the average nominal stress rate) in $\Omega_2-\Omega_1$ vanishes. These *exact* results hold for any W of any arbitrary shape, containing any arbitrary transformation velocity gradient, $\dot{\mathbf{F}}^*(\mathbf{x})$. There are a number of other results of this kind, which are discussed in detail by Nemat-Nasser (1997).

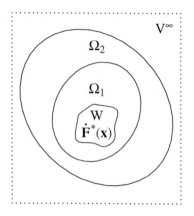

Figure 9. An ellipsoidal Ω_2 in an unbounded uniform V^∞, contains an ellipsoidal Ω_1 which contains an arbitrary region W; arbitrary transformation velocity gradients $\dot{\mathbf{F}}^*(\mathbf{x})$ are distributed within W.

4.4. Overall Properties of Polycrystals

The exact results of the preceding subsections may be used to estimate the effective properties of polycrystals from the properties of the single crystal constituents, at finite deformations and rotations, based on various *models*.[9] Indeed, models used in homogenizing linearly elastic composites (Nemat-Nasser and Hori, 1993), may be generalized by the aid of the exact results presented above, and applied to finite-deformation problems; for a comprehensive account, see Nemat-Nasser (1997). Here, we simply report the effective constitutive relations based on Hill's self-consistent method, as generalized in Nemat-Nasser and Obata (1986) for rate-dependent cases. In this case, the effective pseudo-modulus tensor, $\boldsymbol{\mathcal{F}}$, is used to calculate the concentration tensor, $\boldsymbol{\mathcal{A}}^\Omega$, in (4.3b). Then, from (4.3a), (4.2a), and upon averaging over all crystals, obtain

$$\boldsymbol{\mathcal{F}} = <\boldsymbol{\mathcal{F}}^\Omega : \boldsymbol{\mathcal{A}}^\Omega>, \quad \mathbf{S} = <\mathbf{1}^4 + \boldsymbol{\mathcal{F}}^\Omega : \boldsymbol{\mathcal{A}}^\Omega : \boldsymbol{\mathcal{P}}^\Omega>^{-1} : <(\boldsymbol{\mathcal{F}}^\Omega : \boldsymbol{\mathcal{A}}^\Omega : \boldsymbol{\mathcal{P}}^\Omega + \mathbf{1}^4) : \mathbf{s}>. \quad (4.12a,b)$$

For self-consistency, $<\boldsymbol{\mathcal{A}}^\Omega>$ must equal the fourth-order identity tensor, $\mathbf{1}^4$. To achieve this, the concentration tensor must be normalized as, $\overline{\boldsymbol{\mathcal{A}}}^\Omega = \boldsymbol{\mathcal{A}}^\Omega : <\boldsymbol{\mathcal{A}}^\Omega>^{-1}$; see Walpole (1969), Willis (1981), and Nemat-Nasser and Obata (1986).

---

[9] See, *e.g.*, Budiansky and Wu (1962), Brown (1970), Hutchinson (1970, 1976), Zarka (1972), Beveiller and Zaoui (1979), Weng (1982), Iwakuma and Nemat-Nasser (1984), Nemat-Nasser and Obata (1986), Molinari *et al.* (1987), Lipinski *et al.* (1990, 1992), Havner (1992), and more recently, Bassani (1994).

5. COMPUTATIONAL SIMULATIONS

The rate-dependent constitutive relations in metal plasticity at finite deformations and rotations are highly nonlinear. The approach that has generally been followed in past studies is based on the linearization of the nonlinear constitutive relations which are then evaluated using an implicit time-integration technique. Recently, a semi-explicit and efficient finite-deformation algorithm has been developed, and applied to stiff phenomenological constitutive equations with considerable success; see Nemat-Nasser (1991, 1992), Nemat-Nasser and Chung (1992), Nemat-Nasser and Li (1992), and Rashid and Nemat-Nasser (1992). The method exploits the physical fact that a large part of the finite deformation in a metal is due to plastic flow, with only a very small elastic contribution. Hence, in finite-deformation problems, it is more efficient to assume that the total deformation increment is due to plastic flow (plastic-predictor), and then to correct the results to account for the accompanying elastic deformations (elastic-corrector). This predictor-corrector method has been successfully applied to many large deformation problems, where large time steps have been used without impairing the required accuracy. With such time steps, the forward-gradient method loses both stability and accuracy due to the stiffness of the corresponding system of equations. The method has been further refined by Balendran and Nemat-Nasser (1995) and Fotiu and Nemat-Nasser (1996), and also is combined with the forward-gradient approach for application to crystal-plasticity calculations by Nemat-Nasser and Okinaka (1996). Space limitation does not allow a presentation of an account of this computational technique. Hence, interested readers are referred to the cited publications.

ACKNOWLEDGEMENT

This work has been supported by the Army Research Office under contract No. DAAL03-92-G-0108 with the University of California, San Diego.

REFERENCES

Balendran, B., and S. Nemat-Nasser, "Integration of inelastic constitutive equations for constant velocity gradient with large rotation," Appl. Math. Comp., Vol. 67 (1995) 161-195.

Bassani, J.L., "Plastic flow of crystals," Advances in Applied Mechanics, Vol. 30 (1994) 191-258.

Berveiller, M., and A. Zaoui, "An extension of the self-consistent scheme to plastically flowing polycrystals," J. Mech. Phys. Solids, Vol. 26 (1979) 325-344.

Brown, G.M., "A self-consistent polycrystalline model for creep under combined stress state," J. Mech. Phys. Solids, Vol. 18 (1970) 367-381.

Budiansky, B., and T.T. Wu, "Theoretical prediction of plastic strains of polycrystals," Proc. 4th U.S. Natl. Congr. Appl. Mech., ASME (1962) 1175-1185.

Clifton, R.J., "Dynamic plasticity," J. Appl. Mech., Vol. 50 (1983) 941-952.

Fotiu, P.A., and S. Nemat-Nasser, "A universal integration algorithm for rate-dependent elastoplasticity," Comp. Struc., Vol. 59 (1996) 1173-1184.

Havner, K.S., "The theory of finite plastic deformation of crystalline solids," in *Mechanics of Solids, The Rodney Hill 60th Anniversary Volume,* (ed. H.G. Hopkins and M.J. Sewell) Pergamon Press (1982) 265-302.

Havner, K.S., *Finite Plastic deformation of Crystalline Solids*, Cambridge University Press (1992) 235 pages.

Hill, R.,*The mathematical theory of plasticity,* The Oxford Engineering Science Series,

478

Clarendon Press (1950) ix+355 pages; other editions, (1965, 1960, 1964, 1967, 1971).

Hill, R., "Continuum micro-mechanics of elastoplastic polycrystals," J. Mech. Phys. Solids, Vol. 13 (1965) 89-101.

Hill, R., "On constitutive macro-variables for heterogeneous solids at finite strain," Proc. Roy. Soc. Lond., Vol. 326A (1972) 131-147.

Hill, R., and J.R. Rice, "Constitutive analysis of elastic-plastic crystals at arbitrary strain," J. Mech. Phys. Solids, Vol. 20 (1972) 401-413.

Hill, R., and J.R. Rice,"Elastic potentials and the structure of inelastic constitutive laws," SIAM J. Appl. Math., Vol. 25 (1973) 448-461.

Hill, R., "On macroscopic effects of heterogeneity in elastoplastic media at finite strain," Math. Proc. Camb. Phil. Soc., Vol. 95 (1984) 481-494.

Hutchinson, J.W., "Elastic-plastic behavior of polycrystalline metals and composites," Proc. Roy. Soc. Lond., Vol. 319A (1970) 247-272.

Hutchinson, J.W., "Bounds and self-consistent estimates for creep of polycrystalline materials," Proc. Roy. Soc. Lond., Vol. 325A (1976) 101-127.

Iwakuma, T., and S. Nemat-Nasser, "Finite elastic-plastic deformation of polycrystalline metals," Proc. Roy. Soc. Lond., Vol. 394A (1984) 87-119.

Klepaczko, J.R., and C.Y. Chiem, "On rate sensivity of fcc metals, instantaneous rate sensivity and rate sensivity of strain hardening," J. Mech. Phys. Solids, Vol. 34 (1986) 29-54.

Kocks, U.F., A.S. Argon, and M.F. Ashby, *Thermodynamics and Kinetics of Slip,* Progress in Materials Science, Vol. 19 (Oxford: Pergamon) (1975).

Lipinski, P., J. Krier, and M. Berveiller, "Elastoplasticite des metaux en grandes deformations: comportment global et evolution de la structure interne," Revue Phys. Appl., Vol. 25 (1990) 361-388.

Lipinski, P., A. Naddari, and M. Berveiller, "Recent results concerning the modeling of polycrystalline plasticity at large strains," Inter. J. Solids Struc., Vol. 29 (1992) 1873-1881.

Molinari, A., G.R. Canova, and S. Ahzi, "A self-consistent approach to the large deformation polycrystal viscoplasticity," Acta Metall, Vol. 35 (1987) 2983-2994.

Naghdi, P.M., "Stress-strain relations in plasticity and thermoplasticity," in *Plasticity- Proc. 2nd. Symp. Naval Struc. Mech.* (ed. E.H. Lee and P.S. Symonds) Pergamon Press (1960) 121-169.

Nemat-Nasser, S., "On finite plastic flow of crystalline solids and geomaterials," J. Appl. Mech., Vol. 50 (50th Anniversary Issue, 1983) 1114-1126.

Nemat-Nasser, S., and M. Obata, "Rate-dependent, finite elasto-plastic deformation of polycrystals," Proc. Roy. Soc. Lond., Vol. 407A (1986) 343-375.

Nemat-Nasser, S., "Rate-independent finite-deformation elastoplasticity: a new explicit constitutive algorithm," Mech. of Materials, Vol. 11 (1991) 235-249.

Nemat-Nasser, S., J.B. Isaacs, and J.E. Starrett, "Hopkinson techniques for dynamic recovery experiments," Proc. Roy. Soc., Vol. 435A (1991) 371-391.

Nemat-Nasser, S., "Phenomenological theories of elastoplasticity and strain localization at high strain rates," Appl. Mech. Rev., Vol. 45 (1992) S19-S45.

Nemat-Nasser, S., and D.-T. Chung, "An explicit constitutive algorithm for large-strain, large-strain-rate elastic-viscoplasticity," Comp. Methods Appl. Mech. Eng., Vol. 95 (1992) 205-219.

Nemat-Nasser, S., and Y.-F. Li, "A new explicit algorithm for finite-deformation elasto-plasticity and elastoviscoplasticity: performance evaluation," Comp. Struc., Vol. 44

(1992) 937-963.

Nemat-Nasser, S., and M. Hori, *Micromechanics: Overall Properties of Heterogeneous Solids,* Elsevier Science Publishers (1993) 707 pages.

Nemat-Nasser, S., Y.F. Li, and J.B. Isaacs, "Experimental/computational evaluation of flow stress at high strain rates with application to adiabatic shearbanding," Mech. of Materials, Vol. 17 (1994) 111-134.

Nemat-Nasser, S., and J.B. Isaacs, "Direct measurement of isothermal flow stress of metals at elevated temperatures and high strain rates with application to Ta and Ta-W alloys," Acta Metall. in press (1996).

Nemat-Nasser, S., and T. Okinaka, "A new computational approach to crystal plasticity: fcc single crystal," Mech. of Materials, in press (1996).

Nemat-Nasser, S., *Plasticity: A Treatise on Finite Deformation of Heterogeneous Inelastic Solids,* to appear (1997) approx. 700 pages.

Ono, K., "Temperature dependence of dislocation barrier hardening," J. Appl. Phys., Vol. 39 (1968) 1803-1806 .

Perzyna, P., "Modified theory of visco-plasticity: application to advanced flow and instability phenomena," Arch. Mech., Vol. 32 (1980) 403-420.

Perzyna, P., "Constitutive modeling of dissipative solids for postcritical behaviour and fracture," ASME J. Eng. Mat. Tech. (1984) 410-419.

Rashid, M., and S. Nemat-Nasser, "A constitutive algorithm for rate-dependent crytal plasticity," Comp. Methods Appl. Mech. Eng., Vol. 94 (1992) 201-228.

Regazzoni, G., U.F. Kocks, and P.S. Follansbee, "Dislocation kinetics at high strain rates," Acta Metall. , Vol. 35 (1987) 2865-2875.

Rice, J.R., "Inelastic constitutive relations for solids: an internal-variable theory and its application to metal plasticity," J. Mech. Phys. Solids, Vol. 19 (1971) 433-455.

Senseny, P.E., J. Duffy, and H. Hawley, "Experiments on strain rate history and temperature effects during the plastic deformation of close-packed metals," J. Appl. Mech., Vol. 45 (1978) 60-66.

Tanaka, K., and T. Mori, "Note on volume integrals of the elastic field around an ellipsoidal inclusion," J. Elasticity, Vol. 2 (1972) 199-200.

Walpole, L.J., "On the overall elastic moduli of composite materials," J. Mech. Phys. Solids, Vol. 17 (1969) 235-251.

Weng, G.J., "A unified self-consistent theory for the plastic-creep deformation of metals," J. Appl. Mech., Vol. 49 (1982) 728-734.

Willis, J.R., "Variational and related methods for the overall properties of composites," Adv. Appl. Mech., Vol. 21 (1981) 1-78.

Zarka, J., "Generalization de la theorie du potential plastique multiple en viscoplasticite," J. Mech. Phys. Solids, Vol. 20 (1972) 179-195.

Theoretical and Applied Mechanics 1996
T. Tatsumi, E. Watanabe and T. Kambe (Editors)

Mechanics of saturated-unsaturated porous materials and quantitative solutions

B.A. Schrefler

Dipartimento di Costruzioni e Trasporti, via Marzolo, 9, 35131 Padova, Italy

1. ABSTRACT

A model for the mechanical behaviour of saturated-unsaturated porous media is presented. The necessary balance equations are derived using averaging theories. Constitutive equations and thermodynamic equations for the model closure are introduced. The governing equations are then solved numerically and the numerical properties are discussed. Application examples conclude the paper.

2. INTRODUCTION

Porous materials are materials with an internal structure. They are made of a solid phase and closed and open pores. Attention is here focused on the case where the open pores are filled with one ore more fluids i.e. multiphase media. Porous media are considered as unsaturated if at least one gas phase is present. For instance in case of geomaterials i.e. soil, rock, concrete, the fluids may be water, water vapour and dry air. In the first two materials there may be also gas, oil, water (reservoir engineering). The solid and the fluids usually have different motions. Because of the different motions and of the different material properties there is interaction between the constituents. Further the pore structure has in general a complicated geometry. These facts make a mechanical description of the problem rather difficult. What is in general used for engineering purposes is a substitute model at macroscopic scale, where the interacting constituents are assumed to fill the entire control space. This distribution is obtained by means of the volume fraction concept. Volume fractions are given by the ratio of the volume of the constituents to the volume of the control space. The pores are assumed to be statistically distributed over the control space and the equality of volume and surface porosity follows as a statistical necessity [1,2]. Consequence of the volume fraction concept is that the substitute constituents have reduced densities. These substitute continua can be treated with the methods of continuum mechanics

Two strategies are generally used to arrive at the description of the behaviour of these substitute continua: one starts from macro mechanics and the other from micro mechanics. Phenomenological approaches and the mixture theory integrated by the concept of volume fractions belong to the first strategy. Averaging theories, also called hybrid mixture theories belong to the second one.

The topic is of interest for research workers since considerable time. An extensive review of the history of porous media theories can be found in [1]. Only the more recent contributions are mentioned here. Biot [3,4] developed a phenomenological approach for three-dimensional consolidation of fully saturated porus media. Modern mixture theories were presented by Morland [5], Goodman and Cowin [6], Sampaio and Williams [7] and Bowen [8,9]. Finally averaging theories where developed by Hassanizzadeh and Gray [10-12].

In this paper the governing equations of non-isothermal deforming saturated-unsaturated porous media are established using averaging theories. The equations are then solved numerically by means of a finite element discretization in space and finite differences in time. The numerical properties of these solutions are discussed and several examples are shown.

3. THE MATHEMATICAL MODEL

3.1. Kinematic Equations

As indicated above ,a multiphase medium can be described as the superposition of all π phases, $\pi = 1, 2,...\kappa$, whose material points $\mathbf{X}^\pi$ can be thought of as occupying simultaneously each spatial point $\mathbf{x}$ in the actual configuration. The state of motion of each phase is however described independently. Based on these assumptions, the kinematics of a multiphase medium is dealt with next.

In a Lagrangian or material description of motion, the position of each material point $\mathbf{x}^\pi$ at time t is function of its placement in a chosen reference configuration, $\mathbf{X}^\pi$ and of the current time t

$$\mathbf{x}^\pi = \mathbf{x}^\pi\left(\mathbf{X}^\pi, t\right) \tag{1}$$

To keep this mapping continuous and bijective at all times, the Jacobian J of this transformation must not equal zero and must be strictly positive, since it is equal to the determinant of the deformation gradient tensor $\mathbf{F}^\pi$

$$\mathbf{F}^\pi = \operatorname{Grad} \mathbf{x}^\pi \qquad \left(\mathbf{F}^\pi\right)^{-1} = \operatorname{grad} \mathbf{X}^\pi \tag{2}$$

Because of the non-singularity of the Lagrangian relationship (1), its inverse can be written and the Eulerian or spatial description of motion follows

$$\mathbf{X}^\pi = \mathbf{X}^\pi\left(\mathbf{x}^\pi, t\right) \tag{3}$$

The material time derivative of any differentiable function $f^\pi(\mathbf{x}, t)$ given in its spatial description and referred to a moving particle of the π phase is

$$\frac{\overset{\pi}{D} f^\pi}{Dt} = \frac{\partial f^\pi}{\partial t} + \operatorname{grad} f^\pi \mathbf{v}^\pi \tag{4}$$

If superscript α is used for $\dfrac{D}{Dt}$, the time derivative is taken moving with the α phase.

The deformation process of the solid skeleton is described by the velocity gradient tensor, $\mathbf{L}^s$ which, referred to spatial co-ordinates, is given by

$$\mathbf{L}^s \equiv \operatorname{grad} \mathbf{v}^s = \mathbf{D}^s + \mathbf{W}^s \tag{5}$$

Its symmetric part, $\mathbf{D}^s$, is the Eulerian strain rate tensor, while its skew-symmetric component, $\mathbf{W}^s$, is the spin tensor.

3.2. Microscopic balance equations

The microscopic situation of any π phase is described by the classical equations of continuum mechanics. At the interfaces to other constituents, the material properties and thermodynamic quantities may present step discontinuities.

For a thermodynamic property Ψ the conservation equation within the π phase may be written as

$$\frac{\partial(\rho\psi)}{\partial t} + \operatorname{div}(\rho\psi\dot{\mathbf{r}}) - \operatorname{div}\mathbf{i} - \rho\mathbf{b} = \rho\mathbf{G} \tag{6}$$

where $\dot{\mathbf{r}}$ is the local value of the velocity field of the π phase in a fixed point in space, $\mathbf{i}$ is the flux vector associated with ψ, $\mathbf{b}$ the external supply of ψ and $\mathbf{G}$ is the net production of ψ. The relevant thermodynamic properties ψ are mass, momentum, energy and entropy. The values assumed by $\mathbf{i}$, $\mathbf{b}$ and $\mathbf{G}$ are given in [11,13]. The constituents are assumed to be microscopically non polar, hence the angular momentum balance equation has been omitted. This equation shows however that the stress tensor is symmetric.

3.3. Macroscopic balance equations

In this section the macroscopic balance equations for mass, linear momentum, angular momentum and energy as well as the entropy inequality are given, which have been obtained by systematically applying the averaging procedures to the microscopic balance equations (6) as outlined in [10-12]. It has been shown in [14] that under appropriate assumptions the averaging theory yields the same balance equations as the classical mixture theory. The balance equations have here been specialised for a deforming porous material, where heat transfer and flow of water (liquid and vapour) and of dry air is taking place [13].

The local thermodynamic equilibrium hypothesis is assumed to hold because the time scale of the modelled phenomena is substantially larger than the relaxation time required to reach equilibrium locally. The temperatures of each constituent in a generic point are hence equal. Further the constituents are assumed to be immiscible except for dry air and vapour, and chemically non reacting. All fluids are in contact with the solid phase. Stress is defined as tension positive for the solid phase, while pore pressure is defined as compressive positive for the fluids.

In the averaging procedure the volume fractions η^π appear which are identified as follows: for solid phase $\eta^s = 1 - n$ where $n = \dfrac{dv^w + dv^g}{dv}$ is porosity and dv^π is the volume of constituent π within a R.E.V.; for water $\eta^w = nS_w$ where

$S_w = \dfrac{dv^w}{dv^w + dv^g}$ is the degree of water saturation, and for gas $\eta^g = n\,S^g$ with

$S_g = \dfrac{dv^g}{dv^w + dv^g}$ the degree of gas saturation. It follows immediately that $S_w + S_g = 1$.

The averaged macroscopic mass balance equations are given next. For the solid phase this equation reads

$$\frac{\overset{s}{D}\big[(1-n)\rho^s\big]}{Dt} + \rho^s(1-n)\,\mathrm{div}\,\overline{\mathbf{v}}^s = 0 \tag{7}$$

where ρ^π is the intrinsic phase averaged density and $\overline{\mathbf{v}}^s$ the mass averaged velocity. For water we have

$$\frac{\overset{w}{D}(nS_w\rho^w)}{Dt} + nS_w\rho^w\,\mathrm{div}\,\overline{\mathbf{v}}^w = nS_w\rho^w e^w(\rho) \tag{8}$$

where $nS_w\rho_w e^w(\rho) = -\dot{m}$ is the the quantity of water per unit time and volume, lost through evaporation. For vapour the mass balance equation is

$$\frac{\overset{g}{D}(nS_g\rho^{gw})}{Dt} + \mathrm{div}\mathbf{J}_g^{gw} + nS_g\rho^{gw}\,\mathrm{div}\,\overline{\mathbf{v}}^g = \dot{m} \tag{9}$$

where $\mathbf{J}_g^{gw}$ is the diffusive-dispersive mass flux, and finally for gas (mixture of vapour and dry air) this equation reads

$$\frac{\overset{g}{D}\big(n\,S_g\rho^g\big)}{Dt} + n\,S_g\,\rho^g\,\mathrm{div}\,\overline{\mathbf{v}}^g = \dot{m}. \tag{10}$$

The linear momentum balance equation for the fluid phases is

$$\mathrm{div}\,\mathbf{t}^\pi + \rho_\pi\big(\overline{\mathbf{g}}^\pi - \overline{\mathbf{a}}^\pi\big) + \rho_\pi\big[e^\pi(\rho\,\dot{\mathbf{r}}) + \hat{\mathbf{t}}^\pi\big] = \mathbf{0}. \tag{11}$$

where $\mathbf{t}^\pi$ is the partial stress tensor, $\overline{\mathbf{g}}^\pi$ the external momentum supply, $\rho_\pi\overline{\mathbf{a}}^\pi$ the volume density of the inertia force, ρ_π the phase averaged density, $\rho_\pi e^\pi(\rho\,\dot{\mathbf{r}})$ the sum of the momentum supply due to averaged mass supply and the intrinsic momentum supply due to a change of density and referred to the deviation $\tilde{\mathbf{r}}$ of the velocity of constituent π from its mass averaged velocity, and $\hat{\mathbf{t}}^\pi$ accounts for exchange of momentum due to mechanical interaction with other phases. $\rho_\pi e^\pi(\rho\,\dot{\mathbf{r}})$

is assumed to be different from zero only for fluid phases. For the solid phase the linear momentum balance equation is hence

$$\text{div } \mathbf{t}^s + \rho_s \left(\bar{\mathbf{g}}^s - \bar{\mathbf{a}}^s \right) + \rho_s \, \hat{\mathbf{t}}^s = \mathbf{0} \tag{12}$$

The average angular momentum balance equation shows that for non-polar media the partial stress tensor is symmetric $\mathbf{t}^\pi = \left(\mathbf{t}^\pi \right)^\mathrm{T}$ at macroscopic level also and the sum of the coupling vectors of angular momentum between the phases vanishes.

The macroscopic energy balance equation is

$$\rho_\pi \frac{\overset{\pi}{D} \overline{E}^\pi}{Dt} = \mathbf{t}^\pi : \mathbf{D}^\pi + \rho_\pi h^\pi - \text{div } \tilde{\mathbf{q}}^\pi + \rho_\pi R^\pi \tag{13}$$

where $\overline{E}^\pi$ accounts for averaged specific energy and for averaged kinetic energy related to $\tilde{\mathbf{r}}$, h^π is the sum of averaged heat sources, $\tilde{\mathbf{q}}^\pi$ a macroscopic heat flux vector,

$$\rho_\pi R^\pi = \rho_\pi \left[e^\pi \left(\rho \, \hat{E} \right) - e^\pi(\rho) \, \overline{E}^\pi + Q^\pi \right] \tag{14}$$

with $\rho_\pi e^\pi \left(\rho \hat{E} \right)$ the exchange term of internal energy due to phase change and possible heat exchange between constituents and Q^π represents exchange of energy due to mechanical interaction. The energy balance equations are subject to

$$\sum_\pi \rho_\pi \left[e^\pi \left(\rho \hat{E} \right) + e^\pi \left(\rho \, \tilde{\mathbf{r}} \right) \cdot \overline{\mathbf{v}}^\pi + \frac{1}{2} \, e^\pi(\rho) \, \overline{\mathbf{v}}^\pi \cdot \overline{\mathbf{v}}^\pi + \hat{\mathbf{t}}^\pi \cdot \overline{\mathbf{v}}^\pi + Q^\pi \right] = 0 \tag{15}$$

Finally the entropy inequality for the volume fraction mixture may be obtained as [11,14]

$$\sum_\pi \left[\rho_\pi \frac{\overset{\pi}{D} \overline{\lambda}^\pi}{Dt} + \rho_\pi e^\pi(\rho) \, \overline{\lambda}^\pi + \text{div} \left(\frac{1}{\theta^\pi} \mathbf{q}^\pi \right) - \frac{1}{\theta^\pi} \rho_\pi h^\pi \right] \geq 0. \tag{16}$$

3.4. Constitutive equations

Constitutive models are here selected which are based on quantities currently measurable in laboratory or field experiments and which have been extensively validated. Most of them have been obtained from the entropy inequality, see [12,15].

Moist air (gas) in the pore system is assumed to be a perfect mixture of two ideal gases, dry air and water vapour. The equation of perfect gas is hence valid

$$p^{ga} = \rho^{ga} \theta R / M_a \qquad p^{gw} = \rho^{gw} \theta R / M_w \tag{17}$$

where M_π is the molar mass of constituent π, R the universal gas constant, p^π the macroscopic pressure of the π phase and θ the absolute temperature. Further Dalton's law applies and yields the molar mass of gas

$$\rho^g = \rho^{ga} + \rho^{gw}, \qquad p^g = p^{ga} + p^{gw}, \qquad M_g = \left(\frac{\rho^{gw}}{\rho^g} \frac{1}{M_w} + \frac{\rho^{ga}}{\rho^g} \frac{1}{M_a} \right)^{-1} \tag{18}$$

Water is usually present in the pores as a condensed liquid, separated from its vapour by a concave meniscus because of surface tension. The capillary pressure is defined as $p^c = p^g - p^w$.

The momentum exchange term of the linear momentum balance equation for fluids has the form

$$\rho_\pi \hat{\mathbf{t}}^\pi = -\mathbf{R}^\pi \eta^\pi \overline{\mathbf{v}}^{\pi\alpha}. \tag{19}$$

It is assumed that $\mathbf{R}^\pi$ is invertible and is defined by the following relation

$$\mathbf{K}^\pi = \eta^\pi \left(\mathbf{R}^\pi \right)^{-1} = \frac{\mathbf{k}}{\mu^\pi} \left(\rho^\pi, \eta^\pi, T \right) + p^\pi \operatorname{grad} \eta^\pi \tag{20}$$

where μ^π is the dynamic viscosity, $\mathbf{k}$ the intrinsic permeability and T the temperature. In case of more fluids flowing the intrinsic permeability is modified as

$$\mathbf{k}^\pi = k^{r\pi} \mathbf{k} \tag{21}$$

Diffusive-dispersive mass flux is governed by Fick's law which for a binary gas may be written as

$$\mathbf{J}_g^{ga} = -\rho^g \frac{M_a M_w}{M_g^2} \mathbf{D}_g \operatorname{grad} \left(\frac{p^{ga}}{p^g} \right) = \rho^g \frac{M_a M_w}{M_g^2} \mathbf{D}_g \operatorname{grad} \left(\frac{p^{gw}}{p^g} \right) = -\mathbf{J}_g^{gw} \tag{22}$$

where $\mathbf{D}_g$ is the effective dispersion tensor.

In case of thermodynamic equilibrium the stress tensor in the solid and liquid phases yields the following form of the effective stress principle [13,16], which is employed here

$$\sigma' = \sigma + \mathbf{I} \left(S_w p^w + S_g p^g \right) \tag{23}$$

where σ the total stress and σ' the effective stress tensor, resposible for all major deformations in the skeleton. The expression for the pressure in the solid phase, i.e. the terms in brackets, is also valid under non-equilibrium conditions when the solid grains are incompressible. The effective stress is linked to the strain rate tensor by means of a constitutive relationship

$$\frac{D\sigma'}{Dt} = \boldsymbol{D}_T\left[\left(\boldsymbol{D}^s - \boldsymbol{D}_o^s\right)\right], \qquad \boldsymbol{D}_T = \boldsymbol{D}_T\left(\boldsymbol{D}^s, \sigma', T\right) \tag{24}$$

where $\boldsymbol{D}_o^s$ represents all other strains not directly associated with stress changes and T is the temperature above some datum.

The total heat flux $\mathbf{q}$ in the multiphase medium is governed by Fouriers law

$$\mathbf{q} = -\chi_{\text{eff}}\,\mathrm{grad}\,T \tag{25}$$

The saturation S_w is an invertible function of capillary pressure and temperature. Finally, for the model closure the Kelvin-Laplace law is needed which the equilibrium water vapour pressure p^{gw}

$$R.H. = \frac{p^{gw}}{p^{gws}} = \exp\left(\frac{p^c M_w}{\rho^w R\theta}\right) \tag{26}$$

where the vapour saturation pressure p^{gws} is obtained from the Clausius-Clapeyron equation.

4. GENERAL FIELD EQUATIONS

Introduction of the constitutive equations into the macroscopic balance equations and some rearrangements [13] yield the following general field equations, which will be solved numerically.

The macroscopic mass balance equations are: for the solid phase

$$\frac{(1-n)}{\rho^s}\frac{\overset{s}{D}\rho^s}{Dt} - \frac{\overset{s}{D}n}{Dt} + (1-n)\mathrm{div}\,\mathbf{v}^s = 0 \tag{27}$$

for dry air

$$-n\frac{\overset{s}{D}S_w}{Dt} - \beta_s(1-n)S_g\frac{\overset{s}{D}T}{Dt} + S_g\mathrm{div}\,\mathbf{v}^s + \frac{S_g n}{\rho^{ga}}\frac{\overset{s}{D}}{Dt}\left(\frac{M_a}{\theta R}p^{ga}\right)$$

$$-\frac{1}{\rho^{ga}}\mathrm{div}\left[\rho^g\frac{M_a M_w}{M_g^2}\boldsymbol{D}_g\,\mathrm{grad}\left(\frac{p^{ga}}{p^g}\right)\right] + \frac{1}{\rho^{ga}}\mathrm{div}\left(n\,S_g\rho^{ga}\mathbf{v}^{gs}\right) = 0 \tag{28}$$

for the water species, i.e. liquid water and vapour

$$n\left(\rho^w - \rho^{gw}\right)\frac{\overset{s}{D}S_w}{Dt} - \beta_{swg}\frac{\overset{s}{D}T}{Dt} + \left(\rho^{gw}S_g + \rho^w S_w\right)\mathrm{div}\,\mathbf{v}^s$$

$$+\frac{n\,\rho^{w}S_{w}}{K_{w}}\frac{\overset{s}{D}p^{w}}{Dt}+S_{g}\,n\,\frac{\overset{s}{D}}{Dt}\left(\frac{M_{w}}{\theta R}p^{gw}\right)-\text{div}\left[\rho^{g}\,\frac{M_{a}M_{w}}{M_{g}^{2}}\,\mathbf{D}_{g}\,\text{grad}\left(\frac{p^{gw}}{p^{g}}\right)\right]$$ (29)

$$+\text{div}\left\{\rho^{gw}\,\frac{\mathbf{k}\,k^{rg}}{\mu^{g}}\left[-\text{grad}\,p^{g}+\rho^{g}\left(\mathbf{g}-\mathbf{a}^{s}-\mathbf{a}^{gs}\right)\right]\right\}+\text{div}\left\{\rho^{w}\,\frac{\mathbf{k}\,k^{rw}}{\mu^{w}}\left[-\text{grad}\,p^{w}+\rho^{w}\left(\mathbf{g}-\mathbf{a}^{s}-\mathbf{a}^{ws}\right)\right]\right\}=0$$

where

$$\beta_{swg}=\beta_{s}(1-n)\left(S_{g}\rho^{gw}+\rho^{w}S_{w}\right)+n\,\beta_{w}\rho^{w}S_{w}$$ (30)

with β_{π} the thermal expansion coefficients.
The linear momentum balance equation for fluids is

$$\eta^{\pi}\mathbf{v}^{\pi s}=\frac{\mathbf{k}\,k^{r\pi}}{\mu}\left[-\text{grad}\,p^{\pi}+\rho^{\pi}\left(\mathbf{g}-\mathbf{a}^{s}-\mathbf{a}^{\pi s}\right)\right]$$ (31)

and for the multiphase medium

$$-\rho\,\mathbf{a}^{s}-n\,S_{w}\,\rho^{w}\left[\mathbf{a}^{ws}+\mathbf{v}^{ws}\cdot\text{grad}\,\mathbf{v}^{w}\right]-n\,S_{g}\,\rho^{g}\left[\mathbf{a}^{gs}+\mathbf{v}^{gs}\cdot\text{grad}\,\mathbf{v}^{g}\right]+\text{div}\,\sigma+\rho\,\mathbf{g}=0.$$ (32)

Finally the enthalpy balace for the multiphase medium may be written as

$$\left(\rho C_{p}\right)_{eff}\frac{\partial T}{\partial t}+\left(\rho_{w}C_{p}^{w}\mathbf{v}^{w}+\rho_{g}C_{p}^{g}\,\mathbf{v}^{g}\right)\cdot\text{grad}\,T-\text{div}\left(\chi_{eff}\text{grad}\,T\right)=-\dot{m}\,\Delta H_{vap}$$ (33)

where

$$\left(\rho C_{p}\right)_{eff}=\rho_{s}C_{p}^{s}+\rho_{w}C_{p}^{w}+\rho_{g}C_{p}^{g}\,,\quad \chi_{eff}=\chi^{s}+\chi^{w}+\chi^{g}\,,\quad \Delta H_{vap}=H^{gw}-H^{w}$$ (34)

with H^{π} the specific enthalpy and C_{p}^{π} the heat capacity.

5. INITIAL AND BOUNDARY CONDITIONS

The initial conditions specify the full fields of gas pressure, capillary or water pressure, temperature, displacements and velocities

$$p^{g}=p_{o}^{g},\quad p_{c}=p_{o}^{c},\quad T=T_{o},\quad \mathbf{u}=\mathbf{u}_{o},\quad \dot{\mathbf{u}}=\dot{\mathbf{u}}_{o},\quad p^{\pi}=p_{o}^{\pi},\quad \text{at}\ \ t=t_{0}$$ (35)

The boundary conditions can be imposed values on Γ_{π} or fluxes on Γ_{π}^{q}, where the boundary $\Gamma=\Gamma_{\pi}\cup\Gamma_{\pi}^{q}$. The imposed values on the boundary for gas pressure, capillary or water pressure, temperature and displacements are

$$p^g = \hat{p}^g \quad \text{on } \Gamma_g, \qquad p^c = \hat{p}^c \quad \text{on } \Gamma_c, \qquad T = \hat{T} \quad \text{on } \Gamma_T, \qquad \mathbf{u} = \hat{\mathbf{u}} \quad \text{on } \Gamma_u \qquad (36)$$

The volume averaged flux boundary conditions for water species and dry air conservation equations and the energy equation, to be imposed at the interface between the porous media and the surrounding fluid are as follows

$$\left(\rho^{ga}\overline{\mathbf{v}}^g - \rho^g\overline{\mathbf{v}}^{gw}\right) \cdot \mathbf{n} = q^{ga} \quad \text{on } \Gamma_g^q,$$

$$\left(\rho^{gw}\overline{\mathbf{v}}^g + \rho^w\overline{\mathbf{v}}^w + \rho^g\overline{\mathbf{v}}^{gw}\right)\mathbf{n} = \beta_c\left(\rho^{gw} - \rho_\infty^{gw}\right) + q^{gw} + q^w \quad \text{on } \Gamma_c^q, \qquad (37)$$

$$-\left(\rho^w\overline{\mathbf{v}}^w \Delta h_{vap} - \lambda_{eff}\nabla T\right) \cdot \mathbf{n} = \alpha_c\left(T - T_\infty\right) + q^T \quad \text{on } \Gamma_T^q$$

where $\mathbf{n}$ is the unit vector, perpendicular to the surface of the porous medium, pointing toward the surrounding gas, ρ_∞^{gw} and T_∞ are, respectively, the mass concentration of water vapour and temperature in the undisturbed gas phase distant from the interface, α_c and β_c are convective heat and mass transfer coefficients, while q^{ga}, q^{gw}, q^w and q^T are the imposed dry air flux, imposed vapour flux, imposed liquid flux and imposed heat flux respectively.

Equations (37) are the natural boundary conditions, respectively, for the dry air conservation equation (28), water species conservation equation (29) and energy conservation equation (33), when the solution of these equations is obtained through a weak formulation of the problem, as usually done with the finite element method.

The traction boundary conditions for the displacement field are

$$\sigma \cdot \mathbf{n} = \mathbf{t} \quad \text{on } \Gamma_u^q \qquad (38)$$

where $\mathbf{t}$ is the imposed traction.

6. THE NUMERICAL MODELS

An extensive review of numerical solutions of subsets of the above equations can be found in [17]. Here only a few recent solutions are listed. A model for the quantitative study of static and dynamic isothermal behaviour of fully and partially saturated soils was put forward by Zienkiewicz et al. in two companion papers [18,19]. The model is based on an extension of Biot's theory [3,4], where in unsaturated conditions the air pressure was assumed to remain equal to the ambient air pressure. This model was then extended to finite strain assumptions in [20]. Finite strain effects in fully saturated situations were also considered in [21-23]. Isothermal small strain multiphase solutions for slow phenomena involving a wetting and a non-wetting phase (water and bitumen) were developed by Li [24] and Li and Zienkiewicz [25] and for water and air by Schrefler and Zhan [26]. Finally a non-isothermal three-phase model for slow phenomena with heat transfer through conduction and convection and latent heat transfer, based on averaging theories, was presented in [27]. In the following section the numerical solution of such a model for slow phenomena is shown and in section 6.2 a model for dynamic behaviour of isothermal partially saturated porous media. In the second case the gas phase is either at atmospheric pressure [20] or only water and its vapour (no dry air) are present.

6.1. Coupled heat, water and gas flow in deforming porous media

The model is based on all the equations of section 4, where inertia terms have been neglected and small strains have been assumed. Furthermore the pertinent initial and boundary conditions of section 5 are taken into account. Its numerical solution is carried out by means of the finite element method [28] in space and by finite differences in time. Displacements, capillary pressure, gas pressure and temperature are chosen as macroscopic field variables. The equations of section 4 are rearranged accordingly.

A weak formulation of these governing equations is obtained by applying Galerkin's procedure of weighted residuals [28]. Terms involving second spatial derivatives are transformed by means of Gauss's theorem. Then field variables are approximated in space as is usual in finite element techniques and expressed in terms of their nodal variables.

Discretization in space yields the following system of equations

$$-\int_{\Omega} \mathbf{B}^T \mathbf{\sigma}' \, d\Omega + \mathbf{K}_{ug}\mathbf{p}^g + \mathbf{K}_{uc}\mathbf{p}^c + \mathbf{K}_{ut}\mathbf{T} + \mathbf{f}_u = 0$$

$$\mathbf{C}_{gg}\dot{\mathbf{p}}^g + \mathbf{C}_{gc}\dot{\mathbf{p}}^c + \mathbf{C}_{gt}\dot{\mathbf{T}} + \mathbf{C}_{gu}\dot{\mathbf{u}} + \mathbf{K}_{gg}\mathbf{p}^g + \mathbf{K}_{gc}\mathbf{p}^c + \mathbf{K}_{gt}\mathbf{T} + \mathbf{f}_g = 0$$

$$\mathbf{C}_{cg}\dot{\mathbf{p}}^g + \mathbf{C}_{cc}\dot{\mathbf{p}}^c + \mathbf{C}_{ct}\dot{\mathbf{T}} + \mathbf{C}_{cu}\dot{\mathbf{u}} + \mathbf{K}_{cg}\mathbf{p}^g + \mathbf{K}_{cc}\mathbf{p}^c + \mathbf{K}_{ct}\mathbf{T} + \mathbf{f}_c = 0 \qquad (39)$$

$$\mathbf{C}_{tg}\dot{\mathbf{p}}^g + \mathbf{C}_{tc}\dot{\mathbf{p}}^c + \mathbf{C}_{tt}\dot{\mathbf{T}} + \mathbf{C}_{tu}\dot{\mathbf{u}} + \mathbf{K}_{tg}\mathbf{p}^g + \mathbf{K}_{tc}\mathbf{p}^c + \mathbf{K}_{tt}\mathbf{T} + \mathbf{f}_t = 0$$

The matrices are listed in [13,27]. The effective stresses, $\mathbf{\sigma}'$ appearing in the first equation, are obtained from integration of eq. (24), starting with the known initial values of the problem. The non-symmetric non-linear and coupled system of ordinary differential equations takes the form

$$\mathbf{C}(\mathbf{x})\dot{\mathbf{x}} + \mathbf{K}(\mathbf{x})\mathbf{x} + \mathbf{f}(\mathbf{x}) = 0 \qquad (40)$$

where $\mathbf{x}^T = \{\mathbf{u}, \mathbf{p}^g, \mathbf{p}^c, \mathbf{T}\}$ and $\mathbf{C}$, $\mathbf{K}$ and $\mathbf{f}$ are obtained by assembling the submatrices indicated in eq. (39).

In the following only the direct or monolithic solution of the system of equations ensuing from integration in time domain will be considered. An alternative to this procedure is the staggered procedure, where after operator splitting, iterations between subsets of equations are carried out. The interested reader is referred e.g. to [29-30].

Discretization in time is accomplished through a fully implicit finite difference scheme (backward difference) [28]

$$\mathbf{C}(\mathbf{x}_{n+1})\frac{\mathbf{x}_{n+1} - \mathbf{x}_n}{\Delta t} + \mathbf{K}(\mathbf{x}_{n+1})\mathbf{x}_{n+1} + \mathbf{f}(\mathbf{x}_{n+1}) = 0 \qquad (41)$$

where n is the time step number, Δt is the time step. The solution of eq. (41) is obtained with a Newton-Raphson procedure

$$
\frac{1}{\Delta t}\left[\frac{\partial}{\partial \mathbf{x}}\mathbf{C}\left(\mathbf{x}_{n+1}^{\ell}\right)\left(\mathbf{x}_{n+1}^{\ell}-\mathbf{x}_n\right)+\mathbf{C}\left(\mathbf{x}_{n+1}^{\ell}\right)\right]\Delta\mathbf{x}_{n+1}^{\ell}+\left[\frac{\partial}{\partial \mathbf{x}}\mathbf{K}\left(\mathbf{x}_{n+1}^{\ell}\right)\mathbf{x}_{n+1}^{\ell}+\mathbf{K}\left(\mathbf{x}_{n+1}^{\ell}\right)\right.
$$
$$
\left.+\frac{\partial}{\partial \mathbf{x}}\mathbf{f}\left(\mathbf{x}_{n+1}^{\ell}\right)\right]\Delta\mathbf{x}_{n+1}^{\ell}=-\left[\mathbf{C}\left(\mathbf{x}_{n+1}^{\ell}\right)\frac{\mathbf{x}_{n+1}^{\ell}-\mathbf{x}_n}{\Delta t}+\mathbf{K}\left(\mathbf{x}_{n+1}^{\ell}\right)\mathbf{x}_{n+1}+\mathbf{f}\left(\mathbf{x}_{n+1}^{\ell}\right)\right]
$$

(42)

where ℓ is the iteration index, and at the end of each iteration the primary variables are updated as follows

$$
\mathbf{x}_{n+1}^{\ell+1}=\mathbf{x}_{n+1}^{\ell}+\Delta\mathbf{x}_{n+1}^{\ell}
$$

(43)

For the global convergence analysis we assume consistency and convergence of the finite element discretization in space, which is obtained by a proper choice of elements [28]. A sufficient condition for stability in time of the above procedure can be shown to be [13]

$$
\left\|(\mathbf{C}+\Delta t\ \mathbf{K})^{-1}\mathbf{C}\right\|<1 \qquad \forall n
$$

(44)

where $\|.\|$ is the spectral norm and the matrices are evaluated after convergence of eq.(42) to the exact numerical solution, i.e. for $l\rightarrow\infty$ within each time step. The stability condition and consistency property

$$
\mathbf{r}_{\ell}=\mathbf{0}\left(\Delta t^2\right)
$$

(45)

are sufficient for the above procedure to be globally convergent [29].
For convergence of eq. (42) to a unique solution within one time step Banach's contraction mapping principle states that a sufficient condition for the existence of a unique solution is that

$$
\left\|\mathbf{f}'(\mathbf{x})\right\|<1
$$

(46)

where $\mathbf{f}'(\mathbf{x})$ is the Jacobian matrix of eq. (42).
The transition from partially to fully saturated conditions (and vice versa), is treated here with a formal modification of the relationship between saturation S_w and capillary pressure p^c when saturation S_w is equal to one, the sign of the capillary pressure becomes negative and its value equal to the pressure in the liquid above gas pressure. When such a condition is reached the dry air conservation equation is dropped and gas pressure p^g is set equal to the atmospheric pressure. Capillary and gas pressure oscillations can be avoided by fixing a lower limit for the relative gas permeability.

6.2. Dynamic behaviour of isothermal partially saturated porous media

The model of this section uses all equations of section 4 except the mass balance equation for dry air and the enthalpy balance equation. It is applicable either to fully and partially saturated conditions, where in the unsaturated regions the air pressure remains constant and equal to the athmospheric pressure [20], or to the case of water and vapour flowing (no dry air), with phase change taken into account. The approach is isothermal which means physically that heat is supplied or taken away with infinite rate to maintain constant temperature, thus there are no energetic restrictions for phase change. In case of large deformations and rotations an updated Lagrangian formulation is used, where the reference configuration is that of the last converged step. A hypo-elastic constitutive relationship is adopted for the solid phase. Small strain increments are assumed in each step and the Jaumann stress rate tensor is used as a co-rotational measure associated with the objective Eulerian strain rate tensor [20].

Solid phase displacements and water pressures are assumed as macroscopic field variables. The discretization in space is carried out by means of the finite element method as in the previous section and the Newmark scheme is adopted for time integration, with the lowest allowable order for each variable. The resulting non-linear coupled system is solved by a Newton-Raphson procedure, yielding

$$
\begin{pmatrix}
\mathbf{M} + \mathbf{K}_T \beta_2 \Delta t^2 & -\mathbf{Q}^w \,\Theta \Delta t \\
-\left(\mathbf{Q}^w\right)^T \Theta \Delta t & -\dfrac{\Theta}{\beta_1}\left[\left(\mathbf{H}^w + \mathbf{H}^v\right)\Theta \Delta t + \mathbf{S}^w + \mathbf{S}^v\right]
\end{pmatrix}
\begin{pmatrix}
\Delta \ddot{\mathbf{u}} \\
\Delta \dot{\mathbf{p}}
\end{pmatrix}
=
\begin{pmatrix}
-\Psi^u \\
-\dfrac{\Theta}{\beta_1}\Psi^p + \left(\mathbf{Q}^v\right)^T \Delta \ddot{\mathbf{u}}\,\Theta \Delta t
\end{pmatrix}
\tag{47}
$$

The component matrices can be found in [20], with exception of those listed below, which come from the contribution of the vapour phase.

$$
\mathbf{Q}^v = -\int_\Omega \mathbf{B}^T \frac{\rho^{gw}}{\rho^w}\bar{\alpha}\, S_g\, \mathbf{m}\, N_w\, d\Omega
$$

$$
\mathbf{H}^v = -\int_\Omega \left(\nabla N_w\right)^T \mathbf{k}_g \frac{\rho^{gw}}{\rho^w}\frac{\partial p^{gw}}{\partial p^w}\nabla N_w\, d\Omega
\tag{48}
$$

$$
\mathbf{S}^v = \int_\Omega \mathbf{N}_w^T \frac{\rho^{gw}}{\rho^w}\frac{1}{Q_v}\mathbf{N}_w\, d\Omega
$$

Numerical properties of the scheme can be found in [31].

7. EXAMPLES

Applications of the model of section 6.1 can be found in [13, 26, 27]. Here two examples dealing with slow phenomena are shown. The first refers to drainage from an initially fully saturated sand column. In this case the transition from fully saturated conditions to partial saturation is of importance. A comparison between experimental [32] and numerical results is shown in Figures 1 and 2.

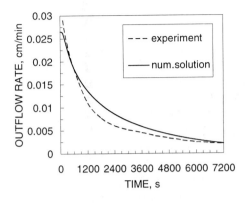

Figure 1. Comparison of the resulting outflow rate history with experimental data [32]

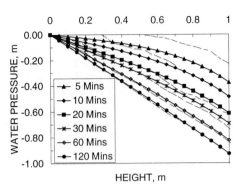

Figure 2. Comparison of the resulting water pressure profiles with experimental data [32]

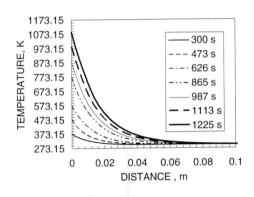

Figure 3. Resulting temperature profiles for heating of concrete

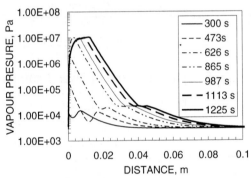

Figure 4. Resulting vapour pressure profiles for heating of concrete

The second example deals with behaviour of concrete at high temperatures. A concrete wall with 95% R.H. at 298,15 K is subjected to an increase of ambient temperature of 80 K/min. In this case phase change is of importance. Temperature and water pressure profiles are shown in Figures 3 and 4.

Applications of the model of section 6.2 to soil mechanics problems can be found in [20] and dynamic strain localisation of dense and medium dense sands in [33-35]. Here an example of strain localisation in an initially fully saturated sample of dense sand is shown. As experimentally observed [36], in such situations strain localisation starts only when relative pore pressures become negative. At the pore pressure values reached cavitation appears and this has been modelled numerically. The sample, effective plastic strain and water pressures are shown in Figures 5, 6, 7 and 8.

494

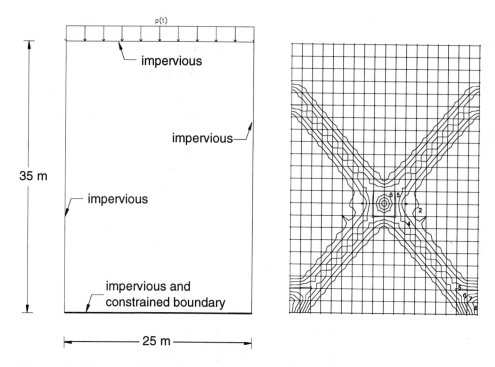

Figure 5. Scheme of the analyzed sample   Figure 6. Effective plastic strain at 0.245 s

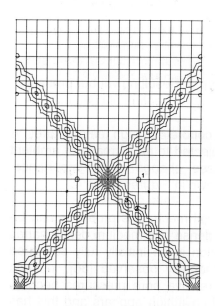

Figure 7. Water pressure at 0.245 s

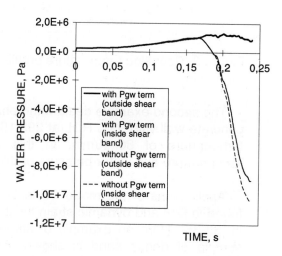

Figure 8. Comparison of two solution for the sand, obtained with or without Pgw term in the same two points: inside and outside shear band

8. CONCLUSIONS

A general model for coupled thermo-hydro-mechanical problems in porous media and their numerical solution has been shown. The few examples quoted give some idea of the variety of problems which can be solved with such a model.

ACKNOWLEDGEMENTS

I wish to thank my co-workers P. Baggio, G. Bolzon, D. Gawin, C.E. Majorana, E.A. Meroi, L. Sanavia, L. Simoni, E. Turska, R. Vitaliani, X. Wang, X. Zhan and H.W. Zhang, who over years have contributed to the work reported in this paper. This work has been partly financed by research funds M.U.R.S.T. 40%.

REFERENCES

1. R. de Boer, Applied Mechanics Reviews, 49 4 (1996) 201.
2. A. Delesse, Annales des mines, 3 series 13 (1848) 379.
3. M.A. Biot, J. Appl. Phys., 12 (1941) 155.
4. M.A. Biot, J. Appl. Mech., 23 (1956), 91.
5. L.W. Morland, J. Geophys. Res. 77 (1972) 890.
6. M.A. Goodman and S.C. Cowin, Arch. Rational Mech. Anal. 44 (1972) 249.
7. R. Sampaio and W.O Williams, Journal de Mechanique, 18 (1979) 19.
8. R.M.Bowen, Int. J. Engng. Sci.,18 (1980) 1129.
9. R.M. Bowen, Int. J. Engng. Sci. 20 (1982) 697.
10. M. Hassanizadeh and W.G. Gray, Adv, Water Resources, 2 (1979) 131.
11. M. Hassanizadeh and W.G. Gray, Adv. Water Resources, 2 (1979) 191.
12. M. Hassanizadeh and W.G. Gray, Adv. Water Ressources, 3 (1980), 25.
13. B.A. Schrefler, Archives of Comp. Methods in Engineering, 2 3 (1995) 1.
14. R. de Boer, W. Ehlers, S. Kowalski and J. Plischka, Forschungsbericht aus dem Fachbereich Bauwesen,54, Universitaet-Gesamthochschule Essen, 1991.
15. W.G. Gray and S.M. Hassanizadeh, Water Resour. Res., 27 (1991) 1855.
16. S.M. Hassanizadeh, Adv. Water Resources, 13 (1990) 169.
17. R.W. Lewis and B.A. Schrefler, John Wiley, Chichester, 1987.
18. O.C. Zienkiewicz, A.H.C. Chan, M. Pastor, D.K. Paul and T. Shiomi, Proc. Roy. Soc. London, A 429 (1990) 285.
19. O.C. Zienkiewicz, Y.M. Xie, B.A. Schrefler, A. Ledesma, N. Bicanic, Proc. Roy. Soc. London, A 429 (1990) 311.
20. E.A. Meroi and B.A. Schrefler, Int. J. Numer. Analytic. Meth. Geomech., 19 (1995) 81.
21. J.P. Carter, J.R. Booker and J.C. Small, Int. J. Numer. Analytic. Meth. Geomech., 3 (1979)
22. O. C. Zienkiewicz and T. Shiomi, Int. J. Numer. Analytic. Meth. Geomech., 8 (1984) 71.
23. S.H. Advani, T.S. Lee, J.H. Lee and C.S. Kim, Int. J. Numer. Methods Eng., 36 (1993) 147.
24. X. Li, Comm. Appl. Numer. Methods, 6 (1990) 125.
25. X. Li and O.C. Zienkiewicz, Comp. Struct., 45 (1992) 211.
26. B.A. Schrefler and X. Zhan, Water Resources Res., 29 (1993) 155.

27. D. Gawin. P. Baggio and B.A. Schrefler, Int. J. Numer. Methods Fluids, 20 (1995) 969.
28. O.C. Zienkiewicz and R.L. Taylor, Mc Graw-Hill, London, 1989
29. E. Turska and B.A. Schrefler, Comp. Methods Appl. Mech. Eng., 106 (1993) 51.
30. E. Turska, K. Wisniewski and B.A. Schrefler, Comp. Methods Appl. Mech. Eng., 114 (1994) 177.
31. A. Gens, P. Jouanna and B.A. Schrefler (eds), Numerical Solutions of Thermo-Hydro-Mechanical Problems, CISM Series, Springer Verlag, Wien, 1995
32. A.C. Liakopoulos, Dr.Eng. Thesis, Univ. of California, Berkeley, 1965.
33. B.A. Schrefler, C.E. Majorana and L. Sanavia, Arch. Mech., 47 1995 577.
34. B.A Schrefler, L. Sanavia and C.E. Majorana, Mech. Cohesive-Frictional Materials, 1 (1995) 95.
35. W. Walther (ed), Strain localisation modelling in saturated sand samples, Forschungsbericht aus dem Fachbereich Bauwesen, Univ. Essen, 1995.
36. M. Mokni, Ph.D. Thesis, Institut de Mecanique de Grenoble, France, 1992.

Theoretical and Applied Mechanics 1996
T. Tatsumi, E. Watanabe and T. Kambe (Editors)
© 1997 Elsevier Science B.V. All rights reserved.

Instability of plastic deformation processes

H. Petryk[a]

[a]Institute of Fundamental Technological Research, Polish Academy of Sciences, Swietokrzyska 21, 00-049 Warsaw, Poland

The energy criterion of instability of equilibrium has been extended to quasi-static deformation processes in time-independent plastic solids under general assumptions that result in a symmetric tangent stiffness matrix. The consistency with the bifurcation approach is demonstrated along with the additional possibility of eliminating unstable paths in non-unique deformation processes. As an application, the computational algorithm for passing bifurcation points with automatic branch switching has been developed and used in finite element computations. Consequences of the material stability requirement are studied. In particular, an analytic formula is derived for the incipient volume fraction of a localization zone in an incrementally nonlinear material.

1. INTRODUCTION

The real behaviour of a slowly loaded plastic body can be in agreement with the simplest theoretical predictions up to some stage of the deformation and deviate from them beyond this stage. Typical examples refer to the conditions which could result in a quasi-static, macroscopically uniform deformation process. Beyond a critical instant, the deformation observed can (A) cease to be quasi-static (as in a dynamic snap-through or total collapse), (B) become nonuniform with overall changes in the body shape (due to buckling, necking, bulging, etc.), or (C) become nonuniform inside the body (e.g. due to shear band formation or internal buckling). In the engineering language, it is common to speak in such cases about instability, although in a more intuitive than rigorous sense. It is convenient to distinguish three kinds of instability: of a dynamic, geometric or material type, corresponding to the examples (A), (B) or (C) above, respectively.

In the framework of time-independent plasticity, the different types of instability have been investigated by using various theoretical approaches, reviewed e.g. in [1-6]. We restrict ourselves here to a few remarks concerning the fundamentals for the class of problems to be discussed below. Instabilities of a dynamic type are usually detected by applying the energy criterion of stability of *equilibrium* [7]. The basic theoretical tool in the analysis of geometric instabilities is the bifurcation theory [7,1], supplemented by studies of post-bifurcation behaviour and imperfection sensitivity [8]. Material instabilities, in both 2D and 3D continua, have been studied by examining the growth of initial imperfections in a band [9], or the possibility of bifurcation within a band [10] connected with the loss of ellipticity of the governing rate-equations and with vanishing speed of propagation of acceleration waves [11].

None of these approaches allows the instabilities of type (A), (B) and (C) to be investigated in a unified manner. This becomes available if the concept of instability of equilibrium is extended to quasi-static deformation *processes*. Under general assumptions that result in a symmetric tangent stiffness matrix, the author showed [12] that the observed distinct forms of plastic instability can be consistently treated as symptoms of the instability of a fundamental process in the unified energy sense. It is the aim of this paper to present this approach along with its justification and selected applications. Several novel contributions are given in subsections 3.5, 5.2, 5.3 and 6.

2. FORMULATION OF THE RATE-PROBLEM

2.1. Constitutive framework

In stability investigations in plasticity, it is essential to incorporate in the material description two effects: of incremental changes in geometry and of vertex formation on the yield surface. The former effect can be of importance when the incremental stiffness moduli are of order of the stresses, no matter whether the strains are small or not [7]. The second effect is invariably predicted by micromechanical models of elastoplastic polycrystals, and is connected with incremental nonlinearity of a plastic constitutive law, cf. [13,14]. For a time-independent (and path-dependent) 'simple' material undergoing isothermal straining, the basic constitutive equation takes the form of a continuous and homogeneous of degree one relationship between the forward rates of stress and strain.[1] Irrespectively of the variables in which it is originally formulated to ensure objectivity, the constitutive rate-equation at finite strain can be expressed as

$$\dot{\mathbf{S}} = \dot{\mathbf{S}}(\dot{\mathbf{F}}, \mathcal{H}) = \mathbf{C}(\dot{\mathbf{F}}, \mathcal{H}) \cdot \dot{\mathbf{F}}, \qquad \mathbf{C} \equiv \frac{\partial \dot{\mathbf{S}}}{\partial \dot{\mathbf{F}}}, \tag{1}$$

where $\mathbf{F}$ is the deformation gradient and $\mathbf{S}$ stands for the first (non-symmetric) Piola-Kirchhoff stress, related to the symmetric Cauchy stress σ and Kirchhoff stress τ by $\mathbf{SF^T} = \det(\mathbf{F})\,\sigma = \tau$, with a superscript 'T' denoting a transpose. A dot over a symbol denotes the *right-hand* material time derivative (with respect to a time-like parameter t), and a dot between two tensor symbols indicates full contraction, in the sense that $(\mathbf{C} \cdot \dot{\mathbf{F}})_{ij} = C_{ijkl}\dot{F}_{kl}$ and $\dot{\mathbf{S}} \cdot \dot{\mathbf{F}} = \dot{S}_{ij}\dot{F}_{ij}$ in Cartesian coordinates with the summation convention. The moduli tensor $\mathbf{C}$ may depend on the *direction* of the strain rate in a discontinuous or nonlinear manner; for the transformation formulae that relate $\mathbf{C}$ to objective moduli associated with any pair $(\mathbf{t}, \mathbf{e})$ of work-conjugate measures of stress and strain, see [1]. The symbolic parameter $\mathcal{H}$ (usually omitted later for simplicity) indicates the influence of the deformation-gradient history. Arbitrary 'hardening' or 'softening' characteristics are allowed, noting that these concepts are not measure-invariant [1].

Following Hill [15,1], we will assume that the relationship (1) admits a potential, $U(\dot{\mathbf{F}}, \mathcal{H})$ say, such that

$$\dot{\mathbf{S}} = \frac{\partial U}{\partial \dot{\mathbf{F}}}, \qquad \mathbf{C} = \frac{\partial^2 U}{\partial \dot{\mathbf{F}} \partial \dot{\mathbf{F}}}, \qquad U = \frac{1}{2}\dot{\mathbf{S}} \cdot \dot{\mathbf{F}}, \tag{2}$$

---

[1]Stresses are assumed here to be independent of the strain-rate, so that rigid-plastic solids are excluded unless the class of admissible deformation modes is appropriately restricted.

U being a homogeneous of degree two, continuously differentiable and piecewise continuously twice differentiable function of $\dot{\mathbf{F}}$. This is equivalent to imposing on (1) the symmetry restriction[2]

$$\mathbf{C} = \mathbf{C}^{\mathrm{T}} \qquad (\Leftrightarrow C_{ijkl} = C_{klij}). \tag{3}$$

The constitutive relationship of the classical elastoplasticity with normality, which at a point on a smooth yield surface can be written as

$$\overset{\triangledown}{\boldsymbol{\tau}} = (\mathbf{L}^e - \frac{\kappa}{g} \boldsymbol{\lambda} \otimes \boldsymbol{\lambda}) \cdot \mathbf{D}, \qquad \kappa = \begin{cases} 1 & \text{if } \boldsymbol{\lambda} \cdot \mathbf{D} \geq 0 \quad \text{(loading)} \\ 0 & \text{if } \boldsymbol{\lambda} \cdot \mathbf{D} \leq 0 \quad \text{(unloading)}, \end{cases} \tag{4}$$

represents a special case of (2) [7]. Here, $\overset{\triangledown}{\boldsymbol{\tau}}$ is the corotational (Zaremba-Jaumann) flux of Kirchhoff stress, $\mathbf{D}$ is the Eulerian strain-rate, $\mathbf{L}^e = \mathbf{L}^{e\mathrm{T}}$ is the current elastic moduli tensor, $\boldsymbol{\lambda}$ is an outward normal to the yield surface in a strain space (with the current configuration as reference), $\otimes$ denotes a tensor product, and g is a positive parameter. More generally, (2) is obtained [16,1] at a point of intersection of N interacting smooth yield surfaces $f_K(\mathbf{e}, \mathcal{H}_L) = 0$, $K, L = 1, ..., N$, with the 'normality structure' for any pair $(\mathbf{t}, \mathbf{e})$ of work-conjugate measures of stress and strain:

$$\dot{\mathbf{t}} = \mathbf{L}^e \cdot \dot{\mathbf{e}} - \sum_K \boldsymbol{\lambda}_K \dot{\gamma}_K, \qquad \dot{f}_K = \boldsymbol{\lambda}_K \cdot \dot{\mathbf{e}} - \sum_L g_{KL} \dot{\gamma}_L, \qquad \boldsymbol{\lambda}_K = \frac{\partial f_K}{\partial \mathbf{e}}, \qquad \mathbf{L}^e = \mathbf{L}^{e\mathrm{T}}, \tag{5}$$

$$\dot{f}_K \leq 0, \qquad \dot{\gamma}_K \geq 0, \qquad \dot{f}_K \dot{\gamma}_K = 0, \tag{6}$$

provided the additional symmetry condition is fulfilled

$$g_{KL} = g_{LK} \tag{7}$$

and that

$$(g_{KL}) \text{ is positive definite} \tag{8}$$

to ensure uniqueness of $\dot{\gamma}_K$'s for prescribed $\dot{\mathbf{e}}$ [17]. If $N = 1$ and $\mathbf{e}$ is the logarithmic strain with the current configuration as reference then (4) is recovered.

The equations (5)-(6) appear to be generally accepted for single metal crystals deformed plastically by multiple slip, at least under ordinary pressures and in the range of strain rates and temperatures in which rate-dependence effects are negligible [17–19]. The question whether (7) is also acceptable is essential for the existence of a velocity-gradient potential, not only for single crystals but also for polycrystals. For, if (5)-(6) are accompanied by (7) at a *micro*-level the the existence of a macroscopic velocity-gradient potential for (1) can be inferred, even without assuming (8) [20]. Although (7) was sometimes criticised as possessing apparently no physical justification, it can be supported by thermodynamic argumentation; this matter is not pursued here.

A constitutive relationship (1) with (3) can be thoroughly nonlinear, and not only piecewise linear as in (4) or (5). For instance, the J_2 corner theory of plasticity, proposed in [21] and used later in a number of numerical calculations, is of that type.

---

[2]If $\mathbf{C}$ in (3) is replaced by stiffness moduli for any work-conjugate variables, this is a measure-invariant property [1].

2.2. The energy functional

We are concerned with isothermal deformation of a continuous solid body of a time-independent material, in general piecewise-smoothly inhomogeneous, subject to a quasi-static loading program parameterized by a scalar loading parameter $\lambda = \lambda(t)$ (possibly a non-monotonic function). Suppose that the body in a fixed reference configuration occupied a spatial domain V bounded by a piecewise smooth surface S, with $\boldsymbol{\xi}$ as a position vector. S_u denotes the part of S where displacements $\mathbf{u} = \bar{\mathbf{u}}(\boldsymbol{\xi}, \lambda)$ are controlled; S_u may only refer to certain components and is regarded as known and independent of λ. Unilateral constraints are not examined here. A class of kinematically admissible deformation processes χ is considered such that the current placement $\mathbf{x} = \chi(\boldsymbol{\xi}, t)$ is a continuous and piecewise sufficiently smooth function of $(\boldsymbol{\xi}, t)$, $\mathbf{F} = \partial\mathbf{x}/\partial\boldsymbol{\xi}$ varies continuously with t, and velocities $\mathbf{v} = \dot{\mathbf{x}}$ are continuous in $V \cup S$ with $\mathbf{v} = \dot{\mathbf{u}}$ on S_u. A tilde over a symbol is used to distinguish a spatial field from its value at a point.

To include elastic supports or fluid-pressure loading, nominal surface tractions $\mathbf{T}$ on $S_T = S \setminus S_u$ are taken to depend not only on λ but also on the local displacements or their gradients; the same is allowed for nominal body forces $\mathbf{b}$ in V (e.g. for 2D problems). The (incremental) loading is assumed to be *conservative* in an overall sense [22,23]. This means that a fixed value of λ, the *total* work done by $\mathbf{b}$ and $\mathbf{T}$ in any virtual motion compatible with the kinematic constraints and leading from a configuration $\tilde{\mathbf{u}}^0$ to any sufficiently close configuration $\tilde{\mathbf{u}}$ is (to second-order) *path-independent*, viz.

$$\int_V \int_{\mathbf{u}^0}^{\mathbf{u}} \mathbf{b}\, d\mathbf{u}\, dV + \int_{S_T} \int_{\mathbf{u}^0}^{\mathbf{u}} \mathbf{T}\, d\mathbf{u}\, dS = \Omega(\tilde{\mathbf{u}}^0, \lambda) - \Omega(\tilde{\mathbf{u}}, \lambda), \quad \lambda = \text{const.} \tag{9}$$

The functional $\Omega(\tilde{\mathbf{u}}, \lambda)$ is defined to within an additive function of λ that may be chosen arbitrarily. With a physically appropriate choice of that function, Ω is identified with the potential energy of the loading device.

The work of deformation within the body can be written as

$$W = \int_V \int \mathbf{S}\, d\mathbf{F}\, dV, \tag{10}$$

where the stresses are determined pointwise by integration of the constitutive rate equation (1) along the deformation path. Define *the energy functional* as [26,12]

$$E = W + \Omega \tag{11}$$

for any kinematically admissible deformation process at varying or fixed λ. In general, E is a functional of the deformation history due to path-dependence of W. An increment of the value of E can be interpreted as the amount of energy which has to be supplied from external sources to the mechanical system consisting of the deformed body *and* the loading device in order to produce quasi-statically a deformation increment, generally with the help of additional perturbing forces. If the mechanical system is imagined to be placed in a heat reservoir then an increment of E can be identified with a change in the internal energy of the resulting compound thermodynamic system (cf. [29]).

2.3. Quasi-static rate boundary value problem

At a given value of λ, every kinematically admissible velocity field $\widetilde{\mathbf{v}} \in \mathcal{V}$ can be decomposed as

$$\widetilde{\mathbf{v}} = \widetilde{\mathbf{v}}^k + \widetilde{\mathbf{w}}, \qquad \widetilde{\mathbf{v}} \in \mathcal{V}, \qquad \widetilde{\mathbf{w}} \in \mathcal{W}, \qquad \mathbf{v}^k = \dot{\mathbf{u}} \text{ on } S_u, \qquad \mathbf{w} = \mathbf{0} \text{ on } S_u, \tag{12}$$

where $\widetilde{\mathbf{v}}^k$ is a fixed field and $\mathcal{W}$ is a linear space. We will assume that all fields in (12) are continuous and piecewise continuously twice differentiable.

From the virtual work principle it follows that a configuration of the body is in mechanical equilibrium if and only if

$$\dot{E}(\widetilde{\mathbf{v}}) = \text{const} \qquad \text{in } \mathcal{V}. \tag{13}$$

It has been shown [12,28] that $\ddot{E}$, the second right-hand time derivative of E, when evaluated in an equilibrium state, does not depend on accelerations and differs from the (Gateaux differentiable) velocity functional appearing in Hill's theory [1] only by a factor two and additive constant terms. Under the assumptions (2) and (9), it follows that a velocity field $\widetilde{\mathbf{v}} \in \mathcal{V}$ is a solution to the first-order rate boundary value problem of continuing equilibrium if and only if it satisfies the variational equality

$$\delta\ddot{E}(\widetilde{\mathbf{v}}, \widetilde{\mathbf{w}}) \equiv \frac{\mathrm{d}}{\mathrm{d}r}\ddot{E}(\widetilde{\mathbf{v}} + r\widetilde{\mathbf{w}})|_{r=0} = 0 \qquad \text{for every } \widetilde{\mathbf{w}} \in \mathcal{W}. \tag{14}$$

If Ω is defined to second order terms for surface loading at prescribed $\dot{\mathbf{b}}$ then the conditions leading to (14) are precisely those adopted in Hill's bifurcation theory [22,1].

The formulation (14) of the exact problem for velocities can be extended to approximate solutions [24]. Consider a spatially discretized problem, in Lagrangian description, with $\widetilde{\mathbf{w}} \in \mathcal{W}$ restricted to the form $\widetilde{\mathbf{w}} = w_\alpha\widetilde{\phi}_\alpha$, $\alpha = 1, ..., M$, where w_α are real numbers and $\widetilde{\phi}_\alpha$ are fixed linearly independent 'shape functions'; the summation convention for repeated Greek subscripts varying from 1 to M is henceforth assumed. A velocity solution $\widetilde{\mathbf{v}} = \widetilde{\mathbf{v}}^k + v_\alpha\widetilde{\phi}_\alpha$ to the discretized problem is still characterized by (14) with $\mathcal{W}$ being now M-dimensional, which leads to a system of nonlinear algebraic equations for v_α, viz.

$$\dot{Q}_\alpha(\widetilde{\mathbf{v}}) = \dot{P}_\alpha(\widetilde{\mathbf{v}}), \qquad Q_\alpha = \frac{\partial\dot{W}}{\partial v_\alpha}, \qquad P_\alpha = -\frac{\partial\dot{\Omega}}{\partial v_\alpha}, \qquad \alpha = 1, ..., M. \tag{15}$$

The quantities Q_α and P_α have the standard interpretation of generalized internal and external forces, respectively. From (2) and (9) it follows that the tangent stiffness matrix for the mechanical *system* {body + loading device}, which appears in the analysis of bifurcation and instability:

$$K_{\alpha\beta}(\widetilde{\mathbf{v}}) \equiv \frac{\partial(\dot{Q}_\alpha - \dot{P}_\alpha)}{\partial v_\beta}(\widetilde{\mathbf{v}}) = \frac{1}{2}\frac{\partial^2\ddot{E}(\widetilde{\mathbf{v}})}{\partial v_\beta\partial v_\alpha}, \qquad \alpha, \beta = 1, ..., M, \tag{16}$$

is symmetric, $[K] = [K]^\mathsf{T}$. Conversely, if this matrix for a rate problem not involving a natural time is found to be always symmetric and satisfies (16) then the second-order energy approach presented here can be applied. This offers a possibility of examining stability of deformation processes without the need of appealing to an exact 3D formulation.

3. INSTABILITY OF A DEFORMATION PROCESS AND THE ENERGY CRITERION

3.1. Path instability in the first approximation

In Hill's theory of uniqueness and stability in elastic-plastic solids [7], the term stability concerns an equilibrium state, while uniqueness and bifurcation refer to a deformation process. In the problem of stability of equilibrium, the effect of small disturbances is examined at constant loading, or more generally, at a fixed value of the loading parameter λ. In the bifurcation problem, no disturbances are considered, and the question of non-uniqueness of an unperturbed solution is examined at varying λ. Another approach, very common in numerical calculations, is to analyze at varying λ the behaviour of a system changed due to small initial imperfections.

The approach to be presented here, complementary to Hill's theory, is to investigate the effect of *small disturbances at varying* λ, that is, to examine *stability* of a deformation *process*. The material system itself remains unchanged, which makes a clear distinction from the initial imperfection approach. The idea can be traced back to Considère, cf. also [25–27]. The motivation comes from the fact that theoretical equations are never satisfied exactly in the course of real deformation. The inaccuracy may be simulated by applying appropriate perturbing forces *during* deformation; in the general theory of stability of motion, such influences are called *persistent disturbances*. There is a vast literature on various theories of stability of motion, in particular, on Lyapunov's theory where *initial* disturbances are considered. Nevertheless, only recently a relation was established [28] between a criterion of path instability applicable to a broad class of plasticity problems and a defined kind of instability of quasi-static motion.

This kind of instability may be called *instability in the first approximation*; it is discussed in [29] in more detail. The essence of the concept is that at each value of λ the incremental problem is *partially* linearized while the incremental nonlinearity present in the constitutive relationship (1) is fully retained. In particular, the rates in (1) are replaced by small increments provided the latter are reached on a direct route in $\mathbf{F}$-space; the approximation is not valid for circuitous paths. Now, if under arbitrarily small perturbing forces in any finite subinterval of λ a finite distance from a fundamental deformation path can be reached (on a direct route) then the fundamental process is regarded as unstable. For a non-discretized continuum, this concept must be made precise by specifying the measures of the magnitude of perturbing forces and of the distance between two deformation paths, cf. [28].

3.2. The energy criterion of instability of a deformation process

Under the conditions that lead to the variational formulation (14) of the basic problem for velocities, the following stability criterion for a process of elastoplastic deformation has been proposed [26,12]: *In a stable process of quasi-static deformation, the increment of E calculated with accuracy to second-order terms is minimized within the class of all kinematically admissible deformation increments.* This may be regarded as a specification of the intuitive engineering hypothesis that a real deformation mode in ductile metals exhibits a tendency to minimize the energy consumption. Justification of the criterion as a necessity condition for stability in the first approximation was given in [28]. The question of sufficiency was discussed in [29] in a thermodynamic framework.

Henceforth, all quantities in a fundamental process χ^0 whose stability is examined are distinguished by a superscript '0'. On account of (13), the above criterion can be reformulated as follows: *In a stable process χ^0 of quasi-static deformation,*

$$\ddot{E}(\widetilde{\mathbf{v}}) \geq \ddot{E}(\widetilde{\mathbf{v}}^0) \qquad \textit{for every } \widetilde{\mathbf{v}} \in \mathcal{V}. \tag{17}$$

By (14), any velocity field $\widetilde{\mathbf{v}}^0$ satisfying (17) represents automatically a quasi-static solution; this is an essential point. If a solution is not unique then (17) can provide a criterion of choice within the solution class. For a discretized problem, it is understood that $\mathcal{V}$ is finite-dimensional and that $\widetilde{\mathbf{v}}^0$ is a solution to (15).

The criterion can be illustrated [28] by the classical example of buckling of the Shanley column with a central two-flange elastic-plastic hinge [30]. The model has two degrees of freedom: the rotation angle Θ of rigid arms and the nondimensional relative vertical displacement u of the end points (Fig. 1a); $\widetilde{\mathbf{v}}$ can thus be identified with $(\dot{\Theta}, \dot{u})$. The graph of $\ddot{E}(\dot{\Theta}, \dot{u})$ consists of four quadrics joined smoothly along the lines $\dot{\Theta} = \pm \dot{u}$. In the uniqueness range $P < P_T$ where P_T is the tangent modulus load, $\ddot{E}$ is minimized by the fundamental mode $\widetilde{\mathbf{v}}^0$ with $\dot{\Theta}^0 = 0$, $\dot{u}^0 = 1$. If the hardening modulus decreases continuously along the fundamental path then the uniqueness range terminates at $P = P_T$; this is the primary bifurcation point at which an absolute minimum of $\ddot{E}(\dot{\Theta}, \dot{u})$ is attained at any $\widetilde{\mathbf{v}}$ from the 'fan' of velocity solutions: $\dot{u} = \dot{u}^0$, $|\dot{\Theta}| \leq \dot{u}^0$ (cf. Fig. 1a). For the straight column within the range $P_T < P < P_R$, where P_R is the reduced modulus load, the fundamental solution $(\dot{\Theta}^0, \dot{u}^0)$ becomes a saddle point of $\ddot{E}$ while an absolute minimum value of $\ddot{E}$ is attained at the secondary solution points $(\dot{\Theta}^*, \dot{u}^*)$ (cf. Fig. 1b). In this range, the fundamental *process* of compression without buckling cannot be stable according to the criterion (17). It follows that the buckling path emanating at the *primary* bifurcation point is the only path for $P > P_T$ not excluded by (17), which is consistent with the conclusion from the initial imperfection approach. For $P > P_R$ in an unbuckled configuration, $\ddot{E}$ is unbounded from below and the equilibrium state is unstable in the dynamic sense.

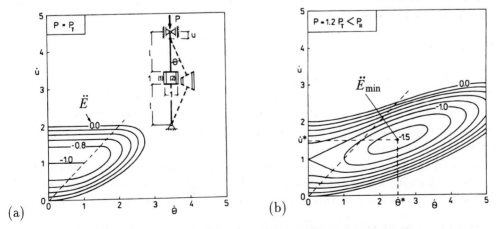

Figure 1. Contours of the functional $\ddot{E}$ for the Shanley column: (a) at the tangent modulus load, (b) between the tangent and reduced modulus loads. From [28].

504

This simple example indicates that the energy criterion of path instability is closely related to other approaches to plastic instability phenomena, and brings the advantages of a unified approach. The connections will now be examined in a general setting.

3.3. Relation to the energy criterion of instability of equilibrium

Hill's [7] condition for stability of equilibrium, in the present notation rewritten as

$$\ddot{E}(\widetilde{\mathbf{w}}) > 0 \quad \text{for every } \widetilde{\mathbf{w}} \in \mathcal{W}, \ \widetilde{\mathbf{w}} \neq \widetilde{\mathbf{0}} \quad \text{at } \lambda = \text{const}, \tag{18}$$

can be interpreted as being sufficient for *directional* stability[3] in the following sense: If (18) holds then a departure from equilibrium at constant λ on a straight (direct) path in a dynamic motion induced by a vanishingly small disturbance is excluded by imposing an energy barrier in all directions. Conversely, *for discretized systems with a symmetric tangent stiffness matrix* $[K](\widetilde{\mathbf{v}})$, *the condition* (18) (*with* $\ddot{E}(\widetilde{\mathbf{w}}) = K_{\alpha\beta}(\widetilde{\mathbf{w}})w_\alpha w_\beta$) *is necessary for stability of equilibrium in the first approximation* [28]. Accordingly, a change in sign of the inequality in (18), typically at a generalized limit point, is expected to be accompanied by an instability of a dynamic type.

It can be shown [12] that (17) implies (18) with > replaced by ≥, and reduces to it if λ is kept constant. The energy criterion (17) represents thus an extension of (18) to deformation processes at varying λ. Since (18) in general does not imply (17), a deformation process can become unstable without loss of stability of the traversed equilibrium states, as illustrated above by the example of the Shanley column. An opposite situation is not possible: as remarked in [28], the instability of a quasi-static process is a wider concept that includes instability of equilibrium as a particular case.

3.4. Relation to the bifurcation theory

It has been shown [12,28,3] that the criterion (17) is consistent with Hill's fundamental theory of bifurcation in elastic-plastic solids.[4] Main conclusions are quoted below in reference to a discretized problem [28,24].

Let $\mathbf{C}^0 = \mathbf{C}(\dot{\mathbf{F}}^0, \mathcal{H})$ denote the current tangent moduli tensor in a given, fundamental deformation process χ^0, and $[K^0] = [K](\widetilde{\mathbf{v}}^0)$ be the respective global tangent stiffness matrix for the system, both assumed to be well defined. Suppose that the following constitutive inequality is satisfied

$$(\dot{\mathbf{S}} - \dot{\mathbf{S}}^0) \cdot (\dot{\mathbf{F}} - \dot{\mathbf{F}}^0) \geq (\dot{\mathbf{F}} - \dot{\mathbf{F}}^0) \cdot \mathbf{C}^0 \cdot (\dot{\mathbf{F}} - \dot{\mathbf{F}}^0) \quad \text{for every } \dot{\mathbf{F}}. \tag{19}$$

This inequality is similar to, but less restrictive than, the 'relative convexity property' with respect to $\mathbf{C}^0$, cf. [1]. Validity of (19) was originally shown [7] for the classical elastic-plastic solids obeying (4), extended later to 'singular' plasticity described by (5)-(8) [16], and further extended to materials obeying (5)-(7) at a *micro*-level [20], in each case for $\dot{\mathbf{F}}^0$ corresponding to the fully active loading.

By the 'comparison theorem' proved in the references just cited, bifurcation in velocities is excluded under the assumption (19) as long as $[K^0]$ is positive definite, and the fulfillment of (17) is ensured in this range. Moreover, (17) holds exactly as long as $[K^0]$

---

[3]Additional assumptions are needed to prove stability for arbitrarily circuitous paths, and for the continuum problem the condition in (18) should be strengthened.

[4]There are also connections with Nguyen's [4] theory of uniqueness and stability; cf. [29,31].

is positive semi-definite, that is, up to the typical instant of primary bifurcation if the lowest eigenvalue of $[K^0]$ is continuously decreasing. At this instant, the energy criterion gives way to the secondary solution, which yields the same value of $\dot{E}$ as the fundamental mode. In those circumstances, the onset and the mode of 'geometric' instability predicted by the energy criterion of path instability coincide with those found from Hill's bifurcation theory. If (19) is not satisfied then the predictions can be distinct since earliest bifurcation modes can correspond to higher values of $\ddot{E}$, without affecting path stability in the energy sense. The energy condition (17) provides *a criterion of choice* of the post-bifurcation branch, which is absent in the bifurcation theory alone.

For finite-dimensional systems, the following theorem has been proved [28,24]:
If (18) *holds for the discretized rate-problem with symmetric* $[K](\widetilde{\mathbf{v}})$ *then:*
(i) *there exists a solution which assigns to* $\ddot{E}(\widetilde{\mathbf{v}})$ *its absolute minimum value in* $\mathcal{V}$,
(ii) *for uniqueness of a solution* $\widetilde{\mathbf{v}}^0$ *it is* necessary *that* (17) *holds with strict inequality for* $\widetilde{\mathbf{v}} \neq \widetilde{\mathbf{v}}^0$.

It follows that there is a bifurcation in velocities at *every* point on the segment of a fundamental path along which the condition (18) of stability of equilibrium holds but (17) of path stability does not. Contrary to incrementally linear (e.g. elastic) solids, the spectrum of bifurcation points, defined by the two *inequalities*, is *continuous* along the path, without the need of singularity of $[K^0]$. For the simplest systems as the Shanley column [30] this result has been known for many years; a general proof was given in [28] and in another version in [24]. The secondary solutions in the range discussed are energetically preferable to $\widetilde{\mathbf{v}}^0$ in the sense of a lower value of $\ddot{E}$. In comparison with the general interpretation of the path instability for continuous systems, a finite deviation from the fundamental path for a discretized system can take place in this range at perturbing forces being just zero and not only arbitrarily small.

3.5. Relation to the initial imperfection approach

In a certain interval of 'time' t with $\lambda = \lambda(t)$, consider a fundamental deformation process χ^0 for a perfect system and a family χ^z of quasi-static solution paths for imperfect systems, parameterized by a number $z \geq 0$ which defines the size of imperfections. The body domain V in the reference configuration is assumed to be independent of z; geometric imperfections can still be be included by allowing a part of the reference volume to be filled by a fictitious material of zero stiffness. All kinematic constraints are also assumed to be not influenced by imperfections, and the classes $\mathcal{V}$ and $\mathcal{W}$ of kinematically admissible velocity fields and their variations at any value of λ are taken to be the same for all z.

$\ddot{E}(\widetilde{\mathbf{v}})$ at some λ depends on z both explicitly (through the initial material properties) and implicitly (through the influence of the changed deformation history in a process χ^z). In a short-hand notation, this complex dependence will be indicated by writing $\ddot{E}^z$, and $\widetilde{\mathbf{v}}^z$ will denote the current velocity solution in a process χ^z. We shall say that the family χ^z converges to χ^0 in the energy sense if

$$\lim_{z \to 0} \ddot{E}^z(\widetilde{\mathbf{v}}) = \ddot{E}^0(\widetilde{\mathbf{v}}) \quad \text{for every } \widetilde{\mathbf{v}} \in \mathcal{V}, \quad \text{and} \quad \lim_{z \to 0} \ddot{E}^z(\widetilde{\mathbf{v}}^z) = \ddot{E}^0(\widetilde{\mathbf{v}}^0) \tag{20}$$

at any value of λ from the interval under consideration. $(20)_1$ means that overall mechanical properties of an imperfect system at given λ tend to those for the perfect system as the imperfection size z tends to zero, and $(20)_2$ can then be regarded as a consequence of convergence of the respective velocity solutions.

The theorem which follows appears to be new.

Suppose that along every quasi-static solution path from a family χ^z for discretized imperfect systems:

(i) *each traversed equilibrium state is stable in the first approximation,*

(ii) *a solution in velocities is uniquely defined at each instant.*

If the family χ^z converges to χ^0 in the energy sense then the stability condition (17) *is satisfied along the path χ^0.*

Proof. From (i) and the theorem from Subsection 3.3 it follows that the condition (18) holds along any path from the family χ^z. If this is so and the condition (ii) above is satisfied then, by the theorem from Subsection 3.4, along each path from the family χ^z we must have

$$\ddot{E}^z(\widetilde{\mathbf{v}}) - \ddot{E}^z(\widetilde{\mathbf{v}}^z) > 0 \qquad \text{for every } \widetilde{\mathbf{v}} \in \mathcal{V}, \quad \widetilde{\mathbf{v}} \neq \widetilde{\mathbf{v}}^z. \tag{21}$$

On determining the limit value of the left-hand expression as $z \to 0$ by using the definition of the convergence in the energy sense, we arrive at

$$\ddot{E}^0(\widetilde{\mathbf{v}}) - \ddot{E}^0(\widetilde{\mathbf{v}}^0) \geq 0 \qquad \text{for every } \widetilde{\mathbf{v}} \in \mathcal{V} \tag{22}$$

along the path χ^0. This is what we set out to prove.

This theorem provides still another justification for the energy criterion of plastic instability, which perhaps can be seen more clearly from the following *corollary: If the condition* (17) *is not satisfied along a deformation path χ^0 for a perfect system then χ^0 cannot be approximated arbitrarily closely by a sequence of solution paths satisfying* (i) *and* (ii) *for imperfect discretized systems.*

4. A COMPUTATIONAL APPROACH BASED ON THE ENERGY CRITERION

As an outcome of the theoretical analysis, especially of the results quoted in Subsection 3.4, an algorithm for crossing bifurcation points in numerical step-by-step analysis was proposed [24] for the problems with a symmetric tangent stiffness matrix $[K](\widetilde{\mathbf{v}})$. The main step is to determine a solution in velocities to the actual, *nonlinear* rate-problem as a *minimizer* $\widetilde{\mathbf{v}}^*$ of the velocity functional $\ddot{E}(\widetilde{\mathbf{v}})$ in $\mathcal{V}$, and not only as a stationarity point according to (14) (Fig. 2). This makes no difference in the range where uniqueness is ensured, but the distinction becomes essential just after the primary bifurcation point. As soon as the tangent stiffness matrix $[K^0]$ ceases to be positive definite and becomes indefinite, the fundamental solution path is no longer followed. This is in accord with the energy criterion (17) of path instability, since $\widetilde{\mathbf{v}}^0$ ceases to assign to $\ddot{E}$ a minimum value in $\mathcal{V}$. The secondary solution in velocities is determined as a minimizer $\widetilde{\mathbf{v}}^*$ whose existence is ensured provided the condition (18) of directional stability of equilibrium still holds. A secondary solution path is initiated in this way *automatically*, irrespectively of the type of primary bifurcation and without changing the numerical technique used along the whole path. Further branching points, if any, are treated analogously. If (18) fails indicating instability of equilibrium, then J decreases unboundedly [12] and the computations are terminated.

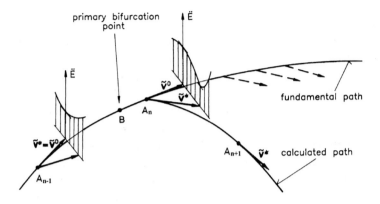

Figure 2. The mechanism of automatic branch switching according to the energy criterion of path instability.

An application of this approach to determining the onset of necking in plane strain tension by using the finite element method was given in [24]. As the material model satisfying (2), the J_2 corner theory of plasticity [21] was employed. It has been shown that the algorithm can be applied without any changes if the primary bifurcation is induced by a discontinuous jump of the current material stiffness so that $[K](\tilde{\mathbf{v}}^0)$ becomes abruptly indefinite without being singular at any instant. Usefulness of the approach in passing multiple bifurcation points was illustrated in [32].

The example analyzed in [33] makes contact with instability of a material type examined in the next section. FEM simulation of the kinematically controlled biaxial stretching of a rectangular sheet of a finite thickness (h_0 initially) was performed under the assumption that all quantities of interest are independent of the in-plane coordinate x_2. The material

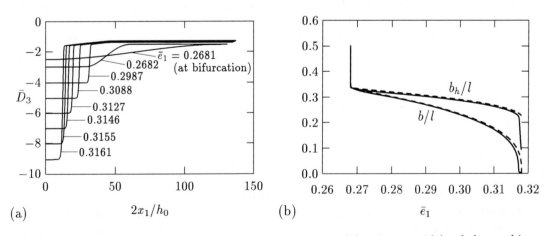

Figure 3. (a) Distribution of thickness strain-rate, and (b) relative width of the necking band vs overall logarithmic strain $\bar{e}_1$, in a rectangular sheet subject to balanced biaxial stretching. From [33].

model and the algorithm of initiation of a post-bifurcation branch were as indicated above. The plot of the thickness strain rate $\bar{D}_3$ vs the in-plane coordinate x_1 had a typical sinusoidal form at the bifurcation point but later underwent rapid redistribution towards that consisting of two zones of practically constant strain rate (Fig. 3a). During subsequent deformation, the necking zone was shrinking at increasing rate of necking within it. That process is illustrated in Fig. 3b, where $b/l = (\bar{D}_3^{\mathrm{mean}} - \bar{D}_3^{\mathrm{max}})/(\bar{D}_3^{\mathrm{min}} - \bar{D}_3^{\mathrm{max}})$ is a conventional width of the current necking band relative to the sheet dimension l in direction x_1, and b_h/l is a relative neck width determined analogously but from the thickness distribution. Dashed lines were determined independently, for an infinitely thin sheet treated as a two-dimensional continuum, the indeterminacy in the post-critical non-elliptic range being removed by using the local stability condition, (24) below. Since the two ways of calculation are different, the agreement observed in Fig. 3b may be treated as an argument supporting the previous justification of the energy criterion of stability of a deformation process.

5. MATERIAL INSTABILITY

Material instability in time-independent plastic solids was understood in the literature in various ways, and no general agreement about it precise meaning has been worked out yet. The well-known concept based on the sign of the second-order work during a virtual *uniform* deformation increment [34] involves an arbitrary choice of the stress-rate measure if the changes in geometry are not disregarded. As explained in [1], the arbitrariness is related to an unspecified notion of what constitutes an appropriate 'passive' environment of a material element.

The influence of boundary conditions can be eliminated if a homogeneous material element is treated, in another scale but under the current uniform Cauchy stress, as an infinite continuum. Material instability has been frequently understood in the sense of quasi-static bifurcation within a planar band, following [35,36,10]. If an incrementally *linear* constitutive relationship in the whole loading branch is assumed then the primary bifurcation takes place at the instant of ellipticity loss. This is accompanied by vanishing speed of propagation of acceleration waves [37,11] and marks simultaneously the onset of instability of equilibrium in the energy sense (*op. cit.*) and in a dynamic sense [38]. The single criterion of ellipticity loss has, however, multiple counterparts when the constitutive law is incrementally *nonlinear* [39].

Conditions for material instability can also be derived as those for local instability of a deformation process in a finite body, without the need of referring to an auxiliary problem for an infinite continuum. In particular, the energy criterion (17) can be specified for a material element *in situ*; implications are discussed below.

5.1. Energy conditions for material stability

Consider a particular class of instability modes that are confined to a neighborhood G of a regular point $\bar{\boldsymbol{\xi}}$ in the reference body domain V (Fig. 4). A continuous perturbation $\widetilde{\mathbf{w}}$ superimposed on the fundamental velocity field $\widetilde{\mathbf{v}}^0$ is taken to vanish over the boundary ∂G of G. Then, as a condition necessary for (17) and thus for stability of the unperturbed process within G, we obtain [39] that the constitutive potential $U(\dot{\mathbf{F}})$ must be *quasiconvex*

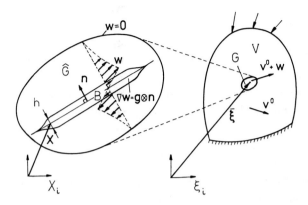

Figure 4. Local perturbations of the fundamental velocity field.

at $\dot{\mathbf{F}}^0$ at given $(\bar{\xi}, \mathcal{H})$, viz.

$$\int_{\hat{G}} U(\dot{\mathbf{F}}^0 + \nabla\mathbf{w}(\mathbf{X}))\,d\mathbf{X} \geq |\hat{G}|\,U(\dot{\mathbf{F}}^0) \qquad \text{for every } \widetilde{\mathbf{w}} : \; \mathbf{w} = \mathbf{0} \text{ over } \partial\hat{G}, \tag{23}$$

where the domain $\hat{G}$ of volume $|\hat{G}|$ may be chosen arbitrarily. In (23), $\mathbf{X}$ is a variable position vector within $\hat{G}$ at fixed $\bar{\xi}$, and $\nabla\mathbf{w} = \partial\mathbf{w}/\partial\mathbf{X}$ is a *non-uniform* perturbation of the fundamental mode $\dot{\mathbf{F}}^0 = \dot{\mathbf{F}}^0(\bar{\xi}, t)$ independent of $\mathbf{X}$ in $\hat{G}$. If $\dot{\mathbf{F}}^0 = \mathbf{0}$ then (23) reduces to the condition for stability of equilibrium within a homogeneous material element with a rigidly constrained boundary.[5]

Let $\nabla\mathbf{w}$ be effectively concentrated in a disk-like volume B of predominantly uniform thickness h which is small compared with the dimensions of $\hat{G}$ (Fig. 4). Then (23) implies [39] that U must be *rank 1 convex* at $\dot{\mathbf{F}}^0$, that is

$$\mathcal{E}(\dot{\mathbf{F}}^0, \dot{\mathbf{F}}^0 + \mathbf{g}\otimes\mathbf{n}) \geq 0 \qquad \text{for all vectors } \mathbf{g}, \mathbf{n}, \tag{24}$$

where $\mathcal{E}$ is the Weierstrass function associated with the constitutive potential $U(\dot{\mathbf{F}})$:

$$\mathcal{E}(\dot{\mathbf{F}}, \dot{\mathbf{F}}^*) = U(\dot{\mathbf{F}}^*) - U(\dot{\mathbf{F}}) - \dot{\mathbf{S}}\cdot(\dot{\mathbf{F}}^* - \dot{\mathbf{F}})\,. \tag{25}$$

If (24) does not hold then the concentration of deformation within a band (or bands) of orientation $\mathbf{n}$ is shown to be energetically preferable to uniform straining with the velocity gradient $\dot{\mathbf{F}}^0$. The requirement of semi-strong ellipticity of the current tangent moduli:

$$(\mathbf{g}\otimes\mathbf{n})\cdot\mathbf{C}(\dot{\mathbf{F}}^0)\cdot(\mathbf{g}\otimes\mathbf{n}) \geq 0 \qquad \text{for every } \mathbf{g}, \mathbf{n} \tag{26}$$

is implied by (24) in the limit as $|\mathbf{g}||\mathbf{n}| \longrightarrow 0$. If $|\mathbf{g}||\mathbf{n}| \longrightarrow \infty$ then as a consequence of (24) we obtain another condition

$$U(\mathbf{g}\otimes\mathbf{n}) \geq 0 \qquad \text{for every } \mathbf{g}, \mathbf{n}, \tag{27}$$

---

[5]Throughout this article, the discussion of instability is based on the assumption that the stress-rate at a given state of a material element is uniquely defined by the velocity gradient. This need not always be true, e.g. in the case of the nouniqueness of a multislip mode in models of single crystals, cf. [19]. As proposed in [40], the stability criterion (17) can then be used to select an 'optimal' solution within a class of *uniform* deformation modes. In particular, (23) may be applied, and the set of $\dot{\gamma}_\kappa$'s in a material law (5)-(7) without (8) may be selected such that it minimizes the value of $(\dot{\mathbf{S}}\cdot\dot{\mathbf{F}})$ at prescribed $\dot{\mathbf{F}}$.

interpreted as a local condition necessary for stability of *equilibrium* [12]. Since the above conditions (discussed in detail in [39], with special reference to propagation of acceleration waves) are derived from the global energy criterion (17), a connection has been established between material instability and the instability in a finite solid body with specified boundary conditions.

If the constitutive law (2) satisfies the inequality (19) then (26) becomes equivalent to (24) (and to (23), but not to (27)). Then, the familiar critical condition [11,10] is recovered:

$$\det(C_{ijkl}(\dot{\mathbf{F}}^0)n_j^*n_l^*) = 0, \qquad \text{or} \qquad C_{ijkl}(\dot{\mathbf{F}}^0)n_j^*n_l^*g_k^* = 0 \quad \text{for some } \mathbf{n}^*, \mathbf{g}^* \neq 0, \qquad (28)$$

which is usually regarded as a condition for the onset of localization of deformation in planar bands (or shear bands). However, fully localized deformations need not occur immediately beyond the point of ellipticity loss of the *linearized* governing equations at which (28) is met, unless the critical points related to (26) and (27) happen to coincide as in classical elastoplasticity obeying (4). If (27) fails then the localization of deformation into a planar band can take place as a *dynamic* process, or as an internal snap-through, at a fixed value of the external loading parameter λ [39].

The situation is different if (27) still holds in a certain range beyond the critical instant (28), that is, when $\mathbf{g}^* \otimes \mathbf{n}^*$ in (28) is directed outside the constitutive domain of $\mathbf{C}(\dot{\mathbf{F}}^0)$. Then, the concentration of deformation can develop gradually. It can also saturate in a post-critical solution with a single band [41]. On the other hand, such a solution violates the stability requirement (26) outside the band if the uniform solution does, and may analogously be regarded as unrealizable in a homogeneous material. The question thus arises: what can be said about the deformation pattern beyond the instant of ellipticity loss, still using the constitutive framework for time-independent materials not possessing any internal length scale. This question is addressed below by applying the energy criterion (17), or rather the local conditions (24) or (26) derived from it, to an indeterminate post-critical deformation process within a material element. This novel approach constitutes an extension of that used recently in [33] in a study of post-critical deformation of a biaxially stretched sheet regarded as a two-dimensional continuum, cf. Section 4.

5.2. Incipient volume fraction of localization zone

Consider a homogeneous and uniformly stressed material element as an infinite continuum deforming with a uniform rate $\dot{\mathbf{F}}^0$. At the critical instant, $t = t^*$ say, at which (28) is met, the current configuration is taken for simplicity as reference (with a position vector denoted by $\mathbf{X}$), so that $\mathbf{F}^0 = \partial\mathbf{x}/\partial\mathbf{X} = 1$ at this instant. In the time-independent 'simple' material without any intrinsic length scale, the width of a single bifurcation band is left arbitrary. Similarly, the number and spatial arrangement of bands are indeterminate. There is thus an inherent indeterminacy of the post-critical deformation. Nevertheless, with the help of the stability condition (24) or (26) applied to a post-critical deformation process, there is a possibility to determine the incipient *volume fraction* of the localization zone in an incrementally nonlinear solid.

A class of velocity solutions in the infinite continuum at the critical instant can be constructed, for a pair of unit vectors $(\mathbf{n}^*, \mathbf{g}^*)$ that satisfy (28), as follows

$$\dot{\mathbf{F}}(\mathbf{X}) = \dot{\mathbf{F}}^0 + s(X_{\mathbf{n}})\,\mathbf{g}^* \otimes \mathbf{n}^*, \qquad X_{\mathbf{n}} = \mathbf{X} \cdot \mathbf{n}^* \in (-\infty, +\infty) \qquad (29)$$

with the restrictions

$$s^- \leq s \leq s^+, \qquad \left| \int_{-\infty}^{+\infty} s \, dX_{\mathbf{n}} \right| < \infty.$$ (30)

The limiting values s^- and s^+ are such that within the resulting range of $\dot{\mathbf{F}}$ there is $\mathbf{C}(\dot{\mathbf{F}}) = \mathbf{C}(\dot{\mathbf{F}}^0)$. The restrictions imposed on s ensure that the difference in the mean velocity gradient between the fundamental and bifurcation solutions in the re-scaled material element vanishes, and that the pointwise difference in velocities is infinitesimal relative to fundamental velocities at the element boundary if $\dot{\mathbf{F}}^0 \neq \mathbf{0}$. The bifurcation solutions represent thus a kind of branching of the material response under macroscopically uniform kinematic boundary data. The sense of $\mathbf{g}^*$ is chosen below such that $\mathbf{D}^* \cdot \mathbf{D}^0 \geq 0$, where $\mathbf{D}^0$ is the fundamental Eulerian strain-rate and $\mathbf{D}^* = \frac{1}{2}(\mathbf{g}^* \otimes \mathbf{n}^* + \mathbf{n}^* \otimes \mathbf{g}^*)$.

In the particular case when $s \neq 0$ only in a finite interval of $X_{\mathbf{n}}$, the solution is obtained which was analyzed in many papers after [10]. This solution violates the path stability condition (24) outside the localization band just beyond the critical instant, assuming that (26) fails along the fundamental path. It is a natural assumption that the rate of the left-hand expression in (26), starting from $\mathbf{g} = \mathbf{g}^*$, $\mathbf{n} = \mathbf{n}^*$, can be made negative at t^* not only for $\dot{\mathbf{F}} = \dot{\mathbf{F}}^0$ but also for all $\dot{\mathbf{F}}$ defined by (29) with (30). By the continuity argument, this is so if the rate of this expression is negative along the fundamental path and if the current constitutive cone is sufficiently sharp. In the example below, validity of this assumption was checked numerically. If this is so then (26) can be satisfied just beyond the critical instant only due to the transition to a more 'stiff' constitutive branch, which is possible if the strain-rate $\mathbf{D} = \mathbf{D}^0 + s\mathbf{D}^*$, wherever defined, lies at t^* precisely *on* the boundary of the constitutive cone containing $\mathbf{D}^0$. This means that $s(X_{\mathbf{n}})$ in (29) can only take two values, s^- or s^+, which define the intersection points, $\mathbf{D}^-$ and $\mathbf{D}^+$, respectively, of a straight line $\mathbf{D}(s) = \mathbf{D}^0 + s\mathbf{D}^*$ in $\mathbf{D}$-space with the cone boundary (Fig. 5a). The bifurcation solution takes the form of a mixture of layers of a constant velocity gradient, $.../\dot{\mathbf{F}}^-/\dot{\mathbf{F}}^+/\dot{\mathbf{F}}^-/\dot{\mathbf{F}}^+/...$ (Fig. 5b), in some analogy to fine phase mixtures discussed in [42]. The initial localization zone, which occupies a volume fraction η^+, say, is defined as the collection of all layers where $\mathbf{D} = \mathbf{D}^+$.

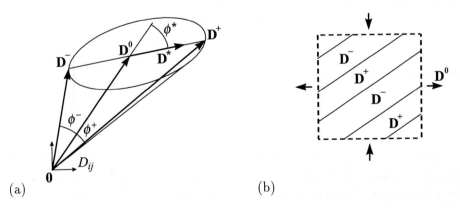

(a) (b)

Figure 5. Schematic sketch of the strain rates (a) and their spatial distribution (b) in a bifurcation solution within a material element.

From the integral condition in (30) we obtain the equation

$$\eta^+ s^+ + (1 - \eta^+)s^- = 0 \qquad \text{or} \qquad \eta^+ \mathbf{D}^+ + (1 - \eta^+)\mathbf{D}^- = \mathbf{D}^0 . \tag{31}$$

On substituting trigonometric relationships within the hyperplane spanned by $(\mathbf{D}^0, \mathbf{D}^*)$ in $\mathbf{D}$-space , the *incipient* volume fraction occupied by the localization zone is determined as

$$\eta^+ = \frac{\cot(\phi^+) - \cot(\phi^*)}{\cot(\phi^+) + \cot(\phi^-)} \qquad \text{for } \phi^+ < \phi^* . \tag{32}$$

ϕ^+, ϕ^- and ϕ^* stand here for positive angles, according to a definition based on a scalar product, between $\mathbf{D}^0$ and $\mathbf{D}^+, \mathbf{D}^-$ and $\mathbf{D}^*$, respectively (Fig. 5a). The formula (32) can be derived, for instance, by substituting the identity (with $|\cdot|$ as an Euclidean norm)

$$\frac{s^+}{-s^-} = \frac{|\mathbf{D}^+| \sin(\phi^+)}{|\mathbf{D}^-| \sin(\phi^-)} = \frac{\sin(\phi^* + \phi^-) \sin(\phi^+)}{\sin(\phi^* - \phi^+) \sin(\phi^-)} \tag{33}$$

into (31) and rearranging in a straightforward manner. In the particular case when $\phi^+ = \phi^-$, (32) reduces to the formula given in [33]. One can check that the value of η^+ is independent of the choice of an angular measure in $\mathbf{D}$-space, on account of the invariance of (s^+/s^-) under affine transformations of this space.

5.3. Post-critical behaviour

Local stability of a *non-uniform* deformation process initiated in the infinite continuum at the critical instant t^* will now be examined. In a preliminary study aimed at here, the post-critical solution is taken in the form (cf. (29))

$$\dot{\mathbf{F}}^0(\mathbf{X}, t) = \dot{\bar{\mathbf{F}}}(t) + \mathbf{g}(X_\mathbf{n}, t) \otimes \mathbf{n}^*, \qquad \left| \int_{-\infty}^{+\infty} \mathbf{g} \, dX_\mathbf{n} \right| < \infty . \tag{34}$$

where $\dot{\bar{\mathbf{F}}}$ stands for previous $\dot{\mathbf{F}}^0$ and denotes the prescribed mean velocity gradient in the material element. The analysis similar to that presented in [33] shows that two zones, (a) and (b), of uniform velocity gradient can still exist in the post-critical range, cf. Fig. 3a, in general accompanied by a developing transitory zone where $\mathbf{g}$ is not uniform. The reference volume fraction $\eta^{(b)}$ of a uniform localization zone (b) is allowed to be fixed or decreasing in time. For simplicity, we assume here that the constitutive inequality (19) is satisfied, so that (26) is used in place of (24). The analysis in [33] was carried out without this assumption.

A condition necessary for stability of a post-critical deformation path is

$$C_{gn}^{(b)} \geq 0 , \qquad C_{gn}^{(b)} \equiv \min_{|\mathbf{g}|=1} (\mathbf{g} \otimes \mathbf{n}^*) \cdot \mathbf{C}(\dot{\mathbf{F}}^{(b)}) \cdot (\mathbf{g} \otimes \mathbf{n}^*), \tag{35}$$

as a consequence of (26) within the zone (b) with $\dot{\mathbf{F}}^0 \equiv \dot{\mathbf{F}}^{(b)}$. As long as $C_{gn}^{(b)} > 0$, $\eta^{(b)}$ may remain constant. If $C_{gn}^{(b)}$ has decreased to zero and further deformation with fixed $\eta^{(b)}$ would make $C_{gn}^{(b)}$ negative, then $\eta^{(b)}$ can start to decrease in accord with the condition for quasi-static propagation of a velocity-gradient discontinuity, cf. [11]. The evolution criterion for $\eta^{(b)}$ takes thus the form

$$C_{gn}^{(b)} \dot{\eta}^{(b)} = 0 , \qquad C_{gn}^{(b)} \geq 0 , \qquad \dot{\eta}^{(b)} \leq 0 \tag{36}$$

at each instant. A similar condition can be established for the zone (a).

(36) represents the additional condition derived from the requirement of path stability. Jointly with the usual kinematic and quasi-static conditions, it can be used to determine *overall* characteristics of post-critical deformation, although a spatial arrangement of different zones at a micro-level is indeterminate.

To illustrate how (36) works, consider plane-strain compression of an (almost) incompressible material. Then, $C_{gn}^{(b)}$ is found as the current modulus for simple shearing superimposed on $\dot{\mathbf{F}}^{(b)}(t)$. Let an incrementally nonlinear constitutive law (2) along a proportional straining path be specified in an inverse form as[6]

$$\mathbf{D} = \frac{\partial \Psi}{\partial \overset{\triangledown}{\tau}}, \qquad \Psi = \frac{1}{2}F(\beta)|\overset{\triangledown}{\tau}'|^2 + \frac{1}{2}\overset{\triangledown}{\tau}\cdot\mathbf{M}^e\cdot\overset{\triangledown}{\tau}, \qquad \cos\beta = \frac{\tau'\cdot\overset{\triangledown}{\tau}'}{|\tau'|\cdot|\overset{\triangledown}{\tau}'|}, \tag{37}$$

$$F(\beta) = \frac{1}{H}\cdot\begin{cases} \pi - 2\bar{\beta} + \sin 2\bar{\beta}\cos 2\beta & \text{for } 0 \le \beta \le \bar{\beta} & \text{(total loading)} \\ \pi - (\beta + \bar{\beta}) + \frac{1}{2}\sin 2(\beta + \bar{\beta}) & \text{for } \bar{\beta} \le \beta \le \pi - \bar{\beta} & \text{(partial unloading)} \\ 0 & \text{for } \pi - \bar{\beta} \le \beta \le \pi & \text{(total unloading)}. \end{cases} \tag{38}$$

τ' is the Kirchhoff stress deviator, and $\mathbf{M}^e$ denotes the tensor of elastic compliances which have a secondary effect on the results. If the range of partial unloading is entered then (37) is no longer valid for an arbitrary stress-rate but may still be used about the *actual* $\overset{\triangledown}{\tau}$ to determine current $\mathbf{C}$, under certain restrictions on the straining path. Changes in the nonlinear constitutive law needed to satisfy (19) may be left unspecified.

Numerical results have been obtained for $H > 0$ calculated in an indirect manner from a power hardening law with an exponent 0.1, and for an angle $\bar{\beta} > \bar{\beta}^{\min} = 55°$; details are omitted here. The plots in Fig. 6 start at t^* and correspond to an initial stage of post-critical deformation during overall plane-strain compression along X_1-axis; $C_{gn}^{(b)}$ is normalized by an initial uniaxial yield stress τ_y. Immediately beyond t^*, the solution returns to the strongly elliptic regime and $\eta^{(b)}$ remains equal to η^+ determined from (32). Somewhat later, $C_{gn}^{(b)}$ falls again to zero and then $\eta^{(b)}$ starts to decrease in accord with (36).

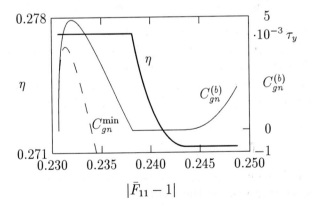

Figure 6. Volume fraction η of the localization zone and the shear modulus $C_{gn}^{(b)}$ vs post-critical overall strain in an example of plane-strain compression.

---

[6]A more detailed discussion of this constitutive model will be given elsewhere.

The actual variations of $\eta^{(b)}$ are not computed, rather, its approximation η is determined from a formula analogous to that for b/l in the example from Section 4, which requires calculations only in the uniform zones (a) and (b). For the material parameters assumed, $C_{gn}^{(b)}$ becomes soon positive and $\eta^{(b)}$ stops again. It can be remarked that there is a competition between several factors that influence the value of $C_{gn}^{(b)}$, and that the example has an illustrative character.

An absolute minimum value, $C_{gn}^{\min}$, of the left-hand expression in (26) for $|\mathbf{g}| = |\mathbf{n}| = 1$ with $\mathbf{n} \neq \mathbf{n}^*$, plotted as a broken line, is shown first to increase and next to fall below zero. This means that the post-critical solution of the assumed form (34) ceases to satisfy the stability condition (26), and a solution of a more complex spatial structure should be considered. This problem requires further study, especially in connection with investigations of internal length scale effects.

6. STABILITY OF HETEROGENEOUS MATERIALS

Stability of the post-critical nonuniform deformation (which produced, in effect, a material of inhomogeneous properties) was examined in the preceding section in a local sense. The energy criterion in its general form can be applied to heterogeneous materials (as metal polycrystals or composites) to investigate not only local but also overall instabilities. Results of an introductory analysis are presented below.

Let curly brackets $\{\cdot\}$ denote an unweighted volume average over a representative sample of a heterogeneous material in a reference configuration. Let a continuous field $\widetilde{\mathbf{w}}$ have the same meaning as above and vanish over the sample boundary. The path stability condition (17) applied to the quasi-static velocity field $\widetilde{\mathbf{v}}^0$ within the sample gives

$$\{U(\nabla \mathbf{v}^0 + \nabla \mathbf{w})\} \geq \{U(\nabla \mathbf{v}^0)\} \qquad \text{for every } \widetilde{\mathbf{w}}. \tag{39}$$

Accordingly, a minimization procedure: $\{U(\nabla \mathbf{v})\} \to \min$ might be used to determine numerically the material response (with the averaged potential $\bar{U}$) in the non-uniqueness range; cf. Section 4. (39) implies (cf. [28])

$$\{\nabla \mathbf{w} \cdot \mathbf{C}(\nabla \mathbf{v}^0) \cdot \nabla \mathbf{w}\} \geq 0 \qquad \text{and} \qquad \{U(\nabla \mathbf{w})\} \geq 0 \qquad \text{for every } \widetilde{\mathbf{w}}, \tag{40}$$

which in turn imply pointwise inequalities (26) and (27), respectively.

If the sample is embedded in an infinite deforming continuum of the same heterogeneous material then the instability mode illustrated in Fig. 4 can be considered at a *macro-level*, with h being of the order of the sample dimensions. By employing macro-variables [43] distinguished by a bar, and proceeding analogously as in [39], the following stability condition can be derived (cf. (24))

$$\bar{U}(\dot{\bar{\mathbf{F}}}^0 + \mathbf{g} \otimes \mathbf{n}) - \bar{U}(\dot{\bar{\mathbf{F}}}^0) - \dot{\bar{\mathbf{S}}}^0 \cdot (\mathbf{g} \otimes \mathbf{n}) \geq 0 \qquad \text{for every } \mathbf{g}, \mathbf{n}. \tag{41}$$

On using the conditions of continuing equilibrium to be satisfied by $\widetilde{\mathbf{v}}^0$, and the divergence theorem, the stability requirement can be expressed in the form

$$\{\mathcal{E}(\nabla \mathbf{v}^0, \nabla \mathbf{v}^0 + \nabla \mathbf{w} + \mathbf{g} \otimes \mathbf{n})\} \geq 0 \qquad \text{for every } \widetilde{\mathbf{w}}, \mathbf{g}, \mathbf{n}, \tag{42}$$

which encompasses both (39) (for $|\mathbf{g}||\mathbf{n}| = 0$) and (41) (for $\widetilde{\mathbf{w}}$ minimizing $\{U(\nabla \mathbf{v}^0 + \nabla \mathbf{w} + \mathbf{g} \otimes \mathbf{n})\}$).

If the macro-homogeneous continuum body surrounding the material sample has finite dimensions then the length scale effect comes into play, and (41) may be too restrictive. This interesting topic deserves further investigations.

7. CONCLUDING REMARKS

It has been shown that various plastic instability phenomena as snap-through, buckling, necking or shear banding can be examined by using a unified approach, based on the energy criterion of instability of a quasi-static deformation process. This may be regarded as an extension of the classical energy method for elastic conservative systems to instabilities in a broad class of incrementally nonlinear solids. The computational algorithm for passing bifurcation points with automatic choice of the post-bifurcation branch has been developed as a natural outcome of the theoretical analysis. In the circumstances specified in the preceding sections, the results of the bifurcation or initial imperfection approaches can be closely reproduced in this way. Novel results are available especially when non-uniqueness persists along a post-critical deformation path, so that a criterion of choice becomes unavoidable. In particular, post-critical solutions describing early stages of localized necking or shear have been calculated with the help of the path instability criterion while otherwise they are indeterminate. There are perspectives of applications to heterogeneous materials. It should be added that the problem of adequate constitutive modeling is of primary importance since the instability predictions are sensitive to the details of the incremental constitutive law.

The applicability of the presented approach is limited by the constitutive assumptions, and in particular by the symmetry restrictions imposed upon the material law and boundary conditions. Other limitations are due to the continuity assumptions (of $\mathbf{v}$ in space and of $\mathbf{F}$ in time) which exclude instability phenomena related to solid-solid phase transformations, cavitation, fracture, etc. Possible extensions have not been discussed here.

Acknowledgment

This work was supported by the State Committee of Research (KBN) in Poland under the project No. 3 P404 035 07.

REFERENCES

1. R. Hill, Adv. Appl. Mech. Vol. 18, p. 1, Acad. Press, New York 1978.
2. A. Needleman and V. Tvergaard, Appl. Mech. Rev. 45, no 3, part 2 (1992) S3.
3. H. Petryk, CISM Lecture Notes No. 327, p. 95, Springer, Wien - New York, 1993.
4. Q.S. Nguyen, Appl. Mech. Rev. 47, no 1, part 1 (1994) 1.
5. Y. Tomita, Appl. Mech. Rev. 47, no 6, part 1 (1994) 171.
6. H. Petryk, Arch. Comp. Meth. Engng 4 (1997) (in press).
7. R. Hill, J. Mech. Phys. Solids 6 (1958) 236.
8. J.W. Hutchinson, Adv. Appl. Mech. Vol. 14, p. 67, Acad. Press, New York 1974.
9. Z. Marciniak and K. Kuczyński, Int. J. Mech. Sci. 9 (1967) 609.
10. J.R. Rice, Theoretical and Applied Mechanics (W.T. Koiter, Ed.), p. 207, North-Holland, Amsterdam 1977.

11. R. Hill, J. Mech. Phys. Solids 10 (1962) 1.

12. H. Petryk, Plastic Instability, Proc. Considère Memorial, p. 215, Ecole Nat. Ponts Chauss., Paris 1985.

13. R. Hill, J. Mech. Phys. Solids 15 (1967) 79.

14. J.W. Hutchinson, Proc. Roy. Soc. Lond. A 319 (1970) 247.

15. R. Hill, J. Mech. Phys. Solids 7 (1959) 209.

16. M.J. Sewell, Stability (H.H.E. Leipholz, Ed.), p. 85, Univ. of Waterloo Press, Ontario 1972.

17. R. Hill and J.R. Rice, J. Mech. Phys. Solids 20 (1972) 401.

18. R.J. Asaro, Adv. Appl. Mech. Vol. 23, p. 1, Acad. Press, New York 1983.

19. K.S. Havner, Finite Plastic Deformation of Crystalline Solids, Cambridge Univ. Press, 1992.

20. H. Petryk, J. Mech. Phys. Solids 37 (1989) 265.

21. J. Christoffersen and J.W. Hutchinson, J. Mech. Phys. Solids 27 (1979) 465.

22. R. Hill, J. Mech. Phys. Solids 10 (1962) 185.

23. M.J. Sewell, Arch. Rat. Mech. Anal. 23 (1967) 327.

24. H. Petryk and K. Thermann, Int. J. Solids Structures 29 (1992) 745.

25. V.D. Klushnikov, Stability of Elastic-Plastic Systems (in Russian), Nauka, Moscow 1980.

26. H. Petryk, Stability in the Mechanics of Continua (F.H. Schroeder, Ed.), p. 262, Springer, Berlin - Heidelberg 1982.

27. Z.P. Bažant and L. Cedolin, Stability of Structures: Elastic, Inelastic, Fracture and Damage Theories, Oxford Univ. Press, New York 1991.

28. H. Petryk, Arch. Mech. 43 (1991) 519.

29. H. Petryk, CISM Lecture Notes No. 336, p. 259, Springer, Wien - New York 1993.

30. F.R. Shanley, J. Aero. Sci. 14 (1947) 261.

31. Q.S. Nguyen, Eur. J. Mech. A/Solids 13 (1994) 485.

32. H. Petryk and K. Thermann, Computational Plasticity: Fundamentals and Applications (D.R.J. Owen and E. Oñate, Eds.), p. 647, Pineridge Press, Swansea 1995.

33. H. Petryk and K. Thermann, Int. J. Solids Structures 33 (1996) 689.

34. D.C. Drucker, Proc. 1 -st U.S. Nat. Congr. Appl. Mech., p. 487, ASME 1951.

35. R. Hill and J.W. Hutchinson, J. Mech. Phys. Solids 23 (1975) 239.

36. J.W. Rudnicki and J.R. Rice, J. Mech. Phys. Solids 23 (1975) 371.

37. J. Hadamard, Leçons sur la Propagation des Ondes et les Equations de l'Hydrodynamique, Hermann, Paris 1903.

38. J. Mandel, Rheology and Soil Mechanics (J. Kravtchenko, Ed.), p. 58, Springer, Berlin 1966.

39. H. Petryk, J. Mech. Phys. Solids 40 (1992) 1227.

40. P. Franciosi and A. Zaoui, Int. J. Plasticity 7 (1991) 295.

41. J.W. Hutchinson and V. Tvergaard, Int. J. Solids Structures 17 (1981) 451.

42. J.M. Ball and R.D. James, Arch. Rat. Mech. Anal. 100 (1987) 13.

43. R. Hill, Proc. Roy. Soc. Lond. A 326 (1972) 131.

Theoretical and Applied Mechanics 1996
T. Tatsumi, E. Watanabe and T. Kambe (Editors)
© 1997 Elsevier Science B.V. All rights reserved.

The Mechanics of Manufacturing Processes

P. Wright, J. Stori and C. King[†]

Department of Mechanical Engineering, The University of California at Berkeley, CA., 94720, U.S.A.

1. ABSTRACT

Economic pressures, particularly related to the quality of manufactured goods and 'time-to-market' are forcing designers to think not only in terms of *product design*, but also in terms of integrated product and *process design*, and finally, in terms of deterministic *manufacturing planning and control*. As a result of these three high level needs, there is now an even greater need for comprehensive simulations that predict material behavior during a manufacturing process, the stresses and/or temperatures on associated tooling, and the final-product integrity. The phrase 'manufacturing processes' of course covers a broad scope; it includes semiconductor manufacturing, injection molding of polymers, metal machining and precision lapping, wood and textile production, and the final assembly of piece-parts into a consumer product. It can be seen from this partial listing that the fields of fluid mechanics, solid mechanics, dynamics and tribology can all play a role. The introduction to the paper will contain a review of manufacturing processes and describe where simulations have been successfully applied, and where simulations are still lacking. The best of the simulations are those where the models accurately fit the physical phenomena, where accurate constitutive equations are available, and where boundary conditions are realistic. Thus, the body of the paper will focus on the results from one of these more successful simulations. It has been used to predict the deflections of tooling and the most appropriate operating conditions for the manufacturing process under study. A new method for manufacturing planning is described. In this method, closed form, somewhat simplified, analytical models are used to determine manufacturing planning parameters and then the results from these simpler models are refined by the fuller simulations. A case study in machining parameter selection for peripheral finish milling operations is developed in which constraints on the allowable form error, D, and the peripheral surface roughness, R_a, drive the process parameter selection for a cutting operation intended to maximize the material removal rate. In four machining scenarios, achieving agreement (within a 5% deviation tolerance) between the simulation and constraint equation predictions required only three simulation execution cycles, demonstrating promise that simulation tools can be efficiently incorporated into parameter selection decisions.

2. INTRODUCTION

How can large scale simulation tools be best exploited and used for manufacturing planning? In this paper, we first briefly review how simulation tools have made an impact on the general understanding of manufacturing processes such as forming and machining. However, there is "some distance to go" before simulations are used on a more regular basis by production personnel in the day-to-day planning of specific production processes. To address this shortcoming, a novel methodology for production planning is

---

† Now with Sandia National Laboratories, Livermore, CA., 94551.

therefore presented that can be seen as a workable compromise between the large scale simulation packages and simpler, closed form analytical models. Such simpler, analytical models provide a relationship between the main inputs and main outputs of a particular manufacturing process. They are usually based on "first-order relationships" with some experimentally adjusted parameters. However by starting with the results from such simpler analytical models, and then by verifying them with the more complex simulation tools in the domain of interest, a satisfactory compromise can be reached between expediency and accuracy. The example that is presented focuses on obtaining very accurate features, with a good surface finish, in a machined component. The overall goal is to enrich a CAD/CAM environment: a) with physically accurate visualizations of the manufacturing process, and b) with process planning modules that allow the detailed selection of manufacturing parameters (e.g. processing speeds and tooling angles). Economically, the aims are to ensure a high quality product, and to reduce "time-to-market" by eliminating ambiguities and "re-work" during CAM (Richmond 1995).

Over the years, the main reference books in the analysis of the high-strain, plastic flow of materials, of course show an increasing emphasis on simulations and numerical methods. For example, each new edition of Rowe (1965-1977) or of Kobayashi et al. (1965-1989) shows an interesting trend: Early editions focus on closed form analytical solutions, often based on slip-line-fields backed up by experimental observations. By contrast, later editions focus much more on detailed simulations often involving the finite element method (Zienkiewicz 1977). Understandably, it is now common to see comprehensive volumes that devote themselves entirely to numerical approaches (Pittman et al. 1982; Shen and Dawson 1995). Simulations are now available for a wide variety of metal processing operations, including: two-dimensional drawing (Lu and Wright 1988); sheet metal forming (Bamman et al. 1995a); forging (Bamman et al. 1995b); machining (Komvopoulos and Erpenbeck 1991); welding (Michaleris et al. 1995); and casting (Byrne et al. 1995; Tortorelli et al. 1994) as well as polymer-process modelling (Smith et al. 1995). In such studies, the stresses (Bagchi and Wright 1987), temperatures (Stevenson et al. 1983), and residual effects (Bamman et al. 1995c) in the formed product are the usual parameters of interest. Constitutive models (Appelby et al. 1984), material properties (Oxley 1962) and boundary conditions (Smith et al. 1995; Wright 1990) are the key, high level issues to be addressed in the search for accurate models and simulations. Such simulations allow a more detailed prediction of the effects of various parameters on manufacturing processes than do closed-form analytical predictions. While purely closed-form relations are valid only within a small subset of the feasible operation space, simulations are often able to provide accurate predictions over a wider range of parameter values.

Unfortunately, increasingly reliable simulation tools do not directly address the planning issue. While the effects of changes in process parameters may be explored, the actual selection of parameters remains an independent task. The intent of the present work is to use available simulation tools as feedback to the parameter selection module. In order to achieve this, an iterative optimization procedure has been developed. Figure 1 schematically depicts the iterative relationship between simulation calls and the optimization module when such a method is employed. The algorithm outlined in Figure 2 illustrates the details of this architecture.

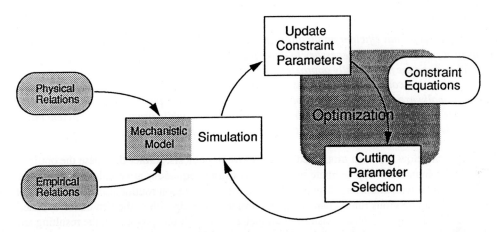

Figure 1. Incorporation of Planning and Simulation

3. THE ITERATIVE OPTIMIZATION APPROACH

3.1. Closed-form Analytical Solutions Versus Simulations

The complexity of machining processes has long challenged attempts to develop reliable and robust models to relate process parameters and state variables (Bagchi and Wright 1987; Elbestawi et al. 1996). As the process geometry and modelled physical interactions become increasingly complex, it is often no longer possible to maintain a closed-form solution (Sullivan et al. 1978; Thangaraj and Wright 1988). In such cases, simulations are implemented. In a simulation, obtaining a prediction of output or state variables requires running a computer program rather than solving an equation or evaluating a function. This introduces new challenges for the parameter selection process. *While a closed-form function can often be inverted to reveal a mapping from outputs to inputs, this is impossible with simulations.* Additionally, gradient information which is useful in optimization is more computationally expensive to obtain. Finally, due to the iterative and computationally demanding nature of simulation tools, a significant expense is typically associated with acquiring simulation results. This effect is exacerbated when a large number of users relies on a single simulation engine, or when the simulation is remotely located, requiring sharing of network bandwidth.

Multivariable optimization requires extensive calls to an objective and constraint functions before a local optimal point can be identified. Depending on the number of variables, constraints, and program structure, these functions may be called hundreds or even thousands of times before arriving at a single solution. When gradient information is not present, (as is the case when simulation tools are employed) the number of function calls is increased further. A significant practical challenge in incorporating simulation results into planning optimization is to make do with a greatly reduced number of calls to the simulation from the optimization algorithm.

3.2. Hypothesis

Analytic models permit efficient application of multivariable optimization techniques, yet rarely possess the accuracy and robustness necessary to make reliable planning decisions. By incorporating simulation feedback to fine-tune functionally approximately correct constraint equations, local convergence between the simulation and constraint predictions may be achieved through a limited number of calls to the computationally expensive simulation tools. The resulting process parameters satisfy the design constraints within the accuracy of the simulation predictions, while providing an efficient balance between the free variables arising from the functional form of the optimization model.

The approach below has been motivated by the above hypothesis. Rather than base the optimization solely on the simulation, approximate parametric constraint equations are employed with a limited number of unknown parameters. The initial call to the optimization routine uses mean values of these parameters from previous runs. Then, at each iteration of the algorithm, the simulation is called, and the results are used to update these constraint parameters. Through this procedure, the resulting analytic model becomes *increasingly locally accurate in the neighborhood of the optimal solution*, and may be expected to posses the correct behavior for local deviations from that point. The optimization routine is executed at each stage of the algorithm, using the improved model. The new "quasi-optimal" solution vector is then passed to the simulations, and the process repeats. Termination occurs when the model predictions agree with the simulation results to within a previously specified tolerance.

The analytic model must incorporate the effects of changes in the control variables. It is very likely, however, that second order effects will exist which are not modelled explicitly. These effects will result in deviations between the model and simulation predictions. By updating the parameters incorporated in the model, the model and simulation can be brought into agreement, at least within a restricted subspace around the optimal point. The role of the models within the optimization is to arrive at a reasonable compromise between control variables such that the simulation predictions satisfy the design requirements.

In Figure 2, the machining parameter selection algorithm is outlined. The procedure is detailed below for the generic case. In the following section, particular constraints and models for the parameter selection in finish periphery milling are developed and used as a case study of the approach. The generic algorithm and notation are detailed below:

Let x be a vector of process variables that must be selected. k is a vector of model parameters. y is an output or state variable that is constrained by the design requirements, such as maximum form error or surface roughness. A series of analytic closed-form expressions estimate the output characteristics, $\hat{y}_i$, given both the process variables and model parameters. The subscript i refers to the iteration stage of the algorithm. Each of the vectors x, k, and y may be expected to change at each iteration.

$$\hat{y}_i = f(x_i, k_i) \qquad (1)$$

Given:
y_j - desired state variables
x_0 - initial guess at solution
k_0 - initial model parameters

Desire:
x_j - parameter values such that
$$y_j = f(x_j, k_j) = SIM(x_j)$$

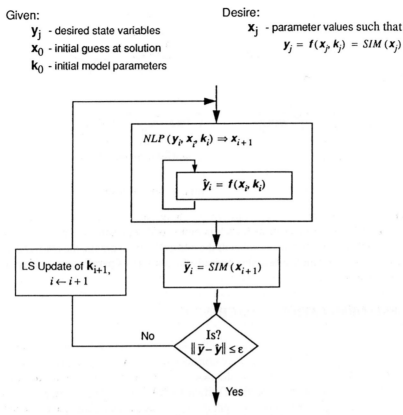

$$NLP(y_i, x_i, k_i) \Rightarrow x_{i+1}$$

$$\hat{y}_i = f(x_i, k_i)$$

LS Update of k_{i+1}, $i \leftarrow i+1$

$$\bar{y}_i = SIM(x_{i+1})$$

No

Is?
$\|\bar{y} - \hat{y}\| \le \varepsilon$

Yes

Figure 2. Algorithm for Machining Parameter Selection

Recognizing the greater range of feasible process space that can be accurately covered using the simulation tools, $\bar{y}_i$ is taken to be the "true" value of the output parameter, given the input vector x. (Note that the model parameters are not passed to the simulations. The simulation itself may possess a number of empirical parameters, but these are not visible to the planning algorithm.)

$$\bar{y}_i = SIM(x_i) \tag{2}$$

The optimization module generates a vector of process variables that are estimated to satisfy the design criteria. When these same process variables are fed into the available simulation tools, a discrepancy is found between the estimated and simulated values. This discrepancy is used to update the model parameters by performing a non-linear least squares fit to the model parameters using the available simulation points. The algorithm terminates when the deviation between the constraint equation and simulation predictions is less than or equal to the termination tolerance, e.

In the first few iterations of the algorithm, there will be more free model parameters than simulation data

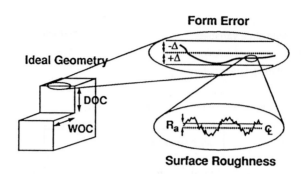

Figure 3. Planning Constraints.

points, making a least squares fit ambiguous. To avoid this situation, only a single additional model parameter is permitted to be updated with each iteration of the algorithm. Thus, for the first several iterations, the number of parameters is equal to the number of data points, the exactly determined case. In subsequent iterations, the number of points exceeds the number of parameters, as is typical in least squares applications.

4. PROBLEM FORMULATION IN MACHINING

To illustrate the above approach, a case study will now be presented in parameter selection for finish milling operations. Two design constraints are considered: a maximum peripheral *form error*, and a maximum peripheral *surface roughness*. In the case of finish milling operations, these tolerance-based constraints tend to drive parameter selection (Wright 1990). For roughing operations, constraints related to spindle power and tool wear/failure drive parameter selection, while tolerance and surface finish concerns are typically neglected. Figure 3 shows the configuration of the milling operation and the planning constraints.

In the parameter selection problem, it is assumed that the cutting tool has already been chosen. Four independent process parameters remain: spindle speed, feed rate, radial depth of cut and axial depth of cut. In this model, the cutting speed has been chosen based on values recommended in the Machining Data Handbook (1980), as tool wear models have not been incorporated into the simulations. The axial depth of cut has been restricted by the cut geometry, leaving two parameters, the radial depth of cut, and the feed rate. These constitute the free optimization variables.

The objective of the optimization is to maximize material removal rate by manipulating the free variables subject to the two primary design constraints: form error and surface roughness. Additional optimization constraints include the non-negativity of the process variables, as well as an upper bound on the spindle speed.

Two simulation tools have been employed which provide estimations of the resultant form error and surface roughness that may be expected in a peripheral milling operation. The EMSIM simulation [http://mtamri.me.uiuc.edu/software.testbed.html] is based on a mechanistic model of the end-milling process (DeVor et al. 1980; Kline et al. 1982a 1982b and 1983; Sutherland and DeVor 1986), and was used to provide the estimates of the form error. The SURF simulation (Melkote 1993; Melkote and Thangaraj 1994; Stark and Thangaraj 1996) has been used to predict the peripheral surface roughness.

Below, expressions for the two primary constraint equations are formulated. In the first case, a closed-form equation for the form error is developed. Although based on the same model as underlies the simulation, the desire to construct a closed-form expression required that a number of simplifying approximations be introduced, such that the constraint equation and simulation will not provide identical results. The constraint equation parameters are used to fit the constraint equation to the simulation data in a local region of the parameter space. The surface roughness constraint equation adopts a much simpler form, which is strikingly different from the geometric model of the surface roughness simulation. The success of the parameter selection algorithm requires only that the models preserve *functionally approximately correct* relationships between input and output variables. Then, through updates to the model parameters, these models can be tuned in-process for local agreement with the simulation results.

4.1. Maximum Form Error

The formulation of a constraint equation related to maximum form error, was derived from the physical models of the end milling process, as cited above.

Using the ideal approximation of the chip thickness derived by Martelotti (1941, 1945), $t_c = f \sin \alpha$, a power-law expression for the specific cutting force, $K'_t = C_t t_c^P$ and Sabberwal's approximation [1961/2] that $(\sin \alpha)^{P+1} \cong \sin \alpha$ for common values of P (i.e. $P @ -0.3$), the elemental force components on a "slice" are:

$$dF_t = C_t f^{(p+1)} \sin(\alpha) \, dz \tag{3}$$

$$dF_R = K_R dF_t \tag{4}$$

The power-law variation of cutting force with chip thickness is included explicitly in the constraint formulation in an attempt to reduce the variability of the remaining empirical constraint parameters. The following simplified form for an infinitesimal axial element shows the elemental force component:

$$dF_y = C_t f^{(p+1)} \sin \alpha \, (\sin \alpha + \cos \alpha) \, dz \tag{5}$$

For the case of finish milling with shallow radial widths of cut, only a single flute of the cutter is engaged with the workpiece. To predict F_y in this orientation, the elemental force components may be integrated along the cutting edge. Using a change of variables,

$$d\alpha = \left(\frac{\tan(\lambda)}{R} \right) dz \tag{6}$$

Equation 5 can be integrated along the length of the engaged flute. Note that the new constant term introduced by the change of variables has been absorbed within the C_t constant.

$$F_y = C f^{(p+1)} \left(\frac{1}{2} sin^2 \alpha + K_R \left(\frac{\alpha}{2} - \frac{1}{4} sin 2\alpha \right) \right) \Bigg|_{\alpha_{bot}}^{\alpha_{top}} \tag{7}$$

The limits of the integration are the orientation angle, a, at the top and the bottom of the cut. To predict surface error at the bottom of the cut, we have $a_{bot} = 0$. At the top of the cut,

$$\alpha_{top} = min \left(DOC \, tan \, (\lambda) \, / R, \alpha_{en} \right) \tag{8}$$

where a_{en} is a function of the radial width of cut and tool radius.

In the climb milling process, tangential forces act to separate the tool and the workpiece regardless of cut geometry. The surface form error is predicted by assuming that the tool deflects as a cantilever beam with a concentrated force applied at the "force center," CF_y. For the purposes of the constraint equation, CF_y is taken to be the midpoint of the cut in the axial direction. The form error is generated by applying this deflection position at the point of contact between the tool and the final surface, termed the surface generation point. As we are interested in the maximum form error, the surface generation point is taken to be the bottom of the cut, which is obtained when $a_{bot} = 0$. As a constraint inequality this results in:

$$\hat{\Delta} = \frac{F_y CF_y^2 (3CF_y - L)}{6EI} \leq \Delta_{max} \tag{9}$$

where L is the effective length of the cutter, as we are interested in deflection at the bottom of the cut. The moment of inertia of the tool is taken to be that of a solid cylinder with an equivalent diameter 0.80 of the tool diameter (Kops and Vo 1990).

When Equation 9 is expanded through substitution of Equations 7 and 8, there remain three empirical coefficients in the final expression, C_f, K_R, and, P.

4.2. Peripheral Surface Roughness

In contrast to the rather involved expression for the maximum form error, a very simple parametric expression has been used for the peripheral surface roughness constraint. Boothroyd and Knight (1989), and others (Shaw 1984; DeVries 1992), have derived expressions relating the feed per tooth and the tool radius to the center-line average roughness, R_a. From purely geometric considerations, the ideal surface roughness is given by:

$$R_a = 0.0321 \, (f^2 / R) \tag{10}$$

In the case of practical machining operations, this geometric ideal will not be met. Nevertheless, the form of the variable relationships should remain valid. For the purposes of the optimization constraint, then, two parameters are used in place of the above constants.

$$R_a = K_a f^b / R \leq R_{amax} \tag{11}$$

These added parameters should provide the flexibility necessary for the model to locally conform to the simulation predictions. For example, in certain run-out situations, only one of the four flutes may be cutting. This is equivalent to an increase in the feed per tooth of a factor of four, or equivalently, assuming b to be close to 2, an increase in K_a by a factor of 16.

5. MODEL IMPLEMENTATION

The algorithm was implemented using the non-linear optimization routines provided in the optimization toolbox available with MATLAB (Grove 1990). For the constrained optimization problem, a sequential quadratic programming (SQP) method was employed, using the BFGS update for the estimate of the Hessian at each iteration. The QP subproblem is solved using an active set method. For the non-linear least squares fit, the Levenberg-Marquardt method was used (Luenberger 1989).

The EMSIM simulation is presently available on the World Wide Web (http://mtamri.me.uiuc.edu/ software.testbed.html). For the results presented here, EMSIM was run manually using the results of the optimization algorithm. The output of the simulation was then manually re-entered into the program before the algorithm could proceed. EMSIM has recently been enabled to provide an automated means of accessing, through the Data Transfer Mechanism (DTM) protocol developed at NCSA (http:// slice.me.uiuc.edu/~venkat/emserv/emserv.html). The SURF simulation was imported from Michigan Technological University and run in house. During the course of the work presented in this paper, a surface roughness prediction capability was added to the EMSIM software. The two simulations are based on very similar models, and results between the two are nearly identical.

6. RESULTS AND DISCUSSION

Four planning scenarios for finish periphery milling were explored using the iterative parameter selection algorithm presented above. Two different cutting tools were selected, and each was considered in both an ideal and runout situation. Scenarios A and B represent the cases for the 1/2" tool, while a 3/4" tool was used in scenarios C and D. Cases A and C were ideal in the sense of no runout, while B and D included rather severe runout conditions. The maximum form error constraint for all four was prescribed to be 0.001", while the surface roughness constraint was 2min for the 1/2" tool scenario and 8min for the 3/4" tool scenario. Table 1 summarizes the fixed parameters, design tolerances, the final variable selections, and the number of simulation iterations and calls to the constraint model during optimization.

Figures 4 and 5, following, present the variable trajectories during the execution of the algorithm. The termination criteria was a 5% maximum deviation between the model and simulation predictions. Of particular note is the very few number of calls that were necessary to the simulation in order to bring the optimization results in line with the simulation results. In all four cases, within three iterations of the algorithm, the termination criteria was satisfied. Also of interest is the final values of the process variables that were selected. The results reveal the significant impact of effects which are often neglected, such as cutter run-out, when making planning decisions. Also revealed through the optimization is a preference for large feeds over large radial widths of cut when planning. The objective of maximum material removal rate always resulted in the feed rate being increased to the bound limited by the surface roughness constraint. Both the surface roughness and form error constraints were active in all of the final outcomes.

The variable space trajectories, as presented in Figures 4 and 5, provide a visual mean of interpreting the behavior of the algorithm. The two horizontal axes represent the two free process variables which are to be selected by the algorithm. The vertical axis represents the percentage deviation between the constraint predictions and the simulation predictions at each stage of the algorithm. Termination is achieved when the trajectory arrives within 5% of the horizontal (z=0) plane. The deviation from the arbitrary initial starting point is provided for reference. As may be noted from the plots, the deviation will in many cases increase in the first optimization iteration. The simulation is run for the first time at the second point of the trajectory. With the addition of the simulation feedback, the trajectory rapidly converges towards the termination criteria in subsequent iterations.

Table 1. Planning Scenarios and Results

	Scenario	A	B	C	D
	Tool Diameter	1/2"	1/2"	3/4"	3/4"
	Free Length	1.3"	1.3"	2.25"	2.25"
Tool	Num Flutes	2	2	4	4
params	PAOR, e	0.000"	0.005"	0.000"	0.002"
	ATR, τ	0.0°	0.0°	0.0°	0.015°
y	Δ_{max}	0.001"	0.001"	0.001"	0.001"
values	R_{amax}	2μin	2μin	8μin	8μin
	DOC	0.4"	0.4"	0.4"	0.4"
x	WOC	0.046"	0.047"	0.024"	0.023"
values	f	0.0039"	0.0019"	0.0100"	0.0024"
	RPM	4584	4584	3056	3056
	Iterations	2	3	2	3
	Function Evaluations	54	55	39	64
ε	Max. Deviation	0.7%	1.9%	2.7%	0.6%

The number of function calls required by the SQP routine for each of the scenarios is reported in the second to last row of Table 1. This provides some indication of the benefits provided by a hybrid analytic/ simulation driven procedure. Although the simulation was called only once or twice in each scenario, obtaining the optimal solutions required a vastly larger number of calls to the model constraint functions. Each of the jogs in the parameter trajectories represents a search direction within the non-linear optimization routine. Each search stage requires an evaluation of the objective function and constraint equations. As is typical in non-linear optimization, the optimal search direction is not known, and a certain amount of exploration takes place en route to a solution. The practical efficiency of the algorithm presented stems from the fact that simulation calls are not required at each intermediate search stage.

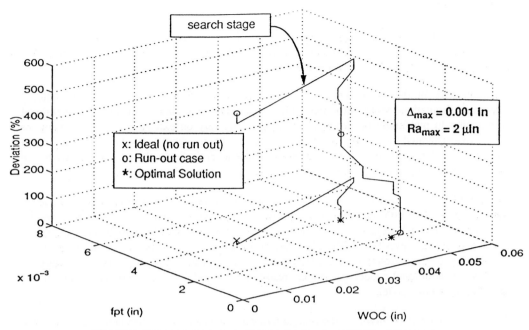

Figure 4. Process Parameter Trajectories for 1/2" Tool Scenario

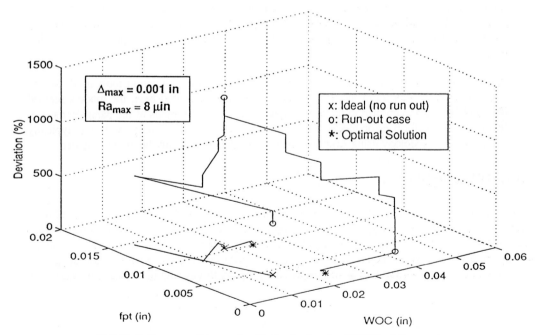

Figure 5. Process Parameter Trajectories for 3/4" Tool Scenario

528

7. CONCLUSIONS

The results presented above provide encouraging support for the initial hypothesis - that process simulation tools may be efficiently incorporated into a multivariable optimization approach to machining parameter selection. Specifically, we note the following:

1. The objective of this work was to provide a methodology for incorporating increasingly accurate simulation tools into machining operation planning decisions. Conventional approaches to machining operation planning typically rely on extensive empirical results or rule-bases generated by human experts. Rarely do such approaches posses the granularity over a broad parameter subspace that simulations are able to provide. In the present approach, the operation planning problem was formulated as a constraint-based optimization problem.

2. By updating the constraint models from simulation feedback, the model was tuned to fit simulation results. With only two or three calls to the relatively expensive simulation, results were obtained which were optimal to within close agreement with the simulation predictions, verifying the stated hypothesis. The efficiency of this approach lay in the use of the optimization model to guide the search through the simulation subspace. Each new simulation point was used to tune an additional model parameter, resulting in rapid local convergence between the simulation and the model.

3. The specific case study involved the selection of process parameters when constrained by form error and surface finish requirements on the final surface of a peripheral milling operation. Results from the iterative planning algorithm were presented for four machining scenarios, which included two tools and two runout cases. In all four scenarios, the algorithm arrived at an optimal solution, (within a 5% agreement of the simulation results) in two or three iterations. This provides encouragement that simulations may be efficiently used in parameter selection decisions.

4. The success of this approach is clearly dependent on both the simulation tools used and the formulation of the constraint models. In the case studies presented here, the models seemed adequately robust to conform to the simulation results in local regions of the parameter space. Nevertheless, the form error model was quite involved and more complex than is typically desirable. A methodology using simpler constraint models is currently being investigated.

8. ACKNOWLEDGMENTS

The authors would like to thank Greg Stark at Michigan Technological University for the use of the SURF simulation program, and his assistance in importing the code to Berkeley. We also would like to thank Prof. R. DeVor and Ganesan Venkatasubramanian at the University of Illinois, Urbana-Champaign for many helpful discussions and access to the EMSIM simulation. The second author was supported by the National Science Foundation Graduate Fellowship program throughout the duration of this research.

9. NOMENCLATURE

$\mathbf{x}$	vector of process variables	E	Young's Modulus of cutter
$\mathbf{y}$	vector of output or state variables	I	moment of inertia of end mill
$\mathbf{k}$	vector of constraint equation parameters	Δ	form error
$\bar{y}$	simulation prediction	Δ_{max}	form error planning tolerance
$\hat{y}$	estimated value	R_a	arithmetic average surface roughness
ε	algorithm termination tolerance	R_{amax}	surface roughness planning tolerance
f	feed per tooth	R	cutter radius
α	angular position of flute element	DOC	axial depth of cut
α_{en}	entry angle of flute	WOC	radial width of cut
α_{ex}	exit angle of flute	K_a	surface roughness constraint coefficient
α_{top}	angular position of flute at top of cut	b	surface roughness constraint exponent
α_{bot}	angular position of flute at bottom of cut	λ	cutter helix angle
δF_t	tangential force element	γ	angular spacing between flutes
δF_R	radial force element	θ	angular orientation of cutter
K_t'	specific cutting force	SG	surface generation point
C_t	cutting force coefficient	e	parallel axis offset magnitude
P	cutting force exponent	β	parallel axis offset orientation direction
D_z	axial disk thickness		
K_R	radial cutting force coefficient	τ	tilt runout magnitude
t_c	true chip thickness	ρ	tilt runout orientation direction
F_y	tangential cutting force		
CF_y	tangential force center		
L	effective length of cutter		

530

REFERENCES

Appelby, E.J, Devenpeck, M.L., Lu, C.Y., Rao, R.S., Richmond, O., and Wright, P.K. (1984) Strip drawing: a theoretical-experimental comparison. *International Journal of Mechanical Science* 26, 5, pp. 351-362.

Bagchi, A., and Wright, P.K. (1987) Stress analysis in machining with the use of sapphire tools. *Proceedings of the Royal Society of London* 409, pp. 99-113.

Bammann, D.J., Chiesa, M.L., and Johnson, J.C. (1995a) Modeling large deformation anisotropy in sheet metal forming. *Simulation of Materials Processing: Theory, Methods and Applications*, pp. 657-660, ed. Shen and Dawson, Balkema, Rotterdam.

Bammann, D.J., Jin, P.S., Johnson, G.C., and Odegard, B.C. (1995b) Modeling 304L stainless steel high temperature forgings using an internal state variable model. *Simulation of Materials Processing: Theory, Methods and Applications*, ed. Shen and Dawson, Balkema, Rotterdam, pp. 215-218.

Bammann, D.J., Prantil, V.C., and Lathrop, J.F. (1995c) A Model of Phase Transformation Plasticity. *Modelling of Casting, Welding and Advanced Solidification Processes* pp. 275-285, ed. M. Cross and J. Campbell, The Minerals, Metals & Materials Society.

Boothroyd, G. and Knight, W.A. (1989) *Fundamentals of Machining and Machine Tools* 2nd. edition, Marcel Dekker, Inc.

Byrne, P.E., Dantzig, J.A., Morthland, T.E., and Tortorelli, D.A. (1995) Optimal Riser Design for Metal Castings. *Metallurgical and Materials Transactions* 26B, pp. 871-885.

DeVor, R.E., Kline, W.A., and Zdeblick, W.J. (1980) A Mechanistic Model for the Force System in End Milling with Application to Machining Airframe Structures. *Eight North American Manufacturing Research Conference Proceedings*, pp. 297-303.

DeVries, W.R. (1992) *Analysis of Material Removal Processes*, Springer-Verlag.

Elbestawi, M.A., El-Wardany, T.I., and Mohammed, E. (1996) Cutting temperature of ceramic tools in high speed machining of difficult-to-cut materials. *International Journal of Machine Tools Manufacturing* 36, 5, pp. 611-634.

Grove, A. (1990) *Optimization Toolbox for use with MATLAB™*, The Mathworks, Inc.

The University of Illinois, Mechanical Engineering, MTAMRI [*http://mtamri.me.uiuc.edu/ software.testbed.html*]

Ibidem [*http://slice.me.uiuc.edu/~venkat/emserv/emserv.html*]

Kline, W.A., DeVor, R.E., and Lindberg, J.R. (1982) The Prediction of Cutting Forces in End Milling with Application to Cornering Cuts. *Int. J. Mach. Tool Des. Res.* 22, 1, pp. 7-22.

Kline, W.A., DeVor, R.E., and Shareef, I.A. (1982) The Prediction of Surface Accuracy in End Milling. *ASME Journal of Engineering for Industry* 104, pp. 272-278.

Kline, W.A., and DeVor, R.E. (1983) The Effect of Runout on Cutting Geometry and Forces in End Milling. *Int. J. Mach. Tool Des. Res.* 23, 2/3, pp. 123-140.

Kobayashi, S., Oh, S., and Altan T. (1989) Metal Forming and The Finite Element Method. *Oxford University Press.*

Komvopoulos, K., and Erpenbeck, S.A. (1991) Finite Element Modelling of Orthogonal Metal Cutting. *ASME Journal of Engineering for Industry* 113, pp. 253.

Kops, L., and Vo, D.T. (1990) Determination of the Equivalent Diameter of and End Mill Based on its Compliance. *Annals. of CIRP* 39/1, p.93.

Lu, S. C-Y., Wright, P.K. (1988) Finite Element Modeling of Plane-Strain Strip Drawing With Interface Friction. *ASME Journal of Engineering for Industry* 110, pp. 101-110.

Luenberger, D.G. (1989) *Linear and Nonlinear Programming*, 2nd. ed., *Addison-Wesley.*

Metcut Research Associates (1980*) Machining Data Handbook*, 3rd Edn.

Martelotti, M.E. (1941) An analysis of the milling process. *Trans. Am. Soc. mech. Engrs.* 63 pp. 667-675.

Martelotti, M.E. (1945) An analysis of the milling process: part II-down milling. *Trans. Am. Soc. mech. Engrs.* 67, pp. 233-245.

Melkote, S.N. (1993) The Modelling of Surface Texture in the End Milling Process. *Ph.D. Dissertation, Michigan Technological University.*

Melkote, S.N. and Thangaraj, A.R. (1994) An Enhanced End Milling Surface Texture Model Including the Effects of Radial Rake and Primary Relief Angles. *ASME Journal of Engineering for Industry* 116, pp. 166-175.

Michaleris, P., Tortorelli, D.A., and Vidal, C.A. (1995) Analysis and optimization of weakly coupled thermo-elasto-plastic systems with applications to weldment design. *Simulation of Materials Processing: Theory, Methods and Applications*, ed. Shen and Daw son, Balkema, Rotterdam, pp. 599-604.

Oxley, P.L.B. (1962) Shear Angle Solutions in Orthogonal Machining. *International Journal of Machine Tool Design and Research* 2, pp. 219-229.

Pittman, J.F.T., Wood, R.D., Alexander, J.M., and Zienkiewicz, O.C. (1982*) Numerical Methods in Industrial Forming Operations*, Pineridge Press, Swansea, UK.

Richmond, O. (1995) Concurrent design of products and their manufacturing processes based upon models of evolving physicoeconomic state. *Simulation of Materials Processing: Theory, Methods and*

Applications, ed. Shen and Dawson, Balkema, Rotterdam, pp. 153-155.

Rowe, G.W. (1965-1968) *Principles of Industrial Metalworking Operations*. In three editions, Arnold.

Sabberwal, A.J.P. (1961/1962) Chip section and cutting force during the milling operation. *Ann. CIRP* 10.

Smith, D.E., Tucker III, C.L., and Tortorelli, D.A. (1995) Optimization and sensitivity analysis in polymer processing: Sheet extrusion die design. *Simulation of Materials Processing: Theory, Methods and Applications*, ed. Shen and Dawson, Balkemá, Rotterdam, pp. 619-624.

Shaw, M.C. (1984) *Metal Cutting Principles*. Oxford University Press.

Stark, G.A., and Thangaraj, A.R. (1996) A 3-D Surface Texture Model for Peripheral Milling with Cutter Runout Using Neural Network Modules and Splines. *Trans. North American Manufacturing Research Institution of SME* 24, pp.57-62.

Stevenson, M.G., Chow, J.G., and Wright, P.K. (1983) Further Developments in Applying the Finite Element Method to the Calculation of Temperature Distributions in Machining and Comparisons with Experiment. *ASME Journal of Engineering for Industry* 105, pp. 149-154.

Sullivan, K.F., Wright, P.K., and Smith, P.D. (1978) Metallurgical appraisal of instabilities arising in machining. *Metals Technology*, pp. 181-188.

Sutherland, J.W., and DeVor, R.E. (1986) An Improved Method for Cutting Force and Surface Error Prediction in Flexible End Milling Systems. *ASME Journal of Engineering for Industry* 108, pp. 269-281.

Thangaraj, A., and Wright, P.K. (1988) Computer-assisted Prediction of Drill-failure Using In-process Measurements of Thrust Force. *Journal of Engineering for Industry* 110, pp. 192-200.

Tortorelli, D.A., Tomasko, J.A., Morthland, T.E., and Dantzig, J.A. (1994) Optimal design of nonlinear parabolic systems. Part II: Variable spatial domain with applications to casting optimization. *Computer Methods in Applied Mechanics and Engineering* 113, pp. 157-172.

Wright, P.K. (1990) Transparent Sapphire Tools. *Journal of Manufacturing Systems* 9, 4, pp. 292-302.

Zienkiewicz, O.C. (1977) *The Finite Element Method*. 3rd Ed., McGraw-Hill Book Company.

Theoretical and Applied Mechanics 1996
T. Tatsumi, E. Watanabe and T. Kambe (Editors)
© 1997 Elsevier Science B.V. All rights reserved.

Multibody Dynamics with Multiple Unilateral Contacts

F. Pfeiffer[a]

[a]Lehrstuhl B für Mechanik, TU-München,
Arcisstr. 21, D–80290 München, Germany

Multibody systems are usually considered to be bilaterally coupled or to possess interconnections in the form of force laws. On the other side, a wide area of mechanical systems and of machine applications can be characterized by unilateral behavior including one-sided contacts with impact and friction phenomena. Events of that kind in multiple contacts of multibody systems may depend on each other so that each transition from stick to slip or vice versa and each impact event changes the complete contact configuration. A mathematical description of these processes conveniently starts with a complementarity law of all contact situations with states, that either magnitudes of relative kinematics in every contact are zero and the corresponding constraint forces are not zero or vice versa. Thus, their product is always zero establishing a linear complementarity problem for plane and a nonlinear complementarity problem for spatial contacts. A theory is presented which covers impulsive and frictional processes in multiple contacts of multibody systems and considers stick-slip transitions. Various practical examples confirm the theoretical ideas principally and by close correspondance of theory and measurements specifically.

1. INTRODUCTION

Contact events appear in dynamical systems very often due to the fact that the world of dynamics usually happens to be as much unilateral as it is bilateral. Walking, grasping, climbing are typically unilateral processes, the operation of machines and mechanisms includes a large variety of unilateral aspects. From this there emerges a need to extend multibody theory by contact phenomena.

All contact processes have some characteristic features in common. If a contact is closed, a motion changes from slip to stick, we come out with some additional constraints generating constraint forces. We then call the contact active. Otherwise it is passive. Obviously transitions in such contacts depend on the dynamics of the system under consideration. The beginning of such a contact event is indicated by kinemtical magnitudes like relative distances or relative velocities, the end by kinetic magnitudes like normal force or friction force surplus. This will deliver a basis for the mathematical formulation to follow.

A large variety of possibilities exist in modelling local contact physics, from Newton's, Poisson's and Coulomb's laws to a discretization of local behavior by FE- or BE-methods. But, simulations of large dynamical systems require compact contact laws. Therefore, we shall concentrate on the first types of laws which inspite of their simple structure still are able to describe realistically a large field of applications.

Literature covers aspects like contact laws, FEM- and BEM-analysis, contact statics, contact dynamics and a large body of various applications. With respect to multibody systems with multiple unilateral contacts most of the mathematical fundamentals, though firstly regarding statical problems only, were laid down by European scientists. First considerations were started by MOREAU and his school [5,11,12], which in the meantime continues his efforts in a remarkable way [18]. MOREAU established the linear and nonlinear complementarity problems of contact mechanics, applied the theory of differential inclusions to contact problems, and he introduced some nice ideas to reduce the computing time problem.

The scientific community includes in addition scientists like PANAGIOTOPOULOS and LÖTSTEDT who developed powerful methods for statical and dynamical problems of contact mechanics [14,15,9,10]. Especially PANAGIOTOPOULOS established in a couple of impressive books [1,14,15] a very general theory on unilateral problems in mechanics. LÖTSTEDT applied some of these ideas to rigid motion [10]. The Swedish school in that field is continued with remarkable results by KLARBRING, who focusses his work to problems of FEM- and BEM-modeling [7].

Research in US significantly is influenced in that field by scientists like KELLER, STRONGE, HOLMES, SHAW, TRINKLE, WANG and BRACH. KELLER extended the available models for impacts with friction by a convincing consideration of compression and expansion phases for one impact [6]. STRONGE developed an extended impact law by taking into account dissipation energy during the impact [20]. SHAW was involved already very early in problems with dry friction [19]. The dynamics of a sinusoidally vibrating was studied by KLOTTER [8], that time at Stanford, and by HOLMES [4]. BRACH is involved in simple impact models and the problem of energy loss during impact [2]. WANG considers similar questions describing impacts with friction [22]. TRINKLE belongs to those few American scientists applying in his research the complementarity idea [21].

At the author's institute research has been performed in that field more than ten years, which is mostly summarized in the book by PFEIFFER, GLOCKER [17]. Some newer results on nonlinear complementarity problems may be found in WÖSLE, PFEIFFER [23]. As the author's institute is one of engineering mechanics, most of the research work deals with a transfer of the demanding mathematical fundamentals to an engineering and application-friendly level. This is also the goal of the contribution.

2. CONTACT KINEMATICS

2.1. General considerations
A set of generalized coordinates

$$\boldsymbol{q}(t) = \begin{pmatrix} q_1(t) \\ \vdots \\ g_f(t) \end{pmatrix} \in \mathbb{R}^f \tag{1}$$

is used for the mathematical descirption of the dynamics of a bilaterally constrained system with f degrees of freedom. In order to take into account additional unilateral constraints like contact or friction constraints, we have to derive kinematic contact conditions. The possible motion of each body in a multibody system, which is compatible to the kinematic conditions, is restricted by conditions for normal distances, relative velocities and relative accelerations in the potential contact points (active unilateral constraints).

We now consider a system with n_A contact points and introduce the four index sets [17]

$$
\begin{array}{llll}
I_A & = & \{1, 2, \ldots, n_A\} & \text{with } n_A \text{ elements} \\
I_C & = & \{i \in I_A : g_{Ni} = 0\} & \text{with } n_C \text{ elements} \\
I_N & = & \{i \in I_C : \dot{g}_{Ni} = 0\} & \text{with } n_N \text{ elements} \\
I_T & = & \{i \in I_N : |\dot{\boldsymbol{g}}_{Ti}| = 0\} & \text{with } n_T \text{ elements}
\end{array}
\tag{2}
$$

which describe the kinematic state of each contact point. The set I_A consists of the n_A indices of all contact points. The elements of the set I_C are the n_C indices of the unilateral constraints with vanishing normal distance $g_{Ni} = 0$, but arbitrary relative velocity in the normal direction. In the index set I_N are the n_N indices of the potentially active normal constraints which fulfill the necessary conditions for continuous contact (vanishing normal distance $g_{Ni} = 0$ and no relative velocity $\dot{g}_{Ni}$ in the normal direction). The n_T elements of the set I_T are the indices of the potentially active tangential constraints. The corresponding normal constraints are closed and the relative velocities $\dot{g}_{Ti}$ in the tangential direction are zero. The numbers of elements of the index sets I_C, I_N and I_T are not constant because there are variable states of constraints due to separation and stick-slip phenomena.

The normal distances and the relative velocities in the tangential direction are determined by means of relative kinematics [3]. In general, the normal distances of all contact points

$$
g_{Ni} = g_{Ni}(\boldsymbol{q}, t) \in \mathbb{R}^1 ; \quad i = 1, \ldots, n_A
\tag{3}
$$

are functions of the generalized coordinates $\boldsymbol{q}$ and the time variable t. The normal distance g_{Ni} shows positive values for separation and negative values for penetration. Therefore, a changing sign of g_{Ni} indicates a transition from separation to contact. The relative velocities in the tangential direction

$$
\dot{\boldsymbol{g}}_{Ti} = \dot{\boldsymbol{g}}_{Ti}(\boldsymbol{q}, \dot{\boldsymbol{q}}, t) \in \mathbb{R}^2 ; \quad i = 1, \ldots, n_A
\tag{4}
$$

are additionally dependent on the generalized velocities $\dot{\boldsymbol{q}}$. The relative velocities in the normal direction

$$
\dot{g}_{Ni} = \dot{g}_{Ni}(\boldsymbol{q}, \dot{\boldsymbol{q}}, t) = \frac{\partial g_{Ni}}{\partial \boldsymbol{q}} \dot{\boldsymbol{q}} + \frac{\partial g_{Ni}}{\partial t}
\tag{5}
$$

are the first derivatives with respect to time of equation (3). The relative velocities in equations (5) and (4) can be written in the form

$$
\dot{g}_{Ni} = \boldsymbol{w}_{Ni}^T \dot{\boldsymbol{q}} + \widetilde{w}_{Ni} ; \quad \dot{\boldsymbol{g}}_{Ti} = \boldsymbol{W}_{Ti}^T \dot{\boldsymbol{q}} + \widetilde{\boldsymbol{w}}_{Ti} .
\tag{6}
$$

with

$$\boldsymbol{w}_{Ni} = \left(\frac{\partial g_{Ni}}{\partial \boldsymbol{q}}\right)^T \;\; ; \;\; \boldsymbol{W}_{Ti} = \left(\frac{\partial \dot{\boldsymbol{g}}_{Ti}}{\partial \dot{\boldsymbol{q}}}\right)^T \;\; ; \;\; \tilde{w}_{Ni} = \frac{\partial g_{Ni}}{\partial t} \;\; . \tag{7}$$

A negative value of the relative velocity $\dot{g}_{Ni}$ corresponds to an approaching process of the bodies. In the case of a continual normal contact with $g_{Ni} = \dot{g}_{Ni} = 0$, we can use the relative velocity $\dot{\boldsymbol{g}}_{Ti}$ to indicate a transition from sliding ($|\dot{\boldsymbol{g}}_{Ti}| \neq 0$) to stiction or rolling ($|\dot{\boldsymbol{g}}_{Ti}| = 0$). The relative accelerations of the contact points

$$\ddot{g}_{Ni} = \boldsymbol{w}_{Ni}^T \ddot{\boldsymbol{q}} + \overline{w}_{Ni} \;\; ; \;\; \ddot{\boldsymbol{g}}_{Ti} = \boldsymbol{W}_{Ti}^T \ddot{\boldsymbol{q}} + \overline{\boldsymbol{w}}_{Ti} \tag{8}$$

with

$$\overline{w}_{Ni} = \dot{\boldsymbol{w}}_{Ni}^T \dot{\boldsymbol{q}} + \dot{\tilde{w}}_{Ni} \;\; ; \;\; \overline{\boldsymbol{w}}_{Ti} = \dot{\boldsymbol{W}}_{Ti}^T \dot{\boldsymbol{q}} + \dot{\tilde{\boldsymbol{w}}}_{Ti} \tag{9}$$

are determined by differentiation of equation (6) with respect to time. Continual contact demands $g_{Ni} = \dot{g}_{Ni} = 0$, while separation is only possible if the relative acceleration $\ddot{g}_{Ni} > 0$. Transition from stiction to sliding occurs for a closed contact if the amount of the relative acceleration $|\ddot{\boldsymbol{g}}_{Ti}| > 0$.

2.2. Contact geometry

Planar contacts may be characterized by a contact line, where the direction of motion is known. A good survey on planar contacts may be found in [3].

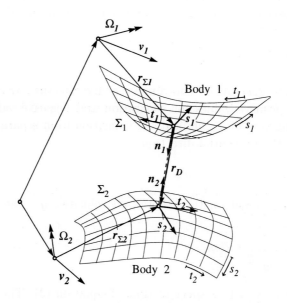

Figure 1. Contact geometry of two surfaces

The situation for spatial contacts is more complicated. We still assume that the two approaching bodies are convex (Figure 1) at least in that area where contact points might occur. The two bodies are moving with $\boldsymbol{v}_i, \boldsymbol{\Omega}_i (i = 1, 2)$.

For the description of a surface Σ we need two parameters s and t: $\boldsymbol{r}_\Sigma = \boldsymbol{r}_\Sigma(s, t)$. The tangents $\boldsymbol{s}$ and $\boldsymbol{t}$, which span the tangent plane at a point of the surface, are defined as:

$$s = \frac{\partial \boldsymbol{r}_\Sigma}{\partial s}, \qquad \boldsymbol{t} = \frac{\partial \boldsymbol{r}_\Sigma}{\partial t} . \tag{10}$$

From these basic vectors the fundamental magnitudes of the first order are calculated:

$$E = \boldsymbol{s}^T \boldsymbol{s}, \qquad F = \boldsymbol{s}^T \boldsymbol{t}, \qquad G = \boldsymbol{t}^T \boldsymbol{t} . \tag{11}$$

The normalized normal vector $\boldsymbol{n}$ is perpendicular to the tangent plane and pointing outwards:

$$\boldsymbol{n} = \frac{\boldsymbol{s} \times \boldsymbol{t}}{\sqrt{EG - F^2}} . \tag{12}$$

We further need the fundamental magnitudes of the second order:

$$L = \boldsymbol{n}^T \frac{\partial^2 \boldsymbol{r}_\Sigma}{\partial s^2}, \qquad M = \boldsymbol{n}^T \frac{\partial^2 \boldsymbol{r}_\Sigma}{\partial s \partial t}, \qquad N = \boldsymbol{n}^T \frac{\partial^2 \boldsymbol{r}_\Sigma}{\partial t^2} . \tag{13}$$

For a potential contact point we demand, that the normal vector of body 1 $(\boldsymbol{n}_1)$ and the distance vector $\boldsymbol{r}_D$ are perpendicular to the tangent vectors of body 2 $(\boldsymbol{s}_2$ and $\boldsymbol{t}_2)$. Thus we obtain four nonlinear equations:

$$\begin{aligned} \boldsymbol{n}_1^T \boldsymbol{s}_2 &= 0, & \boldsymbol{r}_D^T \boldsymbol{s}_2 &= 0, \\ \boldsymbol{n}_1^T \boldsymbol{t}_2 &= 0, & \boldsymbol{r}_D^T \boldsymbol{t}_2 &= 0. \end{aligned} \tag{14}$$

This nonlinear problem has to be solved at every time step of the numerical integration. After the solution is found the distance g_N between the possible contact points can be calculated as

$$g_N = \boldsymbol{n}_1^T \boldsymbol{r}_D = -\boldsymbol{n}_2^T \boldsymbol{r}_D . \tag{15}$$

g_N is used as an indicator for the contact state. Its value is positive for 'no contact' and negative for penetration.

The constraints are again formulated on velocity level, where in the spatial case we have three of them, one in normal direction $\dot{g}_N$ and two in the tangential directions $\dot{g}_S, \dot{g}_T$:

$$\begin{aligned} \dot{g}_N(\boldsymbol{q}, \dot{\boldsymbol{q}}, t) &= \boldsymbol{n}_1^T (\boldsymbol{v}_{\Sigma 2} - \boldsymbol{v}_{\Sigma 1}), \\ \dot{g}_S(\boldsymbol{q}, \dot{\boldsymbol{q}}, t) &= \boldsymbol{s}_1^T (\boldsymbol{v}_{\Sigma 2} - \boldsymbol{v}_{\Sigma 1}), \quad \text{with } \boldsymbol{v}_{\Sigma 1} \text{ and } \boldsymbol{v}_{\Sigma 2} \\ \dot{g}_T(\boldsymbol{q}, \dot{\boldsymbol{q}}, t) &= \boldsymbol{t}_1^T (\boldsymbol{v}_{\Sigma 2} - \boldsymbol{v}_{\Sigma 1}), \end{aligned} \qquad \begin{aligned} \boldsymbol{v}_{\Sigma 1} &= \boldsymbol{J}_{\Sigma 1}(\boldsymbol{q}, t)\dot{\boldsymbol{q}} + \tilde{\boldsymbol{j}}_{\Sigma 1}(\boldsymbol{q}, t) \\ \\ \boldsymbol{v}_{\Sigma 2} &= \boldsymbol{J}_{\Sigma 2}(\boldsymbol{q}, t)\dot{\boldsymbol{q}} + \tilde{\boldsymbol{j}}_{\Sigma 2}(\boldsymbol{q}, t) . \end{aligned} \tag{16}$$

Differentiating this equations with respect to time leads to the constraints on acceleration level:

$$\ddot{g}_N = \boldsymbol{n}_1^T \left(\dot{\boldsymbol{v}}_{\Sigma 2} - \dot{\boldsymbol{v}}_{\Sigma 1} \right) + \dot{\boldsymbol{n}}_1^T \left(\boldsymbol{v}_{\Sigma 2} - \boldsymbol{v}_{\Sigma 1} \right) \ ,$$
$$\ddot{g}_S = \boldsymbol{s}_1^T \left(\dot{\boldsymbol{v}}_{\Sigma 2} - \dot{\boldsymbol{v}}_{\Sigma 1} \right) + \dot{\boldsymbol{s}}_1^T \left(\boldsymbol{v}_{\Sigma 2} - \boldsymbol{v}_{\Sigma 1} \right) \ , \qquad (17)$$
$$\ddot{g}_T = \boldsymbol{t}_1^T \left(\dot{\boldsymbol{v}}_{\Sigma 2} - \dot{\boldsymbol{v}}_{\Sigma 1} \right) + \dot{\boldsymbol{t}}_1^{T} \left(\boldsymbol{v}_{\Sigma 2} - \boldsymbol{v}_{\Sigma 1} \right) \ .$$

The time derivatives of the contact point velocities $\boldsymbol{v}_{\Sigma 1}$ and $\boldsymbol{v}_{\Sigma 2}$ are calculated from eq. (16):

$$\dot{\boldsymbol{v}}_{\Sigma 1} = \boldsymbol{J}_{\Sigma 1}(\boldsymbol{q}, t) \ddot{\boldsymbol{q}} + \bar{\boldsymbol{j}}_{\Sigma 1}(\dot{\boldsymbol{q}}, \boldsymbol{q}, t) \ ,$$
$$\dot{\boldsymbol{v}}_{\Sigma 2} = \boldsymbol{J}_{\Sigma 2}(\boldsymbol{q}, t) \ddot{\boldsymbol{q}} + \bar{\boldsymbol{j}}_{\Sigma 2}(\dot{\boldsymbol{q}}, \boldsymbol{q}, t) \ . \qquad (18)$$

The vecotrs $\dot{\boldsymbol{n}}_1, \dot{\boldsymbol{s}}_1$ and $\dot{\boldsymbol{t}}_1$ can be determined with the formulas of Weingarten and Gauss, which express the derivatives of the normal vector and of the tangent vectors in terms of the basic vectors:

$$\dot{\boldsymbol{n}}_1 = \boldsymbol{\Omega}_1 \times \boldsymbol{n}_1 + \frac{\partial \boldsymbol{n}_1}{\partial s_1} \dot{s}_1 + \frac{\partial \boldsymbol{n}_1}{\partial t_1} \dot{t}_1 \ , \qquad \text{where:}$$
$$\frac{\partial \boldsymbol{n}_1}{\partial s_1} = \underbrace{\frac{M_1 F_1 - L_1 G_1}{E_1 G_1 - F_1^2}}_{\alpha_1} \boldsymbol{s}_1 + \underbrace{\frac{L_1 F_1 - M_1 E_1}{E_1 G_1 - F_1^2}}_{\beta_1} \boldsymbol{t}_1 \ , \qquad (19)$$
$$\frac{\partial \boldsymbol{n}_1}{\partial t_1} = \underbrace{\frac{N_1 F_1 - M_1 G_1}{E_1 G_1 - F_1^2}}_{\alpha_1'} \boldsymbol{s}_1 + \underbrace{\frac{M_1 F_1 - N_1 E_1}{E_1 G_1 - F_1^2}}_{\beta_1'} \boldsymbol{t}_1 \ ,$$

$$\dot{\boldsymbol{s}}_1 = \boldsymbol{\Omega}_1 \times \boldsymbol{s}_1 + \frac{\partial \boldsymbol{s}_1}{\partial s_1} \dot{s}_1 + \frac{\partial \boldsymbol{s}_1}{\partial t_1} \dot{t}_1 \ , \qquad \text{where:}$$
$$\frac{\partial \boldsymbol{s}_1}{\partial s_1} = \Gamma_{11,1}^1 \boldsymbol{s}_1 + \Gamma_{11,1}^2 \boldsymbol{t}_1 + L_1 \boldsymbol{n}_1 \ , \qquad \frac{\partial \boldsymbol{s}_1}{\partial t_1} = \Gamma_{12,1}^1 \boldsymbol{s}_1 + \Gamma_{12,1}^2 \boldsymbol{t}_1 + M_1 \boldsymbol{n}_1 \ , \qquad (20)$$

$$\dot{\boldsymbol{t}}_1 = \boldsymbol{\Omega}_1 \times \boldsymbol{t}_1 + \frac{\partial \boldsymbol{t}_1}{\partial s_1} \dot{s}_1 + \frac{\partial \boldsymbol{t}_1}{\partial t_1} \dot{t}_1 \ , \qquad \text{where:}$$
$$\frac{\partial \boldsymbol{t}_1}{\partial s_1} = \Gamma_{12,1}^1 \boldsymbol{s} + \Gamma_{12,1}^2 \boldsymbol{t} + M_1 \boldsymbol{n}_1 \ , \qquad \frac{\partial \boldsymbol{t}_1}{\partial t_1} = \Gamma_{22,1}^1 \boldsymbol{s} + \Gamma_{22,1}^2 \boldsymbol{t} + N_1 \boldsymbol{n}_1 \ . \qquad (21)$$

The definition of the Christoffel symbols $\Gamma_{\alpha\beta}^\sigma$, $\alpha, \beta, \sigma = 1, 2$, can be found in mathematical handbooks. Inserting eq. (19), (20), (21) in (17) yields the constraint equations:

$$\ddot{g}_N = \boldsymbol{n}_1^T \left(\boldsymbol{J}_{\Sigma 2} - \boldsymbol{J}_{\Sigma 1} \right) \ddot{\boldsymbol{q}} + \boldsymbol{n}_1^T \left(\bar{\boldsymbol{j}}_{\Sigma 2} - \bar{\boldsymbol{j}}_{\Sigma 1} \right) + \left(\boldsymbol{v}_{\Sigma 2} - \boldsymbol{v}_{\Sigma 1} \right)^T \left(\boldsymbol{\Omega}_1 \times \boldsymbol{n}_1 \right) +$$
$$\left(\boldsymbol{v}_{\Sigma 2} - \boldsymbol{v}_{\Sigma 1} \right)^T \left(\left(\alpha_1 \boldsymbol{s}_1 + \beta_1 \boldsymbol{t}_1 \right) \dot{s}_1 + \left(\alpha_1' \boldsymbol{s}_1 + \beta_1' \boldsymbol{t}_1 \right) \dot{t}_1 \right) \ ,$$

$$\ddot{g}_S = \boldsymbol{s}_1^T \left(\boldsymbol{J}_{\Sigma 2} - \boldsymbol{J}_{\Sigma 1} \right) \ddot{\boldsymbol{q}} + \boldsymbol{s}_1^T \left(\bar{\boldsymbol{j}}_{\Sigma 2} - \bar{\boldsymbol{j}}_{\Sigma 1} \right) + \left(\boldsymbol{v}_{\Sigma 2} - \boldsymbol{v}_{\Sigma 1} \right)^T \left(\boldsymbol{\Omega}_1 \times \boldsymbol{s}_1 \right) + \qquad (22)$$
$$\left(\boldsymbol{v}_{\Sigma 2} - \boldsymbol{v}_{\Sigma 1} \right)^T \left(\left(\Gamma_{11,1}^1 \boldsymbol{s}_1 + \Gamma_{11,1}^2 \boldsymbol{t}_1 + L_1 \boldsymbol{n}_1 \right) \dot{s}_1 + \left(\Gamma_{12,1}^1 \boldsymbol{s}_1 + \Gamma_{12,1}^2 \boldsymbol{t}_1 + M_1 \boldsymbol{n}_1 \right) \dot{t}_1 \right) \ ,$$

$$\ddot{g}_T = \boldsymbol{t}_1^T \left(\boldsymbol{J}_{\Sigma 2} - \boldsymbol{J}_{\Sigma 1} \right) \ddot{\boldsymbol{q}} + \boldsymbol{t}_1^T \left(\bar{\boldsymbol{j}}_{\Sigma 2} - \bar{\boldsymbol{j}}_{\Sigma 1} \right) + \left(\boldsymbol{v}_{\Sigma 2} - \boldsymbol{v}_{\Sigma 1} \right)^T \left(\boldsymbol{\Omega}_1 \times \boldsymbol{t}_1 \right) +$$
$$\left(\boldsymbol{v}_{\Sigma 2} - \boldsymbol{v}_{\Sigma 1} \right)^T \left(\left(\Gamma_{12,1}^1 \boldsymbol{s}_1 + \Gamma_{12,1}^2 \boldsymbol{t}_1 + M_1 \boldsymbol{n}_1 \right) \dot{s}_1 + \left(\Gamma_{22,1}^1 \boldsymbol{s}_1 + \Gamma_{22,1}^2 \boldsymbol{t}_1 + N_1 \boldsymbol{n}_1 \right) \dot{t}_1 \right) \ .$$

As can be seen, these equations are only dependent on the Jacobians with respect to the contact points $\boldsymbol{J}_{\Sigma1}, \boldsymbol{J}_{\Sigma2}$, the basic vectors of the surfaces and the time derivatives of the contour parameters $\dot{s}_1, \dot{t}_1, \dot{s}_2, \dot{t}_2$. The Jacobians are known from the rigid body algorithm, the basic vectors from the surface description. The time derivatives of the contour parameters can be calculated by deriving eq. (14) with respect to time:

$$
\begin{aligned}
\left(\boldsymbol{n}_1^T \boldsymbol{s}_2\right)^{\bullet} &= 0, & \left(\boldsymbol{r}_D^T \boldsymbol{s}_2\right)^{\bullet} &= 0, \\
\left(\boldsymbol{n}_1^T \boldsymbol{t}_2\right)^{\bullet} &= 0, & \left(\boldsymbol{r}_D^T \boldsymbol{t}_2\right)^{\bullet} &= 0,
\end{aligned}
\tag{23}
$$

which means, that the conditions for the contact point should not change while the two bodies are moving. Figuring out eq. (23) we obtain a system of equations which are linear in the derivatives of the contour parameters:

$$
\begin{pmatrix}
\boldsymbol{s}_2^T \left(\alpha_1 \boldsymbol{s}_1 + \beta_1 \boldsymbol{t}_1\right) & \boldsymbol{s}_2^T \left(\alpha_1' \boldsymbol{s}_1 + \beta_1' \boldsymbol{t}_1\right) & L_2 & M_2 \\
\boldsymbol{t}_2^T \left(\alpha_1 \boldsymbol{s}_1 + \beta_1 \boldsymbol{t}_1\right) & \boldsymbol{t}_2^T \left(\alpha_1' \boldsymbol{s}_1 + \beta_1' \boldsymbol{t}_1\right) & M_2 & N_2 \\
-\boldsymbol{s}_1^T \boldsymbol{s}_2 & -\boldsymbol{s}_1^T \boldsymbol{s}_2 & \boldsymbol{s}_2^T \boldsymbol{s}_2 & \boldsymbol{s}_2^T \boldsymbol{t}_2 \\
-\boldsymbol{s}_1^T \boldsymbol{t}_2 & -\boldsymbol{s}_1^T \boldsymbol{t}_2 & \boldsymbol{s}_2^T \boldsymbol{t}_2 & \boldsymbol{t}_2^T \boldsymbol{t}_2
\end{pmatrix}
\begin{pmatrix}
\dot{s}_1 \\ \dot{t}_1 \\ \dot{s}_2 \\ \dot{t}_2
\end{pmatrix}
=
\begin{pmatrix}
(\boldsymbol{s}_2 \times \boldsymbol{n}_1)^T (\boldsymbol{\Omega}_2 - \boldsymbol{\Omega}_1) \\
(\boldsymbol{t}_2 \times \boldsymbol{n}_1)^T (\boldsymbol{\Omega}_2 - \boldsymbol{\Omega}_1) \\
\boldsymbol{s}_2^T (\boldsymbol{v}_{\Sigma1} - \boldsymbol{v}_{\Sigma2}) \\
\boldsymbol{t}_2^T (\boldsymbol{v}_{\Sigma1} - \boldsymbol{v}_{\Sigma2})
\end{pmatrix}.
\tag{24}
$$

This linear problem has to be solved at every time step of numerical integration.

Let us summarize the constraint equations in the well known form, by rewriting eq. (22):

$$
\begin{aligned}
\ddot{g}_N &= \boldsymbol{w}_N^T \ddot{\boldsymbol{q}} + \overline{w}_N, \\
\ddot{g}_S &= \boldsymbol{w}_S^T \ddot{\boldsymbol{q}} + \overline{w}_S, \\
\ddot{g}_T &= \boldsymbol{w}_T^T \ddot{\boldsymbol{q}} + \overline{w}_T.
\end{aligned}
\tag{25}
$$

The terms in eq. (22), which are linearly dependent on $\ddot{\boldsymbol{q}}$, are collected in the constraint vectors $\boldsymbol{w}_N, \boldsymbol{w}_S$ and $\boldsymbol{w}_T$, all the rest is included in the scalars $\overline{w}_N, \overline{w}_S, \overline{w}_T$.

3. MULTIPLE CONTACTS

3.1. Unilaterally constrained motion

The equations of motion of any multibody system may be reduced to the standard form [17]

$$
\boldsymbol{M}(\boldsymbol{q}, t)\ddot{\boldsymbol{q}} - \boldsymbol{h}(\boldsymbol{q}, \dot{\boldsymbol{q}}, t) = 0, \quad \boldsymbol{q} \in \mathbb{R}^f,
\tag{26}
$$

with the degrees of freedom f. The mass matrix $\boldsymbol{M}(\boldsymbol{q}, t) \in \mathbb{R}^{f,f}$ is symmetric and positive definite. The vector $\boldsymbol{h}(\boldsymbol{q}, \dot{\boldsymbol{q}}, t) \in \mathbb{R}^f$ contains the gyroscopical accelerations together with the sum of all active forces and moments. In a system with additional unilateral constraints, the number of degrees of freedom is variable. To avoid difficulties with many different sets of minimal coordinates, we take one set of generalized coordinates and consider the active unilateral constraints as additional constraints which necessarily are accompanied by constraint forces. We include these constraint forces, which are in fact the contact forces, into the equations of motion (26) by a Lagrange multiplier technique.

The constraint vectors $\boldsymbol{w}_{Ni}$ and the constraint matrices $\boldsymbol{W}_{Ti}$ in equation (8) are arranged as columns in the constraint matrices

$$
\begin{aligned}
\boldsymbol{W}_N &= [\ldots, \boldsymbol{w}_{Ni}, \ldots] \in \mathbb{R}^{f,n_N} &;& \quad i \in I_N \\
\boldsymbol{W}_T &= [\ldots, \boldsymbol{W}_{Ti}, \ldots] \in \mathbb{R}^{f,2n_T} &;& \quad i \in I_T
\end{aligned}
\tag{27}
$$

for all active constraints. The constraint matrices are transformation matrices from the space of constraints to the configuration space. The transposed matrices are used for the transition from the configuration space to the space of constraints.

The contact forces have the amounts λ_{Ni} (normal forces) and the components λ_{Ti1} and λ_{Ti2} (tangential forces). These elements are combined in the vectors of constraint forces

$$
\boldsymbol{\lambda}_N(t) = \begin{pmatrix} \vdots \\ \lambda_{Ni}(t) \\ \vdots \end{pmatrix} \in \mathbb{R}^{n_N} \quad ; \quad i \in I_N
$$

$$
\boldsymbol{\lambda}_T(t) = \begin{pmatrix} \vdots \\ \boldsymbol{\lambda}_{Ti}(t) \\ \vdots \end{pmatrix} \in \mathbb{R}^{2n_T} \quad ; \quad i \in I_T
\tag{28}
$$

with $\boldsymbol{\lambda}_{Ti}(t) = [\lambda_{Ti1}(t), \lambda_{Ti2}(t)]^T$. In general, the contact forces are time-varying quantities. By the constraint vectors and matrices in equation (8), the contact forces can be expressed in the configuration space. These forces are then added to equation (26),

$$
\boldsymbol{M}\ddot{\boldsymbol{q}} - \boldsymbol{h} - \sum_{i \in I_N} (\boldsymbol{w}_{Ni}\lambda_{Ni} + \boldsymbol{W}_{Ti}\boldsymbol{\lambda}_{Ti}) = 0 \; .
\tag{29}
$$

For the index sets see eq. (2). The contact forces $\boldsymbol{\lambda}_{Ti}$ in equation (29) can be passive forces of sticking contacts or active forces of sliding contacts. We express the tangential forces of the $n_N - n_T$ sliding contacts by the corresponding normal forces using Coulomb's friction law by

$$
\boldsymbol{\lambda}_{Ti} = -\mu_i\left(|\dot{\boldsymbol{g}}_{Ti}|\right) \frac{\dot{\boldsymbol{g}}_{Ti}}{|\dot{\boldsymbol{g}}_{Ti}|} \lambda_{Ni} \; ; \quad i \in I_N \backslash I_T
\tag{30}
$$

where the coefficients $\mu_i(|\dot{\boldsymbol{g}}_{Ti}|)$ of sliding friction may depend on time. The negative sign relates to the opposite direction of relative velocity and friction force. The sliding forces of equation (30) in the configuration space are then

$$
\boldsymbol{W}_{Ti}\boldsymbol{\lambda}_{Ti} = -\mu_i(|\dot{\boldsymbol{g}}_{Ti}|) \boldsymbol{W}_{Ti} \frac{\dot{\boldsymbol{g}}_{Ti}}{|\dot{\boldsymbol{g}}_{Ti}|} \lambda_{Ni} \; ; \quad i \in I_N \backslash I_T \; .
\tag{31}
$$

A substitution of these forces into equation (29) yields the equations of motion

$$
\boldsymbol{M}(\boldsymbol{q},t)\ddot{\boldsymbol{q}}(t) - \boldsymbol{h}(\boldsymbol{q},\dot{\boldsymbol{q}},t) - [\boldsymbol{W}_N + \boldsymbol{H}_R, \; \boldsymbol{W}_T] \begin{pmatrix} \boldsymbol{\lambda}_N(t) \\ \boldsymbol{\lambda}_T(t) \end{pmatrix} = 0 \; ,
\tag{32}
$$

with the additional contact forces as Lagrange multipliers. The matrices $\boldsymbol{W}_N$ and $\boldsymbol{W}_T$ are the constraint matrices of equation (8). The matrix $\boldsymbol{H}_R \in \mathbb{R}^{f,n_N}$ of the sliding contacts has the same dimension as the constraint matrix $\boldsymbol{W}_N$. For $n_T \leq n_N$, $\boldsymbol{H}_R$ consists of the $n_N - n_T$ columns

$$-\frac{\mu_i}{|\dot{\boldsymbol{g}}_{Ti}|}\boldsymbol{W}_{Ti}\dot{\boldsymbol{g}}_{Ti} \; ; \quad i \in I_N \backslash I_T \; ,$$

while the other n_T columns contain only zero-elements.

The relative accelerations of the active normal and tangential constraints in equation (8) can be combined by means of the constraint matrices (27) in the matrix notation. Together with equation (32) we get the system of equations

$$
\begin{aligned}
&\boldsymbol{M}\ddot{\boldsymbol{q}} - \boldsymbol{h} - [\boldsymbol{W}_N + \boldsymbol{H}_R, \; \boldsymbol{W}_T]\begin{pmatrix} \boldsymbol{\lambda}_N \\ \boldsymbol{\lambda}_T \end{pmatrix} = 0 \quad && \in \mathbb{R}^f \\
&\ddot{\boldsymbol{g}}_N = \boldsymbol{W}_N^T\ddot{\boldsymbol{q}} + \overline{\boldsymbol{w}}_N && \in \mathbb{R}^{n_N} \\
&\ddot{\boldsymbol{g}}_T = \boldsymbol{W}_T^T\ddot{\boldsymbol{q}} + \overline{\boldsymbol{w}}_T && \in \mathbb{R}^{n_T}
\end{aligned}
\tag{33}
$$

The unknown quantities are the generalized accelerations $\ddot{\boldsymbol{q}} \in \mathbb{R}^f$, the contact forces in the normal direction $\boldsymbol{\lambda}_N \in \mathbb{R}^{n_N}$ and in the tangential direction $\boldsymbol{\lambda}_T \in \mathbb{R}^{2n_T}$, as well as the corresponding relative accelerations $\ddot{\boldsymbol{g}}_N \in \mathbb{R}^{n_N}$ and $\ddot{\boldsymbol{g}}_T \in \mathbb{R}^{2n_T}$. For the dertermination of the $f + 2(n_N + 2n_T)$ quantities, we have up to now $f + n_N + 2n_T$ equations. In the following the system of equations (33) will be completed by including the missing $n_N + 2n_T$ contact laws. In general, the kinematic equations are dependent on each other if there is more than one contact point per rigid body. The situation results in linearly dependent columns of the constraint matrices $\boldsymbol{W}_N, \boldsymbol{W}_T$ in equation (33). Such constraints are called dependent constraints.

Before discussing the missing contact laws we shall evaluate the equations (33) a bit further. Introducing the magnitudes

$$
\ddot{\boldsymbol{g}} = \begin{pmatrix} \ddot{\boldsymbol{g}}_N \\ \ddot{\boldsymbol{g}}_T \end{pmatrix} \; ; \quad \boldsymbol{\lambda} = \begin{pmatrix} \boldsymbol{\lambda}_N \\ \boldsymbol{\lambda}_T \end{pmatrix} \; ; \quad \overline{\boldsymbol{w}} = \begin{pmatrix} \overline{\boldsymbol{w}}_N \\ \overline{\boldsymbol{w}}_T \end{pmatrix} \; ;
$$
$$
\boldsymbol{W} = (\boldsymbol{W}_N, \; \boldsymbol{W}_T) \; ; \quad \boldsymbol{N}_G = (\boldsymbol{H}_R, \; 0)
\tag{34}
$$

we can write eq. (33)

$$\boldsymbol{M}\ddot{\boldsymbol{g}} - \boldsymbol{h} - (\boldsymbol{W} + \boldsymbol{N}_G)\boldsymbol{\lambda} = 0 \; , \qquad \ddot{\boldsymbol{g}} = \boldsymbol{W}^T\ddot{\boldsymbol{q}} + \overline{\boldsymbol{w}} \; . \tag{35}$$

Outside the transition events and thus for a not changing contact configuration the relative accelerations $\ddot{\boldsymbol{g}}$ are zero. In this case the equations (35) have a solution for $\ddot{\boldsymbol{g}}$ and $\boldsymbol{\lambda}$, which writes

$$\boldsymbol{\lambda} = -\left[\boldsymbol{W}^T\boldsymbol{M}^{-1}(\boldsymbol{W} + \boldsymbol{N}_G)\right]^{-1}(\boldsymbol{W}^T\boldsymbol{M}^{-1}\boldsymbol{h} + \overline{\boldsymbol{w}}) \; , \tag{36}$$

$$\ddot{\boldsymbol{q}} = \boldsymbol{M}^{-1}\left[\boldsymbol{h} + (\boldsymbol{W} + \boldsymbol{N}_G)\boldsymbol{\lambda}\right] \; .$$

The first equation of (36) may also be expressed in the form

$$\boldsymbol{A}\boldsymbol{\lambda} + \boldsymbol{b} = 0 \; , \tag{37}$$

with $\boldsymbol{A}$ and $\boldsymbol{b}$ following from (36). We shall use this form in later considerations.

3.2. Contact laws
3.2.1. Contact laws for normal constraints

Contact situation in normal direction with respect to the plane of contact are charac-
terized by a closure of the contact with vanishing relative distance and velocity ($g_{Ni} = 0, \dot{g}_{Ni} = 0, i \in I_N$) or by a transition to separation. In the first case the normal relative acceleration $\ddot{g}_{Ni}$ also vanishes, and the resulting normal contact force must be compressive.
Therefore

$$\ddot{g}_{Ni} = 0 \;\wedge\; \lambda_{Ni} \geq 0 \,, \;\; i \in I_N \;. \tag{38}$$

Separation of a normal contact can only be achieved by nonnegative relative accelerations and by a vanishing normal force, which gives

$$\ddot{g}_{Ni} \geq 0 \;\wedge\; \lambda_{Ni} = 0 \,; \;\; i \in I_N \;. \tag{39}$$

The two properties (38) and (39) establish a complementary behavior in the sense
$\ddot{g}_{Ni}\lambda_{Ni} = 0$ with $i \in I_N$. We may generalize this for all n_N contacts, which are unambiguously described by the n_N complementarity conditions

$$\ddot{\boldsymbol{g}}_N \geq 0 \;;\; \boldsymbol{\lambda}_N \geq 0 \;;\; \ddot{\boldsymbol{g}}_N^T \boldsymbol{\lambda}_N = 0 \;. \tag{40}$$

The variational inequality

$$-\ddot{\boldsymbol{g}}_N^T(\boldsymbol{\lambda}_N^* - \boldsymbol{\lambda}_N) \leq 0 \;;\; \boldsymbol{\lambda}_N \in C_N \;;\; \forall \boldsymbol{\lambda}_N^* \in C_N \;, \tag{41}$$

is equivalent to the complementary conditions (40). The convex set

$$C_N = \{\boldsymbol{\lambda}_N^* : \boldsymbol{\lambda}_N^* \geq 0\} \tag{42}$$

contains all admissible contact forces λ_{Ni}^* in the normal direction (see for example
[14,23].

The complementarity problem defined in eq. (40) might be interpreted as a corner law
which requires for each contact $\ddot{g}_{Ni} \geq 0$, $\lambda_{Ni} \geq 0$, $\ddot{g}_{Ni}\lambda_{Ni} = 0$. Figure 2 illustrates this
property.

3.2.2. Contact laws for tangential constraints

We shall confine our considerations to the application of Coulomb's friction laws which
in no way means a loss of generality. The complementary behavior is characteristic feature
of all contact phenomena independent of the specific physical law of contact. Furthermore
we assume that within the infinitesimal small time step for a transition from stick to slip
and vice versa the coefficients of static and sliding friction are the same, which may be
expressed by

$$\lim_{\dot{g}_{Ti} \to 0} \mu_i(\dot{g}_{Ti}) = \mu_{0i} \tag{43}$$

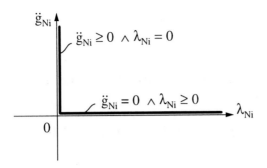

Figure 2. Corner law for normal contacts

For $\dot{g}_{Ti} \neq 0$ any friction law may be applied (see Fig. 3). With this property Coulomb's friction law distinguishes between the two cases

$$\left.\begin{array}{lll} \text{stiction:} & |\boldsymbol{\lambda}_{Ti}| < \mu_{0i}\lambda_{Ni} & \Rightarrow \quad |\dot{\boldsymbol{g}}_{Ti}| = 0 \\ \text{sliding:} & |\boldsymbol{\lambda}_{Ti}| = \mu_{0i}\lambda_{Ni} & \Rightarrow \quad |\dot{\boldsymbol{g}}_{Ti}| > 0 \end{array}\right\} \quad i \in I_N \; . \tag{44}$$

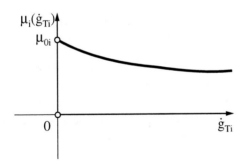

Figure 3. Friction characteristic

For the frictional contact problem, we need a representation of the friction law (44) on the acceleration level in order to determine the tangential relative accelerations $\ddot{\boldsymbol{g}}_T$ in equation (33). The friction forces of the sliding contacts are already taken into account in the first equation (33) by $\boldsymbol{H}_R\boldsymbol{\lambda}_N$ so we only have to transform Coulomb's friction law on acceleration level for the n_T potential sticking contacts. This is possible because the tangential relative velocity and acceleration have the same direction for a transition from $|\dot{\boldsymbol{g}}_{Ti}| = 0$ to $|\dot{\boldsymbol{g}}_{Ti}| > 0$. Thus we get from (44) the conditions

$$\left.\begin{array}{lll} \text{stiction:} & |\boldsymbol{\lambda}_{Ti} < \mu_{0i}\lambda_{Ni} & \Rightarrow \quad \ddot{\boldsymbol{g}}_{Ti} = 0 \\ \text{sliding:} & |\boldsymbol{\lambda}_{Ti} = \mu_{0i}\lambda_{Ni} & \Rightarrow \quad \ddot{\boldsymbol{g}}_{Ti} = \rho_i\boldsymbol{\lambda}_{Ti} \; ; \; \rho_i \geq 0 \end{array}\right\} \quad i \in I_T \; . \tag{45}$$

The inequality

$$\ddot{\boldsymbol{g}}_{Ti}^T \boldsymbol{\lambda}_{Ti} \leq 0 \; ; \; i \in I_T \; . \tag{46}$$

follows from equation (45) and describes the dissipative behavior of Coulomb friction. The variational inequality

$$-\ddot{\boldsymbol{g}}_{Ti}^T (\boldsymbol{\lambda}_{Ti}^* - \boldsymbol{\lambda}_{Ti}) \leq 0 \; ; \; \boldsymbol{\lambda}_{Ti} \in C_{Ti} \; ; \; \forall \boldsymbol{\lambda}_{Ti}^* \in C_{Ti} \; ; \; i \in I_T \; . \tag{47}$$

is equivalent to equations (45) and (46). The convex set

$$C_{Ti}(\lambda_{Ni}) = \{\boldsymbol{\lambda}_{Ti}^* : |\boldsymbol{\lambda}_{Ti}^*| \leq \mu_{0i}\lambda_{Ni}\} \tag{48}$$

contains all friction forces $\boldsymbol{\lambda}_{Ti}^*$ which fulfill Coulomb's friction law.

As in the normal case the complementarity condition (47) can be illustrated by a corner behavior in the $\ddot{g}_{Ti} - \lambda_{Ti}$-plane. Equation (47) writes in more detail

$$\begin{aligned}
|\lambda_{Ti}| &< \mu_{0i}\lambda_{Ni} \wedge \ddot{g}_{Ti} = 0 \quad \} \quad i \in I_T \\
\lambda_{Ti} &= +\mu_{0i}\lambda_{Ni} \wedge \ddot{g}_{Ti} \leq 0 \\
\lambda_{Ti} &= -\mu_{0i}\lambda_{Ni} \wedge \ddot{g}_{Ti} \geq 0
\end{aligned} \left. \begin{aligned} & \\ & \\ & \end{aligned} \right\} \; (i \in I_N) \tag{49}$$

which refers to the cases of stiction and negative, positive sliding, respectively. Figure 4 depicts these properties in the form of a double corner law.

3.3. Equations of motion for multiple contact systems

For establishing all necessary equations for multibody systems with multiple and partly dependent contacts we must combine the results of the two chapters 3.1 and 3.2. Considering in a first step only systems with normal contacts and sliding friction ($i \in I_N$) we come out by a combination of the equations (33,40) with the following set

$$\begin{aligned}
& \boldsymbol{M}\ddot{\boldsymbol{q}} - \boldsymbol{h} - (\boldsymbol{W}_N + \boldsymbol{H}_R)\boldsymbol{\lambda}_N = 0 \; , \\
& \ddot{\boldsymbol{g}}_N = \boldsymbol{W}_N^T \ddot{\boldsymbol{q}} + \overline{\boldsymbol{w}}_N \; , \\
& \ddot{\boldsymbol{g}}_N \geq 0 \; , \; \boldsymbol{\lambda}_N \geq 0 \; , \; \ddot{\boldsymbol{g}}_N^T \boldsymbol{\lambda}_N = 0 \; .
\end{aligned} \tag{50}$$

An elimination of $\ddot{\boldsymbol{q}}$ from the second equation by using the first one results in

$$\ddot{\boldsymbol{g}}_N = \boldsymbol{W}_N^T \boldsymbol{M}^{-1}(\boldsymbol{W}_N + \boldsymbol{H}_R)\boldsymbol{\lambda}_N + (\boldsymbol{W}_N^T \boldsymbol{M}^{-1}\boldsymbol{h} + \overline{\boldsymbol{w}}_N) \; , \tag{51}$$

$$\ddot{\boldsymbol{g}}_N \geq 0 \; , \; \boldsymbol{\lambda}_N \geq 0 \; , \; \ddot{\boldsymbol{g}}_N^T \boldsymbol{\lambda}_N = 0 \; ,$$

which possesses the general structure

$$\boldsymbol{y} = \boldsymbol{A}\boldsymbol{x} + \boldsymbol{b} \; , \; \boldsymbol{y} \geq 0 \; , \; \boldsymbol{x} \geq 0 \; , \; \boldsymbol{y}^T \boldsymbol{x} = 0 \; . \tag{52}$$

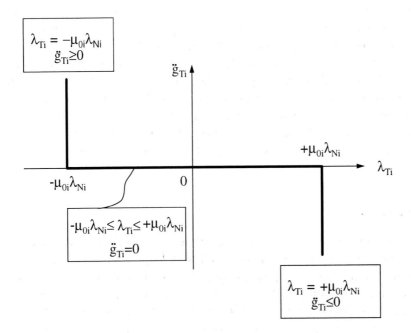

Figure 4. Corner law for tangential contacts

Equation (52) represents a Linear Complementarity Problem (LCP) in its standard form. Numerical solution algorithms for problems of that type are available [13].

Adding to eqs. (50) in a second step also stick-slip phenomena will result in a more complicated situation. As far as all contacts are plane contacts leading to a line contact configuration with known direction we still will be able to reduce the system's equations to a linear complementarity problem. However, in the case of spatial contacts leading to a plane contact configuration with two unknown directions for forces and accelerations we come out with a nonlinear complementarity problem.

The mathematical system consisting of the equations of motion and the constraint equations in equation (33) as well as the variational inequalities (47) and (48)

$$\boldsymbol{M}\ddot{\boldsymbol{q}} - [\boldsymbol{W}_N + \boldsymbol{H}_R, \ \boldsymbol{W}_T] \begin{pmatrix} \boldsymbol{\lambda}_N \\ \boldsymbol{\lambda}_T \end{pmatrix} - \boldsymbol{h}(\boldsymbol{q}, \dot{\boldsymbol{q}}, t) = 0$$

$$\begin{pmatrix} \ddot{\boldsymbol{g}}_N \\ \ddot{\boldsymbol{g}}_T \end{pmatrix} = \begin{pmatrix} \boldsymbol{W}_N^T \\ \boldsymbol{W}_T^T \end{pmatrix} \ddot{\boldsymbol{q}} + \begin{pmatrix} \overline{\boldsymbol{w}}_N \\ \overline{\boldsymbol{w}}_T \end{pmatrix} \tag{53}$$

$$-\ddot{\boldsymbol{g}}_N^T(\boldsymbol{\lambda}_N^* - \boldsymbol{\lambda}_N) \leq 0 \ ; \ \boldsymbol{\lambda}_N \in C_N \ ; \ \forall \boldsymbol{\lambda}_N^* \in C_N \ ; \ I \in I_N$$

$$-\ddot{\boldsymbol{g}}_{Ti}^T(\boldsymbol{\lambda}_{Ti}^* - \boldsymbol{\lambda}_{Ti}) \leq 0 \ ; \ \boldsymbol{\lambda}_{Ti} \in C_{Ti}(\boldsymbol{\lambda}_{Ni}) \ ; \ \forall \boldsymbol{\lambda}_{Ti}^* \in C_{Ti}(\boldsymbol{\lambda}_{Ni}) \ ; \ i \in I_T$$

with the convex sets (42) and (48) completely describes the contact problem. But the system (53) is not solvable in this form. Therefore the variational inequalities are transformed into equalities. From this we get a nonlinear system of equations which can be solved with iterative standard algorithms. This method is shown in [13]. More details with respect to linear and nonlinear complementarity in connection with normal and tangential contact problems may be found in [17,23].

Based on the above theory and applying in addition the impact laws of Newton and Poisson it is also possible to develop a theory of impact without and with friction. The main idea consists in considering a compression and an expansion phase where during compression normal and tangential impulses are stored and where during expansion these stored impulses are partly regained. This theory has been tested by experiments with good results in most cases, though certain configurations still require more research. Details may be found in [3,17].

4. APPLICATIONS

The research on multibody systems at the author's institute was and still is continuously accompanied by investigations of industrial applications. Moreover, most of the fundamental findings like the equations (33,53), the formulation of the complementarity behavior in contact dynamics, the application of complementarity, homotopy or projection methods were directly induced by industrial problems, which could not be dealt with by existing theories. A rather complete survey of this work is presented in [17].

The basic models for an example like a woodpecker toy is principally the same as for gears, turbines, assembly processes and other industrial applications. Therefore, the mathematical models for the example to follow are not presented in detail, only appropriate reference to above formulas will be given. Detailed models may be found in [17].

All research on non-continuous dynamics of multibody systems started at the author's institute with an investigation of the woodpecker toy. At that time a theory on impacts with friction was not available, therefore the friction losses were measured [16]. A woodpecker toy typically consists of a sleeve, a spring and the woodpecker.

The hole of the sleeve is slightly larger than the diameter of the pole thus allowing a kind of pitching motion interupted by impacts with friction. A typical sequence of events is portrayed in Fig. 5.

We start with jamming in a downward position, moving back again due to the deformation of the spring, and including a transition from one to three degrees of freedom between phase 1 and 2. Step 3 is jamming in an upward position (1 DOF) followed by a beak impact which supports a quick reversal of the φ-motion. Steps 5 to 7 are then equivalent to steps 3 to 1. The motion can be described by a limit cycle behavior as illustrated by Fig. 6. The gravitation represents an energy source, the energy of which is transmitted to the woodpecker mass by the y-motion. The woodpecker itself oscillates and possesses a switching function by the beak for quick φ_S reversal and by the jammed sleeve, which transmits energy to the spring by jamming impacts.

The system possesses three degrees of freedom $\boldsymbol{q} = (y, \varphi_M, \varphi_S)^T$, where φ_S and φ_M are the absolute angles of rotation of the woodpecker and the sleeve respectively, and y describes the vertical displacement of the sleeve. Horizontal deviations are negligible. The diameter of the hole in the sleeve is slightly larger than the diamter of the pole. Due to the resulting clearance, the lower or upper edge of the sleeve may come into contact with the pole. A further contact may occur when the beak of the woodpecker hits the

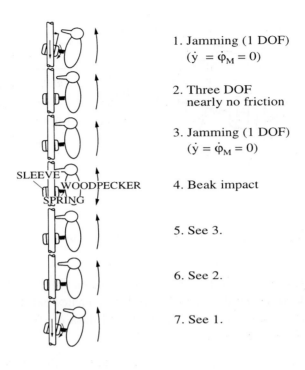

1. Jamming (1 DOF)
   $(\dot{y} = \dot{\varphi}_M = 0)$

2. Three DOF
   nearly no friction

3. Jamming (1 DOF)
   $(\dot{y} = \dot{\varphi}_M = 0)$

4. Beak impact

5. See 3.

6. See 2.

7. See 1.

SLEEVE
WOODPECKER
SPRING

Figure 5. Sequence of events for a woodpecker toy

pole. The derivation of the equations of motion follows the theoretical considerations of the preceding chapter.

As an example we use a woodpecker toy the data set of which may be found in [17]. Some results are shown in Fig. 7, which depicts three limit cycles. We start our discussion at point (6) where the lower edge of the sleeve hits the pole. This completely inelastic frictional impact leads to continual contact of the sleeve with the pole. After a short episod of sliding (6)–(7) we observe a transition of the sleeve to sticking (7). The angle of the woodpecker is now large enough to ensure continual sticking of the sleeve by the self-locking mechanism. In that state the system has only one degree of freedom, and the large 9.10 Hz-oscillation can be observed where the woodpecker swings down and up until it reaches point (1). At (1) the tangent constraint becomes passive and the sleeve slides up to point (2) where contact is lost. It should be noted that the spring is not free of stresses in this situation, thus during the free-flight phase (2)–(3) the high frequency oscillation ($f = 72.91$ Hz) of the unbound system occurs in the phase space plots. In this state the sleeve moves downward (y decreases) and the first part of the falling height Δy at one cycle is achieved. At (3) the upper edge of the sleeve hits the pole with a frictional, completely inelastic impact. Contact, however, is not maintained due to the loaded spring. Point (4) corresponds to a partly elastic impact of the beak against the pole. After that collision the velocity $\dot{\varphi}_S$ is negative and the woodpecker starts to swing downwards. At (5) the upper edge of the sleeve hits the pole a second time with immediate separation.

548

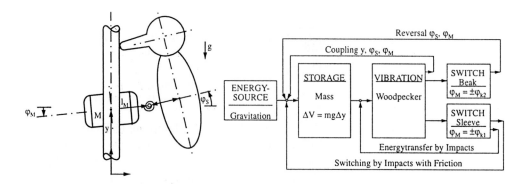

Figure 6. Limit cycle behavior of a woodpecker toy

Then the system is unbound moving downwards (5)–(6), where the second part of the falling height is achieved and the 72.91 Hz frequency can be observed once more.

A comparison of measured and calculated data of the limit cycle fequency f_L and of the falling height Δy in one cycle comes out with:

	Measured	Calculated
f_L [Hz]	9.0	9.2
Δy [mm]	5.3	5.7

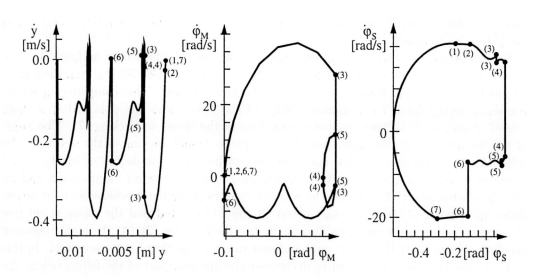

Figure 7. Phase space portraits of a woodpecker toy

5. SUMMARY

The dynamics of multibody systems with multiple unilateral contacts partly depending on each other is essentially characterized by discontinuous transitions like impacts with and without friction and by stick-slip phenomena. To evaluate the relative kinematics of unilateral contact results in costly geometrical developments mainly applying well-known rules of differential geometry. To derive a complete set of the equations of motion including all unilateral constraints affords the application of a basic rule of contact dynamics. It states, that for any contact either magnitudes of relative kinematics are zero and the corresponding constraint forces are not zero, or vice versa. The product of both groups of magnitudes is always zero thus establishing a complementarity problem. Paper presents a survey of the relevant methods and considers multibody systems with multiple contacts, which might be of planar and spatial structure. Much of the theory and the woodpecker example is based on the book [17].

REFERENCES

1. H. Antes, P.D. Panagiotopoulos, The boundary integral approach to static and dynamic contact problems, Birkhäuser Verlag, Basel, Boston, Berlin, 1992.
2. R.M. Brach, Rigid body collisions, Journal of Applied Mechanics, Vol. 56 (1989) 133-138.
3. Ch. Glocker, Dynamik von Starrköpersystemen mit Reibung und Stößen, Fortschrittberichte VDI, Reihe 18, Nr. 182, VDI-Verlag, Düsseldorf, 1995.
4. P.J. Holmes, The dynamics of repeated impact with a sinusoidally vibrating table, J. Sound Vib. Vol. 84 (1982) 173-189.
5. M. Jean and J.J. Moreau, Unilaterality and Dry Friction in the Dynamics of Rigid Body Collections, Proceedings Contact Mechanics International Symposium (ed. Curnier, A.), PPUR (1992) 31-48.
6. J.B. Keller, Impact with friction, Journal of Applied Mechanics, Vol. 53 (1986) 1-4.
7. A. Klarbring, A Mathematical Programming Approach to Three-Dimensional Contact Problems with Friction, Comp. Meth. Appl. Mech. Engng., Vol. 58 (1986) 175-200.
8. K. Klotter, Technische Schwingungslehre, Springer Verlag, Berlin, Heidelberg, New York, 1981 (reprint).
9. P. Lötstedt, Coulomb friction in two-dimensional rigid body systems, ZAMM, Vol. 61 (1981) 605-615.
10. P. Lötstedt, Mechanical systems of rigid bodies subject to unilateral constraints, SIAM J. Appl. Math. Vol. 42, No. 2 (1982) 281-296.
11. J.J. Moreau, On unilateral constraints, friction and plasticity, New Variational Techniques in Mathematical Physics, Edizioni Cremonese, Roma, 1974.
12. J.J. Moreau, Application of convex analysis to some problems of dry friction, Trends in Applications of Pure Mathematics to Mechanics, Vol. 2, London (1979) 263-280.
13. K.G. Murty, Linear Complementarity, Linear and Nonlinear Programming, Sigma Series in Applied Mathematics (ed. White, D.J.), Heldermann Verlag, Berlin, 1988.
14. P.D. Panagiotopoulos, Inequality problems in mechanics and applications, Birkhäuser Verlag, Basel, Boston, Stuttgart, 1985.

550

15. P.D. Panagiotopoulos, Hemivariational inequalities – applications in mechanics and engineering, Springer Verlag, Berlin-Heidelberg-New York-London-Paris-Tokyo-Hong Kong-Barcelona-Budapest, 1993.

16. F. Pfeiffer, Mechanische Systeme mit unstetigen Übergängen, Ing. Arch. Vol. 54 (1984) 232-240.

17. F. Pfeiffer and Ch. Glocker, Multibody Dynamics with Unilateral Contacts, John Wiley, New York, 1996.

18. M. Raous and S. Barbarin, Conjugate gradient for frictional contact, Proc. Contact Mechanics Int. Symp., ed. A. Curnier, PPUR (1992) 423-432.

19. S.W. Shaw, On the dynamic response of a system with dry friction, J. Sound Vib., Vol. 108, No. 2 (1986) 305-325.

20. W.J. Stronge, Rigid body collisions with friction, Prod. R. Soc. Lond., A 421 (1990) 169-181.

21. J.C. Trinkle, S. Sadursky, G. Lo and J.S. Pang, On the numerical solution of dynamic multi-rigid-body contact problems as complementarity systems, to appear.

22. Y. Wang and M.T. Mason, Two-dimensional rigid-body collision with friction, Journal of Applied Mechanics, Vol. 59 (1992) 635-642.

23. M. Wösle and F. Pfeiffer, Dynamics of multibody systems containing dependent unilateral Constraints with friction, Journal of Vibration and Control **2**: 161-192, 1996.

Theoretical and Applied Mechanics 1996
T. Tatsumi, E. Watanabe and T. Kambe (Editors)
© 1997 Elsevier Science B.V. All rights reserved.

The fluid mechanics of natural ventilation

P. F. Linden

Department of Applied Mathematics & Theoretical Physics, University of Cambridge, Silver Street, Cambridge CB3 9EW, United Kingdom

1. INTRODUCTION

Environmental considerations and energy conservation make the use of natural ventilation of buildings in order to provide a comfortable internal environment an attractive proposition. Modern buildings are tightly constructed and made from highly insulating materials. They often include large areas of glazing to maximize the use of natural light. Consequently, these buildings receive relatively high solar gains and, even in temperate climates, ventilation is needed to remove excess heat from these and other internal gains within the building. The ventilation requirements to provide a comfortable temperature are usually far more stringent than the requirements for fresh air for respiration. New building types such as atria and the use of devices such as solar chimneys raise new problems associated with internal stratification and air flow patterns that make traditional design solutions inadequate.

There are two driving forces for natural ventilation: the wind and buoyancy forces associated with temperature variations within a building and between the inside and the outside (the "stack" effect). The aim in designing a natural ventilation system is to harness these forces so that a comfortable indoor environment and good air quality is maintained in the building throughout the year. The crucial part of this problem is the air flow driven by the wind and stack. These couple together in a nonlinear way, and the prediction of the airflow in the complex geometry of a building is an exceedingly difficult problem. Modern designs tend to favour tall spaces, such as atria, which introduce new problems associated with the potential for high levels of thermal stratification. These large open spaces also generate non-standard ventilation patterns and rules of thumb used by architects over the last century are no longer appropriate for these new building designs. This has led to the introduction of air-conditioning into such spaces with consequent large energy costs.

This paper describes the basic ventilation patterns that occur in buildings and describes a method for analyzing these flows. This analysis provides a systematic approach to the physics of these airflows and enables them to be represented in a simple and straightforward way. The theoretical study is supported by a corresponding experimental programme which is used to motivate the theoretical models and to provide data for comparison with their predictions. In addition, the experiments themselves give the information on both the qualitative flow patterns and quantitative estimates of the temperatures and flows within buildings.

The format of this paper is as follows. In Section 2 buoyancy-driven ventilation is discussed and the basic forms of ventilation are examined. Theoretical estimates are given for the ventilation flowrates and it is shown how the arrangement of heat sources within a building is a crucial aspect of the ventilation design. The combination of wind and stack effects is discussed in Section 3 and some general conclusions are given in Section 4. This approach of understanding and quantifying the mechanics of the fluid flow provides a rational basis for making design decisions for natural ventilation systems. It allows new, simplified design rules to be developed that are appropriate to modern buildings. It also shows that natural ventilation is a reasonable option for a wide range of climatic conditions, with consequent savings in energy and other environmental benefits.

2. BUOYANCY-DRIVEN VENTILATION

2.1 Basic ventilation patterns

In the absence of wind, stack-driven ventilation falls into two basic categories called mixing ventilation and displacement ventilation. These two regimes are illustrated in Figure 1.

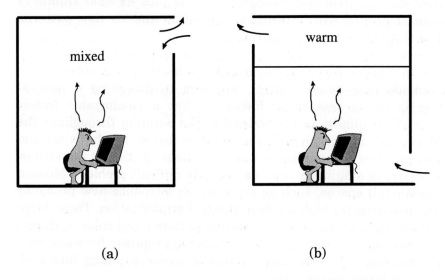

(a)   (b)

Figure 1. Schematics of mixing ventilation (a) and displacement ventilation (b).

In a mixing system the fresh air is introduced in such a way as to provide mixed conditions throughout the ventilated space. The inlets and outlets are arranged so that relatively cool air enters at high levels or relatively warm air enters at low levels within the space so that buoyant convection produces mixing as shown in Figure 1(a). In displacement ventilation the inlets and outlets are arranged so that cool air enters at the bottom of the space and warm air is extracted near the ceiling as shown in Figure 1(b). In this case a stable stratification is produced by the ventilation flow and vertical mixing is minimal. In reality many systems are intermediate between these two, but we will discuss them separately.

2.2 Mixing flows

Mixing flows occur when the inlets are either at high or low levels. If we consider a single opening high in a vertical wall of a space, then the flow into the space will be through a controlled exchange flow at the window as described by Linden & Simpson (1985) and Dalziel (1988). The volume flux Q through the window is given by

$$Q = kA(g'd)^{\frac{1}{2}},\tag{1}$$

where A is the area of the opening of height d and g' is the reduced gravity denoting the strength of the density difference between the inside and outside of the space. The constant k varies with the orientation of the opening. For a vertical window $k = 0.25$ and for a horizontal opening such as a skylight in a flat roof $k = 0.055$ for a circular opening (Epstein 1988) and $k = 0.051$ for a square window (Brown, Wolfson & Selvason 1963). This reduced flow through a horizontal opening reflects the mixing that takes place with the exchange flow in this case. The effect of the orientation of the opening to the vertical has been studied by Davies (1993) and it was found that a non-mixing exchange flow was established for angles greater than 4^0 from the horizontal.

Consider first a space initially filled with light fluid when an inlet is opened. Let g_0' be the initial reduced gravity between the inside and outside of the space. Then mass conservation shows that

$$\frac{g'}{g_0'} = \left(1 + \frac{t}{\tau}\right)^{-2},\tag{2}$$

where

$$\tau = \frac{2V}{kA}\left(g_0'd\right)^{-\frac{1}{2}} \qquad (3)$$

is the timescale associated with ventilating the space of volume V.

When there are sources of buoyancy in the space a steady state will be achieved in which the buoyancy flux through the opening is equal to that produced by the internal sources. The arrangement of sources within a space is unimportant and the steady state buoyancy g_E' is

$$g_E' = \left(\frac{B}{kAd^{\frac{1}{2}}}\right)^{\frac{2}{3}}, \qquad (4)$$

where B is the total buoyancy flux of the internal sources.

2.3 Displacement flows

2.3.1 Single source of buoyancy

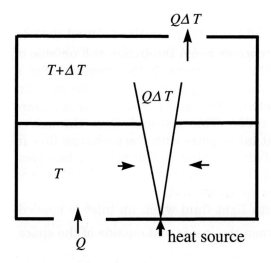

Figure 2. Displacement ventilation with a single source of buoyancy.

Linden, Lane-Serff & Smeed (1990) investigated the flow in an enclosure with high-level and low-level openings generated by a single point source of buoyancy on the floor of the enclosure (see Figure 2). They showed that in this case a very simple stratification develops consisting of two layers separated by a horizontal interface. The lower layer is at uniform ambient temperature and the upper layer is also at a uniform temperature which depends on the buoyancy flux from the source. In an enclosure of height H the dimensionless depth of the cool ambient layer $\xi = h/H$ is given by

$$\frac{A^*}{H^2} = C^{\frac{3}{2}} \left(\frac{\xi^5}{1-\xi} \right)^{\frac{1}{2}}, \tag{5}$$

where A^* is the 'effective' area of the top and bottom openings of the enclosure, and H is the height difference between the top and bottom openings. The constant $C = \frac{6}{5}\alpha \left(\frac{9}{10}\alpha \right)^{\frac{1}{3}} \pi^{\frac{2}{3}}$, where α is the entrainment constant for the plume.

The effective area A^* of the opening is defined as

$$A^* = \frac{c_d a_t a_b}{\left(\frac{1}{2} \left(\frac{c_d^2}{c} a_t^2 + a_b^2 \right) \right)^{\frac{1}{2}}}, \tag{6}$$

where a_t and a_b are the areas of the top and bottom openings, respectively, and c is the pressure loss coefficient associated with the inflow through a sharp-edged opening. A discharge coefficient c_d is used here to account for the vena contracta at the downstream side of the sharp-edged upper vents. If there are n sources of equal strength present on the floor of the enclosure, again a stratification with two uniform layers forms and the non-dimensional height ξ of the interface is given by

$$\frac{1}{n} \frac{A^*}{H^2} = C^{\frac{3}{2}} \left(\frac{\xi^5}{1-\xi} \right)^{\frac{1}{2}}. \tag{7}$$

In the single plume case, or when the sources have equal strength, the height of the interface is independent of the buoyancy fluxes and depends only on the dimensionless vent area A^*/H^2. On the other hand the temperature of the upper layer, which is independent of height, increases as the heat flux of the plumes increases.

These results provide some simple guidelines for the designer. Equations (5) and (7) show that in order to achieve a deep layer at ambient temperature $(\xi \to 1)$ it is necessary to have a very large number of openable vents. In practice this is extremely difficult to achieve, and consequently it is a good idea to have some dead space at the top of an enclosure in which the hot air can accumulate in order to drive the flow. The flow through the system is controlled by the effective area given in (6), and the magnitude of A^* is determined by the smaller vent area. For example, when $a_t \ll a_b$, $A^* \to a_t c_d \sqrt{2}$, and so control of the flow can be achieved by adjusting the smaller openings to the enclosure. A further feature of these flows is that the interface height is independent of the strength of the

556

buoyancy flux from the source, which results from the fact that the position of the interface is governed by the entrainment into the plume.

2.3.2 Multiple sources of buoyancy

Thermal stratification (or stratification of contaminant concentration) in practical situations does not generally exhibit a sharp change in density between two internally well-mixed layers as described in the simple model above. A more gradual change is observed from ambient conditions at the bottom of the enclosure to a maximum temperature at the top (e.g. see Gorton & Sassi 1982; Jacobsen 1988; Cooper & Mak 1991). This type of stratification arises owing to many factors which are not included in the simplified model of a single plume within the space. In practice, heating occurs owing to distributed sources of buoyancy of different strengths, located at various positions within the space.

In an attempt to address this issue the approach of Linden *et al.* (1990) has been extended to cover multiple sources of buoyancy of different strengths. The fluid mechanics are similar to the single source case but the analysis is complicated by the fact that the stronger plumes rise through a stratified region and discharge their buoyancy at higher levels within the space.

2.3.2.1 Two positive sources of buoyancy

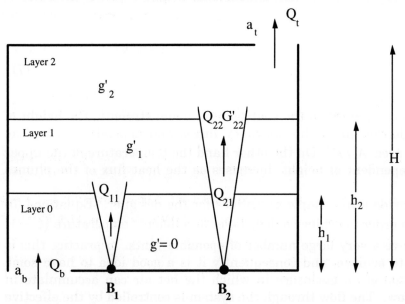

Figure 3. Displacement ventilation with two positive sources of buoyancy.

A schematic of two positive sources of buoyancy is shown in Figure 3. A three-layer stratification in this case and the dimensionless interface heights are given by

$$\frac{A^*}{H^2 C^{\frac{3}{2}}} = \frac{\left(1+\psi^{\frac{1}{3}}\right)^{\frac{3}{2}}}{\left(1+\psi\right)^{\frac{1}{2}}} \left(\frac{\left(h_1/H\right)^5}{1-h_1/H \frac{\left(1-\psi^{\frac{2}{3}}\right)}{\left(1+\psi\right)}\left(\frac{h_2-h_1}{H}\right)}\right)^{\frac{1}{2}} \tag{8}$$

and are shown in Figure 4 (see Cooper & Linden 1996). Here $\psi = B_1 / B_2 \leq 1$ is the ratio of the buoyancy fluxes of the two plumes. This relationship is of a very similar form to that for a single plume (5). The interface heights are independent of the total buoyancy fluxes and depend only on the openable areas of the vents, the height of the enclosure and the ratio of the buoyancy fluxes ψ.

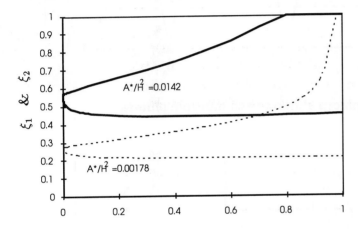

Figure 4. Theoretical prediction of the non-dimensional interface heights ξ_1 and ξ_2 as functions of the ratio B_1/B_2 of the buoyancy fluxes, for two different values of the dimensionless vent area A^*/H^2.

As shown by (8) the interface heights depend only on the dimensionless area A^*/H^2 and the ratio ψ of the buoyancy fluxes. When $\psi=0$ a single interface forms, and for $\psi > 0$ this interface splits into two with the lower interface ξ_2 ascending as ψ increases. The buoyancy of the intermediate layer increases relative to that of the upper layer as ψ increases, and the two are equal when $\psi=1$. At this point the interface ξ_2 disappears and the result for two equal plumes given by (7) with $n = 2$ is obtained.

2.3.2.2 Multiple plumes

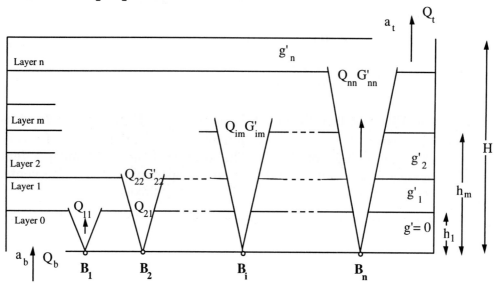

Figure 5. Schematic of displacement flow with multiple plumes.

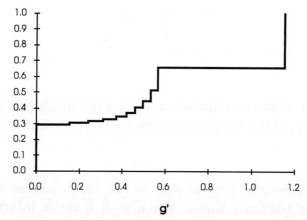

Figure 6. The stratification produced by $n = 10$ plumes, with strengths given by the arithmetic progression (9), plotted against the dimensionless height within the enclosure. The strength of the strongest plume is 20 kW, the dimensionless vent area $A*/H^2 = 0.0167$, H = 5m and $\beta = 0.1$. Note that g'=1 m/s^{-2} corresponds to a temperature difference of about 30°C.

An approximate model, which ignores the effects of the stratification on the plumes has been developed by Linden & Cooper (1996). A schematic for this flow is shown in Figure 5. In this case a layered stratification develops and the strength of the stratification above the ambient zone may be calculated as follows. Consider the case of n plumes where

$$\left. \begin{array}{l} B_i = \dfrac{i\beta}{n} B \; (i = 1, \cdots, n-1, \quad \beta = \text{constant} \le 1) \\[2ex] B_n = B. \end{array} \right\} \qquad (9)$$

The case for 10 plumes with $\beta = 0.1$ is shown in Figure 6. Note that a gradual transition in temperature occurs in the region above the ambient zone.

From a design viewpoint the height of the lowest interface is the critical parameter as this determines the depth of the zone at ambient temperature. The calculations with multiple plumes show that the height of this interface is well approximated by the n-plume result (7) for a wide range of buoyancy fluxes (see Figure 4). The depth of the ambient zone is very sensitive to the number of sources. Thus distributing the buoyancy flux from a single source into 10 equal sources reduces the height of the ambient zone by a factor of about 2. Current design guidelines are generally based on a single source of heat or smoke in a space. The present results show that the height of the smoke-free zone in a naturally ventilated space will decrease in the event of two or more fires being present. However, the presence of other thermal plumes will confine the smoke from a strong fire plume to a thinner region near the ceiling.

2.3.3 Distributed sources

There are many situations, especially involving solar gains, where the sources of buoyancy are distributed over some surface. For a horizontal surface it is possible to consider the plume as arising from a virtual origin at a different height from the floor. This situation enables the results of Section 2.3.2 to be used with some minor modifications (Caulfield 1991). For the case of a distributed source on a vertical wall the situation is more complicated. With displacement ventilation a steady state will form, but if an interface forms at any height then the flux through the space will be the same as the flux in the plume crossing the interface. Since the plume from a distributed source will increase due to the addition of more buoyancy, then the possibility exists that a series of layers will form. At any level where the volume flux in the plume is not equal to the volume flux out of the space there must be a net vertical motion exterior to the plume, and for a steady state fluid elements exterior to the plume must move along surfaces of constant density. The theory described above for point sources can easily be extended to consider this case and it is found that the number of layers is given by

$$N(N+1)^5 = \alpha^3 \pi^2 \frac{H^4}{A^{*2}} . \qquad (10)$$

Experiments by Linden *et al.* (1990) and more recently by Cooper - personal communication (1996) - suggest that intermediate layers do form in this case but considerably more work needs to be done to verify the theoretical prediction (10).

3. COMBINED EFFECTS OF WIND AND BUOYANCY

The relative sizes of the wind-generated and the buoyancy-generated velocities at time t are measured as a Froude number defined by

$$Fr(t) = \frac{U_{wind}}{\sqrt{g'(t)h(t)}}, \tag{11}$$

where U_{wind} is an average velocity of incident wind and $h(t)$ is the depth of the buoyant layer. The wind-induced dynamic pressure drop Δ is related to the square of the wind speed and, hence, the Froude number may be written in the form

$$Fr(t) = \sqrt{\frac{\Delta}{\rho g'(t)h(t)}}. \tag{12}$$

3.1 Reinforcing wind and buoyancy forces

When warm air is escaping from an upper leeward vent and cool air is entering from a lower windward vent, the wind-driven flow reinforces the stack effect. Displacement ventilation is maintained and a stratified interior is established. For a drainage flow, Hunt & Linden (1996), following a similar analysis to that presented by Linden *et al.* (1990), show that the time evolution of the interface height h as a fraction of the total height H over which the buoyancy force acts, may be written in the form

$$\frac{h}{H} = \left(\sqrt{1 + Fr(0)^2} - \frac{t}{te_b} \right)^2 - Fr(0)^2, \tag{13}$$

where $Fr(0) = \sqrt{\Delta/\rho g'H}$ is the 'initial' Froude number. The time taken for the enclosure to empty under the influence of buoyancy forces alone te_b is given by

$$te_b = \frac{2S}{A^*} \left(\frac{H}{g'} \right)^{\frac{1}{2}}, \tag{14}$$

where S denotes the cross-sectional area of the enclosure. The total time taken te for the enclosure to empty with the wind assisting the buoyancy, relative to the time taken to empty under buoyancy alone, is then

$$\frac{te}{te_b} = \sqrt{1 + Fr(0)^2} - Fr(0) .$$

(15)

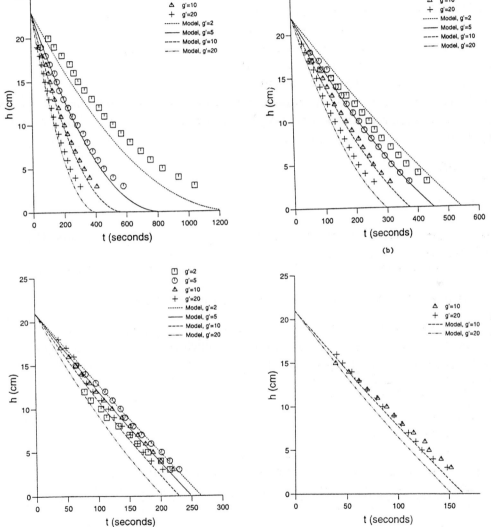

Figure 7. The position of interface as a function of time for $g' = 2, 5, 10$ and $20\, \mathrm{cm s^{-2}}$, (a) $\Delta = 0\, \mathrm{gcm^{-1}s^{-2}}$, (b) $\Delta = 40\, \mathrm{gcm^{-1}s^{-2}}$, (c) $\Delta = 207\, \mathrm{gcm^{-1}s^{-2}}$ and (d) $\Delta = 495\, \mathrm{gcm^{-1}s^{-2}}$.

Figure 7(a) shows the position of the interface as a function of time for the control experiments $(\Delta = 0\,\mathrm{gcm^{-1}s^{-2}})$ in which only the buoyancy force drives the ventilation flow, for the density differences corresponding to $g' = 2$, 5, 10 and $20\,\mathrm{cms^{-2}}$. Also shown in Figure 7(a) are the predicted interface positions deduced from (13). Figures 7(b)-(d) depict the interface position as a function of time for wind-assisted buoyancy-driven ventilation for $\Delta = 40$, 207 and $495\,\mathrm{gcm^{-1}s^{-2}}$, respectively. In each figure the observed and predicted evolution of the interface is shown for $g' = 2$, 5 10 and $20\,\mathrm{cms^{-2}}$. As the buoyant fluid empties from the space, the pressure force exerted by the wind becomes increasingly dominant and the effect of the buoyancy force plays a less significant role in the emptying of the space. In Figure 7(d) the displacement flow is approaching the purely wind-driven limit as only small changes in the total emptying time result from variations in g'. Here the density difference acts merely to keep the fluid stratified and maintain the displacement mode and the wind provides the dominant expelling force. From (12) and (13) the Froude number at time t may be expressed as

$$Fr(t) = \frac{Fr(0)}{\sqrt{h(t)/H}}, \qquad (16)$$

and hence, for a constant wind velocity and density difference, the Froude number increases as the enclosure empties. In other words, wind effects become increasingly more dominant as the thickness h of the lighter layer decreases. In fact, the wind force begins to dominate the buoyancy force as $h < \dfrac{\Delta}{\rho g'}$.

3.2 Opposing wind and buoyancy forces

Experiments to examine the ventilation of an enclosure by the opposing forces of wind and buoyancy were conducted with salt solutions opening both windward (low-level) and leeward (high-level) vents. When the buoyancy force initially exceeds the dynamic pressure force of the oncoming stream, i.e. for $Fr(0) < 1$, three distinct flow regimes are observed. The development of a typical experiment for which $Fr(0) < 1$ is shown in Figure 8 which illustrates the position of the interface (dashed curve) and the average density of the fluid in the enclosure (continuous curve) as a function of time. Initially, an outflow of dense fluid through the low-level, windward opening (i.e. into the oncoming stream) was observed. This fluid was replaced by less dense wind-driven fluid which entered through the high-level, leeward opening, i.e. a displacement flow was set up. As negatively buoyant fluid was displaced from the space the buoyancy force exerted at the low-level, windward opening decreased. The evacuation of the denser fluid by displacement ventilation continued until the buoyancy force was matched by the dynamic pressure force of the wind-driven flow. This displacement flow regime can be seen in Figure 8 as a falling interface and decreasing average density for $t < 300$ s. Due to the unstable nature of the fluid interface at the windward opening, an oscillatory flow then developed in the enclosure.

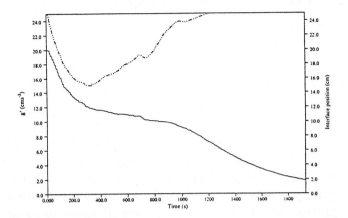

Figure 8. Interface position (dashed curve) and average fluid density (continuous curve) as a function of time, $Fr(0) = 0.78$.

The oscillatory flow observed was one in which there were periods of exchange flow and periods of inflow at the windward opening. Initially, the observed exchange flow was not balanced but was predominantly outflow. The outflow was through the lower section of the opening and the inflow through the upper section. As dense fluid emptied from the enclosure, during the exchange flow, the amount of inflow increased until there was entirely inflow. Fluid entering the enclosure rose as a buoyant plume, which partially mixed with the dense layer of fluid, thereby decreasing its density, and created wave-like disturbances on the fluid interface. The motion of the interface resulted in local pressure variations within the enclosure due to the changes in fluid head. The inflow continued until a sufficiently large amplitude interfacial disturbance, which corresponded to a wave crest vertically above the windward opening, cut-off the inflow. An exchange flow then ensued. The mixing between the incoming fluid and the fluid contained in the enclosure caused the interface to ascend. The non-steady ascent of the interface is shown in Figure 8, as is the relatively small decrease, due to the periods of exchange flow, in the average density of the fluid in the enclosure. The oscillatory flow through the windward opening ceased when the interface reached the top of the enclosure. A mixing type ventilation flow then ensued, with an inflow of light fluid through the windward opening and outflow through the leeward opening. At this stage there is no stratification and the fluid in the enclosure is of an approximately uniform density. The decay in the average density which occurs during the mixing stage can be seen in figure 8 for $t > 1200$ s. Note that if the strength of the wind was now decreased, so that the buoyancy force exerted at the windward opening exceeded the dynamic pressure force of the wind, then the entire ventilation flow sequence described above would be repeated.

For $Fr(0) > 1$, i.e. when initially the dynamic pressure force of the wind exceeds the buoyancy force exerted at the low-level, windward opening, light fluid was observed to enter the space through the windward vent and the outflow was through the leeward vent. A mixing type ventilation flow is established in this case.

4. CONCLUSIONS

In this paper I have described how plume theory may be applied to an analysis of the ventilation flow within a building. I have concentrated on tall spaces with high heat gains, and under these circumstances, displacement ventilation is the preferred mode. It provides a more efficient removal of heat and a greater ventilation rate than mixing ventilation and leads to an occupied zone at ambient temperature. A recognition of the essential dynamical consequences of rising buoyant elements in the form of entraining plumes is an essential feature of this analysis. It shows that the ventilation rate through the building is controlled by the upward air flow in the plumes and, in displacement mode ventilation, this may be controlled by the amount of openable area. The analysis shows that the control of the flow is determined by the smallest opening and this enables designers to make systems which are controllable while allowing occupants considerable individual freedom to regulate their own local environment within a building.

The analysis described here provides simple rules of thumb for designers which make it possible to make fairly simple assessments of the efficiency of likely designs. For example, we observe that the depth of the occupied zone in a building is well predicted by assuming that the major sources of heat are of equal strength and that this depth is independent of the total buoyancy flux in the space. Consequently, a simple analysis of the major heat sources will provide a reasonable estimate of the height of the zone and ambient temperature for a range of vent openings. We also observe that it is difficult to provide an occupied zone which exceeds about 80% of the total height of the space and consequently some dead space is necessary if stack ventilation is to be effective.

The effects of wind have been investigated and here we find that again the nature of the flow can be determined in terms of a simple analysis of likely flow patterns observed in laboratory experiments. We find a range of behaviours depending on whether the stack effect and wind reinforce one another or whether they oppose one another. It is also the case that, while in many circumstances the effect of the wind is potentially much greater than that of the stack, the stack is essential in determining the stratification within the space which, in turn, determines the distribution of the air movement within it. So, while quantitatively the stack effect may appear to be relatively minor, its effect on the interior flow pattern is crucial in determining the ventilation flow.

In this brief survey it has not been possible to consider all the complex flows that can be set up in modern buildings. I have concentrated here on flows within a single space but, of course, many buildings are multiply connected and these require further study. Nevertheless, the approach adopted can be extended to these situations and with a combination of these modelling approaches and experimental work it is possible to obtain quantitative estimates of likely ventilation flows in complex buildings. Another aspect of the problem which I have not addressed is that of transient flow with steady heat sources. When vents are opened or closed, the flow patterns adjust accordingly and the steady states that have been described here are still applicable in these contexts. Most modern buildings respond on relatively short timescales. However, these and other questions such as the possibility of interacting plumes, the effect of radiation from the building fabric and more complex geometries provide a rich field for further study. The work described in this paper suggests that this approach will provide further insights into building ventilation in the future.

Acknowledgements
Much of the research described in this paper has been developed in collaboration with colleagues, particularly P. Cooper, G.R. Hunt, G.F. Lane-Serff, J.E. Simpson and D.A. Smeed, and many of the ideas are theirs.

REFERENCES

Brown, W. G., Wilson, A. G. & Selvason, K. R. 1963 Heat and moisture flow through openings by convection. *J. Am. Soc. Heating Ventilation Air Conditioning Engng,* **5**, 49-54

Caulfield, C. P. 1991 Stratification and buoyancy in geophysical flows. PhD thesis, University of Cambridge, UK

Cooper, P. & Mak, N. 1991 Thermal stratification and ventilation in atria. *Proc. ANZSES (Australian and New Zealand Solar Energy Soc.) Conf., Adelaide, Australia,* 385-391

Cooper, P. & Linden, P. F. 1996 Natural ventilation of an enclosure containing two buoyancy sources. *J. Fluid Mech.,* **311**, 153-176

Dalziel, S. B. 1988 Two-layer hydraulics - maximal exchange flows. PhD thesis, University of Cambridge, UK

Davies, G. M. J. 1993 Buoyancy driven flow through openings. PhD thesis, University of Cambridge, UK

Epstein, M. 1988 Buoyancy-driven exchange flow through openings in horizontal partitions. *Intl Conf. On Cloud Vapor Modelling Nov 1987, Cambridge, MA.*

Gorton, R. L. & Sassi, M. M. 1982 Determination of temperature profiles in a thermally stratified air-conditioned system: Part 2. Program description and comparison of computed and measured results. *Trans. ASHRAE,* **88** (2), paper 2701

566

Hunt, G. R. & Linden, P. F. 1996 The natural ventilation of an enclosure by the combined effects of buoyancy and wind. *Proc. ROOMVENT '96*, **3**, 239-246.

Jacobsen, J. 1988 Thermal climate and air exchange rate in a glass covered atrium without mechanical ventilation related to simulations. *13th Natl Solar Conf. MIT, Cambridge, MA*, **4**, 61-71

Linden, P. F. & Simpson, J. E. 1985 Buoyancy driven flows through an open door. *Air Infiltration Rev.*, **6**, 4-5

Linden, P. F. & Cooper, P. 1996 Multiple sources of buoyancy in a naturally ventilated enclosure. *J. Fluid Mech.*, **311**, 177-192.

Linden, P. F., Lane-Serff, G. F. & Smeed, D.A. 1990 Emptying filling spaces: the fluid mechanics of natural ventilation. *J. Fluid Mech.*, **212**, 300-335.

Theoretical and Applied Mechanics 1996
T. Tatsumi, E. Watanabe and T. Kambe (Editors)
1997 Elsevier Science B.V.

Down-to-earth Temperatures: The Mechanics of the Thermal Environment

R Narasimha[a] *

[a]Centre for Atmospheric Sciences, Indian Institute of Science
Bangalore 560 012, India

This lecture describes some recent attempts at unravelling the mechanics of the temperature distribution near ground, especially during calm, clear nights. In particular, a resolution is offered of the so-called Ramdas paradox, connected with observations of a temperature minimum some decimetres above bare soil on calm clear nights, in apparent defiance of the Rayleigh criterion for instability due to thermal convection. The dynamics of the associated temperature distribution is governed by radiative and convective transport and by thermal conduction, and is characterised by two time constants, involving respectively quick radiative adjustments and slow diffusive relaxation. The theory underlying the work described here suggests that surface parameters like ground emissivity and soil thermal conductivity can exert appreciable influence on the development of nocturnal inversions.

1. INTRODUCTION

In 1979, Lettau[1] listed three micrometeorological paradoxes; this lecture deals with the one that he named after Ramdas. The story goes back to 1932, when Ramdas & Atmanathan published a short paper[2] reporting that, on calm clear nights, air just above bare soil (at heights of up to about a metre) can be cooler than the ground by a few degrees. This report, of what we shall call the lifted temperature minimum or the 'Ramdas effect', was for several reasons received with much scepticism at the time. First of all, conventional wisdom had been (and still largely is) that during night, the temperature is lowest *at* (and not *above*) the ground; this wisdom has apparent support from classical observations of air temperature distributions in the lowest layers of the atmosphere (e.g. [3]). In actual fact, such observations usually do not reach below the standard screen height of 1.2m, and so are not necessarily inconsistent with a lifted minimum below that height. A more serious objection is that such a cold air layer above the ground, resulting in heavier air on top, should be subject to the classical Rayleigh-Benard instability. A calculation of the Rayleigh number, based on the temperature differential over the layer and its thickness, shows it to be often two orders of magnitude higher than the critical value for plane convection between two

---

*Also at Jawaharlal Nehru Centre for Advanced Scientific Research, Bangalore 560 094, India

568

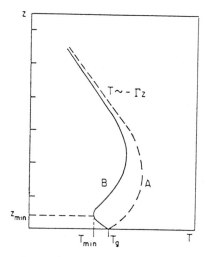

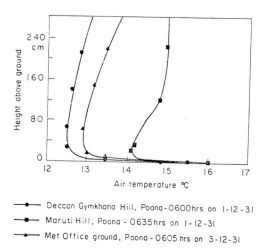

Figure 1. Temperature distribution near ground at night, as generally expected (curve A) and as found by Ramdas under calm, clear conditions (curve B). (From [9]).

Figure 2. Temperature profiles near ground, measured at three sites in Poona (reproduced from Ramdas [6]).

horizontal plates, suggesting that the observed cold layer should be highly unstable. Finally, accurate temperature measurements near the ground present special problems, and it was not clear whether the instruments used by Ramdas were adequate. It is thus no surprise that, as Geiger[4] remarks in his monumental treatise *The Climate near the Ground,* 'These results were at first accepted with some reservations'. In an earlier German edition of the book he had declared that 'the repetition of measurements which lead to such distributions is essential'.

Indeed, the general understanding at the time, as summarized e.g. in the well-known text of David Brunt[5], was that at night the 'normal decrease of temperature [with altitude in the free atmosphere] may be replaced by a condition in which temperature steadily increased from the ground upward, giving what is known as an *inversion* in the lower layers, above which there is a return to the more normal decrease with height'; in other words, according to   Brunt, the temperature distribution is generally like the curve A in Figure 1. The argument for this view is, in the words of [1], that the surface 'is the level of both the source of the outgoing terrestrial radiation and the sink of the two heat fluxes by convection (out of the air) and conduction (out of the soil...)'. What Ramdas & Atmanathan claimed was that on calm clear nights there is a kink below the classical inversion, as in curve B in Figure 1. Figure 2 reproduces observations at three sites in Poona (now Pune) that illustrate the phenomenon; these observations, reported in [6], were said to have been taken 'within a few days of its first detection', and are in fact more typical than the very first data that were published, for reasons that will become clear later.

The authors appear to have seen the phenomenon as special or peculiar to India and the tropics. The generally prevailing view that at night temperatures are lowest at ground is, the authors say, 'no doubt true for temperate latitudes ... But in India ...'. At another point in the paper they write, '... in the tropics the ground heating is so strong during day-time ...'[2]. They finally remark that the data reported by them 'have raised some interesting problems regarding the role of radiation from the ground and the lower layers of the atmosphere'.

2. SOME HISTORY

Lake[7] has suggested that the lifted or elevated temperature minimum may have been observed by Glaisher[8] as far back as 1847. The long tables in this paper include rare occasions when air temperature at a height of 4 ft was lower than on grass and some other surfaces, but it is difficult to decide whether Glaisher had observed a lifted temperature minimum of the same type as Ramdas. Lake also cites a report by Cox in the US, dated 1910, that temperature measured 5 inches above bare soil in Wisconsin was often lower than that at the surface on clear nights. Ramdas and his colleagues, who were clearly unaware of this work, provided not only the first detailed temperature profiles that unambiguously established the existence of an unsuspected minimum, but also the earliest discussion – albeit unsuccessful – of the physical principles that may govern the phenomenon. Some of the early history of work in the area has been traced by Narasimha[9].

The favoured explanation for the phenomenon was that it must be due to advection, i.e. to the flow of colder air from the environs – draining down neighbouring slopes or passing through vegetation. Although Ramdas & Atmanathan admitted that this was a possibility in the case of the Poona measurements they reported in 1932, Ramdas quickly made the other measurements at two hill-tops in Poona that we have shown in Figure 2; advection could not have been a factor here, so one must seek other explanations. In fact during the next two decades he studied the phenomenon extensively, and reviewed at various times the work he had done (e.g. [10]). In the 1951 review he still spoke of phenomena 'which occur rather conspicuously in tropical and sub-tropical micro-climates'.

In the mid-fifties there appears to have been a sudden surge of interest in the subject outside India. Lifted temperature minima were reported by workers in Argentina, USA, UK and several other countries, including the Antarctica, showing that the phenomenon was by no means confined to the tropics ([4], [7], [11]). Interestingly, the most convincing observations came again from India and were published in 1957 by Klaus Raschke[11], a young German agronomist who spent three years with Ramdas in Poona and took up Geiger's suggestion that the earlier measurements must be repeated. He carried out a series of careful and thorough measurements that settled the question once and for all and are a landmark in the field. He made special thermoelectric sensors to measure air temperature, and took care to compensate for the (small) radiation error they were subject to. Based on his observations he distinguished between three different types of temperature distribution near ground: (i) the normal radiative type, with minimum temperature at ground, occurring on clear nights with wind; (ii) the special or singular radiative type (studied by Ramdas), with temperature minimum at a height of some decimeters above ground, occurring on calm, clear nights; and (iii) the advective type, with temperature minimum at some height above ground, occurring on clear or cloudy nights with advection of colder air. To eliminate the possibility of advection, Raschke also made measurements on the remarkably flat top of Chaturshringi Hill in Poona. Reporting on observations made on 183 nights during 1953-54, Raschke found that during the calm winter months, type (ii) (i.e. the Ramdas type) was the one that generally prevailed (98% of the time during November 1954). The existence of the Ramdas type was thus established beyond doubt. In one set of measurements (not on Chaturshringi Hill) where the lifted minimum could have been attributed to advection in the form of a tongue of cold air issuing from a sugarcane field 60 m away, its occurrence even after harvest served to rule out advection as the cause. However with the advent of the monsoons the advective type (iii) became more common (57% of the time in June 1955).

The Ramdas minimum is very sensitive to turbulent transport: Raschke could get rid of it merely by waving a plywood sheet nearby (thereby increasing turbulent transport - we may think of it as 'tripping' the flow in aerodynamic terms). Transitions between different types of temperature distribution were found to occur in times of the order of a minute: Raschke records an instance where the temperature rose 5°C in a 3-minute interval! We shall shortly return to this question.

There had till recently been no completely satisfactory explanation of the phenomenon. As already mentioned, the generally accepted view not long ago (e.g. [3]) was that the cooling of air near the ground was basically a balance between conduction and convection, the radiative heat loss being confined to the surface. Möller[12] asserted that 'radiation processes are meaningless for the behaviour of the air layer near the ground', and considered the formation of fog or advection essential for the occurrence of the lifted minimum. Lettau[1] advocated convection as the mechanism. Nevertheless, following the suggestion of Ramanathan & Ramdas[13], radiation has often been thought to be responsible in some way (raising the immediate question where a length scale of 10-20 cm is relevant in atmospheric radiation). The only quantitative analysis of the problem till recently was due to Zdunkowski[14], who attributed the phenomenon to haze, following a reasoning similar to Möller's. However, the theoretical temperature profiles presented by Zdunkowski were based on values of thermal diffusivity *lower* than the molecular value by a factor of up to 18; no results were presented for realistic values of the parameter. In this respect these results recall a calculation made by Ramdas & Malurkar[15] who also had to assume very low diffusivities (factor of 22) to reproduce the observed steep temperature gradients near the surface (even in the absence of a lifted minimum). There is of course no way that molecular transport can be suppressed, so these unrealistically low diffusivities can only be interpreted as evidence of serious flaws in the theories. In any case, Raschke found no evidence of fog or haze near the ground when he observed a lifted minimum; neither did Oke[16], who made extensive measurements in Canada after the publication of Zdunkowski's paper. (We do often see a thin layer of fog suspended a foot or two above ground; this must therefore be a *consequence* of the lifted minimum, not its cause.)

3. THE VSN MODEL

If in the light of the above observations we infer that radiation plays a key role, it becomes quickly clear that it must be subtle. To see this suppose the medium is optically thick, then a flux-gradient relation for radiative transfer would be valid, and the net effect on the energy balance would be the same as an enhancement of thermal diffusivity. This renders the governing equation for the evolution of the temperature profile parabolic; and it can then be proved that no lifted minimum can be sustained [17]. If on the other hand the medium is optically thin, radiation just escapes across, and cannot affect the medium. Thus in neither of the two obvious limits can radiation explain the phenomenon. If therefore it holds the key, we must seek it in the peculiar properties of atmospheric absorption. A theory that includes a fairly elaborate model for radiation has recently been proposed by Vasudeva Murthy, Srinivasan & Narasimha[18] (VSN henceforth). This theory assumes clear nights with no advective changes, and incorporates humidity and the wind profile (needed to assess the possible effects of any residual turbulence) as parameters. With the air temperature T a function only of time t and the vertical coordinate z, the problem is completely governed by the one-dimensional energy equation, which may be written

$$\rho_a c_p \partial T / \partial t = -\partial Q / \partial z, \tag{1}$$

where ρ_a is the density of air, c_p is the specific heat at constant pressure and Q is the total energy flux, conveniently split into three components representing respectively the contributions of molecular conduction (Q_m), convection (Q_t) and radiation (Q_r). The first two are simply given by

$$Q_m = -k_m \partial T / \partial z, \quad Q_t = -k_t \partial \theta / \partial z \tag{2}$$

where k_m is the thermal conductivity of air, $\theta = T - \Gamma z$ is the potential temperature, Γ is the adiabatic lapse rate (9.86K km^{-1} in dry air), and k_t a suitable eddy conductivity, which may be represented by

$$K_t \equiv k_t / \rho_a c_p = k_* U_* z \phi(\text{Ri}) \tag{3}$$

in a stratified surface layer; here k_* is the von Karman constant, U_* is the friction velocity and

$$\text{Ri} = \frac{k_*^2 g z^2}{U_*^2 \theta} \left(\frac{\partial \theta}{\partial z} \right) \tag{4}$$

is a Richardson number appropriate to turbulent flow in the surface layer, which is governed by the 'wall' variables U_* and z (g is the acceleration due to gravity). Among the many proposals made for the function ϕ, we find it adequate for the present purpose to adopt the relations

$$\phi(\text{Ri}) = \quad 1.35(1 - 9\text{Ri})^{1/2} \quad \text{for } \text{Ri} < 0,$$
$$= \quad 1.35(1 + 6.35\text{Ri})^{-1} \quad \text{for } \text{Ri} > 0;$$

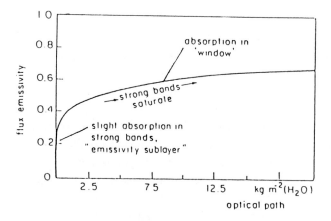

Figure 3. Variation of the flux emissivity of humid air with the optical path, showing rapid variation in an 'emissivity sublayer' near ground, followed by a slow variation where the absorption is due to the wings of the strong lines and the atmospheric window.

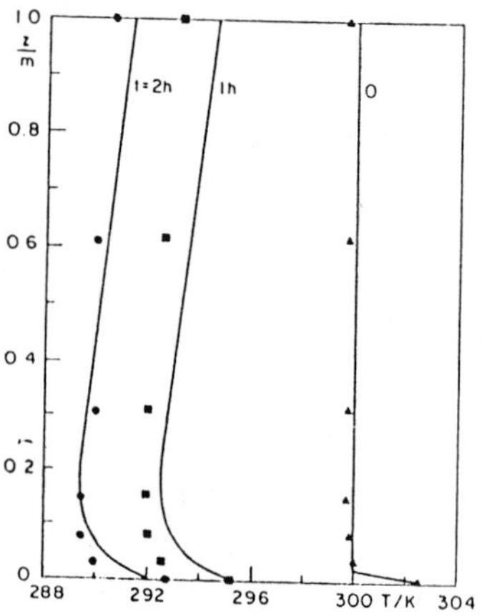

Figure 4. Comparison of observed temperature profiles [13] showing a lifted minimum, with predictions from theoretical model [18].

these are based on the data of [19] and have been widely used (see [20]).

It is a well established fact that during clear nights and in the absence of haze and other particles, the infrared cooling in the lowest 2 km of the atmosphere is chiefly due to water vapour (e.g. [21]). There are conditions (e.g. over snow) when the carbon dioxide content of air is also of some importance; this can when necessary be taken into account by a suitable enhancement of the flux emissivity of air (to be introduced below). The net longwave radiative flux Q_r can be **written** as

$$Q_r = F^\uparrow - F^\downarrow, \tag{5}$$

where $F^\uparrow$, $F^\downarrow$ are the upward and downward fluxes. It is convenient to compute the radiative flux divergence in terms of the differential mass path length Δu of the absorbing gas (water vapour in the present instance) along a differential path Δz; this is given by $\Delta u = \rho_w \Delta z$, where ρ_w is the density of water vapour. Accounting for the effect of pressure variation on the absorption process [22], the corrected water vapour mass path length is taken as

$$u(z) = \int_0^z \rho_w(z') \left\{ \frac{p(z')}{p(0)} \right\}^\delta dz', \tag{6}$$

where $p(z)$ denotes the pressure of air at level z and δ (chosen empirically) lies in the range $0.5 < \delta < 1$. A value of 0.9 is taken for δ, as recommended by [23].

We use the broadband flux emissivity method (e.g. [21]), which expresses the fluxes as

$$F^{\downarrow}(u) = \int_u^{u_\infty} \sigma T^4(u',t) \frac{d\epsilon}{du'}(u'-u)du', \tag{7}$$

$$F^{\uparrow}(u) = \{\epsilon_g \sigma T_g^4(t) + (1-\epsilon_g)F^{\downarrow}(0)\}\{1-\epsilon(u)\} - \int_0^u \sigma T^4(u',t)\frac{d\epsilon}{du'}(u-u')du', \tag{8}$$

where $u_\infty = u(\infty)$ is the total atmospheric path-length, ϵ_g is the emissivity of ground, $T_g(t)$ is the ground temperature at time t and $\epsilon(u)$ is the broadband flux emissivity function of water vapour, given by

$$\begin{aligned} \epsilon(u) &= 0.04902\ln(1+1263.5u) \quad \text{for } u \le 10^{-2}\text{kg m}^{-2}, \tag{9} \\ &= 0.05624\ln(1+875u) \quad \text{for } u > 10^{-2}\text{kg m}^{-2}; \tag{10} \end{aligned}$$

these expressions were originally proposed by [24], and are considered to apply up to $u = 0.5$kg m^{-2}, which is adequate for the present purposes. The relations are displayed in Figure 3.

Note the steep increase in ϵ near $u = 0$ displayed in Figure 3; we shall find it convenient to think of the region $u < 10^{-3}$ kg m^{-2} as constituting an 'emissivity sublayer' because of this rapid variation.

The energy equation should be supplemented with initial and boundary conditions. VSN propose

$$T(z,0) = T_{g0} - \Gamma z, \quad T(0,t) \equiv T_g(t) = T_{g0} - \beta\sqrt{t}, \quad \frac{\partial T}{\partial z}(\infty,t) = -\Gamma, \tag{11}$$

where $T_g(t)$ is conveniently taken to obey the well-known Brunt solution (e.g. [3]), and T_{g0} is the value of T_g at a suitably defined initial instant $t = 0$. This initial instant, which we may call 'nominal' sunset, tends to occur slightly ahead of actual sunset, and is best determined, along with β, by fitting a $\sqrt{t}$ curve to observed ground temperature variation when it is available; in the model β is a specified ground cooling rate parameter. Now VSN actually carried out detailed computations using a coupled air-soil model, and showed that it was sufficient to specify a temperature boundary condition in the form (11), which is in good agreement with observation under the conditions considered. This simplifies our treatment considerably, as the soil thermal conductivity κ_s may then be replaced by the more convenient cooling rate parameter β, which is proportional to $\kappa_s^{-1/2}$.

The basic idea in this approach is to assume a uniform lapse rate Γ in the "free" atmosphere, and discuss the evolution of a thin thermal layer near the ground associated with falling ground temperature.

4. THE SOLUTIONS

The equation (1) is solved numerically by the method of lines, with a suitably graded mesh: details are available in VSN.

Although there are now numerous observations of the lifted minimum, these rarely report the precise initial and boundary conditions; in particular the ground emissivity and the friction velocity at the time of measurement have never been estimated. In these circumstances, the best we can do is to make reasonable estimates for ϵ_g and U_* and offer illustrative comparisons. For this purpose we select the data of [13], which provides good initial and boundary conditions on the temperature under wind conditions stated to have been calm (so we shall assume $U_* \approx 0$ as a first approximation). The temperature profile reported at 1800h (local time) is taken here as the initial condition. Measured surface temperature variation soon after sunset suggests $\beta = 7.5 \mathrm{Kh}^{-1/2}$. In Figure 4 the prediction of the model, with this value of β and assuming $\epsilon_g = 0.8$ and $U_* = 0$, is compared with observations. We find that the numerical simulation is able to predict closely the temperature profile in the lowest metre of the atmosphere. Because of residual uncertainties associated with the values of the various parameters involved we should not perhaps consider the agreement shown in figure 4 as definitive, but there can be no doubt whatever that the theory predicts the right kind of behaviour.

5. PHYSICAL DISCUSSION

In the VSN theory, surface emissivity ϵ_g and the cooling rate parameter β play key roles.

Consider first ϵ_g. In meteorological investigations it is usual to assume that the ground is radiatively black; it is indeed nearly so, but the small departures present do matter for a special reason. The radiative cooling of air near ground may be shown [18] to depend on the product

$$(1 - \epsilon_g)(d\epsilon/dz)_{z=0}, \tag{12}$$

which cannot be ignored even if ϵ_g is close to unity, because the multiplying derivative is huge (Figure 3). It must be emphasized that cooling and heating rates depend on *gradients*, and that near ground the gradient of the emissivity of air often overwhelms that of the temperature: *differential* absorption is therefore the key to understanding why the air cools down so much[25].

The value of the ground emissivity depends very much on the nature of the surface, and can often be rather less than unity. A table compiled by Paltridge & Platt[26] shows that for some surfaces ϵ_g may be lower than 0.8. Indeed these authors point out that the common assumption $\epsilon_g = 1$ can pose problems that are not minor. Furthermore, the values listed in such tables usually refer to vertical emissivities appropriate to remote measurement of temperature by vertically oriented radiometers mounted on spacecraft or similar platforms. The quantity relevant in our discussions is the global or hemispherical emissivity, which for natural surfaces may be rather less because of radiation incident at low angles. It is therefore reasonable to assume that the (global) emissivity of natural surfaces is in the range 0.8 to nearly 1.0.

It has been shown [25] that the energy balance near ground is rather delicate, in the sense that the net cooling rate is the (small) difference between terms one order of magnitude larger – namely upflux emitted from ground and from the air layers just above ground, and a drop in the surface emissivity from unity to 0.8 reverses the net balance from warming to cooling. Any theory that does not take account of the rapid variation with height of the flux emissivity of air cannot therefore be expected to provide a rational explanation of the phenomenon.

It is well known (e.g. [21]) that the flux emissivity variation mentioned above arises from the spectral characteristics of infra-red absorption by water vapour. Thus, the rapid increase in emissivity near ground is due to the strong absorption of radiation in the vibration/rotation bands of the water vapour molecule; and the slow increase at larger heights comes from the wings of the strong absorption lines and the near-transparent atmospheric window in the 8-12μ band. We could thus say that the lifted temperature minimum arises because of the nature of radiative absorptivity in moist air, characterised as it is by the presence of a small nearly transparent window embedded in bands where air is nearly opaque to radiation.

It should perhaps be addded that although the balance is delicate, it is also stable: the temperature minimum will sustain itself from sunset to sunrise in the absence of turbulence.

The second parameter in the VSN model, namely soil conductivity, is important because it determines how fast the ground cools; higher the conductivity lower is the cooling rate, because of greater upwelling heat flux from warm soil underground. Raschke[11] realized the importance of this parameter.

The gist of the proposed theory can now be simply stated. When ground is not perfectly black air above ground can cool radiatively because of the rapid variation (with height) of the absorption of infrared radiation by water vapour. Because of long photon mean free paths (especially in the atmospheric window) a radiative slip at ground will usually be present underneath the classical inversion. The sign of this slip can be such that the ground is warmer if the soil is sufficiently conducting for the upwelling heat to keep ground cooling slow. A touch of diffusion, chiefly only molecular, smears out the slip into the cold layer observed by Ramdas.

There remains the question of how the Rayleigh-Benard instability is circumvented. This must be an effect of radiation, for it is easy to see that radiation is stabilizing. For a given temperature difference the presence of radiation adds to the heat transfer by conduction and so may be thought of as increasing the effective conductivity of the medium. This becomes literally true when the medium is nearly opaque, for then a radiative thermal diffusivity K_r can be formally introduced. The viscosity of air is of course not affected by radiation, so momentum diffusion is much less than total thermal diffusion, i.e. air behaves in this limit like a low Prandtl number fluid. Now it is well known that the critical Rayleigh number for onset of convection is unaffected by Prandtl number. It is therefore reasonable to expect that a modified Rayleigh number, with $K_r + K_m$ replacing the molecular diffusivity K_m, would have the same critical value for instability as in the non-radiative situation, everything else remaining the same. The classical Rayleigh number (defined with only the molecular thermal diffusivity K_m in the denominator) will therefore have, with radiative diffusion, a critical value that is larger by a factor of order $(1 + K_r/K_m)$. This factor can be considerable, so the critical Rayleigh number can go up substantially. When the medium is optically thin the argument is not literally valid, but the additional transfer of heat by radiation should even then have the qualitative effect of stabilizing the flow. A rigorous assessment of the effect of radiation with the semi-open boundary conditions characteristic of the atmospheric problem is still to be carried out, but estimates based on arguments involving the ratio of radiative to diffusive time constants show that the critical Rayleigh number under conditions of interest can go up by a factor that is anywhere between 10 and 150[18]. Furthermore, heat transport by free convection increases only slowly at supercritical Rayleigh numbers: for it to reach four times the conductive flux the Rayleigh number has to be about 60 times higher than critical[18]. We thus see that the lifted minimum can be maintained without completely over-turning either solely due to radiative stabilization or because heat transport is only weakly enhanced even at relatively large super-critical Rayleigh numbers: the heavier fluid is prevented from sinking because viscosity aided by radiation holds it aloft, so to speak. Thus the Ramdas layer is not as unstable as it seems at first sight: and, as we have seen, it often sustains itself from sunset to sunrise.

The theory predicts that the lifted minimum is weaker on rough, dark (radiatively, that is) and insulating ground, and disappears if turbulent transport exceeds some four times the molecular transport. These predictions lend themselves to experimental checks, which therefore now become interesting to carry out.

6. STEADY STATE OR CONTINUOUS EVOLUTION ?

This is a question on which the observational evidence is apparently conflicting. A diagram presented by Ramdas[6] appears to suggest that the height of the lifted minimum, as measured by him at a site in Poona, is almost constant - at about 30cm - throughout the night, beginning about an hour after sunset. However, he could not have detected any change in the location of the minimum in the range 30 - 60 cm above ground, as no thermometers or probes were mounted in this range. Hence, we can only conclude from these observations that the lifted minimum must have remained between 30 and 60 cm, and any evolution that might have taken place must have been very slow.

On the other hand, some kind of evolution is implicit in the observations of Raschke[11], who reports that at his site (also in Poona but different from that of Ramdas), the lowest temperature was almost always recorded by the thermocouple at 10 cm height, but that in the early morning hours the values measured at a height of 1 cm were the lowest. Lake[7] reports that 'the negative difference between the surface temperature and that at any height up to 54 in. increased with time as long as the conditions remained stable', supporting the idea of continuous evolution through the night.

In this context it should be noted that the nocturnal boundary layer itself has been shown to be evolving in several studies (e.g. [27]), and hence it has been suggested that its structure cannot be studied using steady state formulations.

Furthermore, however, we have to contend with Raschke's observation about the extremely rapid response of the temperature distribution to what we have called 'tripping', and to gusts. It is therefore of interest to study the *dynamics* of the phenomenon, and to explore the time constants that govern the physics in the light of the VSN model.

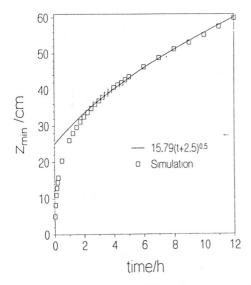

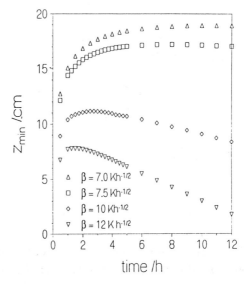

Figure 5. Evolution of the height of the lifted temperature minimum, with $\beta = 2$ Kh$^{-1/2}$, $\epsilon_g = 0.8$, showing parabolic growth beginning 2 hours after sunset.

Figure 6. Evolution of height of lifted minimum at high ground cooling rates, showing near-steady state or collapse.

7. THE TIME CONSTANTS IN THE PROBLEM

A number of simulations based on the VSN model have recently been carried out by Ragothaman, Narasimha & Vasudeva Murthy[28]. Some of the results are displayed in Figures 5, 6. At low β, z_{min} varies like $t^{1/2}$ (Figure 5), suggesting immediately diffusive behaviour as in the analogous, thermal Rayleigh problem. However, it must be noted that the ground continues to cool throughout the night, i.e. in reality the boundary condition is time-dependent unlike in the Rayleigh problem. The analogy thus breaks down for relatively high values of β, indeed to such an extent that eventually the lifted minimum disappears altogether. Thus, from Figure 6, we can see that there are four regimes in the evolution of the Ramdas layer; it can grow, remain relatively steady, collapse or not occur at all depending on surface parameters. The conflicting observations recounted earlier have therefore the simple explanation that they could be in different (site-dependent) regimes.

It can also be seen that z_{min} increases rapidly during the first few minutes after "nominal sunset" ($t = 0$), and much more slowly at larger times; thus z_{min} is already about 10 cm at $t = 6$ minutes and only 60 cm 12 hours later (Figure 5). In general, it is found that, irrespective of surface properties, the short-time evolution of z_{min} is relatively rapid, but that this may be followed by either a slower increase, or if β and ϵ_g are sufficiently high, even a gradual decrease.

The time evolution of ΔT_{min} confirms the above picture.

To probe the question further, it is convenient to parameterise a gust within the framework of the VSN model, on the basis that it is a period of relatively high turbulence

578

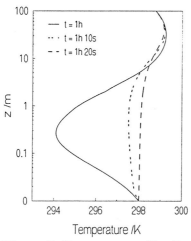

Figure 7. Temperature distributions near ground during gust episode with $U_* = 1$ ms^{-1}, showing disappearance of lifted minimum.

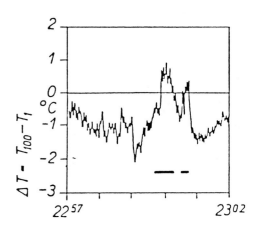

Figure 8. Temperature differential between 100mm and 1mm above ground, as affected by "tripping": a plywood sheet is waved in the neighbourhood during the time intervals marked. (From [11].)

during which the thermal diffusivity is also correspondingly high. This can be done by prescribing a relatively high value (say 1 ms^{-1}) for the friction velocity during the gust, and switching it off to zero during calm periods. In the simulation the friction velocity assumes a top hat profile; the gust is turned on 1 hr after nominal sunset, and turned off 30 s later.

Figure 7 shows the temperature distributions during such a 30s gust episode. It can be seen that, 20s after the gust has commenced, the lifted minimum has disappeared; at this instant the temperature distribution is almost isothermal up to about a meter from the ground, and at greater heights - above 50 m (beyond the inversion) - it follows the prescribed lapse rate. It is also found that after the gust ceases at $t = 1$h 30s, the lifted minimum has already reappeared within 10s; the re-emergence is thus very rapid. This is in qualitative agreement with the observations of Raschke mentioned above; the result of his experiment when a piece of plywood was waved near the site of his instrumentation is reproduced in Figure 8.

Furthermore, it is seen from Figure 9 that the location as well as the intensity of the lifted minimum are still evolving an hour after the gust ceases, approaching the no-gust baseline solution very slowly: both z_{min} (28 cm out of 32 cm) and ΔT_{min} (4.1 K out of 4.4 K) are within 90% of the baseline solution only 1 h after the gust episode although the lifted minimum begins to reappear very quickly after the gust is stopped.

Thus, when calm conditions are restored after the gust, there occurs a quick adjustment leading to the re-emergence of the lifted minimum, followed by a much slower relaxation to the baseline solution.

The simulations we have carried out indicate strongly that the quick adjustment is radiative while the relaxation is diffusive [28].

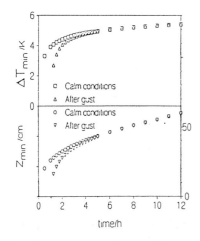

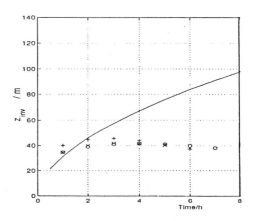

Figure 9. The approach of lifted-minimum parameters to the base-line solution after a gust episode.

Figure 10. Evolution of inversion depth z_{inv} for $\epsilon_g = 0.8$ and $\beta = 2.0$ Kh$^{-1/2}$ (— $U_* = 0$ (showing Taylor's $t^{1/2}$ law); xxx U_*=5 cm s^{-1}; ooo U_*=10 cm s^{-1}; +++ $U_* = 20$cm s^{-1}.)

8. THE NOCTURNAL TEMPERATURE INVERSION

As the VSN model takes into account all the modes of energy transfer that are relevant under the conditions assumed, it is interesting to find out if the model contains the necessary physics to describe the nocturnal temperature inversion.

Among the attempts that have been made to model the nocturnal boundary layer and inversion we may mention Anfossi et al.[29] who take into account the contribution from radiation, and others who have attempted to explain the evolution in terms of turbulent heat transfer alone (e.g. [30]). However, it is now clear from recent studies (e.g. [31]) that both turbulent heat transfer and clear air radiative cooling need to be taken into account for a more precise description. Another interesting study is due to Nieuwstadt[32]. It is interesting to recall that as long ago as 1915 G.I. Taylor[33] proposed a $t^{1/2}$ law for the evolution of the depth of the inversion.

Observational data have also been analysed by several workers [31, 34, 35]; the latter two have examined the prevalence or otherwise of the $t^{1/2}$ law proposed by Taylor.

It has been implicitly assumed in all these studies that surface properties are not directly relevant to a study of the inversion layer, but the present theory would suggest otherwise.

Simulations varying the parameters ϵ_g and β show interesting effects on the inversion height z_{inv} as well as the temperature differential across it. It is found, for example, that when β is increased from 2 Kh$^{-1/2}$ to 5 Kh$^{-1/2}$, the inversion height, say 11 h after sunset, increases from about 120 m to about 140 m; when ϵ_g is increased from 0.7 to 1.0, it decreases from 130 m to 70 m, whereas the temperature differential increases from 2.4 K to 3.7 K. Surface properties thus appear to have a strong influence an inversion parameters.

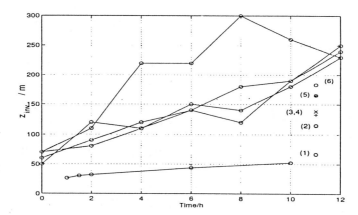

Figure 11. Evolution of z_{inv} as reported by [36] at various sites, compared with predictions shown at $t = 11$ hr to avoid clutter. (1) ϵ_g=1.0 and β =2.0 Kh$^{-1/2}$; (2) ϵ_g=0.8 and $\beta = 2.0$ Kh$^{-1/2}$; (3) ϵ_g=0.8 and β=5.0 Kh$^{-1/2}$; (4) ϵ_g=0.7 and β=2.0 Kh$^{-1/2}$; (5) ϵ_g=0.8 and β=10.0 Kh$^{-1/2}$; (6) ϵ_g=0.8 and β=15.0 Kh$^{-1/2}$.

Eddy diffusion can also have a strong effect (Figure 10). Under calm conditions, z_{inv} follows a Taylor-type $t^{1/2}$ law; but with wind it begins to depart from this law after about 2 h, and to decrease after about 3 h.

The temperature differentials show an earlier and steeper fall with time. They begin to decrease after reaching a maximum at about 2 h, and may disappear altogether at the end of 7 h for the case when U_* is about 20 cm s^{-1}.

A major problem in comparing the theory with observations is that the latter rarely include values of ϵ_g, β and wind, all of which (as we have seen) exert a significant influence on inversion characteristics. Under the circumstances, we can only hope to find out whether the present simulations reproduce observations for some reasonable set of values of the unmeasured parameters.

Figure 11 shows a comparison of the inversion height obtained from the present simulations for calm conditions with those measured by Piggin & Alley[36] and by Ramdas[6]. The former were made under "light winds" (wind speed less than about 2 ms^{-1}); in the latter work there is no direct information. It can be seen that the inversion heights obtained from the simulations with ϵ_g in the range 0.7 - 1.0 and β in the range 2 - 10 Kh$^{-1/2}$ fall around the range of observed values, if allowance is made for the initial conditions, i.e. the time of nominal sunset $t = 0$ is advanced by about 2 hours. The inversion heights obtained from the present simulations are also in qualitative agreement with the range of values compiled by Yu[37] from the Wangara experiment.

9. CONCLUSION

The results reported here have been based largely on computer simulations. It is however possible to construct an approximate asymptotic theory, utilizing the ratio of the radiative to the diffusive time constant (mentioned above) as a small parameter[38]. The difficulty with such a theory is that handling the radiative transfer integrals in

(8, 9) poses certain problems, in spite of the simplifications that are suggested by the analysis of Narasimha & Vasudeva Murthy[25]; so some computation still appears inevitable. Nevertheless, the analysis clearly reveals the structure of the solution, with a thin conduction layer underneath an inversion, both in general evolving continuously and slowly after a short and rapid transient at nominal sunset.

It is incidentally interesting to note that these disparate time constants have been noted in micrometeorological observations during a total solar eclipse [39] – where transitions can be even sharper than during a tropical sunset.

It appears remarkable that unraveling the physics of the very lowest layers of the atmosphere – a region that we explore with our feet all the time! – should have taken so long. Perhaps there is need for special attention from practitioners of mechanics to what may be called, with justification, sub-micro-meteorology[9] – i.e. meteorology beneath the well-known surface layer of the atmospheric boundary layer. The subject is of great interest to agricultural scientists, who have extensively studied it by observation and some modelling[4]. Retrieving correct land-surface parameters from satellite or aircraft remote sensing would also demand a thorough understanding of the subject.

What the present work has shown, however, is that surface parameters can have a large influence on even such apparently larger-scale phenomena like the nocturnal inversion. Understanding what governs such inversions can again be of great importance: recall that the terrible effects of the tragedy that occurred in Bhopal in the early hours of 3 December 1984 were almost certainly attributable in part to the strong nocturnal inversions that prevail during winter in most parts of India.

This brings us back to the question about why these observations, and in particular the discovery of the lifted minimum, were first made in India. We already remarked on the undercurrent in Ramdas & Atmanathan's first report about conditions supposedly peculiar to the tropics. It is now of course known that the effect occurs in all parts of the world, but what may be somewhat unusual in the tropics is the wide prevalence of calm, clear conditions during winter; and quick, sharp sunsets can highlight rapid transients. In the absence of any of these conditions the effect is weak or non-existent.

Acknowledgements

The work reported here would not have been possible without the collaboration of many colleagues and students, including in particular Prof J Srinivasan, Dr A S Vasudeva Murthy, Mr. S Ragothaman, Mr Saji Verghese and Prof A Prabhu, to all of whom I am most indebted. It is a pleasure to acknowledge partial support from the Department of Science & Technology, Government of India, and the U.S. Office of Naval Research (contract no. N00014-94-1-1133).

REFERENCES

1. H. H. Lettau, *Boundary Layer Meteorol.* 17 (1979) 443-464.
2. L. A. Ramdas and S. Atmanathan, *Beitr. Geophys.* 37 (1932) 116-117.
3. O. G. Sutton, *Micrometeorology.* McGraw-Hill, New York, 1953.
4. R. Geiger, *The Climate Near the Ground.* Harvard Univ. Press, 1965.

582

5. D. Brunt, *Physical and Dynamical Meteorology*. Cambridge Univ. Press (1941).

6. L. A. Ramdas, *WMO Tech. Note* 104 (1968) 399-405.

7. J. V. Lake, *Q. J. R. Meteorol. Soc.* 82 (1956) 187-197.

8. J. Glaisher, *Phil. Trans. Roy. Soc.* 37 (1847) 119-216.

9. R. Narasimha, *Current Science* 66 (1994) 16-22.

10. L. A. Ramdas, *Arch. f. Met. Geophys. Bioklim.* 3 (1951) 149-167.

11. K. Raschke, *Met. Rundschau* 10 (1957) 1-11.

12. F. Z. Möller, *Z.f. Meteorol.* 9 (1955) 47.

13. K. R. Ramanathan and L. A. Ramdas, *Proc. Indian Acad. Sci.* A1 (1935) 822-829.

14. W. Zdunkowski, *Beitr. Phys. Atmos.* 39 (1966) 247-253.

15. L. A. Ramdas and S. L. Malurkar, *Ind. J. Phys.* 6(1932) 495-508.

16. T. R. Oke, *Q. J. R. Meteorol. Soc.* 96 (1970) 14-23.

17. A. S. Vasudeva Murthy, J. Srinivasan and R. Narasimha, Report 91AS1, Centre for Atmospheric Sciences, Indian Institute of Science, Bangalore (1991).

18. A. S. Vasudeva Murthy, J. Srinivasan and R. Narasimha, *Phil. Trans. Roy. Soc. (London)* A344 (1993) 183-206.

19. J. A. Businger, J. C. Wyngaard, Y. Izumi and E. F. Bradley *J. Atmos. Sci.* 28 (1971) 181-189.

20. D. A. Haugen, *Workshop on Micrometeorology*. Amer. Met. Soc., Boston (1973).

21. K. N. Liou, *An Introduction to Atmospheric Radiation*. Academic Press, New York. 1980.

22. J. T. Houghton, *The Physics of Atmospheres*. Cambridge Uni. Press, 1986.

23. R. Garratt and R. A. Brost, *J. Atmos. Sci.* 38 (1981) 2730-2746.

24. W. Zdunkowski and F. G. Johnson, *J. Appl. Met.* 4 (1965) 371-377.

25. R. Narasimha and A. S. Vasudeva Murthy, *Boundary-Layer Meteorol.* 76 (1995) 307-321.

26. G. W. Paltridge and C. M. R. Platt, *Radiative Processes in Meteorology and Climatology*. Elsvier, Amsterdam (1976).

27. J. R. Garratt, *The Atmospheric Boundary Layer*. Cambridge Uni. Press (1992).

28. S. Ragothaman, R. Narasimha and A. S. Vasudeva Murthy. Submitted for publication.

29. D. Anfossi, P. Bacci and A. Longhetto, *Q. J. R. Meteorol. Soc.* 102 (1976)

30. J. Kondo, *J. Meteorol. Soc. Japan* 9 (1971) 75-94.

31. J. C. Andre and L. Mahrt, *J. Atmos. Sci.* 39 (1982) 864-878.

32. F. T. M. Nieuwstadt, *J. Appl. Meteorology* 19 (1980) 1445-1447.

33. G. I. Taylor, *Phil. Trans. Roy. Soc.* A215 (1915) 1.

34. A. D. Surridge, *Boundary-Layer Meteorol.* 36 (1986) 295-305.

35. A. D. Surridge and D. J. Swanpoel, *Boundary-Layer Meteorol.* 40 (1987) 87-98.

36. I. G. Piggin and S. K. Alley, *Mausam* 45 (1994) 243-254.

37. T. Yu, *J. Appl. Meteorol.* 17 (1978) 28-33.

38. A. S. Vasudeva Murthy and R. Narasimha. To be published (1996).

39. R. Narasimha, A. Prabhu, K. N. Rao and C. R. Prasad, *Proc. Ind. Natl. Sci. Acad.* 48A (1982) 175-186 (Supp.).

LIST OF CONTRIBUTED PAPERS
PRESENTED AT THE CONGRESS

Abba, A., R. Bucci, C. Cercignani and L. Valdettaro
Anisotropy Effects in Large Eddy Simulation

Abdaimi, Y., C. Licht and G. Michaille
Some Remarks about the Behavior of Heterogeneous Media Capable of Large Elastic Deformations

Abdul-Latif, A. and K. Saanouni
Effect of Some Parameters on the Plastic Fatigue Behavior with Micromechanical Approach

Abe, H.: See Lee, J.H.

Abe, H.: See Saka, M.

Abe, K.: See Mizuno, M.

Abe, O.: See Nohguchi, Y.

Abe, Y.: See Kiya, M.

Adachi, S.: See Ishii, K.

Adachi, T.: See Tomita, Y.

Adachi, T. and Y. Tomita
Trabecular Remodeling Simulation of Stress Regulation Toward Uniform Distribution

Adam, S.: See Schnerr, G.H.

Adamson, R.M.: See Dempsey, J.P.

Adolfsson, E. and P. Gudmundson
Matrix Crack Initiation and Growth in Composite Laminates Subjected to Bending and Extensional Loads

Agrawal, S.C.: See Bhargava, R.R.

Agrawal, S.K.: See Pandey, S.

Ahmad, R. and F.T. Smith
Three-Dimensional Vortex Flows in Distorted Pipes

Aizawa, S.: See Fujita, T.

Aizawa, T.
Aggregate Modeling for Superplastic Analysis

Aizawa, T.: See Iwai, T.

Aizawa, T.: See Tamura, S.

Aizawa, T.: See Tsumori, F.

Akamatsu, T.: See Takahira, H.

Akamatsu, T.: See Tamagawa, M.

Akylas, T.R. and T.-S. Yang
Asymmetric Gravity-Capillary Solitary Waves

Alboussiere, T., Y. Delannoy, A. Kljukin, R. Moreau and V. Uspenski
Two Dimensional Turbulence Developed from a Free Shear Layer

Allin, J.: See Jain, M.

Al-Ostaz, A.: See Ostoja-Starzewski, M.

Alzebdeh, K.: See Ostoja-Starzewski, M.

Ambrosio, J.A.C.: See Silva, M.P.T.

Amielh, M.: See Anselmet, F.

Andersen, J.R. and K.B. Dysthe
Evolution of an Oscillating Instability due to Double-Diffusion

Andersson, L.-E.: See Nygren, T.

Andreasen, J.H. and B.L. Karihaloo
Steady-State Analysis of an Array of Semi-Infinite Edge Cracks in a Transformation Toughening Ceramic

Andreyko, S.S.: See Zhuravkov, M.A.

Anselmet, F., L. Pietri, M. Amielh and L. Fulachier
Experimental Investigation of the Near-Field Region of Variable Density Turbulent Jets using Laser-Doppler Anemometry

Arai, M.: See Song, W.

Arai, Y.: See Takai, S.

Arai, Y.: See Tsuchida, E.

610

618